TOPICAL AREAS

AEROSPACE POWER SYSTEMS AND TECHNOLOGIES

The 2001 IECEC Aerospace Power Systems and Technologies (AT) Sessions highlight many advances in Aerospace power over the past year. Highlights range from the first flight data from the International Space Station power system to development of many innovative new power system components to analysis of possible power technology developments for the next generations of aerospace systems. Session topics include power electronics, photovoltaics and solar arrays, batteries and flywheels for energy storage, and integrated space power systems. Panel discussions have also been expanded this year, highlighted by the Aerospace Plenary Session that will include summaries of the status and directions of each of the six major alternatives for thermal to electric conversion on air and space vehicles. Other panels will highlight issues for solar array alternatives, future spacecraft energy storage systems, and an overview of the US government's power technology research and development programs and plans.

Paper numbers with an asterisk next to them are not in the Proceedings, but will be presented at the Conference with copies provided by the authors.

Coordinators:
Douglas M. Allen
Schafer Corp.

Dr. Valerie Lyons
NASA Glenn Research Center

* * * * * * * * * * * * * * * * * * *

CONVERSION TECHNOLOGIES

Current events highlight the importance of efficient, cost effective thermal to electric energy conversion. The papers in this topical area cover many of the technologies under development today to meet the electricity demand of tomorrow.

Coordinator:
Dr. Mike Schuller
Texas A&M University

RENEWABLE ENERGY

Interest in renewable energy is on the rise due to concerns about global warming, lack of availability of fossil fuels in certain parts of the world, and the vulnerability of energy prices to socio-political factors. Technological advances have resulted in electrical energy generated by modern large wind turbines the lowest cost and environmentally benign option at the present time. Cost of photovoltaic devices is steadily declining and hydropower continues to lead in the amount of electrical energy generated from renewable resources. Biomass and solar thermal energy are being utilized in a variety of ingenious ways to satisfy specific energy needs.

The papers presented in the renewable energy sessions exhibit the richness of renewable energy in its various manifestations and the authors deserve credit and recognition for their efforts in sharing their knowledge and aspirations.

Coordinator:
Dr. Rama Ramakumar
Oklahoma State University

CONTENTS

vii

VOLUME 2

ENERGY SYSTEMS
Building Energy Systems

Thermal and Mechanical Energy Storage

Air Conditioning Systems

Proceedings of IECEC'01
36th Intersociety Energy Conversion Engineering Conference
July 29-August 2,2001,Savanna,Georgia
IECEC2001-AT-20

SPACECRAFT POWER SYSTEMS STUDIES USING PSpice

by W. Billerbeck and Gilbert Lewis, Jr
Veridian Systems Engineering, Chantilly Va.

ABSTRACT

This paper describes development work leading to a new tool for analysis of spacecraft power systems. It uses the very powerful and convenient schematic level mode of ORCad PSpice® to lay out and analyze the system elements as connected functional blocks on a personal computer. This software has proved to be a mature, stable, and well-supported product. It evolved from the Berkeley SPICE, which in its various versions has become the standard software for electronic circuit simulation.

The basic concept of our analysis tool is to simulate each major power system element as an easily configurable block, which can be pulled from a library, and connected up into a system schematic for analysis work. Typical library file blocks include solar cell arrays, rechargeable batteries of various types, buck and boost regulators, constant current and constant power loads, and several types of pulsed TWTA and solid state transmitter loads. Through the use of dynamic analogies, mechanical and thermal features of power devices like inertial energy storage wheels can easily be simulated. Similar blocks describe the functional performance of a family of devices. For instance, solar arrays are divided into silicon, dual junction, triple, and quad junction cell types. Complete battery characteristics including charge and discharge performance are modeled. Cell types include Nickel-cadmium, Nickel-Hydrogen, and Lithium-Ion. Arrays are sized with numerical inputs to each library functional block, and temperature input terminals are provided where they are significant. The blocks are wired up by connecting lines to pins in typical schematic fashion. Utility blocks for mission analysis and energy balance work provide stimuli like programmed on/off commands, solar aspect variations, eclipse periods, and temperature changes. Although this particular effort is aimed at satellite power systems, the analysis approach and many of the models are believed to be applicable to a broad range of power systems.

OrCad PSpice® is a registered trademark of OrCad, Inc

INTRODUCTION

This analysis tool had its origin in a spacecraft power system study performed for the Naval Research Laboratory, Ref 1. That work was done at the detail intensive, netlist level familiar to most SPICE students and professional users. The new approach reported here is based on using the schematic level capabilities of PSpice® with a library of power unit blocks at the next higher level of abstraction, and is therefore an order of magnitude easier to use. The level we chose is the one normally presented in satellite power systems diagrams. Each block simulates the functions of a complete power system element, such as a DC-DC converter, or a series string of battery cells. Parameters are inserted to scale and configure the selected element. For instance, the battery Amp-Hour capacity and number in series, or the converter power capability and its output voltage, is specified by a set of parameters attached to the face of each block. A weight estimate is automatically generated by each of these units.

A complete power system can be assembled quickly at the schematic level by a process similar to the "pick and place" technique used for circuit board design. The system units are selected from the library, placed in the block diagram, sized using the parameter inputs attached to each one, and wired up in conventional manner. Then, the system design is analyzed for each specific mission using stimuli generated by the simulation utilities. This allows very quick trade studies, and analysis of electrical performance under various mission scenarios. Analysis of the effects of single point failures or other anomalies can readily be performed. With sufficient detail in the appropriate portions of the power system diagram, the analysis of system voltage ripple and transient conditions can be performed. Sunlight aspect angles and eclipse periods are selected. Loads can be switched "on" or "off" by a load control utility box. This allows very quick trade studies, and calculation of electrical performance under various mission scenarios. Analysis of single point failures or other anomalies can also be performed. With sufficient detail in the appropriate portions of the power system diagram, the analysis of bus voltage ripple and transient conditions can be performed.

Much of the data used in our simulation models of system level units for power systems has been collected over a number of years of Spice circuit analysis work at our company. These PSpice® netlist models have been derived from electrical device manufacturer's data sheets, as well as various published and unpublished sources. However, it became clear that our quasi-static models did not adequately address device AC performance of batteries, or the relatively new supercapacitor devices. In this study, a number of representative parts were selected for subcontractor testing to address this lack of data. The capacitor testing was quite extensive. Some 32 different samples of production and experimental devices ranging from conventional 100uF wet tantalum capacitors up to 3200 Farad supercapacitors were photographed, weighed, and measured. Electrical data was used to develop Ragoni type power vs energy plots to map out their optimum range of applicability, as shown in Fig. 1, from Ref. 1

A DC bias at the normal working voltage of the device was applied to each capacitor. Typically this was about 2.0 volts (about 80% of max) for the supercapacitors. A low amplitude AC was injected on top of the DC voltage to provide a stimulus for AC impedance measurements. The results were presented in the form of Bode plots of amplitude and phase changes from low frequency up to about 65kHz, which is the range of interest to most power system designers. A typical plot is shown in Fig.2

A five-stage resistor-capacitor network of the form shown in Fig. 3 was fitted to the impedance data measured at several DC bias voltages, to obtain a best fit finalized dynamic model of each capacitor type. Each capacitor model was then tested for fidelity by comparing the pulse response of the model with actual measured data obtained by subjecting the capacitor to a pulse discharge.

A somewhat similar process was used to test and model samples of rechargeable battery cells of interest for spacecraft. These samples included a 21.4 Amp-hr Nickel-Cadmium cell and an 86 Amp-hr Nickel-Hydrogen cell. These were satellite flight quality devices. A 6 Amp-hr cylindrical Lithium-Ion cell, and some small Lithium-Ion commercial cells were tested, as well as a 3/4 Amp-hr Lithium-Polymer cell. These battery cells were charged to about 80% capacity, and at that state an AC signal was injected. The low impedance of the larger cells created a somewhat challenging situation for the experimental work. Once the data was measured, the process used for fitting the simulation model to the impedance data was somewhat similar to that used for the capacitors. Again, a five-stage RC network was used as the basic model. In this case, it was necessary to add an empirical equation of source voltage vs state of charge to obtain good fits to the pulse discharge data. A comparison of the 6 Amp-hr Lithium-Ion model with

measured data under a C/1.5 discharge transient is presented in Fig. 4.

FUNCTIONAL MODEL DEVELOPMENT

It was believed to be important to build the power unit functional blocks to be simple in concept and as generalized as possible. This would provide rapid computation of large power system models, and maximum flexibility of use in analysis work. Development of these power unit functional blocks proved to be a multi-step process, as follows:

- Define unit requirements
- Draw conceptual schematic for the unit
- Write PSPice® netlist as a subcircuit
- Run unit analysis
- Test, debug, and correct
- Convert power unit subcircuit into a Pspice "part"
- Debug system interface problems, as necessary

Initially, the requirements for a power unit block were somewhat unclear. The requirements eventually included the definition of a common set of terminal number designations, and a careful consideration of startup conditions. For instance, a constant power load attempts to draw infinite current at zero source voltage! The Analog Behavioral Modeling features of PSpice®, Ref.2, were essential to this work. The ability to control voltage and current with simple algebraic equations made the compact description of complex unit behavior feasible. These equations for voltage or current sources can utilize net variables as well as parametric inputs, and can include useful logic functions like "Limit", and "Greater or Less Than" statements. A number of books and publications, including Refs. 3, 4, and 5 provided basic concepts for functional simulation of electrical units, as well as methods for developing analog treatments of mechanical devices, and thermal features of devices. Some of these, Refs. 6, and 7, also provide coding details for less publicized math functions and combinations of functions that can be used.

Some of the functional unit blocks are quite simple, while others are more complex. Two simple examples are selected to demonstrate typical power unit blocks. The first is a three-junction solar cell shown in Fig.4 selected as an example from the family of solar cell arrays. A sun intensity input ranging from zero to 1 is input from a utility module, and an equation driving the current generator,G1, calculates the temperature corrected solar current for a one sq. cm. solar cell. This current feeds into a diode especially tailored to produce the voltage-current profile of a presently produced 3-junction cell at 28C. The diode characteristics also represent the reverse bypass diode normally used with these cells, as seen in Fig. 5. A temperature input to an equation-driven voltage generator corrects for cell temperature. The current drawn in the output circuit is scaled down to the one sq. cm. reference by parameters listing the cell area and number in parallel. The cell voltage is scaled up to array output voltage by a parameter representing the number of cells in series. A separate equation not shown in the diagram calculates a solar array weight consistent with current satellite design.

A second example is a block that functions as a main bus shunt limiter, shown in Fig 6. This is a module that smoothly limits the maximum voltage developed by a solar array to the value specified by an input parameter. It can be used to represent several different types of main shunt units using various combinations of digital and analog switching to regulate smoothly, while minimizing thermal dissipation. In the model, a comparator amplifier compares the setpoint voltage specified by an input parameter with the actual bus voltage, and drives a shunt current generator to limit the voltage. A power rating input parameter is used to calculate the weight of the bus limiter, assuming a design consistent with typical flight units.

Rather extensive debugging of the unit simulation blocks was found to be essential. We found that the initial analysis runs to obtain successful results needed to be followed by a set of tests to weed out spurious results when the current source or sink polarities were reversed. Limit functions were quite helpful in solving these problems. A next set of tests was set up incorporating a stressful range of input variables and parameters. These results were scanned for the range all the variables, both voltages and currents after test runs.

To obtain convergence, the double-precision math used in PSpice® requires the maximum range of variables be less than 1e+12. Since the program solution accuracy default values are intended for circuits with voltages up to 10 volts and milliamp currents, higher values may be required for power circuits. For voltages in the kilovolt range, VNTOL, the best accuracy of voltages, may need to be raised from 1uV to 1mV. Similarly, for the Ampere range, ABSTOL of 1nA is recommended, and for kiloamps, ABSTOL should be 1uA. For many applications, the default relative tolerance limit for all variables, RELTOL, is also more accurate than necessary, and can be reset from 0.1% to 1%. These limits make it clear that proper scaling of the variables within the unit simulation modules is essential. A unit module that could be operated successfully alone would not necessarily work smoothly with other modules that had a dramatically different range of variables. After initial trials at the power systems schematic level, we found that revision was needed on some modules to rescale variables to a more compatible range.

SYMBOL GENERATION

After operation of the power unit model was thoroughly verified, a symbol was created for use in system level work. The symbol created has the model's name, input and output pins, and any changeable parameter values that are passed to the subcircuit. Additionally, the revision date of the subcircuit was shown on the symbol. This is mainly a development feature that ensures that the latest model

is in use. All of the symbols use simple block shapes to allow pin names and subcircuit parameters to be visible at all times.

Creating symbols for use in PSpice® simulations from subcircuit libraries has ever been easier. All of the symbols were created from subcircuit netlists. ORCAD Schematics' Symbol Wizard allows easy creation of symbols directly from the subcircuit text file. The Symbol Wizard reads the file and determines how many pins the symbol should have, what parameters are passed to the subcircuit, and defines all other attributes fore the symbol. This includes creating the template for the PSpice subcircuit call, previously a difficult task.

The Symbol Wizard uses only the SUBCKT call line (which defines the interface for the subcircuit) from the subcircuit netlist to define the symbol attributes. The SUBCKT call line is defined as follows:
.SUBCKT <name> [node]*
+[OPTIONAL: <<interface node>=<defaultvalue>>*]
+ [PARAMS: < <name> = <value> >*]

The symbol graphics can be based on previously created symbols or on custom symbols made from scratch. To use previously created symbols the number of pins for the subcircuit must match the number of pins of the symbol. A list of our presently available power unit symbols is in Table 1.

POWER SYSTEMS ANALYSIS PROCESS

The process of analyzing a power system falls into four logical steps. These are: layout of a basic block diagram for the system, detailing of the block diagram, startup and checkout, and analysis of the cases of interest. The assembly of the power system block diagram requires a clear understanding of the basic functional performance of each unit, and selection of a library module, a PSpice "part", with matching function capabilities. These are entered into a computer schematic diagram in the same "pick and place" manner one normally uses for discrete parts in a typical circuit diagram. Once this step is completed, the parameters controlling the scaling of each unit are entered into its box representation. Parameter inputs also control functions like output voltage, undervoltage limits, power capability, etc. Determination of these parameters usually requires data from unit specifications. Utility blocks are added to provide on/off or pulsing control of the various functional blocks, and stimuli for the desired analysis. If a mission analysis is desired, the stimulus utility block will contain parameters like orbit time, eclipse time, solar aspect angles, etc. These system blocks are wired up in conventional manner, as shown in Fig. 8.

One would expect that the next step is to fire up and go for an analysis! However, we have found through experience that one must proceed in a deliberate manner at this stage. The situation is similar to that encountered in a typical circuit analysis using Spice. Debugging of an entire circuit simultaneously can become very complex and frustrating. If one checks out separate sections of the circuit in a stepwise manner, it is easier to proceed confidently toward solution of the entire circuit. In similar fashion, work with these power unit schematics is best performed by starting at the power source, checking out and adding unit blocks step by step. Parameters and connections are adjusted at each step as needed to obtain proper operation. After these checkout steps are performed, an analysis of the complete system can be initiated. Once this first successful analysis is completed, it is often convenient to save the schematic, and then copy, rename, and revise it for analysis of additional cases.

EXAMPLES OF SYSTEMS ANALYSES

Two examples of the types of analysis that can be performed with this new tool are presented here. The first one is a low altitude, sun synchronous satellite with a high power pulsed payload, somewhat similar to the Canadian RADARSAT, Ref. 8. The block diagram layout and schematic of this spacecraft power system is shown in Fig. 8 and 9. One of the interesting investigations run with this power system was the addition of a large bank of high performance supercapacitors on the main bus to handle the pulsating load current generated during radar transmitter operation. The analysis results presented in Fig. 10 provide a tool for further investigation of a properly designed series/parallel supercapacitor bank for use in this application. Please note that this analysis and the one presented in Fig. 13 were run in our "speedup" mode, where one second of problem time equals one hour of real time.

The second power system block diagram layout is a design offering significant possibilities for weight reduction in the near term using high specific energy solar arrays and batteries, as shown in Fig. 11 and 12. This design also includes the application of high energy density supercapacitors on 8 volt input feeds to each of the solid state power amplifiers. This is believed to be a promising application for these lightweight devices, since only a small number are needed in series to keep them within their safe operating voltage. Initial analysis results for this design during a low orbit eclipse mission while operating transmitters in a high power pulsing mode are presented in Fig. 13.

CONCLUSIONS

Block diagram level schematics of several satellite power systems were assembled, and quasi-static mission analyses were performed. It was found that the complete system could be assembled from the element library, connected up in a short time, and analyzed on a typical PC computer with run times of a minute or less. This is satisfyingly convenient for trade study work. The longest time involved proved to be the assembly of a thoroughly complete description of the power system to be simulated!

This work is continuing, with the effort focused on the improvement of functional descriptions in the battery cell models, and the addition of typical dynamic characteristics to the batteries and other appropriate library functional blocks. The latter should provide "first cut" evaluation of the dynamic operation of the power systems. An especially interesting aspect for future applications is that, as they are developed, detailed Spice models of specific electronic circuits could be operated with the overall system model. The basic conceptual approach described here, and indeed many or most of the library blocks could be applied to the prototyping and analysis of almost any type of power system, including the pertinent mechanical and thermal features of all the equipment.

ACKNOWLEDGEMENTS

Our sincere appreciation goes to Mr. Joseph Stockel who provided continuous encouragement in the development of this new simulation tool. The writers would also like to acknowledge the efforts of Dr. John Miller of JME Capacitor for his excellent work in measuring and modeling the supercapacitor and battery dynamic characteristics. We would also like to thank Mr. Jack Brill of Eagle-Picher Corp for his help in arranging the loan of a Nickel-Hydrogen cell to us for testing.

REFERENCES

1. D.T Gallagher and W.J. Billerbeck, System Level Analysis of Spacecraft Power Systems with Pspice", Proceedings of the 27th IECEC, Aug 1992, Vol 2, p2..57 (SAE paper no. 929313)

2. Cadence Orcad Pspice A/D User's Guide, 1998. http://www.orcad.com

3. J. Miller and S. Butler, "Electrochemical Capacitor Pulse Power Performance", Proc. 10th International Seminar on Double Layer Capacitors and Similar Devices, Deerfield Beach, Fla. Dec 4, 2000

4. Wilson, I.M., "Analog Behavioral Modeling using Pspice", Proc. Of the 32nd Midwest Symposium on Circuits and Systems, 1989, IEEE

5. Olson, Harry F. , "Dynamical Analogies", D. Van Nostrand Co, 1948

6. P.W. Tuinenga, Spice: A Guide to Circuit Simulation and Analysis Using Pspice" Prentice Hall 1988, ISBN: 0-13-158775-7

7. A.Vladimirescu, "The Spice Book" John Wiley and Sons 1994, ISBN: 0-471-60926-9

8. J.E. Moore, et al. "RADARSAT: Spacecraft Bus and Solar Array", Canadian Jour. Of Remote Sensing, Special Issue,Vol19, No4, Nov-Dec 1993 ISSN 0703-8992

....

Table 1. LIST OF AVAILABLE LIBRARY POWER UNIT MODELS

1. Simulation Utilities
· Orbitsim - simulates orbital illumination, eclipse, and solar array temperature
· LoadControl - switches vehicle loads on and off
· Pulser - pulses loads as specified by parameters

2. Solar Arrays
· Solar ArrayK47 - Array of Spectrolab K4.702 silicon cells with 13.3% efficiency
· Solar ArrayK67 - Array of Spectrolab K6.7 silicon cells with 13.5% efficiency
· Solar Array2J - Array of Spectrolab dual junction cells with 21.5% efficiency
· Solar Array3J - Array of Spectrolab triple junction cells with 24.7% efficiency
· Solar ArrayTEK2J - Array of TEKSTAR dual junction cells of 20.5% efficiency

3. Batteries
· Nickel-Cadmium Battery
· Nickel-Hydrogen Battery
· Lithium-Ion Battery
· Thin film Lead-Acid Battery

4. Energy Storage Wheel
· Wheel Unit Model

5. Main Bus Electronic units
· MainShuntLim - Main bus shunt voltage limiter type of regulation unit
· ArraySwitcher - Unit simulating solar array string switching for bus voltage limiting

6. Battery Chargers
· VTcurrlim - Charger for Ni-Cd or Ni-H2 with V/T and input current limits
· NiH2_Charger - Battery charger for Ni-H2 with pressure and current limits
· Li_battCharger - Battery charger for Li-Ion with V/T and input current limiting

7. Power Load Units
· Load_ConstPwr - Constant power load
· Load_ConstRes - Constant resistance load unit
· Load_ConstCurr - Constant current load
· Heater - Constant resistance load with temperature cutoff
· Buck_Reg - Buck regulator or DC-DC converter unit
· Boost_Reg - Boost regulator or DC-DC converter unit
· Buck-Boost - Buck- Boost regulator or DC-DC converter unit

8. Payload Units
· TWTA - Traveling Wave Tube Amplifier (including power conditioner)
· TR_module - Solid State Transmit/Receive module

9. Components

· MEPCO_CAP - typ. wet slug tantalum capacitor based on the 100Uf, 75V device
· ELNAcap - 2.5V ELNA supercapacitor based on the one Farad device
· PowerstorCap - 2.xV Powerstor based on the 10F device
· PansonicCap - 2.3V Pansonic based on the 50 F device
· MaxwellCap - 2.5V Maxwell based on the 100F device
· Supercapxx125 - Supercapacitor based on the Australian 125F R&D device
· Inductor - typical power inductor

Energy Density vs. Power Density

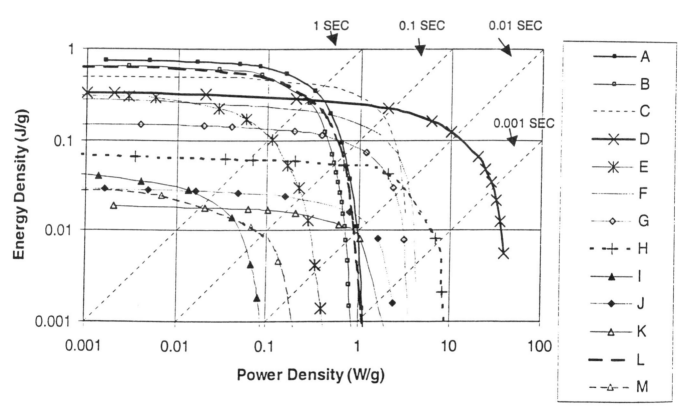

Figure 1: Pulse power Ragone plots for the evaluated capacitors.

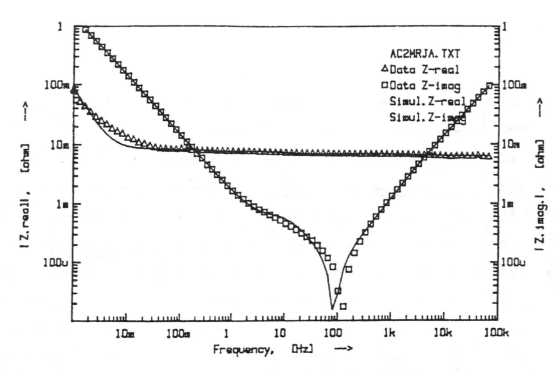

Figure 2: Example of the impedance data from one of the evaluated capacitors.

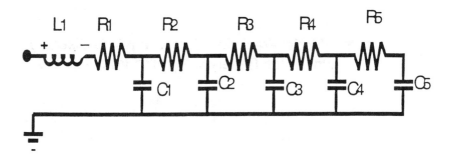

Figure 3: Equivalent circuit model for the capacitor in Figure 2.

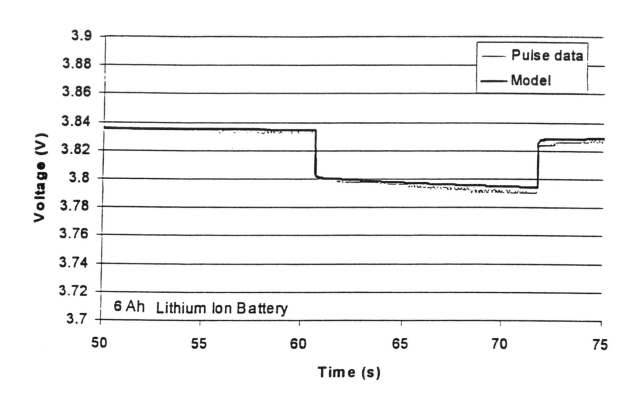

Figure 4 – Model Prediction Compared With Discharge for
Lithium Ion Battery with Pulse from 4 A to 8 A

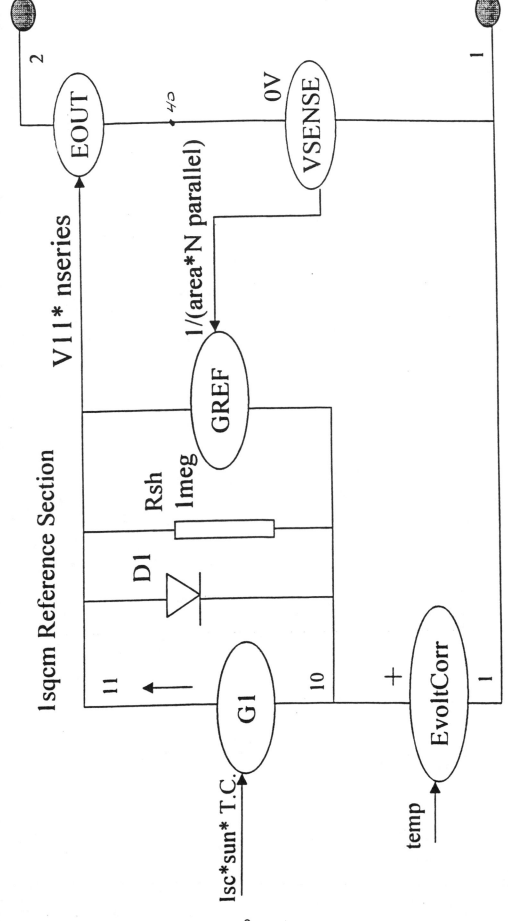

3 J SOLAR ARRAY CIRCUIT

Fig. 5 Solar Array Subcircuit Model

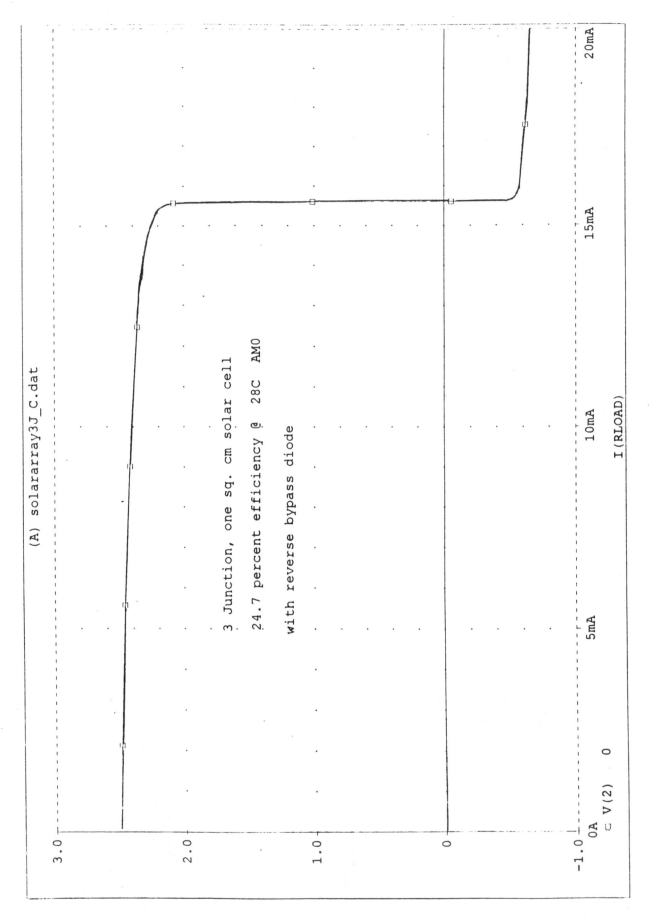

Fig. 6 Solar Array Output Curve

MAIN SHUNT LIMITER CIRCUIT

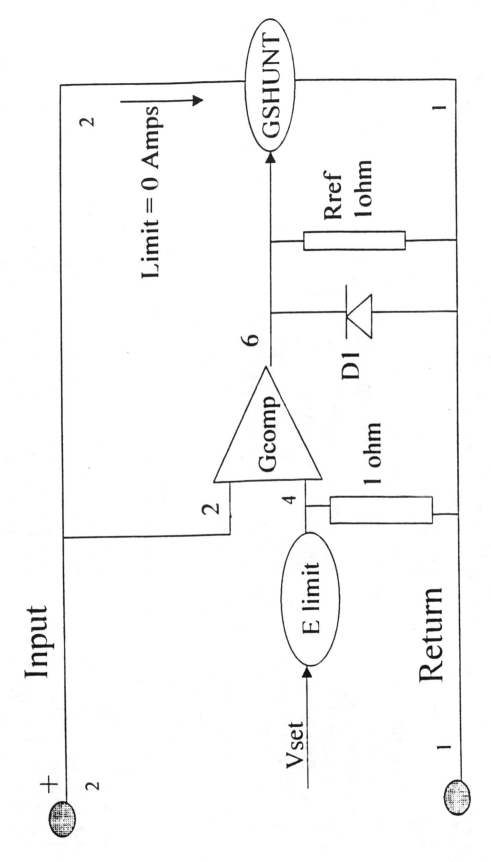

Fig. 7 Main Shunt Limiter Schematic

EXAMPLE #1 SATELLITE POWER SYSTEM

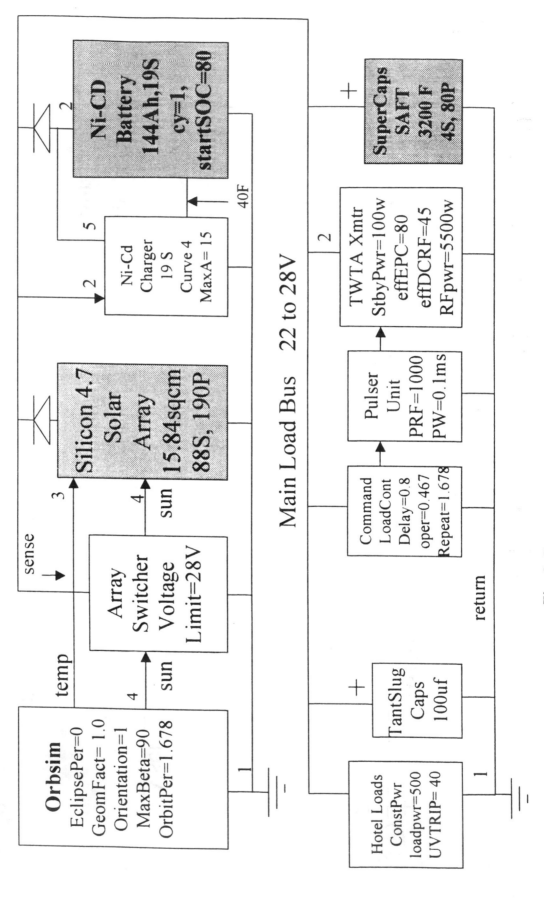

Fig. 8 Example #1 Power System Block Layout

12

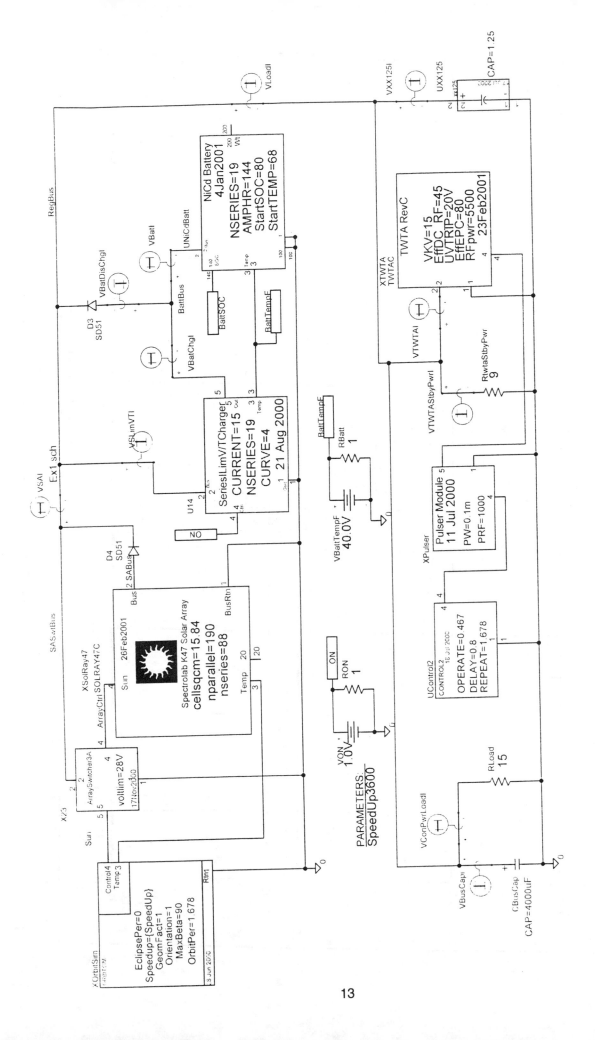

Fig. 9 Schematic #1 – prepared for analysis

13

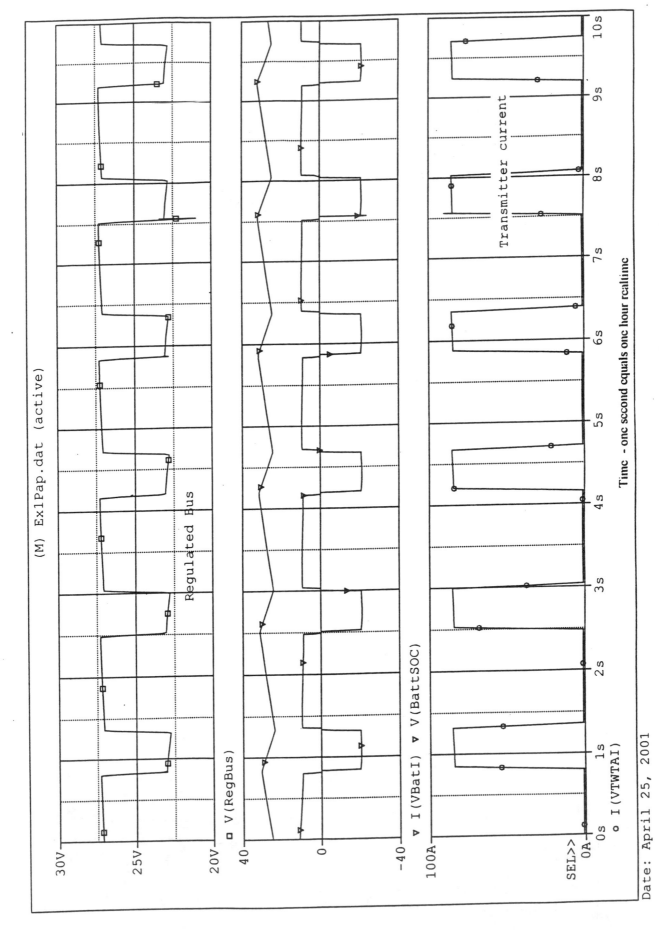

Fig. 10 Analysis results – Power System Schematic #1

Date: April 25, 2001

14

EXAMPLE #2 ADVANCED POWER SYSTEM

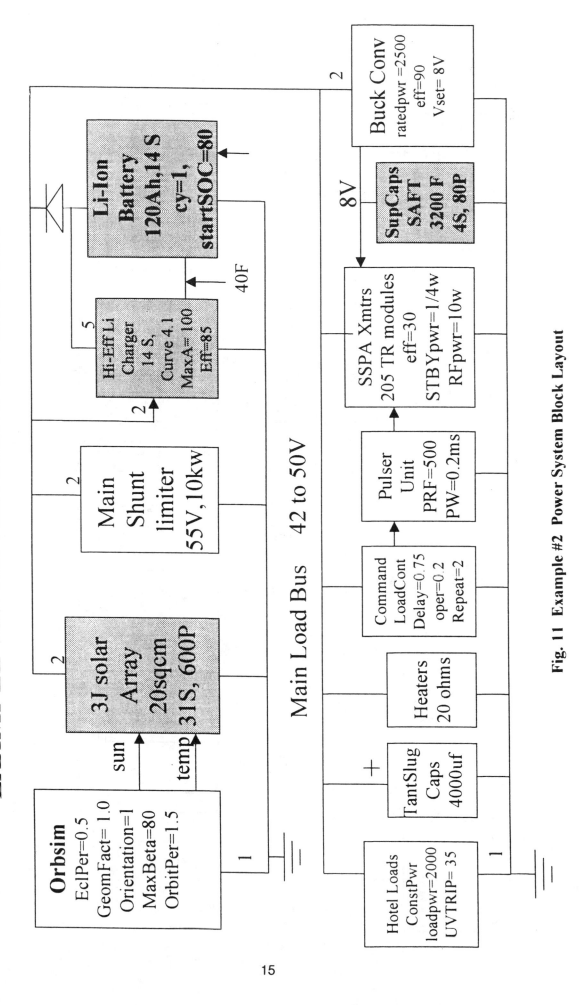

Fig. 11 Example #2 Power System Block Layout

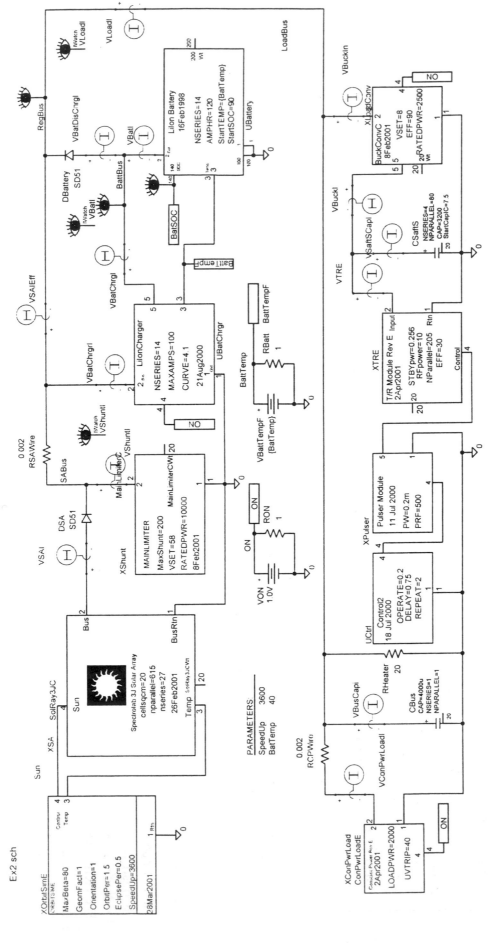

Fig. 12 Schematic #2 – prepared for analysis

16

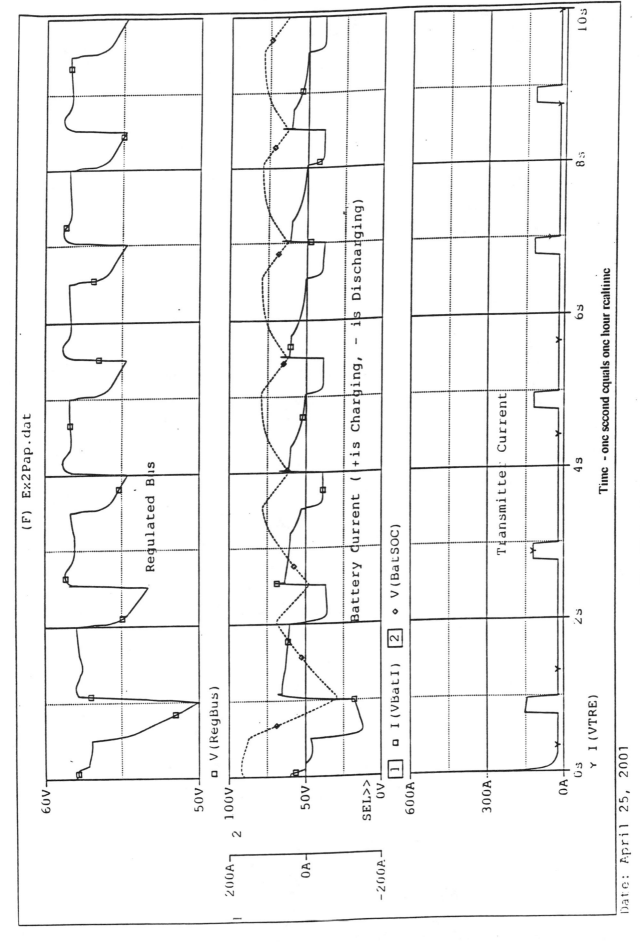

Fig. 13 Analysis results – Power System Schematic #2

Proceedings of IECEC:
36th Intersociety Energy Conversion Engineering Conference
July 29 - August 2, 2001 Savannah, Georgia

2001-AT-18

A NOVEL APPROACH TO SIMULATE POWER ELECTRONIC SYSTEMBY EMBEDDING MATLAB OBJECTS INTO SABER

Zhenhua Jiang and **Roger Dougal**
Department of Electrical Engineering,
University of South Carolina,
Columbia, SC 29208 USA
Email: jiang@engr.sc.edu

ABSTRACT

Matlab/Simlink is a powerful tool for designing control algorithms for power electronic systems. We describe here a method for inserting a matlab-based controller directly into a Saber circuit simulation, which then yields a means to prove the viability of a comprehensive power electronic system design. This approach combines the best aspects of these two powerful tools, making full use of the advantages of each. The major procedure is to convert the MATLAB model to a Saber model. We describe in detail the process for doing that, and then illustrate the process with an example. The results obtained by such mixed model simulation are compared to those obtained by an all-native Saber simulation; numerical results were the same, but the execution time was about 50% longer with the embedded MATLAB objects.

INTRODUCTION

Simulation of power electronic systems presents peculiar challenges due to the need for detailed modeling of both circuitry and control algorithms [1]. Saber is a powerful platform for analyzing electric circuits, while MATLAB is an excellent tool for analyzing control systems. It is often desired to make full use of the advantages of each of these.

Although a MATLAB command window can be invoked in the Saber environment, a Saber template cannot call a function in a MATLAB M-file. Therefore, Saber cannot directly refer to a MATLAB object. Fortunately, a MATLAB M-file can be compiled into a C function and a Saber template can call the foreign C function. This provides an approach to import matlab objects into Saber. A future extension of the work will lead to methods to import an entire Simulink model into Saber.

The process to import a MATLAB model into Saber model involves the following steps:
- Design the controller using MATLAB
- Use the Matlab code generator to produce a C language program

- Compile the C program into a dynamically linked library (DLL) file
- Describe a Saber template that interfaces with the dll file for the controller, and
- Design an icon for the controller in Saber and produce a new model for the controller inside the model library.

This paper is organized as follows. Section 1 introduces the method. Section 2 describes the method of communication between Saber and MATLAB. Details of the process to convert a MATLAB model to a Saber model are presented in Section 3. In Section 4 we illustrate application of the technique in a typical power electronic circuit, and then compare the results against those obtained using only native Saber models. Section 5 concludes the paper.

OVERVIEW OF COMMUNICATION BETWEEN MATLAB AND SABER

Saber is a simulator for electrical/electronic and multi-technology circuits and systems. It can be used at many levels, from design of integrated circuits through to design of systems. It handles analog, mixed-analog/digital, and mixed-technology devices. It is the first single-kernel simulator that allows simultaneous simulation of analog and digital devices, as well as simulation using non-electrical devices. Saber is widely used in the power electronics industry because it has many advantages such as a capability to model networks on a behavior level, good Differential Equation Solver, good Waveform Analysis Capability and a large library of models for electro-mechanical components. But it also has many disadvantages. For example, it has no front end for algorithmic modeling and no links to rapid prototyping or to operation with hardware in the simulation loop.

MATLAB/Simulink, on the other hand, is widely used for development of control algorithms. Its strong points include hardware independent functional modeling and fits into tool chains for rapid prototyping and hardware in the loop. It has a good front end for algorithmic modeling. However, MATLAB has a limited capability to model electrical networks and a limited Differential Equation Solver for electric networks. It is obvious, then, that power electronics simulation needs a combination of the capabilities of both Saber and MATLAB.

Analogy Inc. (now called Avant!) has distributed the SaberLink Analysis Interface, which integrates MATLAB software into the SaberDesigner environment. It allows users to apply Matlab's tools for numerical analysis (including MATLAB's visualization and graphing tools) to Saber simulation data [2, 3, 4]. Despite this efficient interface, it remains difficult to use the Matlab functions during a simulation -- such as one would do to simulate a power electronic system including its controller. Because a Saber template can neither communicate with MATLAB/Simulink models nor call the function in a MATLAB M-file straightforwardly, Saber cannot refer to a MATLAB object directly. Fortunately, a MATLAB M-file can be compiled into a C function and a Saber template can then call the foreign C function. This makes it possible to indirectly use Matlab functions within the Saber simulator. A schematic diagram of the desired hybrid simulation environment is shown in Fig. 1. The two-headed arrow means exchange of simulation data between MATLAB and Saber environments.

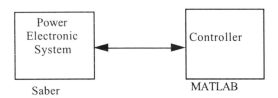

Fig.1 schematic diagram for simulation environment

THE APPROACH TO EMBED MATLAB INTO SABER

The goal of this process is to import MATLAB objects into the Saber simulation environment. Five steps are required, as shown in Fig. 2.

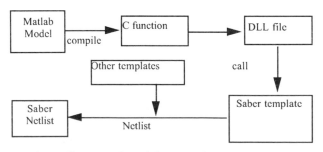

Fig. 2 diagram of model conversion process

The first step is to design the controller in MATLAB. Here, the input and output connections provide the bridge to exchange simulation data between MATLAB and Saber.

Implementation of the controller in MATLAB M-file is performed in the second step. The name of the function that represents the characteristics of the controller is strictly designed to an expected format. This is the MATLAB terminal of data exchange.

Thirdly, the MATLAB M-file must be exported to C code, and then compiled into a DLL file, which can be called by Saber. There are many restrictions on the compilation process. The MATLAB command *Mcc* invokes the MATLAB compiler using the syntax

mcc [-options] mfile1 [mfile2 ... mfileN] [C/C++file1 ... C/C++fileN

One or more MATLAB Compiler option flags can be specified.

If the C programming language is being used to create a foreign routine for use with MAST and the Saber simulator, the routine header must appear exactly as follows:

_declspec(dllexport) void _stdcall CROUTINE(double* inp, long*

ninp, long* ifl, long* nifl, double* out, long* nout, long* ofl, long* nofl, double* aundef, long* ier)

{

}

Note that the *_declspec* statement is important for Windows NT since it indicates that the routine is exported from the Dynamic Link Loader and can be found by the Saber simulator. When using, substitute the foreign routine name for CROUTINE. Note that the CROUTINE string must be entered in upper-case characters.

In the fourth step, a Saber/MAST template for the controller will be created. This template is, in fact, a MAST wrapper for the controller model. The Saber/MAST language makes it available to use foreign functions or foreign subroutines that are outside of all templates. The procedure for doing this includes the following:

- Declaring foreign subroutines in Saber template
- Calling foreign subroutines in Saber template
- Writing foreign subroutines. The foreign subroutines are written in MATLAB M-file. There are some requirements for the foreign routine. This has been explained in detail in the third step.
- Compiling foreign subroutines into a DLL file
- Making sure that the directory of the DLL file is in the SABER_DATA_PATH.

The last step provides an interface for Saber to call the MATLAB object. It requires definition of a symbol to represent the matlab functions, including the input and output terminators, which are of the type to exchange data.

The whole process of model conversion can be considered as the abstraction of model. All the components, which include generated C codes of MATLAB model, generated C code wrapper, differential solver, data logger and MATLAB data structure, are converted to a foreign model. The exported model is represented by two files: a MAST file and a dll file. The procedure of model abstraction is shown in Fig. 3. The entire procedure is involved in a batch file. What the user should do is to push the button, which can call the batch file to produce the dll file and Saber/MAST wrapper. There, however, are still some limitations in the procedure of model conversion. For example, the model conversion procedure is dependent on the manual interface. Some work is under way to improve this procedure.

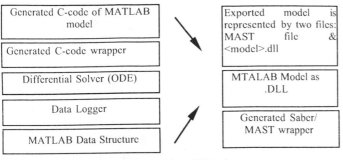

Fig. 3 Generated model blocks

CASE STUDY

We show here a case study that illustrates this technique. The example system consists of a circuit containing a voltage source, a buck converter, and a resistive load as shown in Fig. 4. V_{in} is the input voltage of the buck converter, that is to say,

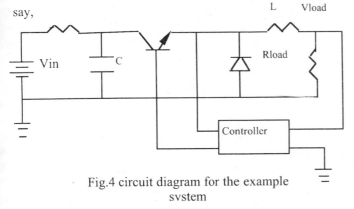

Fig.4 circuit diagram for the example system

the voltage of the voltage source. V_{Load} represents the voltage across two ends of the load. The parameters of the circuit are as follows: L=100mH, C=100 uF, Frequency=20.0 kHz, R_{Load} =10 ohms, V_{in} =30 volts, Simulation time=100 ms, Time step=1 us. The expected load voltage is 15 volts. At t=60 ms, the load resistance is changed to 12 ohms.

The input of the controller is the load voltage and the output is the voltage between the base and emitter terminals of the transistor, which controls the switching state of the transistor. The goal is to define in MATLAB a feedback function that measures the load voltage and controls the duty of the buck converter so that the load voltage is independent of the load resistance. For simplicity, we used a PID control strategy. We first wrote a MATLAB M-file that reads in the load voltage, performs the PID control calculations, and then writes

out the commanded value of switch duty. The MATLAB M-file for the controller is listed below. Note that *in(1)* is the value of the current load voltage while *in(2)* is the value of the load voltage before exactly a time step and *h* is the time step.

```
function mypidcontroller(in,nin,ifl,nifl,out,nout,
                ofl,nofl,aundef,ier)
{
    outi_memory=in(3);      % memory for integration
    h = in(4);              %time step
    Kp=in(5);
    Ki=in(6);
    Kd=in(7);       % coefficients of pid controller
    outp = in(1)*Kp;
    outi = outi_memory + in(1)*h*Ki;
    outd = (in(1)-in(2))*Kd/h;
    out(1) = outp+outi+outd; % output of the controller
    out(2) = outi;                 % to output the current
integration
}
```

Then we converted the M-file to a C function, and then called the C function in a Saber MAST template that represents the PID controller object in a Saber simulation. The controller template is designed to call the foreign MATLAB model. The Saber/MAST template for the controller is listed below.

```
template pid in1 in2 in3 out1 out2 = Kp, Ki, Kd
    input nu in1 in2 in3
    output nu out1 out2
    number Kp=1.0, Ki=1.0, Kd=1.0
    {
        val v work1,work2
        foreign mypidcontroller
        values{
        (work1,work2) =
                mypidcontroller(in1,in2,in3,1u,Kp,Ki,Kd)
        }
        equations {
        out1: out1 = work1
        out2:out2 = work2
    }
}
```

The embedded pid controller was used together with a delay model. The remaining parts of the circuit were placed into the circuit using the standard models from the Saber model library. The parameters for pid controller are as follows: Kp=1.0, Ki=0.1, Kd=0.01. The computed load voltage is shown in Fig. 5.

In order to verify the effectiveness of the embedded MATLAB object, the results were compared to a similar simulation that used a native Saber model for the PID controller. The same circuit and parameters were used.

Simulating the entire circuit in Saber gives the load voltage as shown in Fig. 6. From Fig.5 and 6, it can be seen that the

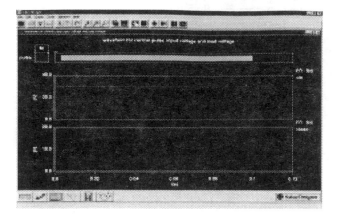

Fig.5 Load voltage with the presented method

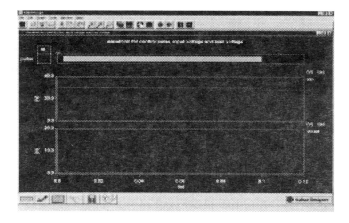

Fig.6 Load voltage in native saber simulation

simulation results with foreign MATLAB objects in Saber simulation are similar to those with all native Saber objects.

The total CPU execution time was 20.2 seconds when using all native Saber models and 31.4 seconds when using the embedded Matlab object.

CONCLUSIONS

There exists a reasonable routine for a Saber simulation to call a foreign MATLAB objects. The simulation results obtained using a foreign MATLAB object as the controller were similar to those obtained using a native Saber controller object. The execution time for simulations using embedded MATLAB code is longer than the time for the native saber function to complete the simulation.

ACKNOWLEDGEMENT

This work was supported by the Office of Naval Research under Grant N00014-00-1-0131.

REFERENCES

[1] R. Dougal, T. Lovett, A. Monti, E. Santi *A Multilanguage Environment for Interactive Simulation and Development of Controls for Power Electronics*, IEEE Power Electronics Specialists Conference, Vancouver, Canada, June 17-22,

[2] SaberLink™ Analysis Interface to MATLAB - Interactive analysis interface for circuit simulation, http://www.mathworks.com/products/connections/product_main.shtml?prod_id=16.

[3] SaberLink Analysis Interface to MATLAB, http://www.spsda.com.sg/analogy.htm

[4] SaberLink™ Analysis Interface to MATLAB, Seamless Integration into SaberDesigner environment, http://www.analogy.com/Products/DataSheets/SaberLink_Datasheet.pdf

Proceedings of IECEC'01
36th Intersociety Energy Conversion Engineering Conference
July 29-August 2, 2001, Savannah, Georgia

IECEC2001-AT-01

SIMULATION AND ANALYSIS OF A THREE-PHASE AUXILIARY RESONANT COMMUTATED POLE INVERTER FOR AN INDUCTION MOTOR DRIVE

Qinghong Yu
Dept. of Electrical & Computer Engineering
152 Broun Hall
Auburn University, AL 36849-5201
Email: qinghong@eng.auburn.edu
Phone:(334) 844-1840
Fax:(334) 844-1809

R. M. Nelms
Dept. of Electrical & Computer Engineering
200 Broun Hall
Auburn University, AL 36849-5201
Email: nelms@eng.auburn.edu
Phone:(334) 844-1830
Fax:(334) 844-1809

ABSTRACT

Two soft switching inverters for AC motor drives, based on auxiliary resonant commuted pole (ARCP) and auxiliary resonant snubber technique, are simulated and analyzed. Both inverters obtain low stress switching of all the switches, and additional control based on load current direction removes the excessive current stress on the primary switches and reduces the loss in auxiliary switches. This soft switching technique is used in a SVPWM driven inverter for a three-phase induction motor drive.

INTRODUCTION

Soft-switching techniques have been widely used in converters and inverters for reduced EMI and switching stresses on power devices [1]. The soft-switching condition is usually created by resonance in an L-C tank circuit, and challenges exist in many aspects. Resonance often creates excessive conduction losses in power devices, and the voltage and current stresses during resonance may require components of much higher rating. The resonant tank, subject to high current and voltage stresses, consumes power and can be difficult to design. Various load conditions may destroy the soft-switching condition in some topologies. Finally, the control complexity of soft-switching topologies tends to be higher and prevents implementation of PWM techniques [1].

The wide range of inductive load currents and various duty cycles of a PWM inverter make the application of soft-switching techniques completely different from that for a DC-DC converter. One soft-switching technique for a motor drive is the resonant DC link [2]. A resonant circuit controlled by auxiliary switches is inserted in the DC link of the inverter where the load current and the input voltage are nearly constant. The resonance in the DC link is controlled to happen before the switching of any switch in the inverter, and the voltage or current in the DC link resonates to zero so that the commutation in the inverter switches can be accomplished under zero voltage or zero current. However, the resonant DC-link topology also has the weakness of excessive current or voltage stress, and many attempts have been made to address this problem [3]. This is at the cost of difficulty in design, control, system complexity and reliability.

Another more recent soft-switching technique is the Auxiliary Resonant Commutated Pole (ARCP) inverter [4]. An LC resonant snubber circuit, triggered by auxiliary switches, is connected to each pole of the inverter to create the low stress switching condition. More auxiliary switches and passive components are used in ARCP inverters than resonant DC link inverters, but the over-current and voltage stress is reduced. The resonance occurs alternatively in several LC tanks and auxiliary switches, thus the average loss in each tank and power device is very small. Various topologies have been derived from the basic ARCP circuit with different numbers of auxiliary switches as well as complexities of control strategy [5]. The ARCP inverter and other soft-switching topologies in this family can be employed in motor drives using either SPWM (Sinusoidal Pulse Width Modulation) or SVPWM (Space Vector Pulse Width Modulation) control [6].

In our study of an electromechanical actuator powered by a bank of ultra-capacitors, two topologies derived from an ARCP cell and a full-bridge resonant snubber inverter (RSI) cell are considered and compared. The commutations are simulated in PSPICE and analyzed under different load conditions. According to the results of the simulation, an improved control strategy is proposed to remove excessive current stress on the primary switches and reduce the loss on auxiliary switches.

NOMENCLATURE

Vdc = Input DC voltage of the inverter
Iload = Load current in one phase of the inverter
Ir (t) = The resonant current in the inductor
Iboost = Ir(t) - Iload
ARCP: Auxiliary Resonant Commutated Pole
RSI: Resonant Snubber Inverter
SPWM: Sinusoidal Pulse Width Modulation
SVPWM: Space Vector Pulse Width Modulation

Inverter based on ARCP cells

Three-phase inverters for an induction motor drive based on ARCP cells and RSI cells are shown in Figure 1. Both inverters have the same number of switches except that the ARCP inverter requires capacitors to maintain a stable midpoint voltage. Switches SW1-6 are primary switches on the inverter, while SW7-12 are auxiliary switches.

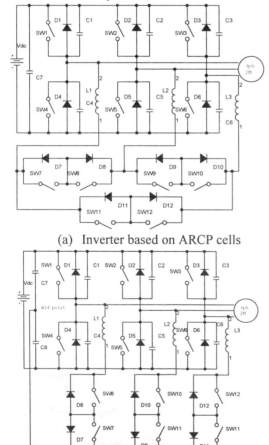

(a) Inverter based on ARCP cells

(b) Inverter based on Full-bridge RSI cells
Figure 1 Soft switching inverters being studied

To illustrate the operation of the soft switching inverter, it is necessary to break down the intricate three-phase inverter into basic single-phase cells, as shown in Figure 2. Since the load current of the induction motor is sinusoidal and of much lower frequency than the commutation, it can be approximated as a DC current in the PSPICE simulation of the inverter commutation. By adjusting the value of this DC current source, the commutation behavior can be simulated as the load current varies sinusoidally during its period.

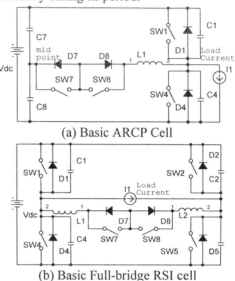

(a) Basic ARCP Cell

(b) Basic Full-bridge RSI cell
Figure 2: Basic Cells of the soft switching inverter

The operation of both inverter cells is the same and the commutation process of basic ARCP cell is illustrated in Figure 3. The circuit operation is divided into by 5 modes.

Mode 0 (t0 to t1)
This is the moment before the commutation. SW1 is off, SW4 is on, load current is leaving the pole through the freewheeling diode of SW4, D4. Auxiliary switches SW7 and SW8 are off.

Mode 1(t1 to t2)
At the moment t1, SW7 is turned on. SW4 is still on and a resonant current begins to build up in L1 and flow through SW7, D8 and SW4. This current increases linearly until it reaches the load current at t2. At this point, there is no current flowing in SW4 or D4.

Mode 2 (t2 to t3)
The resonant current continues to increase after it exceeds the load current. After the gate signal of SW4 falls to zero, D4 remains forward biased, providing a path for the resonant current. At t3, both SW4 and D4 are turned off, but the resonant current in the L1, Ir, exceeds Iload and Iboost [2] is positive.

24

For the existence of constant Iload, Iboost is the portion of resonant current that charges and discharges C1 and C4 after t3 and creates zero voltage across SW1. The value of Iboost is controlled by the gating signal of SW7 and SW4. When both SW7 and SW4 are on (t1 to t3), the resonant current Ir increases linearly at approximate rate of ½ Vdc/L1. The current path for Ir in this mode is SW7-D8-SW4 (or D4).

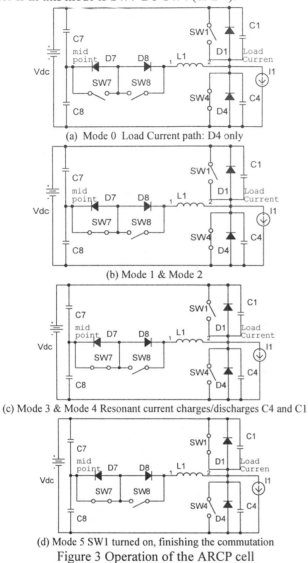

(a) Mode 0 Load Current path: D4 only

(b) Mode 1 & Mode 2

(c) Mode 3 & Mode 4 Resonant current charges/discharges C4 and C1

(d) Mode 5 SW1 turned on, finishing the commutation

Figure 3 Operation of the ARCP cell

Mode 3(t3 to t4)

At t3, Iboost begins to charge and discharge the capacitors C1, C4 and the nonlinear capacitance of the MOSFET. The nonlinear capacitance on the MOSFET makes the current trace not strictly sinusoidal at the start of this phase; with the increasing of voltage on SW4, linear capacitors C1 and C4 dominate the resonance with L1 and the current trace becomes sinusoidal. The resonant current reaches its peak value when the voltage on C4 equals the midpoint voltage, then it swings back. This is a typical resonant cycle and the voltage on C4 increases steadily until it reaches the point when the voltage

across C4 is Vdc and voltage on C1 is zero. The path for the resonant current in this mode is SW7-D8-L1-C1//C4.

Mode 4(t4 to t5)

At t4, the voltage on SW1 and C1 is clamped to zero, and Ir still exceeds the load current and part of Ir is flowing through D1, returning energy to the DC source. In this mode, Ir is decreasing linearly due to the voltage of ½ Vdc on L1. At the t5, SW1 is turned on with zero voltage stress, and this starts mode 5.

Mode 5(t5 to t6)

The resonant current continues to decrease; when it is less than the load current, SW1 starts to provide the load current. Now the load current commutates from D4 to SW1. The resonant current keeps decreasing, and it is blocked by the diode D8 at t6. After t6, the auxiliary switch SW7 is turned off under zero current condition.

The corresponding schematic for PSPICE simulation and waveforms are shown in Figure 4, where the IRFP150 MOSFETs are used as switches. The waveforms of the simulation are provided in Figure 5 for a load current of 6A. It is shown that all the switches operate under zero current or voltage condition, and soft switching does not increase the stress on the main switches of the inverter. The voltage on the auxiliary switch is ½Vdc, but the resonant current peak value can be higher than the load current [4]. Since the auxiliary switches conduct at a very low duty cycle, the current stress and the resulting conduction loss are not significant.

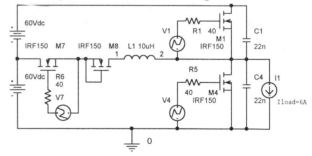

Figure 4 Schematic for Pspice simulation

A parasitic high frequency current oscillation can be observed in Figure 5 (a) and (c) after SW7 is turned off, which is caused by the reverse recovery current of D8 and parasitic capacitance of SW7. This does not count for a major loss or stress but can be avoided by proper selection and application of a fast recovery diode instead of using the body diode of the MOSFET.

The value of the Iload ranges from 0 to a peak value. A soft-switching inverter should always be able to operate with low switching stress. In this simulation, a sweep of the load current from 0 to 8A yields the waveforms in Figure 6. Only

25

the waveform on SW1, which is being turned on, is shown, because the others are not so critical to evaluate the performance of soft switching.

The simulation waveform in Figure 6 shows that when the load is 0A, the voltage on SW1 is clamped to zero very well when SW1 is turning on. But when the load current is 8A, at t5, the voltage on SW1 does not swing to zero, though close to zero. There is also a small current peak in SW1, which is caused by the charge and discharge of C1 and C4, because C1 is not yet discharged to zero and C4 is not charged to Vdc. So when the load is 8A, the circuit operates on the boundary of soft switching. This is because the Iboost is enough for a load current of 0, but marginally enough when the load current is 8A.

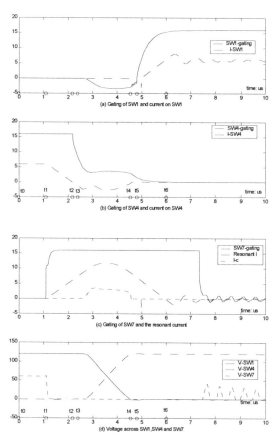

Figure 5 Simulation waveforms with Pspice

It is in a completely different situation when the load current reverses its direction. In this case, the auxiliary resonance will unnecessarily create excess current stress on SW4 as well as SW1. The reversed load current will be flowing into SW4 before the commutation. When SW4 is turned off, the load current will discharge C1 and freewheel through D1, and the voltage on SW1 is clamped to zero and SW1 can turn on with a   low stress. If resonance is created, the resonant current also flows into SW4, adding extra current stress to SW4

when it is turned off. The resonant current also produces more loss on the D1 after SW1 is turned off and on the auxiliary switches too. The simulation waveforms, with and without the resonant current, for a load current of –6A, are show in Figure 7 for comparison. It is very clear that in this case when the load current is entering the pole, the resonance should not be triggered and no auxiliary switch should be turned on.

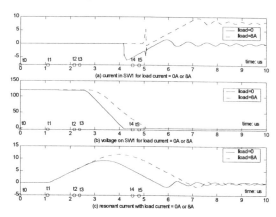

Figure 6 Waveforms on SW1 when Load current is 0A or 8A

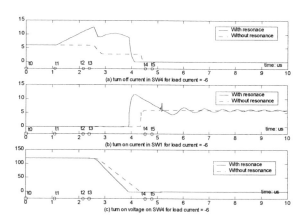

Figure 7 Switching waveforms with negative load current

In all these simulations, only SW7 is turned on and resonance is unidirectional for the commutation from SW4 to SW1. For the case when the commutation is from SW1 to SW4, it is SW8 that must be turned on to create the auxiliary resonance. The control of the auxiliary switch to create resonance is summarized in Table 1. Notice in Table 1, when the load current is "near zero", that one of the auxiliary switches is always turned on, because the load current, even if it helps to create zero voltage for the switch turn on, is not large enough to discharge the capacitor. So resonant current is always needed in this situation.

Table 1 Switching condition of the auxiliary switches

Commutation		Load current	Auxiliary switches	
From	To		SW7	SW8
SW1	SW4	Leaving the pole	Off	Off
SW1	SW4	Enter the pole	Off	On
SW1	SW4	Near zero	Off	On
SW4	SW1	Leaving the pole	On	Off
SW4	SW1	Enter the pole	Off	Off
SW4	SW1	Near zero	On	Off

The control of the gating signal is straightforward. For given resonant tank parameters, Vdc and the range of Iload, the commutation behavior can be predicted. The range of Iboost can be controlled by the interval of t1 to t3 to be large enough to create the zero current condition. A low cost current sensor is enough to detect the low frequency load current direction to judge if the auxiliary switch should be turned on.

Inverter based on Full-bridge RSI cells

The full-bridge RSI cell based soft-switching inverter is very similar to the ARCP cell based inverter. They have the same number of switches and similar control signals. The only difference is another pole takes the position of the capacitors, so there is no "mid point" with a fixed voltage of ½Vdc. When the commutation occurs, more than one pole is involved in the resonance. The simulation schematic of a single-phase inverter is shown in Figure 8. Different from the ARCP cell based inverter, the three-phase inverter based on full-bridge RSI cells cannot be broken down into basic cells, because all three inverter poles are involved for each commutation. The operation principle can be illustrated in the simplified single-phase version.

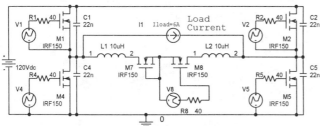

Figure 8 Pspice simulation schematic for RSI cell based inverter

The commutation of the current from SW2 and SW4 to SW1 and SW5 is illustrated in Figure 9. Auxiliary resonance is created by SW8 and the procedure is also explained in Figure 9, and the corresponding waveform is shown in Figure 10.

The simulation shows waveforms similar to the ARCP inverter, and the only difference is that the voltage on the auxiliary switch is now Vdc. This causes more loss, stress and higher parasitic oscillation on the auxiliary switches, but the advantage is that the Vdc voltage is fixed, while in the ARCP inverter the mid point voltage may drift from ½ Vdc when the

input capacitors are not perfectly balanced or large enough. A drifting mid point voltage will change the increasing rate of the resonant current. If the resonant current is not a feedback to the timing control circuit, the zero voltage turn off condition in the ARCP inverter will be undermined. Obvious a timing control with feedback of the high frequency resonant current is very complicated. For an inverter based on full-bridge RSI cells, the fixed Vdc is in the place of the mid point voltage, so it does not have the weakness of ARCP inverters, which typically need bulky capacitors to maintain the midpoint voltage at ½ Vdc.

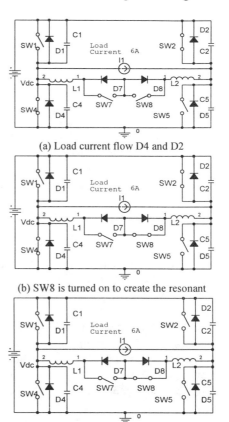

(a) Load current flow D4 and D2

(b) SW8 is turned on to create the resonant

(c) Resonant current exceeds the load current, charge and discharge C1 and C5

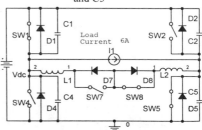

(d) While voltage on SW1 and SW5 are clamped to zero, SW1 and SW5 are turned on with low stress.

Figure 9 The Operation of commutation on RSI cell

The three-phase auxiliary resonant snubber inverter is an extension of the single-phase version. For an example, as in Figure 1, for the same commutation of the load current from

SW4 to SW1, both SW8 and SW12 can be turned on to create resonance so as to turn SW1 on with zero voltage stress.

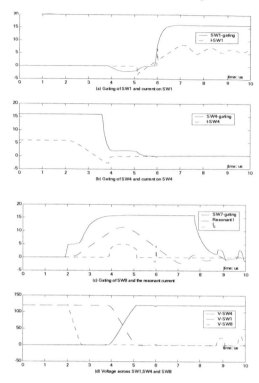

Figure 10 The simulation waveforms of Single phase Resonant Snubber Inverter

Conclusions

Both the ARCP and resonant snubber based soft-switching inverters are capable of operation with low switching stress on all of the switches. The primary switches will turn on with zero voltage, and the capacitors across the switches make the turn off under zero voltage condition, too. The auxiliary switches switch under zero current condition.

There are no excessive stresses on the primary switch, but the peak current in the auxiliary switches is higher than the load current. However, since the auxiliary switches conduct at very low duty cycle, the loss is very small for a single power device. Therefore, the auxiliary switches have a lower power rating.

The control of both soft-switching inverters is the same and low cost current sensors can be used in the control to detect the current direction. The control of the auxiliary switches can be accomplished according to the timing of the gating signal of the primary switches. The primary switches can actually operate independently from the auxiliary switches, and this soft-switching technique can be used on inverters controlled by SPWM or SVPWM for high performance AC motor drivers.

The additional current sensors detect only the direction of the load current and enable or disable auxiliary switches

accordingly. This simple and low cost improvement of control removes the excessive current on the primary switches and one half of the loss on auxiliary switches.

The ARCP inverter has one half of DC input voltage on the auxiliary switches, but the stability of the mid point voltage requires bulky input capacitors. But in the application for an electromechanical actuator powered by a bank of ultra-capacitors, the mid point voltage stability is well maintained.

ACKNOWLEDGMENTS

This work was supported by the Electromechanical Systems Laboratory. The authors would like to take this chance to express our gratefulness to our colleague, Mr. Bill Dillard, for his help in the study with his expertise in PSPICE simulation and his insightful discussions throughout the course of this work.

REFERENCES

[1] Bellar, M.D.; Wu, T.S.; Tchamdjou, A.; Mahdavi, J.; Ehsani, M., 1998, "A Review of Soft-Switched DC-AC Converters", IEEE Transactions on Industry Applications, **34,** Issue: 4, July-Aug. 1998 Page(s): 847 –860.

[2] Divan, D.M, "The resonant DC link converter-a new concept in static power conversion", 1989, IEEE Transactions on Industry Applications, **25,** Issue: 2, March-April 1989 Page(s): 317 -325.

[3] Cho, J.G.; Kim, H.S.; Cho, G.H.; "Novel soft switching PWM converter using a new parallel resonant DC-link", 1991, Power Electronics Specialists Conference,. PESC '91 Record., 22nd Annual IEEE , 1991 Page(s): 241 –247.

[4] R.W. De Doncker and J.P. Lyons,1990, "The Auxiliary Resonant commuted Pole converter", Industry Applications Society Annual Meeting, 1990, Conference Record of the 1990 IEEE, **2,** Page(s): 1228 –1235.

[5] Jih-Sheng Lai, "Fundamentals of a new family of auxiliary resonant snubber inverters", 1997, IECON 97, 23rd International Conference on Industrial Electronics, Control and Instrumentation, **2,** 1997 Page(s): 645 –650.

[6] Jih-Sheng Lai, "Resonant snubber-based soft-switching inverters for electric propulsion drives", 1997, IEEE Transactions on Industrial Electronics, **44,** Issue: 1 , Feb. 1997 Page(s): 71 –80.

IECEC2001-AT-67

INTEGRATED ELECTRO-HYDRAULIC SYSTEM MODELING AND ANALYSIS

S. Johnny Fu, Boeing Phantom Works
Seattle, Washington

Mark Liffring, Boeing Phantom Works
Seattle, Washington

Ishaque S Mehdi, Boeing Phantom Works
Seattle, Washington

ABSTRACT

The approach develop the model of an integrated Electro-Hydraulic System is presented in this paper. In particular, the development of the EASY5 Electric Power Distribution and Control (EPD&C) models, an integral part of the Space Shuttle Electric Auxiliary Power Unit (EAPU) is discussed. EASY5 is a simulation and modeling tool developed and marketed by the Boeing Company. Preliminary modeling and analysis results, model interface issues, EPD&C interactions with the power sources and EPD&C interaction with electro-hydraulic load are discussed. This approach and the model developed for EAPU can be generalized and adapted for the simulation and analysis of other aerospace and commercial electro-mechanical systems.

INTRODUCTION

Extensive research effort has been devoted to the analytical and numerical modeling of components and subsystems in electrical/electronic, mechanical and thermal-hydraulic disciplines and several commercial off-the-shelf simulation packages are available to facilitate these tedious modeling tasks. However, integrating the models from different technical disciplines as an integrated end-to-end system model still presents a major challenge due to the differences in time constant, coupling at model interfaces and the numerical integration methods unique to each technical domain. This paper describes the analytical techniques of modeling integrated electro-hydraulic systems and provides numerical examples of simulation results obtained from the models developed in EASY5[1] software tool. The End-to-End electro-hydraulic system models are developed for the purpose of analyzing the performance and the dynamic response of Space Shuttle Electric Auxiliary Power Unit under development. The integrated model encompasses batteries, EPD&C, motor drive, brushless DC motor, hydraulic pump and associated valves and fluid volumes. Taking advantage of the hierarchical modeling capability in EASY5, related functional component models are grouped as a submodel to accommodate quick model build-up, checkout, validation and updates. Stand-alone submodels representing subsets of the integrated system are also useful for the detailed analyses of individual subsystems in the overall system. The complexity and fidelity of these models represent the first undertaking in industry.

MODEL DEVELOPEMNT

The battery and EPD&C models provide power source to the elctro-hydraulic loads and respond to the load variations. Battery cells can be modeled either by the electrochemical reactions, such as the modeling approach by Dayle[2], Fuller[2] and Newman[2], or by the equivalent circuits representing the electrical characteristics. The equivalent-circuit of a lithium-ion battery cell is modeled as a voltage source in series with the equivalent source impedance shown in Figure 1. The voltage source is the cell no-load output voltage as a function of battery cell state of charge (SOC). The cell resistance, capacitance and induction in Figure 1 are modeled as functions of cell temperature, SOC and aging factors.

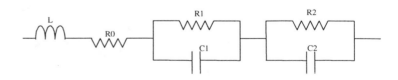

Figure 1. Equivalent-Circuit Battery Cell Model

EPD&C models the electrical characteristics of power distribution and protection hardware including bus bars, cables, connectors, contactors and fueses, and the power control and switching algorithms.

The mechanical load presented to EPD&C is modeled by metering valves that control the flow area of the hydraulic system, resulting in system flow and pressure, which command the pump displacement. The pump torque is coupled to the motor load. The motor isdriven by the voltage output of EPD&C.

Due to the physical and numerical complexity of the integrated model and the flexibility and expandability required by this modeling task, EASY5 is chosen as the software tool. EASY5 is a family of commercial software tools used to model, simulate and analyze dynamic systems that may contain hydraulic, pneumatic, mechanical, thermal, electrical and digital sub-systems. EASY5 also has a complete set of user-friendly modeling, analysis and design features for trade studies, stability evaluation and sensitivity analysis. The block diagram in Figure 2 shows the toolkits and libraries availabel in EASY5. The top-level schematic diagram of the integrated end-to-end model is shown in Figure 3.

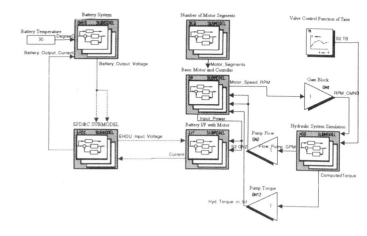

Figure 3. EASY5 Electro-Mechanical Model

EXAMPLES OF SIMULATION AND ANALYSIS

The capability to analyze the source impedance of EPD&C is essential in developing the power quality requirements for electro-mechanical loads and the verification of the end-to-end system stability. An example of the calculated source impedance at the output of the EPD&C submodel in Figure 3 is shown in Figure 4. The EASY5 model can generate the source impedance including battery and EPD&C, with the effects of temperature and battery state-of-charge included in the impedance calculations.

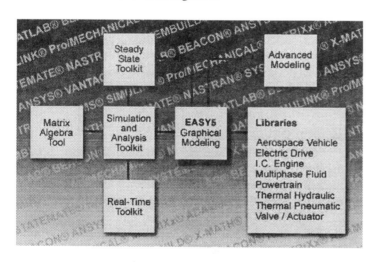

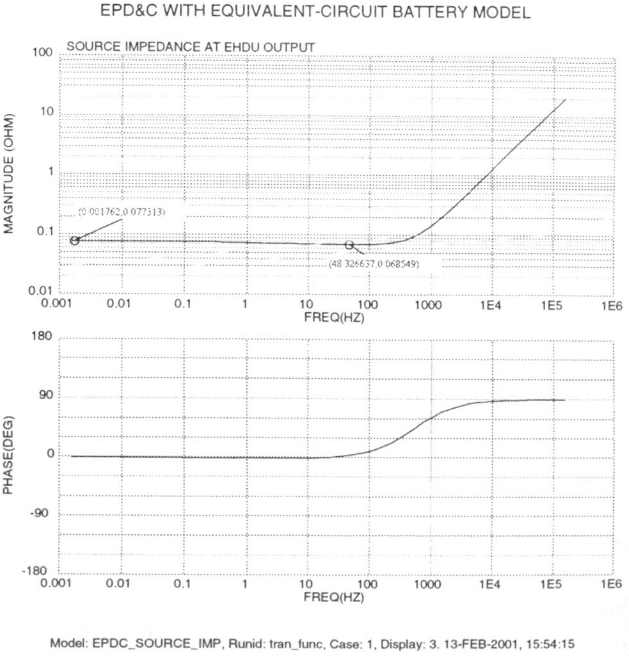

Figure 2. EASY5® Simulation Tool

Model: EPDC_SOURCE_IMP, Runid: tran_func, Case: 1, Display: 3. 13-FEB-2001, 15:54:15

Figure 4. Source Impedance

The capability of the model to evaluate power quality is demonstrated by the example in Figure 5, which shows the voltage transient at three interfaces in the battery and EPD&C in response to hydraulic load fluctuations.

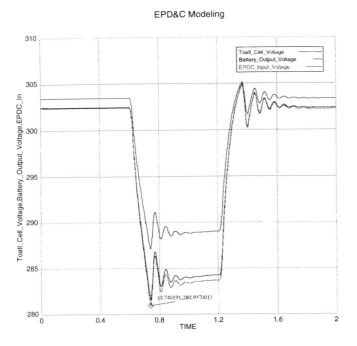

Figure 5. **Transient Voltage Response**

The model is also used to analyze transient and switching conditions of the eletro-hydraulic system. Figure 6 shows the transient current response during power source switching. Power source is switched from battery 1 to battery 2 that higher output voltage (higher SOC) than battery 1, at time =50 ms. The power source ransition from battery 1 to battery 2 takes 50 ms to complete. The results in Figure 6 indicate that battery 1 is charged by battery 2 until the switch to battery 1 is disconnected from EPD&C at time = 100 ms.

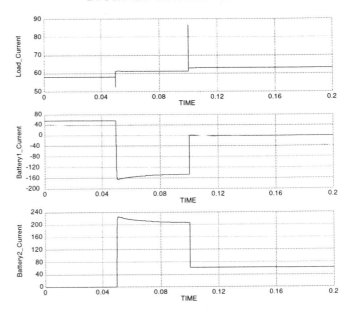

Model: Battery_SWOV, Runid: simulation, Case: 1, Display: 2. 20-FEB-2001, 11:55:07

Figure 6. **Transient Current Under Battery Switching**

CONCLUSION

The approach to model and simulate a complex end-to-end electro-mechanical system is presented in this paper. Detailed modeling the integrated end-to-end system presents significantly more challenge than the conventional approach of modeling components in one specific technical discipline. The comprehensive model using EASY5 component models has demonstrated the unique capabilities of EASY5 in creating large-scale end-to-end system model for simulation and analysis.

REFERENCES

1. Tollefson, Joel, EASY5 Reference Manual, 2000 The Boeing Company.
2. Doyle, M., Fuller, T. F., Newman, J., Modeling of Galvanostatic Charge and Discharge of the Lithium/Polymer/Insertion Cell, J. Electrochem. Soc., Vol. 140, No.6, June 1993.

2001-AT-07

SIMULATION OF FLYWHEEL ENERGY STORAGE SYSTEM CONTROLS

Long V. Truong, Frederick J. Wolff, and Narayan V. Dravid
NASA John H. Glenn Research Center, Cleveland, Ohio

ABSTRACT

This paper presents the progress made in the controller design and operation of a Flywheel Energy Storage System [1]. The switching logic for the converter bridge circuit has been redefined to reduce line current harmonics, even at the highest operating speed of the permanent magnet motor/generator. An electromechanical machine model is utilized to simulate charge/discharge operation of the inertial energy in the flywheel. Controlling the magnitude of phase currents regulates the rate of charge/discharge. Simulation results show the improvements made.

INTRODUCTION

A Flywheel Energy Storage System is under active consideration as a replacement for the traditional Electrochemical Battery System in Spacecraft electrical power systems. This is because the Flywheel system is expected to improve, both the depth of discharge and working life, by a factor of 3 compared to its Battery counterpart [2]. While flywheels have always been used in Spacecraft Navigation and Guidance Systems, their use for energy storage is new. However, the two functions can easily be combined into a single system. The NASA John H. Glenn Research Center [3], in a cooperative activity with industry and academia, have spearheaded this developmental effort which will culminate in an experiment where one battery pack on the International Space Station (ISS) will be replaced by a flywheel unit. Simulation of the flywheel system operation in support of this effort continues.

In the previously reported work [4], the charging/discharging process between the electrical machine and the dc bus, via the converter, was simulated to show power transfer that takes place in the respective modes. However, the electrical machine was a circuit element and not an electromechanical device. Thus, inertial energy transfer could not be simulated as a continuous function of machine speed. Also, the harmonic content of the motor phase currents was much too high to be easily corrected by filters alone.

This paper addresses both of the above problems. Using an electromechanical machine model, inertial energy storage and transfer is simulated as a function of rotational speed. Similarly, the converter switching logic has been redefined to substantially reduce phase current harmonic content to a manageable level.

SYSTEM DESCRIPTION

The simulated Flywheel Energy Storage System (Fig. 1) consists of a flywheel, shaft-coupled to a permanent magnet, 3-phase, synchronous motor/generator unit, powered by a dc source through a 3-phase, bi-directional, half-bridge converter. In the charge mode, energy is transferred from the dc source to the flywheel by increasing the flywheel rotational speed. The reverse operation takes place during discharge mode. Under the current design, the flywheel operating speed will be between 20,000(min) and 60,000(max) RPM. Since the inertial energy stored in a flywheel varies as the square of its RPM, it can discharge 90 percent of its maximum stored energy from maximum to minimum speed limits. The flywheel rotational inertia constant selection is based on energy storage requirements. Reliability and safety considerations govern the maximum speed limit while further lowering the minimum limit does not yield any significant depth of discharge. Friction and winding losses are assumed to be insignificant.

The voltage magnitude and frequency at the machine terminals are directly proportional to the speed of rotation. Therefore, the control of the converter operation determines the motoring (charge) or generating (discharge) action as desired. The phase current magnitude and phase angle determine the amount and direction of power, respectively. This is achieved by pulse width modulated (PWM) switching of the converter switches. The switching action is also affected by the rotor position (actually the position of the magnet pole axis with respect to the phase A winding axis). Such a control is known as 'closed loop control'. A detailed explanation of the controller will be given later.

Nominal filter circuits are connected on the dc side of the converter. These have not yet been tuned to further reduce harmonic contents of the dc bus voltage or current. A series connected inductive filter is utilized to reduce harmonic content of the phase windings.

THE CONTROLLER

The switch controller is the main operating element of this system (Fig. 1). Its output controls the pulse width modulated operation of the converter switches to meet the requirements. PWM switch operation is common knowledge and, therefore, will not be described here. Following is brief description of the requirements that govern the switching action.

Mode Command

The system can be operated under any one of the three modes of Charge, Discharge, and Idle. Presently, the user sets these as a function of time. However, they can be set based upon some other operating criteria. There is also a reset mode to start the simulation.

Switch Current Limit

The absolute value of instantaneous current through each switch is monitored to detect over-current conditions. If one exists, the information is fed to the controller, which, then turns off the PWM signal for that switch. In the simulation, it remains so until the start of the next time step.

Motor Current command

A 3-phase, sinusoidal current is generated as reference for limiting and regulating the phase winding current. The limiting is for protection while regulating winding current to be sinusoidal eliminates substantial amount of harmonic contents. The reference current generator is fed a magnitude and frequency value. Presently, the user sets the magnitude although it could be generated as a function of some system operating

condition. The frequency is computed as a time derivative of sensed rotor position angle.

Sixty-Degree Control

The bi-directional, dc to 3-phase, variable frequency, ac converter has six switches to control its operation. At any time, switches may be ON, OFF, or PWM switched. Since the three windings are connected together in a star point, current in one winding returns through the other two windings. Hence it is necessary that the switches connecting the windings to the dc bus be properly operated. Figure 2 shows the back EMF waveforms for the three phases. A1 denotes the switch connecting phase A to the positive DC bus while A2 is for negative DC bus. "B1, B2" and "C1, C2" are for phases B and C, respectively. The 180 deg. span is subdivided into three 60 deg. spans as shown. In the 0 - 60 deg. span, the phase A voltage is rising towards its positive peak, the phase B voltage is around its negative peak, while phase C voltage is falling away from its positive peak. This would suggest that phases A and C should be connected to the positive dc bus while phase B should be connected to the negative dc bus. Also, to maintain sine wave shape for winding currents, while the phase B switch remains ON, phases A and C switches are PWM controlled. All other switches are OFF. The switch operations are similarly coordinated during subsequent 60 deg. intervals. Figure 3 shows operation of the same switches. The PWM mode is shown as series of pulses while a single, wide pulse denotes ON mode. There are no pulses during OFF mode.

Appropriate logic signals from the above-described operating requirements become input to the Controller, which, in turn, produces commands to operate the converter switches.

MOTOR MODEL

The motor model is Permanent Magnet Synchronous Motor (PMSM) from Saber [5], the same simulation tool used in the previous work. Model data are given below. Motor data are based on a design by Ashman Technologies for a test machine at NASA Glenn.

Motor model (Saber's PMSM library):

Self -inductance of winding, micro Henry	16.7
Back EMF constant per pole pair, V/(radians/s)	0.00828
Torque constant per pole pair, Newton-Meters/A	0.00828
Motor inertia constant, kg-m^2	0.0001
Winding resistance per phase, milli Ohm	14.5
Number of poles	2
First back EMF Fourier coefficient	1

Note: The choice of motor inertia constant value was based on reasonable simulation run time.

SIMULATION PROCEDURE

The basic simulation performed consists of charging the flywheel to a reasonably high speed, coasting at that speed, and then discharging the flywheel to a lower speed.

Operating Modes:

Charge period, seconds _____ 0 - 0.4
Idle period, seconds _____ 0.401 - 0.475
Discharge period, seconds _____ 0.480 - 1

Motor current limits for this simulation:

Charge mode, Amperes _____ 100
Discharge mode, Amperes _____ 100, 75, and 50
DC (bus) power supply, Volts _____ 160
Converter Switch Current limit, Amperes _____ 150
PWM switching frequency, Hertz _____ 8000

Note: The choice of PWM switching frequency is based on reasonable simulation run time.

The simulation run starts at steady state with zero flywheel speed. The inverter circuit outputs voltage and frequency to the motor during the charging process. Motor winding current limit is set at 100 amps. Initially, this constitutes charging at constant motor current. Later, as the motor back EMF builds up, the motor current drops down from the limit value. At a selected point in time, the system is signaled into the IDLE mode. This turns OFF all the switches and power transfer becomes zero. The motor maintains whatever speed it has achieved. This is possible only when the machine back EMF cannot discharge though the switch body diode into the dc bus. At a later time, the circuit is pulled off the IDLE mode while it was already in the DISCHARGE mode; the two modes being independently controlled. Now the machine, acting as generator, begins discharging energy from the flywheel into the dc bus at a constant current that is determined by the winding current limit set by the user. Changing the current limit value can change the discharge power. Some other criteria such as 'constant power' could be used for charging or discharging.

SIMULATION RESULTS

Figure 4 shows the results from the above described scenario. Fig 4(a) denotes the Charge/Discharge and Idle commands. Fig 4(b) denotes the winding current limits set by the user, there being three separate limits during discharge. Fig 4(c) denotes Machine Speed as it increases, coasts, and then decreases. Fig 4(d) denotes machine back EMF as it mirrors the change in the machine speed. Fig 4(e) shows the machine winding current, which is at the limit value in the beginning and then decreases. When the system goes into Idle mode, the machine current becomes zero. During the Discharge

mode, the current follows the three limit values as set by the user.

Figure 5 is a magnified view of the first part of Fig 4(d) and clearly shows the increasing frequency of the back EMF waveform.

Figure 6 shows the comparison of the winding currents with the corresponding reference currents. The winding current is not smooth because of the PWM switching action. For the selected value of PWM frequency, the discharge current, at 487 Hz, is noisier than the charge current at 135 Hz because we have more samples per cycle in the latter.

CONCLUSIONS

An improved version of Flywheel Energy Storage model has been presented. This model incorporates an electromechanical machine model, which is able to simulate energy transfer to/from the flywheel. This operation is shown to be explicitly user controlled but can be performed based on some other system operating criteria. This simulation also incorporates improvements made in the controller design for closer regulation of the machine winding current to a sinusoidal form.

ACKNOWLEDGMENTS

The authors thank James Soeder and Ray Beach from NASA Glenn Research Center, Cleveland, OH, for their advice and support.

REFERENCES

1. Truong, L.V.; Wolff, F.J.; Dravid, N.V.; and Li, P.: Simulation of the Interaction Between Flywheel Energy Storage and Battery Energy Storage on the International Space Station. AIAA-2000_2953, or NASA/TM-2000-210341.

2. Patel, M. R.: Flywheel Energy Storage for Spacecraft Power Systems. Proceedings of the 34th Intersociety Energy Conversion Engineering Conference, SAE Paper 1999-01-2589, 1999

3. NASA/Aerospace Flywheel Development Program. http://space-power.grc.nasa.gov/ppo/project/flywheel/ last modified Nov. 30, 2000, accessed May 15, 2001.

4. Truong, L.V.; Wolff, F.J.; and Dravid, N.V.: Simulation of a Flywheel Electrical System for Aerospace Applications. IAAA-2000-2908, or NASA/TM-2000-210342.

5. Avant! http:www.avanticorp.com copyrighted 2001, accessed May 15, 2001.

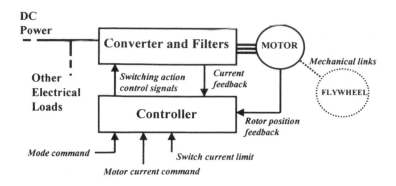

Figure 1: The simulated flywheel energy storage system.

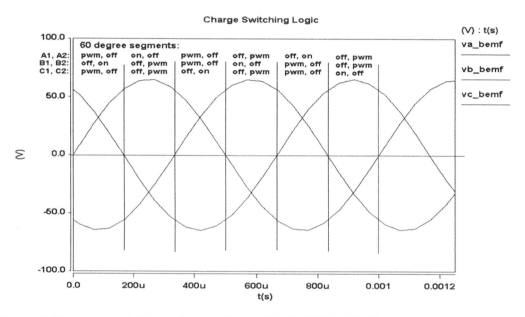

Figure 2: The motor winding induced voltages (Back EMF) with 60-degree segments highlighted.

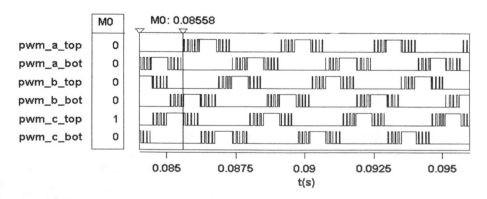

Figure 3: Details of the 60-degree PWM controlled signals. The PWM mode is shown as series of pulses while a single wide pulse denotes ON mode.

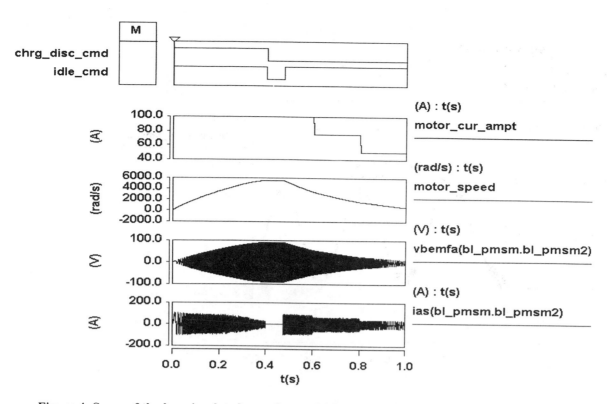

Figure 4: Some of the key simulated waveforms. (a) The Charge, Discharge, and Idle commands. (b) The winding current limit. (c) Machine speed as it increases, coasts, and then decreases. (d) Machine back EMF. (e) Machine winding current.

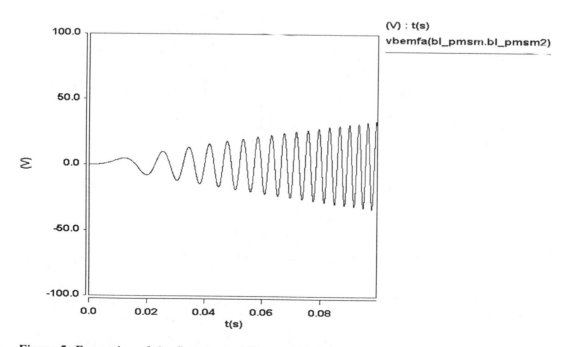

Figure 5: Expansion of the first part of Figure 4(d). It's clearly shown the increasing of both magnitude and frequency of the back EMF waveform.

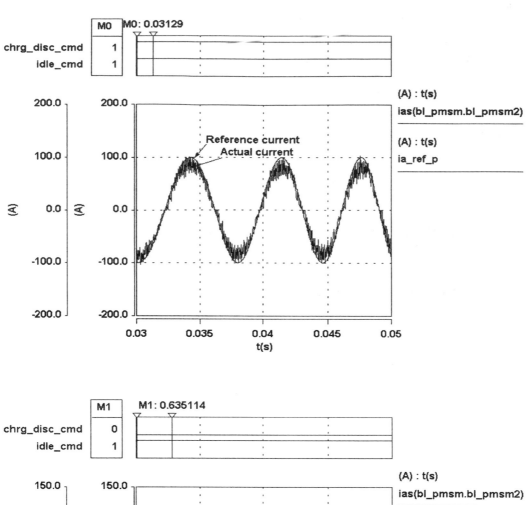

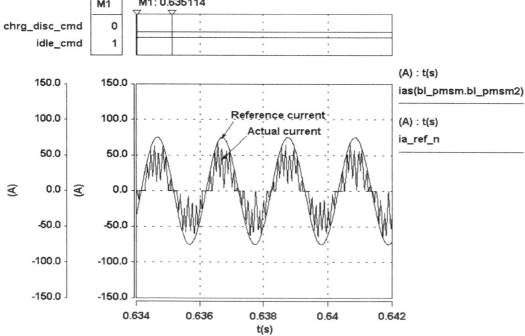

Figure 6: Comparison of the winding actual currents with their corresponding reference currents. (a) During charging from 0.03 to 0.05 second. (b) During discharging from 0.634 to 0.642 second.

2001-AT-04

THE ELEMENTS OF MOTIVE THERMODYNAMICS OF LIGHT FOR NEW SOLAR ENGINEERING SYSTEMS

Anatoly Sukhodolsky

General Physics Institute of Russian Academy of Sciences, Vavilov street 38, Moscow , Russia
7-095-138164 :sukhodol@kapella.gpi.ru

Abstract. The objective of this paper is to introduce the principles of renewable and non-renewable thermodynamics of solar light to use in a new generation of propulsion and extraction and power systems. The principles of renewable thermodynamics were found by laser simulation of sunlight in laboratory conditions to effectively generate vibration and capillary propulsive forces within the renewable volume of the system as a whole. The two branches of such renewable motive forces of light for engineering applications proposed are: a) direct conversion of solar light with high effective acceleration of bodies for elevating and solar vibration propulsion and extraction purposes; b) building new kinds of solar renewable power plants. The principle of non-renewable out-Carnot thermodynamics is introduced for using sunlight as energy for solar propulsion (SP) with non-renewable rocket-like constructions, when acceleration of objects is due to jets induced by sunlight energy in fuels. As far as the second fuel-component, solar radiation that is available for free, for example in space, SP can be competitive with both electrical propulsion and traditional two-fuel technologies. The outline of both renewable and non-renewable thermodynamics of the motive forces of light for new generations of propulsive solar systems is presented in the framework of the laws of energy and momentum conservation to evaluate opportunities for their application in engineering (see: http://www.come.to/LSTcenter)

INTRODUCTION

The thermodynamic principles of direct conversion of sunlight were studied earlier in a treatment of the Carnot principle [Clausius, 1850] by the entropy of Planck in two-member form [Plank, 1945] to include the creation of local heat sources [Sukhodolsky, 1998] powered by light. Such a thermal source becomes a source of mechanical energy under the condition that only a part of thermal energy is available for renewable conversion. The purpose of this paper is to unite the previous thermodynamics of renewable conversion [Sukhodolsky, 1998] with the production of mechanical energy and with non-renewable processes to change the volume of a system to evaluate the application of both for propulsive purposes, in space, [Sukhodolsky, 2001] where solar radiation is available for free. Let us begin with the problem of mechanical energy source in the condition of momentum conservation specific for solar propulsion in the following way.

Suppose one considers a closed energy system, say spacecraft in space, with energy $\Delta G = \Delta E + \Delta H = const$ equals sum of mechanical ΔH and thermal ΔE by the solar excitation $\Delta L = \Delta E = C_V \Delta T$ for propulsive purposes, when initially $\Delta H_0 = 0$. The objective is to find the fundamental restrictions for $thermo \rightarrow dynamic (\Delta E \rightarrow \Delta H)$ of conversion for light ΔL available for the next mechanical propulsion. The problem may be considered in two ways to principally perform conversion: 1. The renewable (RN) mechanical propulsion is due to creation of new energy ΔH by constant mass and volume of a system (such as a spacecraft); 2. The non-renewable (NRN) propulsion induced by sunlight takes place in the case of new synergetic energy, for example kinetic, $\Delta H = \Delta T = \dfrac{m_2 (\Delta u_2)^2}{2}$ is to accelerate a mass of fuel m_2 to put the rest of spacecraft of mass m_1 in motion. To solve the problem of available maximum for conversion efficiency, one can introduce entropy of excitation $\Delta S^{(-)} < 0$ [Sukhodolsky, 1998], which provides sunlight by energy ΔL for propulsion divided by two parts renewable $\Delta S_R^{(-)}$ and non-renewable $\Delta S_{NR}^{(-)}$ to unite the mechanical (synergetic) and thermal (caloric) features of the system in question.

$$\Delta S^{(-)} = \Delta S_R^{(-)} + \Delta S_{NR}^{(-)} \tag{1}$$

To find how RN and NRN contributions are divided between the two principally different propulsion schemes with regard to the law of momentum conservation, one can calculate the entropy excitement for spacecraft $\Delta S_1^{(-)}$ and work substance (fuel) $\Delta S_2^{(-)}$ in the modernized Plank meaning [Plank, 1945] as:

$$\Delta S_1^{(-)} = C_1\left[\ln\left(\frac{T_1}{T_0}\right) - \frac{T_1 - T_0}{T_0}\right] \text{ and } \Delta S_2^{(-)} = C_2\left[\ln\left(\frac{T_2}{T_0}\right) - \frac{T_2 - T_0}{T_0}\right] \qquad (2)$$

with temperatures of space T_0 spacecraft T_1 and excited work substance T_2 and heat capacities $C_0 = \infty$ and $C_1 = c_1 m_1, C_2 = c_2 m_2$ correspondingly. If one supposes the Carnot principle to be valid only for RN-conversion, the summary entropy $\Delta S^{(-)} = \Delta S_1^{(-)} + \Delta S_2^{(-)}$ can be divided on RN and NRN parts of Eq.1 if one introduces average temperature of 1+2 system $T = \frac{C_1 T_1 + C_2 T_2}{C}$ with $C = C_1 + C_2$. If such a closed system as a whole spontaneously relaxes into this thermal equilibrium, its entropy measured with regard to space as surroundings equals:

$$\Delta S_3^{(-)} = C\left[\ln\left(\frac{T}{T_0}\right) - \frac{T - T_0}{T_0}\right] \qquad (3)$$

Therefore, the entropy of excitement as a thermodynamic quantity to describe the following efficiency of conversion with regard to the Carnot principle applicable for RN and not applicable for NRN conversion can divided along the identity:

$$\Delta S^{(-)} = \Delta S_1^{(-)} + \Delta S_2^{(-)} = \ln\left(\frac{T_1^{C_1} T_2^{C_2}}{T^C}\right) + C\left[\ln\left(\frac{T}{T_0}\right) - \frac{T - T_0}{T_0}\right] = \Delta S_R^{(-)} + \Delta S_{NR}^{(-)} \qquad (4),$$

which is derived if one extracts $\Delta S_3^{(-)} = \Delta S^{(-)}{}_{NR}$ as a separate term from initial excitement $\Delta S^{(-)} = \Delta S_1^{(-)} + \Delta S_2^{(-)}$.

THE LAWS OF CONSERVATION

The concept of entropy divided by two principally different parts can be the fundamental basis to formulate both *two laws of motive thermodynamics* and concept of *complex work* necessary to describe high motive forces observed experimentally.

- **Vibration motive forces** were created by direct conversion of continuous CW into mechanical vibrations for propulsive purposes (Sukhodolsky, 1998/2). Thermo-hydraulic actuator (THA) by using process of direct generation is able to accumulate in more than 10 000 times CW energy of light into sequence of strong hydraulic shocks. They, for example, are able with threshold of pumping power 0.1 W to lift bodies 0.1 kg up due to acceleration much more than gravity acceleration g. The test-of-principle experiments with solar radiation provide with a good outline available in engineering.
- **Capillary motive forces** In case of two phase emulsion which consists of the opaque drops (disperse phase DP) in transparent mother phase (isperse matter DM) , the canalled transport particles of DP along laser beam with velocity up to 10 cm/sec toward light has been observed (Sukhodolsky,1999). Gibbs surface as work substance was earlier supposed to describe creation of such motive forces. The method of capillary extraction of water has experimentally proved when motive forces extract DP and DM by phase transitions.

The purpose of motive thermodynamics is to unite the thermal and mechanical properties of light-powered direct systems at an elementary level, when any element of new mechanical energy dH is considered to be the result of the work of two motive forces, $\mathbf{F}^+$ and $\mathbf{F}^-$, which equal in magnitudes and have opposite directions. The inevitable symmetry of motive forces to keep equilibrium between action and reaction in mechanics allows one to introduce two complex numbers, iF and $-iF$, on an axis of their mutual action. Both vectors are non-located vectors with moveable points of applications. In this case, the previous calculation for the motion of state vector [Sukhodolsky, 2000] can be done in keeping the law of momentum conservation. An elementary mechanical source in complex form may avoid the paradox of dynamic boundary conditions [Sukhodolsky, 1999] in hydrodynamics when thermocapillary motive forces put any system into motion by a gradient force at liquid surface, which has no reaction. If work to convert one kind of caloric thermal energy into other synergetic mechanical form is written in such a complex format, anyone who takes the elementary source into consideration will keep the elementary symmetry of motive forces. Any mechanical source can create new energy from a thermal one, by only working in a couple of forces. Another advantage of complex numbers at any elementary thermodynamic calculations may be to better describe different experimental methods of renewable and non-renewable propulsion. In the case of vibration propulsion [Sukhodolsky, 1998/2], which takes place by interaction with a rigid support, there would be an advanced mathematical tool to consider levitation of the center of gravity.

Thermodynamic Laws

The First law is nothing more than the statement that the sum of output mechanical - ΔH and internal thermal energy that is not available for conversion - ΔE and heat ΔQ - available to transfer into surroundings equals the input energy of light:

$$\Delta L = \Delta H + \Delta E + \Delta Q \tag{5}$$

One can also denote energy members in balance Eq. 5 by two principally different schemes for classification:

1. With regard to a kind of energy, one can write the first law as:

$$\Delta L = \Delta H + \Delta J \tag{6}$$

to point out $\Delta J = \Delta E + \Delta Q$ as having chaotic caloric and ΔH as having a synergetic mechanical kinds of energy.

2. With regard to energy interaction with surroundings:

$$\Delta L = \Delta G + \Delta Q \tag{7}$$

to point out the maximum conversion available for adiabatic systems $\Delta Q = 0$ of the direct conversion of light $\Delta L \Rightarrow \Delta G$ as a more desirable and effective way of conversion. Two (6 and 7) treatments of the First Law can be useful along the main form (Eq.5) in different, more specific cases.

The Second Law has a twofold objective: 1. to show the direction of natural processes, and 2. to find fundamentals for their efficiency. With light as a motive power, this law has the following treatment:

1. As far as by excitement (Eq. 4) $\Delta S^{(-)} < 0$, one can use the sign $\Uparrow$ instead of $=$ within any energy balance of the First Law to point out that excitement occurs. The direction of any spontaneous process with the creation of the mechanical energy is exactly within the law of entropy production $dS^{(+)} > 0$. This can be denoted for the left and right sides of energy balance by $\Downarrow$ besides $=$. Again, the entropy $dS^{(\pm)}$ given in two-terms $\dfrac{dE_1}{T_1} + \dfrac{dE_2}{T_2}$. Plank's meaning [Plank, 1945] is to consider both consumption and production entropy (in other words, creation and dissipation of heat sources denoted with $\Uparrow$ and $\Downarrow$). The arrows are to differ processes from the simple energy balance of the First Law. Light as a motive power principally differs the entropy dS with regard to the equilibrium entropy of Clausius $\partial S^C = \dfrac{\partial Q}{T}$ [Clausius, 1850]. This equilibrium consideration is not applicable for the high level of thermal non-equilibrium that can be stimulated with light and for describing the gigantic motive forces. The entropy of light definition in more details can be found in [Sukhodolsky, 1998/1]. Notice that if one involves a source of light (for example, sun) into consideration as a hot body and considers S^Σ as entropy of the universe, any global processes (including local $\Uparrow\Downarrow$) take place strongly by $dS^\Sigma > 0$. The excitement $\Delta L \Uparrow \Delta E$ of internal thermal energy ΔE can be considered as nothing more than a local fluctuation [Landau, 1995] induced by light as a motive power. This is the principal difference of energy transfer from the joining of cold to hot parts of a system [Sukhodolsky, 1998/1] available by light. The only notation is available $\Delta Q \Downarrow \Delta E$ in the case of internal energy changed by heat.

2. The energy balance (Eq.5) of the First Law provides the objective evidence to seek the maximum of conversion between adiabatic systems $\Delta Q = 0$ (Eq.7). One can consider conversion $\Delta L \Uparrow \Delta E^* = \Delta \widetilde{E} + \Delta \overline{E} \Downarrow \Delta H + \Delta \overline{E}$ of internal thermal energy divided by two parts available $\Delta \widetilde{E}$ and non-available for direct conversion $\Delta \overline{E}$ in the condition that only part $\Delta \widetilde{E} \Downarrow \Delta H$ internal energy can be converted by spontaneous entropy production.

The summary for the two points of the Second Law in motive thermodynamics can be shorted as follows:

Light as a motive power provides entropy $S^{(\pm)}$, which does not undergo the principle of superposition. While the Fist Law describes energy balance, entropy indicates energy distribution in the framework of the Second Law. Physically, $dS^{(-)} < 0$ (once again, it can be denoted by $\Uparrow$ instead $=$ in energy balance) means motion from equilibrium, which provides a source of light in the distribution of internal energy by excitement. Entropy production $dS^{(+)} > 0$ ($\Downarrow$ sign instead of $=$) means the tendency of any natural system to spontaneously go into thermal equilibrium. This process can also be stimulated by work production to increase its velocity. The First and Second laws of motive thermodynamics provide a strong methodological meaning of work as a thermodynamic quantity. Work is not a kind of energy. Therefore, it is beyond the two kinds of energy members available for direct measurement: synergetic - ΔH and caloric - ΔJ (Eq. 6). The significance of work is to show an actual element of the path of system dW in time motion, which in case keeping the law of momentum conservation suggested to be in complex form. The motive forces are two equal length with opposite directions presented by two complex numbers iF and $-iF$. In case of the most effective adiabatic systems $G = const$, the problem of available efficiency of conversion $\Box = \dfrac{\Delta H}{\Delta L}$ is to study the motion of the constant vector $\mathbf{G}$ on a complex energy surface driven by actual work entropy production $dS^{(+)} > 0$. Each specific path dW can be given by the specific interaction between synergetic H and chaotic J forms of energy in both phenomena and engines. This suggestion to put mechanical energy into scope elementary thermodynamics was proposed earlier by the consideration of a renewable motive cycle of an actual heat engine instead of the approach of reversible Carnot cycles [Sukhodolsky, 2000].

Work of Motive Forces

Now, one can consider more effective adiabatic system after excitement $\Delta L \Uparrow \Delta G = \Delta E + \Delta H$ in case the mechanical energy grows due to the work of motive forces ($dH > 0$) to introduce complex work for the simplest system of two bodies m_1 and m_2 given after solar excitement. One can suppose that in a point of current time t the complex velocities equal $\mathbf{u}_1 = iu$ and $\mathbf{u}_2 = -iu$ and the momentums $\mathbf{p}_1 = m_1\mathbf{u}_1$ and $\mathbf{p}_2 = m_2\mathbf{u}_2$ are created along a complex axis defined for the given cinematic restriction. The actual direction of the complex mathematic axis in physical space depends on the problem in question. For construction, one can put an imaginary axis along the physical symmetry axis of the cylinder and piston system. For phenomena, the complex axis is perpendicular or tangential to the phase-discontinuity surface for vibration and capillary motive forces, correspondingly. The center of gravity with coordinate $Z = \dfrac{m_1 z_1 + m_2 z_2}{M}$ for system of mass $m = m_1 + m_2$ can be associated with the center of a complex axis to define the law of momentum conservation in the form $\mathbf{p}_1 = \mathbf{p}_2^*$ and to derive the differential of scalar kinetic energy from increments $d\mathbf{p}_1$ and $d\mathbf{p}_2$ according to following relations:

$$dT_1 = \frac{(\mathbf{p}_1 + d\mathbf{p}_1)(\mathbf{p}_1 + d\mathbf{p}_1)^*}{2m_1} - \frac{(\mathbf{p}_1)(\mathbf{p}_1)^*}{2m_1} \text{ and } dT_2 = \frac{(\mathbf{p}_2 + d\mathbf{p}_2)(\mathbf{p}_2 + d\mathbf{p}_2)^*}{2m_2} - \frac{(\mathbf{p}_2)(\mathbf{p}_2)^*}{2m_2} \tag{8}$$

By summarization Eq.8 with using normalized mass $M = \dfrac{m_1 m_2}{m_1 + m_2}$ the increment of mechanical energy is:

$$dT = \frac{(\mathbf{p}_1 d\mathbf{p}_2)(\mathbf{p}_2 d\mathbf{p}_1)}{2M} = \frac{d(\mathbf{p}_1 \mathbf{p}_2)}{2M} \tag{9}$$

42

To find the elementary source, we can substitute $\mathbf{p}_1 = m_1\mathbf{u}_1, \mathbf{p}_2 = m_2\mathbf{u}_2, d\mathbf{p}_1 = \mathbf{F}_1 dt, d\mathbf{p}_2 = \mathbf{F}_2 dt$ into Eq.9:

$$\frac{dT}{dt} = \frac{(m_1+m_2)}{2m_2}(\mathbf{u}_1\mathbf{F}_2) + \frac{(m_1+m_2)}{2m_1}(\mathbf{u}_2\mathbf{F}_1) \tag{10}$$

As far as the distance Z between bodies of masses m_1 and m_2 connected with their positions iz_1 and $-iz_2$ with regard to center of gravity, according to the relation specific for momentum conservation:

$$\mathbf{z}_1 = \frac{izm_2}{m_1+m_2} \text{ and } \mathbf{z}_2 = \frac{-izm_1}{m_1+m_2} \tag{11}$$

one can apply Eq.11 to symmetrize Eq.10 into the following form:

$$\frac{dT}{dt} = \frac{\mathbf{F}_1\mathbf{U}_1^*}{2} + \frac{\mathbf{F}_2\mathbf{U}_2^*}{2} = \frac{\mathbf{F}_2\mathbf{U}_1}{2} + \frac{\mathbf{F}_1\mathbf{U}_2}{2} \tag{12}$$

here $\mathbf{U}_1 = i\dfrac{dZ}{dt}$ and $\mathbf{U}_2 = -i\dfrac{dZ}{dt}$. As a result, to find the power of a mechanical source, one can conjugate a motive force acting on a first body with half of the relative velocity measured with regard to another. Since the work of motive forces has the feature of central symmetry, after it is necessary to add vice versa. Again, the elementary mechanical source in two terms of Eq.12 is due to the fact that the equation is written by keeping the law of momentum conservation. Both non-localized vectors of motive forces $\mathbf{F}_1$ and $\mathbf{F}_2$ have coordinates of application points both $\mathbf{z}_1$ and $\mathbf{z}_2$ moving with velocities both $\mathbf{u}_1$ and $\mathbf{u}_2$ with regard to the center of gravity describe conversion $E \rightarrow H$ available also for more effective direct conversion by $dG = 0$. The mirror symmetry of central motive forces provides one with a principal possibility to describe the path of conversion by $dW = d\mathbf{W}d\mathbf{W}^*$. In spite of the fact that thermal caloric energy has no momentum, one may formally consider thermal properties at the same complex surface to unite mechanical work dW_m of motive forces and thermodynamic dW_t from thermodynamic adiabatic equation $pdV + C_v dT = 0$. By multiplication by dt one can rewrite from Eq.12 for work by any most effective adiabatic processes of direct conversion of light as:

$$dW_m = \frac{\mathbf{F}_1 d\mathbf{z}_2}{2} + \frac{\mathbf{F}_2 d\mathbf{z}_1}{2} = dW_t = \frac{\mathbf{P}_1 d\mathbf{V}_2}{2} + \frac{\mathbf{P}_2 d\mathbf{V}_1}{2} = \frac{\mathbf{C}_1 d\mathbf{T}_2}{2} + \frac{\mathbf{C}_2 d\mathbf{T}_1}{2} \tag{13}$$

The law of momentum conservation involved into thermodynamics of motive forces leads to division of work as a thermodynamic quantity over differentials of both mechanic and thermal energies when indexes 1 and 2 are omitted from the symmetric properties of both:

$$dW = \frac{\mathbf{F}d\mathbf{z}^*}{2} + \frac{\mathbf{C}_V d\mathbf{T}^*}{2} \tag{14}$$

Notice, $\mathbf{C}_V$ and $\mathbf{T}$ both are complex numbers, therefore, (+) in Eq.13 does not contradict $pdV + C_v dT = 0$. The current suggestion to keep strong symmetry (including signs at present instance) is due to following promotion of such complex work into thermodynamics of system with negative energy. Negative energy according common accepted knowledge in theoretical mechanics is specific to describe finite spice motion. As a classical example follows - $H = -T < 0$ from virial theorem (Landau, 1988). The motive forces of light have the main fundamental feature that talking about the "energy of work" alone makes no physical sense, as long as the mechanical energy is as the sole measurable quantity involved in the First Law of Thermodynamics (Eq.5). Work (in Eq.14) describes conversion of one kind of energy E into another H and vice versa. The fundamental symmetry of motive forces to put the new energy with an elementary source by two equal symmetrical parts provides one with the suggestion to associate ½ conversed energy with mechanical and another ½ with thermal energy. In other words, complex work W is a boundary to have the features of both. Now, one can consider the problem of the maximum for propulsive conversion as a previously introduced vector of energy [Sukhodolsky, 2000] but advanced into a complex form $\Delta \mathbf{G} = \Delta \mathbf{E} + i\Delta \mathbf{H}$. By this suggestion, the general energy can be written over the operations with complex numbers:

$$\Delta G = \Delta \mathbf{G} \Delta \mathbf{G}^* \tag{15}$$

Here and below, [*] an asterisk means a complex conjugation.

THE PROPULSIVE PROCESSES

Entropy, according to Eq. 1, is nothing more than availability of thermal energy after excitement to be converted into mechanical synergetic form. This process is available along two principally different schemes of performance for work entropy production. Entropy allows to principally differ the two propulsive scenarios 1) by keeping mass and volume of the system as a whole (renewable), and 2) non-renewable when one part (say spacecraft) can be put in motion by the loss in mass.

Renewable Cycles of Conversion

The path of the system by renewable conversion is closed $\oint_R Z(t)dt = 0$. The actual efficiency of renewable conversion can be found for solving a problem of actual motion of vector $\mathbf{G}$ of energy state driven by the work of motive forces. Obviously, along with kinetic energy (Eq.10) should be the internal (for example potential) return forces to renew the initial volume of the system (for example internal pressure difference). But to introduce the problem of efficiency, one can start from the method of Carnot cycles that is well known for maximum available thermal efficiency. The problem can be to find the principal relation between classical Carnot thermodynamics and proposed motive thermodynamics. While the initial excitement $\Delta L \Uparrow \Delta E^*$ with entropy $\Delta S^{(-)}{}_{NR}$ describes creation of heat source, any actual path down $\Downarrow$ of state vector is supposed to be driven by entropy production $dS^{(+)} > 0$ with a given source. The theorem of Carnot is applicable only for the second process because it does not deal with the problem of heat source at all. The Carnot cycle is not a direct process of energy conversion, as it was shown earlier [Sukhodolsky, 2000]. In order to keep an actual not reversible motion, it should be added by a source of mechanical energy for the preliminary excitement of internal energy ΔE_A for the first adiabatic compression. As a result, if one likes to use any Carnot cycles in motive thermodynamics, there are two actual processes of interaction with an external mechanical source. These adiabatic strokes cannot change the energy of cycle as a whole but should be actually presented in any Carnot cycle. Therefore, in order to keep the actual path for entropy production [Sukhodolsky, 2000] one should keep two accompanied processes in motive thermodynamics:

$$\Delta H_A \Uparrow \Delta E_A = \Delta E_A \Downarrow \Delta H_A \tag{16}$$

The main feature of this addition of energy $\Delta H_A = \Delta E_A$ is its independence on the energy of a cycle. Both an infinitely small dL and a finite ΔL need by conversion the same finite $\Delta H_A = \Delta E_A$ by given temperatures source T_h and sink T_s. Two strokes with mechanical reservoir can be written [Sukhodolsky, 2000] for perfect gas as:

$$\Delta H_A \Uparrow \int_0^1 pdv = R\frac{T_h - T_0}{k-1} = \int_2^3 pdv \Downarrow \Delta H_A \tag{17}$$

This relation can help to find the relation between reversible and renewable processes. Suppose one has a reversible work substance to convert an element of thermal energy $\Box E_h$ from a source of thermal energy by temperature T_h by transfer energy to sink with temperature T_s. Thermal energy that comes to sink $\Box E_s$ becomes less because of the energy conversion $\Box E_s = \Box E_h - \Box W$. The energy exchange with source and sink by actual time processes dE_h and dE_s should have a finite mechanical source with energy ΔH_A does not depend on $dE_h - dE_s$. The motive thermodynamics has full differential notation for any actual processes. Therefore, in order to keep the sign of equality $dH = []W$, as a "bridge" between the thermal cycles of reversible thermodynamics and the actual (irreversible) cycles of motive thermodynamics, one needs to associate with any small element of actual path $dE_h - dE_s \Downarrow dH$ the finite synergetic energy $\Delta H_A \Uparrow \Delta E_A \Downarrow \Delta H_A$. This is necessary to connect two sources, E_h and sink E_s, with temperatures T_h and T_s by actual $dW = dH$ path instead of a reversible $[]W$ one. If one suppose that the reservoir of mechanical energy Eq. 16 is absolutely perfect (reversible), the Carnot theorem [Sukhodolsky, 1998/1] will have the following form:

$$dE_h - dE_s \Updownarrow dH \qquad (18)$$

here the symbol $\Updownarrow$ means the energy balance between left and right occurs by $dS = 0$. In addition, $\Updownarrow$ is associated with internal necessity to interact with mechanical reservoir. The actual Carnot cycle efficiency $dH = \left(\dfrac{T_h - T_s}{T_s} \right) dE_h$ provides

identity $dS = \dfrac{dE_h}{T_h} - \dfrac{dE_h - dH}{T_s} = \dfrac{dE_h}{T_h} - \dfrac{dE_h}{T_s} + \dfrac{dE_h}{T_s} - \dfrac{dE_h}{T_h} = 0$ to prove Eq.18. One has a heat source and a reversible work substance to transfer energy into sink. Entropy as a function depends on both temperature of sink and source simultaneously. It is not a kind of hypothetical fluid. The reversible $[]S^C$ cannot be associated with any actual time derivative $\dfrac{dS^C}{dt}$ because of 1) the finite ΔH_A is necessary to put Carnot cycle in actual motion, and 2) the mathematical paradox occurs described earlier [Sukhodolsky, 1998/1]. Entropy is the quantity to measure the unbalance of energy between specifically both a source and a sink. The theorem of Carnot by renewable conversion means that the maximum of thermal efficiency can be obtained in the case of $dS^{(+)} = 0$. This particular case, denoted by symbol $\Updownarrow$, can serve as a methodological gate to connect classical reversible and proposed motive thermodynamics to calculate efficiency of any renewable cycles in motive thermodynamics. In the case of renewable conversion, the complex numbers for thermal properties can provide the exact difference between the available and non-available heat of Clausius [Clausius, 1850] in motive renewable thermodynamics and its relation with entropy (Eq.1).

Non-renewable propulsion

The actual path of the system by non-renewable conversion is open $Z(t) \rightarrow \infty$. Let us begin with the statement of the problem: What is the maximum light available for non-renewable conversion?

Thermal efficiency

To prove the available thermal efficiency, one can first notice that the internal energy of a work substance, for example, a perfect gas as fuel for propulsion, can be defined by two principally different differentials for adiabatic processes:

$$dE = pdV \quad \text{and} \quad dE = dL \qquad (19)$$

In favor of such a consideration, both definitions of internal energy in motive thermodynamics in parallel can serve at least three circumstances:

1. Adiabatic processes by pdV practically realized by either very fast compression or expansion. dL is able to put energy into a system also very fast, for example, by laser pumping. By simply focusing solar light within a diffraction limited spot for typical thermoconductivity $\dfrac{dL}{dt} \gg \dfrac{dQ}{dt}$, at least, in 10^3. In the case of actual solar excitement, only 0.1% can be lost; therefore, the process of solar conversion can be direct. Both pdV and dL are very similar as finite-time time processes. Both can be put in the definition of internal energy to keep as a sign of equality. And vice

versa dQ- has meaning only as slow heat-transfer in motive thermodynamics due to thermoconductivity.   Notice, the ideal case of equilibrium thermodynamics of heat as motive power means $\dfrac{dQ}{dt} \to 0$.

2. pdV and dL both are "cold" ways to get energy from surroundings; therefore, this second feature differs from dQ the First and Second Laws. The changing of internal energy by dQ is a "hot" way because of a more hot body (heater) with internal energy E_h nearby necessary to change internal energy dE. The excitement $dL \Uparrow dE$ means, again, that internal energy spontaneously comes from more cold surroundings to a more hot body similar to $pdV \Uparrow dE$ by compression. In contrast, by heat transfer $dQ \Downarrow dE$ can be only done as $_h$ energy flows from hot to more cold bodies.

3. pdV and "light" ways dL to change internal energy may be also considered in parallel because

considers light as a flow of particles, the change of internal energy is nothing more than a kind of "inelastic impact."

transfer   rgy after the impact of both initially "cold hammers." This differs
dL from

As it was shown earlier [Sukhodolsky, 2000], the sec
energy, provide thermal efficiency for any process with a fuel for non-

$$_t = \frac{(c_p \quad c_v)\Delta}{\Delta T \,(k-1} \quad 1$$

(21)

The physical sense of out- 100% thermal efficiency for non-renewable motive forces follows from the definition of internal energy (Eq.19). One can suggest consideration of the conversion as a two-stroke process: 1. going away $\Uparrow$ from a state with internal energy E_0 into an excited state $E_0 + \Delta L \Uparrow E_1$ by light, and 2. being in the same adiabatic conditions, the excited system relaxes with making work $E_1 - \Delta W \Downarrow E_2$. Such thermal efficiency simply follows from the definition of internal energy and assumption $E_2 \Downarrow E_0$ (that again means that $E_2 = E_0$ and $S(E_0) > S(E_2)$ simultaneously with the entropy of perfect gas $\Delta S^{(-)}{}_{NR}$ according to Eq 4 $\Delta S^{(-)} = \Delta S^{(-)}{}_{NR} = C_2 \left[\ln\left(\dfrac{T_2}{T_0}\right) - \dfrac{T_2 - T_0}{T_0} \right]$, for an example of fuel adiabatic from another mass of spacecraft [Sukhodolsky, 2001] .

Actual Efficiency

The term of actual efficiency (Sukhodolsky, 2000) was introduced for solar engines to evaluate efficiency that accompanied conversion from an excited by light state into a point of mechanical $p_2 = p_0$ not thermal equilibrium. $T_2 = T_0$. This approach needs to be considered more precisely in non-renewable cases for space application when actual pressure in space $p_0 = 0$. While there is no problem in accepting 100% thermal efficiency as the upper theoretical limit in the case of a non-renewable process, the actual efficiency here is based on the two following factors to not get this figure:

- Any actual source of mechanical energy according to Eq.12 should be accompanied by integration of work of the elementary sources. Therefore, any propulsive jets induced by sunlight to have a 100% theoretical limit will have the incompatible contradiction to keep adiabatic conditions during a long time. The actual conversion should be restricted at least from radiation dissipation of energy.

- The more important problem is the restricted value of classical thermodynamics of perfect gas as a massless work substance. For application in non-renewable solar propulsive systems, the new methods and mathematical tools to unite thermal and mechanical degrees of freedom should be promoted.

CONCLUSIONS

The purpose of this paper was to consider mechanical and thermal processes for solar propulsive systems within a joint elementary consideration. While the conducted calculations have not been applied to concrete technical performances in spice, the attempt to consider the concept of entropy specific for thermodynamics together with the laws of energy and momentum conservation specific for mechanics is expected to be useful in seeking better ways to use solar radiation for different propulsive purposes, for example in space (Sukhodolsky, 2001) for example along with know methods of electrical propulsion (NASA web reference).

REFERENCES

Clausius, R (1850), 'On the motive power of heat, and on the laws which can be deducted from it for the theory of heat," LXXIX, pp. 500-513.

Landau, L.D. and E.M. Lifshitz (1995), Theoretical Physics, Volume V, *Statistical Physics*, Part.1, Nauka-Fiz-mat-lit, Moscow, (in Russian).

Landau, L.D. and E.M. Lifshitz (1999), Theoretical Physics, Volume 1, *Mechanics*, Part.1, Nauka-Fiz-mat-lit, Moscow, (in Russian).

NASA web reference http://trajectory.lerc.nasa.gov/aig7820/projects/electric/

Plank, Max (1945), *Treatise on Thermodynamics*, Dover Publication, 1780 Broadway, New York. (In Max Plank's "A survey of physics, a collection of lectures and essay," New York and E.P Dutton and Company Publisher, Printed in Great Britain, 1920, entropy in two-member format was the sole math relation.)

Sukhodolsky, A.T. (1998/1), "An Introduction to Thermodynamics of Renewable Cycles for Direct Solar Energy Conversion," in proceedings of the *33rd Intersociety Engineering Conference on Energy Conversion, Colorado Springs, August 2-6, 1998/1. (http://www.gpi.ru/~sukhodol/free/h1/u1.htm)*

Sukhodolsky, A.T. (1998/2), "Thermo-hydraulic actuator as a new way for conversion of solar energy in space," 33rd Intersociety Energy Conversion Engineering Conference (IECEC), Colorado Springs, CO. August 2-6, 1998, paper #I390, in proceedings of *the 33rd Intersociety Engineering Conference on Energy Conversion Colorado Springs, August 2-6, 1998/2 (http://www.gpi.ru/~sukhodol/free/h2/u2.htm*

Sukhodolsky, A.T. (1999), "Capillary Motive Forces for Liquid Extraction in Microgravity Conditions by Sunlight," and "Microgravity Fluid Physics and Heat Transfer," in proceedings of the *International Conference on Microgravity Fluid Physics, held at the Turtle Bay Hilton, Oaju, Hawaii, September 19-24, 1999*, pp 1-9. http://www.gpi.ru/~sukhodol/free/h3/u3.htm

Sukhodolsky, A.T. (2000), "Introduction into Applied Renewable Thermodynamics for Direct Solar Energy Conversion Systems," in proceedings of the *35th Intersociety Energy Conversion Engineering Conference Nevada, 24-27 July 2000.* http://www.gpi.ru/~sukhodol/engines.pdf

Sukhodolsky, A.T. (2001), 'On New Generations of Solar Propulsive Systems for Space," in proceedings of *"Space Exploration and Transportation: Journey into the Future, "* February 11-February 14, 2001, Albuquerque, New Mexico. Transfer of Copyright agreement to American Institute of Physics (September 4, 2000).

Proceedings of IECEC'01
36th Intersociety Energy Conversion Engineering Conference
July 29-August 2, 2001, Savannah, Georgia

IECEC2001-AT-46

UPDATE ON DEVELOPMENT OF 5.5 INCH DIAMETER NICKEL-HYDROGEN BATTERY CELLS

Jack N. Brill, Fred L. Sill, and Matthew J. Mahan
Power Subsystems   Eagle-Picher Technologies, LLC
Joplin, Missouri 64801

ABSTRACT

A series of 5.5 inch diameter nickel-hydrogen battery cells is being developed and tested at Eagle-Picher Technologies. These individual pressure vessel (IPV) designs are larger in diameter from the more commonly found 3.5 inch design, but provide capacities which exceed the limits of the smaller diameter cells. The current series of Eagle-Picher designs include cells capable of capacities from 100 Ah to 375 Ah. Life tests at Eagle-Picher are in process, and current results are available, demonstrating nominal performance after 6000 cycles on two designs, and nominal performance after 9200 cycles for another of the designs. The life tests are currently interrupted to conduct storage tests. Several of the 350 Ah cells have been provided to and are being evaluated at a customer's facility. Progress in the area of manufacturing and testing is described in this presentation. Two groups of 4.5 inch, 2.5 volt, CPV cells are also under test and that data is contained herein.

INTRODUCTION

Nickel-hydrogen (NiH_2) cells are the system of choice for both low-earth-orbit (LEO) and geosynchronous-earth-orbit (GEO) communications and surveillance satellites. The NiH_2 system offers unequaled performance and cycle life, as well as extreme abuse tolerance, simplified state-of-charge indication and high reliability levels. Until recent applications cells having diameters of 2.5 to 4.5 inches with capacities of 4 Ah to 200 Ah supported the power needed. Recent applications have power requirements up to 400 Ah which approach the manufacturing limitations of the cell pressure vessel and create a need for cells of a greater diameter to minimize the battery height and weight. To fill this need Eagle-Picher Technologies, LLC began development and qualification of cells having a diameter of 5.5 inches. Cells in this diameter will deliver capacities between 200 Ah to 400 Ah

Another development is underway to offer a more efficient weight and footprint in the 4.5 inch diameter series of cells.

Two groups of dual-cell, 2.5 volt CPV cells are also in development. These cells have capacities of 60 Ah and 100 Ah.

Figure 1 illustrates the progression of cells currently manufactured by Eagle-Picher Technologies, LLC including the 5.5 inch diameter cell under development.

Figure 1, Progression of Cells Having Diameters of 2.5, 3.5, 4.5 And 5.5 Inches

Development

The 5.5 inch diameter development effort is planned to generate a series of cells having capacities between 200 Ah to 400 Ah in the IPV configuration. The 5.5 inch diameter CPV designs will range from 100 Ah to 200 Ah.

The 4.5 inch diameter development is designed to qualify and offer a range of 2.5 volt, CPV cells to complement the current series of 1.25 volt cells. These designs offer a reduced battery footprint and a reduction in battery weight.

All of the cells utilize the vast heritage of Eagle-Picher's experience in supplying NiH₂ cells for the past 20 years. No new technology is introduced other than the increase in diameter. The same materials, electrode designs, electrolyte concentrations, separator material and thermal considerations are carried forth into these cells.

Design

In the development of the cells in the 5.5 inch diameter configuration several basic considerations are followed. The fundamental flight proven design precepts of cells in smaller diameters are maintained in the 5.5 inch configuration. The cells have a common heritage with respect to the electrodes, separator and internal materials. Each cell retains a near common thermal cross-section across the electrode stack from the internal diameter to the outer edge of the stack. All pressure vessels satisfy the safety and life requirements of MIL-STD-1522.

Manufacturing

The development of the 5.5 inch diameter cells has progressed by the manufacture and test of three distinct groups. Each group uses the basic technology present in Eagle-Picher Technologies, LLC's current flight cells. No changes in materials or electrodes, other than size, are introduced. The pressure vessels are of age hardened Inconel 718. The pressure vessels are sized to have a burst pressure of at least 2.5 times the maximum expected operating pressure (MEOP) for each design.

The positive electrodes are the standard aqueous impregnated, slurry sinter used in most Eagle-Picher flight cells. Active material loading is the normal 1.65 grams per cc of void volume. The electrodes are stacked in pairs with the leads extending through the center core. A separator is placed on each side of the positive electrodes. The negative platinum electrodes are placed with the platinum black surface next to the separator material. A plastic diffusion screen is placed on the back side of each negative electrode to facilitate gas diffusion to the back side of this electrode. These components constitute one module of the electrode stack. This arrangement is repeated until the number of modules is reached to provide the required capacity.

Initial Cells

The initial group involved both IPV and CPV designs. The first cells were a 200 Ah IPV (1.25 volt) design. The second were 100 Ah CPV (2.50 volts) cells. These cells are alike in design and assembly. The cells were assembled in a dual stack arrangement (see Figure 2) with the weld ring centrally located in the stack. The 200 Ah IPV cell has the two

stacks connected in series (1.25 volts) while in the CPV the stacks were connected in parallel (2.50 volts).

Figure 2, Dual Stack 200 Ah NiH₂ Cell

The cells were subjected to characterization testing and then placed on a life cycle test with a LEO regime. The characteristics of these two cell designs are shown in Table 1.

Cell Design	200Ah IPV	100 Ah CPV
Cell Voltage (volts)	1.25	2.50
Actual Capacity (Ah) @ C/2, 10°C	205	101
Operating Pressure (psig)	800	825
Cell Length w/o Terminals (in.)	10.5	10.5
Cell Length w/Terminals (in.)	15.5	15.5
Cell Diameter (in.)	5.71	5.71
Weight (grams)	5450	5550
Specific Energy (Wh/kg)	47.8	46.2
Energy Density Wh/liter	72.2	71.1

Table 1

Second Generation

A second series of cells were manufactured having the same diameters. These were 350 Ah IPV designs. Two groups were assembled at different times. They differed only by the use of strain gages in the second group. All of the cells were subjected to characterization tests and then placed on LEO cycling as a life test. Two of the cells having strain gages were supplied to an outside facility for evaluation testing. The remaining cells are scheduled for life testing at Eagle-Picher

Technologies, LLC. The characteristics for the 350 Ah designs are shown in Table 2.

Cell Design	350 Ah	350 Ah w/Strain Gages
Cell Voltage (volts)	1.25	1.25
Actual Capacity (Ah) @ C/2, 10°C	371	367
Operating Pressure (psig)	940	940
Cell Length w/o Terminals (in.)	15.3	15.3
Cell Length w/Terminals (in.)	20.3	20.3
Cell Diameter (in.)	5.71	5.71
Weight (grams)	8592	8592
Specific Energy (Wh/kg)	54.8	54.2
Energy Density (Wh/liter)	83.8	82.9

Table 2

Third Generation

The third phase of development began in 1999. Cells manufactured during this phase are of three sizes and configurations. The first series consists of cells in the CPV (2.50 volt) configuration with a rated capacity of 147 Ah. These cells are made to perform in a LEO environment. They were subjected to characterization testing and at the present time are being prepared for life test at a LEO regime. The cells utilize the same electrodes as the previous groups. Mechanically the design is more optimized for weight and volume.

The second series are 350 Ah IPV (1.25 volt) cells designed for use in a LEO regime. These cells utilize the same electrode designs as previous cells. They will be characterized through test. After characterization the cells will be fitted with a thermal interface and placed on extended life testing for cycling in a LEO regime. These cells, compared to second-generation cells, are also more optimized for weight and volume.

The third series are also 350 Ah IPV (1.25) cells. However, these are designed for use in a GEO mission. The cell stack is the same as the LEO designs with the exception of the separator and the amount of electrolyte in the stack. These cells will be characterized with the LEO design cells and then prepared with thermal interface sleeves for life testing. This group will be placed on a simulated GEO orbital test

The characteristics of the three designs are shown in Table 3.

Acceptance (Characterization) Testing

Each group of cells was characterized through test. The tests consisted of 100% depth of discharge (DOD) electrical cycles stabilized at temperatures of 20°C, 10°C and 0°C. A self-discharge measurement was also made with a 72-hour open circuit storage between charge and discharge and the cells stabilized at 10°C.

Cell Design	147 Ah CPV	350 Ah IPV (LEO)	350 Ah IPV (GEO)
Cell Voltage (volts)	2.50	1.25	1.25
Actual Capacity (Ah) @ C/2, 10°C	155	355*	355*
Operating Pressure (psig)	1000	900	900
Cell Length w/o Terminals (in.)	13.2	15.3	14.6
Cell Length w/Terminals (in.)	18.2	20.3	19.6
Cell Diameter (in.)	5.71	5.71	5.71
Weight (grams)	7041	7764	7420
Specific Energy (Wh/kg)	55.9	58.1	60.8
Energy Density (Wh/liter)	83.0	80.2	84.6

Table 3

Each capacity measurement is made from a 16-hour C/10 charge and a C/2 Discharge to either 1.00 or 2.00 volts (depending on an IPV or CPV). A summary of the tests for each group of cells is listed as in Table 4.

	RNH200-A IPV	RNHC100-A CPV	RNH350-A IPV	RNH350-B IPV	RNHC147-A CPV
20°C (AH)	185	88	323	338	138
10°C (AH)	205	101	371	367	155
10°C, 72 hr Retention (AH)	180	85	326	324	135
0°C (AH)	212	105	388	390	165

Table 4

Typical charge and discharge curves can be seen in Figures 3 through 6. These have the same basic roll over during charge and fall off at the end of discharge as typical IPV or CPV cells. At 10°C the maximum charge voltage at a C/10 rate was 1.57 volts for an IPV cell and 3.1 volts for a CPV cell.

Qualification Testing

A mini qualification test was performed using a 350 Ah cell. The test sequence was a 10°C standard capacity measurement followed by vibration and a second 10°C, standard capacity measurement. The standard capacity measurements were conducted using a 16-hour charge at a C/10 rate and a C/2 discharge.

	Axial	
10 - 80 Hz	+6 db/Oct	
80 - 700 Hz	$0.1g^2$/Hz	
700 - 2000 Hz	-6 db/oct	
Overall 10.5 Grms		
	Radial	
10 - 80 Hz	+4 db/oct	
80 -1000 Hz	$0.05 \ g^2$/hz	
1000 - 2000 Hz	-3 db/oct	
Overall 9 Grms		

Table 5, Vibration Levels

The vibration level selected was a qualification level used on an existing battery program. The levels can be seen in Table 5.

The cell performed throughout the testing without incident. No physical damage was noted. The voltage and capacities were normal before and after vibration.

Life Cycle Testing

Life tests with 5.5-inch diameter cells are ongoing at Eagle-Picher Technologies, LLC. These tests include IPV cells having rated capacities of 200 Ah and 350 AH. Another test involves 100 Ah CPV cells.

All the tests are 90-minute cycles (35-minute discharge/55-minute charge) with a recharge ratio of 1.04. The depth of discharge for all cycles is 40% based on nameplate. The cells are stabilized at 5°C in each test. A summary of the tests is shown in Table 6.

Due to an interest in the effects of activated storage on cycle life, the cells have been removed from cycle and placed on storage for approximately two (2) years. The cells are stored at an ambient of 5°C in a fully discharged condition. At the end of the storage period the cells will be characterized and returned to cycle testing.

Cell	Quantity	Rated Capacity (AH)	DOD %	R.R.	Number of Cycles
RNH 200-A	10	200	40	1.04	9268
RNHC 100-A	10	100	40	1.04	6524
RNH 350-A	10	375	40	1.04	6182

Table 6, Life Tests

4.5 Inch Diameter Cells

Two groups of six (6) cells were assembled in the 4.5 inch diameter pressure vessels. These were 2.5 volt CPV cells having rated capacities of 60 Ah and 100 Ah. The basic characteristics of these cells can be seen in table7.

	RNHC 60-11	RNHC 100-17
Cell Voltage (volts)	2.5	2.5
Rated Capacity (AH)	60	100
Operating Pressure (psig)	900	950
Actual Capacity @ C/2, 20°C (AH)	64	96.8
Actual Capacity @ C/2, 10°C (AH)	69.7	109.5
Actual Capacity @ C/2, 10°C (AH) after 72 hr OCV	63	94.4
Actual Capacity @ C/2, 0°C (AH)	74.5	117.8
Cell Length w/o Terminals (in.)	8.6	12.8
Cell Length w/Terminals (in.)	12.4	16.5
Cell Diameter (in.)	4.64	4.64
Weight (grams)	3350	4357
Specific Energy (WHr/kg)	52	62.8
Energy Density (WHr/liter)	86	85

Table 7, 4.5 inch Diameter, 2.5 volt Cell Designs

The same electrodes and materials previously described for the 5.5 inch cells were utilized. The cells are of the dual stack configuration with the weld ring centrally located and the two stacks connected in series to attain 2.5 volts. Both cells are alike with the number of electrodes varying to acquire the desired capacity. As with the 5.5 inch designs the cells were activated, characterized and placed on cycle test. Prior to being placed on cycle the cells were subjected to a series of capacity measurements at 20°C, 10°C and 0°C. A series of three different rates were applied twice during each discharge. Figures 5 through 7 depict the discharge results for these tests.

The rates were 20 amps, 30 amps and 40 amps for the 60 Ah design. A 10 second, 120 amp pulse was applied between the two series of rates changes. The cells were then completely discharged at a C/2 rate.

The rates used in the 100 Ah test were 40 amps, 50 amps and 60 amps. The high rate test was not applied to these cells due to equipment limitations at the time. These will be performed later.

The cycle tests were performed at 10°C. The regime consisted of a 54-minute charge followed by a 36-minute discharge. The cells were repeatedly discharged to a 40% depth of discharge. Cycles 1 and 200 for each design can be seen in Figures 13 through 16. To date the 60 Ah cells have attained over 400 cycles and the 100 Ah cells have exceeded 350 cycles.

SUMMARY

To date Eagle-Picher Technologies, LLC has manufactured IPV cells having capacities ranging from 200 Ah to 350 Ah. The performance of each cell has shown the same voltages and capacities as would be expected of any NiH₂ cell.

The ability to successfully withstand vibration to a level of 10.5 grms has been demonstrated by a 350 Ah cell.

CPV cells in the 5.5 inch diameter have also been manufactured in the range of 100 Ah to 147 Ah. Performance by these cells also compares well with other NiH₂ CPV cell designs.

Life Cycling at a 40% DOD, 5°C, has been demonstrated to 9258 cycles.

Both 60 Ah and 100 Ah CPV cell designs in the 4.5 in diameter have demonstrated satisfactory performance. Cycling is continuing at 40% DOD without incident. The RNHC 100-17 cells have exceeded 500 cycles while the RNHC 60-11 cells have surpassed 600 cycles.

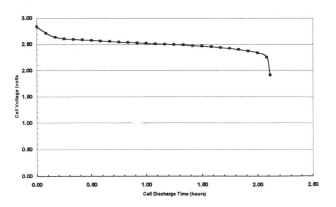

Figure 4, CPV, Typical 10°C Charge

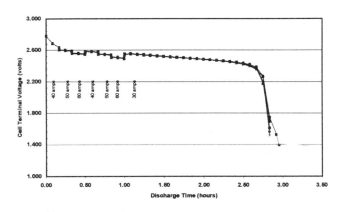

Figure 5, RNHC 100-17, 20°C Discharge

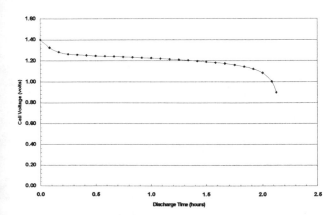

Figure 3, IPV, Typical 10°C Discharge

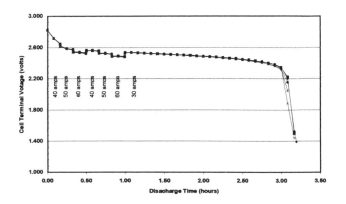

Figure 6, RNHC 100-17, 10°C Discharge

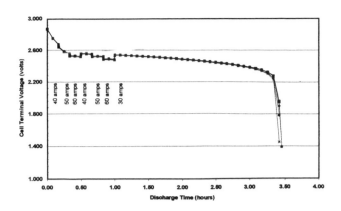

Figure 7, RNHC 100-17, 0°C Discharge

Proceedings of IECEC'01:
36th Intersociety Energy Conversion Engineering Conference
July 29 - August 2, 2001, Savannah, Georgia

IECEC2001-AT-47

UPDATED LIFE CYCLE TEST RESULTS FOR FLIGHT QUALIFIED NICKEL-HYDROGEN CELL DESIGNS

Ron Smith / Eagle-Picher Technologies, LLC
C and Porter Streets
Joplin, Missouri 64801

Jack Brill / Eagle-Picher Technologies, LLC
C and Porter Streets
Joplin, Missouri 64801

ABSTRACT

Nickel-hydrogen batteries continue to be an industry standard for both military and commercial spacecraft power, where high reliability, long life, and abuse tolerance is required. This battery system currently provides power to hundreds of earth-orbiting vehicles where battery failure is an unacceptable option, and where quality performance is required throughout the expected life span of the mission.

A significant portion of satellites currently in operation uses cell and/or battery designs developed and manufactured at Eagle-Picher. The quality of this hardware is extensively proven by flight heritage. For additional confidence and characterization, however, Eagle-Picher has maintained a continuous life cycle test for many of the established cell designs. Both real-time and accelerated cycle testing are represented in the test, including regimes appropriate for both low earth orbit and geosynchronous orbit. In some cases the tests attempt to duplicate the on-orbit charge/discharge profile of the mission as closely as possible, and subject the hardware to the extreme conditions expected during actual use. The test database was established to support early flight programs using nickel-hydrogen technology, and has been expanded over the past 20 years. In addition to the testing performed at Eagle-Picher, some life cycle tests shown in this presentation include results from the testing of EP hardware at customer facilities and independent laboratories. The updated test results include results from a variety of nickel-hydrogen designs, including individual pressure vessels (IPV), common pressure vessels (CPV), and dependent pressure vessel (DPV).

INTRODUCTION

Over the past 25 years, the nickel-hydrogen battery system has become the reliable power source of choice for communications satellites in all types of orbits. As the system replaced the frequently used nickel-cadmium systems, aerospace missions have benefited from increased mission life and improved mass characteristics. Recent information regarding Eagle-Picher nickel-hydrogen hardware indicates that over 300 launches have been supported, and over 400,000,000 cell-hours in space have been accumulated. This flight heritage for nickel-hydrogen, which began in 1977 with the Navy NTS-2 and Air Force Flight Experiment programs, has provided end-users with confidence that sustained missions are both possible and expected using the system. In addition to the countless hours of successful flight time accumulated by the on-orbit applications, Eagle-Picher has maintained a test bed of life tests which further characterize the performance of EPT nickel-hydrogen hardware. Over time, the life test has been modified and/or expanded to include the four varieties of existing nickel-hydrogen technology, including the individual pressure vessel (IPV), the common pressure vessel (CPV), the dependent pressure vessel (DPV), and the single pressure vessel (SPV).

Life tests at both the Eagle-Picher facility in Joplin and those ongoing at various outside facilities represent a range of nickel-hydrogen cell and battery designs. Both geosynchronous and low earth orbits are represented in the test methods. Some tests demonstrate real time testing, and others use an accelerated cycle regime. The information shown here is a recent status of the number of cycles accumulated for the various test items, and a general discussion of the resulting hardware performance.

In the standard set-up for testing individual cells at the Joplin facility, the cells are mounted in aluminum thermal sleeves, and positioned horizontally within a controlled temperature chamber. Unless otherwise required by specified test parameters, the temperature is maintained at either 5 or 10 degrees C. See Figures 1 and 2.

Figure 1. RNHC-10 cells in life test at Eagle-Picher.

Figure 2. 50 amp-hour IPV cells on test at Eagle-Picher.

LIFE TEST INFORMATION

Table 1 describes the cells which are either currently on test or have been on test at the Joplin facility, having been subjected to a charge/discharge regime consistent with a low earth orbit (LEO). A portion of these have been tested using an accelerated LEO regime, a shallow depth of discharge in the range of 15% of the nominal cell/battery capacity. The

majority, however, have been tested using a typical 90 to 98 minute cycle, the depth of discharge (DOD) being approximately 30% of the nominal capacity, and recharge rates sufficient to return approximately 103% of that which had been removed during the discharge.

Table 1. In-House LEO Cycle Test Results

Cell	Type	Qty.	Regime	%DOD	# of Cycles
RNH-50-15	IPV	19	ACC LEO	15	127,770
RNH-80	IPV	1	ACC LEO	15	117,908
RNH-50-15	IPV	1	ACC LEO	15	112,903
RNH-76-3	IPV	5	ACC LEO	15	108,015
RNH-76-9	IPV	1	ACC LEO	15	96,866
RNH-30-1	IPV	10	RT LEO	30	69,806
RNH-50-15	IPV	7	RT LEO	45	56,880
RNH-50-RL	IPV	50	RT LEO	40	3,475
RNH-76-3	IPV	7	RT LEO	45	55,879
RNH-76-3	IPV	10	ACC LEO	15	50,170
RNH-76-3	IPV	4	RT LEO	45	35,179
RNHC-10-1	CPV	11	RT LEO	40	36,062
RNHD-40-1	DPV	3	RT LEO	30	25,523
RNHD-60-1	DPV	2	RT LEO	30	3,385
RNHD-90-1	DPV	3	RT LEO	30	1,121
RNH-56-1	IPV	3	RT LEO	70	13,600
RNH-56-1	IPV	3	RT LEO	70	13,600
RNH-56-1	IPV	1	RT LEO	70	13,600
RNH-56-1	IPV	4	RT LEO	70	13,600
RNH-56-1	IPV	3	RT LEO	70	11,400
RNH-56-1	IPV	4	RT LEO	70	3,700
RNH-100-9	IPV	8	RT LEO	22	13,405
Battery					
SAR-10013	IPV	27	RT LEO	12	29,750

Table 2 describes the cells that are on test or have been on test at the Joplin facility which are representative of geosynchronous orbit cycling (GEO). Typically, these cells are subjected to regimes which resemble actual mission eclipse seasons, 44 day periods where the depth of discharge varies as the season progresses. In most cases, the regime specified will include extended periods where a trickle charge is used to sustain the state of charge. However, in some tests, the trickle charge portion of the normal GEO cycle is omitted so that more cycles can be accumulated in a shorter period of time. These accelerated GEO tests are referred to in the associated charts as ACC GEO. The geosynchronous regime tests at Eagle-Picher have periodically employed some type of reconditioning cycle.

Table 2. Summary of GEO Tests at Eagle-Picher

Cell	Type	Qty.	Regime	%DOD	# of Seasons
RNH-120-H	IPV	4	ACC GEO	75	44
RNH-65-15	IPV	6	ACC GEO	56	42
RNH-120-1	IPV	6	ACC GEO	100	23
RNH-76-11	IPV	6	ACC GEO	57	21
RNH-100-7	IPV	6	ACC GEO	57	20
RNH-65-33	IPV	4	ACC GEO	70	17
Battery					
SAR-10017	IPV	27	ACC GEO	75	9
SAR-10017	IPV	27	ACC GEO	75	24

Table 3 describes the cell/battery tests which are ongoing or have been performed at facilities other than Eagle-Picher (Joplin). A combination of LEO and GEO type cycle results is shown. All units listed are IPV cells.

Table 3. Summary of Off-Site Life Tests

Cell	Qty.	Regime	%DOD	# of Cycles	Facility
RNH-100-1	8	LEO	40	44488	LM
RNH-30-1	14	LEO	13	53200	MSFC
RNH-90-3	4	LEO	7	42700	MSFC
RNH-90-3	4	LEO	7	41600	MSFC
RNH-90-3	132	LEO	7	41500	MSFC
RNH-90-3	22	LEO	7	40900	MSFC
RNH-90-3	4	LEO	7	40860	MSFC
RNH-90-13	8	LEO	30	27100	MSFC
RNH-48-1	5	LEO	35	4700	MSFC
RNH-48-1	5	LEO	45	4700	MSFC
RNH-65-15	6	GEO	55	42 Seasons	TRW
RNH-100-7	6	GEO	65	20 Seasons	TRW
RNH-98-1	2	GEO	65	2 Seasons	TRW
RNH-98-3	4	GEO	65	2 Seasons	TRW

Individual Pressure Vessels

Briefly described, the individual pressure vessel cell (IPV) is a single electrochemical battery cell stack hermetically sealed within a pressure vessel, the capacity of the cell determined by the number of electrode pairs within the stack, the typical terminal voltage being 1.25 V. The IPV is the most mature of the four designs mentioned within this presentation, and represents the largest quantity of cells currently in space, and the largest quantity of cells on ground life test.

The earliest nickel-hydrogen life tests were initiated to provide support data for the flight programs in the early 1980's. The first test began in June of 1983 using RNH-30-1 cells (30 ampere-hour nominal capacity at 10 degrees C). While the test has not operated continuously over that interval of time, this set-up at the Joplin facility has accumulated in excess of 69,000 LEO type cycles (real time), representing over 12 years of mission performance. See Figure 3.

Another IPV test which has been ongoing at the Joplin facility has characterized the longevity and performance of the RNH-50-15 cells. A portion of these cells has been subjected to the previously described accelerated low earth orbit cycling, having accumulated over 127,000 cycles at approximately 15% depth of discharge. These results, examining end-of-charge and end-of-discharge voltages reveal stable performance. See Figure 4.

Similar results are seen reviewing the EOC and EOD voltage data from the 76 ampere-hour cells which have been tested using an accelerated cycle regime. These cells have completed in excess of 108,000 shallow DOD cycles (approximately 15%), and demonstrate stable performance. See Figure 5.

Common Pressure Vessels

The common pressure vessel cell (CPV) is defined as one pressure vessel containing two nickel-hydrogen electrochemical cell stacks, electrically connected to produce a nominal terminal voltage of 2.5. Similar to the IPV design the cell capacity is determined by the number of electrodes included in the length of each stack. The electrochemistry and structure of the internal components is identical to that used in IPV cells. The CPV design, therefore, takes advantage of the flight heritage of the IPV cell, but offers certain volumetric and gravimetric advantages.

The significant CPV life test currently in operation at the Joplin facility is the cycling of 11 RNHC-10-1 cells, currently having accumulated greater than 36,000 cycles, using a real time LEO regime (40% DOD). This test, having been initiated in 1994, was originally intended to demonstrate nominal performance of positive electrodes manufactured on a new production line (Eagle-Picher Joplin - Rangeline facility), and continues to provide favorable results.

Dependent Pressure Vessels

The dependent pressure vessel cell (DPV) is another design concept which attempts to minimize packaging space required by a battery. This "flat-pack" or "pancake" design features a circular cells with flat sides which are restrained by adjacent cells or by an end-panel which supports the end of the battery. These external restraints replace the more voluminous pressure vessels in the baseline design, and provide a more compact footprint. Cells of 40, 60, and 90 ampere-hour nominal capacities have been manufactured and tested at Eagle-Picher's Joplin facility. Currently, the RNHD-40-1 DPV cell has accumulated more than 25,000 cycles using a real time LEO regime (approximately 30% DOD). The results for this test also demonstrate stable EOC and EOD performance.

Figure 3. RNH-30-1 Real Time LEO Cycle Results

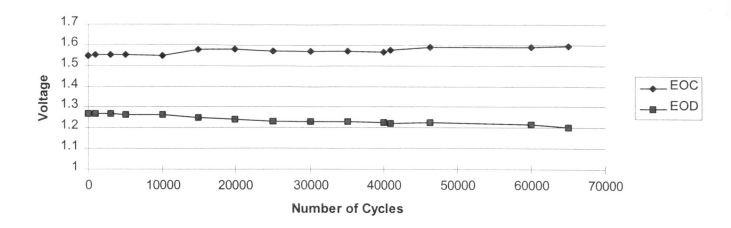

Figure 4. RNH-50-15 Accelerated LEO Cycling Results

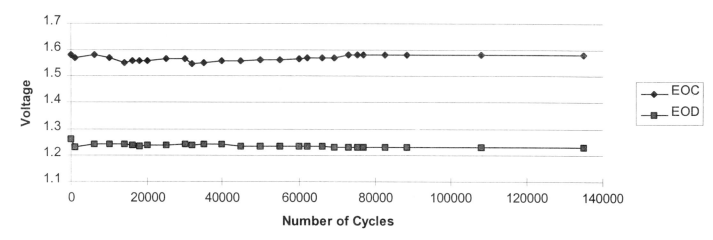

Single Pressure Vessels

Another design which is currently on life test in-house at Eagle-Picher is the single pressure vessel (SPV) battery. This concept, a more recent addition to the family of nickel-hydrogen battery designs, features one pressure vessel as an enclosure for the entire complement of series-connected-cells. Each individual cell within the large pressure vessel has its own gas permeable container, preventing electrolyte sharing between cells, but providing for a common hydrogen reservoir during the charge portion of the regime. This design gained its most prominent on-orbit heritage during the Iridium ® program, providing power for all of the spacecraft in the constellation.

A life test study had been conducted to characterize battery performance early on in the manufacture of Iridium ® flight hardware. A battery was subjected to a 100 minute accelerated LEO regime, to a 70% depth of discharge. During this test, the battery was charged at 36.7 amps for 51 minutes, followed by a 49 minute discharge at 36.7 amps. As the test progressed, the rates were adjusted slightly, as necessary, to maintain an end-of-discharge voltage of 1.0 V per cell, or greater. A total of 4612 cycles were achieved before this portion of the test was terminated. The next phase of the characterization used the same battery from the previous portion, but the rates were changed in order to obtain a 31% depth of discharge during the cycling. Lower rates were used during this 100 minute cycle, and a total of 4812 cycles had been achieved before the test was considered complete and was terminated.

Two SPV batteries are currently on test in Joplin. A 50 ampere-hour (SAR 10065) battery is being subjected to a 100 minute cycle regime at -5 degrees C, and has completed in excess of 25, 287 cycles. The second SPV is an SAR 10081, a

60 ampere-hour unit which is also seeing LEO type cycling, has accumulated in excess of 21,350 cycles.

Special Test Regimes

In addition to the various standard (repetitive cycle) LEO and GEO life tests at Eagle-Picher (Joplin), there are tests ongoing which closely resemble the actual mission regime, where the cycling simulates nominal tasking and housekeeping type loads, as well as occasional worst case conditions. The programmed test mentioned here, which includes the characterization of four 100 ampere hour 8-cell packs (See Figure 5), follows the expected loads for one full year of mission, including all eclipse times and full sun periods, with the schedule repeating itself only after 365 days of the simulation is complete. The oldest 8-cell pack in this test has accumulated over 1600 mission days, has demonstrated satisfactory performance during that interval, and continues to accumulate real time data.

Figure 5. RNH-100-9 Eight-Cell Pack

Conclusions

A variety of nickel-hydrogen battery cell life tests, currently being performed at Eagle-Picher and at other facilities, continues to characterize performance over time, and demonstrate product longevity. The test data generally supports the integrity of Eagle-Picher nickel hydrogen cells and batteries, and provides performance characterization information when mission simulation is required.

Table 4 summarizes a portion of the results acquired to date, as it related to Eagle-Picher nickel hydrogen products. Figure 6 [1] illustrates the cycle data results in chart form.

Table 4. Summary of Data Acquired

Total Spacecraft On-Orbit	314
Total Cells in Orbit	11,225
Total Operational Cell Hours in Orbit	> 437,247,323
Longest Running ACC LEO	> 127,770 Cycles
Longest Running RT LEO	> 69,806 Cycles
Longest Running ACC GEO	> 44 Seasons
Longest Running RT GEO	> 42 Seasons
Longest Running On-Orbit LEO	> 59,529 Cycles
Longest Running On-Orbit GEO	> 30 Seasons

Figure 5. Eagle-Picher Cycle Data

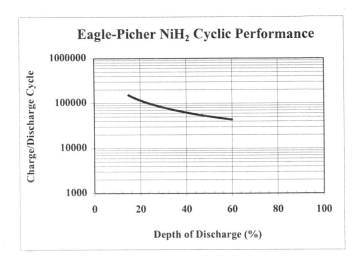

Acknowledgment

1) Information provided by George Datum, Eagle-Picher Technologies, LLC, Joplin, MO

LITHIUM-ION POLYMER BATTERIES FOR SPACE SYSTEMS

Oliver Gross and Chris Fox
Valence Technology Inc.
301 Conestoga Way
Henderson, NV 89015

David P. Roller and Les Shimanek
Alliant Techsystems Inc.
Alliant Precision Fuse Company, Power Sources Center
104 Rock, Road, Horsham, Pa 19044

ABSTRACT

Alliant Techsystems Inc. and Valence Technology, Inc. have designed and delivered a Lithium-Ion Polymer Battery for the USAF MicroSat XSS-10 Program. This program's objective was to determine if the Li-Ion Polymer technology could be developed to meet the rigors of aerospace and space environments.

The current battery technologies presently in use by industry and government have been Silver-Zinc (Ag-Zn) for launch vehicle batteries and Nickel-Cadmium (Ni-Cd), Nickel-Metal Hydride (Ni-MH) and Nickel-Hydrogen (Ni-H2) for satellite systems. Li-Ion Polymer battery technology has the ability to provide an upgrade path to a more modern technology, which can, in turn offer reduced total system cost and weight, as well as improved performance.

INTRODUCTION

In 1998 Alliant Techsystems and Valence Technology received a contract from the Boeing Company, NR99440978, for the design and development of the battery power source for the USAF/DICE XSS-10 MicroSat Program. This battery had to be designed to meet the rigors of space vehicle launch and pass all man rated safety requirements, as this battery was planned for a Space Transport System (shuttle) launch.

The initial design identified the power demands and the physical/ environmental specifications for the battery. In addition to meeting shuttle launch and safety guidelines, the battery design also had to pass safety requirements as stipulated by NASA's Johnson Space Center (JSC).

CELL DESCRIPTION

The power requirements for the battery were constrained by both the maximum allowable volume and weight, thereby requiring a battery with an energy density greater than either Ni-Cd or Ni-MH. Alliant Techsystems Inc. and Valence Technology examined these needs and selected a commercially available cell pack. By selecting a commercial-off-the-shelf (COTS) product, the development cost of the battery was greatly reduced.

PERFORMANCE RQUIREMENTS

The environmental conditions the battery experiences during the vehicle launch are shown in the flowing table:

CONDITION	LEVEL	
Acceleration	20 G constant	
Random Vibration	20 Hz	.04 g^2/Hz
	80 – 500 Hz	0.040 g^2/Hz
	2,000 Hz	0.04 g^2/Hz
	Overall	6.8 Grms
Pyrotechnic Shock	100 Hz	28 G's
	800 Hz	1,500 G's
	5,500 Hz	1,500 G's
	10,000 Hz	3,000 G's
Temperature	Operating	0 - + 40 C
	Non-Operating	-10 to + 50 C
Vacuum	1 x 10^{-5} to 950 torr	

Most importantly, the power storage system must operate in a space vacuum, and at satellite operating temperatures.

BATTERY TESTING

Safety testing was conducted at both cell pack and battery levels. Since Lithium-Ion Polymer technology is relatively new and has no flight heritage, it is mandatory that safety test data specific to normal satellite operation and abuse conditions be evaluated. Evaluation of performance to worst-case power and capacity demands was also conducted in order to establish a baseline for this particular battery technology.

Tests performed were as follows:

TEST	DESCRIPTION
1	Capacity Performance – at the battery module.
2	Cell to Cell and Cell to Battery Box Isolation – at the battery module level.
3	Maximum Discharge Current Profile (External Maximum Sustained Discharge Current) – at the cell pack and battery module.
4	Internal Cell Short – at the cell pack and battery module level
5	Cell Element, Single Cell Unbalance – at the battery module level.
6	Cell Bag Venting – at the cell pack and battery module level.
7	Cell Deformation – at the cell pack level.

1) <u>Capacity Testing</u>. The capacity performance test was conducted to simulate the mission profile and verify the margin of safety during the satellite mission. The tests were performed at ambient and cold operating temperature. The performance of the battery is shown in Figure 1.

2) <u>Isolation Testing</u>. The cell-to-cell isolation test was conducted to ensure that the potential of a short circuit occurring was low and that the structure of the Li-Ion Polymer cell pack (the foil pouch) would not "bleed" voltage and therefore adversely impact the operation of the satellite system.

3) <u>Maximum Discharge Current</u>. Testing of the XSS-10 Battery Module was conducted through the application of a 100A shunt across the module contacts. The testing simulated a sustained 2C discharge rate for a period of 30 minutes.

4) <u>Internal Cell Short.</u> The internal cell short test was conducted by penetrating an active cell pack with an aluminum nail and recording the cell voltage response. The results are shown in Figure 2.

5) <u>Cell Element, Single Cell Unbalance.</u> This test consisted of connecting a fully discharged cell in series with fully charged cells, and then discharging the series string. The discharged cell was driven to 0V and the cell's temperature was monitored during the test.

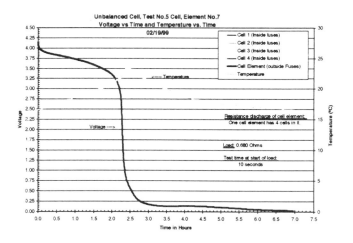

6) <u>Vacuum Testing.</u> A cell pack vacuum test was conducted to ensure that the seal of the cell pack pouch would remain intact throughout the operational life of the satellite. The Boeing Company conducted the cell-pack level test at their Canoga Park facility. Testing at the battery module level was conducted on the engineering models, qualification and flight batteries.

7) <u>Cell Deformation Testing.</u> This test was performed to determine the level of safety of the battery module in the event that the battery was severely deformed in a manned environment (damaged in the shuttle cargo bay). The ability of the cell packs to withstand the deformation shown in the picture below provided a sufficient indication of the robust design of the Valence cell pack.

MECHANICAL DESIGN

The mechanical design of the battery module centered on examining the structural support and the vibration dampening provided by the cell packs. Due to the flat form factor of the cell packs, the volume of active battery components could be maximized. To meet additional safety requirements, "space rated" fuses were incorporated into each cell pack. In addition, thermocouples were installed in order to provide telemetry data to the satellite electronics and ground control stations.

The physical configuration of the battery is shown in the graphic below, the case has been eliminated to show the efficient packaging.

Each cell pack has a fuse ultrasonically welded to the negative lead. Four cell pack/fuse assemblies in parallel form a cell element. Four cell elements in series form the battery module, multiple modules form the satellite battery.

As a result of the inherent strength of the modular approach using individual cell packs, the vibration models of the battery module exceeded the vibration requirements of the USAF. The structural analysis is shown in Figure 3.

The added benefit of the cell packs providing support to the overall assembly led to an extremely light-weight (4.4 pounds) module.

RESULTS AND DISCUSSION

The battery modules delivered to The Boeing Company are currently being integrated into the satellite. Launch of the XSS-10 mission, scheduled for the last quarter of this year, will provide the final demonstration of this battery technology for space applications. A summary of the battery specification is shown in Figure 4.

FURTHER PURSUITS FOR LITHIUM-ION POLYMER BATTERIES

Space uses for batteries apply to four basic applications. These include launch vehicles, Geosynchronous (GEO) Satellites, Low Earth Orbit (LEO) Satellites and Space Platforms. Examining each of these four categories separately shows how the applications can benefit from Li-Ion Polymer technology.

Launch Vehicle Batteries. The main battery type for launch vehicles is currently Ag-Zn. In most launch vehicle applications the three types of batteries are: Main/Core Battery, Flight Termination System Battery and Pyrotechnic battery. Due to the shorter "wet life" of the Ag-Zn electrochemistry, it is possible for the batteries to be removed and discarded in favor of fresh ones in the event that a mission launch experiences extended delays. This can result in large cost overruns and time delays while the battery replacement is taking place. Preliminary designs for this type of battery have been developed and are summarized below:

CONFIGURATION:

- **Capacity:**... 6.0 AHr
- **Rated Capacity:**....................................... 2.0 AHr
- **Battery OCV:**..32.8 VDC
- **Ma Sustained Discharge Rate:**..................... 6.0 A
- **Min Volatge (Worst Case):**.......................29.8 VDC
- **Battery Mass:**...6.9 lbs
- **Battery Impedance:**..................................120 mΩ

GEO and LEO Satellites. Since the first commercial flights using the technology in 1983, the development of batteries for GEO and LEO applications has typically been supported with Ni-H2 batteries. The flight proven heritage of this system and its ability to exhibit proven abuse tolerant performance ensures a future for this type of battery. The largest issue with Ni-H2 batteries is the need for pressure containment. Because of this feature, these batteries are heavy, and the development of new sizes is costly; thus the cost of Ni-H2 batteries far exceeds that of Li-Ion Polymer.

Lithium-Ion Polymer testing at TRW, the Boeing Company has shown the long cycle life demonstrated by this technology. Lab cells tested to 100% Depth-of-Discharge (DoD) have cycled through 3,500 cycles. LEO cycling data is shown in Figure 5.

The Li-Ion Polymer battery does not require a pressure vessel. As a result the need to develop large pressure vessels when resizing the battery is not required. The end result is faster development and lighter weight – both key issues when developing satellite systems.

Space Platforms. High Cycle Life, Long Shelf Life and High Energy Density batteries are required to support numerous systems on the International Space Station. Building upon the LEO/GEO structure, the size of the batteries needed for ISS can be met by the Li-Ion Polymer technology available today.

SCALEABLE TECHNOLOGY

Scaleable technology will focus on two attributes; the type of film and the size of the bi-cell unit and cell pack.

By adjusting the film weight of the bi-cell laminate and the size of the bi-cell the customer demands can be met.

Complete manufacturing capability of the laminate at Valence Technology's Mallusk, Northern Ireland facility enables optimization of the film to each customer's needs.

Alliant Techsystems and Valence Technology have developed a Flexible Automated Bi-Cell and Lamination (FABAL) machine that allows for quick changeover in sizes of bi-cells. This machine allows for production of bi-cells from as small as 2.5 inches to as large as 12 inches. The available ranges are shown in Figure 6.

What these capabilities bring to the aerospace community is the ability to produce cell packs that vary in energy density from 100 Watt-Hours/Kilogram to as high as 190 Watt-Hours/Kilogram.

As a result of this technology, a module approach of has been developed that can meet the needs of the aerospace market. Alliant Techsystems has built 135 AHr and 220 AHr batteries. Alliant Techsystems is working on scaling this technology to a larger level (700 to 1,200 AHr batteries) in order to support undersea systems. A modular approach for satellite applications is shown in the Figure 7 – *SCALEABLE SATELLITE BATTERY SIZE; 50V BATTERY*. This module, with by-pass switches can be connected in a maintainable and producible configuration, in order to provide the satellite's required capacity and voltage level.

**Proposed Lithium-Ion Polymer Battery
Virtual Energy Storage Unit**

The modular approach enables Alliant Techsystems and Valence to rapidly size the Energy Storage Unit to accommodate the needs of the customer. An example of this capability is shown for a 50V modular system in Figure 7.

CONCLUSIONS

Lithium-Ion Polymer technology is continuing to improve by raising the energy density, improving cycle life and lowering cost.

Alliant Techsystems and Valence Technology have developed the design rules and production methodology to mass-produce large capacity cells and deliver energy storage systems to meet the aerospace market.

ACKNOWLEDGEMENTS

The development work for large-scale Lithium-Ion Polymer cells was supported through DARPA funding in the COSSI program in addition to support from the US Navy Office of Naval Research; USN Chief Technology Office, and the USN Deep Submergence Program Office.

Aerospace development for the USAF XSS-10 program was through the USAF Research Lab, Kirkland AFB and The Boeing Company – Rocketdyne Division.

Mr. Chip Potter of Advance Simulation Technology supported mechanical simulation effort for the XSS-10 program and the USN programs.

FIGURES

FIGURE 1 – CAPACITY TESTING

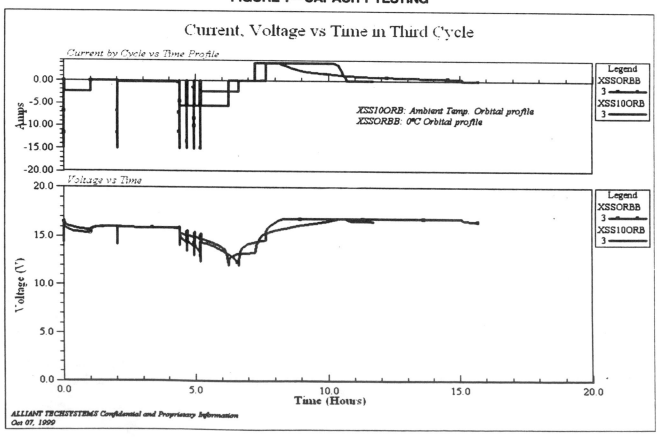

FIGURE 2 – NAIL PENETRATION TEST

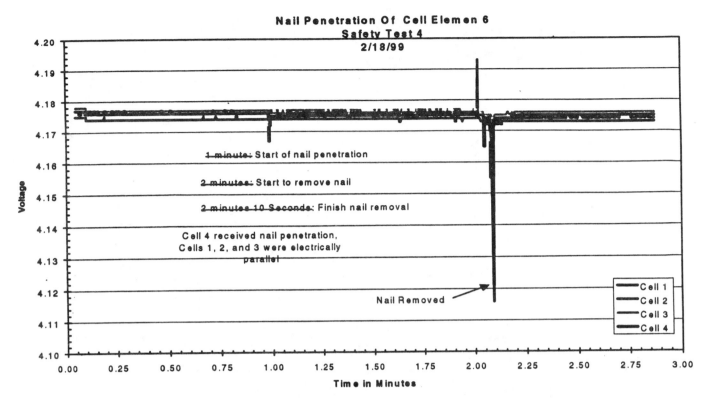

Nail Penetration Of Cell Elemen 6
Safety Test 4
2/18/99

1 minute: Start of nail penetration

2 minutes: Start to remove nail

2 minutes 10 Seconds: Finish nail removal

Cell 4 received nail penetration,
Cells 1, 2, and 3 were electrically
parallel

Nail Removed →

Cell 1
Cell 2
Cell 3
Cell 4

Voltage (y-axis)
Time in Minutes (x-axis)

FIGURE 3 – STRUCTUAL ANALYSIS

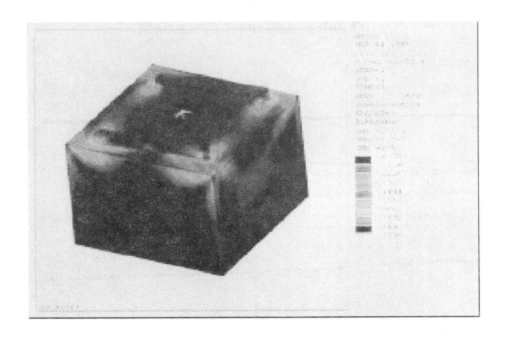

FIGURE 4 – XSS-10 BATTERY DESCRIPTION

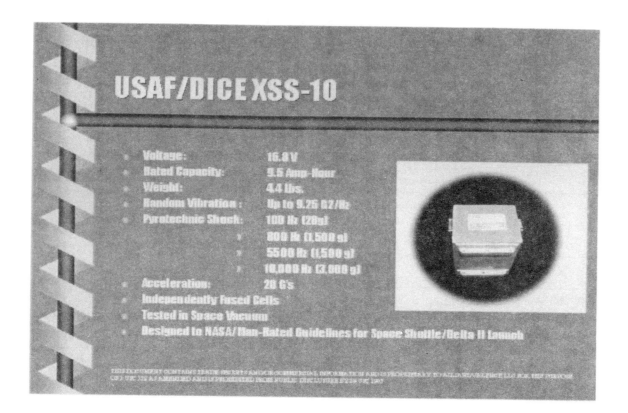

FIGURE 5 – LEO ORBIT CYCLE DATA

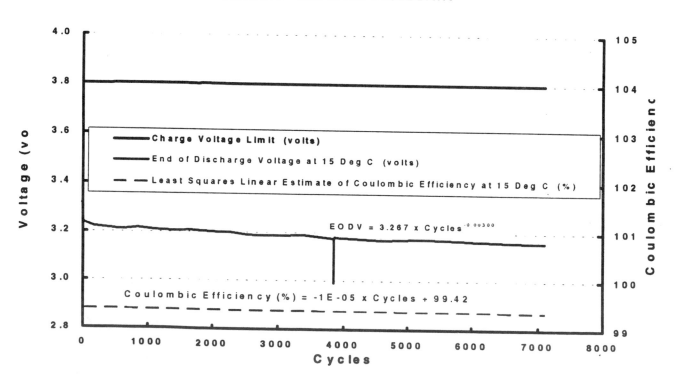

FIGURE 6 – BI-CELL FABRICATION CAPABILITIES

Valence Cell Size Range Summary

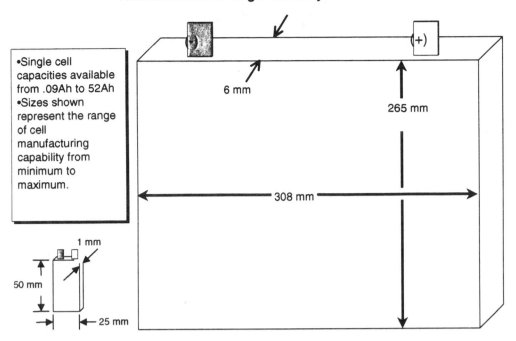

•Single cell capacities available from .09Ah to 52Ah
•Sizes shown represent the range of cell manufacturing capability from minimum to maximum.

6 mm

265 mm

308 mm

1 mm

50 mm

25 mm

Contact Valence Engineering for Capacity Projections for Your Specific Size
Cell Thickness from 1mm to 6mm available for all sizes.
Sketch proportionate however, not to scale.

FIGURE 7 – SCALEABLE SATELLITE BATTERY SIZE; 50V BATTERY

Required Rate (kw)	Calculated Capacity (kWhr)	VESU Voltage	Quantity	Li-ion Battery Configuration Capacity Ahr	Whr	Size Vol. (liter)	Mass (kg)	Virtual Energy Storage Unit (VESU) # of Packs	Config-uration	Ahr	Whr	Width (mm)	Mass (kg)
2	5.4	12 V	8	1,740	6,264	25.6	56.8	15	3 x 5	218	783	97.7	7.1
2.5	6.75	12 V	8	2,088	7,517	30.5	65.0	18	3 x 6	261	940	116.2	8.1
3	8.1	12 V	8	2,436	8,770	35.3	73.1	21	3 x 7	305	1,096	134.8	9.1
3.5	9.45	12 V	8	2,784	10,022	40.2	81.3	24	3 x 8	348	1,253	153.3	10.2
4	10.8	12 V	8	3,132	11,275	45.1	89.4	27	3 x 9	392	1,409	171.9	11.2
4.5	12.15	12 V	8	3,480	12,528	49.9	97.6	30	3 x 10	435	1,566	190.4	12.2
5	13.5	12 V	8	3,828	13,781	54.8	105.8	33	3 x 11	479	1,723	208.9	13.2
5.5	14.85	12 V	8	4,176	15,034	59.7	113.9	36	3 x 12	522	1,879	227.5	14.2
6	16.2	4 V	24	4,872	17,539	72.0	162.2	14	1 x 14	203	731	91.5	6.8
6.5	17.55	4 V	24	5,220	18,792	76.9	170.4	15	1 x 15	218	783	97.7	7.1
7	18.9	4 V	24	5,568	20,045	81.7	178.6	16	1 x 16	232	835	103.9	7.4
7.5	20.25	4 V	24	5,916	21,298	86.6	186.7	17	1 x 17	247	887	110.1	7.8
8	21.6	4 V	24	6,264	22,550	91.5	194.9	18	1 x 18	261	940	116.2	8.1
8.5	22.95	4 V	24	6,612	23,803	96.3	203.0	19	1 x 19	276	992	122.4	8.5
9	24.3	4 V	24	6,960	25,056	101.2	211.2	20	1 x 20	290	1,044	128.6	8.8
9.5	25.65	4 V	24	7,308	26,309	106.0	219.4	21	1 x 21	305	1,096	134.8	9.1
10	27	4 V	24	7,656	27,562	110.9	227.5	22	1 x 22	319	1,148	141.0	9.5
10.5	28.35	4 V	24	8,004	28,814	115.8	235.7	23	1 x 23	334	1,201	147.1	9.8
11	29.7	4 V	24	8,352	30,067	120.6	243.8	24	1 x 24	348	1,253	153.3	10.2
11.5	31.05	4 V	24	8,700	31,320	125.5	252.0	25	1 x 25	363	1,305	159.5	10.5
12	32.4	4 V	24	9,048	32,573	130.3	260.2	26	1 x 26	377	1,357	165.7	10.8
12.5	33.75	4 V	24	9,396	33,826	135.2	268.3	27	1 x 27	392	1,409	171.9	11.2

Proceedings of the IECEC '01
36th **Intersociety Energy Conversion Engineering Conference**
July 29-August 2, 2001, Savannah, Georgia

IECEC2001-AT-37

FLIGHT TEST OF A SOLAR ARRAY CONCENTRATOR ON MIGHTYSAT II.1

Theodore G. Stern
Composite Optics, Inc.
San Diego, CA 92121
Email: tstern@coi-world.com

Patrick Bonebright
Composite Optics, Inc.
San Diego, CA 92121
Email: pbonebright@coi-world.com

ABSTRACT

As part of the experiment suite on Mightysat II.1, a Solar Array Concentrator (SAC) panel was flown. The SAC experiment uses the Light Concentrating Panel (LCP) technology, an orthogrid arrangement of ultra-lightweight composite mirror strips that form an array of rectangular mirror troughs to reflect light onto standard, high-efficiency solar cells at a concentration ratio of approximately 3:1. The solar cells are mounted onto a flat, high-conductivity composite laminate with a co-cured Kapton flex-circuit. The mirror strips also provide structural rigidity, essentially replacing the honeycomb substrate usually used on planar photovoltaic panels with an orthogrid "superstrate." The design and analysis of the SAC module is described, and the data from the first eight months in orbit is analyzed.

INTRODUCTION

Solar concentrating photovoltaic arrays have been developed over the last twenty-five years to a state of flight readiness (Stribling, 2000; Allen et al, 1995; Allen and Piszczor, 1994). The main impetus for their development has been to decrease the cost of prime power generation, in terms of $/watt, by replacing expensive high efficiency space-quality solar cells with less expensive optics. The optics effectively replace solar cell area by collecting insolation (sunlight) and directing it to the now smaller cell areas. Another potential advantage of solar concentrators is the ability to more effectively shield the solar cells against hazardous environments, while maintaining the low mass needed for space solar arrays.

A perception of increased risk to mission success has been a factor preventing even more widespread use of concentrator technology. The primary areas of concern are associated with degradation of optics, thermal control of the solar cell under increased and possibly uneven illumination, and the need to more accurately point the arrays towards the sun. Optical degradation is a particularly difficult risk to manage because the sources of contamination products and their environmental interactions are difficult to predict, model, and test on the ground. The implementation of flight experiments of concentrator modules is a method for retiring these risks. The results from one such flight experiment are described in this paper.

NOMENCLATURE

SAC – Solar Array Concentrator
LCP – Light Concentrating Panel
BOL – Beginning Of Life
EOL – End Of Life
LEO – Low Earth Orbit
AM0 – Air Mass Zero, i.e., space insolation
Vmp – Maximum Power Voltage
Imp – Maximum Power Current
Pmp – Maximum Power Point
Voc – Open Circuit Voltage

Light Concentrating Panel Technology Description

The design and theory of operation of a modular, front-lit reflective solar concentrator, which we call the Light Concentrating Panel (LCP), was described in detail in a previous paper (Stern, 2000). The LCP design is technology transparent, that is, it is similar enough in physical and operational characteristics to be used as a plug-in replacement for a rigid planar solar panel. The basic design, shown in Figure 1, comprises an array of concentrator elements, each having a four-sided mirror trough with each wall having a high solar reflectance, and a solar cell positioned at the bottom aperture.

The mirror trough and cell are mounted upon a base plate consisting of a sheet of lightweight composite material having high thermal conductivity. The solar cells interconnect to a flex-circuit wiring layer which is integral to the base plate using conventional interconnect derived from electronic packaging technology. The LCP objectives are to enable higher performance at lower cost by reducing the required quantity, per unit panel area of the highest efficiency solar cells.

The technology transparency features of the LCP include thickness similar to large rigid planar panels (about 2.5cm), solar cell types, layout and assembly approaches identical to planar solar panels, and required pointing (+/-20 degrees in the most sensitive axis) that allows conventional pointing mechanisms. In the Mightysat II.1 program, the ability to substitute a SAC experiment module for a modular planar experiment module was enabled by these technology transparency features.

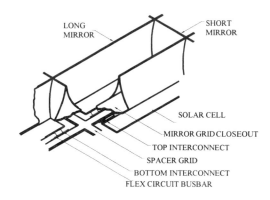

Figure 1. LCP technology used for the SAC experiment

Mightysat II.1 SAC Experiment Design Analysis

One of the unusual aspects of the SAC experiment on Mightysat II.1 is that it formed part of the prime power system for the spacecraft. This required that the SAC, while designed as a technology demonstration experiment, also have sufficient margin to provide the power performance needed to execute the spacecraft mission. In addition, the design reviews had to demonstrate that the SAC experiment would not endanger the entire mission by propagating failures or degradation to the remainder of the solar prime power array.

A complete set of structural, thermal and electrical/photovoltaic analyses was performed on the SAC to assure that it would meet these requirements. The LCP module's structural durability was analyzed by creating a Finite Element Model of the structural components of the design, adding in the non-structural mass, applying launch loading conditions and determining the stresses imposed on the various structural components. The results of the analysis of maximum principal and shear stresses showed a positive margin of safety of greater than 0.19 for all components of the design, above the factor of safety of 1.4. This shows a robust structural response.

A thermal model was used to determine the temperatures of the solar cell and other components in orbit. A detailed or micro-model of the LCP element, was used to determine the detailed temperatures of the components in a steady state and orbital transient thermal environment. The micro-model assumed the SAC was not influenced by the supporting array structure, and so could use the repeatability of the design to allow increased model resolution. A second, less detailed or macro-model was used to determine the thermal effect of the solar panel frame on the SAC. The objective of thermal analysis was to determine cell temperature to use as input to the electrical analysis, and to check for excessive component temperatures or temperature gradients.

The micro thermal model nodal model, shown in Figure 2, was run for the maximum case environment with worst case end-of-life material thermal properties listed in Table 1, in order to evaluate the worst-case temperature conditions. The result of the micro-thermal model under these conditions is shown in Figure 3. The peak solar cell temperature is 104C, and the temperature gradients throughout the concentrator element, i.e. from cell to radiator to mirror, are relatively small, in the range of 3C.

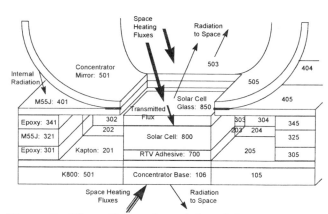

Figure 2. Micro-thermal model nodal arrangement

Table 1. Micro thermal model worst case, end-of-life material properties and environments

Property/Environment	Value
Solar Cell Absorptance	0.88
Solar Cell Efficiency	0.15
Mirror Absorptance	0.15
Backside Absorptance	0.39
Solar Flux – Hot Case	1393 W/m2
Earth IR	236 W/m2
Earth Albedo	42% of Solar

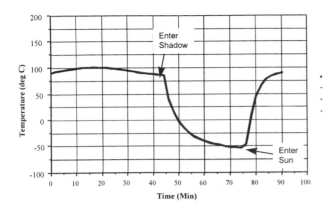

Figure 3. Results of the micro-thermal model

Given a predicted SAC element cell temperature of 104C, we then used the macro thermal model determined the effects of mounting the concentrator on the Mightysat II.1 solar panel frame. Assuming the frame was not thermally connected to the SAC, thermal interaction was caused only by radiation interchange and earth albedo/radiation blockage. The magnitude of this thermal gradient was 4C and results in an estimation of peak solar cell temperature increasing to 108C. This peak solar cell temperature was then used in the electrical analysis for sizing the solar cell strings.

The SAC electrical design was based on the Mightysat II.1 Solar Panel Specification, including a 1-year End-Of-Life (EOL) power output of 27.5W at any solar array bus voltage from 26.7V to 33.7V, and a specific aperture and height envelope for the SAC. The electrical design arranged the cells in series parallel relationships to meet these electrical requirements at EOL, and arranged these strings in row-column sets to create a physical layout compatible with the SAC envelope. The schematic of Figure 4, having two strings of 45 series-connected SAC elements arranged in two row-column arrays of 9x5 SAC elements, was implemented in a flex circuit configuration.

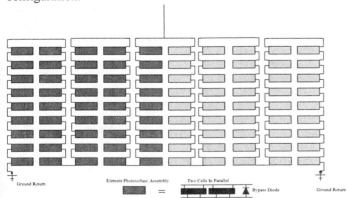

Figure 4. SAC Electrical Schematic

The electrical analysis was divided up into a beginning-of-life (BOL) analysis to determine the voltage and current of the design at peak temperature, and an end-of-life (EOL) analysis to evaluate the effects of contamination, space radiation, micrometeoroids and thermal cycling that can degrade the optics and solar cell strings. The BOL analysis is summarized in the power chain analysis shown in Table 2. The EOL degradation factors are broken up into their effects on voltage and current, separately and summarized in Table 3. Note that both of these analyses consider the operation of the SAC at the peak power point, which is important for correlating these results to the flight telemetry analysis presented at the end of this paper.

Table 2. Beginning of life analysis of LCP module electrical outputs

Item	Efficiency	W/m^2
Insolation	—	1393
Optical Efficiency	0.95	1323
18%GaAs @108C	0.149	197
Wiring Efficiency	0.99	195
Mismatch Loss	0.99	193
Net Efficiency	0.138	193
SAC Power @ $0.18m^2$ = 34W		

Table 3. End of life analysis of LCP module outputs considering environmental degradation factors

Item	Vmp	Imp	Pmp
BOL Outputs at 108C	33.8V	1.01A	34W
Optical Degradation		0.95	
Coverslip Darkening		0.985	
Radiation Degradation	0.99	0.99	
Series Resistance Increase (From Thermal Cycling)	0.98		
EOL Outputs at 108C	32.8V	0.936A	31W
1yr @ 300n.mi.			

SAC Flight Test Article and Results

The SAC flight test article was fabricated and delivered for integration into the Mightysat II.1 solar array. The module used single-junction GaAs solar cells fabricated by Spire Corporation, and incorporated three temperature sensors, one each on a reflector, a solar cell, and a radiator fin area of the base plate. Prior to delivery, the module was tested for electrical isolation, power output under simulated AM0 solar insolation at various incidence angles, and thermal cycling qualification. The mass of the SAC module was measured at 697 grams. With a measured room temperature AM0 output of 40.6W, this translates into a specific power of better than 58 Watts / kilogram.

Figure 5. SAC Flight Test Article module (top), and integrated into the MightysatII.1 array (bottom)

Mightysat II.1 launched on July 19, 2000 and initial checkout showed that all power subsystems, including the SAC, were functioning within normal parameters. Telemetry from the SAC was transmitted from the spacecraft from that day and has continued through the present. The telemetry received from the SAC includes frames for time, the three temperatures, open circuit voltage for the module, solar array bus voltage, SAC current delivered to the solar array bus, and a flag to indicate whether the spacecraft was illuminated.

One of the challenges in interpreting the telemetry from the SAC experiment arises from how the module is used in the power system. The Mightysat II.1 power system uses a battery charge control approach that draws power initially from all of the solar panel modules, and then sequentially opens modules as the battery state-of-charge increases. The SAC, as an experimental solar panel module, was wired to be the second module to be shut off, i.e., to make it less critical to the function of the power system. Because of this, most of the telemetry shows the SAC to be active only for a short time after eclipse exit, when the battery is drawing the most current and needs all of the modules to be turned on. Since this is also the time when the temperatures are low and rapidly varying, and the solar array bus voltage is low, additional analysis was needed to interpret the data.

We evaluated the SAC data in raw form, and also when corrected for seasonal variation, and to a specific operating temperature (the 108C predicted peak temperature). The raw data allowed us to evaluate how many watts are being delivering to the bus and determine if the SAC module met the

power system requirements. These data, shown in Figure 6, indicate a slightly excessive reduction in power output of 10.5% for the first eight months (we predicted a ~10% reduction in one year). The tail-off in the most recent data is attributable to seasonal solar illumination decrease and shorter periods of current draw from the battery charge control. Despite the decrease in power, the SAC has sufficient design margin, so as to allow it to continue to meet mission requirements.

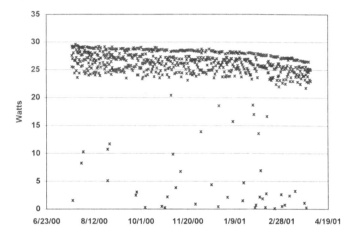

Figure 6. Actual power delivered to bus, not seasonally adjusted

The analysis of power delivered also shows a number of data points below the maximum power capability of the module. The flight data shows that these low-power points occur at low post-eclipse temperatures. At lower temperatures, widening of the GaAs bandgap results in higher red-loss and therefore lower light-generated current. The solar cell temperature telemetry in Figure 7 show a highly condensed view of the temperature cycling experienced by the solar cells. It shows the high occurrence of post-eclipse low temperature illuminated condition. The temperature data also show a consistent high-side temperature of around 104C (some seasonal increase in illumination can be seen in the December time frame), and a relatively benign low-side eclipse temperature of –65C. These temperature extremes are nearly identical to those predicted in the thermal analysis.

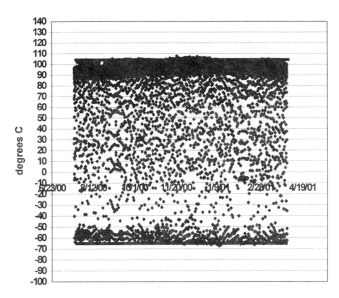

Figure 7. Solar cell temperatures on orbit

The SAC current delivered to the solar array bus is shown in Figure 8. A degradation of 13% was more than predicted and cannot be accounted by solar cell radiation, which was expected to have an effect of less than 2%. We have further evidence that the solar cell has not degraded from the open circuit voltage telemetry in Figure 9, which shows a reduction of less than 0.5% in Voc the same time frame. This current degradation of 13% over eight months exceeded our prediction of 7% for one year. The bulk of the degradation appears to have occurred in the first five months and the current values appear to be reaching an asymptote.

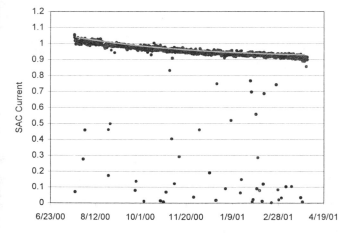

Figure 8. SAC current trends, corrected for seasonal and temperature effects

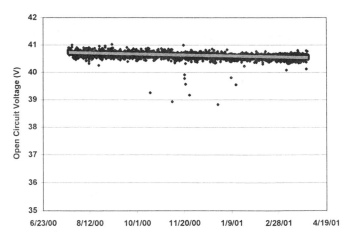

Figure 9. SAC Open Circuit Voltage trends, corrected for seasonal and temperature effects.

Discussion of Flight Test Results

Flight test data for the SAC experiment on Mightysat II.1 showed an excellent correlation to predicted values at beginning of life, both electrically and thermally. It was recognized during the design phase that current decreases from changes in the performance of the reflected optics would be difficult to predict.

Electrical current degradation in reflective concentrators in space is a phenomenon observed on prior LEO flight experiments (Stern, 1988; Severns et al, 1990). There are several hypotheses that can explain this observation, but most revolve around the potential for organic contaminants degrading the mirror reflectance. For first surface protected silver coatings of the kind used here, several factors could exacerbate contamination effects. The transparent protective layer is typically SiO_x/SiO_2 – a good insulator that can build up electrostatic charge and attract outgassing contaminants from the spacecraft or solar array. Once on the surface, ultraviolet radiation darkens organic contaminants and increases their absorptance. In addition, the nature of a reflective optic results in two passes through any contaminant layer, doubling the absorption loss. Finally, our review of the SAC design showed that we could have reduced the potential for self-contamination if we had included a backside vent-path for outgassing of adhesives and other organics used in the design (Katz, 2000).

Next Steps

The use of the International Space Station would provide an ideal way to evaluate LEO orbital environment effects on new technologies such as the LCP. This could easily advance the state-of-the-art of reflective concentrators, as a follow-up experiment to validate the corrective actions needed to achieve durability of mirror surfaces. Such an experiment could be accomplished at a fraction of the cost of an autonomous spacecraft flight experiment, and would have the advantage of post-test direct observation.

Separate economic analysis on the LCP has shown the ability to reduce space panel costs by 40% by eliminating 2/3 of the required solar cell area, even when additional optics degradation is included. The final space qualification of this technology could enable these cost benefits to be realized.

Conclusions

A reflective solar array concentrator experiment flown on Mightysat II.1 has shown the ability of the LCP, a panel of modular reflective concentrator elements, to act as a plug-in replacement for planar photovoltaic arrays. The power and thermal response of the SAC was initially within expected parameters. While we have observed higher-than-expected current degradation during the first 8 months on orbit, the SAC has been able to perform as both a useful experiment and part of the prime power system for Mightysat II.1. The flight experiment has shown that this technology is suitable for near-term application in spacecraft needing higher efficiency arrays at lower cost.

ACKNOWLEDGMENTS

The author gratefully acknowledges the following individuals for their contribution to this work:

1st Lt Jacob A. Freeman of the U.S. Air Force Research Laboratory for his assistance in obtaining, processing and interpreting the SAC telemetry.

Rick Kettner of Spectrum Astro Incorporated for his assistance in integrating the SAC into the Mightysat II.1 solar array.

Ken Qassim of the U.S. Air Force Research Laboratory for his assistance in promoting the SAC experiment and integrating it into the Mightysat II.1 flight test program.

REFERENCES

Allen, D. M. et al, "The SCARLET Light Concentrating Solar Array," Proceedings of the 25th IEEE Photovoltaic Specialists Conference, 1995

Allen, D. M. and Piszczor, M., "Spacecraft Level Impacts of Integrating Solar Concentrator Arrays," Proceedings of the 24th IEEE Photovoltaic Specialists Conference, 1994

Katz, Dr. I., personal communication, 2000

Severns, J. G. et al, "Systematic Anomalies in Experiment Data from LIPSIII," Proceedings of the 21st IEEE Photovoltaic Specialists Conference, 1990

Stern, T., "Interim Results of the SLATS concentrator Experiment on LIPSIII," Proceedings of the 20th IEEE Photovoltaic Specialists Conference, 1988

Stern, T., "Technology Transparent Light Concentrating Panel for Solar Arrays," Proceedings of the 18th AIAA International Communications Satellite Systems Conference, 2000

Stribling, R., "Hughes 702 Concentrator Solar Array," Proceedings of the 28th IEEE Photovoltaic Specialists Conference, 20008

Proceedings of IECEC'01:
36th Intersociety Energy Conversion Engineering Conference
July 29—August 2, 2001, Savannah, Georgia

IECEC2001-AT-38

"SOLARCON" CONCENTRATOR SOLAR ARRAY

Michael A. Brown, PE
Naval Research Laboratory, Washington, DC

James Moore
SRS Technologies, Huntsville, Alabama

ABSTRACT

NRL and SRS are developing a trough (channel) solar concentrator for spacecraft photovoltaic systems. A pair of thin film reflector sheets on either side of the PV array directs additional sunlight onto the array. Structural booms located every two panels along the array support the reflectors and tension them. Catenary-like edges on the reflector sheets distribute the suspension forces into isotropic stress that creates a perfectly flat reflective surface. The reflectors roll up on themselves and stow beside the PV panel stack, leaving the outermost panel exposed to generate power in the stowed position. Solarcon has a maximum geometric concentration ratio of about 2.5:1 which puts 2.47 Suns on the PV panels. This produces approximately a 100% increase in specific power for any photovoltaic panel array, reaching 180W/kg for 30% efficient cells.

INTRODUCTION

Spacecraft designers have been interested in solar concentrators for photovoltaic arrays since the early days of spaceflight, their purpose being to save some combination of cost, mass and stowed volume. Unfortunately, concentrators have proven difficult to implement for a variety of reasons, and they have come into regular use only recently. Of the many types considered, only two remain in the running: linear Fresnels and channels, both using high efficiency polycrystalline cells. "Ultralight" arrays using the lower efficiency amorphous silicon cell but without solar concentration are also in development.

Solarcon is a channel concentrator designed to meet the requirements of "general" spacecraft applications. These include generation of power in the stowed position or if the spacecraft tumbles (i.e., loses attitude control), and the ability to examine the arrays after final acceptance testing. The channel configuration is the only design that meets these requirements. Desirable "marketing" features, helpful in overcoming a generally negative perception of concentrators by spacecraft builders, include a substantial increase in specific power over unaugmented arrays, some decrease in stowed volume, and a minimum of novelty and risk—spacecraft builders are quite conservative!

SOLARCON SYSTEM DESCRIPTION

Figure 1 shows a generic channel system where a pair of flat reflectors directs additional sunlight onto the PV panels. The figure also shows the relation among the geometric concentration ratio (GCR-aperture divided by PV panel width), the ratio of reflector width to PV panel width, and the angle between these two. The practical limit of GCR is seen to be about 2.5:1. A fundamental requirement here is that the reflectors be optically flat: an uneven distribution of sunlight on the PV array decreases efficiency. Previous channel designs failed because the designers used rigid aluminum panels to get the flatness required. The weight penalty proved excessive, and the rigid panels could store only by folding behind the PV panels, limiting their width and therefore the GCR (to 2:1).

Soalrcon's patented (1,2) solution is to use thin film polyimide reflectors with several exclusive features. Figure 2a shows a 2-panel array with the catenary-like edge suspension that provides isotropic tension to create an optically flat reflector. The 4-panel array in Figure 2b shows how the reflectors roll up on themselves for storage, and 2c shows the reflectors stowed beside the folded PV array. This last feature leaves a full panel of cells exposed to generate power in the stowed configuration. A conventional system of pulleys and cables synchronizes the panel deployment and maintains the support booms parallel to each other as the array deploys. Springs in the support booms tension the reflectors at their corners, after the PV panels deploy and lock up. The end booms on geosynchronous satellites can rotate to a position past perpendicular to the PV panels to enable the reflectors to illuminate the entire length of the PV array, even at the extreme diurnal motion of the Sun (see 2a). The booms at the ends of the array have facing cavities, and these booms interlock in the stowed position and enclose the rolled-up reflectors and intermediate booms, to protect them during ground handling and launch. Solarcon requires an array to have an even number of PV panels, with a reflector support boom spaced every two panels. However, the reflectors are continuous sheets for the full length of the array, so there are no shadow lines cast on the cells by discontinuous reflector sections when the array's beta

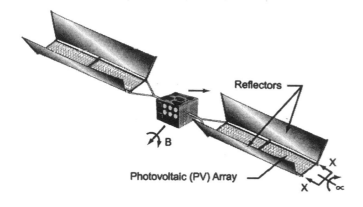

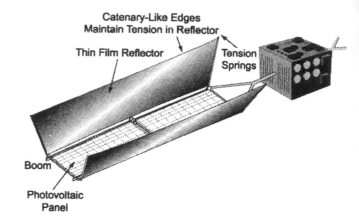

Figure 2a. 2-Panel Array Deployed

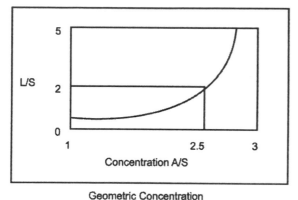

Geometric Concentration
Ratio = A/S

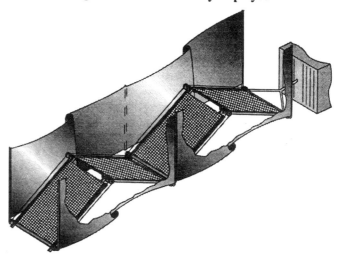

Figure 2b. 4-Panel Array Deploying

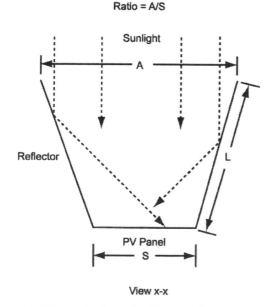

View x-x

Figure 1. Generic Channel System

Figure 2c. Array Stowed

angle is not zero. For maximum packing efficiency a panel's length should be at least twice its width.

Solarcon can be marketed as a reflector kit that bolts to PV panel arrays that require only the addition of tapped inserts at the ends of their side edges (Figure 3). The PV array uses its own springs and hinges for deployment, and the Solarcon system includes the pulley and cable control system. This allows the satellite builders to use their own PV arrays, minimizing the "novelty" of the concentrator system.

Specific Power

This calculation is derived from a cell manufacturer's design of a 100kW array (3). The cells have 23.1% efficiency at one Sun, 28C, and the 408m^2 array has a mass of 1647kg, or

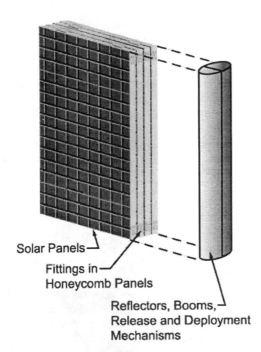

Solar Panels

Fittings in
Honeycomb Panels

Reflectors, Booms,
Release and Deployment
Mechanisms

Figure 3. Retrofit Kit

$4.04 kg/m^2$, with 90% cell packing factor. The efficiency at the 70C operating temperature is
$100kW/(408m^2 \times 1353W/m^2 \times 0.9pf) = .2013$, or 20.13%.
The power from the plane array is then
$1353W/m^2 \times 0.2013 \times 0.9 = 245.1W/m^2$.

The efficiency decrease with temperature is −0.0707/C. By iteration, the operating temperature of a 2.5:1 GCR Solarcon at GEO is found to be 120C, and the operating efficiency is then 17.1% which includes 0.5% increase because of the concentrated sunlight. The BOL power from the Solarcon PV panels augmented by 98% efficient reflectors is
$1353W/m^2 \times .171 \times 0.9 \times (1 + 1.5 \times 0.98) = 514.3W/m^2$
Solarcon's goal has been to provide a 100% increase in specific power for a given array. The allowable mass addition for the Solarcon reflector system is then found from
$514.3W/m^2/(2 \times 245.1W/m^2) = 1.05$,
a 5% increase or $0.05 \times 4.04kg/m^2 = 0.202kg/m^2$ (of PV panel area). Each square meter of PV panel requires about $4.4m^2$ of reflector material, including edge stress-relief area (described below). This area of 0.0127mm CP1 polyimide at 1.434g/cc has a mass of 0.08kg, leaving $0.122kg/m^2$ of PV panel for the support booms and the deployment and tensioning mechanisms. This brings us very close to the 100% specific power increase goal, which can be reached with higher efficiency cells that convert less sunlight to heat and therefore have a higher relative efficiency under concentrated sunlight.

The cell manufacturer has given estimates of 68W/kg for large, high quality PV arrays with 23.1% efficient cells, and this extrapolates to 88W/kg with the new 30% efficient cells. If Solarcon reflectors can reach 98% efficiency, the array should approach 180W/kg. Thin film cell arrays could also use Solarcon reflectors, but the specific power increase would be less, perhaps 50%. Thin film arrays are already very light, and

silicon cells have a sharper decrease in efficiency with increasing temperature.

Reflectors

Solarcon's reflectors are made from SRS's CP1 material, a fluorinated polyimide with exceptional UV resistance, thermal stability, transparency and solubility. NASA's Langley Research Center developed CP1 and SRS holds the sole license to produce the material. CP1 is soluble in a variety of commercially available solvents, enabling it to be cast on many different surfaces and with varying thickness. Innovative cast and cure techniques can produce films that are optically flat and exhibit isotropic material properties in nearly unlimited sizes. These films are far superior to commercially available polyimides, which exhibit residual stresses due to the manufacturing process thereby limiting their flatness.

Reflector panel flatness is critical to achieve maximum performance gain from the concentrators. Any wrinkling that occurs alters the reflected incidence flux on the photovoltaic cells. This effect can generate local hot spots resulting in poor concentrator performance and possible damage. The Solarcon catenary design coupled with CP1 isotropic properties enables the film to be loaded into a state of uniform biaxial stress. This configuration provides a large margin with regard to avoiding local wrinkling. Therefore, the reflector panels can be manufactured with a highly specular and reflective surface to provide the optimum incidence flux gain on the solar panels.

A reflective surface is achieved by sputtering aluminum onto the CP1 substrate. Facilities are available to coat very large sheets of CP1 using this process. The vacuum deposited aluminum coatings provide 88 percent reflectivity over the solar waveband. Advanced processes involving self-metalizing polyimides are currently being developed that promise improved reflectivity up to approximately 92 percent with a goal of 98 percent for silvered reflectors. The self-metalizing process also eliminates the costly vacuum coating procedure.A 5 x 250-mil polyimide bead forms the edges of the reflector sheet, all of which form a parabolic curve to distribute the loading from the reflector's four corners. In the ideal case the stress in the thin reflector sheet at the edge of the bead is always perpendicular to the bead, causing isotropic stress in the reflector. The problem is that the edge bead stretches slightly, putting transverse stress in the reflector that shows up as wrinkles. The solution is to create a stress relief zone by heat sagging strips between the bead and the rectangular active reflector area. This is equivalent to the vertical cables joining a suspension bridge's roadbed and catenary structure.

The first Solarcon reflectors used aluminized Mylar. This material came from a rolling process, and required at least 30psi to eliminate the wrinkles and achieve a flat reflector surface. SRS's cast CP1 is perfectly flat at zero stress, so the design now calls for an order-of-magnitude decrease in support stress to 3psi. The forces in the reflector and support booms are calculated as follows. Consider a 2-panel array, each PV panel 3' x 6' for an overall dimension of 3' x 12'. The reflectors are then 6' x 12'. An isotropic stress of 3psi causes a vertical and horizontal force in the half-mil sheet of 0.216# and 0.108# respectively. At any corner the forces are half this value as

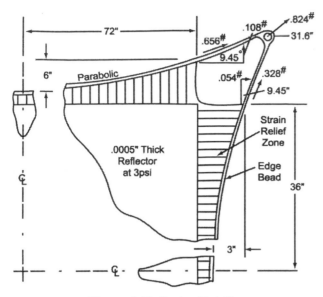

Figure 4. Reflector Details

shown in Figure 4. The top edge bead carries the vertical force, and the magnitude depends on the depth of the parabolic curve which translates into the angle at the tensioned active reflector area. A depth of 6" gives an end angle of 9.45 degrees, so dividing 0.108# by sin 9.45 degrees gives a top edge bead force of 0.656 pounds. The side edge bead has half this force, and the combination makes the force at the boom attach point 0.824 pounds at 31.6 degrees from horizontal.

The distance from the PV panel to the attach point is about 80", so the two booms at each end, at a 22 degree outward cant, subject the PV panel to an out-of-plane torque of

2 x 0.824# x cos31.6 x 80" x cos22 = 104 inch-pounds.

This is countered by 12 inch-pounds from the bottom attach point, which is 9" below the solar panel. The net torque on the 144" solar panel is then 92 inch-pounds. A 0.75" thick composite panel with 20 mil facesheets will deflect only about 0.60" at each end under this load.

Status

The NRO supported Solarcon's development until that organization stopped funding "bus" technology. Its development status is around NASA's TRL 4 or 5; the following have been completed successfully

- Laboratory deployment tests of a 500W-sized array
- Flash power tests at NASA/GSFC
- Thermal cycling reflector in NRL's vacuum chamber
- Micrometeorite impact tests at NASA/MSFC.

In addition, SRS has completed development of the basic design of the Solarcon reflector. The CP1 polyimide material is space qualified and being used on the Hughes 702 Bus concentrator. The Solarcon concentrator is now ready for a flight demonstration, beginning with the development of a full-scale mechanical model to demonstrate deployment and tensioning mechanisms, and deployment in a thermal-vacuum chamber.

Summary

The Solarcon reflector system provides a means for spacecraft builders to reach the maximum possible specific power from a PV panel, without the risks associated with other "advanced" arrays.

- Solarcon's specific power approaches 180W/kg with advanced polycrystalline cells. For a given power requirement, Solarcon has half the mass and cost, fewer than half the cells.
- Stowage requires about half the volume of a plane array of the same power
- Solarcon has a full panel of cells exposed to generate power in the stowed position—for orbit transfer, and for risk reduction
- The large aperture of the channel configuration provides sufficient power if a satellite tumbles
- Heritage and reliability of conventional solar panel structures and deployment systems is a marketing plus.

ACKNOWLEDGEMENTS

Thanks to Dr. Martin Mikulas of the University of Colorado for assistance in developing the reflector suspension system. Thanks also to Mr. Ken Steele at Tecstar, City of Industry, California, for providing technical data.

REFERENCES

1. Michael A. Brown, US Patent 5,885,367, *Retractable Thin Film solar Concentrator for Spacecraft.*
2. Michael A. Brown, US Patent 6,051,775, *Device for Tensioning Sheet Members.*
3. Tecstar, Ref. No. PM99-501.

Proceedings of IECEC'01
36th Intersociety Energy Conversion Engineering Conference
July 29–August 2, 2001, Savannah, Georgia

IECEC2001-AT-39

THE STRETCHED LENS ARRAY (SLA), AN ULTRALIGHT CONCENTRATOR FOR SPACE POWER

Mark J. O'Neill
ENTECH, Inc.
1077 Chisolm Trail
Keller, Texas 76248 USA
Tel: 817-379-0100
Fax: 817-379-0300
E-Mail:
mjoneill@entechsolar.com

Michael F. Piszczor
NASA Glenn
21000 Brookpark Rd.
Cleveland, Ohio 44135 USA
Tel: 216-433-2237
Fax: 216-433-6106
E-Mail:
michael.piszczor@grc.nasa.gov

Michael I. Eskenazi
AEC-ABLE
7200 Hollister Avenue
Goleta, CA 93117 USA
Tel: 805-685-2262
Fax: 805-685-1369
E-Mail:
meskenazi@aec-able.com

A. J. McDanal
ENTECH, Inc.
1077 Chisolm Trail
Keller, Texas 76248 USA
Tel: 817-379-0100
Fax: 817-379-0300
E-Mail:
ajmcdanal@entechsolar.com

Patrick J. George
NASA Glenn
21000 Brookpark Rd.
Cleveland, Ohio 44135 USA
Tel: 216-433-2353
Fax: 216-433-2995
E-Mail:
patrick.j.george@grc.nasa.gov

Matthew M. Botke
AEC-ABLE
7200 Hollister Avenue
Goleta, CA 93117 USA
Tel: 805-685-2262
Fax: 805-685-1369
E-Mail:
mbotke@aec-able.com

Henry W. Brandhorst
Space Power Institute
231 Leach Center
Auburn Univ., AL 36849 USA
Tel: 334-844-5894
Fax: 334-844-5900
E-Mail:
brandhh@mail.auburn.edu

David L. Edwards
NASA Marshall
Huntsville, Alabama 35812 USA
Tel: 256-544-4081
Fax: 256-544-5103
E-Mail:
david.edwards@msfc.nasa.gov

Paul A. Jaster
3M
3M Center
St. Paul, MN 55144 USA
Tel: 651-733-1898
Fax: 651-733-5299
E-Mail:
pajaster1@mmm.com

ABSTRACT

A high-performance, ultralight, photovoltaic concentrator array is being developed for space power. The stretched lens array (SLA) uses stretched-membrane, silicone Fresnel lenses to concentrate sunlight onto triple-junction photovoltaic cells. The cells are mounted to a composite radiator structure. The entire solar array wing, including lenses, photovoltaic cell flex circuits, composite panels, hinges, yoke, wiring harness, and deployment mechanisms, has a mass density of 1.5 kg/sq.m. NASA Glenn has measured 27.4% net SLA panel (Fig. 1) efficiency, or 375 W/sq.m. power density, at room temperature. At GEO operating cell temperature (80C), this power density will be 300 W/sq.m., resulting in 200 W/kg specific power at the full wing level. SLA is a direct ultralight descendent of the successful SCARLET array on NASA's Deep Space 1 spacecraft. This paper describes the evolution from SCARLET to SLA, summarizes the SLA's key features, and provides performance and mass data for this new concentrator array.

Fig. 1 – Stretched Lens Array (SLA) Prototype Panel

INTRODUCTION AND BACKGROUND

Since 1986, ENTECH and NASA have been developing and refining space photovoltaic arrays using refractive concentrator technology [1]. Unlike reflective concentrators, these refractive Fresnel lens concentrators can be configured to minimize the effects of shape errors, enabling straightforward manufacture, assembly, and operation on orbit. By using a unique arch shape, these Fresnel lenses provide more than 100X larger slope error tolerance than either reflective concentrators or conventional flat Fresnel lens concentrators [2].

In the early 1990's, the first refractive concentrator array was developed and flown on the PASP+ mission, which included a number of small advanced arrays (3). The refractive concentrator array used ENTECH mini-dome lenses over Boeing mechanically stacked multi-junction (MJ) cells (GaAs over GaSb). The mini-dome lenses were made by ENTECH from space-qualified silicone (DC 93-500), and coated by Boeing to provide protection against space ultraviolet (UV) radiation and atomic oxygen (AO). This array performed extremely well throughout the year-long mission in a high-radiation elliptical orbit, validating both the high performance and radiation hardness of the refractive concentrator approach (3). In addition, in high-voltage space plasma interaction experiments, the refractive concentrator array was able to withstand cell voltage excursions to 500 V relative to the plasma with minimal environmental interaction [3].

In the middle 1990's, ENTECH, NASA, and 3M developed a new line-focus Fresnel lens concentrator, which is easier to make and more cost-effective than the mini-dome lens concentrator. Using a continuous roll-to-roll process, 3M can now rapidly mass-produce the line-focus silicone lens material in any desired quantity.

In 1994, AEC-ABLE joined the refractive concentrator team and led the development of the SCARLET® (Solar Concentrator Array using Refractive Linear Element Technology) solar array [4]. SCARLET uses a small (8.5 cm wide aperture) silicone Fresnel lens to focus sunlight at about 8X concentration onto radiatively cooled multi-junction cells. Launched in October 1998, a 2.5 kW SCARLET array powers both the spacecraft and the ion engine on the NASA/JPL Deep Space 1 spacecraft, shown in Fig. 2. SCARLET achieved over 200 W/sq.m. areal power and over 45 W/kg specific power. With SCARLET working flawlessly, Deep Space 1 is currently about 180 million miles from earth, on its way to a comet rendezvous later this year [5].

Over the past three years, the team, now including Auburn's Space Power Institute, has developed a new space concentrator array technology, called the stretched lens array (SLA). SLA provides even higher performance than SCARLET at dramatically reduced mass and cost [6 and 7].

Fig. 2 - Deep Space 1 Probe Launched October 1998

In 1999-2000, under NASA's Space Solar Power program, the SLA team designed, developed, fabricated, and tested a fully functional prototype SLA panel (Fig. 1). This panel achieved unprecedented performance, characterized by a net solar-to-electrical conversion efficiency of 27.4% under simulated space sunlight at room temperature. [7]. Furthermore, the same SLA technology provided unprecedented performance under outdoor terrestrial sunlight, characterized by 25-29% net conversion efficiency at operating temperature. [8].

In 2001-2002, under NASA's Advanced Cross-Enterprise Technology Development program, the SLA team is developing an optimized, near-term, robust, SLA solar array wing. This SLA wing builds upon the 15 year heritage of refractive concentrators for space power, including the successful flight heritage of the PASP+ and SCARLET arrays referenced above. The SLA team is also working on a NASA/JPL New Millennium Program Space Technology 6 study which could lead to a near-term flight test for SLA.

EVOLUTION FROM THE SCARLET ARRAY TO THE STRETCHED LENS ARRAY (SLA)

The patented SLA [6 and 9] is an ultralight descendent of SCARLET as shown by comparing Figs. 3 and 4, and as discussed in the following paragraphs. Both SCARLET and SLA use the same small, arch-shaped, line-focus Fresnel lenses (8.5 cm aperture width) to focus sunlight onto high-efficiency multi-junction solar cells (1.0 cm active width). This concentration ratio (8.5X) provides ± 2 degree sun-pointing tolerance about the critical axis, and can be adjusted for specific mission requirements. 3M makes the continuous web of thin lensfilm material from space-qualified DC93-500 silicone.

Fig. 3 shows a SCARLET panel during assembly at ABLE. To support and UV-protect the 200-micron-thick silicone lens material, each lens was laminated to a 75-micron-thick, thermally shaped, ceria-doped glass arch. The laminated lenses were then inserted into a protective frame, made from composite material. The lens-populated lens frame is the upper deck in Fig. 3. The photovoltaic receivers were attached to a high thermal conductivity composite honeycomb panel, which is the lower deck in Fig. 3. For launch, the lens frames were stowed against the honeycomb panels, which were folded together in the same fashion as for a planar solar array. Once on orbit, the SCARLET panels unfolded to form a wing, and the lens frames deployed to their proper position [5].

Fig. 4 shows a prototype SLA panel. The silicone lens material is now supported as a stretched membrane between end arches, so both the glass arches and the lens frame have been eliminated. The composite radiator is now a 190-micron-thick composite sheet, which is more than adequate for excellent thermal performance. Like the silicone lens, the radiator sheet is now supported as a stretched membrane between edge elements, so the honeycomb panel has been eliminated. Without any sacrifice in optical, thermal, or electrical functionality, the SLA panel in Fig. 4 is approximately four times lighter (per square meter of lens aperture) than the SCARLET panel in Fig. 3. Indeed, the SLA performance is higher than SCARLET because the optical losses caused by the glass arches and the lens frame have been eliminated.

Fig. 4 shows the SLA prototype panel under terrestrial sunlight illumination. Note the focal lines on each of the four photovoltaic receivers, which utilize conventional copper-clad polyimide flex circuit construction. Performance measurements for this SLA prototype panel are discussed below.

PROTOTYPE SLA PANEL PERFORMANCE TESTING

Under the NASA Space Solar Power program, the prototype SLA panel shown in Figs. 1 and 4 was designed, developed, fabricated, and tested by the refractive concentrator team. Both Spectrolab and TECSTAR developed monolithic triple junction (GaInP/GaAs/Ge) concentrator cells for the prototype panel. These cells were equipped with ENTECH prism covers to eliminate the normal gridline shadowing loss. Cells were assembled into photovoltaic receivers using polyimide flex circuits, which were attached to the graphite sheet radiator with space-qualified silicone pressure sensitive adhesive (PSA).

The prototype SLA panel was first tested under terrestrial sunlight by ENTECH, then tested in a large area pulsed solar simulator (LAPSS) by ABLE, and finally LAPSS-tested by NASA Glenn. To ensure accuracy, NASA Glenn flew cells from the same production run on the NASA Lear Jet to determine their AM0 short-circuit currents, and then used one

Fig. 3 - Deep Space 1 *SCARLET* Panel Assembly

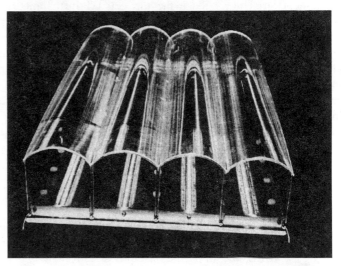

Fig. 4 – Stretched Lens Array (SLA) Prototype Panel

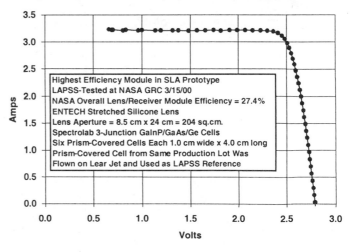

Highest Efficiency Module in SLA Prototype
LAPSS-Tested at NASA GRC 3/15/00
NASA Overall Lens/Receiver Module Efficiency = 27.4%
ENTECH Stretched Silicone Lens
Lens Aperture = 8.5 cm x 24 cm = 204 sq.cm.
Spectrolab 3-Junction GaInP/GaAs/Ge Cells
Six Prism-Covered Cells Each 1.0 cm wide x 4.0 cm long
Prism-Covered Cell from Same Production Lot Was
Flown on Lear Jet and Used as LAPSS Reference

Fig. 5 – NASA-Measured 27.4% Module Efficiency (AM0)

of these cells to set the intensity of the LAPSS lamp to maintain 1 AM0 sun irradiance at the lens aperture. Fig. 5 shows the measured IV curve for the best of the four lens/receiver modules in the prototype panel. These results were measured at room temperature (about 20C), and correspond to 27.4% net module efficiency and 375 W/sq.m. output power density, based on the module aperture area (8.5 cm lens aperture width x 24.0 cm photovoltaic receiver active length = 204 sq.cm.).

OTHER SLA TESTING

A number of other important tests have been performed for the key elements of the stretched lens array (SLA), as summarized in the following paragraphs.

Stretched Lens Optical Efficiency

Numerous stretched membrane lenses have been tested under outdoor terrestrial sunlight by ENTECH, to determine the net optical efficiency. Since the top cell (GaInP) limits the current of the triple-junction stack, the short-circuit current output of a reference GaInP cell is used to determine the lens efficiency. Under a variety of test conditions, the optical efficiency of the stretched silicone lens has typically been measured at 92-93%. Knowing this lens efficiency value, the corresponding cell efficiency for the prototype panel discussed in the previous paragraph is seen to be 30% (27.4% module/92% lens). This cell efficiency value is in very close agreement with Spectrolab's in-house cell efficiency measurement.

Stretched Lens Thermal Cycling

To verify the thermal durability of the new stretched membrane lens, multiple samples were exposed to GEO thermal cycling by ABLE, and all passed the equivalent of more than 20 years in GEO (over 1,830 thermal cycles from −180C to +90C).

Lens Material Space Ultraviolet Exposure Testing

Like the silicone mini-dome lenses on the PASP+ mission, the SLA lenses will be equipped with a thin-film ultraviolet rejection (UVR) coating to protect the silicone. To verify the durability of the coated lens material, NASA Marshall has completed 7,000 equivalent sun hours (ESH) of near ultraviolet (NUV) exposure testing on both coated and uncoated lens material samples made by ENTECH [10]. Figs. 6 and 7 show the NASA Marshall spectral transmittance measurements for coated and uncoated samples, respectively. Interestingly, the NUV-exposed uncoated lens material still has a 93% transmittance, for the current-limiting top cell (GaInP) response spectrum, after 7,000 ESH of NUV. The coated material has degraded to 89% transmittance for the top cell response spectrum after 7,000 ESH of NUV. While these NUV tests are continuing, these early results may indicate that a UVR coating is not critical for the DC 93-500 lens material, perhaps due to the unique ENTECH curing process.

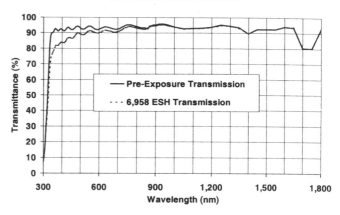

Fig. 6 – NASA Marshall NUV Exposure Test Results for UVR-Coated DC93-500 Lens Material Sample

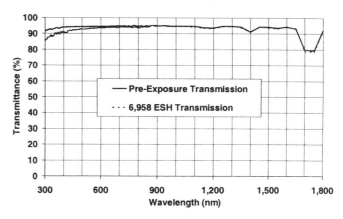

Fig. 7 – NASA Marshall NUV Exposure Test Results for Uncoated DC93-500 Lens Material Sample

Micrometeoroid Impact Testing of Lenses and Cells

Auburn University's Space Power Institute has conducted micrometeoroid impact tests of both lens and cell samples. The lens impact tests showed clean penetrations with no peripheral damage such as tearing. ENTECH mounted the single-cell photovoltaic receivers to composite radiator sheet, and fully encapsulated the receivers to enable high-voltage operation. Auburn then tested these samples in a simulated LEO plasma, with the cells biased to more than 400 volts relative to the plasma. Micrometeoroids were then shot at the samples, causing minor damage to the cover glass over each cell, but no electrical problems (such as arcing) were observed. This high-voltage capability, with very little mass penalty, is one of the key advantages of the SLA approach over conventional planar arrays. The small cells can be super-insulated without adding much mass to the array, due to the small size of the solar cells. For high-power arrays (e.g., 20 kW and larger), this high-voltage capability provide significant savings in wiring mass and cost compared to conventional lower-voltage planar arrays. The added encapsulation can also be designed to provide excellent radiation tolerance for high-radiation missions.

SLA RIGID PANEL WING APPROACH

In 1999-2000, the SLA team thoroughly investigated a flexible blanket platform for the stretched lens array (SLA) [7]. While this blanket approach has an advantage in stowage volume over more conventional rigid panel platforms, it lacks the maturity and flight heritage of the rigid panel wing. This year, under the new NASA Advanced Cross-Enterprise Technology Development program, the SLA team is developing a new ultralight rigid panel wing as the platform of choice for SLA. Figs. 8-10 show the basic rigid panel SLA wing approach in schematic form. The flexible lenses fold down flat against the rigid panels for compact stowing during launch (Fig. 8). On orbit, as the panels unfold, spring driven end arches deploy and tension the individual stretched lenses across the panel's length (Fig. 9). The wing continues to deploy until the panels are all co-planar in their final locked wing position (Fig. 10).

This unfolding rigid panel solar array approach has been widely used for many years for NASA, DOD, and commercial spacecraft. One unique feature in the new SLA rigid panel array relates to the panels themselves, which use only a single face sheet and no honeycomb, except as a support frame around each panel. The use of very lightweight "picture-frame" panels is enabled by the very low mass of the supported cells and lenses, and by snubbing during launch to the inboard and outboard panels which are reinforced honeycomb panels. The

Fig. 8 – Stowed Rigid Panel SLA Wing

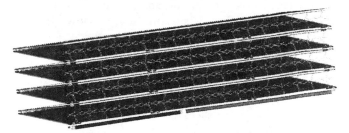

Fig. 9 – Lenses Deploying on Rigid Panel SLA Wing

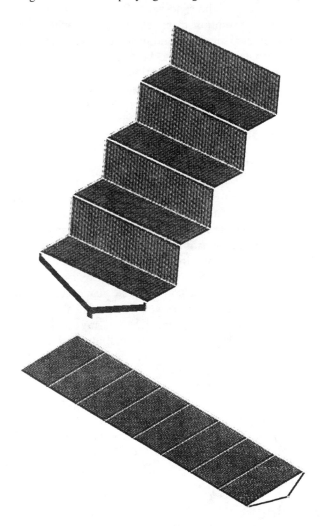

Fig. 10 – Panels Deploying on Rigid Panel SLA Wing

Lens Stowed Against Radiator

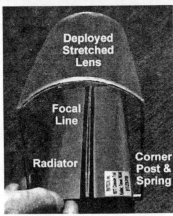

Deployed Stretched Lens

Focal Line

Radiator

Corner Post & Spring

Fig. 11 – Prototype SLA Model with Pop-Up Lens

thin composite face sheet forms the photovoltaic receiver mounting surface and the waste heat radiator, and is stretched in drum-like fashion over the peripheral honeycomb picture-frame structure. The individual pop-up lenses use the same basic deployment and support approach that has been used successfully on numerous SLA prototypes (Fig. 11).

SLA WING-LEVEL MASS AND PERFORMANCE

A detailed mass and performance analysis has been done for the rigid panel SLA wing point design shown in Fig.12. This wing has a total mass of 37 kg and provides a total lens aperture area of 24 sq.m. At beginning of life (BOL), this wing provides over 7 kW of output power and over 190 W/kg of specific power. These values are based on the use of a 190 micron thick composite sheet radiator; with a thinner 130 micron thick composite sheet radiator, the BOL specific power exceeds 200 W/kg.

Key features of the rigid-panel SLA wing are summarized in Table 1. In addition to excellent performance, mass, and stiffness characteristics, the rigid panel SLA wing approach also enables the outermost panel to be populated with planar cells for pre-deployment power generation (e.g., during LEO to GEO orbit transfer), if a specific mission needs this capability. Furthermore, many mission planners and spacecraft program managers prefer the low-risk, rigid panel deployment approach over flexible blanket approaches.

CONCLUSIONS

A new lightweight rigid-panel concentrator array is under development for future space power applications. SLA provides a substantial cost advantage over planar arrays by using only one-eighth as much expensive solar cell material for the same power output. In addition, the thin silicone lensfilm and the thin composite sheet radiator provide a substantial mass/area advantage over conventional planar cell assemblies. The new rigid-panel SLA wing employs a conventional, conservative, well-proven approach to array deployment and support on orbit, while offering outstanding performance, high-voltage capability, and radiation tolerance.

ACKNOWLEDGMENTS

The authors gratefully acknowledge NASA's support of the stretched lens array (SLA) development under the Space Solar Power program, the Advanced Cross-Enterprise Technology Development program, and the New Millennium Space Technology 6 program.

REFERENCES

1. M.F. Piszczor and M.J. O'Neill, "Development of a Dome Fresnel Lens/GaAs Photovoltaic Concentrator for Space Applications," 19th IEEE-PVSC, 1987.
2. M.J. O'Neill, "Silicon Low-Concentration, Line-Focus, Terrestrial Modules," Chapter 10 in **Solar Cells and Their Applications**, John Wiley & Sons, 1995.

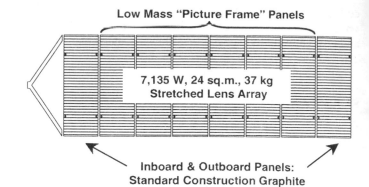

Fig. 12 – Point Design Parameters for SLA Wing

Feature	Value or Characteristic
Point Design Basis	7,135 Watts (BOL)
SLA Implementation	Pop-up lenses
Base Platform Design Maturity	Most components flight proven on DS1
Specific Power (130 micron facesheets)	203 W/kg
Stowed Volume	0.093 m³/kW
Stowed Stiffness	40 Hz
Deployed Stiffness	0.12 Hz
Stowed Power	Easily implemented on outer panel
Ease of Adding Planar Panel	Easily implemented on outer panel
Flatness & Warping	Well understood flat stable platform
Deployment Testing	Can use existing off-loaders
Power Testing	Pop-up lenses allow each panel to be tested as a complete assembly before wing integration
Commercial Appeal	Easier to integrate on commercial spacecraft. Readily accepted configuration.
Self Shadowing	No self shadowing

Table 1 – Key Features of the Rigid Panel SLA Wing

3. H. Curtis and D. Marvin, "Final Results from the PASP Plus Flight Experiment," 25th IEEE-PVSC, 1996.
4. P.A. Jones et al., "The SCARLET Light Concentrating Solar Array," 25th IEEE-PVSC, 1996.
5. D.M. Murphy, "The SCARLET Solar Array: Technology Validation and Flight Results," Deep Space 1 Technology Validation Symposium, Pasadena, 2000.
6. M.J. O'Neill, "Stretched Fresnel Lens Solar Concentrator for Space Power," U.S. Patent 6,075,200, 2000.
7. M.J. O'Neill et al., "The Stretched Lens Ultralight Concentrator Array," 28th IEEE-PVSC, 2000.
8. M.J. O'Neill et al., "Development Of Terrestrial Concentrator Modules Incorporating High-Efficiency Multi-Junction Cells," 28th IEEE-PVSC, 2000.
9. M.J. O'Neill, "Color-Mixing Lens for Solar Concentrator System and Methods of Manufacture and Operation Thereof," U.S. Patent 6,031,179, 2000.
10. D.L. Edwards et al., "Optical Analysis of Transparent Polymeric Material Exposed to Ultraviolet Radiation," with 8th International Symposium on Materials in a Space Environment," Arcachon, France, 2000.

Proceedings of IECEC '01
36th Intersociety Energy Conversion and Engineering Conference
July 29–August 2, 2001, Savannah, Georgia

IECEC2001-AT-08

ENERGY STORAGE FOR AEROSPACE APPLICATIONS

Marla E. Pérez-Davis
NASA Glenn Research Center

Lisa L. Kohout
NASA Glenn Research Center

Patricia L. Loyselle
NASA Glenn Research Center

Michelle A. Manzo
NASA Glenn Research Center

Kenneth A. Burke
NASA Glenn Research Center

Mark A. Hoberecht
NASA Glenn Research Center

Carlos R. Cabrera
University of Puerto Rico

ABSTRACT

The NASA Glenn Research Center (GRC) has long been a major contributor to the development and application of energy storage technologies for NASA's missions and programs. NASA GRC has supported technology efforts for the advancement of batteries and fuel cells. The Electrochemistry Branch at NASA GRC continues to play a critical role in the development and application of energy storage technologies, in collaboration with other NASA centers, government agencies, industry and academia.

This paper describes the work in batteries and fuel cell technologies at the NASA Glenn Research Center. It covers a number of systems required to ensure that NASA's needs for a wide variety of systems are met. Some of the topics covered are lithium-based batteries, proton exchange membrane (PEM) fuel cells, and nanotechnology activities. With the advances of the past years, we begin the 21st century with new technical challenges and opportunities as we develop enabling technologies for batteries and fuel cells for aerospace applications.

INTRODUCTION

Over the past three decades, the Electrochemistry Branch of the NASA Glenn Research Center (GRC) has been a major contributor to the development and application of advanced battery and fuel cell systems for NASA missions and programs. The demand for terrestrial-based systems in the 70's prompted work in the area of battery systems for electric vehicles, alkaline fuel cell technology for the Shuttle, and fuel cells for large-scale powerplant applications. As the work became more focused toward aerospace applications in the 80's and 90's, NASA GRC became involved in the development of nickel-hydrogen batteries for low earth orbit (LEO) and geosynchronous orbit (GEO) spacecraft and PEM fuel cell technology for high altitude aircraft. Today, the Electrochemistry Branch at NASA GRC continues to play a critical role in the development and application of energy storage technologies, in collaboration with other NASA centers, government agencies, industry and academia.

BATTERY TECHNOLOGY

The NASA Glenn Research Center has a long history of contributing to the development of battery storage systems for both aerospace and terrestrial applications. Early efforts focused on the development and support of nickel-cadmium, silver-zinc, silver-hydrogen and hydrogen-oxygen battery systems. Hydrides were investigated as a means to store hydrogen for both battery and fuel cell systems. In the 1970's NASA GRC conducted a Department of Energy (DoE)-funded electric vehicle effort. Work focused on the development and evaluation of nickel-zinc and lead acid battery systems for electric vehicles. The culmination of this effort was the successful demonstration of a GRC-designed battery powered automobile.

As support for the terrestrial related energy programs diminished in the 1980's, efforts became more focused on aerospace-related applications. The battery team at GRC became involved in the development of nickel-hydrogen technology for aerospace applications. Many of the GRC-initiated advances, including the use of 26% KOH and catalyzed wall-wicks, are routinely incorporated into today's aerospace cells [1]. In addition, GRC conducted both in-house and contractual efforts that demonstrated the feasibility of bipolar nickel-hydrogen battery designs with the potential to reduce mass and volume by addressing battery requirements at the system level [2]. Parallel efforts at improving the specific energy and energy density of nickel-hydrogen systems focused on the development of a light-weight nickel electrode. Replacement of the traditional sintered substrate used in nickel electrodes with a light-weight, highly porous nickel felt significantly reduces the weight of a nickel-hydrogen cell [3]. Throughout the 70's and 80's, GRC conducted parallel separator development programs aimed at replacing the asbestos material used in both fuel cells and alkaline battery systems. There were extensive efforts aimed at separator development for zinc based systems as well. This separator expertise was later applied to nickel-hydrogen and nickel-cadmium systems in the 80's and 90's and culminated with the publication of a manual outlining standard test procedures for evaluating separator materials for alkaline cells [4].

Today, the efforts of the Electrochemistry Branch at GRC are focused on supporting NASA's current and future needs for rechargeable batteries for a wide variety of aerospace applications. Current applications cover a wide range of performance requirements. These include planetary missions, such as landers and rovers, that require batteries ranging in capacity from 6-35 ampere-hours, that are capable of operating at temperatures as low as –20°C following the long stand times associated with travel to the mission destination; LEO spacecraft and planetary orbiters that require 30,000-50,000 cycles over a five to ten year period; GEO spacecraft that have an operating life greater than fifteen years; and aircraft applications that require high voltage (28-300V) and high capacity (50-100 AH) batteries that can operate from –40°C to +60°C. Further, these batteries are required to meet the stringent environmental requirements associated with flight applications such as vibration, shock, and high impact.

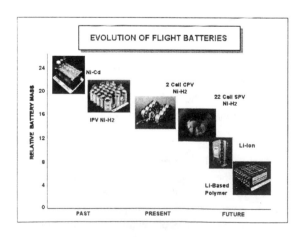

Figure 1: Evolution of Flight Batteries

The evolution of flight batteries over time is illustrated in Figure 1. As technology advances the systems are becoming more lightweight with increased cycle and performance capabilities. Nickel-cadmium batteries have served as the workhorse system since NASA's earliest missions. In recent years, they have been replaced with nickel-hydrogen systems. These systems are capable of operating at deeper depths-of-discharge while providing comparable life to nickel-cadmium systems thus resulting in a lighter-weight energy storage option. Improvements to the existing chemistries and cell designs can yield lighter, more compact and longer-lived systems. For future battery applications, lithium-based battery systems offer the potential of lighter weight, less complexity, and higher performance. Lithium-based batteries offer >2-5X improvement in specific energy over state-of-the-art nickel-hydrogen batteries. Advanced lithium-based batteries will operate over a much wider temperature range and reduce power system mass and volume while decreasing the cost, thus increasing mission capabilities and enabling many future missions. The most challenging goals for the lithium-based battery system are to develop batteries with improved energy density and specific energy that are capable of long calendar life required for GEO and long cycle life required for LEO applications.

The wide range of requirements for future NASA missions cannot presently be met by a single battery chemistry. In order to address NASA's near and far term battery system requirements, the battery-related programs at GRC address multiple systems covering a wide range of technology readiness levels.

Battery Technology Development

The majority of NASA's near-term (<5 yrs) missions will be using nickel-based battery systems. Current efforts at GRC that support the development of nickel-based batteries include the development of a bipolar nickel-metal hydride battery, via a contracted effort with Electro Energy Incorporated [5], and the demonstration of scaled-up light-weight nickel-electrode technology in flight hardware.

In order to address the next generation (3-8 yrs) of secondary batteries for future aerospace missions, a joint Department of Defense (DoD)/NASA program has been established to develop lithium-ion batteries with the capabilities required by future NASA and DoD missions. The objectives of this program are to: 1) develop high specific energy and long cycle life lithium-ion cells and batteries; 2) establish production sources; and 3) demonstrate technology readiness for a variety of applications, such as rovers and landers, LEO and GEO missions, aviation/unmanned aerial vehicles, and military terrestrial applications. The technical approach involves: a) development of advanced electrode materials and electrolytes to achieve improved low temperature performance and cycle life; b) optimization of cell design to achieve high specific energy; c) development of cells (6-100 Ah) and batteries (16-300 V) of various sizes required for various future missions; and d) the development of control electronics for smart battery management. These batteries will be initially used in missions where weight and volume are critical and cycle life requirements are low to moderate (200-1000).

NASA's far-term (6-12 yrs) battery requirements will be met by the development of lithium-based polymer electrolyte batteries. This development is supported by a new battery initiative at GRC known as the Polymer Energy Rechargeable Systems (PERS) Program. These lithium-based batteries have the potential to offer five times the energy of conventional energy storage at 1/3 mass, 1/10 volume, and 1/3 cost. In addition, the system is leak-free and non-toxic. The cells are flexible and conformable, making them ideal candidates for many diverse applications, including spacecraft, unmanned aerial vehicles, portable wireless electronics and electric vehicles. A combination of contracted and in-house efforts has been initiated that presently focuses on the development and evaluation of the various polymer electrolytes as well as cathodes, anodes and related support elements.

The Electrochemistry Branch is also investigating the application of nanotechnology to advanced battery systems and advanced battery concepts that incorporate thin-film solid state devices and conducting polymer electrodes.

NASA Aerospace Flight Battery Program

In addition to the technology development efforts, GRC leads the NASA Aerospace Flight Battery Systems Program, an agency-wide effort aimed at ensuring the quality, safety, reliability, and performance of flight battery systems for NASA missions. The program supports the development of guidelines, documents, and procedures related to diagnostic techniques developed for identifying failure modes. In addition, it supports the establishment and maintenance of a central database to serve as the repository for battery characterization and verification test data that supports the validation of battery technologies for flight use, as well as modeling efforts aimed at predicting performance of aerospace battery systems. The majority of the program resources are dedicated to the testing and validation of aerospace design nickel-cadmium, nickel-hydrogen, nickel-metal hydride, and lithium-ion secondary battery systems. There are also significant efforts characterizing and validating commercial off-the-shelf secondary battery technologies for use in manned missions. The program also addresses the issues related to the safety and reliability of primary lithium battery systems used in manned space operations [6].

Mission Support

In addition to the technology development activities discussed above, the Electrochemistry Branch is supporting the evaluation of flight battery technologies for the International Space Station and the electric auxiliary power unit (EAPU) replacement for the Space Shuttle. GRC's role in support of nickel-hydrogen technology has largely centered on the evaluation of multiple nickel-hydrogen cell design options from various vendors. More than 475 cells have been tested as part of this effort. The majority of the testing has been conducted at the Naval Surface Warfare Center (NSWC), Crane Division with a smaller proportion being conducted in-house at GRC [7]. These efforts culminated with the launch of the first nickel-hydrogen batteries for the ISS in November of 2000.

FUEL CELL TECHNOLOGY

The NASA Glenn Research Center has long been a major contributor to the development and application of fuel cell technologies for NASA's missions and programs. In the 70's, NASA GRC was responsible for advancing the state of fuel cell technology to a level which qualified it for the Shuttle on-board power system. Parallel technology advancement programs on the Gemini proton exchange membrane (PEM) fuel cell and the Apollo alkaline fuel cell were conducted at GRC. These were a combination of GRC in-house and contractual fuel cell R&D efforts. When the alkaline technology was selected for Shuttle, an optimized alkaline cell and stack technology was provided to the NASA Johnson Space Center (JSC) for Shuttle system development. GRC continued to support the JSC system development effort to the mid-80's by working to improve the life and performance of the Shuttle alkaline fuel cell technology [8].

In the 80's, NASA GRC culminated management of the DoE/Gas Research Institute Phosphoric Acid Fuel Cell Program for a 40 kW powerplant field test, stack and the balance-of-plant technology development for 200 kW, 7.5 MW, and 11 MW powerplants [9]. During this period, GRC also examined regenerative fuel cell (RFC) concepts for Lunar/Mars applications in support of the Space Exploration Initiative [10]. On the automotive side, GRC conducted a study of the feasibility of using a PEM fuel cell in an electric vehicle as part of the DoE-funded electric vehicle program. This study was one of the initial guiding elements in what has become the very large PEM fuel cell program for electric vehicles. As a result of this work, NASA GRC was tasked to lead the team to produce DoE's 10-year Fuel Cells for Transportation Plan in the early 1990's [11]. Moreover, GRC continued to address the RFC PEM concept for Space Station, high altitude balloon [12], and high altitude aircraft applications during the 90's.

Today, NASA GRC continues its involvement in the development of PEM fuel cell and regenerative fuel cell systems for a wide variety of applications, including earth-based and planetary aircraft, spacecraft, planetary surface power, and terrestrial use. The GRC fuel cell team is currently participating in the development of technologies to reduce CO_2 emissions from civil transport aircraft. The concepts under investigation involve the conversion of the propulsion systems to hydrogen fuel and the introduction of new energy conversion technologies, i.e. air-breathing fuel cells, to produce an environmentally benign, low cost and durable system for aircraft propulsion. GRC is involved in the system-level design and analysis of the fuel cell systems, which range in power from 100 kW to 90 MW. In addition, NASA GRC is developing and intends to demonstrate revolutionary energy conversion technologies to achieve reduced emissions aircraft operations. The focus of this program is on far-term, breakthrough technologies. The overall approach is a multidisciplinary effort to develop a revolutionary, non-traditional fuel cell power/propulsion system for aircraft applications. Areas under investigation include cell chemistries, advanced materials, and novel cell, stack, component and system designs.

For space vehicle applications, PEM fuel cell technology offers major advantages over existing alkaline fuel cell technology, including enhanced safety, longer life, lower weight, improved reliability and maintainability, higher peak-to-nominal power capability, compatibility with propulsion-grade reactants, and the potential for significantly lower costs. A team comprised of NASA GRC, NASA Marshall Space Flight Center, and Honeywell [13] (formerly AlliedSignal Aerospace) has just completed the development of modular PEM fuel cell stack technology for use in future reusable launch vehicles. Small substacks as well as a prototype 5.25 kW modular PEM fuel cell stack have been successfully built and operated. This 5.25 kW stack is shown in Figure 2.

In addition, NASA GRC, NASA Johnson Space Center, NASA Kennedy Space Center, and NASA Marshall Space Flight Center have just begun a 5-year PEM fuel cell powerplant development program that will culminate in the delivery to NASA of an engineering model PEM fuel cell powerplant for test and evaluation. A modular approach to powerplant design that relies on commonality with commercial hardware will allow NASA to leverage the evolving and highly competitive automotive and residential markets in PEM fuel cell technology, assuring technology transfer and low costs well into the future.

GRC is also involved in the development of passive ancillary component technology to be teamed with a hydrogen-oxygen unitized regenerative fuel cell (URFC) stack to form a revolutionary new RFC storage system for aerospace applications. Replacement of active RFC ancillary components with passive components minimizes parasitic power losses and allows the RFC to operate as a H_2/O_2 battery. The goal of this program is to demonstrate an integrated passive 1 kW URFC system.

NANOTECHNOLOGIES
Recently, advances in the area of nanotechnology have developed materials and techniques which could significantly improve the performance, mass, and volume of energy conversion/storage devices.

Carbon nanotubes have exhibited many interesting properties including electrical properties that range from conducting to insulating, exceptionally high mechanical strength, and a potential to store large amounts of hydrogen and other atoms and molecules within the tubes and tube bundles. Because of these unique properties, energy conversion/storage devices incorporating nanotubes have the potential to display significant improvements in performance and energy density over the current state-of-the-art [14, 15, 16, 17].

Figure 2: Honeywell Prototype 5.25 kW PEM Fuel Cell Stack for Reusable Launch Vehicle Applications

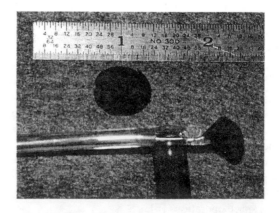

Figure 3: GRC Research Electrodes Made From Carbon Nanotubes

Currently, the NASA Glenn Research Center is assessing the technical feasibility of utilizing carbon nanotubes and nanotechnologies in future energy storage systems designs with initial investigations focusing on their capacity to reversibly store hydrogen under various conditions. An electrode made at GRC from carbon nanotubes is shown in Figure 3. This electrode is being used to evaluate the electrochemical properties of nanotubes with respect to hydrogen absorption.

CONCLUDING REMARKS
In summary, the Electrochemistry Branch at the NASA Glenn Research Center continues to play a critical role in the development and application of battery and fuel cell technologies, in collaboration with other NASA centers, government agencies, industry and academia. With the advances of the past years, we begin the 21st century with new technical challenges and opportunities as we develop enabling battery and fuel cell-based technologies for aerospace power and propulsion systems.

ACKNOWLEDGMENTS
The authors would like to acknowledge Mr. Paul Prokopius and Dr. Lawrence Thaller for their help in providing information on the historical background of fuel cell and battery development at the NASA Glenn Research Center.

REFERENCES
[1] Stadnick, S., and Rogers, H., 1994, "Use of 26% KOH Electrolyte for Long Term Nickel-Hydrogen Geosynchronous Missions", Proceedings of the 29th Intersociety Energy Conversion and Engineering Conference.

[2] Manzo, M., Lenhart, S., Hall, A., 1989, "Bipolar Nickel-Hydrogen Battery Development – A Program Review", Proceedings of the 24th Intersociety Energy and Engineering Conference.

[3] Britton, D., 1999, "Progress in the Development of Lightweight Nickel Electrode for Nickel-Hydrogen Cell", NASA TM 1999-209430.

[4] Guasp, E. and Manzo, M., 1997, "Test Procedures for Characterizing, Evaluating, and Managing Separator Materials Used in Secondary Alkaline Batteries", NASA TM 107292.

[5] Ralston, P., 2001, "Nickel-Hydride and Bipolar Battery Development - Flight Battery Option", Final Report, NAS3-27787, Contract Period October 1995-January 2001.

[6] Manzo, M., O'Donnell, P., 1996, "NASA Aerospace Flight Battery Systems Program Update", Proceedings of the NASA Battery Workshop.

[7] Moore, B., Brown, H., and Miller, T., 1997, "International Space Station Nickel-Hydrogen Battery Cell testing at NAVSURWARCENDIV Crane", Proceedings of the 32nd Intersociety Energy and Engineering Conference.

[8] Warshay, M., Prokopius, P., Le, M., Voecks, G., 1996, "NASA Fuel Cell Upgrade Program for the Space Shuttle Orbiter", Proceedings of the 31st Intersociety Energy and Engineering Conference.

[9] Prokopius, P., Warshay, M., Simons, S., King, R., 1979, "Commercial Phosphoric Acid Fuel Cell System Technology Development", NASA TM-79169.

[10] Kohout, L., 1989, "Cryogenic Reactant Storage for Lunar Base Regenerative Fuel Cells", NASA TM 101980.

[11] "National Program Plan Fuel Cells In Transportation", 1993, U.S. Department of Energy, Office of Propulsion Transportation Technologies, DOE/CH-9301a.

[12] Loyselle, P., Maloney, T., Cathey, H., Jr., 1999, "Design, Fabrication, and Testing of a 10 kW-hr H_2-O_2 PEM Fuel Cell Power System for High Altitude Balloon Applications, Proceedings of the 34th Intersociety Energy and Engineering Conference.

[13] Rehg, T., Birschbach, M., Loda, R., Simpson, S. Tourbier, D., Weng, D., Woodcock, G., 2000, "PEM Fuel Cell System Development at Honeywell", Proceedings of the 2000 Fuel Cell Seminar.

[14] Lee, S., Lee, Y., 2000, "Hydrogen storage in single-walled carbon nanotubes" Applied Physics Letters 76 2877.

[15] Nutzenadel, C., Zuttel, A., Chartouni, D. Schlapbach, L., 1999, "Electrochemical Storage of Hydrogen in Nanotube Materials", Electrochemical and Solid State Letters 2 30.

[16] Qin, X., Gao, X., Liu, H., Yuan, H., Yan, D., Gong, W., Song, D., 2000, "Electrochemical Hydrogen Storage of Multiwalled Carbon Nanotubes", Electrochemical and Solid State Letters 3 532.

[17] Rajalakshmi, N., Dhathathreyan, K., Govindaraj, A., Satishkumar, B., 2000, "Electrochemical Investigation of Single-walled Carbon Nanotubes for Hydrogen Storage", Electrochimica Acta 45 4511.

IECEC2001-AT-11

ADVANCED MOTOR CONTROL TEST FACILITY FOR NASA GRC FLYWHEEL ENERGY STORAGE SYSTEM TECHNOLOGY DEVELOPMENT UNIT

Barbara H. Kenny[1]

Michael Mackin[1]

[1]NASA Glenn Research Center
21000 Brookpark Road
Cleveland, Ohio 44135

Peter E. Kascak[2]

Walter Santiago[1]

[2]Ohio Aerospace Institute
Cedar Point Road
Brookpark, Ohio 44135

Heath Hofmann[3]

Ralph Jansen[2]

[3]Department of Electrical Engineering
Pennsylvania State University
University Park, PA 16802

ABSTRACT

This paper describes the flywheel test facility developed at the NASA Glenn Research Center with particular emphasis on the motor drive components and control. A 4-pole permanent magnet synchronous machine, suspended on magnetic bearings, is controlled with a field orientation algorithm. A discussion of the estimation of the rotor position and speed from a "once around signal" is given. The elimination of small dc currents by using a concurrent stationary frame current regulator is discussed and demonstrated. Initial experimental results are presented showing the successful operation and control of the unit at speeds up to 20,000 rpm.

INTRODUCTION

One of the key components of the flywheel energy storage system is the electric motor and its control. Energy storage and recovery are achieved by using the motor to increase or decrease the flywheel rotor speed as necessary. Good control of the motor is thus very important for the proper operation of the flywheel system. As part of the flywheel technology development effort, NASA Glenn Research Center has built a test facility with the capability to rapidly test and evaluate advanced motor control algorithms. It is the purpose of this paper to describe the test facility (with particular emphasis on the motor control portion), the basic motor control algorithms developed and to present initial experimental results.

NOMENCLATURE

L_q is the q-axis machine inductance, henries.

L_d is the d-axis machine inductance, henries.

f_q^r q-axis voltage or current in the rotor reference frame.

f_d^r d-axis voltage or current in the rotor reference frame.

f_q^s q-axis voltage or current in the stator reference frame.

f_d^s d-axis voltage or current in the stator reference frame.

p is the derivative operator, d/dt.

θ_r is the angle between the stator q-axis and the rotor q-axis, radians.

λ_{af} is the flux linkage due to the rotor magnets, volt-sec.

ω_r is the electrical rotor speed, radians/second.

FLYWHEEL TEST FACILITY

The flywheel system test facility consists of a test cell and a control room. The test cell is physically separated from the control room for safety purposes. The present flywheel system under test consists of a 4 pole permanent magnet synchronous machine, a high strength composite rotor, magnetic bearings and a housing structure which is sealed and pumped to a low vacuum with water cooling capability. The flywheel system itself is housed in a containment structure that is closed during operation. Power for the motor is derived from a standard six switch three phase inverter with a dc power supply source. The inverter is a commercial off the shelf intelligent power module (gate drive circuitry included in the unit) rated at 600 volts, 200 amps and uses IGBTs for the power switches.

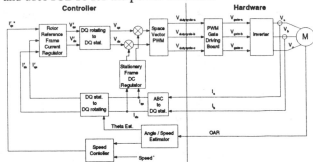

Figure 1: Block Diagram of Flywheel Motor Control

The control room has independent stations for various functions: magnetic bearing control, motor control, dc power supply control and temperature monitoring. The motor control function is implemented digitally using a commercially available microprocessor board. The control algorithms are written in a high level block diagram simulation language and converted to the necessary microprocessor code via software provided by the manufacturer of the microprocessor board. The motor current is measured in the test cell and fed back to an analog to digital (A/D) converter board in the control room. The inverter power switch duty cycles are calculated digitally as part of the control algorithm and output as analog signals using a digital to analog (D/A) converter. The duty cycle analog signals are sent to a pulse width modulation (PWM) generation board in the test cell. The output of the PWM generation board is a set of gate drive signals for the inverter unit. A block diagram of the motor control is shown in Figure 1 and described extensively in the next section.

MOTOR CONTROL ALGORITHM

The motor control algorithm is based on the principles of field orientation control or "vector control" [1]. The basic idea behind this control technique is to orient the applied stator currents to the rotor magnetic field. When this is done, the motor control is simplified because the control variables become dc quantities in steady state. The operating point of the machine can then be accurately and dynamically controlled so that high efficiency and fast response are obtained. A brief mathematical description for this technique follows.

A three phase machine, without a neutral connection, can be equivalently described as a two phase machine through a transformation from abc coordinates to dq coordinates as follows [2].

$$f_q = f_a \tag{1}$$

$$f_d = -\frac{1}{\sqrt{3}} f_a - \frac{2}{\sqrt{3}} f_b \tag{2}$$

The reverse transformation is

$$f_a = f_q \tag{3}$$

$$f_b = -\frac{1}{2} f_q - \frac{\sqrt{3}}{2} f_d \tag{4}$$

$$f_c = -\frac{1}{2} f_q + \frac{\sqrt{3}}{2} f_d \tag{5}$$

The d and q variables described in (1) and (2) are in the stator reference frame. From (1) it can be seen that the q-axis is aligned with the 'a' phase. This means that the 'a' phase current is equal to the q-axis current and the 'a' phase voltage (V_{an}) is equal to the q-axis voltage.

In a permanent magnetic machine, it is convenient to transform these variables to a reference frame that is rotating synchronously with the rotor magnetic field. In the rotor reference frame, the d-axis is defined to be co-linear with the rotor magnetic field axis. The transformation from the stator frame to the rotor frame is given by (6) and the inverse transformation is given by (7) [2].

$$\begin{bmatrix} f_q^r \\ f_d^r \end{bmatrix} = \begin{bmatrix} \cos\theta_r & -\sin\theta_r \\ \sin\theta_r & \cos\theta_r \end{bmatrix} \begin{bmatrix} f_q^s \\ f_d^s \end{bmatrix} \tag{6}$$

$$\begin{bmatrix} f_q^s \\ f_d^s \end{bmatrix} = \begin{bmatrix} \cos\theta_r & \sin\theta_r \\ -\sin\theta_r & \cos\theta_r \end{bmatrix} \begin{bmatrix} f_q^r \\ f_d^r \end{bmatrix} \tag{7}$$

The permanent magnetic synchronous machine can then be modeled in the rotor reference frame as follows [1]. Equation (9) gives the stator flux linkages in the rotor reference frame and (10) is the torque expression.

$$\begin{bmatrix} V_{qs}^r \\ V_{ds}^r \end{bmatrix} = \begin{bmatrix} R_s + L_q p & \omega_r L_d \\ -\omega_r L_q & R_s + L_d p \end{bmatrix} \begin{bmatrix} i_{qs}^r \\ i_{ds}^r \end{bmatrix} + \begin{bmatrix} \omega_r \lambda_{af} \\ 0 \end{bmatrix} \tag{8}$$

$$\begin{bmatrix} \lambda_{qs}^r \\ \lambda_{ds}^r \end{bmatrix} = \begin{bmatrix} L_q & 0 \\ 0 & L_d \end{bmatrix} \begin{bmatrix} i_{qs}^r \\ i_{ds}^r \end{bmatrix} + \begin{bmatrix} 0 \\ \lambda_{af} \end{bmatrix} \tag{9}$$

$$T_e = \frac{3P}{4} [\lambda_{ds}^r i_{qs}^r - \lambda_{qs}^r i_{ds}^r] \tag{10}$$

In field orientation control the d-axis current, i_{ds}^r, is commanded to 0. From (9) and (10) it can be seen that this results in a simplified expression for torque, given in (11).

$$T_e = \frac{3P}{4} [\lambda_{af} i_{qs}^r] \tag{11}$$

Because the magnitude of λ_{af} is constant due to the rotor magnets, the torque of the machine is proportional to the q-axis current. This result is similar to the dc motor with a constant field winding current where the torque is equal to the armature current times the torque constant. Thus control of the torque is achieved by properly controlling the rotor reference frame currents, i_{qs}^r and i_{ds}^r.

The torque command to the machine is determined from the outer loop control. In the results presented here, the outer loop is a speed controller, that is, errors in speed result in a correcting torque command. In an actual flywheel energy storage system, the torque command is derived from dc bus current or voltage commands. This is described in a companion paper [3].

To control the currents in the machine to be the commanded values, a current regulated voltage source inverter is used. This means that current errors result in voltage commands to the inverter which increase or decrease the applied voltage to increase or decrease the current respectively. This is accomplished through a synchronous frame current regulator [4] that is basically a PI controller operating on the rotor reference frame currents. The output of the controller is considered to be a voltage command in the rotor reference frame that is then transformed to the stator reference frame. The stator frame voltage commands are then used to calculate the inverter switch duty cycles.

There are several methods to find the duty cycles from the commanded stator frame voltages [5]. In the implementation used here, space vector modulation is used. Space vector modulation is a digital technique to calculate the duty cycles directly from the stator reference frame d and q voltages. One advantage of space vector modulation is that it increases the dc bus utilization to the maximum value. This means that for a given dc bus voltage, the maximum fundamental phase voltage is achievable by using space vector modulation. This is in contrast to the more common sine-triangle modulation which results in a lower dc bus utilization.

In this implementation, the duty cycles are output as analog signals which are sent to a PWM generation board. The analog signals are compared to a carrier signal triangle wave to create the gate drive signals to the inverter. This method allows the switching frequency to be separate from the sample frequency which is limited by the calculation time of the control algorithm. In this implementation, the sample time is set to 100 usec and the PWM switching frequency is set to twice that value at 20 kHz which is the maximum switching frequency of the inverter.

It can be seen by the rotor reference frame transformations of (6) and (7) that the rotor reference frame quantities are based on the rotor angle, θ_r. Thus it is necessary to know what this angle is. In industrial implementations of field orientation, it is common to use an encoder or resolver to feed back the rotor position. The high speed nature of this application makes it difficult to use the traditional feedback devices. Instead, the rotor position is estimated based on knowledge of a "once around" (OAR) position marker. In this machine, one half of the rotor hub is coated with a dark oxide, the other half is polished aluminum. The once around signal is based on the optical detection of the dark or light surface and thus a square wave at the same frequency as the machine speed is generated with the rising edge corresponding to the change from dark oxide to polished aluminum. Due to the large inertia of the flywheel rotor, the speed is reasonably assumed to be constant between once around signals and from this, the position can be estimated. This will be discussed further in the next section.

INITIAL IMPLEMENTATION ISSUES

Back EMF Test

As explained previously, to properly control the motor using the field orientation technique, the instantaneous position of the rotor q-axis with respect to the stator q-axis (θ_r) must be known. The once around signal indicates the position of the color change (dark to light) on the rotor but this does not necessarily coincide with the position of the rotor q-axis. What needs to be determined is the angle between the rotor q-axis and the stator q-axis as a function of the once around signal. This can be determined experimentally by measuring both the phase voltages of the machine and the once around signal as the rotor freely decelerates ("spins down"). For a freely decelerating machine, the stator current is 0 so the phase voltages are entirely due to the back emf of the machine. Stated mathematically, (for 0 stator current) [1]:

$$V_{qs}^s = p\lambda_{qs}^s \text{ where } \lambda_{qs}^s = \lambda_{af}\sin(\theta_r) \qquad (12)$$
$$V_{ds}^s = p\lambda_{ds}^s \text{ where } \lambda_{ds}^s = \lambda_{af}\cos(\theta_r) \qquad (13)$$

By definition, $V_{qs}^s = V_{an}$ because the q-axis in the transformation of (1) is aligned with the stator a-phase. Thus in a freely decelerating machine, the relationship between θ_r and V_{an} is given in (14).

$$V_{an} = \lambda_{af}\cos(\theta_r) \qquad (14)$$

V_{an} can be measured if there is a neutral terminal on the machine or V_{ab} can be measured and V_{an} calculated. In this case, the motor neutral was available for measurement purposes (not normally connected). Figure 2 shows a plot of the once around signal and the V_{an} voltage.

From (14) it can be seen that when $\theta_r = 0$, V_{an} will be at a maximum. Figure 2 shows that this maximum occurs twice in one cycle of the OAR signal at approximately the middle of each pulse. The machine is a 4 pole machine so between every rising edge of the OAR signal, 360 mechanical and 720 electrical degrees will have passed. Figure 2 shows that the V_{an} maximum occurs approximately 90 mechanical degrees or 180 electrical degrees behind the rising edge of the OAR signal. This means that the θ_r transformation angle is equal to the electrical angle determined by the OAR signal minus 180°.

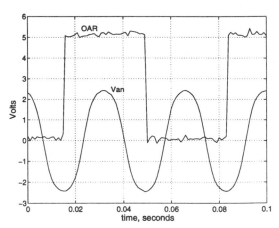

Figure 2: Phase voltage and OAR signal for freely decelerating rotor.

Once the relationship between θ_r and the once around signal is known, the transformation from the stator to the rotor reference frame can be done. From (8), it can be seen that for a freely decelerating rotor, when $i_{qs}^r = i_{ds}^r = 0$, the back emf voltage appears entirely in the rotor reference frame q-axis and $V_{ds}^r = 0$. This means that the transformation angle θ_r can be fine-tuned. The voltages measured in the back emf test can be transformed to the rotor reference frame using the relationship

$$\theta_r = \text{OAR electrical angle} - 180°. \qquad (15)$$

If the transformation angle θ_r given by (15) is not exactly correct, then V_{ds}^r will not be exactly zero. Figure 3 shows a plot of V_{ds}^r where θ_r lags the OAR electrical angle by 180°. It can be seen that V_{ds}^r is slightly less than zero. Figure 3 also shows the case where θ_r lags the OAR electrical angle by 170°. It can be seen that this average value of V_{ds}^r is now more nearly equal to zero. Thus the actual transformation angle used in the control algorithm is given by (16).

$$\theta_r = \text{OAR electrical angle} - 170° \text{ degrees} \qquad (16)$$

The ac components of the q and d axis voltages in Fig. 3 are due to the spatial harmonics of the motor windings and are not considered in this control.

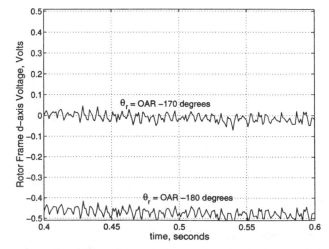

Figure 3: V_{ds}^r for Variations in θ_r.

Starting the machine

To operate the machine using field orientation, the rotor position angle must be known. However, the machine must be spinning to estimate the position from the once around position. Initially, this dilemma was resolved by starting the machine in "open loop" operation. A fixed voltage and frequency were applied to the machine directly until the rotor started to move. Once a OAR signal was established, the control of the machine was shifted to closed loop speed control.

It was additionally observed that the machine could start under closed loop control if a low speed command was given.

EXPERIMENTAL RESULTS

Closed Speed Loop Operation

Closed loop speed control was achieved by using a PI loop around the torque control as shown in Fig. 1 and the unit was tested at speeds up to 20,000 rpm. Using the once around signal, two methods were developed to estimate the speed and position. The first method works well at low speeds and degrades as the speed increases due to quantization errors. The second method is limited at low speeds due to saturation of the digital counter register but works very well at high speeds.

Speed Estimate: Method One

In this method, the speed was determined by counting the number of sample periods, t_p, between once around signals and multiplying by the sample rate, T_s as shown in (17). The speed was assumed to be constant between OAR signals leading to the position estimate shown in (18) where t_{pz} is the number of sample periods since the last rising edge of the OAR signal. The position estimate was reset to zero at every rising edge of the OAR signal.

$$Speed_{rpm} = \frac{60}{t_p T_s}. \tag{17}$$

$$\theta_{electrical}{}^{rad} = Speed_{rpm} \frac{2\pi}{30} t_{pz} \tag{18}$$

The sample rate of the control algorithm is 100 usec. This leads to the possibility of error shown in Table 1. The quantization errors of this method led to speed oscillations which became progressively worse as the speed increased. Figure 4 shows the speed estimate for a commanded speed of 5000 rpm. Oscillations on the order of 12 rpm are clearly evident. Figure 5 shows the impact of the speed oscillations on the phase current.

Actual Machine Speed (rpm)	Maximum Possible Speed Estimate Error, Method One, rpm	Maximum Possible Speed Estimate Error, Method Two, rpm
1,000	1.66	.03
5,000	41.32	.83
10,000	163.93	3.33
20,000	645.16	13.32
60,000	5454.54	119.76

Table 1: Possible Speed Errors for Alternate Methods

Speed Estimate: Method Two

This method was based on a commercially available timer board which contains two 16 bit counters. Counter 1 is used to

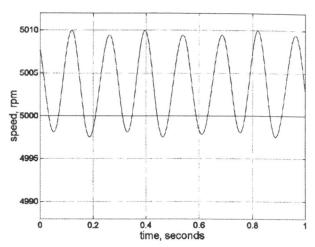

Figure 4: Speed Oscillation due to Quantization Error with Method One Speed Estimate

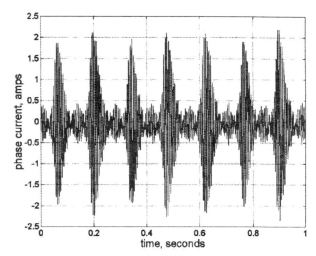

Figure 5: Oscillations in Phase Current due to Speed Estimate Oscillations for Method Two

determine the OAR frequency and counter 2 is used to determine the time that has elapsed since the most recent rising edge of the OAR signal. The counter clocks are scaled so that a signal in the range of 7.64 Hz to 500 kHz may be detected. (The clock rates may be scaled higher in software as necessary for higher rotor speeds.) The speed is determined from counter 1 and (19) and the OAR electrical angle is determined from (20) where 'f' is the output of counter 1 and 't' is the output of counter 2.

$$Speed_{rpm} = 60 \frac{f}{2} \tag{19}$$

$$\theta_{electrical}{}^{\circ} = 4\pi ft \tag{20}$$

The motor speed using speed estimate method two for feedback is shown in Fig. 6. The oscillations in both the speed estimate and the phase current (Fig. 7) are eliminated.

DC Offset Currents

Figure 8 shows the frequency spectrum of the phase current at an operating speed of 10,000 rpm. The fundamental component, at 333 Hz, clearly dominates. However, there is a significant dc component also present. Although not predicted theoretically, the dc component may be caused by slight

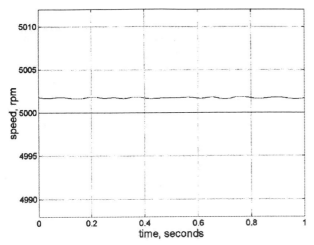

Figure 6: Speed oscillation with Method Two Speed Estimate

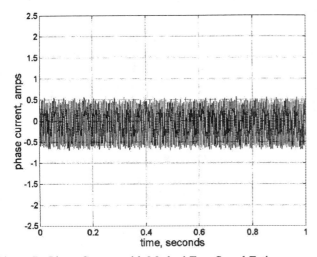

Figure 7: Phase Current with Method Two Speed Estimate

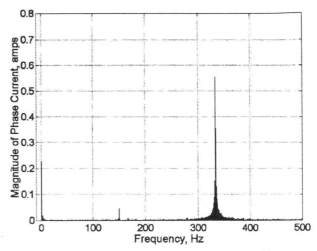

Figure 8: Phase Current Frequency Spectrum at 10,000 rpm

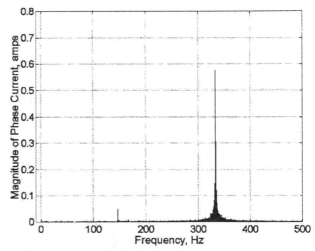

Figure 10: Phase Current Frequency Spectrum at 10,000 rpm with stationary frame current regulator

mismatches or offsets in the phase voltages applied to the machine.

The synchronous frame current regulator was tuned for a bandwidth of 200 Hz. The synchronous electrical speed at 10,000 rpm is 333 Hz. This means that currents in the frequency range of 133 Hz to 533 Hz are within the bandwidth of the current regulator. Thus the current regulator will not effectively eliminate the dc currents. One approach to this problem is to introduce an additional current regulator which operates in the stationary reference frame. This current regulator is a PI controller tuned to have a very low bandwidth so it essentially only regulates the dc current as shown in Fig. 9. Due to the low bandwidth of the PI, the fundamental current passes through the regulator unchanged while the dc current is regulated to zero. Using the stationary frame current regulator in addition to the synchronous frame current regulator resulted in the elimination of the dc component as shown in Fig. 10.

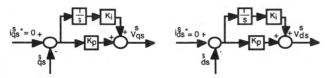

Figure 9: Block Diagrams of d- and q-axis Stationary Frame DC Current Regulators

Sensitivity to Error in Position Estimate

In this implementation of field orientation control the commanded d-axis current, i_{ds}^f, is set equal to zero. In this strategy, the direct axis stator flux linkage, λ_{ds}^f, is constant and due entirely to the rotor magnet with no contribution from the stator current. The torque is produced by the interaction of this flux with the q-axis current according to (11). However, if the rotor reference frame transformation angle, θ_r, is not correct, then there will be d-axis current in addition to the commanded q-axis current. For small errors in angle estimate, this additional current does not contribute significantly to the total phase current in the machine. However, as the error approaches 90 electrical degrees, the theoretical maximum, the total stator current increases rapidly. This is shown in Fig. 11. The data points in Fig. 11 were taken at a speed of 7,000 rpm.

20000 RPM OPERATION

The speed of the motor was limited to 20,000 rpm due to limits on the particular flywheel rotor used in the system. The following results show steady state operation at 20,000 rpm. The small magnitude of the phase current (rms value just over 1 amp) indicates the very low losses of the machine when suspended on magnetic bearings and operated in a vacuum. Fig. 12 shows the motor phase current. The current looks somewhat distorted however an examination of the spectrum shown in Fig. 13 shows the major frequency component at 667 Hz as expected. Also, the current shown in Fig. 12 has an extremely small magnitude which may contribute to the

perceived distortion.

Finally, Fig. 14 shows the duty cycle for the phase A high switch at 20,000 rpm with a 120 volt dc bus for both sine-triangle modulation and space vector modulation. This plot shows that the space vector modulation technique needs less dc bus voltage than the sine-triangle technique to generate the same phase voltage. The space vector duty cycle is seen to be well within the duty cycle range of zero to one for operation at this speed from this voltage. However, the duty cycle for the sine-triangle technique is clamped at one at some points and at zero at others. This duty cycle which would result in a higher distortion of the synthesized phase voltage.

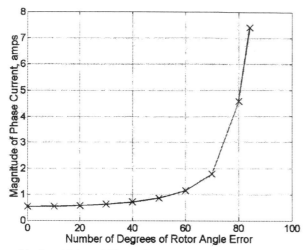

Figure 11: Sensitivity of Phase Current to Rotor Angle Error

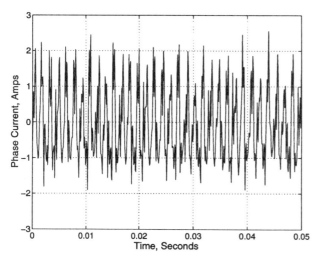

Figure 12: Motor Phase Current at 20,000 rpm Operation

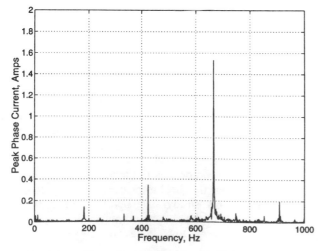

Figure 13: Motor Phase Current Spectra for 20,000 rpm Operation

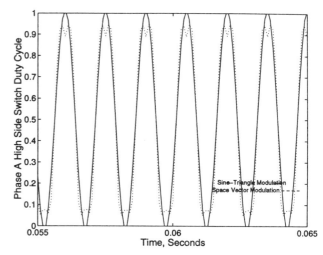

Figure 14: Phase A High Switch Duty Cycle for Space Vector Modulation and for Sine-Triangle Modulation

CONCLUSIONS

This paper has presented the initial experimental results of the flywheel motor test facility at the NASA Glenn Research Center. The once around signal was successfully used for speed and position feedback. The back emf test and the subsequent fine-tuning was shown to accurately determine the relationship between the OAR and the rotor q-axis. The dc offset current was eliminated with the addition of the concurrent stationary frame regulator. Space vector modulation was shown to increase the dc bus utilization over sine-triangle modulation.

REFERENCES

[1] Krishnan, Ramu, *Permanent Magnet Synchronous and Brushless DC Motor Drives: Theory, Operation, Performance, Modeling, Simulation, Analysis and Design*, Virginia Tech., Blacksburg, Virginia, 1999.

[2] Krause, Paul, *Analysis of Electric Machinery*, McGraw-Hill Book Company, New York, New York, 1986.

[3] Kascak, P., B. Kenny, W. Santiago, R. Jansen, T. Dever "International Space Station (ISS) Bus Regulation Studies with NASA Glenn Research Center (GRC) Flywheel Energy Storage System Development Unit", *to be published at 2001 IECEC*, Savannah, Georgia.

[4] Rowan, T. and R. Kerkman, "A New Synchronous Current Regulator and an Analysis of Current-Regulated PWM Inverters", *IEEE Transactions on Industry Applications*, Vol IA-22, No. 4, July/August 1986, pp. 678-690.

[5] Holtz, Joachim, "Pulsewidth Modulation for Electronic Power Conversion", *Proceedings of the IEEE*, Volume 82, No. 8, August, 1994, pp. 1194-1214.

Proceedings of IECEC'01
36th Intersociety Energy Conversion Engineering Conference
July 29-August 2, 2001, Savannah, Georgia

IECEC2001-AT-81

MAGNETIC SUSPENSIONS FOR FLYWHEEL BATTERIES

Professor Alan Palazzolo
Yeonkyu Kim
Andrew Kenny
Dr. Uhn Joo Na
Shuliang Lei
Erwin Thomas
Texas A&M University
Mechanical Engineering

Raymond Beach
NASA Glenn

Albert Kascak and Gerald Montague
U.S. Army at NASA Glenn

ABSTRACT

This paper provides a comprehensive overview of test and analysis procedures for designing and implementing a high speed flywheel magnetic suspension. These techniques were successfully employed to spin a 11.4 kg flywheel at 60,000 rpm

INTRODUCTION

High-speed flywheels are presently being developed for application to energy storage (ES) and attitude control (AC) on satellites and the ISS. This NASA funded effort is driven by potentially large savings in electrochemical battery replacement costs and the dual role capability of flywheels for ES and AC. Conventional rolling element bearing suspensions for flywheels are not acceptable due to high parasitic drag torque losses and consequent loss in energy conversion efficiency. Electromagnetic bearings (EMB) provide suspension of the high speed flywheel with a far reduced drag torque and lower levels of transmitted vibratory force, reducing disturbances to micro-gravity experiments. The cost for this gain in efficiency and isolation is a sophisticated design and commissioning task, typical for feedback control of a large order, spinning, flexible system with strong gyroscopic torques, inherent sensor noise, dominant high frequency (1000 hz) poles, and high speed (60,000 rpm). The Texas A&M Vibration Control and Electromechanics Lab (VCEL) was contracted by NASA Glenn to meet this challenge on the ACESE (Attitude Control and Energy Storage) and FESU (Flywheel Energy Storage Unit) programs. This team effort with NASA Glenn and U.S. Army personnel was funded to lower long-term operation costs and increase reliability on the ISS and LEO satellites.

This paper addresses the major technical tasks required for the successful commissioning of two prototype, ACESE class flywheels. A parallel approach for a FESU class flywheel is now ongoing with commissioning planned for early 2001. The literature has various references for magnetic bearing supported flywheels. Noteable among these is the works of Ahrens, et. al (1994) and Okada (1992).

COMPONENT MODELING

Rule #1 is successful magnetic levitation of a high-speed flywheel will rarely be achieved unless it can be performed first "on paper". Violation of this rule has led to numerous failures, incurring losses of resources and time far exceeding that required for the modeling task. This step though necessary, is not sufficient for success, due to the unavoidable presence of unmodelled dynamics and disturbances. Nevertheless, the modeling effort may still provide good engineering insights into addressing these problems if encountered during commissioning.

The flywheel module system dynamics model is composed of seven major elements: (a) Wheel and Shaft, (b) Shaft Position Sensors, (c) Electromagnetic Actuators, (d) Feedback Controller, (e) Power Amplifiers, (f) Module Casing and Support Structure and (g) Auxiliary Bearings. These are illustrated in the flow diagram of Figure 1.

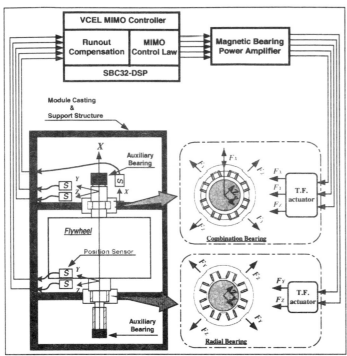

Figure 1. Flywheel Module and Control Flow Diagram

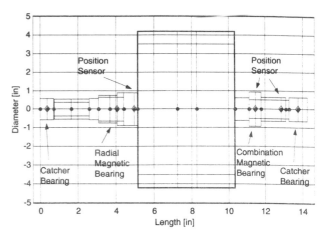

Figure 2. FE Model of Flywheel and Shaft

Flywheel and Shaft

The flywheel is typically press-fit onto a single shaft or bolted onto two shaft stubs, and is made from high strength composite materials with hoop strength near 1,000,000 psi. The energy storage density of the rotating assembly increases as the ratio of its polar to transverse moments of inertia (I_P/I_T) increases. This also has an effect of yielding larger changes in momentum for more effective attitude control. Typically I_P/I_T ratios from 0.5 to 1.5 are employed with speeds ranging from 35,000 to 60,000 rpm. Flexible hub designs are also desirable, concentrating mass at larger radii thereby increasing the energy density while maintaining a constant total weight. Here again, a trade off in performance and design complexity appears, in that as I_P/I_T increases a lightly damped, dominant pole also increases in frequency, requiring wider control bandwidth and higher susceptibility to noise. The control of this pole, in a 1-2 kW-hr, 60,000 rpm wheel, presents an interesting challenge!

High speed, high bandwidth operation mandates modeling of the rotating assembly as a flexible body, a task performed by the VCEL with its suite of finite element based, closed loop feedback, electromechanical modeling codes. These codes have been developed with the support of NASA and other sponsors over the past 15 years, for tasks ranging from piezoelectric actuator based active vibration control to magnetic bearing development for gas turbine-aircraft engines and machine tools. Figure 2 shows the F.E. model for an I_P/I_T = 0.8 NASA flywheel along with the locations for position feedback sensors, electromagnetic actuators and auxiliary mechanical bearings.

The number of states present from the flexible shaft model ranges from 100 on an ACESE class flywheel, to as much as 600 on a locomotive flywheel. The TAMU-VECL was contracted by U.S. Flywheel Systems to develop a magnetic suspension controller for the former flywheel and by the UT Austin CEM for the latter flywheel. Conversion of these physical coordinate states to modal coordinate states reduces their numbers tenfold, providing a practical sized model for simulation.

Position Feedback Sensors (PFS)

An important task in magnetic suspension design is for the shaft to accurately track designated target positions with magnetic bearings. Typically these targets are static and correspond to the geometric or magnetic centers of the EMB actuators. Dynamic target applications have been investigated by the VCEL for NASA and the Navy, with aim of the former sponsor to control rotating stall in an aircraft engine. The PFS array provides instantaneous shaft positions to the controller which adjusts the EMB forces to reduce the error between the actual and target positions. Non-concentricity of the rotating shaft deceives the position sensor into transducing false "run-out", which masquerades as dynamic motions of the shaft centerline. Non-uniformity in reflectance, or conductivity and permeability have similar effects to out-of-roundness, for optical and eddy current/reluctance position sensors, respectively. Tight machinery tolerances (.0002") and careful shaft surface polishing is required to reduce this run-out, pseudo-noise effect. Sophisticated filtering and feedforward compensation schemes are also used at the controller level to combat run-out before it enters the power amplifier's current servo yielding wasteful and deleterious currents and vibration. Significant current levels have been detected as high as the 12th-15th harmonic of shaft speed, due to run-out. Simple low pass filtering may reduce this range, but must be designed to maintain a high sensor bandwidth (typically 5khz-10khz), for improved feedback control.

Extensive research is now ongoing at NASA and the VCEL to develop sensors with high immunity to run-out and EMI. Issues being addressed include radiation effects on circuitry and fiber optic cable, motor and switching power amplifier induced EMI, long term outgassing and reliability.

Electromagnetic Actuators

High frequency, dominant poles of the flywheel/magnetic suspension system requires high bandwidth actuation to provide damping and improve stability. Minimum weight requirements also demand occasional actuator operation with high flux densities that approach saturation during levitation or upset events. These factors mandate an actuator construction with thin laminates (6-14 mils) made from a cobalt iron alloy such as Hyperco (super mender). Flux fields encountered by the shaft material may vary from zero to a peak value, or from a negative to a positive peak value depending on whether the EMB construction is a homopolar or heteropolar construction, respectively. This results in eddy current and hysteresis losses, and drag torque. Proper design and operation of the EMB actuator will reduce these negative side effects, while maximizing the force/current gain, bandwidth, force slew rate, linearity and load capacity. Both magnetic circuit and 3D FE models for flux, force and power loss are employed by the VCEL for this purpose. Figure 3 shows a typical flux map for an 8 pole heteropolar magnetic bearing. These models often include saturation and eddy current effects. Pole geometry and extent, number of turns, radii and material are varied during the simulation/design process. Of particular importance for flywheel applications on satellites is parasitic drag torque arising from eddy currents in the magnetic bearings. For instance, a drag torque as small as 0.12 in lbs. will cause a power loss of 85 watts at a speed of 60,000 rpm.

Component: BMOD
0.000494048   0.553265   1.106035

Figure 3: Flux/Force Model for an 8 Pole Heteropolar Bearing

Efforts to reduce drag torque are being sponsored by NASA at the VCEL, including extensive 3D, FE model simulations and development of a unique, high speed, vacuum environment test rig for direct magnetic bearing drag torque measurement.

Feedback Controller

If the sensor and actuator acts as the eyes and brawn of the suspension, the controller acts as the brain. Streams of coding continuously execute to command actuator forces that move the shaft nearer to its target positions. A digital signal processing board (DSP) converts analog position signals into digital words, evaluates compensation instructions in terms of difference equations then outputs analog control voltages to the power amplifiers. This process typically occurs at a rate of 5000 to 20,000 times/second. Increasing the sampling frequency raises the controller's bandwidth but reduces the number of coding commands (control order) that may be successfully executed during each cycle. An alternate approach for compensation is an analog realization of the control algorithm. This permits use of a much larger pool of rad hardened components than are currently available for high performance DSP's or microprocessors, but is less easily adapted during the control.

The controller developed and implemented by the VCEL for the ACESE/FESU flywheel program utilizes a MIMO (multiple input-multiple output) control approach, which increases the phase and gain margins for the wheel's dominant nutation and precession poles. Initial attempts at employing a SISO strategy were fruitless since they underestimated the gyroscopic torque effect on these large I_P/I_T ratio (0.8-1.25) wheels.

The VCEL's MIMO controller was successfully employed for magnetic suspension of a 40,000 and a 60,000 rpm, ACESE class flywheels, at U.S. Flywheel Systems. The compensation stages were especially designed to maintain high phase and amplitude margins while rejecting motor drive related EMI and shaft sensor run-out.

Power Amplifiers (PA)

Current controlled power amplifiers are servo devices, yielding an output current that tracks a controller output voltage. Desired bandwidths, supply voltages and maximum currents for flywheel suspension range over 2-3kHz, 80-180V and 5-15 Amps, respectively. The model should replicate the servo action along with both current and current slew rate limits. Switching power amplifiers with pulse width modulation (PWM) are employed to improve efficiency for coil current generation. This benefit is somewhat offset by their EMI production that interferes with the position sensors. Extensive testing has been performed at the VCEL to insure reliable model bandwidths and nonlinear response characteristics for these components.

Module Casing and Support Structure

Dynamic force isolation of the flywheel module from the ISS or LEO satellite is a key concern for limiting disturbances to micro-gravity experiments and sensitive optics and electronics. Soft mounting is employed for this purpose but is constrained by the thermal management objective of providing a good heat conduction path from the module to the host frame. The lightweight casing though possessing many flexible body vibration modes may be modeled as a rigid body if the magnetic suspension is properly designed to isolate the casing from the rotor, and to reject run-out, imbalance and EMI noise related disturbances.

Auxiliary (Catcher) Bearings (AB)

Rub of the high-speed shaft against the stator is restricted by passive, auxiliary (catcher) bearings during an EMB power outage or extreme load condition. The AB's are typically rolling element or sleeve type bearings with a compliant outer liner or support, to reduce impulse loading and to provide damping. In place measurement of the AB stiffness is an important step to include all series or parallel compliances in the AB component model.

SYSTEM SIMULATION

Simulation results of importance to robust and efficient magnetic suspension design may now be extracted from the assembled model. Linear stability is of paramount concern due to the risk of severe rig damage in the event of experiencing an unstable pole event at high speed. Pole locations must be tracked vs. rpm due to their strong dependence on gyroscopic torque. The four dominant pole mode shapes of primary concern are the forward and backward conical modes, the cylindrical mode and the first bending mode. "Forward" and "backward" here describe the whirl direction of the mode relative to the spin direction. These modes are illustrated in Figure 4, for an ACESE class rotor. Root locus and Campbell diagrams clearly show the expected stability margins and pole's speed dependence, as shown in Figures 5a and 5b, respectively.

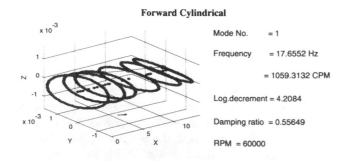

Figure 4a. Dominant Pole Mode Shapes

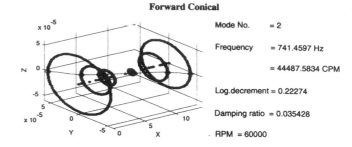

Figure 4b. Dominant Pole Mode Shapes

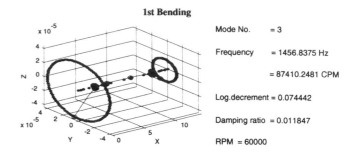

Figure 4c. Dominant Pole Mode Shapes

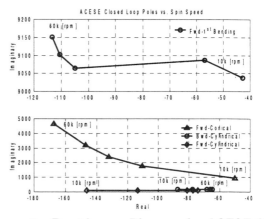

Figure 5a. Root Locus Diagram for ACESE Class Rotor

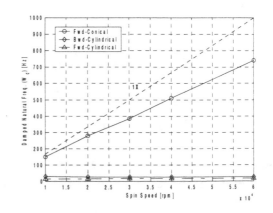

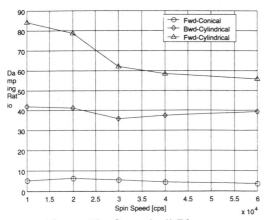

Figure 5b. Campbell Diagram

Effective magnetic stiffness and damping vs. frequency plots lend additional insights into closed loop stability and are shown in Figure 6. Critical speeds are identified from synchronous response plots utilizing either an imbalance or run-out disturbance as illustrated by Figure 7. Transient response plots are also very useful due to the nonlinearities in the component models and the multitude of disturbance harmonics from sensor run-out. Typical plots of shaft displacement, coil current and controller output voltages vs. time are shown in Figure 8. A "waterfall" plot of power amplifier effective voltage vs. frequency and speed can indicate a saturation source and is shown in Figure 9. This plot is obtained by taking the FFT of the steady state portion of a transient response simulation. Finally, upset events such as power loss, or composite wheel cracking may also be simulated as shown by the shaft displacement plots in Figures 10a and 10b, respectively. Figure 11 shows the catcher bearing forces that result from the power loss upset event in Figure 10a. Additional simulations may be obtained from the VCEL simulation codes, however, the types shown in Figures 4-11 are the most useful.

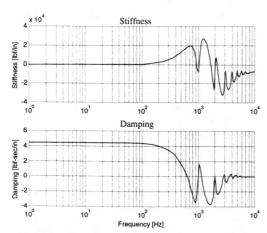

Figure 6. Effective Magnetic Stiffness and Damping vs. Frequency

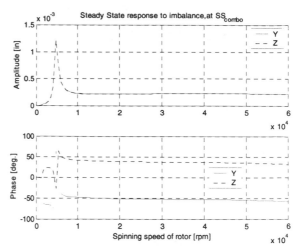

Figure 7. Synchronous Response Plot for ACESE Class Flywheel

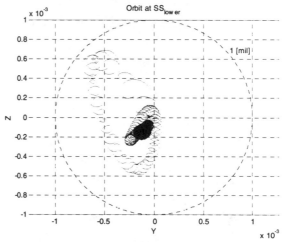

Figure 8a. Transient Imbalance, Run-out, and Sensor Noise Response Plots

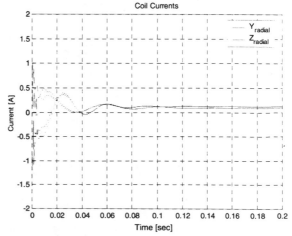

Figure 8b. Transient Imbalance, Run-out, and Sensor Noise Response Plots

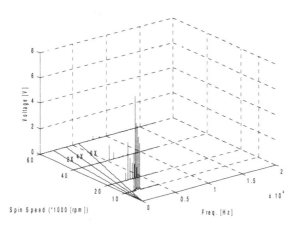

Figure 9. Waterfall Plot for ACESE Class Flywheel

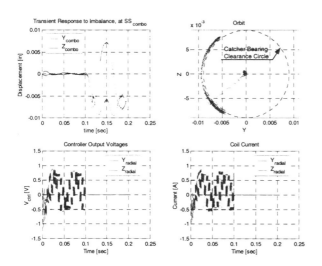

Figure 10a. Power Loss Response of ACESE Class Flywheel

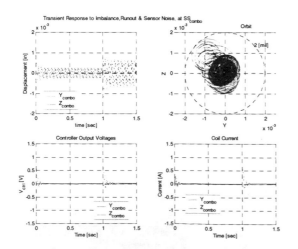

Figure 10b. Transient Response of ACESE Class Flywheel (Simulated Micro-crack)

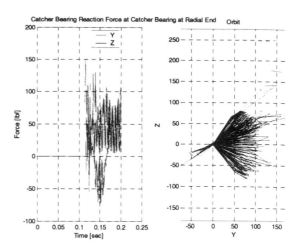

Figure 11. Catcher Bearing Forces of ACESE Class Flywheel

The preceding types of plots provide the tools to judge proposed feedback compensation stages in the controller. These stages are iteratively designed utilizing PID, lead, lag and special filtering blocks. Mode shapes, pole plots and transient responses are checked to guide selection of the compensation stages leading to a control law design. This design then receives final in-situ tuning during the test, when unmodeled disturbances and dynamic effects often appear.

FLEXIBLE HUB FLYWHEEL CONTROL

Control of a flexible hub flywheel poses the additional challenge for maintaining stability of a second forward conical mode that involves relative motion between the flywheel and shaft. Figure 12 shows the rotor FE model of a flexible hub flywheel system. Tables 1 and 2 show how the frequency and damping of the closed loops second conical mode varies with hub stiffness and damping at a constant spin frequency of 667 Hz. It is clearly seen that hub damping stabilizes this mode if it is above the spin frequency and destabilizes it when is is below the spin frequency.

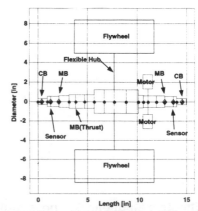

Figure 12. FE Model of Flexible Hub Flywheel System

Table 1. 2nd Forward Conical Mode Frequency [Hz] vs. Hub Stiffness and Damping

Stiffness Damping [%]	Base	0.5 Base	0.25Base	0.1Base
0.1	1191.3	1109.7	934.8	634.75
0.5	1191.5	1109.7	935.31	634.68
1	1191.6	1109.7	935.87	634.62
3	1191.0	1109.3	937.32	634.62
5	1188.4	1108.4	937.47	634.98
10	1172.1	1104.6	931.28	637.2

Table 2. 2nd Forward Conical Mode Damping[%] vs. Hub Stiffness and Damping

Stiffness Damping [%]	Base	0.5 Base	0.25Base	0.1Base
0.1	-0.32063	-0.2355	-2.9853	0.40946
0.5	-0.03926	-0.1512	-2.6364	0.33503
1	0.31347	-0.04539	-2.1997	0.24143
3	1.7369	0.38165	-0.4451	-0.13478
5	3.1833	0.81005	1.3295	-0.5033
10	6.9266	1.8381	5.9021	-1.3129

TESTING STEPS FOR FLYWHEEL COMMISSIONING

Failure at high speed may cost tens of thousands of dollars and months delay in flywheel commissioning. The approach employed by the VCEL and its NASA and Army colleagues consists of three test stages:

    (a)   Component Model Verification Tests (CMVT)
    (b)   System Model Verification Tests
    (c)   Run-up to Full Speed Tests

CMVT is an essential step due to the "garbage in = garbage out" principle of running system models, i.e. if component models are poor then the system model results will also be poor. Consider the components in the order previously presented, beginning with the flywheel and shaft. The rigid body inertias of the rotating assembly should be measured by the pendulum method for both I_P and I_T. Free-free mode shape testing must also be performed to validate the flexible shaft finite element model. Position feedback sensors should be tested for DC voltage/displacement gain, bandwidth, sensor sensitivity and run-out, including the effects of all signal conditioning in the tests. The major properties required to test for on the actuators are DC force/current gain, bandwidth and position stiffness (force/position). The feedback controller possesses filter characteristics that should also be measured to improve the model. Likewise, measurement of the DC

current/voltage gain, and frequency response characteristics of the power amplifiers is essential. Casing modes and inertias are finally measured for system model enhancement.

The non-spinning model characteristics of the system are measured once the rotor has been levitated. Pole frequencies and dampings are obtained through forced response excitations. This procedure is repeated at various speeds during the startup providing data to plot experimental Root Locus and Campbell diagrams. Resonance speeds are also noted and compared with the model predictions.

SUMMARY

Flywheels for energy storage and attitude control will provide significant cost savings for the ISS and LEO satellites. The use of magnetic bearings on these flywheels provides a low power consumption path to reliable, spinning shaft suspension. System design, component selection and testing, system testing and final commissioning are each important steps to successful magnetic suspension of the high speed flywheel. In particular, sensor selection, actuator design and controller realization present "make or break" challenges for the program.

The tools highlighted in this text were employed for the landmark runs of two ACESE class flywheels, to 40,000 rpm and 60,000 rpm, respectively. Terrestrial applications for magnetic suspensions will greatly benefit from NASA's ISS and LEO Satellite "flywheels in space" efforts. This is already evidenced in the VCEL's magnetic suspension control design commissionings with U.S. Flywheel Systems, Optimal Energy Systems, the Center for Electromechanics at UT-Austin and the U.S. Navy-ONR, each receiving technical advancement spinoff's from the VCEL's NASA work. The urgency of the NASA program to incorporate energy storage flywheels on the ISS has been recently made obvious by the space shuttle Atlantis' flight to replace 4 of the 163 lb., $252,000/unit Russian made electrochemical batteries that failed prematurely.

ACKNOWLEDGMENTS
The authors acknowledge the support for this work through NASA Glenn (Grant NRA 99-GRC-2) and from Tom Calvert, Lyn Peterson, and Glenn Bell of the Navy NSWC (grant ONR BAA 97-030).

REFERENCES
Ahrens, M., Kucera, L., and Larsonneur, R., 1996, "Performance of a Magnetically Suspended Flywheel Energy Storage Device," *IEEE Transactions on Control and System Technology*, Vol.4, No. 5, pp.494-502.

Okada, Y., Nagai, B. and Shimane, T., 1992, "Cross Feedback Stabilization of the Digitally Controlled Magnetic Bearing," *ASME Journal of Vibrations and Acoustics*, Vol. 114, pp.54-59.

Proceedings of IECEC'01:
36th Intersociety Energy Conversion Engineering Conference
July 29-August 2, 2001, Savannah, Georgia

IECEC2001-AT-82

AEROSPACE FLYWHEEL TECHNOLOGY DEVELOPMENT
FOR IPACS APPLICATIONS

Kerry L. McLallin
NASA Glenn Research Center

Dr. Jerry Fausz
Air Force Research Laboratory

Ralph H. Jansen
Ohio Aerospace Institute

Robert D. Bauer
Lockheed Martin

ABSTRACT

The National Aeronautics and Space Administration and the Air Force Research Laboratory are cooperating under a space act agreement to sponsor the research and development of aerospace flywheel technologies to address mutual future mission needs. Flywheel technology offers significantly enhanced capability or is an enabling technology. Generally these missions are for energy storage and/or integrated power and attitude control systems (IPACS) for mid-to-large satellites in low earth orbit. These missions require significant energy storage as well as a CMG or reaction wheel function for attitude control. A summary description of the NASA and AFRL flywheel technology development programs is provided, followed by specific descriptions of the development plans for integrated flywheel system tests for IPACS applications utilizing both fixed and actuated flywheel units. These flywheel system development tests will be conducted at facilities at AFRL and NASA GRC and include participation by industry participants Honeywell and Lockheed Martin.

Keywords: Flywheel, IPACS, Energy Storage, Attitude Control, Satellite

INTRODUCTION

The National Aeronautics and Space Administration and the Air Force Research Laboratory are cooperating under a space act agreement to sponsor the research and development of aerospace flywheel technologies to address mutual future mission needs. For some missions, flywheel technology offers significantly enhanced capability or is an enabling technology. Generally these missions are for energy storage and/or integrated power and attitude control systems (IPACS) for mid-to-large satellites in low earth orbit. These missions require significant energy storage as well as a CMG or reaction wheel function for attitude control.

In order to address these mission needs and demonstrate the readiness of flywheel technology, integrated flywheel system development and operational demonstrations are needed. Three major development and test activities that will begin to address the system integration questions are in progress within the NASA and AFRL flywheel programs. These development activities are the subject this paper. A summary description of the NASA and AFRL flywheel technology development programs is provided, followed by specific descriptions of the development plans for integrated flywheel system tests for IPACS applications.

PROGRAM CONTENT DESCRIPTION

A summary description of the NASA and AFRL flywheel technology development programs is provided below.

NASA Program

The Aerospace Flywheel Technology Program is funded by the NASA Headquarters Code R Cross-Enterprise Technology Development Program/Space Base Program and managed by Glenn Research Center. The objectives of this program are to develop advanced aerospace flywheel component and system technologies to meet NASA's long term mission needs. Flywheel technology addresses mission needs for energy storage, integrated power and attitude control, and power peaking.

The near term focus of the program is on "Century" class flywheels with energy storage capacity in the hundreds of watt-hours (300-700) for application to mid-sized satellites. In addition, longer term development of flywheels for small satellite applications is also in progress at energy capacities less than 100 watt-hours. Flywheel technology goals are defined by the metrics in Table 1. Flywheel technology offers significant performance advantages for energy storage and IPACS in low Earth orbit applications such as ISS and Earth Sciences. Other NASA applications that require power peaking such as advanced launch vehicles and launch systems are also areas where flywheels can offer significant performance advantages.

Table 1. Flywheel Technology Program Metrics

Metric	Goal
Usable System Specific Energy	Near term > 45 Whr/Kg
	Long Term >200 Whr/Kg
Cycle Life	> 75,000 cycles
Round Trip Efficiency	> 90%
System Cost Reductions	> 25%

While flywheel technology development is ongoing at NASA GRC, there is also a system prototype development project at GRC funded by NASA Headquarters, Code M for the International Space Station (ISS) called the Flywheel Energy Storage System (FESS) Project. This project is specifically developing a prototype flywheel battery for possible use as replacements for the ISS. If development is successful, a flywheel battery could fly on ISS as early as 2005.

The flywheel technology program supports research and development in three areas: flywheel systems, component technologies, and rotor safe-life technologies. In addition, this program leverages other programs such as the AFRL FACETS Program, the FESS Project, the Commercial Space Centers at Auburn and Texas A&M Universities, NASA NRA's and SBIR's, support from NASA Code Q and GRC/Army Research Lab aeronautics and internal programs. The NASA CETDP NRA contract with Lockheed Martin for the development of the COMET Flywheel System™, which is described below, is a key flywheel system integration development task.

The flywheel technology program conducts research over a broad spectrum of component technologies. Table 2 summarizes these activities, including leveraged program tasks.

Table 2. Flywheel Technology Program Research

Component	Task	Source
Magnetic Bearings	Advanced Bearing Control	GRC, Texas A&M
	Fault Tolerant Actuators and Optimized Design	Texas A&M
	Health Monitoring	Texas A&M, UT-CEM
	Passive Bearings	GRC, Foster-Miller
Power Train	Optimized Motor/Gen Control	GRC, Penn State
	Advanced Motor/Generator	GRC
	High Speed Concepts	GRC, Penn State
Composite Rotors	Rims and hubs	Auburn U., UT-CEM
	Rotor design optimization	GRC, Auburn U., UT-CEM
	Material Characterization	GRC, UT-CEM, FESI
	Rotor Fatigue Testing	GRC, CNRC
	Rotor NDE	GRC
	Standardized Certification Process	GRC, AFRL, Aerospace Corp.

A focal point for the flywheel component and system research and development is the flywheel testbed at GRC. This will allow the demonstration in a systems environment of the ability of advanced component and system technologies developed by the flywheel program to meet performance metrics and NASA mission needs. The flywheel testbed facility is operational and a flywheel module development unit is currently under test. This test program will be described below.

AFRL Program

The capstone of flywheel development at the Air Force Research Lab (AFRL) Space Vehicles Directorate will be the Flywheel Attitude Control, Energy Transmission and Storage ground demonstration on the Advanced STRuctures Experiment (ASTREX) test-bed at the AFRL facility in Albuquerque, NM (Kirtland AFB), Figure 1. This testing represents the first three degree-of-freedom spacecraft simulator demonstration of simultaneous energy storage and attitude control using flywheels.

Figure 1 ASTREX Test-Bed at AFRL

The main difference between the FACETS ground demonstration and the other flywheel-based IPACS demonstrations mentioned in this paper is that the IPACS units in this case will be gimbaled. It is well-known that gimbaled wheels, or control moment gyros (CMGs), are used for attitude control in spacecraft applications requiring large control torques. Using gimbaled wheels provides the maximum flexibility in testing since the flywheel units can be operated in either a locked-gimbal or controlled-gimbal mode to represent both reaction wheel or CMG attitude control actuation. The mounting scheme on ASTREX will also be designed to maximize flexibility in the testing of multiple wheel configurations.

The primary purpose of this effort is to verify that full exploitation of the enormous benefits of flywheels for spacecraft applications is realized in the combined functionality of energy storage and momentum management/attitude control. The FACETS ground demonstration will actually be the culmination of four related efforts:

1. A multi-phase development effort with Honeywell, Inc., Tempe, AZ, to design, develop and individually test IPACS units, Figure 2, for use in FACETS,
2. In-house basic controls research at AFRL focused on the development of practical control algorithms for performing simultaneous energy storage and attitude control with gimbaled wheels as well as control of the magnetic bearings,
3. An international cooperative effort with the Canadian Space Agency to verify that the composite rotors in the Honeywell IPACS units have sufficient cycle life for the demonstration, and
4. Facility development at AFRL, including modification of the ASTREX structure, upgrades to the ASTREX test-bed and setting up the ASTREX high-bay and control room to control and monitor the FACETS demonstration testing.

Figure 2 Honeywell IPACS Unit

The development activities at Honeywell will continue under a Phase II effort beginning in Spring, 2001, aimed at completing the full operational testing of the Phase I IPACS unit and upgrading the unit with housing and electronics suitable for testing on ASTREX. This upgrade will also insure that the Phase II IPACS units have a 'path-to-flight' so that this development will lead to flight-quality flywheel hardware for use in satellite applications.

Development Program Description

There are three development activities currently in progress at NASA and AFRL to demonstrate integrated flywheel operation in IPACS configurations. The AFRL activity is focused on demonstrating flywheels for a gimbaled CMG and energy storage type application. The NASA activities at GRC and Lockheed Martin address the lower momentum requirements more typical of NASA missions and are focused on the use of fixed flywheel configurations for

energy storage and momentum control. These development activities are described below.

AFRL FACETS Development

The ASTREX facility, located in a dedicated high-bay facility within the Space Vehicles directorate at the AFRL, Kirtland AFB, New Mexico Phillips Research Site, will be the focal point of FACETS development. ASTREX is a spacecraft structure simulator mounted on a spherical air bearing, which gives it the 3 degree-of-freedom motion desired for attitude control testing. ASTREX also has additional sensors and actuators (accelerometers, angular encoders, thrusters, etc.) that will aid in the testing and verification of the attitude control function of FACETS. This will be performed simultaneously with the energy storage and power transmission capability. The ASTREX test-bed was originally designed to investigate structural disturbances caused by attitude control actuation and explore methods for mitigating these effects. Because of this, ASTREX also provides the opportunity to investigate and control structural disturbances influenced by the simultaneous energy storage and attitude control functionality of FACETS.

The test plan for the FACETS ground demonstration is built around the primary objective of demonstrating simultaneous energy storage and attitude control capability for both gimbal-locked and gimbal-controlled modes of operation. The testing will be conducted in three main phases:

1. Integration/Calibration Testing
2. Closed-loop Testing
3. Advanced Control Testing

Integration/Calibration Testing. This phase of the testing will focus on integrating the IPACS hardware with the mechanical, electrical and control systems of ASTREX. It will also include calibration of the sensors and actuators (ASTREX and IPACS) to compensate for bias, misalignments and noise. This phase will rely heavily on the system simulation model, using it for hardware-in-loop testing of hardware in place on ASTREX. Test sequences in this phase will include open loop slewing of the ASTREX structure to verify sensor, actuator and system-level operation.

Closed-Loop Testing.. This phase represents the crux of the FACETS ground demonstration. It is in this phase that the loop will be closed on the system level simultaneous energy storage and attitude control algorithm. Test sequences will consist of attitude hold, long duration attitude hold, IPACS hardware desaturation (returning gimbal angles to 'zero' position without introducing torque disturbances), characterization of attitude stability and long term accuracy, and target tracking maneuvers. Energy storage functionality will be tested under several scenarios including average (housekeeping) satellite power levels and worst case peak loading representative of high power, short duty cycle applications.

Advanced Control Testing. The objective of this phase is to improve performance of nominal closed-loop control tested in the previous phase by taking into account lessons learned during that testing. This may involve hardware and software reconfiguration and use of more advanced control methodologies, such as incorporation of singularity avoidance. The test sequence is open-ended since potential areas of improvement will be unknown prior to the nominal closed-loop testing.

NASA Testbed Development

The flywheel testbed at NASA GRC will be used to demonstrate a single axis attitude control and energy storage system (ACESE) and will be used to test prototype electronics for the Flywheel Energy Storage System (FESS) on the International Space Station (ISS). The layout of the test cell, control room, and facilities support equipment room is shown in Figure 3. A picture of the control room is shown in Figure 4.

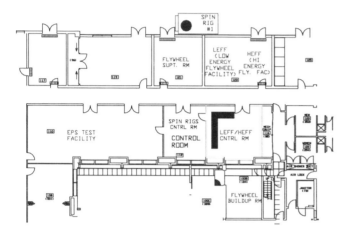

Figure 3. Flywheel Testbed Facilities – Building 333, NASA GRC

Figure 4. Flywheel Testbed Control Room

The hardware configuration for the ACESE experiment consists of two flywheel modules with parallel spin axis vertically mounted on an air bearing table. The electronics and controls to support the magnetic bearing and motor/generator systems and provided overall system control are a combination of COTS and brassboard hardware with rapid prototype software written in Simulink. A system layout is shown in Figure 5.

Each flywheel module stores 350 W·hrs and has a rotor inertia of .066 kg·m². The motor/generator can charge or discharge at 3 kW with a torque of 1.43 N·m. Detailed

specifications are shown in Table 3. The rotor is suspended with active magnetic bearings and has a rolling element touchdown bearing for off nominal conditions. The flywheel module housing provides the vacuum enclosure and mounting locations for the stator components. The two flywheel modules are shown in Figure 6.

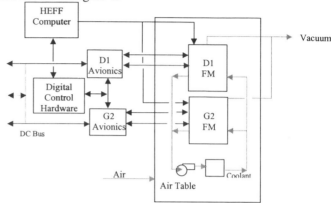
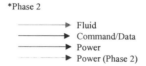

*Phase 2

————→ Fluid
————→ Command/Data
————→ Power
————→ Power (Phase 2)

Figure 5. ACESE System Schematic

Table 3. ACESE Flywheel Module Characteristics

Characteristic	Specification
Useable Energy Storage	320 W-Hrs
Charge/Dischg Power	3000 W
Operating Speed Ratio	3/1
Maximum Speed	60,000 RPM
Motor/Generator	PM Sync, 2-pole, 3-phase Y connected
DC Bus Voltage	130 V
Magnetic bearings	Homopolar PM bias, 4-pole

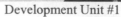

Development Unit #1   Cross-section of new GRC Design, G2

Figure 6. Development Units at GRC

The ACESE experiment has three phases:
1. Demonstrate two flywheel modules operating on a lightly constrained mounting system (air bearing table),
2. Demonstrate momentum and energy control in charge mode with the flywheels on separate power buses, and
3. Demonstrate full momentum and energy control in charge and discharge on a single power bus.

In the first phase, the operational demonstration of a magnetically suspended flywheel module functioning on an air table will be used to verify an analytical modeling effort which is exploring the effect of mount stiffness on magnetic bearing stability. In the second phase of the ACESE experiment the combined charge rate of and net torque of the two flywheel modules will be controlled. This demonstration will be limited to charge mode only, since the flywheels are on separate power buses. Step response, overshoot and regulation band will be compared to analytical models. In the final phase, both flywheel modules will be run on one power bus. Charge mode will operate in the same manner as in the second phase. In discharge the controller will regulate the bus voltage and the net torque of the two-wheel system.

After completing the ACESE work, the same test hardware will support the FESS program. Brassboard and prototype avionics for the space station flywheel will be tested and debugged prior to the delivery of the FESS flywheel modules. FESS testing will focus on efficiency measurement and control verification. The ACESE flywheel modules store 1/10 of the energy of the FESS modules. Since the normal orbit cycle can not be demonstrated with the smaller wheels, two extremes will be tested. First, the system will be run at normal charge and discharge power levels, completing an orbit cycle in nine minutes. Second, the system will be run the full ninety minutes at 1/10 power.

The synergy between the ACESE and FESS efforts have allowed NASA GRC to focus resources and provided value to each program at reduced cost.

Lockheed Martin COMET Flywheel System™ Development

Lockheed Martin in Newtown, Pennsylvania is under contract with NASA Glenn Research Center to design, build, and test a demonstration of an innovative IPACS called the COMET Flywheel System. "COMET" is an acronym for *"coordinated momentum and energy transfer"*. Figure 7. shows the COMET Flywheel System integrated with the satellite power and attitude control subsystems. Typically a flywheel based IPACS uses either gimballed flywheels, or fixed-axis flywheels arrange in counter rotating pairs. In the case of counter rotating pairs, a minimum of six flywheels are required to achieve full attitude torque authority. The COMET Flywheel System uses as few as four fixed-axis flywheels arranged in a pyramid.

At the heart of system is the COMET Flywheel Logic™ used to coordinate the momentum and energy transfer from the flywheels. Each flywheel is used as a torque actuation device. The sum of torques times respective wheel speeds results in

power transfer to/from the power bus (energy transfer). The vector sum of the torques results in a net torque on the spacecraft body (momentum transfer). The individual torques are coordinated so that the momentum and energy transfer meet the needs of the attitude control and electrical power subsystems.

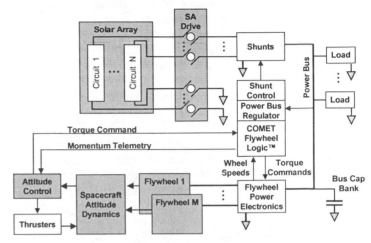

Figure 7. Integrated COMET Flywheel System™

The basic approach (Ref. 1) is to apply a torque allocation matrix to a vector composed of a power command and three-axis torque command to generate individual torque commands to the flywheels. The torque allocation matrix is derived from knowledge of the flywheel alignment and wheel speeds. The work of Ref. 1 has been augmented to include null space control to equalized the wheel speeds to the extent possible in the case that five or more wheels are used to achieve the energy and momentum requirements. Also, and more importantly, a fast wheel speed loop (Ref. 2) is incorporated to reject torque disturbances resulting from load power fluctuation coupling through mismatches between the individual flywheel power electronics.

The objective of contract is to demonstrate feasibility of combining momentum and energy storage/transfer capabilities using the COMET Flywheel System. Three flywheels will be mounted on a force table with their spin axis in the plane of the table arranged at 120 degree intervals. Force transducers will measure the net reaction torque in two axes. Load power and attitude torque profiles will be applied and the power bus voltage and net reaction torque will be observed to validate the system operation. An important part of the program is the generation of a flywheel subsystem specification and a scalability report that shows how the designs used in the demonstration must be modified for a flight program.

The contract began in April, 2001 and will end in March, 2004. The first year is dedicated to trade studies and preliminary design. The second year will be dedicated to detailed design and build-up of the first of the three flywheels. During the third year the other two flywheels will be built and tested. Finally the flywheels will be integrated into the demonstration system and tested.

SUMMARY

The National Aeronautics and Space Administration and the Air Force Research Laboratory are cooperating under a space act agreement to sponsor the research and development of aerospace flywheel technologies to address mutual future mission needs. For some missions, flywheel technology offers significantly enhanced capability or is an enabling technology. Generally these missions are for energy storage and/or integrated power and attitude control systems (IPACS) for mid-to-large satellites in low earth orbit. These missions require significant energy storage as well as a CMG or reaction wheel function for attitude control.

Three major development and test activities that will begin to address the system integration questions are in progress within the NASA and AFRL flywheel programs. Development tests at Honeywell and the AFRL ASTREX Facility will address flywheel system development for CMG/energy storage type IPACS applications. Development activities at NASA GRC and Lockheed Martin will address flywheel system development for fixed flywheel IPACS applications. These development test programs, being conducted as integrated flywheel systems, will significantly advance the state of the art in flywheel applications for IPACS missions.

REFERENCES

1. Tsiotras, P., Shen, H., and Hall, C., "Satellite attitude Control and Power Tracking with Energy/Momentum Wheels", Journal of Guidance, Control, and Dynamics, Vol. 24, No. 1, January-February 2001, pp. 23-34.

2. Goodzeit, N. E., Paluszek, M. A., and Cohen, W. J., "Attitude Control System wiht Reaction Wheel Friction Compensation", U.S. Patent No. 5,201,833, Apr. 13, 1993.

Proceedings of IECEC'01
36th Intersociety Energy Conversion Engineering Conference
July 29-August 2, 2001, Savannah, Georgia

IECEC2001-AT-84

NOVEL ACTUATOR FOR MAGNETIC SUSPENSIONS OF FLYWHEEL BATTERIES

Andrew Kenny
Texas A&M University

Andrew Provenza
NASA Glenn Research Center

Uhn Joo Na
Texas A&M University

Alan Palazzolo
Texas A&M University

Raymond Beach
NASA Glenn Research Center

Albert Kascak
U.S. Army at NASA Glenn

Gerald Montague
U.S. Army at NASA Glenn

Yeonkyu Kim
Texas A&M University

Shuliang Lei
Texas A&M University

ABSTRACT

Magnetically suspended high speed, energy storage flywheels are presently being developed for terrestrial, satellite, and International Space Station applications. A key component in the magnetic suspension is the actuator that applies control forces to the spinning rotor in response to electric currents. The actuator is typically selected using three criteria: efficiency, size (weight), and fault tolerance.

This paper examines the relationship between a bearing's geometric dimensions and its performance. Specifically it examines a magnetic bearing biased with permanent magnet poles. In this paper a method is presented for choosing the dimensions of the rotor, stator, and coils. The method generates families of curves relating bias flux density to the maximum bearing force, power loss in each coil, coil inductance, peak current, number of coils, permanent magnet length, and the bearing outer radius. With the curves the bearing with the highest force and lowest power loss, that still has reasonable inductance, and size can be found.

INTRODUCTION

Since the beginning of the last decade magnetic bearings biased with permanent magnets have seen fruitful research, such as that by Allaire, Maslen, et. al. Lea, Hsiao, Fan, and Ko derived detailed equations for sizing the components to achieve certain a force, stiffness. Fukata, Yutani, and Kouya advanced this work. Flowers, Overstreet, and Szase improved the flux ring to reduce the leakage around the permanent magnet.

As covered by Nataraj and Calvert in detail, electrically biased hetereopolar bearings consume biasing power in the coils due to the wire resistance. Not only does this power loss reduce the efficiency of the bearing, but the heat produced is a major consideration. This has spurred the development of energy efficient heteropolar bearings by Meeker and Maslen and permanent magnet biased homopolar bearings such as that by Grbesa. But homopolar bearings require two planes of laminate stacks. Permanent magnet biased homopolar bearings are long and heavy. Furthermore, the axial flow of bias flux in a high speed rotor is potentially a significant source of parasitic drag. This paper examines a permanent magnet biased bearing with the simplicity of a single laminate stack. It has been shown by Kenny and Palazzolo to have an excellent force capacity to power loss ratio. It may be the best choice for space based flywheel energy storage depending on the very high speed power losses (Kasarda, Allaire, et. al.).

NOMENCLATURE

A_a area of active pole gap

A_{bi} area of between coil slot and outer radias

A_c	area of coil cross section
A_m	area of magnet
A_p	area of active pole cross section
B_{ba}	bias flux density in active gap
B_{bm}	bias flux density in bias gap
i_c	peak control current
l_{ga}	length of active gap
l_{gm}	length of bias gap
l_m	length of magnet
N	number of turns in coil
r_i	radius of coil slot
r_o	outer radius of bearing
r_r	outer radius of rotor
R_{acbi}	reluctance for control flux in stator circumference
R_{acr}	reluctance for control flux in rotor circumference
R_{ba}	reluctance of active pole
R_{bcr}	reluctance for bias flux in rotor circumference
R_{ga}	reluctance of active gap
R_{gm}	reluctance of bias gap
R_m	reluctance of magnet
R_{pa}	reluctance of total control flux path
R_{pb}	reluctance of total bias flux path
t	thickness of laminate stack
Θ_a	angle of active pole
Θ_m	angle of bias pole
Θ_s	angle of separation between pole edges

GEOMETRY AND PERFORMANCE EQUATIONS

Performance curves are calculated from equations that will be presented. Initial parameters for the calculations include material properties and certain dimensions. The rotor outer radius is set by the centrifugal stress at the maximum operating speed. The active pole gap length is set by the quality of the catcher bearing tolerance. For these calculations a few dimensions are set arbitrarily. These are the angle between the edges of the active and bias poles, and the gap length between the rotor and the bias poles. The initial parameters and are shown in Table 1.

Table 1. Known Parameters

Variable	Numerical Value	Definition
r_r	22.12 mm (.871 in)	Outer diameter of rotor laminate stack
l_{ga}	.495 mm (.0195 in)	Length of active pole air gap
l_{gm}	.99 mm (.039 in)	Length of bias pole air gap
Θ_s	2 degrees	Angle separating pole tip edges
d_w	.909 mm (.0358 in)	Diameter of 20 gage triple insulated wire
ρ_w	.0332 ohm/mm (.010228 ohm/ft)	Resistance of 20 gage wire
B_m	1.2 T	Residual flux density of magnet
μ_m	1.05	Relative permeability of magnet
μ_r	2000	Relative permeability of the laminate stack
f	.97	Stacking factor of laminate

Figure 1 shows the bias pole bearing. In addition to the parameters in Table 1, two more dimensions need to be specified to calculate the rest of the bearing dimensions. One of these must be angular, and one must be radial. For these calculations the angular dimension chosen is the width of the active pole, and the radial dimension is the magnet thickness.

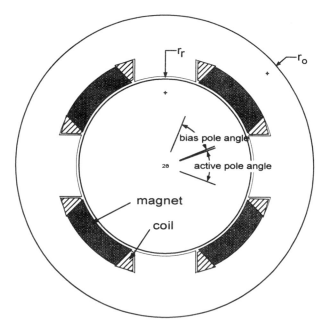

Figure 1. Bias pole bearing dimensions.

The bias pole angle was calculated from the active pole angle using Eq. (1).

$$\Theta_m = 90 \cdot \deg - \Theta_a - \Theta_s \qquad (1)$$

The stack thickness was also calculated using the active pole angle. Square active poles have the largest ratio of pole cross section area to pole perimeter. They give the best ratio of force to coil power loss. The laminate stack thickness was calculated to make the active poles square as given by Eq. (2).

$$t = \Theta_a r_r \tag{2}$$

The flux path cross sections could be calculated as shown by Eq. (3) to Eq. (6). The active gap area is the whole area of the rotor covered by the active pole, where as the active pole area is just the pole cross section. The return path area, A_{bi}, is only seventy-five percent of the active gap area because the pole carries the bias flux and control flux, but the back iron carries only half the bias flux to each of the bias poles on either side of the active pole.

$$A_a = \Theta_a R_r t \tag{3}$$

$$A_p = 2R_r \sin\left(\frac{\Theta_a}{2}\right) t \tag{4}$$

$$A_m = \Theta_m R_r t \tag{5}$$

$$A_{bi} = .75 A_a \tag{6}$$

The radial dimensions were calculated after specifying the magnet length. Equation (7) gave the radius of the coil slots and Eq. (8) gave the outer radius of the bearing. This completed the calculations of the bearing dimensions.

$$r_i = r_r + l_{gm} + l_m \tag{7}$$

$$r_o = r_i + \frac{A_{bi}}{t} \tag{8}$$

After the bearing dimensions were known, the performance parameters could be calculated. Equation (9) to Eq. (16) gives the reluctance of each path in the bearing.

$$R_m = \frac{l_m}{\mu_o A_m} \tag{9}$$

$$R_{ga} = \frac{l_{ga}}{\mu_o A_m} \tag{10}$$

$$R_{gm} = \frac{l_{gm}}{\mu_o A_m} \tag{11}$$

$$R_{bcr} = \frac{(.5\Theta_a + .5\Theta_m + \Theta_s)\dfrac{r_r}{\mu_r}}{\mu_o A_{bi}} \tag{12}$$

$$R_{bcbi} = \frac{.5\Theta_a + .5\Theta_m + \Theta_s}{\mu_o A_{bi}} \frac{r_o + r_i}{2\mu_r} \tag{13}$$

$$R_{ba} = \frac{l_m}{\mu_o \mu_r A_a} \tag{14}$$

$$R_{acr} = \frac{\pi \dfrac{r_r}{\mu_r}}{\mu_o A_{bi}} \tag{15}$$

$$R_{acbi} = \pi \frac{(r_o + r_i)}{2\mu_r \mu_o A_{bi}} \tag{16}$$

The reluctance of the bias flux path is then given by Eq. (17), and the reluctance of the control flux path is given by Eq. (18).

$$R_{pb} = R_{ba} + R_{bcbi} + R_{bcr} + R_{gm} + R_{ga} + R_m \tag{17}$$

$$R_{pa} = R_{acr} + R_{acbi} + 2R_{ba} + 2R_{ga} \tag{18}$$

The bias flux density through the bias pole gap can be calculated using Eq. (19). Here β_m is magnet leakage estimation coefficient that was estimated to be 0.7 by finite element analysis. The bias flux density in the active pole gap depends on the ratio of the active pole angle and the bias pole angle because of flux conservation, as in Eq. (20).

$$B_{bm} = \frac{\dfrac{1}{\mu_o} B_m l_m \beta_m}{R_{pb} A_m} \tag{19}$$

$$B_{ba} = B_{bm} \frac{\Theta_m}{\Theta_a} \tag{20}$$

As Eq. (21) shows, the maximum control flux density through the active pole was set at ninety percent of the bias flux density. This prevented the control flux from ever reversing the bias flux.

$$B_{ca} = .9 B_{ba} \tag{21}$$

The number of ampere-turns needed by each coil can be calculated from the peak control flux density, the active gap area, and the active flux path reluctance as given by Eq. (22).

113

$$Ni_c = \frac{1}{2}\left(B_{ca}A_a R_{pa}\right) \tag{22}$$

To calculate the number of coil turns and peak current, the area available for the coil must be calculated. To reduce the area that the coils take away from the poles, during assembly the coils can first be put over the active poles, and then the magnets can put in place. The only area that needs to be dedicated to the coil is the that required for the .76 mm thick (.030 in) bobbin and the coil itself. The coil packing efficiency used to calculate how many turns would fit in this area, ζ, is $\pi/4$. This is the ratio of the area of a circle to that of a square. This assumes square wire could be packed without any wasted space. The maximum number of coil turns that will fit in the area between the active and the bias poles is calculated from Eq. (23) to Eq. (25). The minimum peak control current required to achieve the previously calculated peak control flux density is calculated from Eq. (26).

$$A_c = \frac{22.5}{180}\left[\pi l_m^2\right] + \Theta_s r_r l_m \tag{23}$$

$$A_{bob} = .76 \cdot mm \cdot l_m \tag{24}$$

$$N = \frac{A_c - A_{bob}}{A_w}\zeta \tag{25}$$

$$i_c = \frac{Ni_c}{N} \tag{26}$$

The length of wire in the coil can be calculated as shown in Eq. (27). The coil resistance can then be calculated, as in Eq. (28), and it follows that the resistance power loss due to the harmonic control current in each coil is given by Eq. (29).

$$l_w = N \cdot \left[\left(\Theta_a + \Theta_s\right) \cdot r_i + t\right] \cdot 2 \tag{27}$$

$$res_c = l_w \cdot \rho_w \tag{28}$$

$$P_c = .5i_c^2 res_c \tag{29}$$

The coil inductance is the ratio of the control flux through the coil to the control current that produces it as shown in Eq. (30).

$$L_c = \frac{NB_{ca}A_a}{i_c} \tag{30}$$

Finally, and importantly, the peak force the bearing can produce in the horizontal and vertical direction is given by Eq. (31). The value of the gap flux fringe factor, α, is estimated at 0.8.

$$F = \frac{2B_{ba}B_{ca}A_a\alpha^2}{\mu_o} \tag{31}$$

GEOMETRY AND PERFORMANCE CURVES

Plots of calculations using all these equations give the performance curves shown in Figures 2 to 8. The bias flux density through the active pole is shown on the abscissa of each plot. The reason for this is so that the curves show the bearing performance that can be achieved with available alloys. The saturation flux density of the laminate stack is between 1.5 T and 2.2 T depending on the alloy chosen. The peak control flux is ninety percent of the bias flux, so flux densities in the active pole should be between .79 T and 1.1 T. Each plot shows several curves. The individual curves are for different active pole angles.

The first plot, Figure 2, shows the maximum force of attraction the bearing can exert on the rotor. The second plot shows the power consumed in the coils. The third plot shows the coil inductance. These three plots are the main set of performance curves. Generally higher force capability and lower power consumed are better. However these objectives are limited by the coil inductance since a high coil inductance may lower the frequency response bandwidth of the bearing-power amplifier system below the controllable limit.

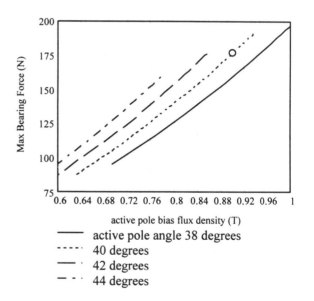

active pole bias flux density (T)

—— active pole angle 38 degrees
----- 40 degrees
—·— 42 degrees
— ·· — 44 degrees

Figure 2. Bearing force versus active gap bias flux density.

114

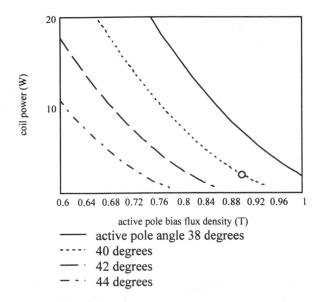

Figure 3. Resistance power loss versus active gap bias flux density.

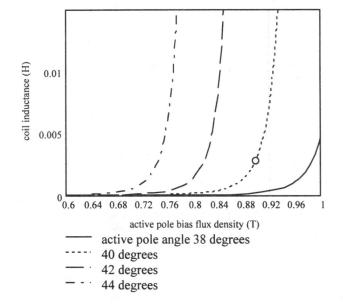

Figure 4. Coil inductance versus active gap bias flux density.

It seems unusual that the force capability in Figure 2 would go up when power consumed in the coil shown in Figure 3 goes down. The reason for this is way the coil is packed into the available space in the slot. As the magnet length increases the bias flux density increases which also increases the force capacity. Increasing the magnet length also increases the length of the bias pole and the length of the slot between the poles.

More coil turns can fit in the longer slot which reduces the coil current and the power consumed by the coil.

The other plots in Figure 5 through Figure 8 show how the physical parameters of the bearing are affected by one's choice of the bearing performance parameters. These physical parameters include the number of turns and the peak current in the coil, the magnet length, and the outer radius of the stator laminate stack. The outer radius of the stator laminate stack is also the outer radius of the bearing.

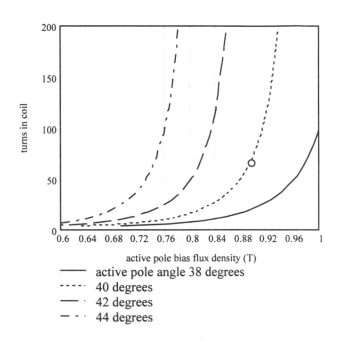

Figure 5. Turns in coil versus bias flux density.

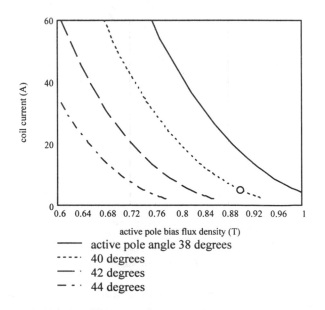

Figure 6. Control current versus bias flux density.

115

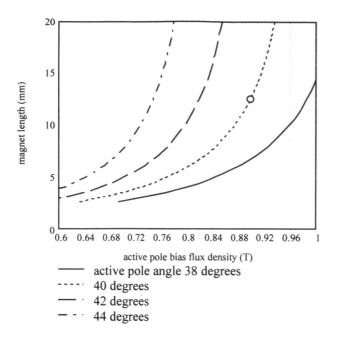

Figure 7. Magnet length versus active gap bias flux density.

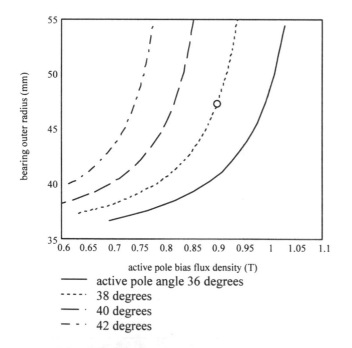

Figure 8. Outer radius of bearing versus active gap bias flux density.

From all these curves a compromise of force capacity, power loss, and inductance was chosen. The choice is indicated on all the figures by a circle on the curve corresponding to an active pole angle of forty degrees and a bias flux density of 0.9 T. The numerical values of the circles are shown in Table 2. An important starting value for the calculations used to generate these curves was the outer radius of the rotor. In this case it was 22.12 mm (.871 in). At 60000 rpm, a rotor of this radius has a maximum tangential stress of 137 MPa (20 ksi). The yield stress of alloy Hyperco50hs is 413 Mpa (60 ksi) in it lowest core loss heat treatment. Table 2 then shows the parameters of a bearing that would run with a factor of three safety at 60000 rpm. The steady state running power loss should be considerably less than the peak power loss shown in the Table. The peak power is only required for start up levitation and to handle sudden bumps to the system.

Table 2. Parameters Chosen from the Performance Curves

Variable	Numerical Value	Definition
F	177 N (39.9 Lb)	Maximum force capacity
P_c	2.3 W	Resistance power per coil
L_c	3 mH	Inductance per coil
N	77 turns	Number of turns per coil
I	4.9 A	Peak coil current
ρ_w	12.5 mm (.5 in)	Magnet length
r_o	47.2 mm (1.86 in)	Bearing outer radius

SUMMARY

Equations were developed for generating a set of performance curves for a bias pole bearing. The starting values needed for the equations were the rotor outer radius, the gap length, and the material properties. The curves were generated by varying the active pole angle and the magnet length. Using the curves a compromise of bearing force capacity, coil power loss, and coil inductance was chosen. This choice caused the bearing and coil dimensions to be specified.

Future research could address some of the starting values needed for the equations, especially the best ratio of active pole to bias pole gap length. Calculations and plots comparing the force to power loss ratios for this type of bearing to other types would also be informative

ACKNOWLEDGMENTS
The authors acknowledge the support for this work through NASA Glenn (Grant NRA 99-GRC-2) and from Tom Calvert, Lyn Peterson, and Glenn Bell of the Navy NSWC (Grant ONR BAA 97-030).

REFERENCES

Fan, Y., Lee, A., Hsiao, F., "Design of a Permanent/Electromagnetic Magnetic Bearing-controlled Rotor System", J. Franklin Inst. Vol. 334B. No. 3. Pp 337-356, 1997.

Fukata, S., Yutani, K., and Kouya, Y., "Characteristics of Magnetic Bearings Biased with Permanent Magnets in the Stator," JSME International Journal, Series C, Vol. 41, No. 2., pp 206-213, 1998.

Grbesa, B., "Low Loss and Low Cost Active Radial Homopolar Magnetic Bearing," *Proceedings of the Sixth International Symposium on Magnetic Bearings*, Cambridge, MA, pp. 286-295, 1998.

Kasarda, M.E.F.; Allaire, P.E.; Norris, P.M.; Mastrangelo, C.; Maslen, E.H., "Experimentally Determined Rotor Power Losses in Homopolar and Heteropolar Magnetic Bearings," *Proceedings of the 1998 International Gas Turbine & Aeroengine Congress & Exhibition,* Stockholm, Paper : 98-GT-317, 1998.

Kenny, A. and Palazzolo, A.B., "Single Plane, Radial Magnetic Bearings Biased with Poles Containing Permanent Magnets," submitted to ASME Journal of Mechanical Design, 2001.

Lee, A., Hsaio, F., Ko, D., "Analysis and Testing of Magnetic Bearing with Permanent Magnets for Bias," JSME International Journal, Series C, Vol. 37, No. 4, pp. 774-782, 1994.

Meeker, D.C. and Maslen,E.H., "Power Optimal Solution of the Inverse Problem for Heteropolar Magnetic Bearings," Proceedings of the Fifth International Symposium on Magnetic Bearings, Kanazawa, Japan, pp. 283-288, 1996.

Nataraj, C. and Calvert, T.E., "Optimal Design of Radial Magnetic Bearings," *Proceedings of the Sixth International Symposium on Magnetic Bearings*, pp. 296-305, Cambridge, MA, August 1998.

Overstreet, R.W., Flowers, G.T., Szasz, G., "Design and Testing of a Permanent Magnet Biased Magnetic Bearing," Symposium on Vibration of Rotating Structures and Systems for the 1999 ASME Design Technical Conferences (17th Biennial Conference on Mechanical Vibration and Noise), Las Vegas, 1999.

Sortore, C.K., Allaire, P.E., Maslen, E.H., Humphris, R.R., Studer, P.A., "Permanent Magnetic Biased Bearings – Design, Construction, and Testing," *Proceedings of the Second International Symposium on Magnetic Bearings*, pp. 175-182, Tokyo, 1990.

Proceedings of IECEC'01:
36th Intersociety Energy Conversion Engineering Conference
July 29-Augist 2, 2001, Savannah, Georgia

2001-AT-83

REVIEW OF OES FLYWHEEL ENERGY STORAGE DEVELOPMENT FOR SPACE POWER

Dr. Dwight W. Swett
Optimal Energy Systems, Inc.
2720 Monterey Street, Suite 401
Torrance, CA 90503

John G. Blanche IV
Optimal Energy Systems, Inc.
2720 Monterey Street, Suite 401
Torrance, CA 90503

ABSTRACT

Researchers have long sought to develop a high-speed flywheel that is capable of storing energy on a sufficient scale to advantageously replace existing electrochemical battery technologies on space systems by offering reduced weights and extended operating life. Previously, no such breakthrough has existed, until now. Optimal Energy Systems, Inc. has developed and tested a flywheel invention that will make flywheel energy storage a reality for the high performance space satellite energy storage community. This paper summarizes some of the recent flywheel energy storage and high-speed electrical machine development activities at Optimal Energy Systems.

The Optimal Flywheel

Optimal Energy Systems has developed a very high specific energy flywheel that is lighter weight and smaller, for a given energy, than any other flywheel ever developed. The flywheel has demonstrated very low stresses at extremely high speeds and can internally "growth match" itself when spinning. The design uses a filament wound graphite composite rim which is attached to a shaft via the patented Optimal Flywheel Hub. The Hub is a self-expanding design consisting of a hollow steel "double cone" structure combined with a filament wound composite flange (see Fig 1). The hub is designed to expand with speed and match the growth rate of the rim throughout the speed range. Using finite element modeling, the hub design can be changed to adjust the growth rate to match any rim design. Figure 2 shows the radial stress distribution from the finite element analysis of one flywheel design simulating a rotational speed of 62,000 rpm. This model only shows the composite flange and rim. The radial stress distribution can be seen to be zero at the flange/rim interface indicating growth matching has occurred. Since the stress in the flywheel is proportional to the square of the speed, the amplitude of the stresses changes with

speed but the distribution does not. Therefore, the flange/rim interface remains as a zero radial stress point regardless of the speed of the flywheel.

Figure 1: OES Patented Growth Matching Hub

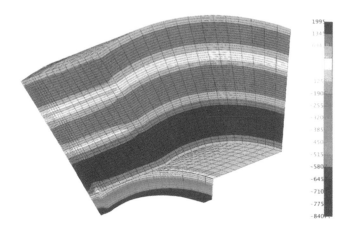

Figure 2: Radial Stress Distribution in OES Phase IID Flywheel Rim and Flange at 62,000 RPM (steel hub not shown)

High speed testing of multiple prototypes has shown excellent correlation with the analytical predictions. The most recent spin test involved testing a 12.25 inch diameter, 24 lb rotor to 60,000 rpm safely thus demonstrating energy storage of 891 W-hr total energy or 82 W-hr/kg specific energy at the flywheel rotor (rim and hub) level (see Fig. 3).

Figure 3: OES Phase II D Flywheel Rotor after 60,000 rpm Spin Test (12.25 inch Dia., 24 lbs)

The Swett Motor/Generator

The second key technology component of a flywheel energy storage system is the motor/generator. Optimal Energy Systems has developed an ultra-high-speed electrical machine that current test data indicates has drastically outdistanced all electromechanical power conversion devices, specifically in the areas of power density, efficiency and weight. The Swett motor/generator is a three phase, 6 pole, axial gap machine,

meaning that the magnets are polarized parallel to the axis of the rotor shaft (see Fig 4). The rotor consists of a titanium rotor constraining 18 rare earth magnets that are arrayed in groups of 3 to make a 6 pole machine. The rotor is attached to the shaft utilizing a modified version of the OES growth-matching hub.

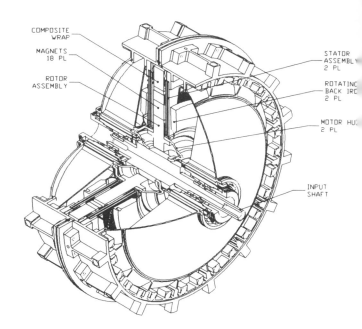

Figure 4: OES Power Conversion Module

For the 50 horsepower prototype machine, a composite wrap is bonded to the outside diameter of the rotor for added strength. The windings are embedded in two stator assemblies, one on each side of the rotor, that are attached to the motor housing. In order to minimize eddy current losses, the stator structure is fabricated from a material which is non magnetic, non electrically conductive with a high thermal conductivity. The magnetic flux path is closed by a backiron structure that rotates with the rotor. The breakthrough technology in the Swett motor/generator is the method by which the magnets are captured in the rotor (see fig 5). The rare earth material which the magnets are made from is very strong in compression and very weak in tension. The patent pending technology in the Swett motor/generator involves the captivation method for the magnets in the titanium rotor. As the rotor spins to high speeds, the magnets are constrained in a proprietary method such that they experience primarily compressive stresses with very minimal tension in the magnets.

Figures 6 & 7 shows stress distribution from the finite element model of the rotor and magnet simulating a rotational speed of 35000 rpm. The plot of the stress in the rotor shows a maximum of 61 ksi in the titanium which has an ultimate strength of 160 ksi. The plot of the magnet stress shows the result in terms of the Hoffman criterion which looks at the total stress condition and accounts for the significant difference between the tensile strength and the compressive strength of the material. Theoretically failure occurs when the Hoffman criterion reaches 1.0, anything less has margin. This plot shows an interactive failure criterion value with a peak of 0.47 at the intermediate 35 krpm case.

Figure 5: Swett Motor/Generator prototype Rotor with backiron

This motor/generator technology has been demonstrated through several prototypes. The rotor has been tested in a spin tank where it was taken to 40,000 rpm with no damage. In addition, OES has built several complete motor/generators and, using an OES designed motor controller, demonstrated controlled speeds of as high as 43,000 rpm with no damage (1.27 Mach magnet array velocity). Figure 8 is a photo of this prototype motor/generator.

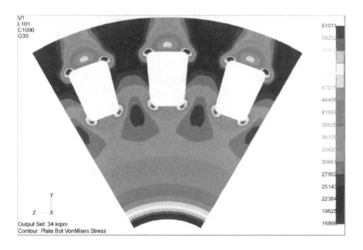

Figure 6: FEM Stress Analysis Results, Titanium Rotor

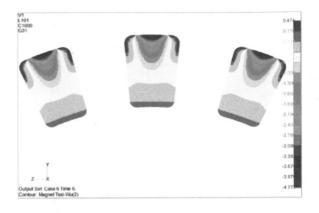

Figure 7: FEM Interactive Stress Analysis Results, Permanent Magnets

The Swett motor/generator can be compared to other electrical machines by comparing the parameter $K_{inductive}$ which is derived using the following machine characteristic:

$$K_{inductive} = N_p E^2 / (\omega L) \tag{1}$$

where N_p is the number of winding phases in the electrical machine, E is the back EMF phase voltage of the electrical machine, ω is the electrical frequency of the voltage, and L is the phase inductance of the machine windings. This parameter is an indicator of the maximum theoretical power available from a given machine. In addition, the net power available, after accounting for resistive losses in the windings, can be compared using the parameter $K'_{inductive}$ which is found from:

$$K'_{inductive} = [N_p E^2/(\omega L)] [1 - R/(\omega L)] \tag{2}$$

Where R is the phase resistance of the windings. In both cases, the higher the $K_{inductive}$ parameter the more peak power available from a given machine. Table 1 shows a comparison of several commercially available electrical machines with three OES machines. From this table it is obvious that the OES electrical machine is a breakthrough in terms of power available from an inductive machine.

121

Figure 8: Prototype Swett Motor/Generator
(50 horsepower machine)

	Phases	EMF (rms)	Freq	Phase L	Phase R	K'$_{inductive}$
No. 1	3	70 V	2000 Hz	78 µH	0.075 Ω	13,849
No. 2	2	22 V	400 Hz	60 µH	0.098 Ω	2,247
No. 3	3	40 V	1000 Hz	27 µH	0.044 Ω	20,956
No. 4 [4]	3	360 V	1333 Hz	850 µH	0.038 Ω	54,309
Optimal FESM [1]	3	130 V	2000 Hz	70 µH	0.100 Ω	51,085
Optimal PCM [2]	3	165 V	1375 Hz	8 µH	0.018 Ω	873,962
Optimal FPoM [3]	3	300 V	693 Hz	21 µH	0.008 Ω	2,704,934

Table 1: Inductive Electrical Machine Comparison

Note: (1) Optimal Flywheel Energy Storage System (satellite application)
      (2) Optimal Power Conversion Module (aviation and terrestrial applications)
      (3) Optimal Flywheel Power Module (pulse & short duration power cycling)
      (4) PNGV Concept Design

Flywheel Energy Storage Module (FESM) Design:

With the two key technologies that have been developed and demonstrated, Optimal Energy Systems has combined them into a high specific energy Flywheel Energy Storage Module (FESM). The OES FESM, shown in Figure 9, was sized for a potential higher-power commercial LEO application (6 kW satellite). This module weighs 125 lbs and will store a total of 2320 W-hr of energy at 41,000 rpm. The operating speed range is 15,000 to 41,000 rpm with a steady-state discharge of 3 kW for 37 minutes. The total roundtrip efficiency at the module level (not counting losses in the electronics) is projected to be 94% based on testing that has been performed on the Swett motor/generator prototypes.

The module consists of a flywheel rim weighing 65 lbs which is attached to the shaft using the OES hub, weighing 9.2 lbs. The flywheel rotor is on a common shaft with the motor/generator, which is a subscale version (approximately 0.75X) of the motor/generator prototype discussed in the previous section. The rotating mass is supported on three homopolar magnetic bearings, two radial and one axial. The magnetic bearing control loop is closed using optical probes looking at shaft targets to determine it's position in the gap. The overall size of the module is 19 inches in diameter by 15.5 inches length.

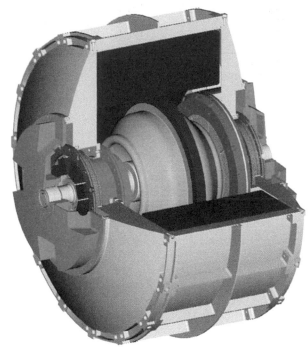

Figure 9: OES Flywheel Energy Storage Module (FESM)

Energy and Power:

Figure 10 is a plot of a typical charge/discharge power cycle for a LEO orbit showing a constant 3 kW of discharge for 37 minutes followed by 58 minutes of charge.

Since the back EMF in the generator is proportional to speed, as the FESM loses speed, the back emf decreases requiring more current to supply the same power. The losses in the FESM will include I^2R losses in the windings of the motor/generator as well as the losses in the magnetic bearing windings and rotors. However, the losses in the electronics are not included, such as losses in the motor drive electronics, and power conditioning electronics. Figure 11 shows those Module losses as a function of speed for the same cycle.

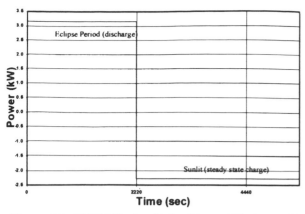

Figure 10: FESM Design-Point Charge/Discharge Cycle

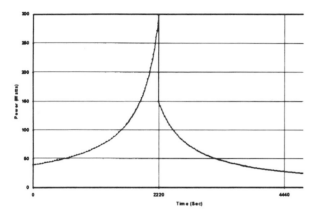

Figure 11: Power Losses in the FESM, Design-Point Charge/Discharge Cycle

Integrating the results yields the results shown in Table 2. At 41,000 rpm the total amount of energy stored in the FESM is 2320 W-hrs. During discharge, the FESM supplies 1943 W-Hrs to the Load. During the charge portion of the cycle, the FESM draws 2054 W-Hrs from the bus. Since it requires 2054 W-hr to get 1943 W-hr, the round trip efficiency is 95%.

E_{total} =	2320 W-hr
$E_{useable}$ =	1943 W-hr (3.15 kW)
$E_{recharge}$ =	2054 W-hr
Efficiency = (roundtrip)	94.55%

Table 2: Round Trip Efficiency (22 W axial magnetic bearing, 15 W each radial magnetic bearing)

Hardware Status:

All of the components for the FESM prototype have been fabricated (see Fig 12). Assembly and Integration has been started. Integration with magnetic bearings will begin in July.

Figure 12: Status of Optimal Energy Systems Flywheel Energy Storage Module Prototype hardware.

Summary

Optimal Energy Systems has developed and demonstrated breakthrough technology in the area of very high specific energy flywheel rotor designs as well as ultra-high-speed power conversion technology. These technologies will be combined in the Flywheel Energy Storage Module that has applications in the space satellite energy storage industry. OES has developed these technologies utilizing state of the art structural and thermodynamics analysis tools and CAD solid modeling tools as well as analytical electromagnetic modeling tools. Prototype hardware has been built and tested with excellent correlation to the analytical models. OES now has the analytical tools and hardware background to scale these designs for a wide range of applications throughout the aerospace energy, commercial and government high-speed power conversion, and commercial power quality management industries.

ACKNOWLEDGMENTS

We would like to acknowledge the technical support of the NASA Glenn Research Center for the use of equipment and technical consultation.

Proceedings of IECEC'01
36th Intersociety Energy Conversion Engineering Conference
July 29–August 2, 2001, Savannah, Georgia

IECEC2001-AT-89

A PASSIVE MAGNETIC BEARING FLYWHEEL

Mark Siebert
Ohio Aerospace Institute

Ben Ebihara
Ohio Aerospace Institute

Ralph Jansen
Ohio Aerospace Institute

Robert L. Fusaro
NASA Glenn Research Center

Albert Kascak
U.S. Army at NASA Glenn

Andrew Kenny
Texas A&M University

ABSTRACT

A 100 percent passive magnetic bearing flywheel rig was designed, constructed and tested without using any active control components. The suspension of the rotor was provided by two sets of radial permanent magnetic bearings operating in the repulsive mode. The axial support was provided by jewel bearings on both ends of the rotor. The rig was successfully operated to speeds of 5500 rpm, which was above the first critical speed of 3336 rpm. It was not continued beyond this point because of the excessive noise generated by the air impeller and because of inadequate containment. Radial and axial stiffnesses of the permanent magnetic bearings were experimentally measured and then compared to finite element results. The natural damping of the rotor was measured and a damping coefficient was calculated.

INTRODUCTION

Actively controlled magnetic bearings are used by industry for rotor levitation. Active magnetic bearings have the disadvantage of requiring complicated control hardware, such as digital signal processors, amplifiers, digital-to-analog converters, analog-to-digital converters, and software. Passive magnetic bearings do not require this hardware; thereby they have the potential to increase system efficiency and reliability. Another advantage is that larger rotor-stator gaps may be possible since the current for active magnetic bearings is proportional to the square of the gap.

The disadvantages of passive magnetic bearing are that they typically have lower stiffness and lower damping than similar size active magnetic bearings. The stiffness and damping are not adjustable "on-the-fly", therefore active vibration control is not used with passive magnetic bearings.

Passive magnetic bearing systems must have sufficient stiffness and damping built in to allow them to perform over their entire operating range. There have been previous studies on passive bearing systems. Passive magnetic levitation of at least one DOF has been achieved using (1) permanent magnets (Fremerey, Ohji et al., and Jansen and DiRusso); (2) passive electrodynamic effects (Ting and Tichy, Post et al., and Post and Ryutov); (3) superconductors (Hull et al., Mulcahy et al., and Hull and Turner); and, (4) diamagnetic materials (Geim et al.). Flywheels have been built and tested using several of these types of passive magnetic levitation systems. They seem well suited since high positional accuracy of the rotor and high stiffness values are not typically required in flywheels.

A few studies have also been published on the stiffness and damping of passive magnetic bearing systems. (1) Ohji et al. investigated the radial disturbance attenuation characteristics of horizontal- and vertical-shaft machines supported radially by permanent magnetic bearings. (2) Fremerey used permanent magnetic bearings with eddy-current damping to levitate a

500 Whr energy storage flywheel for an uninterruptible power supply. (3) Satoh et al. used a mechanical damper attached to the stator magnets in a four passive-axis system. (4) Jansen and DiRusso measured the stiffness and damping of a ferrofluid stabilizer and the stiffness of permanent magnetic bearings.

Jewel bearings have been widely used in electrical measuring instruments such as galvanometers, ammeters, and energy measuring meters. A jewel bearing consists of a ball that spins perpendicular to the contact area. Jewel bearings have been treated analytically and experimentally. Sankar and Tzenov modeled a jewel bearing and determined its steady-state dynamics using a free ball and its motion due to friction.

Spinning ball contact of jewel bearings was experimentally studied at the Glenn Research Center in the late 60's and early 70's. A series of papers were written on the spinning ball contact. The first paper was by Townsend, Dietrich and Zaretsky and included speed effects on ball spinning torque. The second paper by Allen and Zaretsky proposed a microasperity model for the elastohydrodynamic lubrication of a spinning ball on a flat surface. The third paper by Dietrich, Parker, and Zaretsky studied the effect of ball-race conformity on spinning friction. The fourth paper by Townsend and Zaretsky studied the effects on ball spinning torque of antiwear and extreme-pressure additives in a synthetic paraffinic lubricant. The fifth paper by Allen, Townsend, and Zaretsky proposed a new generalized rheological model for lubrication of a ball spinning in a nonconforming groove.

Magnetic bearing systems that are currently being considered for flywheel applications use electromagnets that require active control. This makes them very complex. The objective of this study was to demonstrate the potential of a simple system that uses only passive components that could be used for flywheels on unmanned satellites. The approach was to use passive magnetic bearings for radial bearing support and jewel bearings for axial support.

BACKGROUND

Earnshaw's Theorem states that if inverse-square-law forces govern a group of charged particles, they can never be in stable equilibrium. The theorem is based on the consequence that the inverse-square-law forces follow the Laplace partial differential equation. Since the solution of this equation does not have any local maxima or minima and only saddle-type equilibrium points exit, there can be no equilibrium. The theorem usually applies to charged particles and magnetic dipoles but also can be extended to solid magnets. Thus, this theorem explains why it is impossible to have a completely 3-D stable passive magnetic bearing using only the forces of static fields of permanent magnets.

There are five known cases where Earnshaw's Theorem does not apply: (1) time-varying fields, (2) active-feedback systems, (3) diamagnetic systems, (4) ferrofluids, and (5) superconductors. Active feedback control is the most commonly used method to circumvent Earnshaw's Theorem.

Another method used by Jansen and DiRusso was to use a ferrofluid stabilizer. A ferrofluid is a fluid that contains chemically suspended iron-oxide-magnetic particles. The addition of these particles causes the fluid to have magnetic properties. A ferrofluid stabilizer is based upon the interaction between a permanent magnet and a magnetic fluid. A magnet immersed in a ferrofluid will seek an equilibrium position within the magnetic fluid. If the magnet is displaced from its equilibrium position, a restoring force will result. This type of system is stable in all axes, however the load capacity is minimal. To increase the load capacity, a ferrofluid stabilizer was constructed using a restricted cavity for the magnetic fluid. The magnet geometry was optimized in order to increase the restoring force in one axis relative to the other two translational axes.

To stabilize the passive radial magnetic bearing that was built by Jansen and DiRusso, two permanent magnetic disks were mounted on a rotor and two annular magnetic rings were mounted on a stator. The disks were located concentrically within the rings. The outer face of the disk and the inner face of the ring have the same magnetic polarity, which results in a repulsive magnetic force and suspends the rotor. These disks were coupled with a ferrofluid stabilizer on each end of the shaft. This arrangement resulted in a very stable passive magnetic bearing (Figure 1). However, because of the high viscosity of the ferrofluid, there was considerable drag on rotation. Therefore, the bearing could not be used for high-speed applications. However, the bearing does show promise for use with small oscillations and for use as a vibration isolation device.

Jansen and Fusaro initiated the second design. The same type of radial support was used, but a jewel bearing provided the axial support. A photo of this bearing is shown in Figure 2. The bearing was an improvement over the first design, but there was still considerable torque loss on rotation.

For concept to work a more precise way of aligning the radial magnets would have to be developed. A redesign program was initiated which resulted in the design described in this paper.

Figure 1. Passive magnetic bearing using ferrofluid stabilizers for the axial stabilization.

Figure 2. Passive magnetic suspension using jewel bearings for axial stabilization.

DESCRIPTION OF REDESIGNED RIG

The redesigned rig, shown in Figure 3, is an improvement over the previous two rigs in that it incorporates a precise method of aligning the position of the radial magnetic bearings to minimize the axial force. In the two previous designs, the magnetic bearing stators were held by slide fits on the base and alignment was performed by moving the magnetic bearing stators by hand (Figures 1, 2). In the redesign, load cells were positioned behind each jewel

bearing to measure the axial load whereas in the previous two designs no load cells were incorporated. To accomplish this precise alignment, the entire rotor axis of the support base is internally threaded. This allowed the externally threaded stator magnets to be moved along the axis of the rotor during assembly and during fine adjustment of the axial force. Since the threads have little backlash, and the support base and the rotor are stiff, precise alignment is possible.

In the ferrofluidic rig (Figure 1), each radial magnetic bearing was formed by a single magnet on the stator and on the rotor. In the first jewel bearing rig (Figure 2), each magnetic bearing is formed by a stack of two magnets on the stator and two on the rotor. There was a problem aligning the magnetic forces of the rig in Figure 2 so the rotor was never properly positioned. In the redesigned rig, each magnetic bearing is formed by a stack of four magnets on the stator and four on the rotor.

The redesigned rig has a higher radial bearing stiffness than either two previous designs, because of the different radial magnetic bearing design and the use of magnets with a higher energy product. The redesigned rig also has a means of turning the rotor (an air impeller), whereas the previous two did not.

Figure 3. Photographic image of the redesigned jewel bearing rig showing improvements over previous designs.

A schematic of the passive magnetic bearing rig is shown in Figure 4. The support base is made of a single piece of aluminum alloy. The rest of the rig consists of two passive magnetic bearings, a wheel, jewel bearings on both ends of the rotor, and an air impeller. The two sets of permanent magnetic bearings operating in the repulsion mode provide levitation in the radial direction. The axial movement of the rotor is constrained by a jewel bearing. The axial force is in one direction at a time. Jewel bearings were placed on both ends to allow for force reversal.

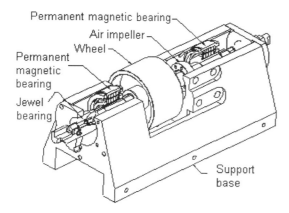

Figure 4. Schematic view of passive magnetic bearing rig. The permanent magnetic bearings and jewel bearing assembly at the near end were sectioned for clarity.

The rotor shaft was made of nonmagnetic stainless steel. The wheel, permanent magnetic bearings and an air impeller were slip fit over the rotor shaft and restrained axially by retaining rings. Each end of the rotor has a ball retainer to hold the 3.175 mm (0.125 inch) diameter ball used in the jewel bearing. The ball retainer is held to the rotor by threads on the rotor bore. The arrangement of components on the rotor is shown in Figure 5.

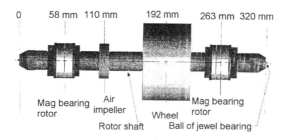

Figure 5. Schematic of rotor that shows the location of components. The total mass of the rotor is 2.26 kg. The diameter of the wheel is 102 mm.

Figure 6 depicts the permanent magnet bearing assembly. The magnets were axially magnetized, meaning the poles are on the flat sides the magnets. A separate magnet stack was attached to the rotor and to the stator. Four

magnet rings comprise each magnet stack. The poles of the magnets were oriented so that the stationary magnets and the rotating magnets repel each other when the non-rotating and the rotating sleeves are axially aligned.

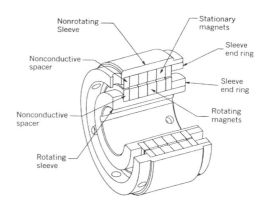

Figure 6. Sectioned view of permanent magnetic bearing assembly. There are two of these assemblies attached to the rotor.

Neodymium-iron-boron (NdFeB) magnets were used, having a nominal BH product of 35 MGOe (Gauss-Oersted x10^6). The dimensions of the rotating and stationary magnets are as follows. The OD of the rotating magnet is 45.7 mm (1.800 in) and the ID is 31.8 mm (1.250 in). The stationary magnet OD is 61.0 mm (2.400 in) and the ID is 48.3 mm (1.900 in). The thickness of magnet rings used in both stacks is 5.08 mm (0.200 inch). Therefore, the total thickness of the magnets in each stack is nominally 20.3 mm (0.800 inch).

Both the non-rotating and the rotating sleeves were made of nonmagnetic stainless steel. The stacks of magnets and the nonconductive spacers were connected together with epoxy in their mounting sleeve. The magnets were stacked with like poles adjacent to each other (NS-SN-NS-SN) and permanently locked in place with epoxy. During the assembly process each magnet was coated with epoxy and put in the sleeve along with the nonconductive spacers and the sleeve end ring. An arbor press was used to provide the force to hold the magnets in the sleeve while press pins could be installed in the sleeve end ring. Once the sleeves were assembled, a thin coat of epoxy was also applied to the free surface of the magnets to prevent the magnets from oxidizing.

The OD of the nonrotating magnet sleeve was threaded. These threads fit the threading along the rotor axis of the support base, allowing the stator magnets to be moved axially. The stator magnet sleeve was rotated by inserting a Ø4.75mm rod into a hole on the end of this sleeve. The rod was turned until the stationary sleeve was positioned over the rotating sleeve. The stator magnets were adjusted to minimize the axial force before operation.

The rotating and stationary magnets had a nominal 1.27 mm (0.05 inch) gap. The nonconductive spacers on the rotor and the stator were axially aligned and have a 0.76 mm (0.03 inch) nominal gap. This smaller clearance was to prevent the rotor and stator magnets from contacting during installation and operation.

A schematic of the jewel bearing assembly is shown in Figure 7. The jewel bearing on each end of the rotor consisted of a removable 3.175 mm (0.125 inch) diameter Si_3N_4 ball held in the ball retainer in the center of the rotor shaft. About 0.50 mm (0.02 inch) of the ball extends beyond the surface of the ball retainer. The ball spins against the stationary 440C stainless steel disk of the jewel bearing. The surface of the disk was perpendicular to the axis of the rotor. A piezoelectric load cell measures the load of the jewel bearing contact.

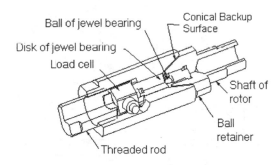

Figure 7. Jewel bearing assembly applied to each end of the rotor. The ball of the jewel bearing rotates against the stationary disk of the jewel bearing during operation.

An air impeller drives the shaft. The air impeller is a disk with blind holes drilled on both sides. One air stream is directed to either side of the disk to minimize the axial force on the rotor. The blind holes catch the air, and rotate the rotor. A pressure regulator controls the air pressure.

RESULTS AND DISCUSSION

SPIN TESTING

The rotor was spun by using an air impeller. A 120-psig-air line was connected to a pressure regulator leading to the air impeller. Adjusting the flow rate of the compressed air against the impeller could control the rotational speed of the bearing. Use of a stroboscope monitored the rotational speed.

Testing consisted of slowly increasing the speed. The speed was gradually increased to 5500 rpm and the bearing ran very smoothly over the entire speed range. It was stopped at this point only because of the loud noise generated by the compressed air and also, for safety precautions (there was not any shielding around the bearing). Other than these two reasons there was nothing observed that would have prevented the bearing from being run at much higher speeds.

The rotor had passed through the critical speed at 3336 rpm, but no vibrations were observed. We repeated the test and still no vibrations were observed.

RADIAL AND AXIAL STIFFNESSES OF BEARINGS

To determine the radial stiffness of the permanent magnetic bearings, weights were statically loaded on the rotor and the displacement of the rotor was measured. The displacement of the rotor was measured with a 0.025mm (0.001 inch) resolution dial indicator. Because of the limitation of space under the rotor, loads of up to 100 N could be applied. This was enough to give a good representation of the stiffness.

To determine the stiffness of the permanent magnetic bearing by the finite element method, a finite element model was made of the permanent magnet rings. The model had 77760 linear brick elements as shown in Figure 8. A magnetostatic solver was used to compute the distribution of the magnetic field. The Maxwell stress tensor was integrated over the surface of the four inner ring magnets to calculate the force on them.

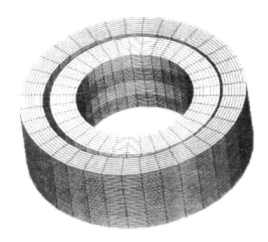

Figure 8. Finite element model of ring magnets.

Figure 9 shows the displacement as a function of radial load for the finite element and experimental data. In order to compare the experimental and the finite element results, linear regressions curve fits were performed. The slope of the experimental curve is 1.776×10^5 N/m (1015 lb$_f$/in). The slope of the finite element curve is 1.87×10^5 N/m (1068 lb$_f$/in). The radial stiffness found by the experimental method is about 5.2 percent less than the finite element prediction, which is considered very good agreement.

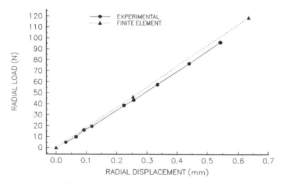

Figure 9. Radial displacement as a function of radial load for one permanent magnetic bearing as determined by experimental testing and finite element analysis.

The radial stiffness per unit length of the magnet stack of each bearing was calculated by dividing the experimental radial stiffness, 1.776×10^5 N/m (1015 lb$_f$/in) by the length of

each magnet stack, 20.3mm (0.8 in). This value is 8.76×10^6 (N/m)/m (1269 (lb$_f$/in)/in).

To measure the axial stiffness the non-rotating magnetic sleeve of one magnetic bearing was rotated relative to the rotating sleeve. This caused the magnet stacks to become axially displaced relative to each other. This axial displacement was measured with a dial indicator. The axial load was measured with the load cell that was behind the disk of the jewel bearing. This procedure was repeated for a number of different displacements and the data plotted in Figure 10. This figure shows the axial load as a function of displacement for one permanent magnetic bearings. A linear regression fit of the curve gave a value of 3.52×10^5 N/m (2008 lb$_f$/in) for the axial stiffness.

Application of Earnshaw's theorem implies that the axial stiffness should be exactly twice the radial stiffness. Comparing the experimental axial stiffness of Figure 10 [3.52×10^5 N/m (2008 lb$_f$/in)] to twice the experimental radial stiffness of Figure 9 [1.776×10^5 N/m (1015 lb$_f$/in)] showed that the 2 to 1 ratio predicted by Earnshaw's theorem is within 1.1 percent. This is considered very good agreement.

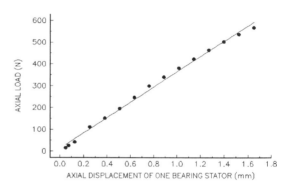

Figure 10. Axial displacement as a function of axial load for one magnetic bearing stator of the permanent magnetic bearing.

DAMPING

The procedure for measuring the free vibration history of the rotor was as follows. The axial force is in one direction. The disk of the jewel bearing at the unloaded end of the rotor was removed so that the rotor was free to move at this end. An accelerometer was attached to the rotor 25 mm from free end. To excite the rotor the wheel was struck with an impact hammer.

Figure 11 gives the free vibration history of the rotor as a function of time.

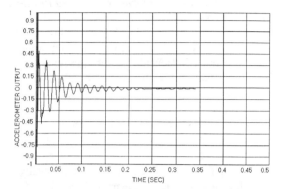

Figure 11. Free vibration history of the rotor from an impact test. The vertical axis is the output of the accelerometer attached to free end of the rotor.

A damping coefficient was calculated from the logarithmic decrement and from the half-power bandwidth methods. The natural logarithm of consecutive amplitude ratios is called the logarithmic decrement, δ, equaling $\ln(x_1/x_2)$, where x_1 and x_2 are consecutive peak amplitudes (Tse et al.). Taking the log decrement of the first five peaks in the time domain gives $\xi=0.0649$, or about 6.5 percent damping.

The frequency response of the rotor is shown in Figure 12. The first fundamental occurred at 55.6 Hz (3336 rpm). Using the frequency response, the damping coefficient of the first fundamental was calculated by the half-power bandwidth method. The damping coefficient of the first fundamental was found to be 0.069, close to the damping coefficient as found by the logarithmic decrement method. Since no vibrations occurred on running the bearing through the critical speed, 6.5 percent damping was adequate.

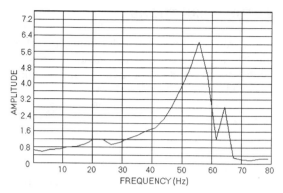

Figure 12. Frequency response of the rotor that occurred from impact test.

SUMMARY OF RESULTS

A completely passive magnetic bearing rig was designed, and constructed. Support of the rotor was accomplished in the radial direction by two sets of permanent magnetic bearings and in the axial direction by two sets of jewel bearings. The following results were obtained.

1. The rig was successfully operated to speeds of 5500 rpm, which was 65 percent above the first critical speed.
2. The radial stiffness of one magnetic bearing was measured to be 1.776×10^5 N/m (1015 lb_f/in) which was within 5 percent of a finite element prediction. The measured stiffness was adequate for the currently designed rotor.
3. The fact that experimental and finite element methods predictions agree demonstrate that finite element methods can be used to accurately design the size and number of the magnets for the radial support of the bearings.
4. The damping was measured to be 6.5 percent of critical damping. This was adequate damping to pass through the first critical speed.
5. The axial stiffness was measured and was shown to be within 1 percent of twice the measured radial stiffness. This is consistent with Earnshaw's theorem.

The results are only preliminary, but they are very encouraging in that we think these bearings can be developed for use with small high-speed flywheels.

FUTURE DIRECTIONS

Since the air drive was shown to be noisy and hard to control, it will be replaced with an electric motor drive that can rotate the rotor to speeds of 30,000 RPM. Proximity probes will be installed to measure vibrations and displacements. Vibration data will be compared to rotordynamics analysis.

RERERENCES

Allen, C. W., Zaretsky, E. V., "Microasperity Model for Elastohydrodynamic Lubrication of a Spinning Ball on a Flat Surface," NASA TN D-6009, 1970.

Allen, C. W., Townsend, D. P., and Zaretsky, E. V., "New Generalized Rheological Model for Lubrication of a Ball Spinning in a Nonconforming Groove," NASA TN D-7280, 1973.

Dietrich, M. W., Parker, R. J., Zaretsky, E. V., "Effect of Ball-Race Conformity on Spinning Friction," NASA TN D-4669, 1968.

Earnshaw, S., "On the Nature of the Molecular Forces which Regulate the Constitution of the Luminiferous Ether," *Trans. Of the Cambridge Philosophical Society*, Vol. 7, Part I, 1842, pp. 97-112.

Fremerey, J. K., "A 500-Wh Power Flywheel on Permanent Magnet Bearings," Fifth International Symposium on Magnetic Suspension Technology, July 2000, pp. 287-295.

Geim, A. K., Simon, M. D., Boamfa, M. I., and Heflinger, L. O., "Magnetic Levitation at Your Fingertips," *Nature*, Vol. 400, July 22, 1999, pp. 232-324.

Hull, J. R., Mulcahy, T. M., Uherka, K. L., Erck, R.A., and Abboud, R. G., "Flywheel Energy Storage Using Superconducting Magnetic Bearings," *Appl. Supercond.*, Vol. 2, pp. 449-455, July/Aug. 1994.

Hull, J. R., and Turner, L. R., "Magnetomechanics of Internal-Dipole, Halbach-Array Motor/Generators," *IEEE Transactions on Magnetics*, Vol. 36, No. 4, July 2000, pp. 2004-2011.

Jansen, R., DiRusso, E., "Passive Magnetic Bearing with Ferrofluid Stabilization," NASA-TM-107154, 1996.

Mulcahy, T. M., Hull, J. R., Uherka, K. L., Niemann, R. C., Abboud, R. G., Juna, J., and Lockwood, T. A., "Flywheel Energy Storage Advances Using HTS Bearings," *IEEE Trans. Appl. Supercond.*, Vol. 9, June 1999, pp. 297-300.

Ohji, T., Mukhopadhyay, S. C., Iwahara, and Yamada, S., "Permanent Magnet Bearings for Horizontal- and Vertical-Shaft Machines: A Comparative Study," *Journal of Applied Physics*, Vol. 31, 8, pp. 4648-4650.

Post, R. F., Ryutov, D. D., Smith, J. R., and Tung, L. S., "Research on Ambient-Temperature Passive Magnetic Bearings at the Lawrence Livermore National Laboratory," LLNL Pub. #231803, Apr. 1997.

Post, R. F., Ryutov, D. D., "Ambient-Temperature Passive Magnetic Bearings: Theory and Design Equations," LLNL Pub. #232382, Dec. 1997.

Sankar, T. S., and Tzenov, P. I., "The Steady-State Dynamics of Jewel Bearing with a Free Ball," *Tribology Transactions*, Vol. 37, 2, pp. 403-409.

Sankar, T. S., and Tzenov, P. I., "Friction and Motion Accuracy in Jewel Bearing," *Tribology Transactions*, Vol. 37, 2, pp. 269-276.

Satoh, I., Shirao, Y., and Kanemitsu, Y., "Dynamics and Vibration of a Single Axis Active Magnetic Bearing System for Small-Sized Rotating Machinery," Fifth International Symposium on Magnetic Bearings, August 1996, pp. 497-502.

Ting, L., Tichy, J., "Stiffness and Damping of an Eddy Current Magnetic Bearing", *Journal of Tribology*, Vol. 114, July 1992, pp. 600-604.

Townsend, D. T., and Zaretsky, E. V., "Effects of Antiwear and Extreme-Pressure Additives in a Synthetic Paraffinic Lubricant on Ball Spinning Torque," NASA TN D-5820, 1970.

Townsend, D. P., Dietrich, M. W., Zaretsky, E. V., "Speed Effects on Ball Spinning Torque," NASA TN D-5527, 1969.

Tse, F. S., Morse, I. E., and Hinkle, R. T., Mechanical Vibrations: Theory and Applications, 2nd Edition, Allyn and Bacon, Inc., 1978, pp. 78-79.

Proceedings of IECEC '01
36th Intersociety Energy Conversion Engineering Conference, 2001
July 29 - August 2, 2001 Westin Hotel Savannah Georgia

IECEC2001-AT-63

Advanced Radioisotope Power System Technology Development for NASA Missions 2011 and Beyond

Jack F. Mondt
California Institute of Technology
Jet Propulsion Laboratory

ABSTRACT

NASA's Office of Space Science requested JPL to lead an assessment of advanced power source and energy storage technologies that will enable future (beyond 2006) NASA Space Science missions and prepare technology road maps and investment strategies. This paper summarizes the result of reviewing the power requirements for future NASA deep space and Mars science missions and providing a technical assessment of the radioisotope power conversion technologies being considered for these future NASA missions.

There are uncertainties in the lifetime performance predictions of Advanced Radioisotope Power Systems (ARPS) conversion technologies, uncertainties in the future supply of Pu-238 and uncertainties in the requirements of NASA missions beyond 2011. Therefore several ARPS conversion technologies need to be further developed before selecting one. The ARPS assessment team recommends that the Advanced Stirling Engine Converter (ASEC), Alkali Metal Thermal to Electric Converter (AMTEC) and Segmented Thermoelectric Converter (STEC) technologies be funded with detailed technical progress reviews yearly to decide to continue or not. The ARPS selected conversion technology would be based on demonstrated technical progress towards meeting technology readiness gates based on NASA's future mission requirements for deep space science missions beyond 2011. Advanced technology funding would be used to develop the ASEC, AMTEC and STEC technologies in parallel for missions 2011 and beyond. A Formal Review Board would review each technology at NASA Technology Readiness Levels (TRL) 3, 4 and 5. The Board would recommend continuing the technology or not based on technical progress in meeting their technology readiness gates and meeting future NASA mission requirements.

INTRODUCTION

All spacecraft require electrical power in order to accomplish their mission. Power is provided either by a photovoltaic (PV) array with batteries or by radioisotope power systems (RPS)s. Over the years the efficiency, specific power and lifetime of PV arrays with batteries have steadily improved. PV arrays with batteries are the power source of choice for most space missions within 2 au of the sun because of their high specific power, efficiency and reliability.

However, there are missions for which PV arrays with batteries are unsatisfactory. These include missions where the solar flux is too low due to large distances from the sun or variable due to eclipses, shadows, dust and changing distances from the sun. Examples include missions to the outer planets, Jupiter and beyond, and missions on Mars that require (a) operation in shadows, (b) extended lifetimes where seasonal variations and settling of dust on PV arrays would be deleterious or (c) where power throughout the 24.66-hour diurnal cycle is essential.

OBJECTIVES

The assessment team goal was to determine the ARPS conversion technologies that best satisfy the future deep space science and Mars mission requirements.

The assessment team objectives were to:

1) Review NASA future missions needs for advanced radioisotope power systems (ARPS).

2) Assess the status and potential performance of ARPS technologies.

3) Estimate resources required to advance ARPS technologies to NASA TRL 5.

4) Prepare road maps for promising technologies that satisfy future mission requirement.

5) Recommend to NASA and DOE investment strategies for developing ARPS technologies.

APPROACH

JPL established the following technical assessment team to assess the conversion technologies and accomplish the above objectives within three months.

Rao Surampudi, NASA JPL, Chairperson
Bob Carpenter, OSC (DOE support contractor)
Mohamed El-Genk UNM
Lisa Herrera, DOE
Lee Mason, NASA GRC
Jack Mondt, NASA JPL
Bill Nesmith, NASA JPL
Donald Rapp, NASA JPL
Robert Wiley, BA&H (DOE support contractor)

The reasons to develop ARPS technologies are:

- To increase the specific power by about a factor of 2 (9 to 10 watts/kg)
- To increase the System efficiency by a factor of 2 to 4 (13% to 25%)
- To reduce the recurring cost of RPSs
- To reduce the flight RPS fabrication time from project start to delivery to the launch site.

ARPS technologies that accomplish all of the above are desired.

The system efficiency of the state-of-art (SOA) SiGe RTG is 6.5% and the specific power is 4.5 watts/kg. The SOA SiGe RTG has a proven long lifetime > 20 years. A more efficient ARPS converter reduces the amount of Pu-238 fuel and the cost for any power level. The team used a scaled-down 100-watt version of the Cassini 285-watt SiGe RTGs as a baseline against which ARPS technologies were compared. Based on most future deep space science and Mars mission the power requirements can be met with an RPS that is sized to deliver 100 watts electric at the beginning of mission (BOM). Some missions may require 2 or 3 such RPS. In this assessment, the team assumed that a RPS uses a specific number of GPHS modules at an assumed BOM thermal power of 240 watts/module. The number of GPHS modules was chosen so that each ARPS technology BOM power output is 100 watts or greater.

CRITERIA

The parameters used to evaluate the advanced technologies are:
1) Safety, 2) Lifetime and Fault Tolerant,
3) Specific Power, 4) Conversion Efficiency,
5) Applicability to a wide range of mission requirements, 6) Development Risk, 7) Spacecraft Interface Issues, 8) Converter/GPHS Interface Issues, 9) Feasibility of Validating the Lifetime Performance for 15-year missions

The team's interpretation of the NASA Technology Readiness Levels (TRL) is described in Figure ES-1. TRL 1 and 2 refer to new technologies that are in the early stages of emergence. NASA TRL 3, 4 and 5 include some demonstrated lifetime performance. NASA Technology development efforts for NASA space missions involve advancing a technology from NASA TRL 3 to TRL 5.

When a technology reaches the NASA TRL 5, successful prototype flight configuration and accelerated lifetime tests of major subassemblies and components have been tested on the ground in a relevant space environment. Subsequently, the technology is transferred to a system integrator to develop a flight RPS that meet the requirements of a specific space science mission. NASA-JPL projects require that a technology reach TRL 6 by spacecraft preliminary design review (PDR) in order to be used on a NASA primary mission

The ARPS assessment team estimated the resources needed to advance selected candidate ARPS from TRL 2, 3 or 4 to TRL 5. The team qualitatively described the development risk that the technology will reach TRL 5 with the estimated resources, as High, Medium or Low. A critical element of technology development is assessing the lifetime of the ARPS. This requires accelerated testing of components, subassemblies and systems to validate lifetime prediction codes. In-depth analysis of failure modes and accelerated tests are required to validate ARPS lifetime performance prior to launch.

FUTURE NASA MISSION REQUIRING RPSs

NASA Mars Exploration Program (MEP) enterprise's future plans includes Mars surface lander missions every four years beyond 2011. RPSs would provide the longevity and versatility required to accomplish the scientific objectives for these missions. Mission duration is not known but is expected to be 3 to 4 Earth years. However, with long life RPS power, Rover missions could possibly extend to 10 years. NASA Solar System Exploration (SSE) enterprise's future plans beyond 2011 includes Pluto-Kuiper Express, Europa Lander, Titan Explorer and Neptune/Triton Orbiter missions.

TRL	Accomplishment
1-2	Concept and application are formulated. Basic phenomena are observed in a laboratory environment.
3	Critical functions are tested in a laboratory environment of breadboard configuration to validate proof-of-concept's potential performance and lifetime.
4	A breadboard system (or at least the major components of the system) are tested in the laboratory and verify that components will work together effectively in a system. At this level, preliminary analytical and experimental efforts are made to calculate lifetime performance of critical components.
5	A realistic breadboard portion of the system is thoroughly tested in a relevant environment that demonstrates the flight system design. Lifetime performance predictions of critical components are validated with accelerated tests.
6	System engineering model with approximate "form fit and function" of a flight system or prototype demonstration tested in a relevant environment on ground or in space. System lifetime performance prediction validated based on accelerated life testing of components, subassemblies and systems.

Figure ES-1. The NASA Technology Readiness Levels for ARPS

These are 6 to 15 years missions that appear to require radioisotope power systems.

NASA Sun-Earth Connection (SEC) enterprise's plans include Solar Probe, Interstellar Probe, Interstellar Trailblazer and the Outer Heliosphere Radio Imager missions, each of which requires radioisotope power sources. These missions are in the early stages of planning and the projected power levels are 200 to 300 watts with lifetimes up to 30 years.

ADVANCED TECHNOLOGIES EVALUATED

Technologies evaluated in this study are given in table ES-1 and described after the table. Advanced Stirling Engine Converter (ASEC), Alkali Metal Thermal to Electric Converter (AMTEC), Segmented Thermoelectric Converter (STEC), Low Temperature Thermionic (LTI), Thermo-Acoustics (TA) and Thermal PhotoVoltaics (TPV) technologies as applied to ARPS were reviewed and evaluated to satisfy future potential NASA Space Missions. Estimated system masses and efficiencies were made for each technology and compared to a scaled-down design of a 100-watt SiGe RTG.

Advanced Stirling Engine Converter (ASEC)-ARPS technology development approach uses a reciprocating free-piston Stirling heat engine or a Thermo-Acoustic heat engine with a linear alternator that are low-mass version of the Stirling engine alternator now under development by DOE and NASA. The ASEC-ARPS has the principal advantage of increased conversion efficiency to almost four times the system efficiency of the SOA SiGe RTG. Advanced versions of the Stirling engine converter ARPS may double the specific power over the SOA SiGe RTG.

The ASEC-ARPS major technical issues are:

- 1) Validating the system lifetime for 15 year missions
- 2) Developing an efficient, low-mass, long life ASEC-ARPS
- 3) Reducing the EMI for space missions that measure very small magnetic fields
- 4) Reducing the Stirling engine alternator vibration for very sensitive seismic instruments

Alkali Metal Thermal to Electric Converter (AMTEC)-ARPS produces electric power by the flow of sodium ions through a Beta-Alumina Solid Electrolyte (BASE) that produces DC current and voltage. AMTEC delivers DC power with no vibration and very small EMI. AMTEC is a young technology with potential system efficiency as high as 20%, which is three times the system efficiency of the SOA SiGe RTG. AMTEC ARPS has the potential for doubling the specific power over the SOA SiGe RTG to 9 watts/kg.

The AMTEC-ARPS major technical issues are:

Table ES-1 Technologies Evaluated

Technology	Specific Technology	Comments
Thermoelectric	SiGe RTGs	Used on Voyager, Ulysses, Galileo and Cassini Missions
	PbTe-TAGS	Used on Viking and Pioneer Missions
Stirling Engine Converter	Version 1.0	Present design; efficiency is very good but mass is high; lifetime is not certain; most mature of the ARPS technologies; only one with a reasonable chance of being made ready for Mars 2007
	Version 1.1	Advanced Technology Low Mass Alternator
Advanced Stirling Engine Converter	Version 2.0 or Thermo-acoustic	Low Mass Stirling Engine, Alternator, Radiator and Controller Potential for high efficiency long life Stirling Engine. Considered as an advanced form of Stirling Technology.
AMTEC	Refractory Metal Chimney	Potential for Low Mass and Medium Efficiency. Being developed by NASA and DOE under existing DOE contract
Segmented Thermoelectric	Advanced Materials/ Segmented Unicouple	Potential for low mass and medium efficiency. Solid state device Radioisotope Power System configuration, operations and handling similar to SiGe RTG
Thermionic	Cesiated triode	Low Temperature 1300K Thermionic early stage of technology development
	Micro-miniature	Early stage of research
Thermo-photovoltaic	Thermal PhotoVoltaic Converter	Early stage of research

- 1) Developing a BASE to metal ceramic seal
- 2) Developing a converter refractory metal containment material fabrication process
- 3) Developing a reproducible wick-evaporator fabrication process
- 4) Developing an electrical feed-through fabrication process.

Segmented-Thermoelectric Converter (STEC)-ARPS contains thermoelectric materials that produce a current and a voltage when placed in a temperature gradient. Each thermoelectric material, whether n-type or p-type, exhibits a maximum figure-of-merit at some temperature. If a single material is used in each leg of the unicouple, the effective efficiency will be an average over the temperature range, which is less than the maximum. If each leg of the unicouple is segmented so that a thermal gradient is established down the leg, the temperature gradient over each segment will be small. Thermoelectric materials developed with a high efficiency over the small thermal gradient for each segment will achieve a higher efficiency

over the entire thermal gradient. The STEC-ARPS could double the efficiency of the SiGe RTG.

The major technical issues are:

- 1) Developing a compatible high temperature (973K to 1273K) thermoelectric material
- 2) Developing joints between the segments with very small thermal and electric resistance
- 3) Developing barriers that prevent inter-diffusion between segments
- 4) Developing joints between the high temperature thermoelectric materials and a hot shoe.

The team assessments of these advanced converter technologies for ARPS are as follows:

ASEC-ARPS has the potential to quadruple the conversion efficiency over the SOA SiGe RTG. Advanced ASEC-ARPS systems may double the specific power over the SOA SiGe RTG. A method to accelerate and validate the ASEC-ARPS lifetime needs to be developed. Thermo-Acoustic Stirling engine technology may offer less vibration and longer lifetime over conventional Stirling engines but it is at an early stage of development.

AMTEC-ARPS has the potential to double the specific power and efficiency over the SOA SiGe RTG. The lifetime of AMTEC ARPS basic conversion components (BASE, Electrodes and current collectors) have demonstrated >20 years. It is planned that accelerated testing will validate the lifetime performance of components, converter and system. There are no EMI or vibration problems.

STEC-ARPS has the potential to double the specific power and efficiency over the SOA SiGe RTG. STEC-ARPS converters are amenable to accelerated lifetime testing as they are being developed. The STEC-ARPS lifetime is well known once the converter lifetimes has been validated. There are no EMI or vibration problems.

Low Temperature Thermionics and Thermal PhotoVoltaics technologies are at NASA TRL1-2. Therefore the team could not assess the efficiency, specific mass or lifetime for an ARPS for these conversion technologies. The team recommends NASA's crosscutting technology program; the DOE PRDA program; SBIRS and STTRs fund these two technologies to NASA TRL 3 so a realistic estimate of system mass, efficiency and lifetime can be prepared.

Table ES-2 summarizes the major characteristics, provides estimated system data and compares the team selected candidate technologies, ASEC, AMTEC and STEC to the SOA SiGe RTG. Thermo-Acoustic technology is considered as part of the advanced Stirling technology. In each case, a GPHS module delivers 240 watts thermal at BOM and the number of GPHS modules was chosen to make the BOM power 100 watts electric or greater.

RESULTS AND RECOMMENDATIONS

The assessment team recommends that ASEC, AMTEC and STEC technologies be funded and developed by NASA in accordance with a technology plan that includes technology readiness gates for each technology. The progress towards meeting these technologies readiness gates for each technology should be reviewed yearly by the same independent Formal Review Board. Two of these technologies should be selected in two or three years based on the progress in meeting their technology gates and meeting the requirements for the greatest number of future NASA ESS, SEC and MEP missions.

The two selected technologies should then be developed to TRL 5 under a joint NASA/DOE technology program. When the technologies reach NASA TRL 5 a NASA flight project and DOE would jointly select and develop the technology that best satisfies the requirements of that project's specific mission.

A top level recommended ARPS technology roadmap is shown in Figure ES-2. This roadmap assumes that NASA and DOE would develop a near term Stirling RPS and/or a 100-watt class RTG, either SiGe or PbTe/TAGS), for potential NASA deep space and Mars science missions that are launched prior to 2011.

ACKNOWLEDGMENTS

The author acknowledges and thanks the dedicated efforts of Rao Surampudi, Bob Carpenter, Mohamed El-Genk, Lisa Herrera, Lee Mason, Bill Nesmith, Donald Rapp and Robert Wiley who made this paper possible. Also acknowledges Joe Parrish for requesting and guiding this assessment for NASA and the guidance and support of Doug Stetson and Greg Vane at JPL.

The research described in this paper was performed at the Jet Propulsion Laboratory, California Institute of Technology under contract with the National Aeronautics and Space Administration.

REFERENCE

Surampudi, Rao, et al, Advanced Radioisotope Power System (ARPS) Team 2001Technology Assessment and Recommended Roadmap For Potential NASA Deep Space Missions Beyond 2011, JPL Report, April, 2001

Table ES-2. Characteristics of Candidate ARPS technologies

Technology	NASA TRL	BOM Watts	System Mass kg	Spec Pwr W/kg	Sys Eff %	GPHS Modules (d)	Resources for TRL5	Dev't Risk	Life Issues	S/C IF Issues (e)	Resiliency to Partial Failure
Small SiGe RTG (a)	8	139	31.2	4.5	6.5%	9	None	None	None	None	Highly modular
Stirling 1.0 (b)	4	110	27	4.1	23%	2	$4.5M 2yr	Low	Engine & Control ElecTronics	AC/DC Control electronics	Failure of one converter may lead to generator failure
Stirling 1.1	4	120	20	6.0	25%	2	$8M 3yr	Med	Helium leakage	Radiator Vib, EMI	"
Stirling 2.0	2	120	16	7.5	25%	2	$13.5M 6yr	High	"	"	"
AMTEC (LMA) (c)	3	139	25	5.6	14.5%	4			Seals, wick Evaporator	Launch vehicle accelera tion	Failure of converter results in generator partial power loss
AMTEC Chimney	3	120	13.6	8.8	16.7%	3	$15M 6yr	High	Containment, matls, fab Process	"	
Low T Segmented TE	2	125	14	8.9	13%	4			Joint bonding, barriers	None	Highly modular
High T Segmented TE	2	144	14	10.2	15%	4	$13.5M 6yr	High	New Material Joint bonding Barriers	None	"

(a) LMA 9 GPHS Vacuum RTG for Europa Orbiter 04 or 05 launch
(b) LMA Stirling RPS study concept for Europa Orbiter
(c) LMA AMTEC preliminary design for Europa Orbiter
(d) Each GPHS module assumed at 240 thermal watts at BOM
(e) Potential spacecraft interface issues with some mission

Technology	FY01	FY02	FY03	FY04	FY05	FY06
Stirling 1.0	Mtls Charact ($3.5M) TRL4	Life Testing ($1.0M) TRL5	System Integration	System Integration	Flight Development (EO, Mars '07) →	
Stirling 1.1	TDC Mod Plan ($0.5M)	Lt Wt Alt/Cntl ($3.5M)	Demow/ElcHS ($4.0M) TRL5	System Integration		
EO & Mars '07, $/FY	($4.0M)	$(4.5 M)	($4.0M)			
Stirling 2.0	Concept Feasibility $0.5M	Advanced Concepts RFP $2.0 M TRL3	Lt Wt Conv Rad, Cntl. $2.0M	Conv, Rad, Cntl, Brdbds $3.0M TRL4	Engr Models Life Tests $3.0M	Engr Models Life Eval $3.0M TRL5
Thermoacoustics NRA (CETDP) Conv Demo	($650K)	($650K)				
AMTEC	Matls Dev Chimney Cell ($2.0M EO) $1.0M	Chimney Cell, Life Validate ($1.0M DOE) $2.0M TRL3	ReproduceCell Life Prediction ($1.0M DOE) $2.0M	Integ 4 Conv Demonstration $4.0M TRL4	Engr Model Life Eval $3.0M	Engr Model Life Eval $3.0M TRL5
Segmented TE	Matls Eval $0.5M	Bond/Barrier Unicouple Fab $2.0 M TRL3	1000KUnicouple Demo $2.0M	4-CpleModule AccelLifeTest $3.0M TRL4	3,18-Couple Engr Models $3.0 M	18-Couple Engr Models $3.0 M TRL5
Segmented TE NRA (CETDP)Unicouple Dv	($350K)	($350K)	($350K)			
New Tech $/FY	$2.0M	$6.0M	$6.0M	$10.0M	$9.0M	$9.0M

TRL

ES 2 RECOMMENDED ARPS TEHNOLOGY ROADMAP

Notes: Stirling 1.0 & 1.1, System Integration & Flight Development costs, EO, DOE & Mars Funds & CETDP costs not included in totals Independent Review Board evaluates technology progress at TRL 3, 4 & 5 to determine to proceed or not.

Proceedings of IECEC'01
36th Intersociety Energy Conversion Engineering Conference
July 29 - August 2, Savannah, Georgia
IECEC2001-AT-72

STRESS ANALYSIS OF DC-DC POWER CONVERTERS

Siamak Abedinpour, and Krishna Shenai
Department of Electrical Engineering and Computer Science
University of Illinois at Chicago, Chicago, IL 60607
Email: shenai@eecs.uic.edu, Tel.: 312 996 2633, Fax.: 312 996 0763

ABSTRACT

Over 2% of the power supplies in the computer and telecommunication industries fail in the first year of operation. Cost and size reduction drive the DC-DC converters used in the power supplies to operate at higher switching frequencies and as a result the power MOSFET switches used in these converters undergo excessive electrical and thermal stress. This paper presents the results obtained from a study conducted to evaluate the various electrical stresses applied to a power MOSFET in a 350V - 3.5V, 225A full-bridge phase-shifted ZVS DC-DC converter. For the converter under study MOSFETs were found to fail under no-load converter operation.

I. INTRODUCTION

Cost and portability specially in computer and telecommunication power supplies are driving the power electronics designers to increase the power density by use of innovative topologies, and power semiconductor devices. In spite of slightly inferior on-state performance compared to bipolar devices, silicon power MOSFETs are extensively used in medium-voltage power supplies because of their simplicity of operation, and faster switching. In conventional hard-switched full bridge (FB) DC-DC converter topology higher switching frequency results in smaller reactive components size, but increases MOSFET switching losses, heat sink requirement or number of paralleled devices, and over-voltage stress as a result of parasitic inductance and body diode reverse recovery.

The performance of hard-switched converters has been improved by introduction of various soft-switching schemes. Zero voltage switching (ZVS) is one of the possible solutions for operation at higher frequencies in order to increase the converter power density. In a full bridge topology, ZVS can be achieved by use of the parasitic intrinsic MOSFET body diode, which reduces the switching losses, and heat sink requirements. The two legs of the bridge are operated with a phase shift, which allows zero voltage turn on of the

MOSFETs by use of the energy stored in the transformer leakage inductance to discharge the output capacitance of the MOSFETs before turning them on. Recently, power MOSFET failures have been reported in the ZVS FB DC-DC converter topology [1,2]. This paper presents the results obtained from a study conducted to evaluate the various electrical stresses applied to a power MOSFET in a 350V - 3.5V, 225A FB ZVS DC-DC converter. For the converter under study MOSFETs were found to fail under no-load converter operation. Normal operation of the FB ZVS DC-DC converter creates conditions that cause forward and reverse recovery of the MOSFET body diode. Body diode recovery introduces an additional stress on the MOSFET. The intrinsic body diode of similar rated (500V-20A) MOSFETs from different manufacturers were tested for forward recovery and reverse recovery performance. It is observed that the peak reverse current of these diodes is much greater than that of the conventional diodes for the same operating conditions. Lifetime control is used in power diodes to improve the switching performance, while no such attempt is made in case of commercial power MOSFETs.

II. CIRCUIT OPERATION

A FB ZVS DC-DC converter, which uses circuit parasitics to achieve resonant switching is shown in Fig. 1a [3]. Various electrical stresses are applied to the MOSFET during a switching cycle. The two legs of the converter are operated with a phase-shift, which causes a resonant discharge of the output capacitance of the MOSFETs by utilizing the energy stored in the transformer leakage inductance (L_{lk}) and determines the converter duty cycle. Larger leakage inductance increases the load range at which the converter operates with ZVS but decreases the effective duty cycle at the secondary of the transformer. Energy transfer occurs during active states, when diagonally opposite MOSFETs conduct for example M_1 and M_4. For approximately half of the switching period M_3 and M_2 will be under a forward blocking voltage stress with the input voltage applied across the drain-source, with gate-source shorted.

When M_4 is turned off, the energy stored in the output inductors charges the output capacitance of M_4 and discharges the output capacitance of M_3, which causes the anti-parallel body diode of M_3 to conduct. A passive state begins and no energy is transferred to the load. M_3 can now be turned on under zero voltage.

III. FORWARD RECOVERY STRESS

When the body diode of M_3 starts to conduct its forward voltage rises to peak forward recovery voltage, which is higher than the final steady state voltage drop. The excess charge in the drift region requires time to build up to its final value and before this time period the resistance of the diode is very high. High level injection increases the conductivity of the lightly doped drift region of the body diode and greatly enhances the current handling capability [4]. To determine the static forward voltage drop the transition from low level injection to high level injection is taken into account. The on-state voltage is expressed as:

$$V_F = \frac{J_F(W_D - x_o)}{q\mu_n N_D} + \frac{W_D x_o}{\tau \mu_a} + V_{p+} + V_{n+} \quad (1)$$

$$x_o = 2L_a \left(\ln \left(\frac{J_F L_a}{2qN_D \sqrt{D_N D_P}} \right) \right) \quad (2)$$

where x_0 is the extent of the drift region that is in high level injection at a give current density J_F, and V_{p+} and V_{n+} are the junction drops at the anode and cathode ends of the drift region respectively. The first term in Eq. (1) is dropped when the entire drift region is in high level injection. During the period in which the excess carrier distribution grows in the drift region, there is no conductivity modulation of the drift region until the space charge is discharged to its thermal equilibrium value. Therefore there is a voltage overshoot before the drift region becomes shorted by the large amount of carrier injection. The duration of the space charge layer discharge and the growth of the excess-carrier distribution in the drift region is governed by both the intrinsic properties of the body diode and the external circuit in which the MOSFET is used. A large value of di/dt minimizes the time needed to discharge the space charge layer. On the other hand a large value of forward current and of carrier lifetime in the drift region will lengthen the time needed for the excess-carrier portion of the transient to be completed. This establishes an inherent trade-off between shorter turn-on transient and higher on-state losses. During the forward recovery time t_{fr}, the power dissipation in the body diode is much higher than predicted from its static characteristics, which is an additional stress on the MOSFETs to be used in the converter. The

current flow through M_3 during this transition is shown in Fig. 6a. As is shown in Fig. 1b after the diode forward recovers and a steady state voltage of V_F is reached across the device ,M_3 is turned on under ZVS and the current is shared between the channel and body diode. When M_1 is turned off similar situation happens for M_2 and its body diode goes under forward recovery and M_2 is turned on under ZVS. This causes the current to decrease and change direction and after reverse recovery of the body diodes of M_2 and M_3 another active state continues to transfer energy to the transformer secondary.

Similar rated devices (500V-20A) from different manufacturers were tested in the forward recovery test circuit of Fig. 2a, where M_3 is the device under test. By turning on M_4 current ramps up in the inductor in parallel with M_3. By adjusting the duration of the gate pulse the forward current was adjusted for five different values as shown in Fig. 2b. The measurements were performed at four different temperatures. Measurement results showing variation of peak forward voltage drop vs. forward current at different temperatures are shown in Fig. 3 for a power MOSFET (C). Test results which indicate the variation of forward recovery voltage drop and forward recovery energy loss as a function of forward current for similarly rated MOSFETs (A-E) and IGBT(V) are shown in Fig. 4a at 50°C. Test results which indicate the variation of forward recovery energy loss are shown in Fig. 4b at different temperatures. The results show that IGBT with its anti-parallel diode, which is optimized for reverse recovery shows the worst performance under forward recovery.

Rigorous two-dimensional (2-D) device simulations were performed to understand the internal plasma dynamics during transient switching conditions. The MOSFET structure obtained from simulation of device performance under static conditions as is shown in Fig. 5a, was used in a 2-D mixed device and circuit simulator, in which semiconductor equations are solved within the device under boundary conditions imposed by external circuit elements [5]. Three similar structures were used in order to study the dynamics of the device under forward recovery in FB ZVS DC-DC converter. The finite element simulation results are shown in Fig. 6b, c, and d at three different time instants indicated in Fig. 1b. Current flow through the channel and body diode of MOSFETs M_3 and M_4 are shown at t_1, which corresponds to the beginning of the forward recovery. Instant t_2 is during the conduction of the body diode and t_3 is when the MOSFET is also conducting the reverse current in its third quadrant of operation. Figure 5b shows the 2-D simulation results of the current through MOSFETs M_3 and M_4 and the drain-source voltage of M_3 during forward recovery of its body diode, by considering the effect of parasitic drain and source lead inductances.

IV. REVERSE RECOVERY STRESS

Under normal operating conditions during a passive to active transition reverse recovery of the body diode occurs, when the drain-source voltage is low due to conduction of the channel. Therefore the reverse recovery energy loss is significantly reduced, and therefore this is not a stressful reverse recovery. Under no-load condition the phase shifted full bridge topology has to operate with extremely short active intervals [6]. For very narrow pulses the normal current-mode control can not be maintained and is replaced by the duty-ratio mode control. This means that the current symmetry in the transformer primary is no longer ensured and some dc current component may build up. This current can be caused by unsymmetry in duration of positive and negative voltage pulses across primary, which is a result of unequal delays in the driving signals. A net dc voltage attempts to appear across the primary winding, which causes a unidirectional current.

Similar rated devices (500V-20A) from different manufacturers were tested in the reverse recovery test circuit of Fig. 7a, where M_1 is the device under test. By turning on M_2 current ramps up in the inductor in parallel with M_1. By adjusting the duration of the gate pulse the forward current was adjusted for two different values as shown in Fig. 7b. The measurements were performed at four different temperatures. Test results, which indicate the variation of peak reverse recovery current and reverse recovery energy of various devices as a function of forward current, are shown in Fig. 8a, and 8b at 60°C. The results show that IGBT (device V) with its anti-parallel diode, which is optimized for reverse recovery shows the best performance under reverse recovery. MOSFET B failed under this testing condition as is shown in Fig. 7c.

Rigorous two-dimensional (2-D) device simulations were performed to understand the internal plasma dynamics during transient switching conditions. The MOSFET structure obtained from simulation of device performance under static conditions was used in a 2-D mixed device and circuit simulator, in which semiconductor equations are solved within the device under boundary conditions imposed by external circuit elements [3]. Two similar structures were used in the reverse recovery test circuit of Fig. 7a. A good match was obtained between the measured and simulated results as is shown in Fig 8c.

The 350V - 3.5V, 225 A FB ZVS DC-DC converter measurement results in Fig. 9b show the variation of peak reverse recovery current as a function of load current, where by increasing the output load, the magnitude of peak reverse recovery current decreases.

V. CONCLUSIONS

This paper presents various electrical stresses in FB ZVS DC-DC converters with a detailed study of the MOSFET body diode recovery stress under normal and low load operating conditions. Experimental results show the variation of peak reverse recovery current, and reverse recovery energy loss at different current levels. The duration of the reverse recovery and the peak reverse recovery current is governed by both the intrinsic properties of the body diode and the external circuit in which the MOSFET is used. Two-dimensional finite element simulation results showed the dynamics of the device under reverse recovery in the FB ZVS DC-DC converter. The experimental results are supported by detailed two-dimensional numerical simulation results. It is shown that reverse recovery under no load conditions is stressful as opposed to the reverse recovery under normal operating condition.

REFERENCES

[1] L. Sao, R. Redl, K. Dierberger, "High-voltage MOSFET behavior in soft-switching converter: analysis and reliability improvements," in *Proc. IEEE Int. Telecommunications Energy Conf. (INTELEC)*, 1998, pp. 30-40.

[2] H. Aigner, K. Dierberger, and D. Grafham, "Improving the full-bridge phase-shift ZVT converter for failure-free operation under extreme conditions in welding and similar applications," in *Proc. IEEE Industrial Applications Society Annual Meeting*, 1998, vol. 2, pp. 1341-1348.

[3] J. A. Sabate et al., "Design considerations for high-voltage high-power full-bridge zero-voltage-switched PWM converter," in *Proc. IEEE Applied Power Electronics Conf. (APEC)*, 1990, pp. 275-284.

[4] B. J. Baliga, *Power semiconductor devices*, PWS Co., 1996.

[5] *Atlas users Manual*, Silvaco International, Santa Clara, CA.

[6] A. Pietkiewiz, and D. Tolik, "Operation of high power soft-switched phase-shifted full-bridge dc-dc converter under extreme conditions," in *Proc. IEEE Int. Telecommunications Energy Conf. (INTELEC)*, 1994, pp. 142-147.

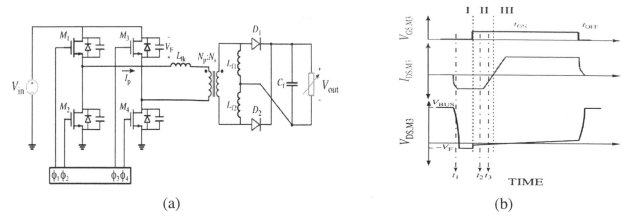

(a)   (b)

Fig. 1 Full bridge ZVS DC-DC converter (a) circuit, and (b) waveforms.

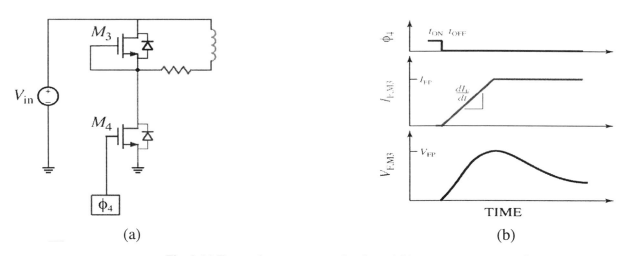

(a)   (b)

Fig. 2 (a) Forward recovery test circuit, and (b) measurement procedure.

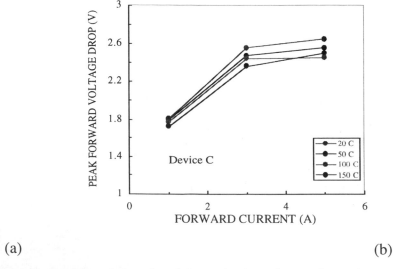

(a)   (b)

Fig. 3 Measurement results showing variation of peak forward voltage drop vs. forward current at different temperatures for a

power MOSFET (C).

144

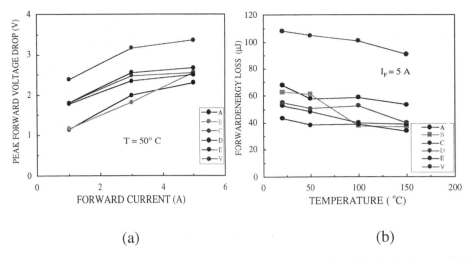

(a)   (b)

Fig. 4(a) Variation of peak forward voltage drop and (b) forward recovery energy loss for similarly rated MOSFETs (A-E), and IGBT (V).

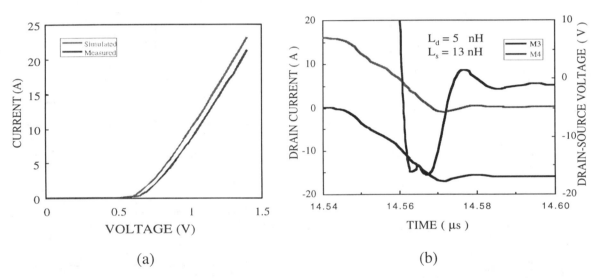

(a)   (b)

Fig. 5 2-D simulation results (a) body diode static iv and (b) during forward recovery transition.

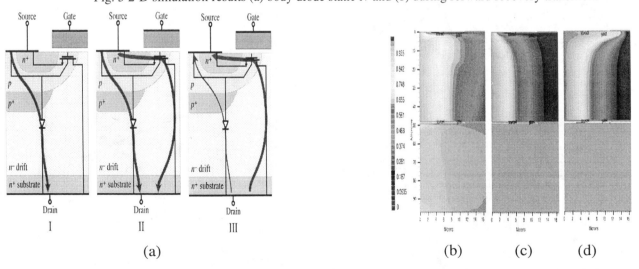

(a)   (b)   (c)   (d)

Fig. 6 (a) Current flow through M_3 and (b) finite element simulation results @t_1, (c) @t_2, and (d)@t_3.

145

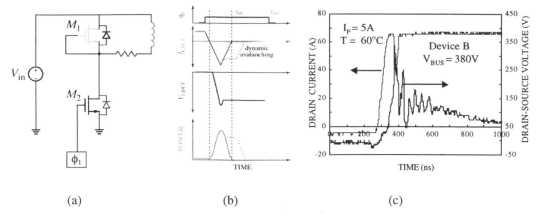

(a)   (b)   (c)

Fig. 7 Reverse recovery (a) test circuit, (b) waveform, and (c) device failure.

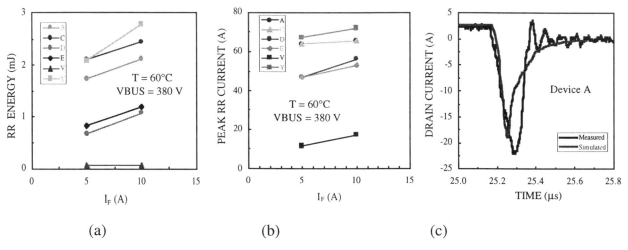

(a)   (b)   (c)

Fig. 8(a) Variation of peak reverse recovery (RR) current, (b) RR energy loss vs. forward current for similarly rated MOSFETs
(A-E) and IGBT (V), and (c) measurement and 2-D simulation results for MOSFET A.

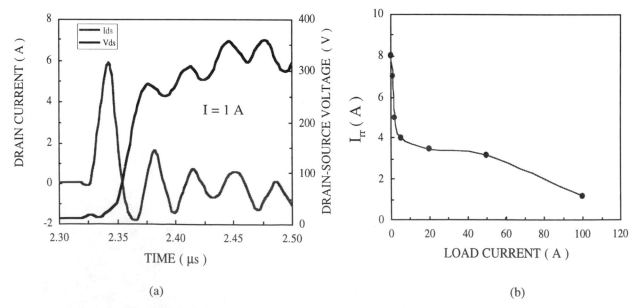

(a)   (b)

Fig. 9 (a) and (b) Converter measurement results under low load operation.

Proceedings of IECEC'01
36th Intersociety Energy Conversion Engineering Conference
July 29–August 2, 2001, Savannah, Georgia

IECEC2001-AT-32

A SURVEY OF POWER ELECTRONICS APPLICATIONS IN AEROSPACE TECHNOLOGIES

M. David Kankam
Power and On-Board Propulsion
Technology Division
NASA Glenn Research Center
Cleveland, OH 44135

Malik E. Elbuluk
Dept. of Electrical Engineering
University of Akron
Akron, OH 44325

ABSTRACT

The insertion of power electronics in aerospace technologies is becoming widespread. The potential and existing application areas summarized in this paper include the International Space Station (ISS), aircraft and satellite power systems, and motor drives in 'more electric' technology (MET) as applied to aircraft and Reusable Launch Vehicles (RLVs), starter/generators (S/G) and flywheel technology, servo systems embodying electromechanical actuation (EMA), and spacecraft on-board propulsion. Continued inroads of power electronics depends on eliminating variable frequency-fixed frequency incompatibility among aircraft conventional utility equipment. Dual-use electronic modules should reduce development cost.

I. THE POWER ELECTRONIC SYSTEM (PES) [1-2]

A power electronic system (PES) can comprise a modular power electronic subsystem (PESS) connected to a source and load at its input and output power ports, respectively. The third port of PESS is connected to the system control, as shown in Fig. 1. PESS has been described as a power device module, intelligent power module, smart power device, power control module and, more recently, power electronic building block (PEBB) of the Office of Naval Research (ONR)-initiated program [1-2].

The next three sections discuss the commonly used power electronic converters, their constituent semiconductor devices, and electric motor drive technologies which are enablers in the 'more electric' technology (MET) for space vehicle upgrades.

I-1. Power Semiconductor Devices

The semiconductor devices, Metal-Oxide-Semiconductor Field Effect Transistor (MOSFET), Insulated Gate Bipolar Transistor (IGBT), MOS-Controlled Thyristor (MCT) and Gate-Turn-Off

Thyristor (GTO), represent the cornerstone of modern power electronic converters. They are fully controllable. Their applications depend on their power and frequency characteristics.

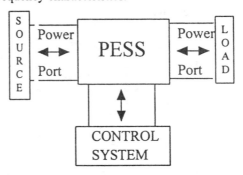

Fig. 1 Three-Port Power Electronic System Structure

The unipolar, voltage-controlled **MOSFET** is relatively very fast, requires only minimal snubbing due to its low switching losses, and is protected from breakdown by its inherent positive temperature coefficient of resistance. Its drain resistance increases with temperature. The device is used in applications of few MHz and few watts to few kilowatts, such as in voltage-source pulse-width-modulated inverter (PWMI) and a zero-voltage switching (ZVS) converter rated at 0.1-2kVA and 0.5-2MHz.

The **IGBT**, a hybrid of a MOSFET and a turn/off Bipolar Junction Transistor (BJT), is a MOS-gated device. When used in 0-300kVA, 1.5-5kHz voltage-source PWMI and 20-150kVA, 40kHz voltage-source converter drives, the IGBT requires minimum cooling, is more reliable, experiences reduced voltage spikes at turn-off, and exhibits improved thermal life.

The **GTO** is turned-on/off by gate current pulses. Turn-off-induced spike in the anode voltage can cause hot spots in, and breakdown of the device. A snubber can protect the GTO the high switching losses of which limit its use to 1.0-2.0kHz PWM. Example application areas are 0.75-1.5kVA, 500-600Hz voltage-source PWMI and 2-12MVA, 700-1200Hz ZVS inverter.

The **MCT** is a MOS-gated device with a high turn-off current gain. With switching speed comparable to that of IGBT, the MCT shows promise for application in snubbered voltage-source inverters at 0-100kVA and 2kHz, zero-current switching (ZCS) at 150kVA and 40kHz, and ZVS converter at 20-150kVA and 25kHz. The above devices are constituent parts of electronic converters which permeate aerospace power systems.

I-2. Power Electronic Converters [3-4]

Aerospace power systems have a considerable real estate of DC power usage. Over the past decade, AC power has emerged as a 'driver' for developing METs. This has increased the use of power electronic converters to condition and control power in the related systems. The high frequency converters of interest and their devices are noted in Fig. 2 [3].

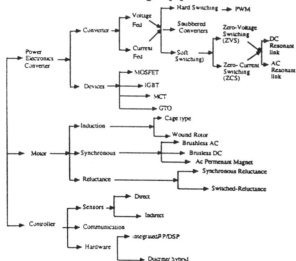

Fig.2 Options for 'More Electric' Technology (MET)

The semiconductor devices in a hard-switching **PWMI** are stressed during turn-on/off modes. Capacitive and inductive stray effects compound the resulting switching losses. The PWMI AC voltage output is harmonic-rich, variable-frequency (VF), and commercially used in motor drives. The PWMI is characterized by limited power range and low switching frequencies, with attendant acoustic noise and reduced efficiency. However, it has low Volt-Ampere rating, and is relatively easy to control.

The **snubbered converter** employs switching similar to that of the PWMI. It incorporates a series inductive snubber to limit the inrush-current though its devices. A parallel capacitive snubber limits the device voltage, and reduces device stress. High switching losses are dissipated in the snubber.

The **zero-current switching (ZCS) converter** uses an inductive snubber for device turn-off without current flow. Similarly, the **zero-voltage switching (ZVS) converter** employs a capacitive snubber for

device turn-on, with an anti-parallel diode conducting. The converters have high switching frequency, lossless devices, high efficiency and reliability.

The **Resonant DC-Link (RDCL)** in Fig. 3 and **Resonant AC Link (RACL) converters** overcome the limitations of PWMI. The RDCL converter has low switching losses, heat dissipation and acoustic noise, higher operating frequency and reliability, and reduced dv/dt and di/dt, resulting in low EMI. Eliminating the Resonant Tank from Fig. 3 yields the conventional hard-switched, voltage-source PWMI [3].

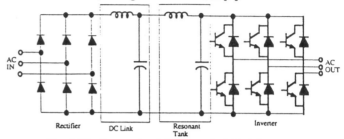

Fig. 3 The Conventional Resonant DC Link Inverter

DC-DC converters have been used in aerospace power systems to provide the required voltage for the secondary distribution network. The basic topologies are the **step-down (buck)** and **step-up (boost) converters** for various load requirements [4]. For a given input voltage, the average output voltage is obtained by switching the electronic devices at constant frequency, while adjusting the device 'ON' duration.

Electronic converters constitute the heart of motor drives which are essential for electromechanical actuation (EMA) in MET.

I-3. Motor Drive Technologies [5-12]

The electric motor is the workhorse in a drive system. Drive characteristics depend on the motor, the power circuit, electronic devices and the controller. Power rating, operating speed range, environment, fault tolerance, reliability, performance, thermal capability and cost affect motor selection for an application [5]. In the past, the excellent drive performance and low initial cost of DC machines made them the primary choice for servo applications. Their built-in commutators, high maintenance and spark-inducing brushes hinder DC machine use in drives. "Brushless" motors have emerged from coupling DC, AC synchronous and induction motors with electronic controllers. The resulting maintenance- and spark-free brushless DC machine (BLDCM) or permanent magnet synchronous motor (PMSM), switched reluctance motor (SRM) and induction motor (IM) have higher torque/inertia ratio, peak torque capability, power density and reliability than the DC brush motor.

The permanent magnet in the rotor of a BLDCM produces an armature current-independent field. The commutatorless BLDCM has laterally stiff rotor which permits higher speed, especially in servo applications. The intra-stator placement of the rotor improves heat conduction which increases electric loading, and yields a higher torque/amp, better effective power factor and higher efficiency. BLDCM-based dives are popular, due to their performance and price improvements. Disadvantages include the need for shaft position sensing and more complex electronic controller.

The IM has been the traditional workhorse for fixed and variable speed drive applications. It is rugged, relatively inexpensive and almost maintenance-free. The rotor slip-dependent torque production worsens the performance and decreases the efficiency of the motor. In the fractional and low integral hp range requiring dynamic performance, high efficiency and a wide speed range, the complexity of the induction motor drive favours the BLDCM.

The gear-less, direct-application SRM drives are widely accepted. Their applicability to aircraft engine starter/generator has been demonstrated [6-7]. The motor design and converter topologies of the drive have undergone significant research and development in over two decades [8-12]. The series connection of the converter phase-leg switches to the motor phase winding prevents shoot-through fault by the converter switches. The motor is economical, compact in construction, and has high torque-to-inertia ratio, high torque output at low-to-moderate speeds, and faster response in servo systems. Its drawbacks are higher torque ripple and acoustic noise, complex control, and the need for an absolute rotor position sensor for the controller to establish the phase current pulses.

II. AEROSPACE POWER SYSTEMS [2]

The advent of MET for aerospace systems has focused attention on AC-, and hybrid DC- and AC-PMAD-based systems. The schematics in Figs. 4 and 5 show a commonality in the use of electronic converters, photovoltaic (PV) solar arrays and batteries [2]. In aerospace systems, PEBB-related integration issues are the level of power and frequency range, application- and mission-dependent extreme temperature range, weight and size, electromagnetic interference (EMI) and performance. Resolution of these issues is expected to promote expeditious insertion of electronic modules in aerospace technologies.

III. POWER ELECTRONICS APPLICATION IN AEROSPACE TECHNOLOGIES

The International Space Station (ISS), satellite power systems, MET, starter/generator system, reusable launch vehicles, flywheel technology and on-board electric propulsion are discussed to highlight the important role of power electronics in these systems.

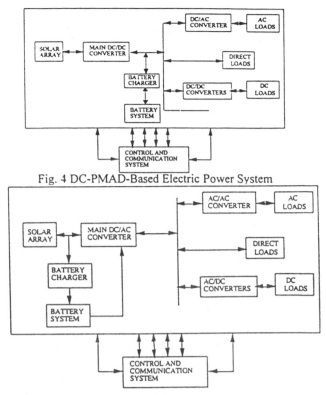

Fig. 4 DC-PMAD-Based Electric Power System

Fig. 5 AC-PMAD-Based Electric Power System

III-1. Space Station Power System [13-14]

A single channel diagram, Fig. 6, of the ISS electric power system (EPS) shows a DC network of PV solar arrays, batteries, power converters, switches and user loads [13]. The networks of 120V American and 28V Russian can exchange bi-directional power flow via American-to-Russian Converter (ARCU) and Russian-to-American Converter (RACU) units. The primary distribution system (PDS) comprises the PV arrays, batteries and the network up to the DC-DC converter units (DDCUs), for 160V-to-120V step-down to the secondary distribution system (SDS). The sequential shunt unit (SSU) regulates the voltage output of the PV array. The DDCUs isolate the PDS and SDS from each other, and condition the source power for the SDS. The batteries store energy during insolation periods, and supply load power during orbital eclipse. The battery charge/discharge units (BCDUs) isolate the battery from the primary bus. Remote power controller modules (RPCMs), or switchgears, distribute power to the load converters. The BCDUs, DDCUs and the load converter contain semiconductor switches in their circuitry, thus underscoring the importance of power electronics in the ISS EPS.

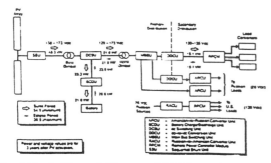

Fig. 6 Single Channel Diagram of ISS Power System

In their modular form, power electronics are expected to facilitate control of PMAD and diagnosis of system malfunctions, to yield reliability improvement [14]. Some level of electronic modularity has already been built into some satellite EPS, to permit adaptability of the EPS to various and future programs, with minimal re-design.

III-2. Satellite Power Systems [15]

A partial block diagram of a satellite modular EPS is shown in Fig. 7. It depicts only the north portion of a 'Dual Bus' power system for a geosynchronous satellite [15]. The primary-side elements of the EPS are the PV arrays, battery and power control unit (PCU). On the secondary side, the PCU embodies the SSU, battery charge and discharge converter modules (BCCM, BDCM), and low voltage converter module (LVCM) of redundant, main bus-connected DDCUs which feed the spacecraft loads via the power distribution unit (PDU). The operation of the sarellite

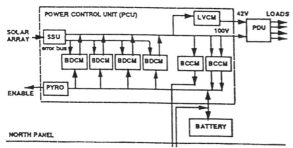

Fig. 7 Block Diagram of Satellite Power System

EPS is similar to that of the ISS, regarding sunlight and eclipse portions of a mission. The built-in modularity makes it possible to vary battery voltage and output power levels, by adding and removing converter module(s). This feature renders the EPS configurable for future missions. Also, the modularity facilitates power electronics packaging, equipment deployment into space, and needed on-orbit EPS modifications.

The use of power electronics-based motion control systems in selected aerospace systems is discussed next.

III-3. Motor Drive Applications [5, 14, 16-33]

The key elements in electric actuation (EA) for MET are the electric motor, its power electronics, the control system and the actuator load(s) [5]. Lower costs and advances in power electronics and high-speed electric machines have fuelled the interest of technologists, developers and researchers in industry [16-19], Government Agencies [20-23], and academia [24-27], in aerospace motion control systems. A key premise of the MET is to replace the traditionally mounted auxiliary drives and bleed air extraction with integral engine starter/generators (S/Gs), electrical-driven actuators and engine-gearbox-driven fuel pumps. The replacement eliminates hydraulic, pneumatic and mechanical power, and minimizes and/or eliminates their associated costs, as well as high pre-flight operation, maintenance and refurbishment of hydrazine-driven auxiliary power units (APUs).

III-3.1. More Electric Technology

Figure 8 shows a conceptual diagram of the Air Force's 'more electric aircraft' (MEA) subsystems [20]. The hydraulic-driven flight control actuators, the engine-gearbox driven fuel pump and air-driven environmental control system (ECS) are electrically powered by electric motor drives. A S/G supplies electric power to a fault-tolerant PMAD subsystem, which feeds power to the EA, engine starting, braking, ECS, fuel pump and anti-icing. Uninterrupted power from an integrated APU and battery system provides redundancy and engine start-up.

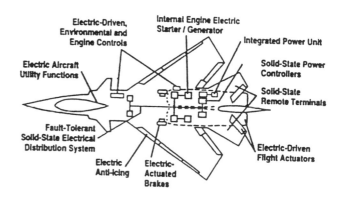

Fig. 8 Concept of More Electric Aircraft

The candidate electric motor drives for the MET are the IM, BLDCM and SRM drives. For each selected drive in the MET, depending on the application, the EA and its electronic controller must suitably match the safety and reliability of hydraulic actuation. This may require motor drive redundancy, for assured flight and landing. Thus, the built-in redundancy in its independent motor windings due to magnetic isolation, and in power switching circuits by electrical isolation, makes the SRM an attractive choice for fault-tolerant EA. This 'limp home' ability of the SRM has been one of the key factors in its selection by the Air Force in their MEA development [21-22].

III-3.2. Starter/Generators

A general variable speed constant frequency (VSCF) S/G system is shown in Fig. 9 [23]. The machine may be any of the three candidates in Section III-3. During motoring, constant frequency (CF) electrical power from the main AC bus is converted to VF by the bi-directional power converter, and fed to the machine to start the load such as an aircraft engine. In the generating mode, the variable speed load

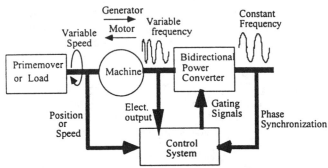

Fig. 9 Typical VSCF Starter/Generator System

provides mechanical power to run the machine the variable frequency of which is converted to a constant frequency for the main bus. The control system receives inputs from the VSCF sources, and provides gating signals for the converter to maintain proper interface between VF and CF requirements.

III-3.3. Reusable Launch Vehicle

EMA has been under consideration for replacing hydraulic systems used on RLVs for thrust vector control (TVC) gimbaling of engines and aerosurface control [28-29]. The projected benefits are as stated in Section III-3. In the early 90's, NASA Glenn (GRC) (formerly Lewis) Research Center and General Dynamics Space Systems Division (GDSS) cooperated on demonstrating EMA technology readiness to meet the hydraulic TVC requirements for the Atlas Expendable Launch Vehicle (ELV) [28]. Concurrently, a study by the NASA Kennedy Space Center indicated that the MET would save nearly sixty-six percent of man-hours needed for hydraulic TVC processing of the Shuttle Solid Rocket Booster [30]. The EMA in Ref. [28] embodied IM drive with a field-oriented control

(FOC) [5, 31-32] for independent control of torque and flux, and a Pulse Population Modulation (PPM) technique [28] for independent control of voltage and frequency.

A more advanced motor drive is in use for on-going Government-Industry development of flywheel technology.

III-3.4. Flywheel Technology

US Government Agencies, Industry and academia are jointly developing advanced flywheel technologies to provide high performance and reliability, and reduced losses in a high-speed, light weight flywheel energy storage system (FESS), peak power and load leveling in spacecraft and aircraft applications [14]. The FESS embodies advanced composite materials-based rotor, low-loss magnetic bearings, high-speed motor-generator set and electronic converter drive. The above efforts and concurrent component technology developments represent an advancement of prior work by the collaborators [33]. NASA GRC is currently leading an effort for a combined FES and attitude control system (ACS), namely, integrated power and control system (IPACS), to enable development of a low cost, lightweight and higher specific energy spacecraft.

Additionally to conditioning power and enabling bi-directional power and stored energy flow in flywheel systems, motor drives feature in electric upgrade of aircraft pumps for the same MET benefits.

III-3.5. Servo System Applications

The MET proposes the use of VF motor drives to operate hydraulic and fuel pumps on aircraft. Using high-density motor drives can eliminate the usual size and weight limitations of drives. However, issues of VF incompatibility with 400Hz-operated aircraft equipment such as fuel and hydraulic pumps [19], attendant increase in motor weight to achieve the required torque at high frequencies, and potentially high upgrade cost must be resolved.

By comparison with fixed frequency power, VF motor controllers can reduce transient inrush current at motor start. Furthermore, a variable speed motor-driven fuel pump can provide only the required amount of fuel. Also, such a fuel pump can improve aircraft performance by reducing engine gearbox weight and enabling direct integration with the aircraft electronic propulsion and flight control [16].

Besides aerospace power systems, electronic converters play an important function in the on-board electric propulsion of spacecraft.

III-4. On-Board Electric Propulsion [34-37]

Power electronics reside in the power processing unit (PPU) of spacecraft electric propulsion (EP) which is credited with reducing launch vehicle requirements, notably for north-south station keeping of commercial geosynchronous (GEO) satellites [34]. A PPU comprises one or more electronic converters. It provides electric power for the spacecraft thruster, and commands and telemetry interface to the electric propulsion system, as shown in Fig. 10 [35]. The converters may be current-controlled and voltage-fed, to rapidly supply constant current to offset thruster voltage variations, typically during start-up period.

Small-sized PPUs with lightweight, and high efficiency via reduced number of high frequency, soft-switching converters, yield increased payload and power [36]. A highly efficient PPU can generate high voltage start pulse to ignite as many as four arcjet thrusters for north/south station keeping orbit maneuvers [37], thus reducing propulsion system mass.

Several challenges must be overcome for continued penetration of power electronics into aerospace systems.

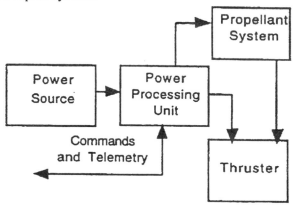

Fig. 10 Electric Propulsion System

IV. FUTURE TRENDS [14,19,24,38]

VF is currently used in turbo-prop and business jets. The cost and savings attraction of VF is tempered by the potential high cost of VF-upgrade for conventional, 400Hz-operating equipment on aircraft. Reference [19] points out judicious use of power electronics, via a hybrid hydraulic/pneumatic/motor drive design, to circumvent the VF-fixed frequency equipment incompatibility.

Continued improvements in power electronic devices and their switching schemes, advances in magnetic materials and capacitors, and better design of motors and electronic controls are expected to ameliorate weight, size and reliability issues of MET application to aerospace systems.

The need for bi-directional converters for battery charge/discharge functions and fixed frequency power and voltage, and expected varying requirements of multiple loads in aerospace systems suggest future use of hybrid AC and DC multi-converters with multi-voltage levels [24].

Increasing use of power electronic modules will require consideration of device ratings, bi-directionality or otherwise of power flow, power density requirements and degree of integration, when developing aerospace systems. Hardware commonality will promote dual-use application of the modules, and decrease system development cost [14]. For instance, NASA-planned development of 2-3kW power processor/thruster is expected to provide modular elements for various mission requirements [38].

V. CONCLUSIONS

This paper presents a survey of power electronics applications in aerospace technologies. It encompasses the International Space Station, satellite and aircraft power systems, flywheel technology, spacecraft on-board propulsion, and the MET insertion in spacecraft, aircraft and launch vehicles.

Power electronic converters are central to the performance of aerospace power systems and spacecraft on-board electric propulsion. Resolution of incompatibility between conventional, 400Hz operating equipment and the VF of MET should promote increased penetration of power electronics into aerospace systems. Future multi-voltage needs and varied load requirements will necessitate the use of multi-voltage level converters. The use of electronic modules with dual-use options and hardware commonality for aircraft and spacecraft should reduce development cost and maximize system re-use, while improving system reliability and performance.

ACKNOWLEDGEMENTS

The authors sincerely thank Dr. Narayan V. Dravid, Eric J. Pencil and Linda M. Taylor for their review and constructive comments. We acknowledge Marie DiNovo for her assistance in processing the diagrams.

REFERENCES

[1] "Power Electronics Building Blocks (PEBBs)", Project Report by Virginia Power Electronics Center, April '96.

[2] Elbuluk, M.E., Kankam, M.D., "Power Electronics Building Blocks (PEBBs) in Aerospace Power Electronics Systems", Presented at the '34th Intersociety Energy Conversion Engineering Conf. (IECEC)', Vancouver, BC, Canada, Aug. 1-5, '99.

[3] Elbuluk, M.E., Kankam, M.D., "Motor Drive Technologies for the Power-By-Wire (PBW) Program: Options, Trends and Tradeoffs, Part II: Power Electronic

Converters and Devices", IEEE Aerospace and Electronics Systems Magazine, Dec. '95, pp. 31-36.

[4] Mohan, N., Underland, T.M., Robbins, W.P., "Power Electronics: Converters, Applications, and Design", Book, John Wiley & Sons, 1989.

[5] Elbuluk, M.E., Kankam, M.D., "Motor Drive Technologies for the Power-By-Wire (PBW) Program: Options, Trends and Tradeoffs, Part I: Motors and Controllers", IEEE AES Magazine, Nov. '95, pp 37-42

[6] MacMinn S.R., Sember J.W., "Control of a Switched-Reluctance Aircraft Engine Starter/Generator Over a Very Wide Speed Range", IEEE/IECEC Record, '89, pp. 631-638.

[7] MacMinn S.R., Jones W.D., "A very High Speed Switched-Reluctance Starter/Generator for Aircraft Engine Application", IEEE/IECEC Record, '89, pp. 1758-1764.

[8] Ray, W.F., et.al, "High Performance Switched Reluctance Brushless Drives", IEEE Trans. on Ind. Electronics, Vol. 22, #4, July/Aug. '86.

[9] Miller, T.J., "Brushless Permanent Magnet and Reluctance Motor Drives", Oxford Univ. Press, '89.

[10] Acarnley, P., Hill, R., Cooper, C., "Detection of Rotor Position in Stepping and Switched Reluctance Motors by Monitoring of Current Waveforms", IEEE Trans. on Ind. Electronics, Vol. 32, # 3, Aug. '85, pp. 215-222.

[11] Harris, W.D., Lang, J., "A Simple Motion Estimator for Variable Reluctance Motor", IEEE Trans. on Ind. Application (IA), Vol. 26, #2, Mar'90, pp. 237-243.

[12] MacMinn, S.R., Szczesny, Rzesos, W.J., Jahn, T.M, "Application of Sensor Integration Technique to Switched Reluctance Motor Drives", IEEE Trans. on IA, Vol. 28, #6, Nov./Dec. '92, pp. 1339-1344.

[13] Gholdston, E.W., Karimi, K., Lee, F.C., Rajagopalan, J., Panov, Y., Manners, B., "Stability of Large DC Power Systems Using Switching Converters, with Application to the International Space Station", Proc. of 31st IECEC, Aug. '96, pp. 166-171.

[14] Kankam, M.D., Lyons, V.J., Hoberecht, M.A., Tacina, R.R., Hepp, A.F., "Recent GRC Aerospace Technologies Applicable to Terrestrial Energy Systems", Proc. of 35th IECEC, Las Vegas, NV, July 24-28, '00, pp. 865-875.

[15] Canzano, S.M., Webber, H.F., Applewhite, A.Z., Hosick, D.K., Pollard, H.E., "A Modular Multi-Mission Electrical Power Subsystem for Geosynchronous Satellites", Proc. of 30th IECEC, Aug. '95, pp. 369-374.

[16] Radun, A.V., "High-Power Density Switched Reluctance Motor Drive for Aerospace Applications", IEEE Trans. on IA, Vol. 28, # 1, Jan./Feb., '92, pp. 113-119.

[17] Mohamed, F., "Use of a Variable Frequency Motor Controller to Drive AC Motor Pumps on Aircraft Hydraulic Systems", IECEC, # AP-17, ASME, '95.

[18] Blanding, D.E., "An Assessment of Developing Dual Use Electrical Actuation Technologies for Military Aircraft and Commercial Application", Proc. of '95 IECEC, pp. 716-721.

[19] Lazarovich, D., et. al., "Variable Frequency Use in Aerospace Electrical Power Systems", IECEC Paper # 1999-01-2498.

[20] Reinhardt, K.C., Marciniak, M.A., "Wide-Bandgap Power Electronics for the More Electric Aircraft", Proc. of 31st IECEC, Aug. '96, pp. 127-132.

[21] Fronista, G.L., Bradbury G., "An Electro- Mechanical Actuator for a Transport Aircraft Spoiler Surface", Proc. of 32nd IECEC, Aug. '97, pp. 694-698.

[22] Cloyd, J.S., "Status of the United States Air Force's More Electric Aircraft Initiative", IEEE AES Magazine, April '98, pp. 17-22.

[23] Elbuluk, M.E., Kankam, M.D., "Potential Starter/Generator Technologies for Future Aerospace Applications", IEEE AES Magazine, Vol. 11, #10, Oct. '96, pp. 17-24.

[24] Emadi, A., Ehsani, M., "Electrical System Architectures for Future Aircraft", IECEC Paper # 1999-01-2645.

[25] Emadi, A., Fahimi, B., Ehsani, M., "On the Concept of Negative Impedance Instability in the More Electric Aircraft Power Systems with Constant Power Loads", IECEC Paper # 1999-01-2545.

[26] Sul, S.K, Lipo, T.A., "Design and Performance of a High Frequency Link Induction Motor drive Operating at Unity Power Factor", IEEE Trans. on IA, Vol. 26, #3, May/June, '90, pp. 434-440.

[27] Sul, S.K, Alan, I., Lipo, T.A., "Performance Testing of a High Frequency Link Converter for Space Distribution System", Proc. of IECEC, '89, pp. 617-623.

[28] Burrows, L.M., Roth, M.E., "An Electromechanical Actuation System for an Expendable Launch Vehicle", Proc. of 27th IECEC, Aug. '92, pp. 1.251-1.255.

[29] Hall, D.K., Merryman, S.A., "Hybrid Electrical Power Source for Thrust Vector Control Electromechanical Actuation", Proc. of 30th IECEC, Aug. '95, pp. 393-397.

[30] McCleskey, "Interim Report of the Launch Site Electric Actuation Study Team", Jan. '92.

[31] Divan, D.M., Lipo, T.A., Lorenz, R.D., Novotny, D.W., "Field-Orientation and High Performance Motion Control", Summary of Publications: '81-'88, WEMPEC.

[32] Bose, B.K., "Power Electronics: An Emerging Technology", IEEE Trans. on Ind. Electronics, Vol. 36, #3, Aug. '89.

33] Christopher, D., Lt. Donet, C., "Flywheel Technology and Potential Benefits for Aerospace Applications", Proc. of IEEE Aerospace Conf., '98, pp. 159-166.

[34] Curran, F.M., Schreiber, J.G., Callahan, L.W., "Electric Power and Propulsion: The Future", Proc. of 30th IECEC, Aug. '95, pp. 437-441.

[35] Hamley, J.A., "Direct Drive Options for Electric Propulsion Systems", 30th IECEC, July-Aug. '95

[36] Pinero, L.R., et.al., "Development Status of a Processing Unit for Low Power Ion Thrusters", Presented at 36th Joint Propulsion Conference and Exhibit, Huntsville, AL, July '00.

[37] Hamley, et. al., "The Design and Performance Characteristics of the NSTAR PPU and DCIU", Paper # AIAA 98-3938.

[38] Sovey, J.S., et. al., Development of an Ion Thruster and Power Processor for New Millennium's Deep Space 1 Mission", Presented at 33rd Joint Propulsion Conference and Exhibit, Seattle, WA, July '97.

IECEC2001-AT-09

DESIGNING ELECTRONIC POWER CONVERTERS FOR SPACE APPLICATIONS

Biswajit Ray
Department of Physics and Engineering Technology
Bloomsburg University
Bloomsburg, PA 17815
Email: bray@bloomu.edu

ABSTRACT

Issues related to the design of electronic power converters for space applications are addressed in this paper. Meeting the customer's specifications for a space level design requires a detailed worst case circuit analysis taking into account the component parameter variations due to initial, temperature, life, and radiation conditions. Part selection becomes heavily dependent on the availability of total ionizing dose (TID) and single-event effects (SEE) radiation data. A simple primary-side controlled PWM dc-dc power converter is considered as an example to illustrate the importance of detailed analysis.

INTRODUCTION

Designing a power converter for space applications includes several engineering areas including electronics design and worst case circuit analysis, radiation hardened part selection and part derating, radiation analysis, thermal and packaging design, reliability analysis, and failure-mode effects and criticality analysis. Part selection becomes an important part of the design process due to space radiation environments' impact on various part parameters. A preliminary radiation analysis is essential to estimate a worst case total ionizing dose at the part level. Single-event effects also play an important role in the part selection process. Once the initial part level temperature and radiation levels are known, a detailed worst case circuit analysis is warranted to make sure that the customer's specifications are met under all electrical and environmental conditions. A detailed part degradation database needs to be developed to account for initial, temperature, life, and radiation tolerances for parameters of interest for various components. In case of unavailability of radiation data for parts, the designer becomes the lead person in making sure the devices get tested for TID and/or SEE.

The space radiation environment is discussed next as it relates to the electronics design. A primary-side current-mode controlled PWM dc-dc power converter is considered in the following sections to illustrate the worst case circuit analysis process. The SEE and enhanced low dose rate sensitivity issues related to power converter design are discussed as well. Importance of a detailed analysis to ensure performance specifications are met under worst case conditions before the design is released to manufacturing can not be overstated in terms of its impact on project schedule and cost.

SPACE RADIATION ENVIRONMENT

The space radiation environment poses a risk to all earth orbiting satellites and missions to other planets. Charged particles in this environment consist primarily of high-energy electrons, protons, alpha particles, and heavy ions (cosmic rays). The radiation effects of these charged particles are dominated by ionization in electronic devices and materials. Energy deposited in a material by ionizing radiation is expressed in "rads", with one rad equal to 100 ergs/gram. However, energy loss per unit mass differs from one material to another because of the atomic differences in various materials. For semiconductor devices, the unit of absorbed dose is rads (Si). There are two types of radiation damage induced by charged particle ionization in the natural space environment: total ionizing dose[1,2] (TID) and single-event effects[3,4] (SEE). The TID effects are cumulative ionization damage caused by the charged particles passing through a semiconductor device[1,2]. For MOS devices, this ionization traps positive charges in the gate oxide and produces interface states in silicon at the $Si\text{-}SiO_2$ interfaces. These effects cause threshold voltage shifts and decreased channel carrier mobility, resulting in increased leakage current and power supply current, and possible loss of device functionality. For bipolar devices, ionization adversely affects current gain and junction leakage currents, causing significant degradation in device performance. This leads to increased offset voltage and bias current in op-amps and comparators, and loss of accuracy and functionality in analog-to-digital and digital-to-analog converters.

Single event effects[3,4] (SEE) are caused by a high-energy single ion (heavy ion or energetic proton) passing through a device. SEE include single event upsets (SEU), single event latchup (SEL), single event burnout (SEB), and single event gate rupture (SEGR). While SEU are non-destructive and do not cause permanent damage to the device, the other single event effects can be destructive. SEU occur due to either the deposition or depletion of charge by a single ion at a circuit node, causing a change of state in the memory cell (bit upset). In very sensitive devices, a single ion hit can cause multiple bit upsets in adjacent memory cells. However, these SEUs cause no permanent damage, and auto-correcting circuits can be implemented in many designs. SEL can occur in any semiconductor device that has a parasitic n-p-n-p path. A single heavy ion or high-energy proton can initiate regenerative action. This leads to excessive power supply current and loss of device functionality. Device burnout may occur unless the current is limited or the power to the device is recycled. SEL is of most concern in bulk CMOS devices. The SEB and SEGR may occur in MOSFETs, however, they are avoidable by design as long as the applied drain and gate voltages are properly derated.

A radiation risk assessment for any electronic device includes the determination of TID and SEE susceptibility of the device caused by the projected radiation environment of the spacecraft. It should be noted that the TID on a device can vary significantly with the amount of shielding interposed between the device and the outside environment, however, the SEE susceptibility do not change significantly with shielding[3,4].

Figures 1 and 2 below show typical TID and SEE environments for a geo-stationary orbit.

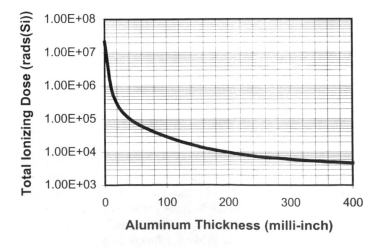

Fig. 1 Typical total ionizing dose for a geostationary orbit

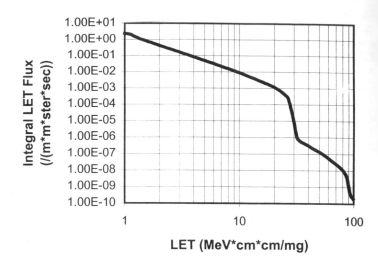

Fig. 2 Typical single-event effects environment for a geostationary orbit

TID testing of devices is generally performed by exposing devices to gamma rays from a Co-60 source with a dose rate of typically 50-300 Rads/sec, per MIL-STD-883. However, dose rates in the natural space environment are very low (0.1 to 10 mrads/sec). It is not feasible to simulate the low dose rate of space environment during ground testing. However, more and more bipolar linear devices such as op-amps, comparators, and linear regulators are tested at low dose rates since the recent discovery of enhance low dose rate sensitivity[5,6] (ELDRS). As a designer, either we select ELDRS tested parts or design very conservatively especially with input offset voltage and bias currents for linear bipolar ICs.

SWITCHING POWER SUPPLY DESIGN

A primary-side electronic power converter (EPC) is considered here for space level design. Block diagram representation of a current-mode controlled PWM power converter is shown in Fig. 3. The output of the EMI filter feeds the startup regulator as well as the main power stage. At startup, the startup regulator supplies power to the PWM controller. Once the switching regulator is up and running, the startup regulator is disabled due to diode or-ing with the boot winding output. The main output of the converter is linearly post-regulated. This control scheme is acceptable for low power application. Higher power applications will require secondary-side controller necessitating gate-drive isolation through either magnetic or optical technology.

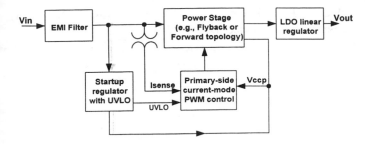

Fig. 3   Block diagram representation of a primary-side
controlled switching power converter

At the onset of the design, thermal as well as radiation environments need to be known. Through preliminary power loss estimate and the packaging design, a worst case board temperature shall be established. The total dose radiation level at the part level and the SEE environment need to be known for proper part selection and circuit design. Mission life becomes an important part as well for estimating end-of-life (EOL) performance specifications. Part derating and reliability analyses are heavily dependent on thermal design of the unit, making the EPC design an engineering team effort.

The start-up design issues are discussed here for a current-mode controlled primary-side PWM flyback converter as shown in Fig. 4.

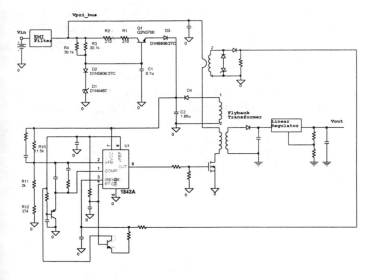

Fig. 4 Primary-side controlled flyback dc-dc converter

The following conditions must be met under worst-case conditions to achieve a successful startup as well as continuous operation:

- The minimum output voltage of the startup regulator shall be greater than the maximum UVLO of the PWM IC,
- The maximum output of the startup regulator shall be less than the minimum output voltage of the flyback boot winding, and

- The dc current gain of the startup regulator transistor (Q1) shall be enough to support the approximately 25 mA of current needed by the PWM IC for successful startup of the power converter.

Part Degradation Database

For the above mentioned analysis, a part degradation database needs to be prepared that takes into account part parameter tolerances due to initial, temperature, life, and radiation effects. Tolerances due to initial and temperature can be obtained form component datasheet. For this example, the minimum case/junction temperature is assumed to be −35°C, and maximum case/junction temperature is assumed to be 75°C for passive components and diodes. For the bipolar transistor and the PWM IC, the maximum junction temperature is assumed to be 90°C.

Life data is usually available from military and commercial databases[7]. Since life data is valid only for a given temperature and mission life, available data needs to be adjusted based on operating temperature of the component and mission life. For this example, a mission life of 18 years is assumed. The temperature effect on life data can be evaluated based on Arrhenius aging model given below.

$$Aging\_factor_{T_2} = Aging\_factor_{T_1} \bullet e^{\left[\frac{E_A}{K} \left(\frac{1}{T_1+273} - \frac{1}{T_2+273} \right) \right]} \quad (1)$$

where, K = Blotzman's constant = 8.62e-5 eV/°K,
E_A = activation energy (e.g., 1.1 eV for semiconductors), and
T_1 and T_2 are temperatures in °C.

Next, the radiation data for semiconductors needs to be taken into account knowing the TID level at the parts. For the analysis presented, a worst-case TID level of 50 krads is assumed. Several radiation databases are available including ERRIC. For this example, radiation induced parameter degradation for 2N3700[8] and UC1843A[9] are considered.

For 2N3700, the dc current gain degradation is given by,

$$h_{FE,post\_rad} = \frac{1}{\dfrac{1}{h_{FE,pre\_rad}} + \Delta\left(\dfrac{1}{h_{FE}}\right)} \quad (2)$$

where,

$$\Delta\left(\frac{1}{h_{FE}}\right) = \frac{1}{h_{FE,post\_rad}} - \frac{1}{h_{FE,pre\_rad}} \quad (3)$$

Table I in the next page summarizes the dc current gain radiation data[10] for 2N3700.

Table I Post-radiation dc current gain of 2N3700
(Ic=10mA; Vce=10V; ERRIC Device ID: 2N3700 17)

Device #	Initial	50 kRads	Delta (1/hFE)
3	133	106	0.001915165
4	133	108	0.001740462
5	130	102	0.002111614
6	129	100	0.002248062
7	128	102	0.001991422
Mean	130.6	103.6	0.002001345
StdDev	2.302173	3.28633535	0.000192752
Mean+3.4*Sigma	137.5065	113.459006	0.002579602
Mean-3.4*Sigma	123.6935	93.740994	0.001423088
Minimum initial	90 (from MIL-PRF-19500/391E data sheet)		
HFE post-rad	73.04222		

Next, the radiation data[10] for UC1843A reference voltage and error amplifier bias current are shown in Tables II and III, respectively.

Table II Radiation data for UC1843A reference voltage
(ERRIC Device ID: 1843 12)

Device #	Initial (V)	75 kRads (V)	Delta (V)
11	5.03E+00	5.03E+00	1.39E-03
12	5.01E+00	5.01E+00	1.17E-03
13	5.02E+00	5.02E+00	1.19E-03
14	5.01E+00	5.01E+00	1.13E-03
15	4.99E+00	4.98E+00	-5.16E-03
16	5.01E+00	5.01E+00	1.07E-03
17	5.02E+00	5.03E+00	7.72E-03
18	5.01E+00	5.01E+00	9.33E-04
19	5.01E+00	5.00E+00	-5.53E-03
Mean	5.01222222	5.012657	0.00043478
StdDev	0.01092906	0.013699806	0.0039229
Mean+3.0*Sigma	5.04500941	5.053756418	0.01220347
Mean-3.0*Sigma	4.97943503	4.971557582	-0.0113339

Tolerance   ± 0.24347 %

Table III Radiation data for UC1843A error amplifier bias current (ERRIC Device ID: 1843 12)

Device #	Initial (uA)	75 kRads (uA)	Delta (uA)
11	-1.70E-01	-2.49E-01	-7.86E-02
12	-2.42E-01	-2.72E-01	-3.00E-02
13	-1.94E-01	-2.42E-01	-4.82E-02
14	-1.82E-01	-2.48E-01	-6.64E-02
15	-2.55E-01	-2.49E-01	6.40E-03
16	-1.94E-01	-2.48E-01	-5.43E-02
17	-1.33E-01	-2.30E-01	-9.68E-02
18	-1.76E-01	-2.49E-01	-7.25E-02
19	-2.55E-01	-2.73E-01	-1.79E-02
Mean	-0.20011111	-0.251033333	-0.050922222
StdDev	0.042093481	0.013598254	0.03242078
Mean+3.0*Sigma	-0.07383067	-0.210238573	0.046340116
Mean-3.0*Sigma	-0.32639155	-0.291828094	-0.148184561

Tolerance   ± 74.05114 %

Once the components initial, temperature, life, and radiation tolerances are known, a worst case tolerance needs to be established. Depending on the directional or random nature of the tolerances, the worst case tolerance can be computed based on one of following three methods: root-sum-square (RSS), extreme value analysis (EVA), and combination approach. EVA is the most rigorous while the RSS yields the least robust design of the three approaches. The combination approach is based on the concept that some part parameter variations are random and some are directional. Table IV in the next page presents the part degradation database used in worst case analysis of the example converter.

Table IV Part degradation database for the example power converter

Part	Ref Des.	Parameter	Nom. Value	25C Max Value	25C Min Value	Max Temp (deg C)	Min Temp (deg C)	Temp Coeff.	Temp Coeff. Type	Aging Max (%)	Aging Min (%)	RADIATION Value	RADIATION Rad Level	DESIGN Max Limit	DESIGN Min Limit
M55342K06B11E5S	R10	Resistance (Ω)	11500	11615	11385	75	-35	0.01	Linear (%/°C)	0.03414	0	0		11676.43	11316
M55342K06B2E00S	R11	Resistance (Ω)	2000	2020	1980	75	-35	0.01	Linear (%/°C)	0.03414	0	0		2030.683	1968
M55342K06B374DS	R12	Resistance (Ω)	374	377.74	370.26	75	-35	0.01	Linear (%/°C)	0.03414	0	0		379.7377	368.016
JANS1N4467	D1	Zener voltage (V) at Iz2 = 2 mA	11.93	12.5265	11.334	75	-35	0.076	Linear (%/°C)	0.12618	0.1262	0.5	%; (50 kRads)	12.983	10.78633
JANS1N5806	D2, D3	Vf (V) at If = 25 mA	0.62	0.682	0.558	75	-35	-1.8	Linear (mV/°C)	0.18926	0	0		0.791291	0.468
		Vf (V) at If = 2 mA	0.5	0.55	0.45	75	-35	-2.5	Linear (mV/°C)	0.18926	0	0		0.701041	0.325
JANS2N3700	Q1	hFE (Ic=10mA; Vce=10V)			90	90	-35	0.9133	Exp %/°C	0	11.152	0.00258	Δ(1/hFE); (50 kRads)		37.42328
		VBE (V); (Ic=25mA, Vce=1V)		0.8911	0.4489	90	-35	-2.00E-03	Linear (V/°C)	2.87093	0	1	%; (50 kRads)	1.05E+00	3.19E-01
UC1843A	U1	Vref (V)	5	5.05	4.95	90	-35	0.0004	Linear (V/°C)	1	1	0.24	%; (75 kRads)	5.13862	4.86462
		Ibias_EA (uA)		1		90	-35	0.00	Exp %/°C	2.87093	2.8709	74.1	%; (75 kRads)	1.769709	

Worst case circuit analysis for the example converter

The minimum and maximum output voltages of the startup linear regulator are given by the following equations:

$$V_{out,lin\_reg\_min} = V_{D1,min} + V_{D2,min} - V_{BE,Q1,max} - V_{D3,max} \quad (4)$$

$$V_{out,lin\_reg\_max} = V_{D1,max} + V_{D2,max} - V_{BE,Q1,min} - V_{D3,min} \quad (5)$$

Based on Table IV, the V_{out,lin_reg_min} and V_{out,lin_reg_max} are calculated to be 9.275 V and 12.89 V, respectively.

Since this voltage is greater than the specified maximum startup threshold of 9 V for UC1843A, the design guarantees a safe startup. Moreover, it does show the role of worst case analysis to ensure that the converter will successfully startup under worst case conditions. This type of analysis shall be done before the design is released to the manufacturing to avoid negatively impacting the schedule and cost.

Next, the dc output range of the boot winding needs to be calculated to make sure that the startup regulator will shutdown once the switching regulator is up and running. The minimum and maximum boot winding voltages are given by:

$$V_{boot,min} = \frac{V_{ref,min}}{2} \bullet \left(1 + \frac{R_{10,min}}{R_{11,max} + R_{12,max}}\right) - R_{10,max} \bullet I_{bias,max} \quad (6)$$

$$V_{boot,max} = \frac{V_{ref,max}}{2} \bullet \left(1 + \frac{R_{10,max}}{R_{11,min} + R_{12,min}}\right) - R_{10,min} \bullet I_{bias,min} \quad (7)$$

Again, using the *part degradation database* above, $V_{boot,min}$ and $V_{boot,max}$ are computed to be 13.83 V and 15.02 V, respectively. This corresponds to a static dc regulation of ±4.1%.

Therefore, this design guarantees that the maximum output voltage of the startup regulator (12.89 V) is less than the minimum output voltage of the boot winding (13.83 V). This ensures that the startup regulator will shutdown once the switching regulator is up and running.

This brings up the gate voltage derating issue. Per the above analysis, the maximum gate voltage applied is 15.02 V, slightly above the acceptable 15 V per 75% derating. Thus the design may need some tweaking in order to meet the maximum gate voltage requirement of less than 15 V. Using 0.1% resistors in place of 1% ones might do the trick in this case. Reducing the boot winding set point will also help. However, it will require adjusting the startup regulator setpoint as well. Using a 2% zener instead of a 5% one will definitely help albeit with added part cost.

Finally, we need to verify that the worst case dc gain of 2N3700 is enough to support 25 mA of current needed by 1843A. Per the *part degradation database*, the worst case dc gain is 37.4, implying a maximum base current need of 0.65 mA. Now, the minimum current through the zener (D1N4467) can be computed using the following equation.

$$I_{zener\_min} = \frac{V_{pri\_bus} - V_{zener\_max}}{\dfrac{R_3 \cdot R_4}{R_3 + R_4}} - I_{base,max} \quad (8)$$

Using Table IV, I_{zener_min} is computed to be 1.81 mA, enough to maintain its intended operation. Therefore, the startup regulator shall be able to support 25 mA of current into 1843A in this design.

SEE AND ELDRS ISSUES

Other important issues not addressed in the above power converter design example are single event effects (SEE) and enhanced low dose rate sensitivity (ELDRS). Single event upsets (SEU), a type of SEE, are an important part of the analog circuit design incorporating op-amps, voltage comparators, and PWM ICs. SEU include single-event transients such as changed voltage and/or frequency level, missing a few pulses, or a wrong logic state. This single event pulse can be of a short duration (5-10 μs), however, its impact on the power converter operation can be benign to catastrophic. For example, a single-event pulse in a latch-type circuitry can cause a false shutdown of the power converter. Only way to get out of this false shutdown is to recycle power to the unit. Depending on application, it may or may not be possible to add filtering to eliminate the effect of single-event pulses altogether. If a single-event pulse results in a catastrophic failure, system reliability requirement may dictate adding redundant unit. In case of a possible unit failure due to single-event effects, it is necessary to estimate an upset rate based on the device cross-section versus LET characteristics and the SEE environment. As an example, SEU test data[11] for LM139 comparator with a ΔV of 0.5 V and biased at +5 V is shown in Fig. 5. Based on SEE test data and the SEE environment, an upset rate can be estimated for all sensitive components leading to a unit level upset rate that feeds into the system level upset rate estimation. A typical acceptable upset rate at the unit level is in the order of 1 in 200 years.

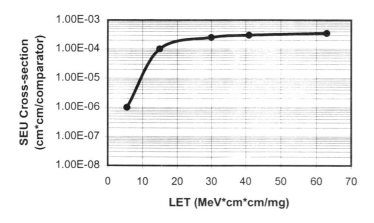

Figure 5   SEU test data for LM139 biased at +5V
with a ΔV=0.5V

Another relatively recent finding is the effect of dose rate on the radiation performance of semiconductor devices. Typical TID testing is done at a dose rate of 50-300 rads/s, way too high compared to a space radiation environment. Recent low dose rate testing has shown that the performance of linear bipolar ICs, a mainstay in power converter design, detoriates much more compared to the historically available high dose

rate data. In particular, the bias current and offset voltages change as much as 10 times more at low-dose rate compared to at high dose-rate. As a power converter designer, the low-dose rate radiation data for bipolar linear ICs such as comparators, op-amps, PWM and linear regulator ICs is critical in ensuring that the performance specifications of the power converter are met under worst case conditions.

SUMMARY

Space level design methodology for electronic power converters is presented. Electronics related space radiation environment is reviewed along with the part degradation database development. A power converter example is presented to illustrate the importance of detailed analysis in ensuring that the unit performance requirements are met under all conditions.

REFERENCES

1. A. H. Johnston, "Radiation effects in advanced microelectronics technologies", IEEE Trans. Nuclear Science, Vol. 45, No. 3, pp. 1339-1354, Jun. 1998.
2. R. L. Pease, "Total dose issues for microelectronics in space systems", IEEE Trans. Nuclear Science, Vol. 43, No. 2, pp. 442-451, Apr. 1996.
3. J. C. Pickel, "Single-event effects rate prediction", IEEE Trans. Nuclear Science, Vol. 43, No. 2, pp. 483-495, Apr. 1996.
4. T. L. Turflinger, "Single event effects in analog and mixed-signal integrated circuits", IEEE Trans. Nuclear Science, Vol. 43, No. 2, pp. 594-602, Apr. 1996.
5. A. H. Johnston, C. I. Lee, and B. J. Rax, "Enhanced damage in bipolar devices at low dose rates: Effects at very low dose rates", IEEE Trans. Nuclear Science, Vol. 43, No. 6, pp. 3049-3059, Dec. 1996.
6. M. R. Shaneyfelt, et. al., "Thermal stress effects and enhanced low dose rate sensitivity in linear bipolar circuits", IEEE Trans. Nuclear Science, Vol. 47, 2000.
7. US Department of Defense, "Electronic Parts, Materials, and Processes for Space and Launch Vehicles", MIL-HDBK-1547A, July 1998.
8. Semicoa Corporation, 2N3700 datasheet, http://www.semicoa.com.
9. Unitrode Corporation, UC1843A datasheet, http://www.unitrode.com.
10. Defense Threat Reduction Agency, ERRIC Database, http://erric.dasiac.com/.
11. R. Koga, et al., "Single event upset sensitivity dependence of linear integrated circuits on bias conditions," IEEE Trans. Nuclear Science, Vol. 44, pp. 2325-2332, 1997.

IECEC2001-AT-16

Low Temperature Performance of High Power Density DC/DC Converter Modules

Malik E. Elbuluk
Electrical Engineering Department
University of Akron
Akron, OH 44325-3904

Ahmad Hammoud
QSS Group, Inc.
NASA Glenn Research Center
Cleveland, OH 44135

Scott Gerber
ZIN Technologies
3000 Aerospace Parkway
Brook Park, OH 44142

Richard L. Patterson and Eric Overton
NASA Glenn Research Center
Power Technology Division, MS 301-5
Cleveland, OH 44135

ABSTRACT

In this paper, two second-generation high power density DC/DC converter modules have been evaluated at low operating temperatures. The power rating of one converter (Module 1) was specified at 150 W with an input voltage range of 36-75 V and output voltage of 12 V. The other converter (Module 2) was specified at 100 W with the same input voltage range and an output voltage of 3.3 V. The converter modules were evaluated in terms of their performance as a function of operating temperature in the range of 25 °C to -140 °C. The experimental procedures along with the experimental data obtained are presented and discussed in this paper.

INTRODUCTION

Presently, spacecraft operating in the cold environment of deep space carry on-board a large number of radioisotope heating units to maintain an operating temperature for the electronics of approximately 25 °C. To reduce system development and launch costs, improve reliability and lifetime, and increase energy densities, electronic components and systems capable of low-temperature operation will be required for many future space missions. These improvements will require proper system and circuit design, component selection and the development of components better suited for operation at extremely low temperatures. At the NASA Glenn Research Center, the Low Temperature Electronics Program focuses on research and development of "low temperature" electrical components and systems suitable for deep space mission.

Past and present activities of the Low Temperature Electronics Program have included characterization and evaluation of components such as semiconductor switching devices, integrated circuits, resistors, magnetic components, and capacitors [1-3]. In addition, low temperature device development activities have been conducted primarily in the area of semiconductor switching devices through Phase I Small Business Innovation Research Grants. The program has also been extensively involved in investigating DC/DC converters for low temperature space missions.

In addition to deep space applications, low temperature electronics have potential uses in terrestrial applications that include magnetic levitation transportation systems, medical diagnostics, cryogenic instrumentation, and super-conducting magnetic energy storage systems.

ADVANCES IN DC/DC CONVERTERS

Recently, there has been a great progress in the design of DC/DC converters with high power density. Converters that operate at power densities of 50% or more greater than the available standard conventional converter designs have been developed. This increase in power density is achieved using new designs, advanced devices and components, and packaging techniques. Today's leading edge DC/DC converter modules use synchronous rectifiers with multi-layer thick film hybrid packaging. This provides more usable output power without the use of a heat sink than do the conventional DC/DC converters that use Schottky diode with a heat sink and thick-film single-layer packaging. The first generation of DC/DC converter modules has a typical power rating of 10-15 W. The second-generation of DC/DC converters has a power rating of about 100-150 W with packaging and thermal management that provide high power density with small temperature gradients. Extensive use of silicon integration resulted in about one third the part count of a first generation converter.

LOW TEMPERATURE POWER CONVERTERS

Most aerospace power management systems are DC-based, and they require DC/DC power converters that operate with different inputs and outputs at various power levels. However, most of the existing DC/DC converter systems are specified to operate at low temperatures between -40 °C and -55 °C.

The Low Temperature Electronics Program has gone through a number of stages in building, testing and evaluating DC/DC converters for potential low temperature missions. During the first stage, a number of DC/DC converters were designed or modified to operate from room temperature to -196 °C using commercially available discrete components such as CMOS-type devices and MOSFET switches. These converters had output power in the range of 5 W to 1 kW and switching frequencies of 50 kHz to 200 kHz. Pulse-width modulation technique was implemented in most of these systems with open as well as closed-loop control. The topologies included buck, boost, multi-resonant, push-pull and full-bridge configuration [5-9].

The second stage followed the recent advancement in design and manufacturing of low power DC/DC converter modules. Several commercial-off-the-shelf (COTS) DC/DC converters have been characterized in terms of their performance as a function of temperature in the range of 20 °C to - 180 °C. These converters ranged in electrical power from 8 W to 13 W, input voltage from 9 V to 75 V and an output voltage of 3.3 V. Test results showed that they operated as expected within the manufacturer's specified temperature range, but at low temperature results varied. For some converters performance degraded rapidly, with others, reasonably good performance was seen down to temperatures between -80 °C and -100 °C. For temperatures below -100 °C, performance was either out of range, erratic, or non-existent for most of the converters [10-12].

The third evaluation stage is presented in this paper. Two of the second-generation high power density DC/DC converter modules have been evaluated at low temperature. The converters had power rating of 100 W and 150 W, input voltage range of 36V to 75 V and output voltages of 3.3 V to 12 V, respectively. The converters were evaluated in terms of their performance as a function of temperature in the range of 25 °C to -140 °C. The experimental procedure along with the experimental data obtained are presented and discussed.

EXPERIMENTAL PROCEDURE

The steady-state performances of the DC/DC converters were characterized as a function of temperature from 25 °C to −140 °C in terms of output voltage and efficiency. In addition, output voltage ripple, input current ripple and output current ripple waveforms were obtained. At a given temperature, these properties were obtained at various input voltages and at different load

levels: from light-load to full-load conditions. The tests were performed as a function of temperature using a environmental chamber cooled by liquid nitrogen. A temperature rate of change of 10 °C/min was used throughout this work. At every test temperature, the module under test was allowed to soak at that temperature for a period of 30 minutes before any measurements were made. After the last measurement was taken at the lowest temperature, the converters were allowed to stabilize to room temperature and then the measurements were repeated at room temperature to determine the effect of low temperature exposure on the converters.

RESULTS AND DISCUSSIONS

Figure 1 shows the output voltage and efficiency of Module 1, normalized to their respective room temperature values, at different load currents and as a function of temperature. The output voltage and efficiency were normalized to compensate for the voltage drop and power loss resulting from long wires connecting the module, inside the chamber, to the equipment and measuring instruments outside the chamber.

At a given load current, the normalized output voltage (Figure 1a) maintains a steady value from room temperature down to -120 °C. The effect of temperature on the efficiency of Module 1 (Figure 1b) under different load conditions showed that, in general, the normalized efficiency drops as the temperature is lowered with temperature having the least effect on the heavy load condition. Figure 1 represents data taken at a nominal input voltage of 49V. Additional, sets of data were taken with input voltages of 36V, 62V and 75V. The data obtained at these input voltages is very similar to that shown in Figure 1.

Figure 2 shows the output voltage and efficiency of Module 2, measured at different loads, input voltage and as a function of temperature. Similar to Module 1, the output voltage and efficiency were normalized to their values at room temperature. In Figure 2a, the normalized output voltage maintains a steady value from room temperature down to -20 °C. For temperatures below -20 °C the normalized output voltage increased with heavy loads values being higher than light load values. This effect was observed only at the lower input voltage of 36 V. In Figure 2c, for an input voltage of 48 V, the normalized output voltage remained constant, similar to that of Module 1.

The effect of temperature on the efficiency of Module 2 under different input voltage and load conditions is shown in Figures 2b and 2d. Opposite to that of Module 1, the normalized efficiency increases as the temperature is lowered. The effect of temperature is greater with the heavy load condition than that of the light load. This effect is greater at the lower input voltage (Figure 2b) than at the higher input voltage (Figure 2d). Additional, sets of data were taken with input voltages of 60V and 72V. The data obtained at these input voltages is very similar to that obtained at 48V input (Figures 2c and 2d).

Figures 3 and 4 show the waveforms of the output voltage ripple, the output current ripple and the input current ripple at room temperature (25 °C) and at a low temperature (-100 °C) for Module 1, respectively. The corresponding ripple in output voltage, input and output current increases with decreasing temperature. Also, instabilities in the input current are observed with the heavy load conditions being more prominent.

Figures 5 and 6 show the waveforms of the output voltage ripple and the input current ripple at room temperature (25 °C) and at the low temperature (-80 °C) for Module 2, for light load and heavy load conditions. The output voltage ripple changes slightly between room temperature and the low temperature, with very high spikes occurring during heavy loading conditions. Opposite to Module 1, the input current ripple at room temperature is higher than that of the low temperature (i.e. the ripple decreases with decreasing temperature). The output current was not recorded because the data under heavy loading could not be taken due to the need of high current probe. Similar to Module 1, instabilities in the input current are observed with the heavy load conditions being more prominent.

CONCLUSIONS

Two commercially available DC/DC converters were characterized in terms of their performance as a function of temperature in the range of 25 °C to -140 °C. The converters were evaluated with respect to their output voltage regulation, efficiency, output voltage ripple, input current ripple and output current ripple in response to environmental temperature. The two converters generally displayed somehow similar behavior with change in temperature. The intensity of any occurring changes, however, varied with the converter type and the test temperature. This work represents only a preliminary investigation into the steady-state effects of low temperature on these two second-generation high power density DC/DC converter modules. To fully characterize their performance at low temperature, further testing and analysis is required.

ACKNOWLEDGEMENT

This work is supported by a grant from the Low Temperature Electronics Group at NASA Glenn Research Center.

REFERENCES

1. Ray, B., Gerber, S.S., Patterson, R.L. and Myers, I.T., "Power Control Electronics for Cryogenic Instrumentation," *Advances in Inst.* and *Control,* Vol. 50, Part 1, Int. Soc. for Measurement and Control, 1995, pp.131-139.

2. Patterson, R.L., Dickman, J.E., Hammoud, A. and Gerber S.S., "Low Temperature Power Electronics Program," NASA EEE Links, Electronic Packaging and Space Parts News, Vol. 4, No. 1, January 1998.

3. Patterson, R.L., Hammoud, A. and Gerber, S.S., "Evaluation of Capacitors at Cryogenic Temperatures for Space Applications," IEEE International Conference on Electrical Insulation, Washington DC, June 7-10, 1998.

4. Ray, B., Gerber, S.S., Patterson, R.L. and Myers, I.T., "77K Operation of a Multi-Resonant Power Converter," IEEE *PESC Record,* Vol. 1, 1995, pp.55-60.

5. Ray, B., Gerber, S.S., Patterson, R.L. and Myers, I.T., "Liquid Nitrogen Temperature Operation of a Switching Power Converter," *Symp. on Low Temp. Electronics* & *High Temperature Superconductivity,* The Electrochemical Society, Vol. 9, 1995, pp.345-352.

6. Ray, B., Gerber, S.S., Patterson, R.L. and Dickman, J., "Low Temperature Performance of a Boost Converter with MPP and HTS Inductor," *IEEE APEC 96 Conference,* Vol. 2, 1996, pp.883-888.

7. Gerber, S.S., Patterson, R.L., Ray, B. and Stell, C., "Performance of a Spacecraft DC-DC Converter Breadboard Modified for Low Temperature Operation," *IECEC 96,* Vol.1, 1996, pp. 592-598.

8. Ray, B., Gerber, S.S. and Patterson, R.L., "Low Temperature Performance of a Full-Bridge DC-DC Converter," IECEC '96, Vol. 1, 1996, pp. 553-559.

9. Gerber, S.S. Miller, T., Patterson, R.L. and Hammoud, A., "Performance of a Closed-Loop Controlled High Voltage DC/DC Converter at Cryogenic Temperature," IECEC '98, Vol. 1, 1998.

10. M. Elbuluk, S. Gerber, A. Hammoud and R. Patterson, "Evaluation of Low Power DC/DC Converter Modules at Low Temperatures," Proceedings of IEEE PESC'00, Galway, Ireland.

11. Elbuluk, S. Gerber, A. Hammoud and R. Patterson, "Efficiency and Regulation of Low Power DC/DC Converter Modules at Cryogenic Temperatures," Proceedings of IEEE IECEC'00, Las Vegas.

12. M. Elbuluk, S. Gerber, A. Hammoud and R. Patterson, "Characterization of Low Power DC/DC Converter Modules at Low Temperatures," Proceedings of IEEE IAS'00, Rome, Italy.

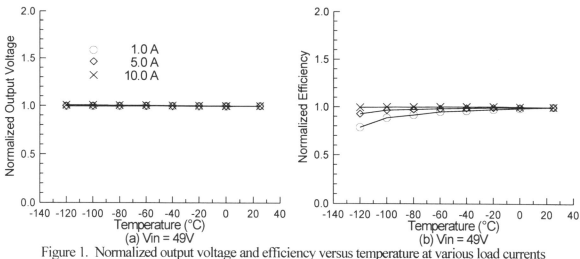

Figure 1. Normalized output voltage and efficiency versus temperature at various load currents for Module#1.

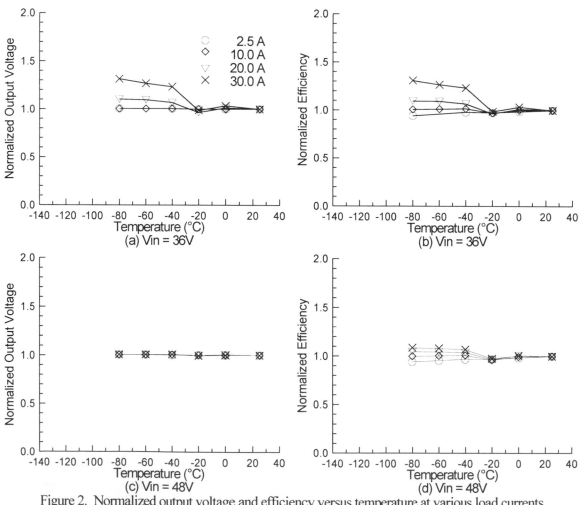

Figure 2. Normalized output voltage and efficiency versus temperature at various load currents for Module #2.

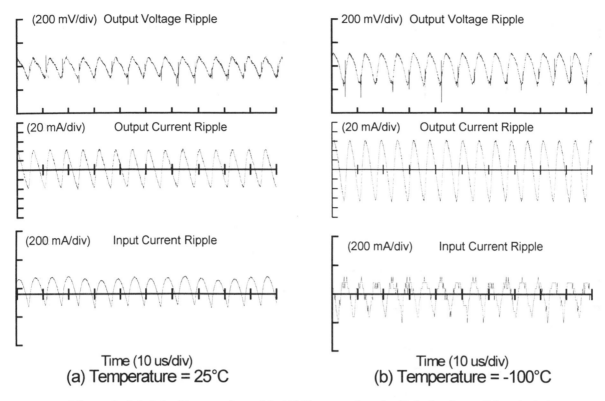

Figure 3. Module #1 operation with 49V input and under light load condition (1.0A).

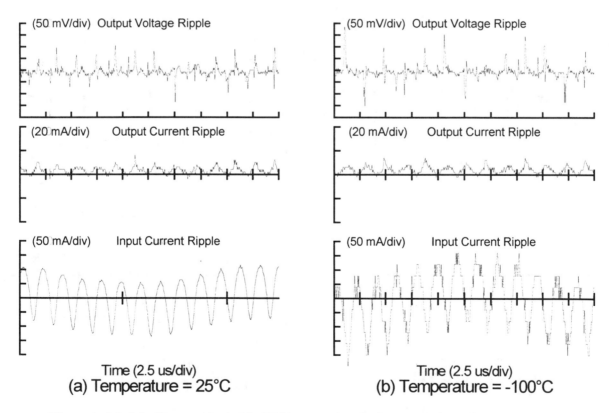

Figure 4. Module #1 operation with 49V input and under heavy load condition (12.5A).

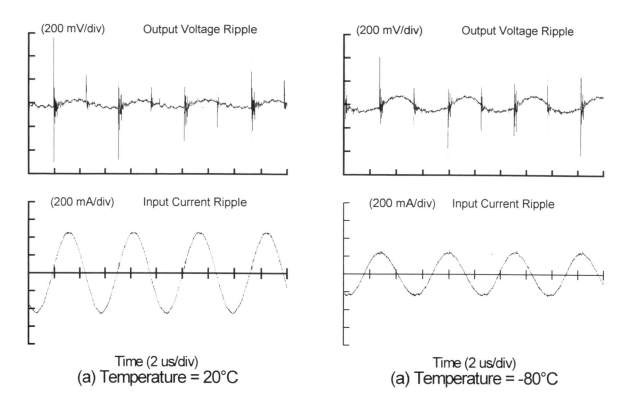

Figure 5. Module #2 operation with 48V input and under light load (5.0A).

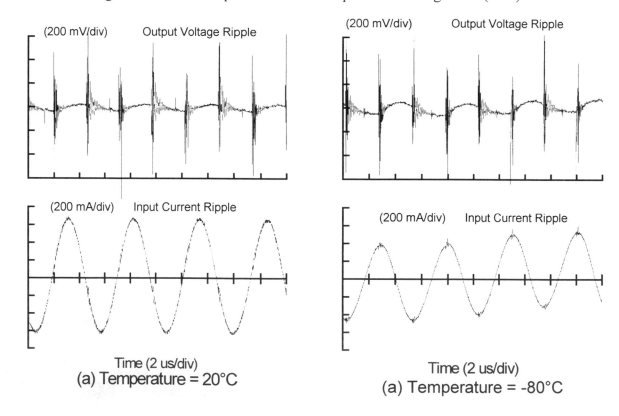

Figure 6. Module #2 operation with 48V input and under heavy load (30A).

Proceedings of IECEC'01
36th Intersociety Energy Conversion Engineering Conference
July 29–August 2, 2001, Savannah, Georgia

IECEC2001-AT-31

LOW TEMPERATURE TESTING OF A RADIATION HARDENED CMOS 8-BIT FLASH ANALOG-TO-DIGITAL (A/D) CONVERTER

Scott S. Gerber
ZIN Technologies, Inc.
and Ahmad Hammoud
QSS Group, Inc.
3000 Aerospace Parkway
Brook Park, OH 44142

Richard L. Patterson
and Eric Overton
NASA Glenn Research Center
21000 Brookpark Road
Cleveland, OH 44135

Malik E. Elbuluk
Electrical Engineering Department
University of Akron
Akron, OH 44325

Reza Ghaffarian,
Rajeshuni Ramesham,
and Shri G. Agarwal
Jet Propulsion Laboratory
4800 Oak Grove Drive
Pasadena, CA 91109

ABSTRACT

Power processing electronic systems, data acquiring probes, and signal conditioning circuits are required to operate reliably under harsh environments in many of NASA's missions. The environment of the space mission as well as the operational requirements of some of the electronic systems, such as infrared-based satellite or telescopic observation stations where cryogenics are involved, dictate the utilization of electronics that can operate efficiently and reliably at low temperatures. In this work, radiation-hard CMOS 8-bit flash A/D converters were characterized in terms of voltage conversion and offset in the temperature range of +25 °C to − 190 °C. Static and dynamic supply currents, ladder resistance, and gain and offset errors were also obtained in the temperature range of +125 °C to − 190 °C. The effect of thermal cycling on these properties for a total of ten cycles between +80 °C and −150 °C was also determined. The experimental procedure along with the data obtained are reported and discussed in this paper.

INTRODUCTION

Certain NASA deep space missions may require electronic components that encounter temperatures well below their minimum manufacturer's specified operating temperature (-55°C for military parts). Presently, spacecraft operating in the cold environment of deep space carry radioisotope heating units to maintain the on-board electronics at a temperature of approximately 25 °C.[1] The radioisotope heating units, however, require active thermal control system management, are expensive, and require elaborate containment structures. In addition to the environment of deep space and planetary exploration, extreme low temperatures are encountered in applications such as infrared-based satellite systems and Next Generation Space Telescope (NGST).[2] In some of these systems, electronic sensors and detector arrays are required to operate at cryogenic temperatures in order to reduce background thermal noise. In addition to detectors and sensors, these electronic systems employ a vast variety of integrated circuits and devices such as A/D and D/A converters, multiplexing and de-multiplexing chips, and solar-based circuits. Electronics circuits and systems, which are capable of operation at cryogenic temperatures, are, therefore, needed in order to meet the requirements of NASA space missions and commercial ventures. Low temperature electronics will not only tolerate the hostile environment of deep space but also reduce system size and weight by eliminating radioisotope heating units or any thermal control measures and associated structures; thereby reducing system development and launch costs, improving reliability and lifetime, and increasing energy densities.

A radiation hardened CMOS 8-bit flash A/D converter is an attractive candidate device for future NASA deep space missions that require a significant level of radiation tolerance. This particular device has been considered by the Jet Propulsion Laboratory (JPL) for potential use on the MUSES CN (Mu Space Engineering Spacecraft) NASA/Japanese mission to Asteroid Nereus.[3] The purpose of this investigation was to determine the A/D converter suitability for use in a low temperature environment. In this work, radiation-hard CMOS 8-bit flash A/D converters were characterized in terms of their performance in the temperature range of +25 °C to -190 °C. The testing activities were performed at the Low Temperature Electronics Facility at the NASA Glenn Research Center and at a contractor facility.

The in-house test activities included voltage conversion, offset voltage, and ladder resistance measurements as a function of temperature. The 8-bit flash A/D converter was tested at a clock rate of 1 MHz. For simplification, only DC values from 0 to 4V (test vector) were applied to the analog input of the converter. For each DC analog input value, the A/D converter produced an effective output bit pattern in binary format that was monitored by a logic analyzer. The experimental test setup and the device layout are shown in Figure 1. Testing of the A/D converter was performed in a chamber whose temperature was controlled using liquid nitrogen as the coolant at a ramp rate of 10 °C per minute. The test started at room temperature and was taken in steps down to -190 °C. At each temperature the device was allowed to soak for 20 minutes in order to reach thermal stability or equilibrium. At each temperature, a test vector was applied to the input of the A/D and the corresponding digital output was recorded.

The static and dynamic supply currents, ladder resistance, gain and offset error measurements were performed in the temperature range of +125 °C to -190 °C. Thermal cycling activities were also performed between +80 °C to -150 °C. All these tests, which comprised characterization of a total of four devices, were carried out at a contractor facility. The results are presented and discussed in this paper.

EXPERIMENTAL RESULTS

The CMOS 8-bit A/D converter was designed for space applications where relatively low power, exceptional accuracy, and very fast conversion speeds were necessary. This radiation-hard A/D converter was tested at a clock rate of 1 MHz. For simplification, the device's enable pins CE1 and CE2 were set for valid data on B8 through B1. In addition, only DC input values were applied to the input (Vin). For each DC input value, the converter produced an effective output pattern (B8 through B1) in binary format. Table I shows the corresponding binary value for each of the 8-bits. With a reference voltage of 4.012V, the A/D converter has a 15.3 mV per bit measurement resolution.

The output voltage of the converter was obtained as a function of the input voltage in the temperature range of +25 °C to -190 °C. The values of the output voltage, in binary as well as decimal format, corresponding to an input voltage variation from about zero to 4 volts are listed at the two extreme temperatures, i.e. +25 °C and -190 °C, in Tables II and III, respectively. In general, the measured output voltage tracked the input voltage quite well. For example, the average offset voltage at room temperature (25 °C) and at -190 °C was 34 and 24mV, respectively. This means that the 8-bit flash A/D converter is able to track the input voltage within about 2 bits. The offset voltage exhibited by the device over the temperature range is plotted in Figure 2. The deviations of about 20 to 40 mV within this temperature range are well within the overall error of the inaccuracies of the test setup. These errors can be attributed to factors inherent to the device, instrumentation error, and circuit layout. It is important to note that although the data presented in Tables II and III pertain to only the extreme test temperatures, the performance of the device was similar throughout the test temperature range of +25 °C to -190 °C.

The performance of four A/D converter devices was obtained over both hot and cold temperature ranges, i.e. from +25 °C to +125 °C and from +25 °C to -190 °C. The properties investigated included: dynamic supply current (IDDD), clock low static supply current (IDDS), clock high static supply current (IDDH), ladder resistance (Lad Res), and gain and offset (Vos) errors.

Table IV lists the properties of the four devices investigated at temperatures of 25, 80, 100, and

125 °C. It can be seen that, at a given test temperature, all investigated properties of the four devices remained within their specifications.

Test results for six parameters measured at low temperatures from −55 °C to −190 °C are listed in Table V. The Vos and gain errors were measured only at −55 °C. Testing of the converter at lower temperatures necessitated the development of a special test fixture, which utilized long wires to provide connections between the instrumentation and the device under test. These wires introduced excessive noise that had detrimental effects on the accurate measurement of low-level signals, such as Vos, Gain Error, and to determine linearity behavior.

Table V shows that, while the dynamic supply current (IDDD) was the only parameter that remained within its specification to −190 °C even though its drift increased with decrease in temperature, the other parameters, i.e. IDDH and IDDS (static supply currents) and ladder resistance moved outside their specification limits at and below −120 °C. The out-of-specification values are shaded in Table V. Test results for four devices were slightly different, but they showed very similar trends with decrease in temperature.

A number of functional tests were performed in-house to determine if the drift in the parameters was real and whether the drift influenced the functionality of the converter. These tests involved measurement of the ladder resistance in the temperature range of +25 °C to −190 °C, while the voltage conversion was also monitored. As an example, the values obtained for the ladder resistance and the converted output for a 2.75 V input are listed in Table VI for this temperature range. It can be clearly seen that the trend of decreasing ladder resistance with temperature was similar to that obtained for devices evaluated at the contractor facilities (see Table V). As temperature was varied, the voltage conversion remained consistent. The 8-bit flash A/D converter functioned well even with a decrease in operating temperature, although ladder resistance drifted to values outside of manufacturer's specifications.

The effect of thermal cycling on the A/D converter was investigated by exposing the A/D converter to a total of ten thermal cycles. Two of the four devices were thermally cycled between −150 °C and +80 °C at a rate of 5 °C per minute with dwell times of 20 minutes. After completion of ten thermal cycles, parameters were measured at the two extreme cycling temperatures, i.e. −150 °C and +80 °C. The device parameter values at these two temperatures prior to and after ten cycles are shown in Table VII. As before, only four parameters were measured. At both low and high temperatures, the four parameters were in good agreement before and after ten thermal cycle exposures.

CONCLUSIONS

A radiation hardened CMOS 8-bit flash A/D converter, which is an attractive candidate device for future NASA deep space missions requiring a significant level of radiation tolerance, has been investigated for suitability for use in a low temperature environment. The A/D converters were characterized in terms of their performance in the temperature range of +125 °C to − 190 °C. Properties investigated included static and dynamic supply currents, ladder resistance, and gain and offset errors. The effect of thermal cycling for a total of ten cycles on these properties was also determined.

Room temperature results have shown that the 8-bit flash A/D converter was able to track the input voltage with an average of 34 mV or about 2 bits. This error, which was well within the overall error of the inaccuracies of the test setup, was the largest error measured for the converter over the temperature range of +25 °C to -190 °C. These preliminary test results have also shown that the A/D converter may be considered as a potential candidate for operation at low temperatures down to -190 °C, well below the manufacturer's minimum specified operating temperature. Although, the converter survived and maintained good operation after a total of ten thermal cycles at extreme temperatures, more comprehensive testing is, however, required to fully-characterize its performance and to determine its suitability for low temperature space missions and commercial applications.

ACKNOWLEDGEMENTS

This work was supported by the NASA Glenn Research Center, Contract # 98008, SETAR Task 0021 and by JPL's NASA Electronic Parts and Packaging Program (NEPP). The authors acknowledge Bell Technologies, Inc. for performing some of the testing activity on this A/D converter. They also acknowledge the MUSES CN Project Office that has identified the A/D converter for their application.

REFERENCES

[1]. S. Gerber, T. Miller, R. Patterson and A. Hammoud, "Performance of a Closed-Loop Controlled High Voltage DC-DC Converter at Cryogenic Temperatures," 33rd IECEC, Colorado Springs, Aug 2 - 6, 1998.

[2]. R. Patterson, A. Hammoud, J. Dickman, S. Gerber and E. Overton, "Development of Electronics for Low Temperature Space Missions," 4th European Workshop on Low Temperature Electronics, Noordwijk, the Netherlands, 21-23 June 2000.

[3]. Discussions with JPL/NASA mission and project groups.

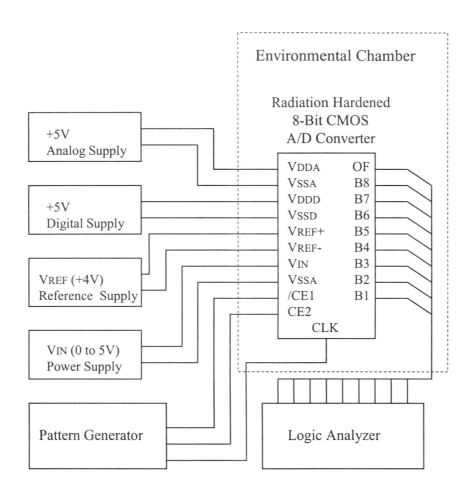

Figure 1. Test Setup for Radiation Hardened 8-Bit CMOS A/D Converter.

Table I. Binary coding.

A/D outputs	B8	B7	B6	B5	B4	B3	B2	B1	
Binary value	128	64	32	16	8	4	2	1	Total 255 bits

170

Table II. Voltage conversion at test temperature of 25 °C

Vin	OF	B8	B7	B6	B5	B4	B3	B2	B1	Binary Vout	Decimal Vout (V)	Δ Vout -Vin (offset) (V)
-0.06	0	0	0	0	0	0	0	0	0	0	0	0.06
1.00	0	0	1	0	0	0	0	0	1	65	1.02	0.02
2.00	0	1	0	0	0	0	0	0	1	129	2.03	0.03
2.50	0	1	0	1	0	0	0	0	1	161	2.53	0.03
2.75	0	1	0	1	1	0	0	0	1	177	2.78	0.03
2.99	0	1	1	0	0	0	0	0	0	192	3.02	0.03
3.97	0	1	1	1	1	1	1	1	1	255	4.01	0.04
4.00	1	1	1	1	1	1	1	1	1	overflow	overflow	average offset = 0.034

Table III. Voltage conversion at test temperature of -190 °C

Vin	OF	B8	B7	B6	B5	B4	B3	B2	B1	Binary Vout	Decimal Vout (V)	Δ Vout -Vin (offset) (V)
-0.04	0	0	0	0	0	0	0	0	0	0	0	0.00
0.21	0	0	0	0	0	1	1	1	0	14	0.22	0.01
1.00	0	0	1	0	0	0	0	0	1	65	1.02	0.02
2.00	0	1	0	0	0	0	0	0	1	129	2.03	0.03
2.41	0	1	0	0	1	1	0	1	1	155	2.44	0.03
2.42	0	1	0	0	1	1	1	0	0	156	2.45	0.03
2.75	0	1	0	1	1	0	0	0	1	177	2.78	0.03
2.99	0	1	1	0	0	0	0	0	0	192	3.02	0.03
3.97	0	1	1	1	1	1	1	1	1	255	4.01	0.04
3.99	1	1	1	1	1	1	1	1	1	overflow	overflow	average offset = 0.024

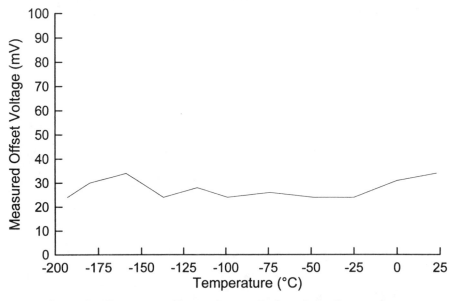

Figure 2. Converter offset voltage as a function of temperature

Table IV. Converter properties at high temperatures

Parameter	Specifications	25 °C	80 °C	100 °C	125 °C
Device #1					
IDDD	0.05 to 135 mA	61	55	53	51
IDDH	0.05 to 135 mA	87	78	74	71
IDDS	0.05 to 80 mA	34	37	35	33
Lad Res	300 to 900 Ohms	577	614	693	675
Vos	-1.25 to 1.25 LSB	0.06	-0.00	-0.04	-0.10
Gain Error	-2.25 to 2.25 LSB	0.16	0.12	0.09	0.09
Device #2					
IDDD	0.05 to 135 mA	62	62	53	51
IDDH	0.05 to 135 mA	90	90	75	71
IDDS	0.05 to 80 mA	34	34	34	33
Lad Res	300 to 900 Ohms	593	571	689	682
Vos	-1.25 to 1.25 LSB	0.08	0.07	0.01	-0.03
Gain Error	-2.25 to 2.25 LSB	0.27	0.05	0.15	0.11
Device #3					
IDDD	0.05 to 135 mA	62	55	53	50
IDDH	0.05 to 135 mA	90	79	75	70
IDDS	0.05 to 80 mA	34	37	36	33
Lad Res	300 to 900 Ohms	590	636	654	685
Vos	-1.25 to 1.25 LSB	0.11	0.10	0.04	-0.00
Gain Error	-2.25 to 2.25 LSB	0.38	0.13	0.09	0.17
Device #4					
IDDD	0.05 to 135 mA	60	53	51	49
IDDH	0.05 to 135 mA	87	76	73	69
IDDS	0.05 to 80 mA	34	34	32	31
Lad Res	300 to 900 Ohms	582	617	704	685
Vos	-1.25 to 1.25 LSB	-0.00	-0.07	-0.13	-0.11
Gain Error	-2.25 to 2.25 LSB	0.30	0.15	0.17	0.08

IDDD: Dynamic supply current
IDDH: Clock high static supply current
IDDS: Clock low static supply current

Lad Res: Ladder resistance
Vos: Offset voltage

Table V. Converter properties at low temperatures

Parameter	Specifications	-190 °C	-180 °C	-170 °C	-150 °C	-120 °C	-80 °C	-55 °C	25 °C
Device #1									
IDDD	0.05 to 135 mA	133	133	133	124	118	110	75	61
IDDH	0.05 to 135 mA	174	174	174	160	151	139	112	87
IDDS	0.05 to 80 mA	100	92	92	84	76	73	39	34
Lad Res	300 to 900 Ohms	199	204	204	284	312	363	466	577
Vos	-1.25 to 1.25 LSB							0.28	0.06
Gain Error	-2.25 to 2.25 LSB							0.23	0.16
Device #2									
IDDD	0.05 to 135 mA	131	128	125	122	113	102	76	62
IDDH	0.05 to 135 mA	171	166	161	156	142	129	113	90
IDDS	0.05 to 80 mA	91	88	84	83	74	68	39	34
Lad Res	300 to 900 Ohms	216	252	273	296	340	385	458	593
Vos	-1.25 to 1.25 LSB							0.28	0.08
Gain Error	-2.25 to 2.25 LSB							0.12	0.27
Device #3									
IDDD	0.05 to 135 mA	132	132	127	122	114	115	77	62
IDDH	0.05 to 135 mA	175	174	166	158	146	127	114	90
IDDS	0.05 to 80 mA	92	89	86	82	80	78	40	34
Lad Res	300 to 900 Ohms	199	207	254	298	328	328	438	590
Vos	-1.25 to 1.25 LSB							0.36	0.11
Gain Error	-2.25 to 2.25 LSB							0.11	0.38
Device #4									
IDDD	0.05 to 135 mA	124	120	120	115	107	97	74	60
IDDH	0.05 to 135 mA	169	160	159	152	140	125	110	87
IDDS	0.05 to 80 mA	83	79	81	75	70	63	38	34
Lad Res	300 to 900 Ohms	204	262	263	297	344	385	447	582
Vos	-1.25 to 1.25 LSB							0.13	-0.00
Gain Error	-2.25 to 2.25 LSB							0.19	0.30

Table VI. Converter output voltage and ladder resistance at low temperatures

Parameter	-190 °C	-180 °C	-170 °C	-150 °C	-120 °C	+25 °C
Ladder Resistance (Ω)	222	229	250	286	333	571
Vin (V)	2.75	2.75	2.75	2.75	2.75	2.75
Converted Vout (V)	2.78	2.79	2.79	2.78	2.78	2.78

Table VII. Summary of Thermal Cycling Data (10 Cycles from +80 °C to -150 °C)

Device #	Parameter	Specifications	Before Thermal Cycling -150 °C	After Thermal Cycling -150 °C	Before Thermal Cycling 80 °C	After Thermal Cycling 80 °C
1	IDDD	0.05 to 135 mA	124	123	55	55
	IDDH	0.05 to 135 mA	160	156	78	78
	IDDS	0.05 to 80 mA	84	78	37	37
	Lad Res	300 to 900 Ohms	284	298	614	617
	Vos	-1.25 to 1.25 LSB			-0.00	-0.00
	Gain Error	-2.25 to 2.25 LSB			0.12	0.03
2	IDDD	0.05 to 135 mA	115	113	53	53
	IDDH	0.05 to 135 mA	152	150	76	76
	IDDS	0.05 to 80 mA	75	71	34	34
	Lad Res	300 to 900 Ohms	297	302	617	623
	Vos	-1.25 to 1.25 LSB			-0.07	-0.09
	Gain Error	-2.25 to 2.25 LSB			0.15	0.09

Shaded areas are outside of specifications.   Blank cells: invalid/questionable data (not reported).

IECEC2001-AT-68

A LOW-COST DRIVER FOR PIEZOCERAMIC FLIGHT CONTROL SURFACES

William C. Dillard
ECE Department
Auburn University
Auburn, AL 36849-5201
dillard@eng.auburn.edu

R. M. Nelms
ECE Department
Auburn University
Auburn, AL 36849-5201
nelms@eng.auburn.edu

ABSTRACT

A low-cost drive circuit for piezoceramic flight control surfaces for unmanned aircraft and guided munitions has been developed. Electrically, the control surfaces are capacitive in nature and actuated by applying dc voltages, generally between ± 75 V. A Zener-limited half-bridge boosting topology is employed to convert a battery voltage of 3 V to the desired output voltage, which is controlled by four MOSFETS. Additionally, a constant-t_{ON}, variable- frequency scheme is used to vary the output voltage rate of change. Measured performance is in agreement with theoretical treatments.

INTRODUCTION

Since 1989, significant progress has been made in the use of smart materials, such as piezoceramics, for active flight control [1-3]. These materials, which deform under a voltage stimulus, have eliminated the need for traditional actuator systems, greatly reducing the weight and size of unmanned aircraft and guided munitions.

Piezoceramic flight control surfaces are capacitive, requiring as much as ±200V for adequate deflection. The actual voltage range scales with the surface's area, which is dependent on the aerodynamic specifications of the aircraft. Continued miniaturization of remotely guided craft will require smaller, lighter avionics. In this work, a low-cost piezoceramic driver circuit for compact applications is presented. The driver, operating from a 3V power source, employs a half-bridge, boosting configuration similar to a dc-dc boost converter to deliver voltages between ±60V to the piezoceramic. To keep the component count low, the driver operates open-loop, relying on a remote guidance system for stable flight and a Zener diode to limit the maximum/minimum voltage. A variable frequency scheme allows control of the output voltage rate of change.

NOMENCLATURE

C_B	= boost capacitance (F)
C_{FIN}	= piezoceramic capacitance (F)
D	= duty cycle
$D_5 - D_0$	= digital rate multiplier input bits
F	= digital rate multiplier prescaler
I_D	= MOSFET drain current (A)
I_P	= Inductor peak current (A)
K_N	= MOSFET transconductance parameter (A/V^2)
L	= boost inductance (H)
P_P	= average transfer losses per switching cycle (W)
R	= resistance (Ω)
R_{EQ}	= time constant equivalent resistance (Ω)
T	= switching period (s)
V_{DS}	= MOSFET drain-source voltage (V)
V_{FIN}	= piezoceramic voltage (V)
V_{DS}	= MOSFET gate-source voltage (V)
V_O	= initial value of output voltage (V)
V_{On}	= normalized output voltage (V)
V_{OUT}	= output voltage (V)
V_T	= MOSFET threshold voltage (V)
V_Z	= Zener voltage (V)
f	= switching frequency (Hz)
n	= switch cycle index
r_{ds}	= MOSFET drain-source resistance (Ω)
t	= time (s)
t_{OFF}	= switch off-time (s)
t_{ON}	= switch on-time (s)
ΔV	= output voltage resolution (V)
ΔV_n	= output voltage resolution (V)
λ	= channel length modulation parameter (V^{-1})

THE DRIVER CIRCUIT

The driver, shown in Figure 1, is a capacitor charger with variable, bipolar output voltage [4 - 6]. A Zener-limited boost topology, consisting of MOSFET M1, inductor L, Zener diode Z and capacitor C_B, is employed in the power supply section to convert a battery voltage of 3V to as much as 120V at V_{OUT}. As in dc-dc boost converters, gating M_1 will boost V_{OUT}. To decrease V_{OUT}, C_B is discharged through M_2. Mutually exclusive switches S_1 and S_2, implemented with MOSFETs and an optocoupler as shown in Fig. 2, determine the polarity of the voltage across the piezoceramic, which is modeled as C_{FIN}. The switches, along with the resistive voltage divider, produce the following relationship for V_{FIN},

$$V_{FIN} \approx \pm \frac{V_{OUT}}{2} \qquad (1)$$

where the sign depends on which polarity switch, S_1 or S_2, is closed. Thus, for ±60V at V_{FIN}, V_{OUT} must be controllable over the range ± 120V.

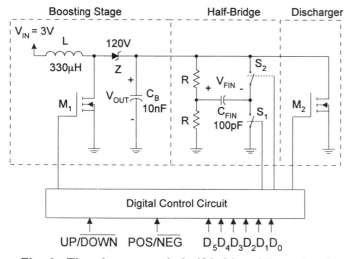

Fig. 1. The piezoceramic half-bridge driver circuit.

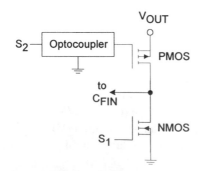

Fig. 2 Implementation of switches S_1 and S_2 using complementary MOSFETs and an optocoupler.

The driver circuit itself is an open-loop system. Flight data is transmitted serially from a remote guidance system, which provides closed loop control through the Digital Control Circuit in Fig, 1. Each control word is eight bits in length, allowing a high bandwidth control scheme. Two bits control the output voltage by enabling the appropriate MOSFETs and switches. The remaining six bits control the rate of change of V_{OUT} by modifying the duty cycle of the MOSFET gating signal. The details of this circuit will be discussed in the next section.

THEORY OF OPERATION

A per-cycle energy balance model can predict the performance of the power supply when boosting the output voltage. Within each switching cycle the inductor energy must be released to the load capacitance or lost in parasitics. Also, for discontinuous operation, the inductor current returns to zero in each switching cycle, as shown in Fig. 3. Thus, the energy balance requirement can be expressed as

$$\frac{LI_P^2}{2} = P_P T + \frac{C_B \left(V_{OUT}^2 - V_0^2 \right)}{2} \qquad (2)$$

where C_B is assumed to be much greater than C_{FIN}. Starting at $t = 0$ with no initial output voltage, a general expression for V_{OUT} can be developed. Relating I_P to V_{IN} yields

$$I_P = \frac{V_{IN} t_{ON}}{L} \qquad (3)$$

After n switching cycles, (2) becomes

$$\frac{V_{IN}^2 t_{ON}^2}{2L} n = P_P nT + \frac{C_{FIN}}{2} V_{OUT}^2 (nT) \qquad (4)$$

For large n, $nT \to t$ and (4) can be written as

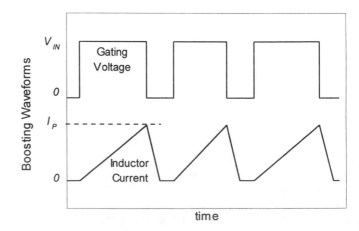

Fig. 3 Standard boost waveforms for gating voltage and inductor current for discontinuous operation.

$$V_{OUT}(t) = \sqrt{\left(\frac{V_{IN}^2 D^2}{LC_{FIN}f} - \frac{2P_P}{C_{FIN}} \right)} \sqrt{t} \qquad (5)$$

where $D = t_{ON}/T$.

The model in (2 - 5) reveals two consequences of the boost topology. First, the output voltage is proportional to the square root of time, reaching a maximum when parasitic losses balance the input power. If P_P is a constant, independent of V_{OUT}, then the output increases without bound. Since in reality the output voltage is limited, losses must increase as V_{OUT} increases. Second, the rate of change of V_{OUT}, or slew rate, depends on the instantaneous value of V_{OUT},

$$\frac{dV_{OUT}(t)}{dt} = \frac{\frac{V_{IN}^2 D^2}{2LC_B f} - \frac{P_P}{C_B}}{V_{OUT}(t)} \qquad (6)$$

with an additional implicit dependence through $P_P(V_{OUT})$. Since the deflection angle of the piezoceramic is linearly dependent on V_{OUT}, flight dynamics are slower at larger deflection angles. Our solution is to control the slew rate by varying the duty cycle in a constant on-time, variable frequency scheme implemented in the digital control circuit in Fig. 4.

The digital rate multiplier masks oscillator pulses based on the value of multiplier inputs $D_5 - D_0$

$$\frac{RateMultiplierPulses}{OscillatorPulses} = \frac{D_5 D_4 D_3 D_2 D_1 D_0}{64} = F \qquad (7)$$

where F is the pulse prescaler. Figure 5 shows the digital rate multiplier operation for $D_5 - D_0 = 000101_2 = 5_{10}$, or $F = 5/64$. Note that while the on-time of the rate multiplier output is the same as that of the oscillator waveform, the off-times are very different and dependent on F. Thus, the gating signals for the MOSFETs are indeed constant on-time, variable frequency signals.

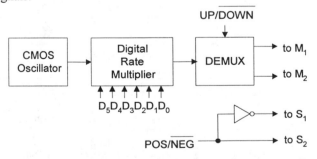

Fig. 4 The digital control circuit consists of an oscillator that sets t_{ON} for M$_1$/M$_2$, a digital rate multiplier for output slope control and routing circuitry to MOSFET gates.

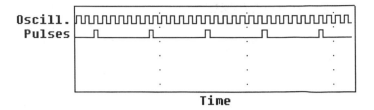

Fig. 5 Simulation of the digital rate multiplier's performance for $F = 5/64$.

To decrease V_{OUT} the discharge MOSFET, M$_2$, is gated. For most output voltage values, the MOSFET will be in saturation and accurately modeled as a current source, as shown in Fig. 6. The change in V_{OUT} per switching cycle is

$$\Delta V_{OUT} = \frac{I_D t_{ON}}{C_B} \qquad (8)$$

Over many switching cycles, (8) can be expressed as

$$V_{OUT}(t) = V_O - \frac{I_D D}{C_B} t \qquad (9)$$

Here, the rate of change of V_{OUT} is, to first order, independent of V_{OUT} and the need for slope control is not as great as when boosting. However, including the rate multiplier in the discharge circuitry adds flexibility to the discharge profile without additional hardware complexity. The rate multiplier modifies (5) and (9) as

$$V_{OUT}(t) = \sqrt{\left(\frac{V_{IN}^2 FD^2}{LC_B f} - \frac{2FP_P}{C_B} \right)} \sqrt{t} \qquad (10)$$

and

$$V_{OUT}(t) = V_O - \frac{I_D FD}{C_B} t \qquad (11)$$

where D and f are still defined in terms of the oscillator period.

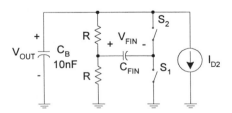

Fig. 6 Circuit model for discharging the output voltage through M$_2$, which is in saturation.

EXPERIMENTAL RESULTS

Measurements for the Zener-limited variable-output piezoelectric driver have been conducted for positive output voltages for a family of prescaler values. Data for the boosting mode are shown in Figure 7 and, as predicted by (5), the profiles generally fit a square root relationship, limited by the Zener diode. The effect of the prescaler is also in agreement with (10). Figure 8 shows the results for the output voltage discharge. Note that the discharge of C_{FIN} is nearly linear, as modeled in (9). Also, the slope is linearly dependent on F, as predicted in (11). Discharge profiles for negative voltage are similar. Boosting mode measurements for negative voltages, shown in Fig. 9, again agree well with (9).

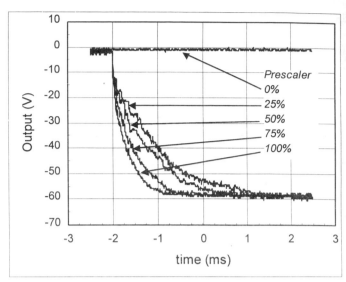

Fig. 9 Measurements of negative boosting again shows the effect of the prescaler on the slope the Zener diode.

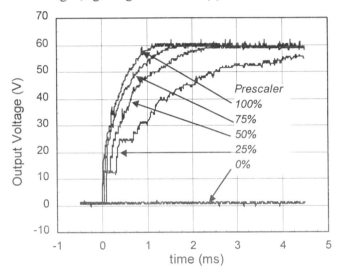

Fig. 7 Measurements of positive boosting clearly shows the effect of the prescaler on the slope and the voltage limit imposed by the Zener diode.

NON-IDEALITIES

A collection of parasitic mechanisms that lead to non-ideal performance in the driver is listed in Table 1 along with parameters that quantify each mechanism. Losses degrade the efficiency of the driver and can be related to the inductor current, I_P, or the output voltage, V_{OUT}. Those losses that are part of the average power lost in transferring energy from the inductor to the boost capacitor, P_P in (2), are shown in boldface type. In general, each component can be selected such that its losses can be minimized. However, for compact aircraft/munitions there are two other concerns that can severely restrict the choices in component selection – circuit size and voltage rating.

For the inductors, miniaturization is the major concern, limiting inductor choices to the smallest possible SMT devices that will not saturate at I_P as defined in (3). Similarly,

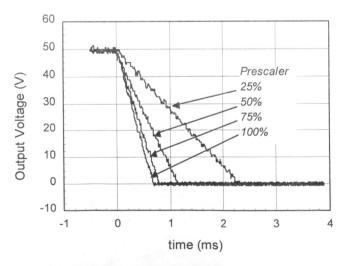

Fig. 8 Measurements for discharging positive voltages for a family of prescaler values.

Table 1. Non-ideality Issues in the Piezoceramic Driver

Mechanism	Parameters	Mode
Inductor core	f, material, I_P	Boosting
Inductor copper	R_{wire}, I_P	Boosting
Zener diode	V_D, I_P	Boosting
M_1	R_{ON}, I_P	Boosting
M_2	V_{OUT}, I_{D2}, r_{ds2}	Discharge
C_B	ESR_B	Both
C_{FIN}	ESR_{FIN}	Both
Voltage divider	R, V_{OUT}, RC_B	Both
S_1	R_{ON}, I_{FIN}	Both
S_2	R_{ON}, I_{FIN}	Both

the MOSFETs must maintain reasonable on-resistances at I_P current levels, and must also block voltages up to 120V. (Signal level, 200V MOSFETs were used in this study.) The capacitor C_B must also be rated at the maximum output voltage.

The resistance of the voltage divider resistors affects performance in three ways. First, when boosting the output, some energy is expended in the divider. This loss is part of P_P in (2) and depends on $V_{OUT}(t)$. Second, since the divider resistors discharge C_B exponentially, it is important that $C_B R_{EQ}$ be large compared to the period of gating signals at the rate multiplier output. This issue is even more critical for small values of F where the time between gating pulses can be long. For example, given a 128kHz oscillator signal at 50% duty cycle and $F = 1$, the gating signal will have the following characteristics.

$$t_{ON} = \frac{1}{2f} = 3.9\,\mu s \tag{12}$$

$$t_{OFF} = \frac{64}{f} - \frac{1}{2f} = 496\,\mu s \tag{13}$$

To maintain the output voltage during t_{OFF},

$$R_{EQ}C_B = 2RC_B \gg 496\,\mu s \tag{14}$$

Given $C_B = 10$ nF, R_{EQ} should be much greater than 25kΩ. In this study, 4MΩ resistors were used.

Finally, during discharge, I_{D2} is a weak function of V_{OUT} and the discharge rate is not strictly linear. This effect is accurately modeled by the output resistance of M_2, r_{ds2}, which itself is a function of V_{OUT} and the gate drive voltage.

$$I_{D2} = K_{N2}[V_{GS2} - V_{T2}]^2[1 + \lambda_2 V_{DS2}] \tag{15}$$

$$r_{ds2} = \frac{\partial V_{DS2}}{\partial I_{D2}} = \frac{V_{DS2} + 1/\lambda_2}{I_{D2}} = \frac{V_{OUT} + 1/\lambda_2}{I_{D2}} \tag{16}$$

As a general rule of thumb, $1/\lambda$ is at least twice the voltage rating of the MOSFET. Thus, from (12), a 200-V device operating at 120V will conduct no worse than 12% more current than that modeled by the simple square law relationship.

VOLTAGE RESOLUTION

In each switching cycle, energy is transferred from inductor to C_B, as described in (2), producing a change in the output voltage, ΔV, which is the resolution of the driver. Equation (2) can be written

$$\frac{LI_P^2}{2} = P_P T + \frac{C_B\left((V_0 + \Delta V)^2 - V_0^2\right)}{2} \tag{17}$$

Ignoring the parasitic power in (17) provides a conservative estimate for the resolution, which decreases as V_0 increases. Normalizing both the resolution and output voltage to the resolution at $V_0 = 0$V yields the general expression

$$\Delta V_n = \sqrt{V_{0n}^2 + 1} - V_{0n} \tag{19}$$

where the resolution at $V_0 = 0$V is

$$\Delta V\big|_{V_0 = 0} = \frac{V_0 t_{ON}}{\sqrt{LC_B}} \tag{20}$$

Equation (19) is plotted in Fig. 10 for the parameter values used in this work over the output voltage range 0 to 120V. The resolution changes by a factor of 30 across the full output voltage span.

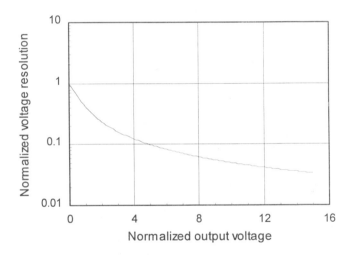

Fig. 10 Normalized resolution is a strong function of normalized output voltage, a direct consequence of a constant on-time, variable-frequency gating scheme.

CONCLUSIONS

A low-cost, compact driver circuit that produces a variable bipolar output voltage suitable for piezoceramic flight control surfaces in miniature aircraft/munitions has been developed. Flight control data, transmitted serially, varies the output voltage value, polarity and slew rate. Measurements conducted for various charging/discharging profiles agree with theoretical treatments.

ACKNOWLEDGMENTS

The authors gratefully acknowledge our colleague Mr. Qinghong Yu for his insightful discussions and encouragement throughout the course of this work.

REFERENCES

[1] Crawley, E.F., and Anderson, E.H., 1989, "Detailed Models of Piezoceramic Actuation of Beams," *Proceedings of the AIAA/ASME/ASCE/AHS 30th Structures, Structural Dynamics, and Materials Conference*, AIAA, Washington, DC, pp. 2000-2010.

[2] Barrett, R.M., 1994, "Active Plate and Missile Wing Development Using Directionally Attached Piezoelectric Elements," AIAA Journal, **32**, pp. 601-609.

[3] Stutts, J.C., and Barrett, R.M., 1998, "Development and Experimental Validation of a Barrel-Launched Adaptive Munition," *Proceedings of the AIAA/ASME/ASCE/AHS Structures, Structural Dynamics, and Materials Conference*, AIAA, Washington, DC, **4**, pp. 2834-2843.

[4] Sokal, N.O., and Redl, R., "Control Algorithms and Circuit Designs for Optimally Flyback-Charging an Energy Storage Capacitor (e.g., for a Flash Lamp)," *Conference Proc. 5th Annual IEEE Applied Power Electronics Conference*, IEEE, Washington, DC, pp. 295-302.

[5] Sokal, N.O. and Redl, R., "Control Algorithms and Circuit Designs for Optimally Flyback-Charging an Energy Storage Capacitor (e.g., for a Flash Lamp of Defibrillator)," IEEE Transactions on Power Electronics, **12**, pp. 885-894.

[6] Newsom, R.L., Dillard, W.C., Nelms, R.M., 1999, "A Capacitor Charging Power Supply Utilizing Digital Logic for Power Factor Correction," *Applied Power Electronics Conference and Exposition*, IEEE, Washington, DC, **2**, pp. 1115-1122.

IECEC2001-AT-02

A COMPARISON OF DC-DC CONVERTER TOPOLOGIES
FOR HIGH VOLTAGE SSP SYSTEMS

William Dillard, Qinghong Yu, R. M. Nelms
Auburn University
ECE Department
200 Broun Hall
Auburn University, AL
36849-5201
(334) 844-1840
dillard@eng.auburn.edu

ABSTRACT

A collection of medium power dc-dc converters topologies; series resonant, phase shift converter and isolated boost [1], have been investigated for high-voltage space solar power (SSP) applications in the range of 5 – 10 kW at 500 - 1000 V output. Several performance criteria are isolated for comparison, namely: efficiency, semiconductor device stresses (switches and diodes), transformer complexity and control system complexity, where all quantifiable criteria were extracted from PSPICE simulations. To provide a common point of reference, all topologies have the same switching frequency, input voltage, output voltage and output power with identical switch, diode and transformer models. Based on simulation results, the phase shift converter is recommended for this application.

INTRODUCTION

The feasibility of directing electrical energy to earth from space-based solar collectors has been under investigation for some time [2]. While various schemes have been proposed, each has common features, however, as seen in the concept schematic in Fig. 1. This work is an investigation, by PSPICE simulation, of the dc-dc converter to determine the best topology candidate based on the criteria listed in Table 1.

Three dc-dc converter topologies were selected for comparison, series resonant, phase shift and isolated boost (or Clarke). A hard-switched converter was included as a comparative baseline. Simulations for steady-state performance were executed at the same switching frequency (20 kHz), input voltage (300 V), output voltage (800 V), and output power (7.5 kW), with identical switch, diode and transformer models.

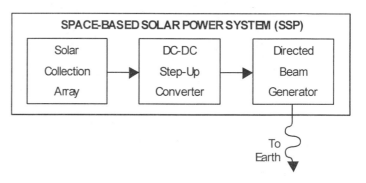

Fig. 1. Concept diagram of a generic SSP system.

Table 1. Converter Comparison Criteria

Converter	Efficiency
Switches/Rectifiers	Voltage Stresses
	Current Stresses
	Conduction Losses
	Switching Losses
Transformer	Saturation
	Complexity
Control Complexity	
Environment Immunity	

NOMENCLATURE

L_{MAG}	Magnetizing inductance (H)
L_{SEC}	Secondary inductance (F)
m	Catch winding transformer turns ratio
n	Transformer turns ratio

A GENERAL TOPOLOGY

For the power and voltage levels under consideration, in addition to the possibility of hostile environments, a full bridge inverter–full bridge rectifier configuration, shown in Fig. 2a, was selected as the best candidate. Several features of this topology also simplify the design of the transformer.

- There is no inherent dc component in the transformer current, exploiting the full excursion of the B-H characteristic.
- As compared to a half bridge inverter, the output voltage is doubled, reducing the transformer turns ratio.
- A full bridge rectifier eliminates the need for a split secondary winding, again reducing the turns ratio and further simplifying transformer design.
- A full bridge rectifier output allows converters to be stacked serially to produce a much higher dc output voltage ($\approx$40 kV) as required by the SSP system shown in Fig. 2b.

The efficiency of the converter has significant impact on the cost and success of the SSP system. Efficiency is affected by all loss mechanisms within the converter. Principle among these are switch and rectifier losses – both conduction and switching. For given input/output conditions, conduction losses depend directly on the on-resistances and forward voltage drops of switches and rectifiers, respectively. On the other hand, switching losses are more manageable and can be minimized by a variety of zero voltage switching (ZVS) and zero current switching (ZCS) schemes [3].

Within the general class of full bridge inverter – full bridge rectifier configurations, the following topologies were selected for simulation: series resonant, phase shift converter and isolated boost converter. Simulations for a hard-switched converter are included as a comparative baseline.

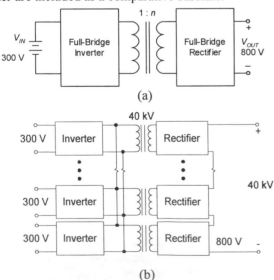

(a)

(b)

Fig. 2. The full bridge inverter – full bridge rectifier dc-dc converter: (a) single and (b) stacked.

TRANSFORMER MODEL

Since the converter outputs in this application are to be connected serially, the transformers must have large isolation voltages. In general, this reduces the coupling between primary and secondary and is modeled as a leakage inductance. The transformer model shown in Figure 3 is used all simulations. Winding L_{LEAK} is the leakage inductance (5 µH), L_{MAG} is the magnetizing inductance (1.8 mH) and L_{SEC} sets the turns ratio is set by the relationship

$$n = \sqrt{\frac{L_{SEC}}{L_{MAG}}} \tag{1}$$

All inductor losses, core and copper, are modeled by the resistor, R_{LOSS}. Its value was selected such that inductor losses are 2% of the 7.5 kW output power, namely 240 mΩ. Interwinding capacitance is modeled by C_{WIND}, set at 50 pF.

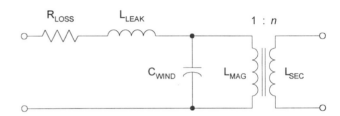

Figure 3. The transformer model used in PSPICE simulations.

MOSFET AND RECTIFIER MODEL

Ten parallel-connected IRFP150 HEXFETs are used for each switch. This creates a lower effective on-resistance, significantly reducing conduction losses. The SPICE model for the IRFP150 is modified, as shown in Table 2 where κ is the number of paralleled devices, such that a single device has an on-resistance of 1/10 and terminal capacitances 10 times those in the IRFP150 datasheet. In Table 2, κ is the number of paralleled devices.

Table 2. Scaling PSPICE MOSFET Model Parameters

SPICE Parameter	Symbol	Unit	Scale Factor
Transconductance Parameter	KP	A/V^2	1
Gate Resistance	RG	Ω	1/κ
Source Resistance	RS	Ω	1/κ
Drain Resistance	RD	Ω	1/κ
Gate Width	W	m	κ
Drain-Source Resistance	RDS	Ω	1/κ
Bulk-Drain Capacitance	CBD	F	κ
Bulk-Source Capacitance	CBS	F	κ

HARD-SWITCHED CONVERTER

A hard-switched converter, shown in Figure 4a, was included as a baseline for comparative purposes. Of particular importance are the switching losses during commutation,

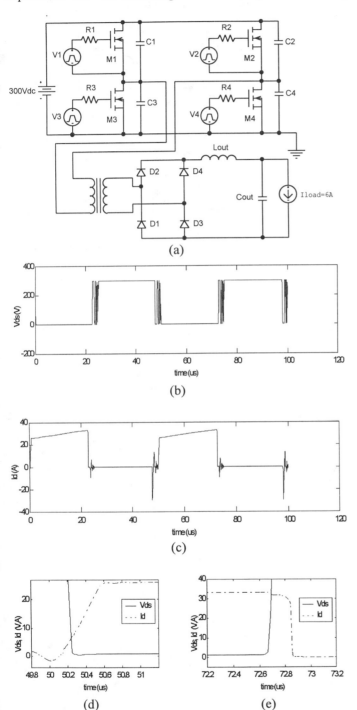

(a)

(b)

(c)

(d)   (e)

Fig. 4. The hard-switched converter: (a) schematic, (b) simulated switch voltage and (c) current waveforms. Switch waveforms emphasizing losses during commutation at (a) turn-on, and (b) turn-off.

characterized by the V_{DS} -I_D waveforms shown in Fig. 4d for turn-on and Fig. 4e for turn- off, and overall efficiency. From Fig. 4e the switching losses at turn-off are significantly larger than at turn-on and should be reduced by some soft switching scheme. Simulations predict total switch losses (both conduction and switching) of 88 W and a converter efficiency of 96%. It is the paralleling of ten switches, and a relatively low switching frequency that produces such a high efficiency.

SERIES RESONANT CONVERTER

Switch voltages and currents waveforms for the series resonant converter, seen in Fig. 5, are given in Fig. 6. Of particular interest are the switching losses. Figure 6c shows strong ZVS in the turn-on transient. While the turn-off transient in Fig. 6d is not zero-switched, it is significantly better than in the hard-switched case.

Simulation results also show that of the three soft-switched configurations, the series resonant converter has the highest switch currents (50 A) and the largest power loss in the transformer (262 W). Another disadvantage of the series resonant converter is the sensitivity of the output voltage to switching frequency, which is rather nonlinear and dependent on the Q of the L_rC_r tank. In the case of stacked outputs, failure of some converters requires adjustment of the frequency in some or all of the other converters, altering the EMI emissions of the system.

Series resonant performance can also be adversely affected by load conditions. Changes in the load current reflect back to primary, affecting the performance of the LC series tank. This can degrade the ZVS effectiveness and increase switching losses. Thus, a converter whose ZVS is optimized at a particular load current may have a limited load range. This is particularly troublesome this application since, should one converter fail, the others must increase their output voltages and currents.

PHASE SHIFT CONVERTER

The Phase Shift converter, shown in Fig. 7, has two features that distinguish it from the hard-switched converter.

- The transformer leakage inductance and the parasitic switch capacitance work in unison to provide some degree of ZVS. Both the inductance and capacitance can be augmented with external elements as necessary.
- The gating signals are independent, allowing the commutation of the poles to overlap. This produces a unique control scenario where duty cycle and frequency are fixed and the amount of overlap regulates the output voltage.

Also, the phase shift converter is more robust than the series resonant converter in that it can maintain zero voltage switching over a wider load range.

Switch voltage and current waveforms in Figure 8 show the piecewise linear nature of the switch current and the ZVS at turn-on. As in the series resonant case, the turn-off transient displays neither ZCS nor ZVS, but is an improvement over the hard-switched baseline. From the simulations, the phase shift converter has the lowest switch and rectifier losses (18.8 W and 9 W, respectively). However, it has the largest rectifier stress (910 V), greater than the output voltage (800 V), due to its LC output filter. Finally, with constant frequency and duty cycle, better control of EMI emissions is possible.

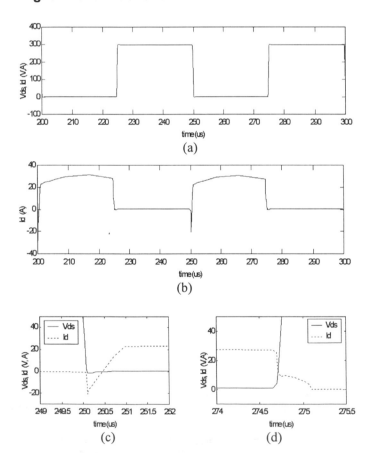

Fig. 5. Schematic for the series resonant converter.

Fig. 6. PSPICE simulations for series resonant converter: (a) switch voltage, (b) switch current, (c). Switch current and voltage simulations in commutation: (c) turn-on, and (d) turn-off.

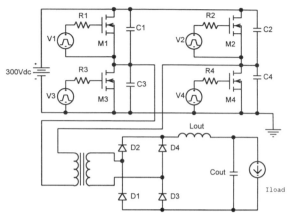

Fig. 7. Schematic for the phase shift converter.

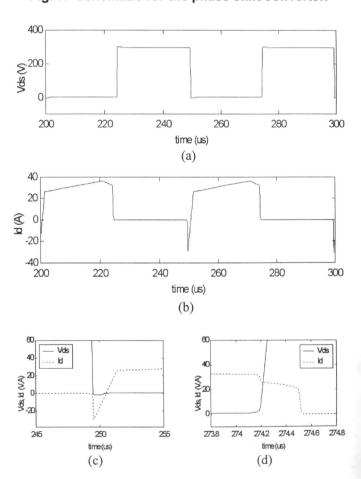

Fig. 8. PSPICE simulations for phase shift converter: (a) switch voltage, (b) switch current, (c). Switch current and voltage simulations in commutation: (c) turn-on, and (d) turn-off.

ISOLATED BOOST CONVERTER

In the isolated boost converter, shown in Fig. 9, the switches alternate between energizing the input inductor, L_{in}, and delivering that energy to the load. Details of switch conditions for a full switching cycle are outlined in Table 4.

During the delivery phase, defined in Table 4, D_{CATCH} is on and the secondary winding voltage equals V_{IN}. This protects the primary side winding from excessive voltages. When energy is delivered from the inductor to the load, the catch winding voltage is clamped by the input voltage and the turns ratio, m.

$$\left|V_{L_{in}}\right| = V_{in}\left(1 + \frac{1}{m}\right) \qquad (2)$$

The value of m strongly affects the currents in the catch winding transformer. For $m > 1$, the current become increasing large, decreasing the efficiency of the converter. For $m < 1$, the currents decrease but the primary voltage in (2) increases. In this investigation, $m = 1$ for all simulations, which sets the maximum voltage in (2) at 600 V.

Simulation waveforms for the switches in the isolated boost, shown in Fig. 10, suggest significantly worse switching losses than the other topologies. This is verified in the data where switch and rectifier losses are 59 W and 14 W, respectively. Also, loss in the catch winding diode, D_{CATCH}, is 31 W. However, the transformer losses are much less, only 91 W, which brings the overall efficiency (95%) in line with the other converters. Based on Figs. 10c and d, operating the isolated boost at higher switching frequencies will degrade the efficiency much more than in the other topologies.

Since the converter is current fed, the input inductance will limit the current rate of change should the switches temporarily short-circuit, perhaps due to radiation-induced triggering within the control circuitry. This simplifies the design of protection circuits. Finally, since there is no LC output filter, the rectifier stress is limited to the output voltage value, while switch voltages can exceed the input voltage.

Table 4. Switching Sequence in the Isolated Boost Converter

Phase	Boost 1	Delivery 1	Boost 2	Delivery 2
On Devices	All	M_1, M_4	All	M_2, M_3

CONCLUSIONS

Critical simulation results for all four converters under study are listed in Table 5. While all topologies have essentially equal efficiencies at the simulation frequency, the switching losses in the isolated boost will degrade significantly at higher switching frequencies highest efficiency, eliminating it as a

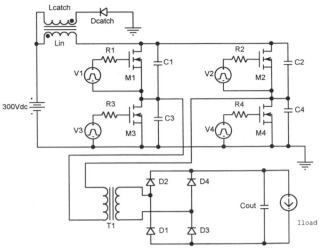

Fig. 9. Schematic for the isolated boost converter.

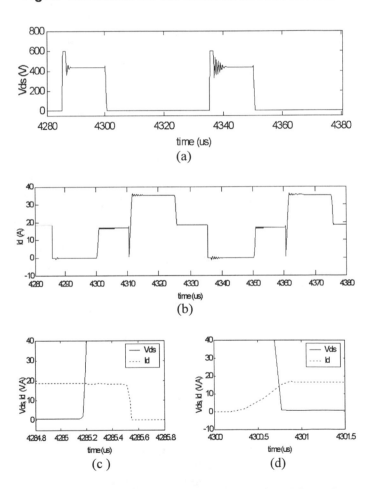

Fig. 10. PSPICE simulations for isolated boost converter: (a) switch voltage, (b) switch current, (c). Switch current and voltage simulations in commutation: (c) turn-on, and (d) turn-off.

viable candidate. The phase shift converter has switching losses comparable to the series resonant converter but has other advantages.

- Lower switch and rectifier losses (switching plus conduction).
- Lower transformer losses
- Constant frequency – constant duty cycle control produces tighter control over EMI emissions.

The sole disadvantage is the rectifier stress, which is higher due to the output filter inductor. For these reasons, the phase shift converter is recommended for the specified medium power SSP converter.

Table 5. Comparison of Converter Performance

Parameter	Full Bridge	Phase Shift	Series Resonant	Isolated Boost	Units
Output Voltage	805	800	805	805	V
Switch Stress[1]	299	299	299	439/600 (2)	V
Switch Current[1]	34.5	33	50	35	A
Rectifier Stress[1]	908	910	805	800	V
Rectifier Current[1]	10.4	10	15.6	19	A
Switch Loss[1]	22	18.8	21	59	W
Rectifier Loss[1]	8	9	10.4	14	W
Duty Cycle	45	NA	NA	70	%
Frequency	20	20	21	20	kHz
Phase Shift	NA	44.2	NA	NA	%
Inductor Current	28.5[3]	29.6[3]	33.3[4]	30.3[5]	Arms
XFMR Losses	176.2	215	262	91	W
Catch Diode Power	NA	NA	NA	31	W
Efficiency (%)	96	96	95	95	%

[1]Per device   [2] V See Fig. 10.a.
[3]Output filter inductor   [4]Resonant tank inductor   [5]XFMR primary

REFERENCES

[1] Clarke, P.W., "Converter Regulation by Controlled Conduction Overlap", U.S. Patent 3,938,024, Feb. 10, 1976.
[2] Glaser, P.E., "Power form the Sun: Its Future," Science, **162**, pp 857-886.
[3] Steigerwald, R.L., 1995, "A Review of Soft-switch Techniques in High Performance dc Power Supplies, *Proceedings of the 1995 IEEE IECON 21st International Conference*, IEEE, Piscataway, NJ, **1**, pp. 1 -7.

Proceedings of IECEC'01
36th Intersociety Energy Conversion Engineering Conference
July 29-August 2, 2001, Savannah, Georgia

IECEC2001-AT-03

DESIGN AND IMPLEMENTATION OF A DIGITAL PID CONTROLLER FOR A BUCK CONVERTER

Liping Guo, John Y. Hung and R. M. Nelms
Department of Electrical & Computer Engineering
200 Broun Hall
Auburn University, Auburn, AL 36849-5201
Phone (334) 844-1830, Fax (334) 844-1809
guolp@eng.auburn.edu

ABSTRACT

Issues in the design and implementation of a digital controller for a buck converter are reported in this paper. An analog PID controller is designed for a generic buck converter. The controller is then transformed into a digital controller, which is implemented on a TI DSP. Three algorithm modifications are implemented to improve the performance of the digital PID controller. The modifications are dead zone, averaging digital filter and two sets of gains. The digitally-controlled buck converter is able to obtain both fast transient response and stable, accurate steady state response.

INTRODUCTION

The focus of this paper is the design and implementation of a digital controller for a buck converter. The circuit diagram of a generic buck converter is shown in Fig. 1. In the diagram, Vin is the input voltage, Vo is the output voltage, L is the filter

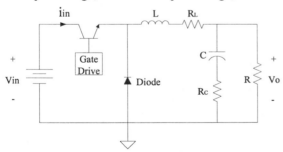

Fig. 1. Circuit diagram of the generic buck converter

inductance and C refers to the output filter capacitance. The parasitic elements are R_L and R_C, which are the winding resistance of the inductor and the equivalent series resistance (ESR) of the capacitance, respectively. The digital controller is

implemented using a TMS320F240 EVM from Texas Instruments, which features a 16-bit fixed point DSP controller with flash memory, three 16-bit, 6-mode, general purpose timers, 12 PWM channels and dual 10-bit, 8-channel ADCs. The control specific peripherals are most beneficial for the realization of a digital controller. The DSP operates at 20 MIPS with an instruction cycle time of 50 μs. The primary advantages of digital control over analog control are higher immunity to environmental changes such as temperature and aging of components, increased flexibility by changing the software, more advanced control techniques, and shorter design cycles [6-8].

Presented in the first section of this paper is the buck converter's small signal model derived using standard state-space averaging techniques. This is followed by a discussion of the control design for the buck converter. Modifications of the PID algorithm to improve converter performance are described in the third section. Experimental results are presented in the fourth section.

I.BUCK CONVERTER'S MODEL

The first step in the design of the controller is to obtain the mathematical model of the plant. The control-to-output transfer function of the buck converter, derived using standard state-space averaging technique, is [1]:

$$\frac{\hat{v}_0(s)}{\hat{d}(s)} = \left(\frac{Vo}{D}\right)\left[\frac{1+sRcC}{1+s\left(RcC+[R/\!/RL]C+\frac{L}{R+RL}\right)+s^2LC\left(\frac{R+Rc}{R+RL}\right)}\right] \quad (1)$$

The input voltage of the prototype buck converter Vin is 20 V, and the desired output voltage Vo is 12 V. L is 150 μH,

C is 1000 μF, and R is 10 Ω. The parasitic elements R$_C$ and R$_L$ are estimated to be 30 mΩ and 10 mΩ, respectively [7,8]. Figure 2 is the Bode plot for the small signal model of the buck converter at its nominal operating point. Frequency response data for this prototype converter was measured using a Model 102B analog network analyzer by AP Instruments, and compares favorably to the results predicted by Eq. (1).

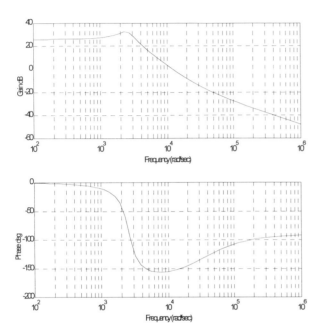

Fig. 2. Bode plot of the small signal model of the buck converter

The control-to-output transfer function is a common two-pole low pass filter, with a left-half-plane zero introduced by the ESR of the filter capacitance [1]. The cutoff frequency of the low pass filter is $\dfrac{1}{\sqrt{LC}}$. The magnitude falls with a slope of - 40dB/decade at the cutoff frequency. The phase associated with it is -180 degrees. The zero is at $\dfrac{1}{R_C C}$. There is a 20dB/decade magnitude rise at that frequency and the phase shift is 90 degrees. The magnitude of the transfer function depends on the duty cycle D. When D increases, the magnitude decreases; when D decreases, the magnitude increases. However, changes in D do not change the shape of the Bode plot. It only shifts the magnitude upward or downward.

II. CONTROL DESIGN OF THE BUCK CONVERTER

The transfer function of a proportional-integral-derivative (PID) controller is given by:

$$Gc(s) = K_P + \frac{K_I}{s} + K_D s \qquad (2)$$

K$_P$ is the proportional gain, K$_I$ is the integral gain, and K$_D$ is the derivative gain. The PID controller has one pole at zero and two zeros. The gains of the controller are selected using standard frequency response techniques [5].

The PID controller is designed based on the transfer function in Eq. (1). Both the sampling frequency and PWM frequency were selected to be 20 kHz. The digital controller acquires a sample once every sampling period (50 μs), utilizes a PID algorithm to calculate a new duty cycle, and updates the new duty cycle at the start of the next switching period. So the duty cycle is only able to be updated after one PWM period delay. This delay doesn't affect the magnitude plot in Fig. 2, but produces a phase delay which increases exponentially with frequency. The transfer function of the delay is e^{-Ts}, where T is the sampling period, and the Bode plot is shown in Fig. 3. Combining the plots in Fig. 2 and Fig. 3 yields the Bode plot of Fig. 4, which is the model used in the design of the PID controller.

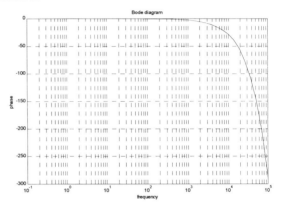

Fig. 3. Bode plot of the time delay

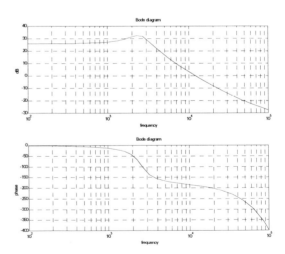

Fig. 4. Bode plot of the model to be compensated

The desired characteristics of the compensated system are that the gain at low frequencies should be high to minimize the steady state error and the crossover frequency should be as high

188

as possible but about an order of magnitude below the switching frequency to allow the power supply to respond to transients quickly. As a result, the phase margin should be in the range of 45 to 60 degrees to satisfy the transient response requirements [2].

A. PID Design I – Fast Transient Response

In design I, one zero is placed an octave below the cutoff frequency (approximately 260 radians/second) and the other one at the cutoff frequency. The Bode plot for the compensated system is shown in Fig. 5. As can be seen in this plot, the gain at low frequency is high, the phase margin for the compensated system is 55 degrees, and the bandwidth is 22,000 radians/second.

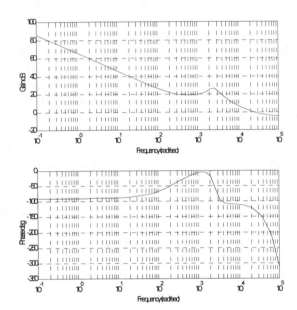

Fig. 5. Bode plot of compensated system by PID design I

B. PID Design II – Slow Transient Response and Higher Phase Margin

PID design II places a set of complex conjugate zeros close to the complex conjugate poles of the buck converter model. The complex conjugate zeros approximately cancel out the phase shift from the complex conjugate poles at the cutoff frequency. The Bode plot for the system compensated with this controller design is displayed in Fig. 6. This PID design yields a bandwidth of 8000 radians/second and a phase margin of 80 degrees. The bandwidth is lower than the one in PID design I, which slows down the system. The phase margin is larger than design I, which makes the system more stable.

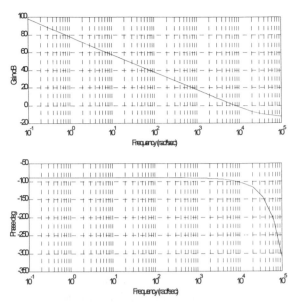

Fig. 6. Bode plot of compensated system by PID design II

C. Transformation Into the Digital Controller

The design in the continuous-time domain is converted to the discrete-time domain [4]. In the backward integration method (Euler rule), s is replaced with $\frac{1 - z^{-1}}{T}$ in Eq. (2) to yield:

$$Gc(z) = K_P + \frac{K_I Tz}{z - 1} + \frac{K_D(z - 1)}{Tz} \tag{3}$$

Equation (3) is converted to a difference equation, which is utilized to calculate a new duty cycle:

$$u(k) = K_P e(k) + K_I T \sum_{i=0}^{k} e(i) + \frac{K_D}{T}[e(k) - e(k - 1)] \tag{4}$$

In the difference equation, u(k) is the controller output for the kth sample, and e(k) is the error of the kth sample. The error e(k) is calculated as e(k) = Ref-ADC(k), where ADC(k) is the converted digital value of the kth sample, and Ref is the digital value corresponding to the desired output voltage. $\sum_{i=0}^{k} e(i)$ is the sum of the errors and e(k)-e(k-1) is the difference between the error of the kth sample and the error of the (k-1)th sample. The block diagram of the algorithm for the difference equation is shown in Fig. 7.

189

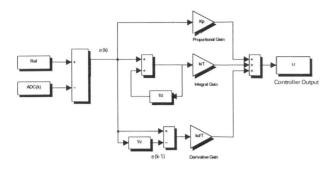

Fig. 7. Block diagram of the regular digital PID algorithm

III. ALGORITHM MODIFICATION

When the regular PID algorithm represented by Eq. (4) is implemented using design I, the duty cycle oscillates very much during steady state. The PID algorithm is modified to reduce this oscillation. The modifications include dead zone, averaging digital filter and two sets of gains.

A. Dead zone

The dead zone is a nonlinear function. The equation of the dead zone is:

$$p(k) = \begin{cases} e(k), when |e(k)| > \varepsilon \\ 0, \quad when |e(k)| \leq \varepsilon \end{cases} \quad (5)$$

p(k) is then supplied to the PID controller as the kth sample's error.

If the value of ε is too small, the transition between $|e(k)| > \varepsilon$ and $|e(k)| < \varepsilon$ will be very frequent, and the goal to stabilize the digital control system can not be reached. On the other hand, if the value of ε is too large, it will produce a large lag for the response of the plant. When ε is zero, the algorithm is a common PID algorithm.

B. Averaging digital filter

The averaging digital filter is used in the digital control system for the buck converter. The flowchart of the averaging digital filter is shown in Fig. 8. When the error between the

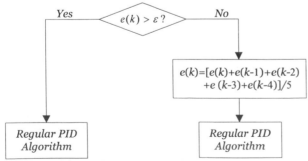

Fig. 8. Flowchart of averaging digital filter

reference and the output voltage is under ε, the averaging digital filter algorithm is applied. The samples taken in the last five sampling periods are added and divided by 5. The average value is used as the sample in the current sampling period to execute the regular PID algorithm, which is represented by the difference equation in (4). When the error is over ε, the regular PID algorithm is directly executed. Therefore, during the transient, the system is able to respond fast according to the current sampled data; while during steady state, the system will remain stable and obtain high output accuracy because the output is filtered to get a smoother value.

C. Two sets of gains

In Section II, there are two PID control designs for the buck converter. Design I has a higher bandwidth and smaller phase margin than design II, which makes the system respond faster and be less stable than the system with design II. As a result, design I gives fast and accurate transient response, but the steady state response contains too much oscillation. Design II improves the steady state response, but the transient response is slower. A digital controller permits the use of two sets of controller gains. The gains which produce a better transient response can be utilized when the input voltage or load changes very rapidly. When the system is operating in steady state, the second set of gains can be employed because of the improved steady state stability. Whether the system is operating in transient or steady state is decided by comparing e(k) with a certain value ε. When e(k)>ε, the system is considered in transient, and when e(k)<ε, it is considered in steady state. This is more difficult to accomplish with an analog controller and illustrates an advantage of digital controllers for DC-DC converters. The flowchart for the two sets of gains algorithm is shown in Fig. 9.

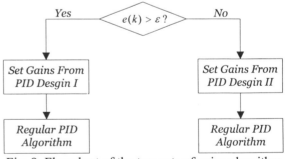

Fig. 9. Flowchart of the two sets of gains algorithm

IV EXPERIMENTAL RESULT

In all three algorithm modifications, ε is an adjustable variable, which can be determined experimentally. Different values of ε were tested on a buck converter. Table 1 compares the system performance for different values of ε: 200mV, 150mV and 100mV.

Table 1. Comparison of the system performance with different values of ε

$\varepsilon = 200$ mV	PWM oscillation	Ripple	Steady State Error
Dead Zone	59%~60%	100 mV	3 mV
Average Filter	56.9%~62%	150 mV	-10 mV
Two Sets of Gains	56.6%~63.6%	100 mV	-6 mV

$\varepsilon = 150$mV	PWM oscillation	Ripple	Steady State Error
Dead Zone	58% ~ 60%	60 mV	7 mV
Average Filter	56% ~ 63%	175 mV	20 mV
Two sets of gains	56.5% ~ 62.2%	100 mV	37 mV

$\varepsilon = 100$ mV	PWM oscillation	Ripple	Steady State Error
Dead Zone	Not Stable		
Average Filter	56% ~ 63%	175 mV	20 mV
Two sets of gains	56.5% ~ 60%	100 mV	20 mV

For the dead zone, the controller is stable only when $\varepsilon = 200$ mV. The smallest oscillation of the PWM signal and steady state error are obtained. When $\varepsilon = 150$ mV, the ripple is smaller. The duty cycle is stable most of the time, but occasionally oscillates for a very short time. This may be because of the frequent transitions between the regular PID algorithm and the dead zone. When ε is 100 mV, the system becomes unstable. Therefore, an optimal value of $\varepsilon = 200$ mV is chosen for the dead zone.

For the averaging digital filter, the oscillation of the PWM signal is very similar when ε varies from 200 mV to 100 mV. However, the ripple and the steady state error is the smallest when ε is 200 mV, so $\varepsilon = 200$ mV is also chosen for the averaging digital filter.

For the two sets of gains algorithm modification, the PWM oscillation and the ripple are very similar as ε varies, but the steady state error is smallest when $\varepsilon = 200$ mV, so this value of ε is chosen.

Based on the experimental results in Table 1, ε is selected to be 200 mV for all algorithm modifications.

The comparison of the start up transient response of the three algorithm modifications is shown in Fig. 10. R1 is the waveform from the dead zone, R2 is the waveform from the averaging digital filter and R3 is the two sets of gains algorithm. Since the dead zone takes any error under 200 mV to be zero, there is some oscillation of the output voltage after it goes to steady state, but after 200 ms the oscillation disappears. The algorithm modification of two sets of gains has a little higher overshoot. It may be because the switch of the gains slows the system down a little.

The ripple of the output voltage of the buck converter operating open loop is shown in Fig. 11. The magnitude of the ripple is under 80 mV. The ripple of the output voltage of the buck converter under closed loop with different algorithm modifications is shown in Figs. 12, 13, and 14, respectively.

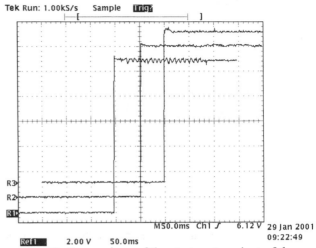

Fig. 10. Comparison of the start up transient of three algorithm modifications

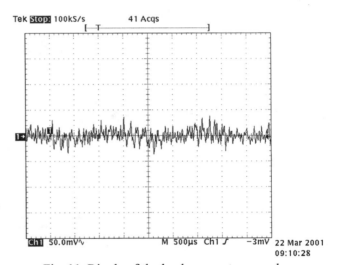

Fig. 11. Ripple of the buck converter open loop

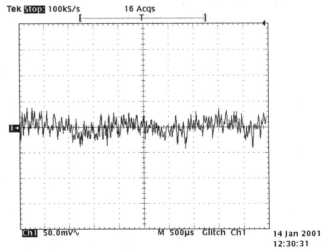

Fig. 12. Ripple of the buck converter with the dead zone

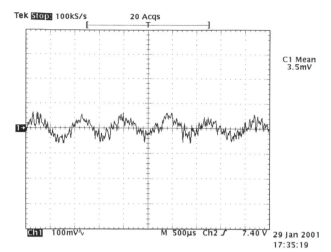

Fig. 13. Ripple of the buck converter with the averaging digital filter

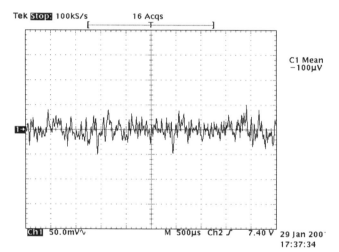

Fig. 14. Ripple of the buck converter with the two sets of gains

The magnitude of the ripple is under 100 mV for both the dead zone and two sets of gains, and is under 150 mV for the averaging digital filter. The ripple for the averaging digital filter has a frequency of 1kHz. However, the ripple with the other two techniques is random. In comparison, the magnitude of the ripple with the averaging digital filter algorithm modification is the highest among the three PID algorithm modifications. The ripple with the other two algorithm modifications is only a little higher than the open loop ripple.

CONCLUSION

Issues in the design and implementation of a digital controller for a buck converter are discussed in this paper. Two designs of the PID controller are presented: one with faster transient response, the other with slower transient response but higher phase margin. The digital PID controller is implemented on a DSP from Texas Instruments. The regular PID algorithm is modified to improve the controller's stability while maintaining the fast transient response. The algorithm

modifications are: the dead zone, the averaging digital filter and the two sets of gains. A table of experimental results compares the difference among the algorithm modifications and when different values of ε are selected. The difference of the results obtained with the three algorithm modifications is very small. ε is chosen to be 200 mV for all algorithm modifications based on the experimental results. Satisfactory results are obtained with the algorithm modifications. The output voltage becomes stable after 1ms with no overshoot at the start up. The steady state error is very small, and the ripple is comparable to the open loop ripple.

ACKNOWLEDGMENTS

This research was supported by the Center for Space Power and Advanced Electronics with funds from NASA grant NCC3-511, Auburn University, and the Centers' industrial partners.

REFERENCES

[1] Severns, R. P., and Bloom, G., 1985, *Modern DC-to-DC Switchmode Power Converter Circuits*, Van Nostrand Reinhold Company, New York.

[2] Mohan, N., Undeland, T. M., Robbins, W. P., 1995, *Power Electronics: Converters, Applications, and Design*, John Wiley&Sons, Inc.

[3] Krein, P. T., 1998, *Elements of Power Electronics*. Oxford University Press, New York, Oxford.

[4] Ogata, K., 1987, *Discrete-Time Control Systems*. Prentice-Hall, Inc., New Jersey.

[5] Phillips, C. L., and Harbor, R. D., 1996, *Feedback Control Systems*, Prentice-Hall, Inc., New Jersey.

[6] Vallittu, P., Suntio, T., and Ovaska, S. J., 1998, "Digital Control of Power Supplies-Opportunities and Constraints", Proceedings of the thirty third Intersociety Energy Conversion Engineering Conference, Vol. 1, pp. 562-567.

[7] Boudreaux, R. R., Nelms, R. M. and Hung, J. Y., 1996, "Digital Control of DC-DC Converters: Microcontroller Implementation Issues," Combined Proceedings of HFPC Power Conversion & Advanced Power Electronics Technology, Power systems World '96, pp. 168-180.

[8] Boudreaux, R. R., Nelms, R. M., Hung, J. Y., and Mathison, L.C., 1995, "Digital Control of a Buck Converter Using an 8-bit Microcontroller," Technical Papers of the Tenth International High Frequency Power Conversion 1995 Conference, pp. 238-251.

Proceedings of IECEC'01
36th Intersociety Energy Conversion Engineering Conference
July 29-August 2, 2001, Savannah, Georgia

IECEC2001-AT-40

INTELLIGENT SYSTEMS FOR POWER MANAGEMENT AND DISTRIBUTION

Robert M. Button
NASA Glenn Research Center
Cleveland, OH 44135

ABSTRACT

The motivation behind an advanced technology program to develop intelligent power management and distribution (PMAD) systems is described. The program concentrates on developing digital control and distributed processing algorithms for PMAD components and systems to improve their size, weight, efficiency, and reliability. Specific areas of research in developing intelligent DC-DC converters and distributed switchgear are described. Results from recent development efforts are presented along with expected future benefits to the overall PMAD system performance.

INTRODUCTION

As power electronics technology has advanced over the past decade, there has been increasing interest in replacing high maintenance mechanical and hydraulic systems in automobiles, aircraft, and aerospace vehicles with all electric systems. To achieve this goal the NASA Glenn Research Center is researching technologies that will enable more modular, reliable, and robust power management and distribution (PMAD) systems.

A PMAD system is defined as the power components necessary to "connect" electrical sources (generators, solar arrays, batteries) to the loads. In DC power distribution, the PMAD system is comprised of source regulators, storage charge/discharge regulators, voltage converters, and protective switchgear. Even the cable itself is part of the PMAD system.

To achieve this vision of reliable and robust PMAD systems, NASA Glenn is researching the benefits of using adaptive, robust, and distributed digital controllers in place of conventional analog controllers in devices such as DC-DC converters and protective switchgear.

Until recently the performance of available microprocessors and digital signal processors (DSPs) were too slow to be used in fast-acting, high frequency PMAD components. However, recent advances in the speed and size of these processors and periphery, such as analog-to-digital converters, has made their use now possible in all kinds of PMAD components and systems.

Just as the fuel efficiency and emission controls of the modern automobile would have been impossible without the advent of engine control computers, we expect that similar use of intelligent, digital control will have many benefits that will allow PMAD systems to achieve

- higher efficiency and lower weight
- better power quality and transient response
- higher fault tolerance and reliability, and
- lower integration costs.

It is really the last two bullets that will make or break electrical power systems in automotive and aerospace vehicles. While higher efficiency, lower mass, and better power quality can be achieved without digital control, there are many several reliability and integration problems that cannot be solved without digital control.

POWER CONVERTERS

Many advantages of intelligent controls are expected to be found in the power converter function. These devices, typically DC-DC converters for spacecraft systems, are some of the most complicated and pervasive devices in the

PMAD system. Very few loads can use the electrical distribution voltage directly, especially in high power spacecraft, so DC-DC converters are required to change voltage levels needed for the loads.

The most important characteristics of a DC-DC converter are its size, weight, efficiency, power quality, fault tolerance, and reliability. These characteristics are heavily inter-related. An improvement in one area may have additional benefits in another area. For example, a converter with high efficiency generates less heat, thereby reducing or eliminating any active or passive cooling. This results in a reduction in the size and weight of the converter.

Today's DC-DC converters rely solely on specialized analog circuits to perform control, regulation, and protective functions of the converter. However the linear, analog design limits their ability to adjust and adapt to the wide range of operating conditions. A digital controller, on the other hand, can easily make adjustments, thereby optimizing the control and operation of the DC-DC converter at all times.

POWER EFFICIENCY

There are several methods for increasing the power efficiency of DC-DC converters using intelligent control. The first method involves non-linear control algorithms to improve the efficiency of a single DC-DC converter. The second method involves using digital controllers and a communication bus to coordinate the function of modular converters.

Variable Switching Frequency

In many conventional DC-DC converters the switching frequency of the input bridge is fixed during the design to achieve a balance between overall efficiency and the size of passive components. As the switching frequency is increased, the size of the passive components is decreased (input capacitors, isolation transformer, and output capacitors) while the switching losses in the primary switch and rectifier diodes is increased. These trade-offs between increased losses and decreased size leads to a fixed frequency operation at one desired optimal point.

The addition of intelligent control can improve the "overall" efficiency of the converter by varying the switching frequency versus load.

By being able to lower the switching frequency of a converter at lower power, the switching losses can be reduced, thereby increasing the low-power efficiency of the converter. In experimental testing, a 1kW full-bridge DC-DC converter with a nominal frequency of 20kHz was switched at frequencies down to 5kHz at different power

levels [1]. The data in Figure 1 shows a marked improvement in efficiency at lower frequencies, especially at low power levels.

It should be noted that the current and voltage ripple in the converter increased, as expected, when the switching frequency was reduced. However, the ripple magnitudes at low power were still well below that seen at full power loads.

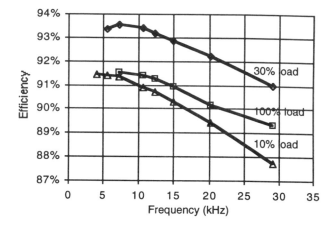

Figure 1 - Converter Efficiency vs. Switching Frequency

Cooperation with Paralleled Units

In systems using on-line spare modules to increase reliability, intelligent control can again improve the overall system efficiency. These on-line spares used to meet N+1 redundancy requirements are excess modules above what the nominal power levels require. Therefore, if one of the modules fail, the excess module will allow continuous operation at rated power levels.

If the number of modules used is small (< 4) then the N+1 redundancy leads to converters operating at lower power levels with lower efficiency. For example, a 2kW load being fed by a converter made up of 1kW modules uses three converters to meet N+1 redundancy. If all converters are operational, then at 75% load (1.5kW) all three converters are operating at 500W each where efficiency is less than ideal. An intelligent controller would be able to coordinate the three paralleled converters so that at 75% load only two modules would be operational, resulting in higher overall efficiency.

POWER QUALITY

Another area that can benefit from intelligent control is the size and power quality of the PMAD system. Power quality specifications for PMAD components and systems are written to ensure that voltage and current ripple in the system is kept to a minimum. These specifications have a profound impact on the design and sizing of passive components in DC-DC converters.

An approach to input filter reduction that has been used in custom DC-DC converters in the past is the technique of phase-staggering paralleled switch-mode power supplies (SMPS). This method coordinates the switching phases of the DC-DC converter input bridges so that the input current ripple from one converter is cancelled by other converters. By doing this, two benefits are accrued. First, the input current ripple magnitude is greatly reduced since the input currents cancel each other out when phase-staggered at $360°÷N$ phase increments, where N is the number of converters operating in parallel or series.

Second, the input current ripple frequency is increased resulting in a ripple frequency that is $f_s•N$ where f_s is the baseline switching frequency. These two factors combine to significantly reduce the capacitive and inductive filtering which makes up a large part of all SMPS.

With current technologies, implementation of the phase-stagger between converters is coordinated by a single controller that accepts the timing signals from a "master" device and generates the staggered "slave" signals used to drive the other devices. This method was used in a recent investigation of phase-stagger benefits (Figure 2) and currently Vicor, Inc. [2] has developed a custom chip to provide phase-stagger functions for up to 12 units in parallel.

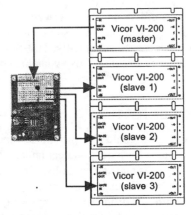

Figure 2 - Diagram of Vicor Phase Shift Test Setup

Using the test setup shown in Figure 2 it was found that phase-stagger techniques reduced the passive component sizes of the input filters by a remarkable 75-95% [3]. Test data is shown in Figures 3 and 4. As you can see, the unfiltered input current ripple has been greatly reduced from 6.75 A_{p-p} to about 2 A_{p-p} while the ripple frequency has been multiplied from 50 kHz to 200kHz. These improvements greatly reduce the amount of passive filtering needed to meet power quality and conducted electromagnetic interference (EMI) specifications.

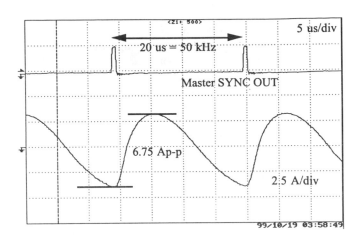

Figure 3 - Converters in-phase. Load = 3.1 A

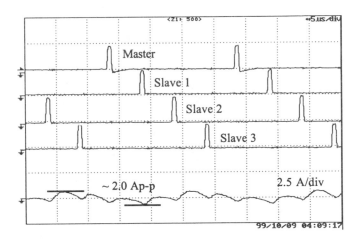

Figure 4 - Phase shifted converters. Load = 3.1 A

It is clear that the phase-stagger control is an ideal technology to improve PMAD system size and power quality. However, the dependence on a centralized controller results in a serious single-point failure that could bring down the entire converter. We can address the single-point failure problem and take advantage of the phase-stagger benefits by using an intelligent, digital controller in each converter

There are two methods we are pursuing to allow this distributed cooperation with paralleled converters. Initially, we'll be investigating a high-speed data bus to synchronize and delay the switching phase of paralleled converters. For fixed frequency converters this adjustment does not need to be done on a cycle-by-cycle basis, so bandwidth requirements are not too critical. However, the communication protocol chosen may present latency problems that can interfere with the critical timing required for proper synchronization.

Ultimately, we want to develop an intelligent controller that can "listen" to the noise ripple on the system and adjust its switching phase to minimize it. This is expected to require very fast digital signal processors (DSPs) running fast Fourier transform (FFT) algorithms with closed loop control on minimizing harmonic amplitudes. If this technique is successfully developed, then each converter would have the means for independent operation, yet be able to negotiate with other devices to achieve minimized ripple currents and result in smaller, more compact DC-DC converters.

COMPONENT HEALTH MONITORING

An intelligent controller can also monitor and track the health of some key components in the DC-DC converter. By monitoring the health of key components, impeding failures may be predicted and allow for corrective actions to be taken before catastrophic failure. The corrective actions may include reducing load current to extend lifetime, activating on-line spares, or even replacing failing converters or devices on manned missions.

For DC-DC converters, the most likely source of failure occurs in the active devices - the input switches, the switch drive electronics, and the output rectifiers. These devices can fail instantaneously or can degrade gradually over time. Instantaneous failures cannot be predicted or, in many cases, prevented, so these failures will temporarily affect system performance. An intelligent controller does, however, have the ability to record sensor data during a failure offering important historical information to determine the cause of failure.

More importantly, an intelligent controller has the ability to "remember" past performance of the converter and can monitor trends in sensor data and control inputs over time that may signal the degradation of the active components. For example, we know that as MOSFETs degrade, their on resistance, $R_{ds}(on)$, increases resulting in higher conduction losses and higher operating temperatures that lead to further accelerated degradation. In a typical full-bridge DC-DC converter, as $R_{ds}(on)$ increases, the pulse width needed to maintain regulation would also increase. By monitoring the load current vs. pulse width data over time, trends may be identified that portend the eventual degradation and failure of active switches.

Successful health monitoring requires confidence in the sensors used to measure key information. If sensor data suggest impending converter failure, it is important to be able to trust that a faulty sensor is not the cause for alarm. Intelligent systems have the added benefit of "sanity checking" sensor data in many components to rule-out sensor drift or failure. A good example would be the cross-checking of a DC-DC converter output current sensor against the input current sensor of a downstream device.

SWITCHGEAR

Another area that we expect to benefit from intelligent control is in the application of remote power controllers (RPCs) in a PMAD system. An RPC is a semiconductor device that can provide load on/off control, soft-start/current limit, fault protection, and load shedding.

The functionality of the RPC is straightforward and can be implemented in analog control circuits. However, the unpredictable cable and load impedance characteristics can lead to stability problems with an analog controller. For example, coordinating the switch's soft start algorithms with its protective function can lead to oscillations between the switch and the load. To combat these potential oscillations, many systems have in-rush current limit specifications levied upon the power converters. This forces the converter manufacturers to include series switch elements to limit the in-rush current, resulting in a system with one semiconductor switch followed by another semiconductor switch in series with the power flow. The result is redundant components, excess power losses, and additional failure modes.

An intelligent switch would be better suited to detect the amount of protection necessary, and would be able to coordinate with downstream devices to eliminate any instability. For example, an intelligent switch would "expect" the current to rise above trip levels when it is turned on into a load and could arbitrarily extend the time limit to prevent oscillations. Once at steady state, another control algorithm would take over shortening the time limit to respond to true faults quickly. This type of flexibility is difficult to achieve using analog control circuits.

SOFT FAULT DETECTION

Intelligent switchgear could also provide advanced fault detection algorithms for hard-to-detect faults like continuous soft-faults and arcing faults. These faults, typically caused by failures in the wire insulation, cause low amplitude currents or intermittent sparking between hot and ground or hot and return. These faults do not draw large amounts of current and therefore go undetected by conventional protection devices. While the faults do not cause system disturbances, they pose a serious fire threat as "hot spots" develop that can ignite combustible materials. In fact, a recent report has concluded that arcing faults played major roles in two recent airline disasters, TWA flight 800 in 1996 and the Swissair III crash off Nova Scotia in 1998 [4]

Analog detectors can do a good job of detecting continuous soft faults. The common household ground fault circuit interrupter (GFCI) is a good example. This device simply detects a ground fault by sensing an imbalance between line and neutral currents using a common core.

The continuous line to neutral fault is more difficult to detect in that there is no imbalance between line and neutral. The only way to detect this fault is to measure an imbalance at each end of the cable. This is where intelligent power systems can help. By using the data gathered from the distributed sensors included in each component, hot to return faults can be detected.

Unfortunately, intermittent arcing faults may escape all conventional analog detection methods due to their very high frequency content and short time duration. One possible detection method is to use high-speed signal analysis to detect the high frequency signature of an arcing fault in the system. Again, distributed digital signal processors will be able to perform this signal analysis and may even be able to coordinate their analysis results to determine an approximate location of the arcing fault. The biggest challenge to this method of arc fault detection is the ability to ignore electrical arcs produced normally by motors, relays, and manual switches. The NASA Glenn Research Center along with Howard University has begun a program to characterize arc faults in high voltage DC systems and to develop advanced algorithms to detect and locate the faults.

SYSTEM BENEFITS

While the concepts presented so far could have positive impacts on almost all PMAD systems, the real benefits are to be found in large, distributed power systems such as the International Space Station and possible future manned missions to Mars. In these systems, not only will components have better performance and higher reliability, but the intelligent controllers will also provide unparalleled levels of flexibility and reconfiguration that are possible in manned systems.

Some of these "system benefits" have already been discussed in the sections above such as the coordination between switchgear and converters to ensure stability and help detect faults.

However, intelligent PMAD components are expected to have the greatest impact on the system design and integration costs associated with large, complex, distributed systems. Two of the goals we are currently working toward for intelligent PMAD systems is "Plug and Play Configuration" and active stability control.

PLUG AND PLAY

The phrase "Plug and Play" is borrowed from the mid-90's marketing of easy-to-integrate personal computer hardware and peripherals. This improvement in hardware integration came from the adoption of hardware identification and programming technologies. In essence, the previously complex process of determining the hardware configuration, manually setting the proper jumpers to work with your system, and knowing what software drivers to install was reworked so that the computer and hardware could do all the work. This is our goal for future PMAD systems and components.

Specifically, we are developing technologies that will allow switchgear to "sense" the source and the load that it is connected between so that key setpoints will be automatically set. To do this, we have to develop technologies that would allow an intelligent component to "know" what device was feeding power to it and what loads were connected to it at all times. Knowing this information and the basic characteristics of the devices attached to it, the component would be able to make independent decisions about how to best handle any anomalies that would occur. Parameters such as maximum current levels or load prioritization could be autonomously set without having an army of ground-based engineers determine if a new configuration would cause any system instabilities.

The implications to having this amount of flexibility coupled with modular system design are great. It could mean that the amount of spare components required for long duration manned missions could be greatly reduced. For example, a failed component in a critical system could be easily swapped-out with a similar component in a non-critical system if they had the ability to reprogram themselves based on their location.

ACTIVE STABILITY

One of the key problems with integrating large, distributed electrical power systems is guaranteeing stable operation between active control loops operating in series. Some examples of possible unstable operation include large DC-DC converters feeding smaller DC-DC converters, each with their own control loops. Also, the problem described previously with active switchgear feeding DC-DC converters causing relaxation oscillators.

Another goal of the intelligent PMAD system is to greatly reduce integration time and cost by developing technologies that will guarantee stability at all times. These will probably include advanced detection algorithms that will look for signs of instability and actively adjust controller gains to return to stable operation. By being able to guarantee stable operation at all times the daunting job of system stability design and analysis is eliminated. This should greatly reduce the time and cost associated with integrating large, complex PMAD systems in the future.

COMMUNICATIONS

Finally, a critical element in developing an intelligent PMAD system is the inter-device communication. Without communication, intelligent devices would be very limited in their ability to detect and isolate faults, verify sensor data, and cooperate with other devices to improve power quality and reliability. By developing a high bandwidth, intra-device communication medium, the promise of an intelligent PMAD system can be achieved.

There are several technologies that we are considering for intra-device communication. Power-line communication technology uses the power cables to transmit digital information. This technology is very attractive since the PMAD devices would not require extra cabling for communication. However, it may not support the high bandwidth required for some functions, and transmitting signals across transformer isolated converters and switches in the off state remain a challenge.

Dedicated, high-speed serial busses like IEEE-1394 (a.k.a. Firewire or iLink) would likely be very capable in meeting the bandwidth requirements. However, the technology also defines a transport protocol that may not be as sensitive to latency as would be required. Also, each device would require extra data ports and dedicated data cables would be necessary for this implementation.

The most attractive option is a wireless communication technology. This would have the benefit of being high bandwidth and wouldn't require any extra cabling. The biggest issue will probably be the acceptance of on-board wireless communication in a space environment that is accustomed to being very noise-free.

CONCLUSIONS

It is clear that future improvements to power management and distribution (PMAD) systems will be found in the flexible and robust control of intelligent, digital systems. They will be able to improve the performance of individual components using non-linear control optimization, and will open up vast possibilities of improving coordination and autonomy at the system level.

CONTACT

Robert M. Button
NASA Glenn Research Center
Mail Stop 333-2
21000 Brookpark Rd.
Cleveland OH 44135
(216) 433-8010
robert.button@grc.nasa.gov

REFERENCES

1. Sustersic, John, et. al., "Design and Implementation of a Digital Controller for DC-DC Power Converters", 2000 SAE Power Systems Conference, San Diego, CA, Oct. 31,2000
2. Vicor Corporation, 25 Frontage Road, Andover, MA 01810, http://www.vicr.com
3. Button, R.M., et. al. "Digital Control Technologies for Modular DC-DC Converters", 2000 IEEE Aerospace Conference, Big Sky, MT, March 18-24, 2000.
4. Furse, Cynthia and Haupt, Randy, "Down to the Wire", IEEE Spectrum, February 2001.
5. Ground Fault Circuit Interrupter, http://www.codecheck.com/gfci_principal.htm

IECEC 2001-AT-45

Using Half-Bridge Topologies as Power Electronic Building Blocks (PEBBS)

Martin Beck
TRW Space and Technology Group
Redondo Beach, Ca 90278

D. Kent Decker
TRW Space and Technology Group
Redondo Beach, Ca 90278

Frederick J. Wolff
NASA John Glenn Research Center at Lewis Field
Cleveland, OH 44135

ABSTRACT

The emphasis on reduced cost in space electronics has led to the investigation of common circuitry across multiple applications. This paper explores the use of a common half-bridge power circuit for use in a space Flywheel Energy Storage System (FESS). The half-bridge circuit is used for magnetic bearing control and bi-directional power flow in conjunction with a synchronous motor. The paper discusses aspects of the circuit design as well as the associated control systems

INTRODUCTION

The emphasis on reduced cost in space power electronics has led to the investigation of common circuitry called power electronic building blocks (PEBBS) that can be used with multiple applications. The broad range of space applications includes solar power conversion, bi-directional processing of power for energy storage devices such as nickel-hydrogen and lithium ion batteries as well as high speed flywheels, and control of the various actuators on a spacecraft. The diverse nature of these applications challenges the task of developing a high performance, low cost, common building block. This paper explores the use of a common half-bridge power circuit as a potential PEBBS candidate for a Flywheel Energy Storage System (FESS).

A Flywheel Energy Storage System (FESS) provides an excellent application to use the PEBBs. The FESS is a technology that uses a high speed spinning rotor to store energy mechanically. A permanent magnet motor/generator with a three-phase stator is used to add energy to or take energy from the rotor, which increases or decreases the speed of the rotor. The stator of the motor/generator can be driven from a DC source bus using three of the half bridge PEBBs in parallel and the appropriate control to manage the current into or out of each phase. For the FESS machine, these currents are ideally sinusoidal for each phase. The motor/generator is rated at approximately 3.6 kW over its entire speed range so each PEBB is processing 1/3 of this power, or 1.2 kW.

The FESS rotor is levitated in a magnetic field to eliminate any losses due to friction of a mechanical bearing. This "magnetic bearing system" provides another application for the PEBB. To suspend the FESS rotor, the magnetic bearing has to apply forces on the rotor in five axes: the x-axis and y-axis or radial axes at each end of the rotor and the z-axis or axial direction of the rotor. To apply force in any axis, current in a series of coils, one on either side controls magnetic flux on each side of the rotor. This controlled electro-magnetic flux adds to or subtracts from the bias flux generated by a permanent magnet. Bi-directional control of the current in the coils is required for each axis of the magnetic bearing so that force can be applied in a plus or minus direction. For example, to apply a force in the "plus X" direction, current is sent to the coils such that it adds to the bias flux on the "plus X" side of the rotor and subtracts from the bias flux on the "minus X" side of the rotor. This control can be implemented with two of the half-bridge PEBBs for each axis. These PEBBs are connected to the same DC source as the motor/generator PEBBs. For the magnetic bearing, fine control of the current is of primary importance to accurately control the force on/position of the rotor. With a good control algorithm running on a fast Digital Signal Processor, currents of less than 1 amp are sufficient to keep the rotor centered.

THE DESIGN VARIABLES

In the general sense, if one is to develop a common building block, the range of options to be considered includes: High voltage vs. low voltage, isolated vs. non-isolated drive, high current vs. low current, current control vs. voltage control, fast switching vs. slow switching, H-bridges vs. ½ bridges vs. ¼ bridges, and current feedback vs. open loop. To add to the complexity, power requirements vary from watts to kilowatts. There often are several options within each choice. For example, when considering isolation, one can use optical, transformer, and differential drivers. Optical is marginal regarding radiation, transformers are more complicated to design and differential drivers are limited to 15 Volt transients from chassis ground. Placement of the isolation is also a variable. Extra secondary power sources are needed if the isolation is too far toward the input. Placement at the last stage has problems with leakage inductance for driving gates.

Power control options include buck and boost converters with output filters and without output filters. These circuits may require current feedback for improved bandwidth, bi-directionality to control in all four quadrants, or operation near zero current (for magnetic bearings and low disturbance position control systems). In the past, it has been necessary to have a multitude of circuits to cover the range of new power processing requirements.

THE COMMON DENOMINATOR

All of the potential applications use a serialized approach where an input signal drives a low power circuit that drives a medium power circuit that drives a high power circuit. This serialized power stage must be isolated to fill all the requirements. A circuit configuration that meets the serialized isolated power processing requirements is represented in block diagram form in Figure 1.

The requirement of the low power circuitry is to take the TTL level input and drive a transformer with pulses. The underlap of the "on" and "off" pulses needs to be selectable up to 5 uSec. in order to accommodate shoot-through timing of the output transistors.

The medium power circuitry must be able to source 20 Amp pulses on the order of 50 nSec from a 15V bus when it is used to drive several paralleled power MOSFETs. This stage must be able to receive bus voltages as power inputs when it is used as a half-bridge drive for motors in a three phase configuration. It must also be possible to tie the power in a +/- configuration and get bi-directional currents in a motor from a half-bridge. The isolation circuit must be designed to allow a train of all "on" or all "off" pulses as well as the normal sequence of alternating "on" and "off" pulses. The output of the medium power stage has to connect to bus in some applications and the gates of the next stage in some applications.

If the high power circuit is configured as a half-bridge and packaged on a single circuit card capable of delivering 200 watts, then it can be used for many applications in a modular fashion. For example, this serialized power stage can be used in buck and boost converters if the lower half of the bridge is used as a synchronous rectifier to replace the usual freewheeling diode. Figure 2 shows two of the serialized power stage channels connected in a full-bridge configuration.

Five full-bridge circuits are used to drive the FESS magnetic bearings where one full bridge circuit is used for each control axis. The full-bridge circuit is also useful for cryo-cooler motors and 2 or 3 phase actuators. Each half-bridge stage uses current feedback to extend bandwidth. The current sense also provides for fault detection that can be used to remove a failed load or shut down the complementary transistor in the half bridge in the case of a semiconductor failure.

This same stage can also be laid out for high frequency, high pulse current operation and serve to drive up to 4 paralleled MOSFETs to get into the KW arena as shown in Figure 3. In this configuration, the current feedback is wrapped around the MOSFET output stage. Figure 4 shows a modular implementation of a multi-kilowatt half bridge configuration using two of the circuits in Figure 3. The circuit of Figure 4 could also be used for fault isolation if the bottom output switch is reversed such that the body diodes face each other. In this manner, the circuit could block or pass current flow in either direction.

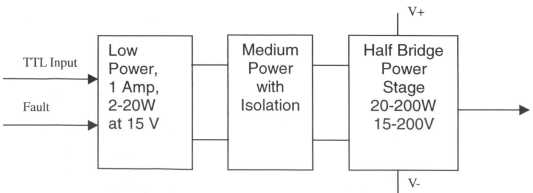

Fiugre 1 Power Electronic Building Block (PEBB)

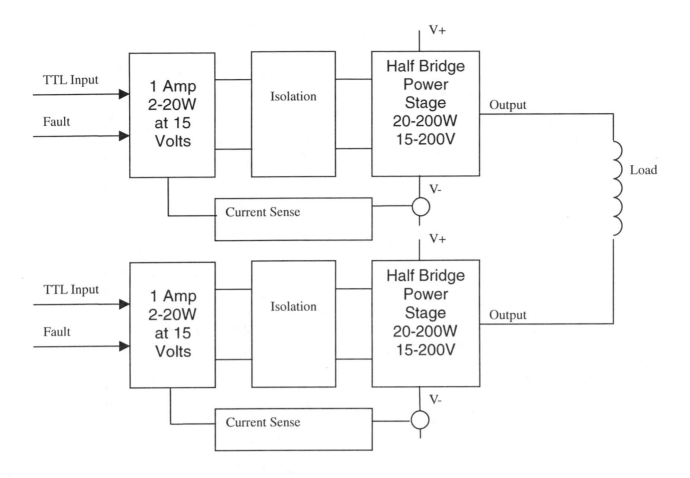

Figure 2: Full Bridge Power Circuit

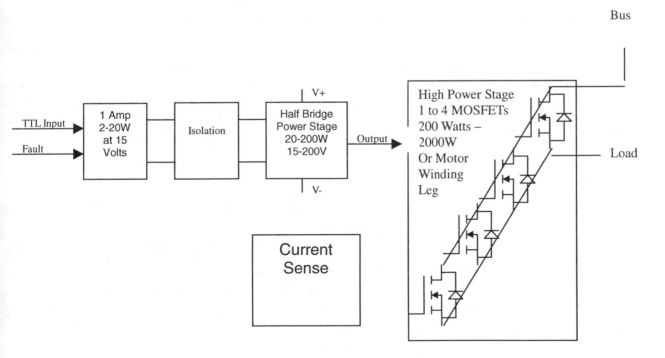

Figure 3: Two Kilowatt Power Switch

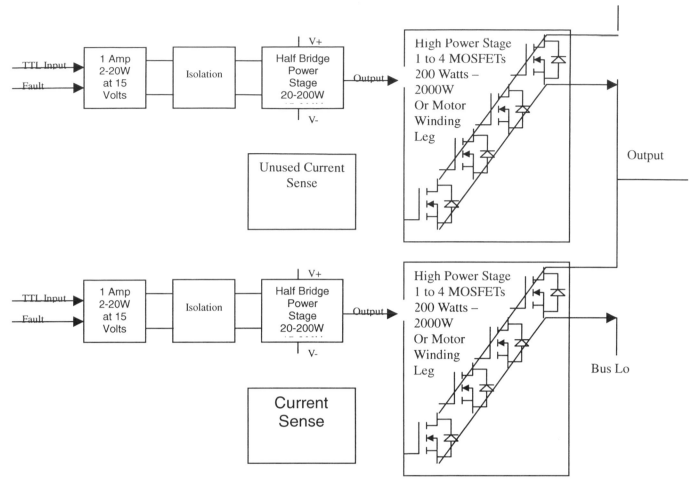

Figure 4: 2KW Power Control-Four Quadrant Motor Drives with Current Feedback

Three of the half-bridge circuits are used in the FESS to control bi-directional power flow through the three-phase motor/generator. During motor operation when the flywheel speed is increasing the half-bridge circuits act as buck converters with the lower switching element functioning as the synchronous rectifier. During generator operation when current is flowing to the dc bus, the half-bridge acts as a boost converter with the upper switch functioning as a synchronous rectifier. Current and voltage regulation as well as motor commutation is implemented with control loops running through a digital signal processor.

PACKAGING CONSIDERATIONS

The goal of this circuit development is to customize the resulting electrical and mechanical architecture without a cost penalty and provide cost savings for repeated applications. One approach to allow reusability of power processing circuitry is to have the silicon on one card and the input, conversion, and output magnetics on another card. This approach invites the use

of a back-plane with the silicon on one card and the power filter components on another card with signals routed between the cards. A cost effective solution under investigation is to have power processing building block circuit cards that stay constant and are plugged into a fairly simple back-plane that changes from program to program to provide the required flexibility. Thermal dissipations and EMI reduction are under investigation with this approach.

PROGRESS TO DATE

The most demanding circuit application, that of driving several paralleled output FETs has already been built and tested and produced notable efficiency results at 1500 Watts on a 150 Volt bus. A design that uses a single circuit card and meets the flexibility of being able to drive either a motor or an output stage of paralleled MOSFETS has been demonstrated in a prototype circuit card.

Local programmability to select configuration is accomplished with the use of several resistors that can range from a few ohms when driving the gates of paralleled power transistors or can be zero ohms when driving a motor.

SUMMARY:

The requirements of a common power processing element have been outlined and an overall architecture within which to develop the circuits has been discussed. The assertion is that a half bridge building block with current feedback is the basic building block. The design is serialized in that the stage that drives a motor can also drive a next stage of paralleled power MOSFETs for KW applications. The assertion is that this could be extended to hundreds of KW.

In general, this building block takes advantage of the fact that flyback diodes in power converters can be replaced with transistors in half bridge building blocks. Often the transistor provides circuit advantages regarding efficiency or playing the role of a synchronous rectifier.

The prototype design meets most of the stated goals but requires further work regarding the product design to allow interfacing to either motors or power MOSFETs without having some inefficiency regarding board area when considering thermal and EMI issues.

The next effort will be directed toward testing for efficiency and fault management for the myriad of other applications where this common power processing block is useful. The evolution of this design element and this architecture is presently mature enough to propose for a flight system.

ACKNOWLEDGMENTS

This work was performed under NASA Contract NAS3-27811, Power Management Control and Distribution for Electrical Power System Testbeds.

Proceedings of IECEC'01
36th Intersociety Energy Conversion Engineering Conference
July 29–August 2, 2001, Savannah, Georgia

IECEC2001-AT-64

NONLINEAR DIGITAL CONTROL IMPLEMENTATION FOR A DC-TO-DC POWER CONVERTER

Jack Zeller, Minshao Zhu, Tom Stimac and Zhiqiang Gao

The Advanced Engineering Research Laboratory
ECE Department, Cleveland State University
Cleveland, OH, USA 44115

ABSTRACT

Closed loop control studies of a DSP-based H-bridge power converter are discussed. The experimental test facility and the analytical development tools being used are described. Open loop modeling results for the NASA-provided power converter test unit are summarized. The performance benefits of nonlinear control algorithms, readily implemented in DSP software, are discussed. Technology issues, specific to the CSU digital control structure are identified and their ongoing development studies are discussed.

INTRODUCTION

Cleveland State University is involved with research to study how the application of direct digital control to spacecraft power converters could enhance their performance and reliability as well as that of complete power management and distribution (PMAD) systems [1]. The work is being conducted by a team of faculty members and students (both graduate and undergraduate) from CSU's Electrical and Computer Engineering (ECE) department. This paper is intended to provide an overview of the entire research activity while focusing on early closed loop control performance results that have been obtained using nonlinear controller algorithms.

To provide background information for readers, the paper begins with a discussion of the objectives for conducting this specific research. This is followed by a brief description of the experimental facility being used to conduct the research. A more detailed description can be found in [2]. Next there will be a description of how the facility was used to experimentally determine linear model representations of the power converter provided by NASA for this program. These linear models [3] serve as the basis for analytically studying a variety of closed loop voltage regulation control strategies. These strategies involve nonlinear control laws that depend upon a digital controller's computational capabilities for their implementation. The main section of the paper will present initial performance results obtained with these digital control strategies. Both simulation and hardware test results will be included and discussed. The final section will: 1) discuss power converter digital control technology areas which warrant further study, and 2) describe DSP control hardware development activities intended to provide tools for broader PMAD control investigations.

RESEARCH OBJECTIVES

Reliable, efficient, well-regulated DC-to-DC power conversion equipment is critical for mission success on most space platforms. As platforms, especially manned spacecraft, become more sophisticated, reliable operation of complete power management and distribution (PMAD) systems becomes a must. As a result there is much interest in determining how and in what areas a more intelligent and robust control structure might be of value. Replacing the present analog control solution with a digital computer based control is one approach toward satisfying this need.

Much work in digital control of DC-to-DC power converters has already been accomplished and documented ([10]-[16]). Either microcontroller-based or DSP-based approaches have been used to realize sophisticated and/or flexible control algorithms, such as PID, Fuzzy Logic, Adaptive Fuzzy, and Feedforward Control. The versatility provided by software programmable digital controllers is well suited to the increasing control performance and reliability demands being placed on new space borne power converters and PMAD systems. Applying previous experience by CSU researchers on highly nonlinear control strategies [5] is the focus of the work to be reported in this paper. One objective of the CSU research will be to evaluate the closed loop performance benefits that these new nonlinear algorithms can bring to DC-to-DC power converters. In addition our DSP-based research will be conducted so as to evaluate a multitude of control opportunities that can only be accomplished digitally. One example will be the ability to use variable PWM frequency as a method of improving low power converter efficiency. Early results of our studies as well as details of the multitude of ongoing efforts will be discussed in the following sections.

CSU RESEARCH FACILITY

In order to provide an effective research environment, CSU's ECE department allocated one of its laboratories to this project to function as a combined laboratory and office in which to conduct this research. This facility has been designated as the Advanced Engineering Research Laboratory (AERL). In order to conduct realistic experimental research, NASA provided to CSU a Westinghouse-designed 1 KW "brassboard" power converter. This SMPS unit was designed to accept an input voltage between 100 and 160 volts DC and provide a regulated and isolated output DC voltage of 28 volts for loads up to 36 Amps. Galvanic voltage isolation was obtained with a stepdown (3:1) transformer whose primary winding was pulse-width-modulated (PWM) with an H-bridge switching configuration of power MOSFET transistors. The lower voltage secondary winding was rectified and filtered to provide the 28 volt DC output. Pulse-width-modulation (PWM) of the switching devices was used to accomplish closed loop voltage regulation. The analog PWM generation circuitry and analog controller circuitry were removed, since the intent of the research is to accomplish these two functions digitally.

It was decided that a DSP-based digital system would be used rather than a microcontroller approach. Equipment needed to support this approach was put into place and configured to realize a versatile research environment. The DSP development system selected was dSpace Inc.'s [4] rapid-prototyping development system. This system is equipped with a high-performance TI DSP chip, A/D conversion capability as well as digital I/O circuitry. To expedite the development and evaluation of digital control strategies, Mathwork's Matlab/Simulink/Real-Time Workshop toolbox software was selected. Simulink provides the ability to model and accurately simulate the transient performance of dynamic processes to arrive at a set of acceptable closed loop control strategies. Mathworks' Real-Time Workshop will convert a controller, modeled in Simulink, into 'C" code which will run on dSpace's DSP processor to control actual experimental hardware (in this case the 1 KW Westinghouse power converter). This is termed hardware-in-the-loop simulation. The control laws can also be programmed in native "C" code. Then dSpace's compiler and libraries will be used to generate the code for the TI DSP chip on dSpace's processor board. This second approach has been found to generate faster operating real-time control code.

A decision was made early in the program to generate the two-phase PWM signals needed by the H-bridge outside of the DSP by using a programmable CPLD chip. This will off-load a potentially heavy computational burden from the DSP controller. A block diagram of this experimental configuration is shown in Figure 1.

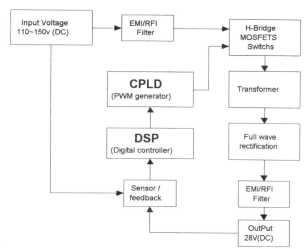

Figure 1: Experimental Facility Block Diagram Description

To complete the experimental research facility, appropriate test equipment was acquired. This included: power supplies, signal generators, digital voltmeters, digital oscilloscopes, and an electronic load bank. The photograph in Figure 2 shows how this array of equipment is configured in our facility. A more detailed description of all of this equipment is included in [2].

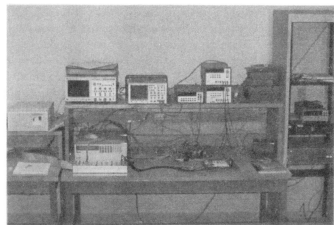

Figure 2. Photograph of AERL Experimental Equipment
Linear Model Development

To expedite the analytical control development studies, a linear (transfer function) model of the power converter process was developed. Obtaining the data for this model was the first research activity which used the AERL experimental hardware-in-the-loop configuration. A methodology for using a CPLD device to generate the PWM signals needed to drive the switching converter's gate circuitry was developed. The DSP's algorithmic logic needed to accept a variable pulse width control input and compute the outputs for the CPLD's input registers was configured for evaluation in Simulink. The RTW toolbox was used to convert the simulated algorithmic logic into dSpace's DSP "C" code equivalent [6].

It *should be noted that the initial design of the algorithm chose an eight (8)bit quantization level for each phase of the CPLD's PWM output. Thus the 28 volt DC output could only be resolved to 0.156 volts, (at 120 volts of input). This quantization has proven to be a performance limitation to the control studies and improvements are being evaluated. A brief discussion of the early results of these improvements will be presented later in this paper.*

Using the hardware-in-the-loop experimental configuration of Figure 1 and the just-described PWM generation software, linear model data was obtained. By varying the input pulse count, the converter's open loop steady-state performance under varying output current (I_l) load levels and for a range of input supply voltages (V_{in}) was determined. The detailed results of this steady-state mapping can be found in [3]. Using those results, an equation was determined [3] which analytically defines the converter's steady-state output voltage over a range of conditions.

$$V_o = \frac{V_{in}}{(3*256)}(PulseCount) - 0.8 - (0.075*I_L) \quad (1)$$

In (1) the division of the input voltage by 3 accounts for the 3:1 turns ratio of the isolation step-down transformer. The 256 factor is the maximum pulse count due to the eight bit quantization used in the initial design. The 0.8 volts accounts for the rectifier's diode drop while the 0.075 is the approximate output impedance of the converter under load. Eq.(1) can be rearranged to yield a pulse count value which would be needed to produce a particular output voltage knowing the input DC voltage and the load current. This relationship is defined as (2) below: \quad (2)

$$PulseCount = [V_o + 0.8 + 0.075*I_L]*(3*256)/V_{in}$$

The next experimental modeling activity of [3] was to determine the transient behavior of the converter process when subjected to disturbance inputs in: 1) pulse count, 2) load current, and 3) input DC supply voltage. Step inputs in each of these three parameters were used to produce time response data. Curve fit approximation's to this data were used to determine linear transfer function models. The details of the testing activity are included in [3]. The result of this activity was the open loop process transfer function block diagram of Figure 3.

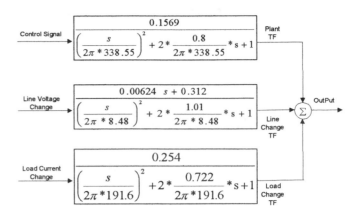

Figure 3 Linear Model Block Diagram

CLOSED LOOP PERFORMANCE RESULTS

Simulink Setup -The results of the simulation studies were obtained using a detailed Simulink model of the digitally controlled converter. The simulation includes the open loop converter model of Figure 3. A comparison of a traditional linear PID control and a nonlinear control (NPID) was performed. Figure 4 shows this model and includes blocks for the two control laws as well as a soft-start feature. Figures 4a-4d are block descriptions of the Simulink subsystems of Figure 4.

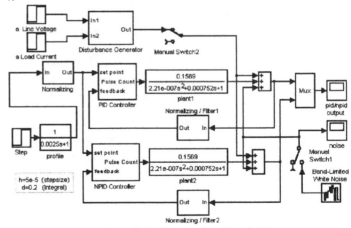

Figure 4 Simulink Simulation Block Diagram

We use a zero-order hold with a sampling period of 50us. The quantizer is used to mimic the dSpace's 12 bitA/D converter. It is set 0.0048828125.

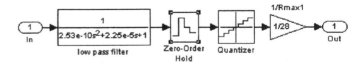

Figure 4a Normalizing and Filtering Subsystem

After comparing the setpoint and feedback signals, the control algorithm is executed and a control signal is produced. It is then converted to a pulse count and sent to the PWM generator to create real PWM control signals for the switching MOSFET's gate drivers.

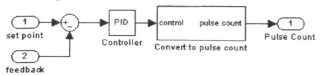

Figure 4b PID Controller Subsystem

The disturbance block in Figure 4c is used to simulate the effects of the Line voltage change and Load current change on the output voltage. It allows us to observe the disturbance rejection performance for each controller. It comes from Figure 3.

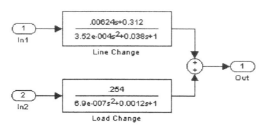

Figure 4c Disturbance Generator Subsystem

The Conversion to pulse count block is shown in Figure 4d, where the saturation limits are set at 0 and 1, respectively. This is because our PWM generator range is 0 - 240 pulse counts. The quantization level in the Quantizer is set at 1.

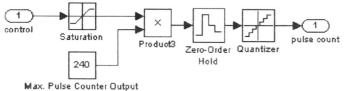

Figure 4d Conversion to Pulse Count Subsystem

NONLINEAR PROPORTIONAL AND INTEGRAL CONTROL

The initial results of the digitally controlled converter using standard Proportional-integral (PI) controller, which is primarily an integral control, have been reported in [2]. Recently, a set of high performance Nonlinear PID control algorithms have been reported [5] and some of them are used here as shown in Figure 4e, which shows the details of the Nonlinear PI block in Figure 4.

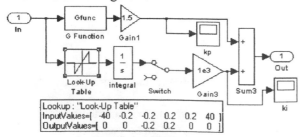

Figure 4e Nonlinear PI Control subsystem

The G-function is Figure 4e is a nonlinear gain function shown in Figure 5, where the green line is normal linear gain and the blue line represents the nonlinear G function. The design philosophy is fully explained in [5]. Here, the intuition is that the gain should be higher when the error is smaller, which makes the controller "more stiff". That is the proportional control is made more sensitive to the small errors. This will also reduce the reliance on the integral control to eliminate steady state errors. Note that the instability is often caused by the 90 degree phase lag in the integral control.

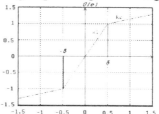

Figure 5 G-Function

It is mathematically expressed as:

$$G(e) = \begin{cases} k_2 * e + (k_1 - k_2) * \delta * \text{sgn}(e) & |e| > \delta \\ k_1 * e & |e| \le \delta \end{cases}$$

(3)

Although the use of this nonlinear gain provides good disturbance rejection and stability robustness, it may make the controller too sensitive to noise. Therefore, a compromise is made between the nonlinear proportional control and a limited nonlinear integral control. In particular, the integral term is reformulated as

$$k_i * \int G_i(e) dt \qquad G_i(e) = \begin{cases} 0 & |e| > \delta_i \\ e & |e| \le \delta_i \end{cases}$$

(4)

That is, the integrator only integrates when the error is "small", typically when the output is within 10% of the set point.. This design strategy allows the control to effectively avoid undesirable overshoots and the integrator wind-up during large disturbances.

Simulation : Transient Results – Figure 5 show a comparison of the transient performance simulation results obtained for a well-tuned linear PID versus a nonlinear PI control. Figure 5(a) contains results for the application of a 20 Amp load while Figure 5(b) shows results for PWM pulse Count (control variable) respectively. The blue curves are for the PID and the green traces are for the nonlinear PI. The nonlinear controller shows a much smaller deviation from steady-state than the linear PID. Also the nonlinear algorithm is faster.

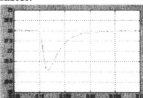
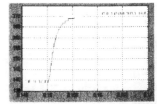

Figure 5(a).Load Transient   Figure 5(b)PWM pulse Count

Experimental Controller Setup –The two control algorithms were then coded in native "C" code , compiled and down loaded to the DSP system. This code could then operate the converter hardware. After extensive experimentation and tuning activity, transient performance comparison results were obtained. The use of dSpace's Control Desk software helped expedite this tuning activity. Figure 6 is a sample of what the computer screen looks like when Control Desk is employed. The designer has a great deal of critical parameter information available at a glance along with the captured transient data. Even though the actual values may not be readable in the paper, the figure is included to show the capability of the Control Desk software for enhancing productivity.

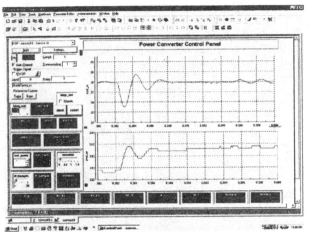

Figure 6 Sample Control Desk Screen

Experimental Transient Results – The transients caused by a sudden change in the load current were captured as was done during the simulation studies.

We used the Control Desk to assist the controller tuning and transient response monitoring. In the following figures, the top trace is Output Voltage, the lower trace is CPLD PWM pulse Count (control signal).

In the hardware test, the load current was changed from 3A to 20A. The lowest load is set 3A so that the inductor in the converter is in continuous conduction.

1) Linear PI (LPI) Controller results

The parameter for Linear PI Controller setting are :
$$K_p = 0.2, K_i = 423$$

And the response is shown in Figure 7, which indicates a 15.2ms recovery time and 3.4V peak-to-peak voltage variation.

2) Two-slope Nonlinear PI (NPI) Controller results

According to (3) and (4), the parameters for the NPI controller setting are set as
$$K_1 = 0.256, K_2 = 0.024, \delta = 0.4$$
$$K_i = 400, \delta_i = 0.8$$

and the response is shown in Figure.8, which yields a 5.7 ms recovery time and a 3.25V peak-to-peak voltage variation.

Comparing to Figure 7 the NPI transient response performs almost 2 times better on the recovery time.

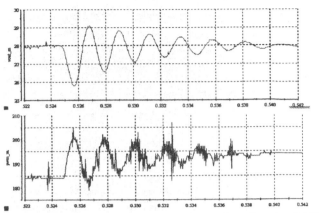

Figure 7 Transient response with Load application(LPI)

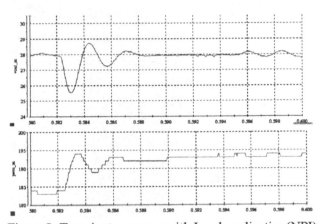

Figure 8 Transient response with Load application(NPI)

The above results show the benefits of using the NPI controller as:

1.Much cleaner control output

2.Much less ringing during load application.

3.Much faster load application recovery time during load application.

ONGOING AND FUTURE RESEARCH PURSUITS

As the AERL team undertook R & D activity to replace the traditional analog SMPS controller with a direct digital solution, a number of technology and system issues became evident. Several of these issues will now be briefly addressed in terms of each one's ongoing design and development activities.

Signal Conditioning – Critical voltage and current variables, which define the converter's performance, must be measured accurately, isolated, and conditioned for sampling by the digital controller. One important aspect of the signal conditioning is the selection of adequate anti-aliasing filters to remove (to filter out) unneeded high frequency information in the measurements. To accommodate these requirements the signal conditioning circuitry was breadboarded for the initial experimental studies. For future control studies, a ruggedized printed circuit version of this circuitry is being designed.

PWM Generation - As stated earlier, CSU's approach to PWM generation is to generate the pulse width gate driving signals with a programmable logic device (CPLD). The present performance limiting eight-bit PWM CPLD will soon be replaced with an alternative CPLD design which provides higher resolution (finer quantization) and will have the ability to vary the PWM frequency directly through commands from the DSP software. Closed loop control testing of the higher resolution CPLD is now underway. Performance studies using the variable frequency feature will start soon. Results will be reported at a later time.

Control Mode Selection-As was shown in the results section, the new nonlinear control strategies show benefits over linear, more traditional, control modes. It must be noted at this point, however, that the AERL team has not yet implemented a current-mode inner loop. Because we generate the PWM signals digitally, a strategy for effectively using sensed transformer primary current in an inner current loop control has not yet been determined. Resolving this control design issue is a major priority.

DSP Control Development Platform –The dSpace rapid-prototype development equipment has played an invaluable role in our controls research. However, at CSU we are designing an easier-to-use DSP development platform to study converter control in a broader PMAD system context. A major feature of this platform design is the inclusion two high-speed IEEE-1394 (Firewire) data communication ports.

SUMMARY

A research program on direct digital control of power converters has been described. Analytical and experimental results for a new nonlinear control strategy are discussed and compared against traditional linear control modes. The results encourage continued study into nonlinear approaches to converter voltage regulation. Finally some of the technology issues related to digital converter control are identified and efforts for their resolution discussed.

ACKNOWLEDGMENTS

The work discussed in this paper is sponsored by the NASA Glenn Research Center under grant #NCC3-699 and by Cleveland State University. The authors would like to thank other members of the AERL team including Charles Alexander, Dave Gerdeman, Dave Wladyka, Ivan Jercic, Greg Tollis, Marcelo Gonzalez, and John Sustersic for their active participation in the development activities.

REFERENCES

1. Robert M. Button," intelligent systems for power management and distribution", NASA Glenn Research Center, Cleveland, OH 44135

2. John Sustersic," Design and Implementation of a Digital Controller for DC-DC Power Converters", 2000 SAE Power Systems Conference, San Diego, CA, Oct. 31, 2000

3. Stimac, Tom," Digital Control of a 1-kW DC-DC switching Power converter", Master's Thesis, Department of Electrical and Computer Engineering, Cleveland State University, December 2000

4. dSPACE Inc. 22260 Haggerty Road - Suite 120 Northville, MI 48167 Tel.: (248) 344-0096 Fax: (248) 344-2060.

5. Zhiqiang Gao, "Linear to Nonlinear Control Means: A practice Progression", To appear in ISA Transactions.

6. David M. Wladyka, " Investigation of C Code Execution on dSpace ", AERL Report, Cleveland State University, Cleveland,OH,44115

7. R.R. Boudreaux, R.M. Nelms, and John Y. Hung, "Simulation and Modeling of a DC-DC Converter Controller by an 8-bit Microcontroller", 0-7803-3704-2/97 IEEE

8. W.C. So, C.K. Tse and Y.S. Lee, "A Fuzzy Controller for DC-DC Converters", 0-7803-1859-5/94 IEEE

9. Keyue Ma Smedley and Slobodan Cuk, "One-Cycle Control of Switching Converters", 0-7803-0090-4/91 IEEE

10. T. Gupta, R.R. Boudreaux, R.M. Nelms, and John Y. Hung, "Implementation of a Fuzzy Controller for DC-DC Converters Using and Inexpensive 8-bit Microcontroller", IEEE Transactions on Industrial Electronics, Jan 1997.

11. R. Vinsant, J. DiFiore, and R. Clarke, "Digital Control Converts Power Supply into Intelligent Power System Peripheral," Ninth International High Frequency Power Conversion Conference, April 1994, pp. 2-6.

12. R.R. Boudreaux, R.M. Nelms, and John Y. Hung, "Digital Control of DC-DC Converters: Microcontroller Implementation Issues," Combined Proceedings of HFP Power Conversion & Advanced Power Electronics Technology, Powersystems World'96. September 1996, pp. 168-180.

13. Barry Arbetter, Dragan Maksimovic," DC-DC Converter with Fast Transient Response and High Efficiency for Low-Voltage Microprocessor Loads", Department of Electrical and Computer Engineering at the University of Colorado, 1997.

14. Barry Arbetter, Dragan Maksimovic, " Feedforward Pulse Width Modulators for Switching Power Converters", IEEE Transanctions on Power Electronics, Volume. 12, No. 2, March 1997.

15. David J. Caldwell, "Digital Power Flexibility: Applications and Advantages", HFPC '99 Proceedings, November 1999

16. Sun-Hoe Huh,Gwi-Tae Park ,"An Adaptive Fuzzy Controller for Power Converters", University of Hong Kong, 1998.

IECEC2001-AT-87

LARGE-SIGNAL ANALYSIS OF THE PEAK-POWER-TRACKING SYSTEM

*Yoon-Jay Cho

The Power Electronic System Laboratory
The School of Electrical Engineering #043 (301-617)
Seoul National University
San 56-1. Shilim-Dong. Kwanak-Ku
Seoul, Korea
*E-mail : feelso@snu.ac.kr

B.H. Cho

The Power Electronic System Laboratory
The School of Electrical Engineering #043 (301-617)
Seoul National University
San 56-1. Shilim-Dong. Kwanak-Ku
Seoul, Korea

ABSTRACT

Large-signal stability analysis of the Peak-Power-Tracking system is performed to facilitate design and analysis of Low-Earth-Orbit satellite power system. Possible instability due to nonlinear dynamic interactions between Solar-Array-Regulator and solar array nonlinear characteristic is identified by simplified system model. Based on Circle criterion, the effect of the slope variation and the sharpness of solar array on the system stability are investigated and the way to avoid the instability by the gain adjustment and attaching the limitations of the PPT controller is presented.

INTRODUCTION

Solar array, as the primary power source for a spacecraft power system, has a highly nonlinear output I-V characteristic as shown in Fig. 1. It has a current source region and a voltage source region separated by the maximum power point (MPP). The location of this MPP varies according to various factors such as the illumination level and the temperature. Especially for the Low Earth Orbit (LEO) Satellite that undergoes the wide temperature variations, extracting maximum power from a solar array requires an adaptive and intelligent Peak-Power-Tacking (PPT) control. The most popular and commonly used method is based on the power slope, which is positive on current source region, negative on voltage source region, and zero at MPP [1]. As suggested in [1], this algorithm is implemented in the PPT controller employing a microprocessor and Fig.2 depicts this PPT system.

In this system, PPT controller tracks the MPP by adjusting a voltage reference, V_{ref}, to the Solar-Array-Regulator (SAR). Then, the inner voltage loop of SAR regulates the solar-array-output-voltage to the V_{ref} set by PPT controller. PPT controller calculates the output power by multiplying the sensed ouput voltage and current of solar array, generating the corresponding power slope, and changes V_{ref} accroding to this information. This algorithm can be presented as the following equation:

$$V_{ref}(k+1) = V_{ref}(k) + M \cdot \frac{\Delta P_{sa}}{\Delta V_{sa}} \quad (1)$$

In designing the PPT system, various factors must be considered such as nonlinearity of solar array coupled with nonlinear switching characteristic of solar array regulator, the dynamic interactions between PPT loop and inner voltage loop. Without understanding these factors and accurate prediction of system operating mode under large-signal perturbation, PPT controller cannot be designed to achieve a desired performance or even causes instability. However the analysis and design of PPT system was performed and presented in [1], it produces a less accurate result since its analysis is based on the small-signal analysis. Its results cannot predict the system stability under large-signal perturbations.

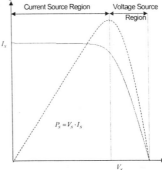

Fig. 1 Solar Array I-V characteristic

Thus, in this paper, a comprehensive, large-signal stability analysis of PPT system is presented. In Section II, employing a simplified PPT system model, the possible instability from the interaction between the nonlinearity of the solar array and the

Solar Array Regulator (SAR) is identified. It is verified that the conventional state-plane analysis based on the small-signal concept has a severe limitations on predicting the large-signal behavior of PPT system.

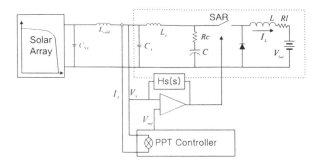

Fig. 2 The Peak-Power-Tracking System

In section III, concentrated on the nonlinear characteristics of the solar array, the design method for the large signally stable PPT system is provided employing Circle criterion. In this procedure, the nonlinear characteristic of the solar array is represented as sectors in Circle criterion.

LIMITATIONS OF THE CONVENTIONAL SMALL-SIGNAL ANALYSIS

In this section, the possible instability from the nonlinearity of the solar array is identified employing the simplified system model as in Fig. 3. In this model, solar array regulator with inner voltage loop closed is represented as a voltage load. Unlike the Direct-Energy-Transfer (DET) system in [2] in which the load line is static, the load of PPT system is varying according to the solar array output power, thus the load must be treated as a variable state rather than as a static load-line as in [1,2]. In this model, C represnets a solar array capacitance, C_{sa} in Fig. 2, and L represents a cable inductance, L_{Cable} in Fig. 2.

The system equation of Fig. 3 can be obtained as:

$$\frac{dv_{sa}}{dt} = \frac{1}{C}(i_{sa} - i_L) \tag{2}$$

$$\frac{di_L}{dt} = \frac{1}{L}(v_{sa} - (v_{ref} + i_L R_l)) \tag{3}$$

$$v_{ref}(k+1) = v_{ref}(k) + M \cdot \frac{\Delta p_{sa}}{\Delta v_{sa}} \tag{4}$$

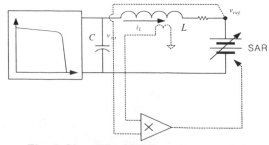

Fig. 3 Simplified PPT system model

Introducing perturbation, $v_{sa} = V_{sa} + \hat{v}_{sa}$, $i_{sa} = I_{sa} + \hat{i}_{sa}$, $v_{ref} = V_{ref} + \hat{v}_{ref}$ and using first order Tylor series expansion, then Eq. (2) and Eq. (3) can be derived as;

$$\frac{d\hat{v}_{sa}}{dt} = \frac{1}{C}(\alpha \cdot \hat{v}_{sa} - \hat{i}_L) \tag{5}$$

$$\frac{d\hat{i}_L}{dt} = \frac{1}{L}(\hat{v}_{sa} - (\hat{v}_{ref} + \hat{i}_L R_l)) \tag{6}$$

where,

$$I_{sa} + \hat{i}_{sa} = f(V_{sa} + \hat{v}_{sa}) = f(V_{sa}) - \alpha \cdot \hat{v}_{sa} \ (\alpha > 0) \tag{7}$$

$f(x)$: The solar array IV equation

α : The absolute value of the slope of solar array at V_{sa}

The equivalent differential equation of Eq. (4) can be derived as the following sequence;

$$\frac{\hat{v}_{ref}(k+1) - \hat{v}_{ref}(k)}{T_S} = \frac{M}{T_S} \cdot \frac{\Delta \hat{p}_{sa}}{\Delta \hat{v}_{sa}}$$

$$\frac{d\hat{v}_{ref}}{dt} = Mg \cdot \frac{d\hat{p}_{sa}}{d\hat{v}_{sa}} \tag{8}$$

where , $Mg = \dfrac{M}{T_S}$: Effective gain of PPT controller

The term ($\frac{d\hat{p}_{sa}}{d\hat{v}_{sa}}$) can be derived as a function of $\hat{v}_{sa}$ as;

$$\frac{d\hat{p}_{sa}}{d\hat{v}_{sa}} = \frac{d}{dv_{sa}}\left[(I_{sa} + \hat{i}_{sa}) \cdot (V_{sa} + \hat{v}_{sa})\right]$$

$$= \frac{d}{dv_{sa}}\left[f(V_{sa}) - \alpha\hat{v}_{sa}\right] \cdot (V_{sa} + \hat{v}_{sa})$$

$$= -2\alpha \cdot \hat{v}_{sa} + \left[f(V_{sa}) - \alpha V_{sa}\right]$$

Thus,

$$\frac{d\hat{v}_{ref}}{dt} = Mg \cdot \frac{d\hat{p}_{sa}}{d\hat{v}_{sa}} = Mg \cdot \left\{-2\alpha \cdot \hat{v}_{sa} + \left[f(V_{sa}) - \alpha V_{sa}\right]\right\} \tag{9}$$

For the system stability analysis, combining equations (5),(6),(9) results in the following state equation.

$$\dot{x} = \begin{bmatrix} -\dfrac{\alpha}{C} & \dfrac{1}{C} & 0 \\ \dfrac{1}{L} & -\dfrac{R_l}{L} & -\dfrac{1}{L} \\ -2Mg \cdot \alpha & 0 & 0 \end{bmatrix} \cdot x$$

$$+ \begin{bmatrix} 0 \\ 0 \\ Mg \cdot (I_{sa} - \alpha \cdot V_{sa}) \end{bmatrix} \tag{10}$$

$$\dot{x} = A_{jacobian} \cdot x + u \tag{11}$$

Note that the first order approximation under small-sigal assumtion is employed in these derivations, thus the only slope information of solar array is considered.

The conventional analysis of the system stability can be divided into two steps. First, search the whole operation region to find the point where the condition, $\dot{x} = 0$, is satisfied. Then this point becomes a eqilibrium point of the system. Secondly, obtain the eigenvalues of the system matrix, $A_{jacobian}$ to analyze the equilibrium point to be stable or not.

For the DET system in [2], the connection point of the solar array source line and the load line becomes the equilibrium point. However, in this PPT system, the point where the following equation is satisfied becomes the equilibrium point.

$$I_{sa} - \alpha V_{sa} = 0 \tag{12}$$

The point where the above condition is satisfied, (I_{sa}, V_{sa}), can be interpreted as a connection point of the solar array source line and the line equation;

$$I - \alpha V = 0 \tag{13}$$

Thus, the equation (13) becomes the load-line representing the variable load according to the Peak-Power-Tracking control. Since the α in this equation is dependent on the operating point of the solar array, V_{sa}, the load-line is varying according to the operating point as shown in Fig. 4. The equilibrium point is moved according to the operating point location. When the operating point exists in the current source region, α becomes small, thus the load-line is drawn as in Fig. 4(a) and the equilibrium point exists in the voltage source region. In case that this equilibrium point is stable, the operating point moves for this point. However since this point is not a real equilibrium point but a temporary equilibrium point that must be moved according to the operating point. To distinguish the real equilibrium point, this point is named as a 'virtual equilibrium point'.

When the operating point is located in the voltage source region, the slope of solar array, α, becomes larger, then the load-line changes as in Fig. 4 (b). Then, a virtual equilibrium point of solar array exists in the current source region. Thus, the operating point moves back into the current source region for this virtual equilibrium point. Near MPP of solar array, the virtual equilibrium point exists in MPP of solar array, the operating point converges to MPP since this point is designed to be stable equilibrium point.

This analysis of PPT system may seem to be perfect, and guarantees the global stability. However, the simulation results in Fig. 5 shows that this analysis also produces the less accurate results under large-signal perturbation. In this simulation, the system is designed to be stable entire operating region. In Case 1, the initial point is set to the near of MPP of solar array and results in the stable behavior. But in case 2 and case3, when initial points are set wide from the MPP of solar array, the current source region and voltage source region respectively,

results in the oscillatory behavior. Thus, this analysis cannot be used to predict the large-signal stability of PPT system.

This inaccuracy is originated from the fact, the stability of PPT system is not only affected by the slope of solar array, but also by sharpness of the solar array, the rate of the variation of the solar array slope. Under the above conventional analysis, this cannot be considered in design process of PPT controller since this analysis is based on the local property of the solar array. Thus, the more advanced methodology must be developed for the proper design of PPT controller.

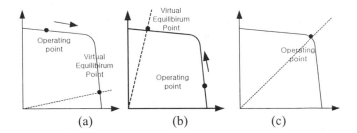

Fig. 4 The state-plane analysis of the PPT system

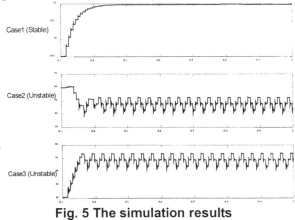

Fig. 5 The simulation results

LARGE SIGNAL ANALYSIS AND DESIGN OF PPT CONTROLLER

The inaccuracy of the above analysis is originated from the small-signal assumption employed in Eq. (7). To avoid the limitations, the function, $f(x)$ can be expressed as;

$$f(V_{sa} + \hat{v}_{sa}) = f(V_{sa}) - g(V_{sa}, \hat{v}_{sa})$$
$$= I_{sa} - g(V_{sa}, \hat{v}_{sa}) \tag{14}$$

The minus sign in above equation is added for the negative slope of solar array to change posituve, thus $g(x, y)$ can be illustrated as in Fig. 8 (b). And the constants, $\alpha_1, \alpha_2, \alpha_3$, represents the slope in the current source region, MPP region and the voltage source region, respectively. The application of Eq. (15) instead of Eq. (7), the equivalent nonlinear differential equation of Eq. (4) can be derived.

213

$$\frac{dp_{sa}}{dv_{sa}} = \frac{d}{dv_{sa}}\Big[(I_{sa} + \hat{i}_{sa})\cdot(V_{sa} + \hat{v}_{sa})\Big]$$

$$= \frac{d}{dv_{sa}}\big[f(V_{sa}) - g(V_{sa}, \hat{v}_{sa})\big]\cdot(V_{sa} + \hat{v}_{sa})$$

$$= -\left[\frac{d}{d\hat{v}_{sa}}g(V_{sa}, \hat{v}_{sa}) - \alpha_2\right]\cdot V_{sa}$$

$$\quad -\left[\frac{d}{d\hat{v}_{sa}}g(V_{sa}, \hat{v}_{sa})\cdot\hat{v}_{sa} + g(V_{sa}, \hat{v}_{sa})\right] + \big[f(V_{sa}) - \alpha_2 V_{sa}\big]$$

$$\overset{\Delta}{=} -l_1(\hat{v}_{sa}) - l_2(\hat{v}_{sa}) + \big[f(V_{sa}) - \alpha_2 V_{sa}\big] \quad (15)$$

The perturbation used in this derivation can be regarded as shifting the center of the axis in the above equation to the point, (I_{sa}, V_{sa}), thus, if this point is set to MPP point of solar array, then, the term ($\big[f(V_{sa}) - \alpha_2 V_{sa}\big]$) becomes zero. Thus;

$$\frac{dp_{sa}}{dv_{sa}} = -l_1(\hat{v}_{sa}) - l_2(\hat{v}_{sa}) = -l(\hat{v}_{sa}) \quad (16)$$

Since $l_2(\hat{v}_{sa})$ in Eq. (16) is not linear equation link Eq. (7), the conventional design method based on the Nyquist theorem cannot be directly applied. Therefore, in this paper, Circle Criterion based on the sector property of the nonlinear equation is applied to analysis of PPT system. With Circle criterion, the powerful graphical analysis like Nyquist plot can be used.

Definition

A memoryless nonlinearity $\varphi : [0, \infty) \times R \to R$ is said to satisfy a sector property or to have sector condition if there exist real α and β such that

$$\alpha x^2 \le x\varphi(t, x) \le \beta x^2, \forall t \ge 0, \forall x \in R$$

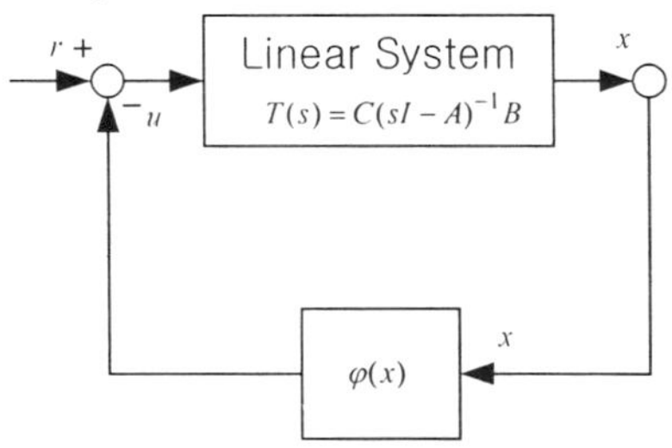

Fig. 6 Circle Criterion

Circle Criterion

Consider a single loop scalar interconnected system in Fig. 6 and the nonlinear function $\varphi(x)$, which satisfies the sector condition as;

$$\alpha \cdot x^2 \le x\varphi(x) \le \beta \cdot x^2$$

then, the system is absolutely stable if one of the following condition is satisfied.

① If $0 \le \alpha \le \beta$, the Nyquist plot of $Tv(jw)$ does not enter the disk $D(\alpha, \beta)$ and encircle it m-times in the counterclockwise direction, where m us the number of poles of $Tv(s)$ with positive real parts.

② If $0 = \alpha \le \beta$, $Tv(s)$ is stable and the Nyquist plot of $Tv(jw)$ lies to the right of the vertical line defined by $Re[s] = -\dfrac{1}{\beta}$.

③ If $\alpha \le 0 \le \beta$, $Tv(s)$ is Hurwitz and the Nyquist plot of $Tv(jw)$ lies in the interior of the disk $D(\alpha, \beta)$.

$D(\alpha, \beta)$ is the closed disk in the complex plane whose diameter is the line segment connecting the points $-\dfrac{1}{\alpha} + j0$ and $-\dfrac{1}{\beta} + j0$.

In this analysis, the first condition in above Circle Criterion is applied to the PPT system. The system equation (5,6) and the PPT controller (7) or (16) can be expressed as the feedback system as in Fig. 7(a). Under small-signal assumption, Eq. (7) is employed and the system stability condition can be obtained with Nyquist Theorem. In this case, Nyquist plot of T(s) must not encircle the point $-\dfrac{1}{\alpha}$. However, under large-signal assumption, Eq. (16) and Circle Criterion are employed. And Nyquist plot of T(s) must not enter the $D(\gamma_1, \gamma_2)$ where γ_1 and γ_1 are sector constants of Eq. (16). Figure 7(b) shows the difference of the stability conditions between small-signal analysis and large-signal analysis. The large-signal analysis presents the more tight condition for the system stability.

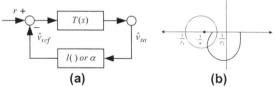

(a)　　　　　**(b)**

Fig. 7 Nyquist Theorem .vs. Circle Criterion

To apply circle criterion to the PPT system, the sector constants of Eq. (16) can be obtained according to the following procedure.

The sector property of $l_2(\hat{v}_{sa})$

$$l_2(\hat{v}_{sa}) = \frac{d}{d\hat{v}_{sa}}g(V_{sa}, \hat{v}_{sa})\cdot\hat{v}_{sa} + g(V_{sa}, \hat{v}_{sa}) \quad (17)$$

As shown in Fig. 8 (b), $g(V_{sa}, \hat{v}_{sa})$ is satisfied with the sector condition as;

$$\alpha_1 \cdot \hat{v}_{sa}^2 \le g(V_{sa}, \hat{v}_{sa})\cdot\hat{v}_{sa} \le \alpha_3 \cdot \hat{v}_{sa}^2 \quad (18)$$

and $\dfrac{d}{d\hat{v}_{sa}}g(V_{sa}, \hat{v}_{sa})\cdot\hat{v}_{sa}$ is also satisfied the following condition;

$$\alpha_1 \cdot \hat{v}_{sa}^2 \le \left[\frac{d}{d\hat{v}_{sa}}g(V_{sa}, \hat{v}_{sa})\cdot\hat{v}_{sa}\right]\hat{v}_{sa} \le \alpha_3 \cdot \hat{v}_{sa}^2 \quad (19)$$

Thus, the sector property of $l_2(\hat{v}_{sa})$ can be obtained as;

$$2\alpha_1 \cdot \hat{v}_s^2 \le l_2(\hat{v}_s)\cdot\hat{v}_s \le 2\alpha_3 \cdot \hat{v}_s^2 \quad (20)$$

The sector property of $l_1(\hat{v}_{sa})$

$$l_1(\hat{v}_{sa}) = \left[\frac{d}{d\hat{v}_{sa}}g(V_{sa},\hat{v}_{sa})-\alpha_2\right]V_{sa} \qquad (21)$$

The term $\left[\dfrac{d}{d\hat{v}_{sa}}g(V_{sa},\hat{v}_{sa})-\alpha_2\right]V_{sa}$ is the first derivatives of the solar array curve as shown in Fig. 8 (c), and the following sector property can be obtained;

$$\beta_1 \cdot \hat{v}_{sa}^{\,2} \le l_1(\hat{v}_{sa}) \cdot \hat{v}_{sa} \le \beta_2 \cdot \hat{v}_{sa}^{\,2} \qquad (22)$$

Thus, the combined sector condition, $l(\hat{v}_c) = l_1(\hat{v}_c)+l_2(\hat{v}_c)$, as shown in Fig. 8 (d) is;

$$(2\alpha_1 + \beta_1) \cdot \hat{v}_s^{\,2} \le l(\hat{v}_s) \cdot \hat{v}_s \le (2\alpha_3 + \beta_2) \cdot \hat{v}_s^{\,2} \qquad (23)$$

$$\gamma_1 \cdot \hat{v}_s^{\,2} \le l(\hat{v}_s) \cdot \hat{v}_s \le \gamma_2 \cdot \hat{v}_s^{\,2} \qquad (24)$$

As shown in Fig. 8 and Eq. (23), the sector property of the dynamic characteristics of solar array is represented by the sum of two properties. One is related to the term, $l_2(\hat{v}_c)$, representing the difference of slope in the voltage source region and the current source region, respectively. And the other is determined from the term, $l_1(\hat{v}_c)$, representing the sharpness of the solar array curve. A solar array whose slope is more sharply decreasing near MPP point has the larger sector constants in $l_1(\hat{v}_c)$, increasing the overall sector constants of $l(\hat{v}_c)$. In this case, the effect of the nonlinear characteristics of solar array on the system stability is increased.

The block-diagram of the interconnected system of PPT controller and Solar-Array-Regulator is presented in Fig. 9(a). Considering the sampling effect, Zero-Order-Hold (ZOH) effect, the discrete equation can be derived and is presented. To apply the Circle criterion, the over-all system can be divided into two blocks as the non-linear part and the linear part as in Fig. 9(b). According to Circle criterion, the Nyquist plot of the transfer function of linear part must avoid to entering the forbidden region, $D(\gamma_1,\gamma_2)$, as shown in Fig. 10(a). The gain of PPT controller of the simulation in above section is set to 0.5 and its Nyquist plot enters into the forbidden region, thus resulting in large-signal instability.

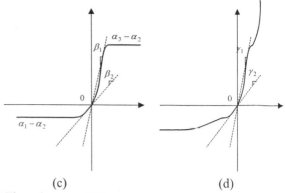

(c)   (d)

(a) The solar-array V-I curve

(b) The shifted solar array curve ($g(V_{sa},\hat{v}_{sa})$)

$$\alpha_1 \cdot \hat{v}_{sa}^{\,2} \le g(V_{sa},\hat{v}_{sa}) \cdot \hat{v}_{sa} \le \alpha_3 \cdot \hat{v}_{sa}^{\,2}$$

(c) The sector property of $l_s(\hat{v}_{sa})$

$$\beta_1 \cdot \hat{v}_{sa}^{\,2} \le l_1(\hat{v}_{sa}) \cdot \hat{v}_{sa} \le \beta_2 \cdot \hat{v}_{sa}^{\,2}$$

(d) The combined sector Property of $l(\hat{v}_{sa})$

$$\gamma_1 \cdot \hat{v}_s^{\,2} \le l(\hat{v}_s) \cdot \hat{v}_s \le \gamma_2 \cdot \hat{v}_s^{\,2}$$

Fig. 8 The sector property

The conventional analysis with Eq. (7) cannot predicts this instability since the Nyquist plot of the transfer function of linear part has only not to encircle the point $-\frac{1}{\alpha}+j0$ under Nyquist thorem.

To avoid this large-signal instability, two different ways can be used. First, by reducing the effective gain of controller from adjusting the gain or the sampling time, the Nyquist plot of PPT controller can be designed not to enter the forbidden region. The PPT controller, whose gain is changed to be 0.3, can be expected to be large-signally stable as in Fig. 10 (a). Thus it can tracks the MPP of solar array without oscillatory behavor as in Fig. 12(a). In this case, to avoid the instability, the performance of the controller must be decreased. The time to arrive MPP point is increased.

As an alternative method, attaching the limit function in front of the output of PPT controller can stabilize the system by making the forbidden region be narrow. As shown in Figure 10(b) and Fig. 11, the limit can reduce the absolute value of the sector constants in (24). With this limitation, the PPT controller can tracks MPP of solar array without reducing the gain of the PPT controller, thus avoid the performance decrease in terms of the speed of PPT controller as in Fig. 12 (b). The limiting function provides the more feasible and effective solution in the practical point of view. Since the system can be suffered from the undesired operation of the digital circuit at Start-Up time or in electrical noisy environments, the variation of the outputs must be limited. By adjusting the limiting function, the large-signal instability can be avoided without additional circuit or the performance-degrade.

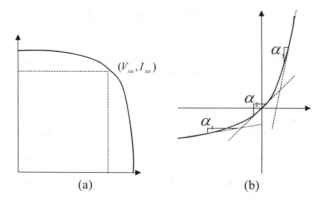

(a)   (b)

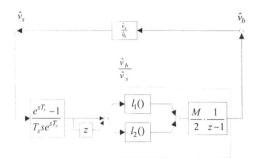

(a) The block diagram of PPT system

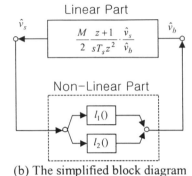

(b) The simplified block diagram

Fig. 9 The nonlinear control block diagram

CONCLUSION

Large-signal stability analysis of the Peak-Power-Tracking system is performed to facilitate design and analysis of Low-Earth-Orbit satellite power system. Possible instability due to nonlinear dynamic interactions between Solar-Array-Regulator and solar array nonlinear characteristic is identified. The effect of the slope variation and the sharpness of solar array on the system stability is investigated and evaluated employing the Circle criterion. The analysis and design method is proposed to avoid the system instability from large-signal nonlinear interactions.

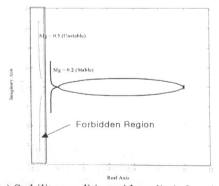

(a) Stability condition without limit function

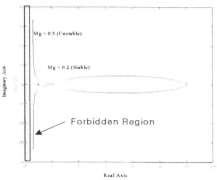

(b) Stability condition with limit function

Fig. 10 Large-signal stability Criterion

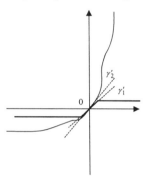

Fig. 11 Effect of the limitation

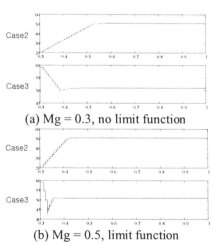

(a) Mg = 0.3, no limit function

(b) Mg = 0.5, limit function

Fig. 12 Simulation results

REFERENCES

[1]   Huynh, P., and Bo.H. Cho ,"Analysis and Design of a Microprocessor-Controlled Peak-Power Tracking System", Intersociety Energy Conversion Engineering Conference, 1992, pp. 1.63-1.78

[2]   Bo.H. Cho, and F.C. Lee, "Large Signal Analysis of Spacecraft Power Systems", Transactions on Power Electronics in Institute of Electrical and Electronics Engineers, Volume 5, No.1, Jan, 1990, pp 110-115

[3]   Huynh, P., and Bo.H. Cho ,"A New Methodology for the Stability Analysis of Large-Scale Power Electronics Systems", IEEE Transactions on Circuit And Systems- I: Fundamental Theory and Applications, VOL. 45, NO. 4, April, 1998, pp. 377-383

[4]   Hassan K. Khalil, "Nonlinear Systems", Second Edition, Prentice Hall

IECEC2001-AT-19

MIR RETURNED SOLAR ARRAY

Robert J. Pinkerton
Spectrum Astro Inc.

ABSTRACT

In November1997 a segment of a solar array wing from the Mir Space Station was removed from the Core Module and stored on the exterior of the Russian space station. NASA mission STS-89 returned it to Earth for analysis in January 1998. The panel spent a total of 10.5 years on-orbit attached to the Core Module. This paper describes the construction of the wing and panel, the measured performance, the performance losses and probable causes of the performance losses. Contamination was a major concern with the array since it was on a manned vehicle. The losses due to contamination, radiation, micrometeoroid and orbital debris (M&OD) have been assessed. The actual loss from each degradation factor is discussed for comparison and reference for future LEO missions.

INTRODUCTION

An eight-panel wing segment from the Core Module of the Mir Space Station was returned to Earth by the Space Shuttle in January 1998. This wing had spent 10.5 years on orbit. A panel of experts was assembled to examine and perform tests on it. Upon arrival at NASA Kennedy Space Center (KSC) the wing was unpacked and inspected by a joint US Russian team. After the initial one-week inspection one panel, #8,

Figure 2 Mir Wing Segment

was removed from the segment and left with NASA. The remaining 7 panels were returned to Russia for analysis. The segment, shown attached to Mir in Figure 1, was removed from the wing and stored on the Docking Module exterior for about two months prior to being returned by the Orbiter. Panel #8, which is representative of the rest of the panels, was then shipped to NASA Langley Research Center (LaRC) for analysis. Figure 2 shows the backside of the entire returned wing. The panel farthest to the left in the photo is panel #8.

Upon arrival at LaRC a detailed examination was performed prior to the removal of five sample sections of the panel for further investigation by various NASA centers. The roles of the NASA centers were: MSFC, JSC and GRC contamination; LaRC and JSC M&OD; GRC electrical performance. The author was added to the team to analyze system performance impacts.

Figure 1 Location of Segment on Mir

Figure 3 Silicone Contamination

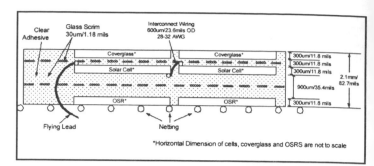

Figure 5 Cross Sectional View

VISUAL INSPECTION

The panel coloration is very non-uniform. The cell colors range from deep maroon to light blue. Some have a milky appearance, indicative of coverglass to cell delamination or poor wetting of bonding surfaces by the adhesive during assembly. Several cells on the periphery of the panel were very dark, with indications of a dark colored contamination. There were puffs of off-white contamination all over both the front and rear sides of the laminate surfaces, which can easily be seen in Figure 3. They appeared to originate from the holes in the laminate used to stitch the laminate to the netting. The silicone contamination apparently resulted from out-gassing from the silicone adhesive inside the laminate.

All uncovered surfaces of the panel had a thick layer of contamination. The frame had originally been painted white, but any exposed area was now brownish in color. There was a coating of contamination on the coverslides that had the appearance of oil on water when viewed through a microscope but did not appear to block much light.

There were multiple M&OD impact sites. These sites ranged from holes where pieces of coverglass and solar cell had been blown away, to tiny surface abrasions visible only under magnification. The Russians assessed that 34% of the panel had M&OD abrasion [1]. The edges of the

laminate were a dark tan in color, as were any exposed areas of adhesive, i.e., not under coverglass. The adhesive between the coverglass and the solar cells was very clear, rendering the coverglass invisible in most cases. The optical solar reflectors (OSR) on the rear of the panel had the same white puffs around the stitching holes. There was also a pattern, matching that of the netting, which was uncontaminated as shown in Figure 4. The OSRs additionally exhibited a speckling of brown contamination pretty evenly across their surface. See Figures 9 and 13.

PANEL CONSTRUCTION

The panel has 409-11% efficiency Silicon solar cells wired into four parallel strings. Three of the four strings are made up of 4.0 cm X 4.8 cm cells and the fourth uses 4.0 cm X 2.4 cm cells. Each string of large cells has 102 cells in series. The string of small cells has 103 cells in series. The panel is 0.62 m x 1.19 m in size and weighs 2.31 kgs.

The panel design is very unique. It is made up of a laminated sandwich of coverglass, scrim, silicon solar cells, scrim, and OSRs, as shown in Figure 5. The laminate is then stitched to a net that has been stretched over a tubular aluminum frame. The frame and netting can be seen in Figure 12. This panel uses the solar cell stack as the substrate, rather than mounting it onto a substrate.

The area between coverglass edges, OSR edges and the area between cells and coverglasses and between cells and OSRs was all filled with silicone adhesive. The laminate thickness is about 2.1 mm. This design transforms layers of glass into a very tough laminate.

The scrim between the solar cells and the coverglass covers about 80% of the surface area of the cell, but is virtually invisible to the naked

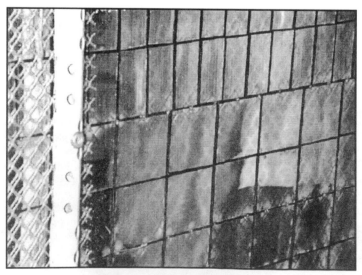

Figure 4 OSR Contamination

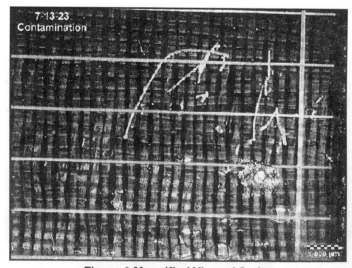

Figure 6 Magnified View of Scrim

Figure 7 Separation of Cell and Coverglass

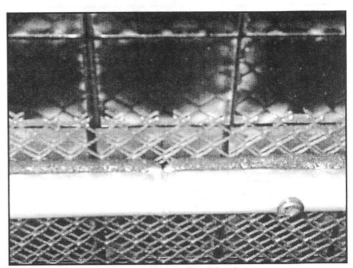

Figure 9 Foam Cushioning

eye, yet a magnified view of the cell clearly shows the scrim.

Figure 6 clearly shows the scrim under magnification. The scrim has little impact on panel efficiency. Solar cells primarily convert visible light into electrical energy, so if something in front of the cell is transparent, it has very little effect on cell efficiency. One or two cells had a milky appearance indicative of coverglass separation. The cell in the upper right hand corner of Figure 7 has the appearance of delamination. On these cells the coverslide could clearly be seen. On a well-bonded assembly it is almost impossible to detect a coverslide. Upon inspection with a microscope the scrim could immediately seen. The scrim could be compared to Fiberglas. The glass fiber is clearly visible before it has been wetted with the binder. However after being wetted, the fiber becomes transparent. The scrim on this cell looked like it had never been wetted with adhesive.

The laminate assembly is attached to the panel frame, shown pictorially in Figure 8. The frame itself is made of tubular aluminum about 1/2 inch in diameter. It has a rectangular shape 0.695m X 1.22m. There are tubular supports running across the middle of the length and the width intersecting in a junction in the center of the frame. The frame has a hole pattern drilled into it, presumably to save weight and ensure venting.

The laminate facing side of the frame has a foam rubber strip adhered to it the same width as the frame and about 1/4 inch thick. It can be seen Figure 9. This frame has netting stretched across it, held in place by a set of pins affixed along the outer edges of the frame. These pins penetrate every other hole in the net and are held to the frame by a series of metal loops.

The edges of the netting material are captured in what appears to be a metal closeout mounted on the edges of the frame.

The laminate is sewn on to the net with cotton string. This stitching was done between every other row of cells starting at the outside of the cells, running down the long dimension of the panel. It appears that the stitching penetrates the laminate from the cell side about every three inches, and picks up a strand of the netting then runs three more inches to pick up another strand of the netting, at another location. It also appears that there are two sets of stitching, their penetrations offset by about 3/4 inch. The string is visible on the front of the array and there are several locations where it is severed and frayed. The laminate is, however, still very firmly attached to the frame. The stitching can clearly be seen in Figure 3. Figure 10 shows an area where the string has been damaged.

ELECTRICAL LAYOUT

The solar cells are soldered together in lengths equal to the width of the panel. There are not a sufficient number of cells along the length of the panel for a complete string. Only 24 of the 4.0 cm X 4.8 cm cells, or 46 of the 4.0 cm X 2.4 cm cells, can fit across the length of the panel and 102 large or 103 small cells are required to make up a string. Figure 11 depicts

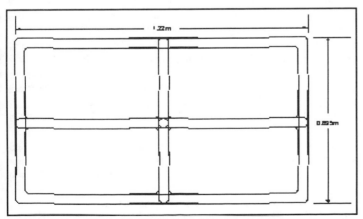

Figure 8 Panel Frame

Figure 10 Damaged Stitching

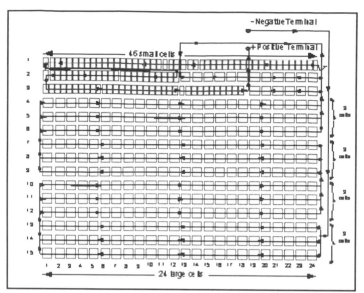

Figure 11 Solar Cell Layout

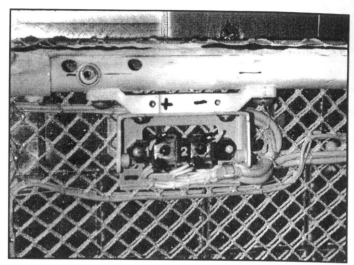

Figure 13 Terminal Block

the lay-down pattern for solar cells on the panel and how they are wired. To complete the string 3 1/3 panel lengths need to be wired in series. This is accomplished by external wiring on the backside of the panel. Each panel length of cells has a pair of 20 AWG equivalent flying leads coming off each end. These leads are fed through the back piece of scrim during the laminate assembly process.

Once the laminate has been mounted to the net, lengths of wire are attached across the ends of the panels. Bussing wires can be seen in the upper right hand quadrant of Figure 12 collecting the positive ends of the strings. This length of 18 AWG equivalent wire is used as a bus to interconnect the three paralleled strings to the next set of three paralleled strings. The bussing wire is shown schematically in Figure 11, on either side of the cells. This bussing is accomplished by striping a space about 1/4 to 1/2 inch long at six equally spaced intervals on the busing wire. The flying leads are then soldered onto this bussing wire at those locations. The bussing wire is tied to the netting using cable lacing. The ends of the wire are bent over and laced tight. There was no coating noted on any of these connections, or on the ends of the bussing wire, again see Figure 12. Bussing wires are installed at several locations on the back of the panel.

Figure 12 Rear Side Wiring

The two ends of the circuits are then bussed into two sets of wires, one positive and one negative. These wires are then routed into a junction box and soldered onto a terminal block, see Figure 13. From there the panel is connected in parallel with the other panels to form up a segment. Interestingly there are no blocking or bypass diodes on the panel. The Russians stated that they were mounted inside the power conditioning hardware. Bypass diodes are used to prevent "hot-spotting". This condition is described in reference 2.

The back of the frame then had a metal grid riveted to it. This panel is of open grid construction, again probably to save weight. The grid has openings of about 2-inch square and about 10% coverage. Only the two end panels of the wing segment, #1 and #8, have these metal grids. The purpose of these panels may be to provide some protection to the folded wing segment since they are on the outside when the wing segment is folded. They can be seen in Figure 2.

The cell interconnect wiring appears to be done by hand on the Mir panel. Each clear insulated wire, equivalent to about a 26 AWG which is no more than a half-inch long, is striped on each end, formed to provide strain relief and is then soldered by hand to each solar cell. There are three interconnects between each of the larger cells and two between each of the smaller cells. This interconnect wiring is depicted in Figure 5.

PANEL ELECTRICAL PERFORMANCE

Some initial testing was done on all eight panels at KSC during the one-week inspection period. A camera light was used to illuminate each panel while a current and voltage measurement was taken. The testing resulted in similar measurements from each panel. Panel #8 was sent to GRC for detailed panel level testing. David Brinker performed tests at all levels that could be achieved without any destruction to the panel.

This complete panel was measured to produce 48.7 W at maximum power. It had a measured efficiency of 4.81%. Its short circuit current was measured to be 1.595 A and its open circuit voltage was measured to be 59 V. The maximum power point was found to be at 42.4 V and 1.153 A. The fill factor was 0.561. NASA GRC performed the power measurements under ambient conditions. Figure 14 is the Panel IV curve. Unflown Russian solar cells, of the same vintage and assembly techniques as were used on the Mir panel, were sent to NASA for comparison. These cells were measured to be between 11.43 and 11.8% efficient.

Table 1 is a summary of the data taken at GRC. It is broken down into three segments. The first segment is the panel level data. The second is the

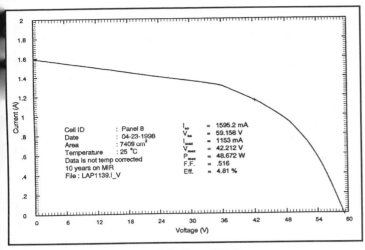

Figure 14 Mir Panel IV Curve

Cell ID : Panel 8
Date : 04-23-1998
Area : 7409 cm²
Temperature : 25 °C
Data is not temp corrected
10 years on MIR
File : LAP1139.I_V

I_{sc} = 1595.2 mA
V_{oc} = 59.158 V
I_{max} = 1153 mA
V_{max} = 42.212 V
P_{max} = 48.672 W
F.F. = .516
Eff. = 4.81 %

Table 1 Performance Data

Circuit	Isc Amps	Imp Amps	Vmp Volts	Voc Volts	Pmax Watts	FF %	η %
Panel	1.59	1.15	42.21	59.11	48.67	.52	4.81
Row 1	0.22	0.17	20.13	24.44	3.40	.62	5.04
Rows 2&3 SC	0.21	0.16	26.86	32.26	4.29	.63	5.08
Rows 2&3 BC	1.89	1.75	2.77	3.55	4.85	.73	9.59
Rows 4, 5 & 6	1.29	0.99	9.32	13.96	9.21	.51	4.55
Rows 7, 8 & 9	1.15	0.81	8.81	13.13	7.16	.48	3.53
Rows 10, 11 & 12	1.82	1.29	10.59	13.74	13.72	.55	6.77
Rows 13, 14 & 15	1.92	1.67	10.87	13.97	18.12	.68	8.94
Sec. 8-6-16	0.62	0.55	0.44	0.57	0.24	.69	9.27
Sec. 8-5-24	0.34	0.22	0.36	0.56	0.08	.42	3.01
Sec. 8-10-7	0.68	0.59	0.39	0.56	0.23	.59	8.67
Sec. 8-15-21	0.23	0.11	0.28	0.53	0.03	.27	1.21
Sec. 8-3-1	0.17	0.13	0.70	1.09	0.09	.51	3.53

data taken on one or more rows of cells. This data is taken by flashing the panel and reading the current and voltage it produced. Connections were made to the bare solder joints where the cell row flying leads attach to the bussing wires on the back of the panel. The third segment is the data taken on the sections of the panel that were removed. Row and column numbers defines the locations. Figure 11 identifies the row numbers down the left side of the figure and the column numbers across the bottom of the figure. Three numbers identify removed sections. The first number is the panel number, all are panel 8, the second number is the row number and the third number is the column number. Sections 8-6-16, 8-5-24, 8-10-7 and 8-15-21 are all single large cells. Section 8-3-1 is two small cells wired in series. Some of the data is suspect, particularly at the section level. This is because the wires were buried in the adhesive making it very hard to get a good connection. No attempt was made to disrupt the section to get better contact so it would be left intact for the first rotation cycle between researchers.

M&OD DAMAGE

Bill Kinard and Don Humes, of LaRC, found the panel damage to be in line with the current M&OD model predictions, both in the particle size and fluence level. The panel had been on-orbit ten years. It had many small divots and about 16 holes in the coverglass. Of these, five completely penetrated the panel and three were complete penetrations that passed light through them. Both Don and Bill believed that there was very little chance of shorting resulting from M&OD damage on the panel. They had found no signs of either.

The panel has less than 1% of its surface area damaged by impact craters. Of this 1% surface area [4], less than 0.15% of the solar cell area was damaged. While about 1% of the coverglass area was shattered, the rest was still optical quality. A much smaller area of solar cells was damaged because the panel construction absorbed the shock and protected the solar cells. A rigid planar array would most likely suffer a much greater amount of damage because it would not absorb the shock damage to the solar cell.

CONTAMINATION

Studies at MSFC, LaRC and GRC determined that the make up of the main component of the contamination found on the panel was SiOx, resulting from outgassing of the silicone adhesive used to bond the laminate

together. There were other contaminates such as urine and detergent, but in much smaller quantities. Performing a Large Area Pulsed Solar Simulation (LAPSS) test on a covered cell, then removing the cover and retesting the cell determined the effect of the contamination on electrical performance to be about 0.72% [3]. Mir was in approximately a 380 km orbit at a 51.6° inclination so it experienced a high flux of incident atomic oxygen. The outgassed silicone adhesive was deposited on the coverglass and the OSRs, and hardened in place by the atomic oxygen forming a white cloudy contaminate film. In orbits above the atomic oxygen rich altitude, it is expected that this contamination would have retained its chemical identity and been darkened by solar UV radiation resulting in a much greater power loss. The contamination depth on the panel coverglass and OSRs ranged from between 0.2 and 5.0 microns. The allowable contamination depth for the International Space Station over an equivalent time period is 0.137 microns.

ARCING DAMAGE

There was little evidence of large scale arcing on this panel. However, after receiving the first section, 8-15-21, there is a clearly visible burned spot near the edge of the cell. There were some other areas examined, under magnification, which showed signs of sputtering and possibly arcing. Evi-

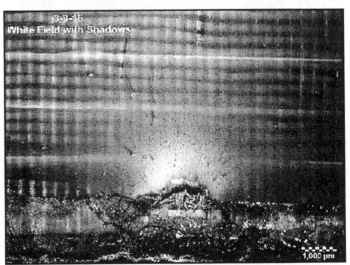

Figure 15 View of Possible Arcing Site

dence of possible arcing can be seen in Figure 15. It appears that the adhesive in these areas may have been darkened or charred but some of the people who looked at all eight of the panels thought there was arcing on some of the panels which were returned to Russia. The Russian engineers believed that there was no arcing observed, but only hot spotting. As previously stated there are no bypass diodes on this array to protect individual cells from reverse bias under shadowed conditions. This condition will cause solar cells to run very hot. The Mir arrays were routinely shadowed during orbital operations, so there would have been hot spotting. The Russians believe all the charring and darkening seen was resulting from this hot spotting.

THERMAL DATA

The absorptivity of the new panel was 0.77 and the emissivity was 0.85. The a/e ratio was 0.906 new. The measured absorptivity of the returned panel was 0.81 and its emissivity was 0.90 for a a/e ratio of 0.900. This would mean the array maximum operating temperature increased by about 7° C over its life, resulting in a 0.41% increase in current, a 3.98% decrease in voltage and a 1.63% power loss at the maximum power point at temperature.

HOTSPOTTING

This panel had no shadow diodes installed on it, but rather the cells were specially screened to ensure they could handle the full reverse current. In the final assessment, it appears that the increased temperature resulting from shadowed operation caused an increased resistance in the cell interconnections. This increased resistance, typically accounted for under thermal cycling loss, made up 38.44% of the 43% loss measured at the panel level. The Russian team is still analyzing the root cause of this phenomenon.

SUMMARY

The main loss factor on this panel appears to be hot spotting due to not placing bypass diodes on the panel. The other factors were quite small and cannot account for the 43% loss the panel incurred. The loss resulting from contamination, even though it was heavily contaminated, was only 0.72%. The M&OD loss was less than 1%. The temperature increase over life was only 7°C and accounted for a 1.63% power loss at the peak power point. The Russians reported that there was less than 1% loss due to natural radiation. These factors account for 4.35% of the total measured loss of 4.91% on several cells that did not suffer hot spotting.

ACKNOWLEDGMENTS

The author would like to acknowledge the work of the other Mir Returned Array Experiment Team members, in particular, David Brinker, Jim Visentine, Gayle Harvey, Don Humes, Bill Kinard and Keith Albyn without which this work could not have been done, and NASA and RSA for making the panel available to be studied. The author would also like to acknowledge the photographic skills and insights of Joe Chott, without whom all but two pictures would be missing, and Jerry Stephenson for the thermal analysis.

REFERENCES

1. Unknown, April 2, 1999, "Russian Study Results," Technical Report SS8352/TTI/AM/Bl

2. Pinkerton, R. J., 2000, "Solar Array String Characteristics in Strange Places," Proceedings, 35th Intersociety Energy Conversion Engineering Conference

3. Visentine, J. T., 2000, "Mir Solar Array Return Experiment Executive Summary of Visual Inspections & Laboratory Analysis Results"

4. Harvey, G. A., Humes, D. H., Kinard, W. H., 1999, "Optical Characterization of Returned Mir Solar Cells", Proceedings, Space Technology Conference & Exposition.

5. D.J. Brinker and D.A. Scheiman, NASA Glenn Research Center, Cleveland, OH; and J.T. Visentine, The Boeing Company, Houston, TX, 2000, "Power Degradation Studies of the Mir Solar Array Return Experiment," Proceedings of the 28th IEEE Photovoltaic Specialists Conference

IECEC2001-AT-33

IN-FLIGHT SOLAR ARRAY PERFORMANCE OF THE CLUSTER II SATELLITES

Roy Crabb,
European Space Agency,
European Space Technology
Centre,
Keplerlaan 1,
P. O. Box 299,
2200 AG,
NOORDWIJK,
The Netherlands.
E-Mail: roy.crabb@esa.int

Antonio Garutti,
European Space Agency,
European Space Technology
Centre,
Keplerlaan 1,
P. O. Box 299,
2200 AG,
NOORDWIJK,
The Netherlands.
E-Mail: antonio.garutti@esa.int

Jim Haines,
European Space Agency,
European Space Technology
Centre,
Keplerlaan 1,
P. O. Box 299,
2200 AG,
NOORDWIJK,
The Netherlands.
E-Mail: jim.haines@esa.int

ABSTRACT

The Cluster-II mission which is an element of the international Solar Terrestrial Science Programme (STSP), is an in-situ investigation of the Earth's magnetosphere using four spacecraft simultaneously. The paper briefly describes the electrical power system of the CLUSTER II satellites and presents their initial in-orbit solar array performances of the four spacecraft

INTRODUCTION

The Cluster-II scientific satellite programme is a mission using four accurately spaced spacecraft, allowing in-situ investigation of the Earth's magnetosphere. The mission allows for the accurate determination of three-dimensional and time varying magnetospheric phenomena and thus can distinguish between spatial and temporal variations.

Following the earlier launcher failure of Cluster I mission, a subsequent build of the satellites as depicted in Figure 1 was commissioned by the European Space Agency, the spacecraft prime contractor being Astrium GmbH of Friedrichshafen, Germany.

This second build was successfully placed into orbit by launching the spacecraft in pairs aboard two Soyuz-Fregate rockets in July and August of 2000. The four Cluster-II satellites were placed in nearly identical, eccentric polar orbits, having a nominal apogee of 22 earth radii and nominal perigee of 4 earth radii.

Figure 1. Three of the Cluster Satellites during Integration

The four spacecraft which are principally of identical design, are of a spin-stabilised configuration. During the scientific mission the spacecraft spin axis is pointed towards the ecliptic north pole, with a solar aspect (on the cylindrically mounted solar array) that ranges from 92 degrees to 96 degrees. Within the mission orbit, periodic eclipses do occur, which having a maximum duration of 4 hours.

POWER SYSTEM DESCRIPTION

The Cluster II spacecraft power demands are provided by the cylindrical solar array and/or five non-magnetic silver/cadmium batteries. The electrical power is conditioned and distributed via a voltage-regulated bus of 28 V +/- 1 % and redundant current-limiting switches. As a result full protection against load short-circuit or overloads is accomplished by limiting the maximum current in any supply line.

Excess solar array output power is automatically routed to Internal Power Dumpers (IPDs) or is shunted by commandable switching to the External Power Dumpers (EPDs). The power bus operates on a linear shunt regulation approach, rendering the main bus voltage extremely " clean " during the payload measurement periods. This significantly reduces any potential electromagnetic disturbances due to the power subsystem.

The Cluster-II Electrical Power System (EPS) conditions the power from the body mounted solar array and/or the Silver Cadmium batteries. The main features of the Cluster EPS are therefore as follows :-

- The provisioning of electrical power via a DC power bus of 28 volts ± 1% at the regulation point and 28 volts + 1%, -2 % at the power distribution interface.

- Distribution of the required power via ON/OFF switchable, Latching Current Limiters (LCLs) and protecting Foldback Current Limiters (FCLs) for the spacecraft essential equipments.

- Distribution of a low power 'keep alive' voltage line to those units requiring a permanent power interface to memorise their operational status during switch-off.

- If required by the spacecraft thermal design, the provisioning of heater power via Latching Current Limiters (LCLs).

The Cluster-II Solar Array Subsystem consists of six identical panels, each panel representing a 60 degree segment cladding of the overall spacecraft cylinder. The solar array output is provided as a single 'bulk' power source by the connection of 48 independent solar cell strings, each string comprised of 57 series connected, 3.21cm x 6.11cm x 210 micron BSR solar cells, covered by a 200 micron CMX glass.

EARLY MISSION PHASE

At first acquisition of signal an initial oscillation by approximately 100W of solar array power had been noticed on all four spacecraft. This was caused by a rather large difference in temperature (+ 60°C to – 40°C) between the sun illuminated part of the solar array and the opposite part of the array that had been in shadow as a result of the three-axis stabilised period of the Fregat flight. This solar array power fluctuation which started with spin up of the spacecraft, displayed an initial power variation of approximately 260W to 150W. It was noted that the power fluctuation decreased with progressive reduction of the temperature difference between the hot and cold part of the array.

After the temperature of the solar array has stabilised under nominal spin conditions, a permanent fluctuation of the solar array power was noticed on spacecraft 1 and 2. The oscillation on spacecraft 1 of approx. ± 3 watts could be explained by the build standard of this solar array since it used 2 panels from the Cluster I production and 4 panels from Cluster II production. The fluctuation on spacecraft 2 amounted to ± 8 watts, which was later declared to be a single solar array string failure.

With respect to the location of this failure a large amount of telemetry data from several days was reviewed in order to correlate the sun reference pulse events coming from the X-Beam Sun Sensor (XBSS) and the observed minimum array current readings. Data from four mission days which included also data for four different spin rates finally gave the consistent picture that the area affected was located about 54 degrees clockwise from the XBSS sensor. This location was the first string on the left hand side of panel 1 which had been the approximate location facing directly to the sun during the three axis stabilised Fregat flight prior to spacecraft 2 separation.

Although this array location had therefore experienced a high temperature during the launcher attached flight period, since the panel design had seen similar elevated temperatures during thermal cycling testing in the qualification programme, it was considered unlikely that this was the cause of the failure. However the panels do have triple string welds at the bottom of the panel and it was postulated that these could have been affected by some physical damage during handling.

The initial solar array power output after stabilisation of the solar array temperature and for given solar aspect angles (SAA) is presented in Table1.

	S/C 1	S/C 2	S/C 3	S/C 4
Solar array power (W)	268.5 ±2.7	268.6 ±8.1	275.2 ±1.5	271.9 ±0.7
SAA	82.83	87.22	88.52	84.10
SA temp. (°C)	3.2	2.3	1.8	1.8

Table 1. Initial Solar Array Power Output

BEGINNING-OF-LIFE (BOL) POWER BUDGET

The in-flight monitored BOL power budgets for all four spacecraft on the 3rd of December 2000 is presented in Table 2. This data which has been subsequently used as the reference for the end-of-life (EOL) budget computations, are based on the following conditions:

- The solar array power has been derived from the telemetry data of the main bus voltage and solar array current, including all the offset correction of the telemetry readings.

- The main bus load power (meaning all the loads connected to the main bus, including experiments and platform subsystems) has been computed by the telemetry data of the main bus voltage and main bus current, including all the offset correction of the telemetry readings.

- The IPD (Internal Power Dumper) power represents the excess array power dumped inside the spacecraft and thus contributing to the heating of the Main Equipment Platform (MEP). It has been computed by the telemetry data of the main bus voltage and IPD current including all the offset correction of the telemetry readings.

- The EPD (External Power Dumpers) power represents all the excess of power dumped outside the spacecraft and has been computed by the difference between the solar array power and the main bus load power.

- The MEP average temperature is based on the telemetry readings of the relevant thermistors.

	S/C 1 RUMBA	S/C 2 SALSA	S/C 3 SAMBA	S/C 4 TANGO
	Mission Oper. Mode	Mission Oper. Mode	Mission Oper. Mode	Mission Oper. Mode
MB Load+PCU+BDR Pwr W]	168.34	167.49	157.18	157.75
IPD Power [W]	51.28	70.74	55.76	82.38
EPD Power [W]	58.81	40.42	69.21	39.51
MEP temp. [C]	18.69	17.65	17.16	18.38
S.A. Power [W]	278.42	278.65	282.14	279.64
Needed power [W]	219.62	238.23	212.94	240.13
Operational Margin [W]	58.81	40.42	69.21	39.51
Thermal Margin [W]	13.45	8.25	5.80	11.90
Total Margin [W]	72.26	48.67	75.01	51.41
Lifetime	BOL	BOL	BOL	BOL

Table 2. Beginning-of-Life Power Budget

RADIATION ASSESSMENT

The overall radiation environment comprises the trapped particle belts and solar proton events. The trapped particle belts are a permanent structure, although the extent and intensity of the electron belt can vary considerably. Estimating the scope of the short-term effects of the radiation belts on a spacecraft is best achieved with in-situ measurements due to their dynamism and the locality of the structures. Without in-situ measurements,

the long term average AP8/AE8 trapped particle models can provide a rough estimate. On the other hand solar proton events are relatively rare and last for several days. This requires a statistical treatment for long-term predictions, but due to the non-localised extent of the effects, also permits the use of remote measurements for estimating the effects of an event.

The trapped particle effects were characterised in the "Cluster-II Radiation Analysis" for the respective transfer orbit and mission duration. As a particle flux measurement instrument was not manifested for the Cluster satellites, it has to be assumed that the radiation received by the spacecraft is the sum of the predicted transfer trajectory component and the operational dose. The operational dose fraction is the ratio of the commissioning time to the operational time (138 days:736 days) of the mission.

Since the beginning of the Cluster II mission in July 2000, two large proton events have occurred and several smaller ones. The first large event in July did not significantly impact that mission as the launch occurred in the declining phase of the event and that the geomagnetic field effectively shielded the initial transfer orbits. The second large event in November was approximately 80% the size of the July event and this dominated the solar proton fluence for the commissioning phase. Using the Radiation Environment Monitor from the GOES-10 spacecraft, the solar proton accumulated radiation fluence over the commissioning phase was calculated for the Cluster-II spacecraft, this accounting for about 1.22E13 equivalent 1 MeV electron fluence .

SOLAR ARRAY DEGREDATION

The solar array performance degradation is mainly function of the radiation effect on the cell efficiency. Beside this, the solar array power production is a function of other factors depending on the environmental conditions experienced by the spacecraft and the solar array itself, such as :

- Solar Array Aspect angle
- Array Temperature
- Orbital conditions (season, albedo influence, manoeuvres)

To have a clear picture of the progressive degradation of any solar array, the power output readings from a spacecraft solar array have to be normalised to a standard condition, which is typically 1 solar constant, sun normal and 23°C. In adopting this approach it is possible to identify the degradation independently from other effects like spacecraft attitude, season and temperature.

Figure 2 shows the basic Cluster II solar array output power (calculated from voltage and current telemetry readings) for each of the four spacecraft during the period from launch to the 3rd of December 2000. The values are referred to Cluster II apogee passage and scaled to eliminate the known effects of telemetry offset.

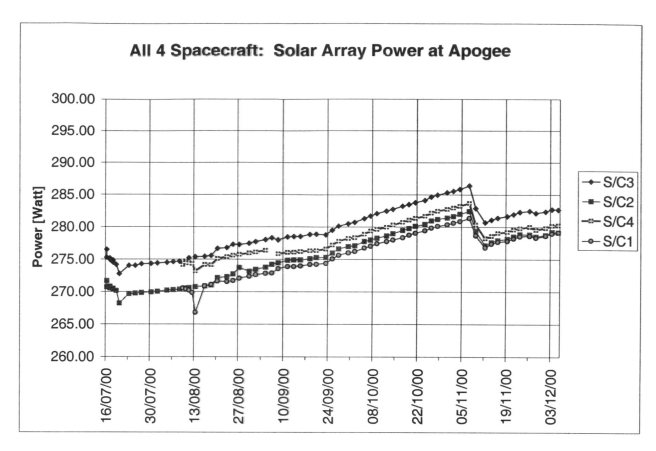

Figure 2. Solar array power evolution of all four Cluster II spacecraft from launch to 3rd December 2000.

Figure 2 shows the evolution of the solar array power for all four Cluster II spacecraft since launch. As can be seen from launch up to December 2000 the solar array output power of all four spacecraft shows a common trend. The values increase steadily with the seasonal change of the solar flux. All traces stay within a 5 watt wide band throughout the observation period as defined by spacecraft 3 (upper graph) and spacecraft 1 (lower graph). The fact that even small deviations (±0.5 watt) in the plots can be observed on all four spacecraft simultaneously gives strong confidence that the data is sound.

Due to telemetry the absolute accuracy of the data presented should not be considered to be better than ±4 watts. Additionally it should be noted that that a 1-bit resolution of the solar array current telemetry at constant main bus voltage, corresponds to 1.3 Watts.

The only difference in the gradual development of the individual S/C solar array performances can be seen from the solar array power history of spacecraft 4. As can be seen from figure 2 and in particular from figure 3, (which depicts the intensity, solar aspect and temperature normalised solar array power for all spacecraft), the solar array power level of spacecraft 4 which was close to the level of spacecraft 3 just

after launch has since reduced down to the power level seen on spacecraft 1 and spacecraft 2.

This performance could be related to one or a combination of the following three reasons:

- the observed trend is within the overall power reading accuracy of +/- 4 watts

- a telemetry offset correction should also have been applied for the solar array power of S/C 4.

- since there is a noticeable trend in the power readings it could be the effect of a deterioration/drift of the solar array cell/string/component performance.

However in summary, it is believed that the solar array data available so far is of acceptable accuracy level in order to perform the mission End-of-Life (EOL) projection. It is expected that a more precise assessment of the power level of each of the spacecraft will certainly become possible after a longer period of mission lifetime has elapsed.

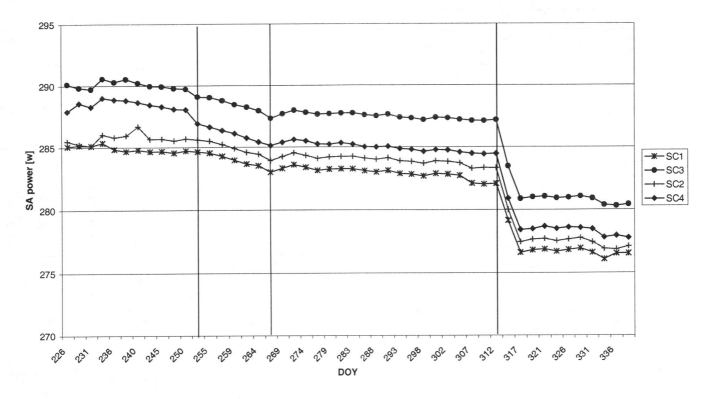

Normalised SA power @ SAA 90, 1AU, 23 C

Figure 3. Normalised Solar Array Output of all four Cluster II spacecraft from launch to 3rd December 2000.

This will be particularly the case when the lifetime will be reached for which a direct comparison of equivalent seasonal data can be made, such as comparing winter solstice 2000 with winter solstice 2001.

From figure 2 it can be seen that up to December 2000, three radiation related degradation events had been experienced on the four Cluster II spacecraft.

The first event took place in the period between the 6th of September 2000 (DOY 250) and the 26th of October 2000 (DOY 270). From available data source it has been verified, that during that period several low magnitude flare events have occurred. The total degradation in solar array power for all four spacecraft due to the accumulated equivalent fluence in this period was less than 1%.

The second event, which was of very high magnitude occurred in the second week of November 2000, with the peak occurring on the 9th of November. This flare which resulted in a 2% power degradation within a relatively short time was experienced by each of the four Cluster II spacecraft and as can be seen caused a visible step in all the power curves.

The last radiation degradation occurred on 26 November 2000. This was a flare of a medium order of magnitude, which caused a loss in power of about 0.5%.

Using the data from figure 3 it can be stated that the degradation in the Cluster II solar array power output, experienced up to December 2000, corresponds to 1.35E13 equivalent 1 MeV electron fluence and results in about 3 % power loss.

PREDICTION OF THE END-OF-LIFE ARRAY POWER

In order to determine the End-of-Life solar array power for each of the Cluster II spacecraft, the following computation has been subsequently performed:

- The starting point is the actual reading of the solar array current telemetry on 3rd Dec. 2000.
- On the basis of this data the WS BOL solar array current is computed for 21st Dec. 2000.
- The WS EOL current is computed applying the expected degradation from the radiation cell test results. All the other boundary conditions like solar aspect angle and solar array temperature remain unchanged.

The End-of-Life (EOL), winter and summer solstice SA power outputs thus predicted are presented in Table 3 and Table 4.

	S/C 1	S/C 2	S/C 3	S/C 4
BOL [3 Dec. 00] S.A. Current	9.99	9.93	10.07	9.98
BOL WS S.A. Current	10.02	9.96	10.10	10.01
EOL WS S.A. Current	8.67	8.61	8.74	8.66
EOL WS S.A. Power	241.55	241.75	244.78	242.61

Table 3. SA Power–Winter Solstice-21 December 2002

	S/C 1	S/C 2	S/C 3	S/C 4
BOL [3 Dec. 00] S.A. Current	9.99	9.92	10.06	9.98
BOL WS S.A. Current	10.02	9.95	10.09	10.01
EOL + 0.5 Y SS S.A. Current	8.06	8.01	8.12	8.05
EOL + 0.5 Y SS S.A. Power	224.64	224.83	227.64	225.63

Table 4. SA Power–Summer Solstice-21 June 2003

END-OF-LIFE SPACECRAFT POWER BUDGETS

Starting from the overall S/C power budget as listed in Table 2 an end of life extrapolation has been performed taking the following assumption and boundary conditions into account.

- The budgets have been determined for the 21/12/02 Winter Solstice environmental condition. It is assumed that the heating power necessary to maintain the Main Experiment Platform (MEP) temperature is the same as at BOL.

- Effects of EOL degradations of the solar array.
  The actual SA degradation taken into account in the computation is the remaining dose (about 95% of the fluence budget considered in the Cluster radiation analysis)

- The MLI thermo-optical property degradation has to be considered and the practical effect of this degradation is an increase of the MEP temperature. The estimated value of the EOL temperature increase is about 3°C which is known to be equivalent to 15 watts of heating power.

The EOL computations have been performed for the same mission worst case spacecraft mode (Mission Operational Mode), in both Winter (Table 5) and Summer Solstice (Table 6) seasonal conditions and to meet the minimum specified MEP temperature of 16 °C.

	S/C 1	S/C 2	S/C 3	S/C 4
	Mission Oper. Mode	Mission Oper. Mode	Mission Oper. Mode	Mission Oper. Mode
MB Load+PCU+BDR Pwr W	168.34	167.49	157.18	157.75
IPD Power [W]	51.28	70.74	55.76	82.38
EPD Power [W]	21.93	3.52	31.85	2.48
MEP temp. [C]	18.69	17.65	17.16	18.38
S.A. Power [W]	241.55	241.75	244.78	242.61
Needed power [W]	219.62	238.23	212.94	240.13
Operational Margin [W]	21.93	3.52	31.85	2.48
Thermal Margin [W]	28.45	23.25	20.80	26.90
Total Margin [W]	50.38	26.77	52.65	29.38
Lifetime	EOL	EOL	EOL	EOL
Season	WS	WS	WS	WS

Table 5. S/C Budget–Winter Solstice-21 December 2002

	S/C 1	S/C 2	S/C 3	S/C 4
	Mission Oper. Mode	Mission Oper. Mode	Mission Oper. Mode	Mission Oper. Mode
MB Load+PCU+BDR Pwr W	168.34	167.49	157.18	157.75
IPD Power [W]	52.83	77.49	64.96	85.48
MEP temp. [C]	16	16	16	16
SA Power [W]	224.64	224.83	227.65	225.62
Needed power [W]	221.17	244.98	222.14	243.23
Operational Margin [W]	3.48	-20.15	5.51	-17.61
Thermal Margin [W]	15.00	15.00	15.00	15.00
Total Margin [W]	18.48	-5.15	20.51	-2.61
Total Margin [W]	43.48	19.85	45.51	22.39
Lifetime	EOL	EOL	EOL	EOL
Season	SS	SS	SS	SS

Table 6. S/C Budget–Summer Solstice-21 June 2002

CONCLUSION

The power budgets as presented in Tables 5 and 6 confirms that the available power margin is sufficiently large for the nominal operation of the Cluster spacecraft and a mission lifetime of two years starting in January 2001 and ending in December 2002.

IECEC 2001-AT-34

CALIBRATION AND MEASUREMENT OF SOLAR CELLS
ON THE INTERNATIONAL SPACE STATION: A NEW TEST FACILITY

Geoffrey A. Landis and Sheila G. Bailey
NASA John Glenn Research Center

Phillip Jenkins, J. Andrew Sexton, and David Scheiman
Ohio Aerospace Institute

John Bowen, Robert Christie, James Charpie, and Scott S. Gerber
Zin Technology

D. Bruce Johnson
Dynacs Engineering

ABSTRACT

The Photovoltaic Engineering Testbed ("PET") is a proposed facility designed for flight on the International Space Station to perform calibration, measurement, and qualification of solar cells in the space environment and then return the cells to Earth for laboratory use. The goal of PET is to allow rapid-turnaround testing of new photovoltaic technology under actual space (true AM0 spectrum) conditions. PET is also designed to allow long-duration exposure tests of cells to the space environment, with regular measurement of changes in cell properties, and to measure the temperature coefficient of the current-voltage (I-V) characteristic of photovoltaic cells under space conditions. PET is designed for mounting on the Japanese Experiment Module (JEM) Exposure Facility. The sample change-out unit is exchanged through the airlock using the Japanese robotic "small fine arm", permitting rapid testing and return of samples.

NOMENCLATURE

AM0 (Air Mass Zero): sunlight under space conditions
FRAM (Flight-releasable Attach Module): a fixture for externally attached cargo to be launched to the space station in the space shuttle cargo bay
JEM: Japanese Experiment Module on the International Space Station
JEM-EF: Japanese Experiment Module Exposed Facility
PET: Photovoltaic Engineering Testbed

INTRODUCTION

Transission of new solar cell technology from the laboratory to flight requires calibration, measurement and qualification of solar cells in the space environment. Historically, the limited ability to test new cell types in space has been a bottleneck to the transition of new technologies to flight readiness. The Photovoltaic Engineering Testbed ("PET") is intended to meet the need for space flight experience on new cell types on an ongoing basis.

PET is a proposed facility designed for flight on the International Space Station to test solar cells in the space environment and then return the cells to Earth for laboratory use [ref. 1-2]. The goal of PET is to allow rapid-turnaround testing of new photovoltaic technology under actual space (true AM0 spectrum) conditions.

The solar spectrum in space is referred to as "Air-Mass Zero" (or "AM0"). This indicates that the mass of air between the solar cell and the sun is zero, or true space conditions. At the surface of the Earth, for light passing perpendicularly through the atmosphere, the mass of air between the cell and the sun is exactly one times the atmospheric thickness; this is thus known as "Air-Mass One" (or AM1). Solar cell performance at air-mass 1 is considerably different from air-mass zero performance, due to the selective absorption of the atmosphere. The facility will be used for three primary functions: calibration, measurement, and qualification.

Calibration is used to create a space-measured reference cell that can be used to calibrate the performance of cells on the ground, allowing ground measurements to be referenced to an actual space-flown standard of the same type.

Measurement consists of measurements of the performance of cells in space, in particular measuring values of parameters that are sensitive to the space solar spectrum or environment, such as the temperature coefficient.

Qualification consists of verifying that the performance of an interconnected solar-cell coupon does not degrade over time in the space environment.

The PET facility is more capable and does more than existing calibration facilities. PET makes solar cell measurements under true space conditions (true air-mass zero). This test will be in space, not in "near" space. No correction factors are required to produce absolute calibration. Furthermore, PET allows long duration testing in the space environment. This will allow in-space measurement of degradation such as Staebler-Wronski effect, radiation (including low-energy electrons), UV degradation, and atomic oxygen and plasma.

Finally, PET is designed to perform temperature coefficient measurements under the AM0 solar spectrum. In-space measurement of solar cell temperature coefficient is becoming of increasing importance for NASA missions such as Solar Probe and MESSENGER, which will operate in high-temperature environments close to the sun; and Mars, asteroid, and Jupiter missions, which operate in low-temperature environments far from the sun. Temperature coefficient is extremely sensitive to the solar spectrum [ref. 4]. The existence of emission lines in the xenon-simulated solar spectrum results in temperature coefficients of current that can incorrect by a factor of five; and in some cases can even have the incorrect sign. It is of increasing importance to measure temperature coefficient under true space AM0 spectrum, rather than simulated sunlight.

Finally, the qualification function of the testbed will be used to verify that the performance of an interconnected coupon does not degrade in the space environment, which is the final critical step leading to flight qualification and acceptance of a new technology. By doing flight exposure on the space station testbed, the qualified samples can be returned for examination after the test. This will give us the ability to diagnose failure mechanisms (if any), allowing a technology to be fine-tuned as required to pass performance specifications.

The space exposure facility of PET can also be used to measure space performance of things other than solar cells, including electronics and sensor packages. It can also, if desired, be used as an exposure platform to measure the effect of the space environment on materials.

ENGINEERING DESIGN

A FRAM (Flight-releasable Attach Module) fixture is used as the attach point for carrying the PET facility into space in the space shuttle cargo bay; this is mounted on the side of the PET. Once brought to the space station in the shuttle cargo bay, the space station robotic arm docks the PET to its permanent mounting position on the Japanese Exposed Facility. The exposed facility provides an attach point which includes an interface for data and power as well as optionally fluid cooling.

Figure 1 shows the PET facility. Two sample change-out trays are shown. Each change-out tray has four sample holders. The top tray is shown with the light shielding in place to eliminate Earth albedo illumination on the cells and reduce the reflected light from space station; the sun shield configuration also incorporates doors that can be opened or shut on orbit to control the duration of space exposure, allowing the cells to be exposed to space only during the periods when the test is occurring. The bottom tray is shown here with a variety of experimental samples mounted without sun shields for exposure testing.

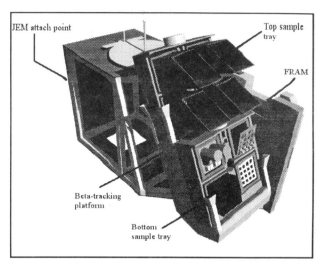

Figure 1: PET facility

Each of the four sample holders per sample tray can hold up to 28 2 cm by 2 cm solar cells, or a smaller number of larger area cells, that can be separately measured. Each sample holder has an independent temperature control,. PET will be capable of controlling each solar cell to the standard reference temperature of 25°C during the I-V measurement, or to allow heating of the sample holder from 20 to 80 °C for temperature coefficient measurement.

PET incorporates single-axis rotation for sun tracking. This allows the solar beta angle to be adjusted. Normal sunlight (0° incidence angle, plus or minus 1°) is achieved for a period of 30 seconds per orbit. A complete I-V characteristic can be taken for each cell during this time. The 92 minute orbital period of the space station allows slightly over 15 measurement periods per day. Cell sample trays can be changed out every time the space shuttle visits the space station, a minimum of once every 90 days, allowing rapid turnaround of samples.

A significant engineering design issue for PET has been the design to allow robotic change-out of samples with no requirement for astronaut EVA [ref. 3]. The JEM Exposed Facility (JEM-EF) facility is ideal for this purpose, since it has a small airlock and a robotic arm with an attached "small fine arm" manipulator tool which can take packages from the airlock and move them to the experiment packages mounted on the exposed facility. The sample trays are designed to be able to fit in the airlock, and have a robotic grapple fixture designed to interface with the JEM robotic arm. The beta-tracking platform on the PET facility includes coarse- and fine-alignment guides to allow the robotic arm to slide the sample trays into position and mate the electrical and data connections to the PET electronics.

Figure 2 shows the location of the PET on the Japanese Exposed facility, showing the Japanese robotic arm in position preparing to change out an experiment tray. (Note that in this figure, the other payloads on the JEM-EF are not shown.)

Figure 2. Location of PET on the Japanese Exposure Facility on the International Space Station.

PET STATUS

The PET facility has finished the conceptual design stage, and is preparing for the preliminary design review. Engineering test hardware for the electronics package has been fabricated and tested, verifying that the I-V measurement can be made under simulated space conditions.

PET is intended to fly on space station assembly flight 2J/A, now scheduled for January, 2005 [ref. 5]. However, the current status of the PET project is uncertain. The ERT program office at NASA Johnson has been shut down, and there is no commitment to fly the facility. Several alternative paths to develoment and flight of PET are now under consideration.

CONCLUSIONS

It has historically been extremely difficult to get new technologies from the lab into flight. Users require space experience before baselining a technology for use in flight. PET is a facility designed to do this.

The best possible way to calibrate solar cells is to actually fly the cells in space and calibrate them to a true space conditions. With the PET facility, we have an opportunity to make this measurement routine.

ACKNOWLEDGMENTS

Engineering development of the PET project has been supported under the Engineering Research and Technology (ERT) program at NASA Johnson Space Center. The authors would also like to acknowledge their colleagues at the Photovoltaics and Space Environmental Effects branch at NASA Glenn for assistance with the project, and would like to acknowledge the support of Carlos Parra, of the Engineering Research and Technology program at NASA Johnson Space Center.

REFERENCES

1. G. Landis and A. Sexton, "An Engineering Research Testbed for Photovoltaics," *AIP Conference Proceedings Volume 458*, pp. 438-441, presented at Space Technology & Applications International Forum 1999 (STAIF-1999), Conference on Space Station Utilization, Albuquerque NM, Jan. 30 - Feb. 4 1999.
2. G. Landis, S. Bailey, P. Jenkins, A. Sexton, D. Scheiman, *et al.*, "Photovoltaic Engineering Testbed," presented at the IEEE Photovoltaic Specialists Conference, Sept. 17-22 2000, Anchorage AK.
3. G. Landis, A. Sexton, R. Abramczyk, J. Francz, D. Johnson, L. Yang, D. Minjares, and J. Myers, "The Photovoltaic Engineering Testbed: Design Options and Trade-offs," *AIP Conference Proceedings Volume 504*, presented at Space Technology and Applications International Forum (STAIF-2000), Jan. 30 - Feb. 3, 2000, Albuquerque NM.
4. G. Virshup *et al.*, "Temperature Coefficients of Multijunction Solar Cells," *Proceedings of the 21st IEEE Photovoltaic Specialists Conference, Vol. 1*, pp. 336-338 (1990).
5. http://spaceflight.nasa.gov/station/assembly/flights/chron.html

Proceedings Of IECEC'01:
36th Intersociety Energy Conversion Conference
July 29–August 2, 2001, Savannah, Georgia
IECEC2001–AT–35

STATUS OF PHOTOVOLTAIC CALIBRATION AND MEASUREMENT STANDARDS

Cosmo Baraona, Sheila Bailey, Henry Curtis, and David Brinker
National Aeronautics and Space Administration
Glenn Research Center
Cleveland, Ohio 44135

Phillip Jenkins and David Scheiman
Ohio Aerospace Institute
22800 Cedar Point Road
Brook Park, Ohio 44142

ABSTRACT

The 7th International Workshop on Space Solar Cell Calibration and Measurement was held on September 25–27, 2000 in Girdwood, Alaska. Representatives from eight countries discussed international standards for single and multijunction solar cell measurement and calibration methods, round robin intercomparisons, and irradiation test methods for space solar cells. Progress toward adoption of an ISO standard on single junction cells was made. Agreement was reached to begin work on new standards for multijunction cells and irradiation testing. Progress on present single junction round robin measurements was discussed and future multijunction round robins were planned. The next workshop will be held in Germany in October 2001.

INTRODUCTION

Photovoltaic (PV) systems (cells and arrays) for spacecraft power have become an international market. This market demands accurate prediction of the solar array power output in space throughout the mission life. Since the beginning of space flight, space-faring nations have independently developed methods to calibrate solar cells for power output in low Earth orbit (LEO). These methods rely on terrestrial, laboratory, or extraterrestrial light sources to simulate or approximate the air mass zero (AM0) solar intensity and spectrum. Beginning in 1994, the PV community (national space agencies and industry) in Asia, Europe, and the Americas have been working together to develop an international standard for single junction cell AM0 short circuit current measurement and calibration. Annual workshops have been held since 1994 to discuss and compare balloon, aircraft, direct terrestrial sunlight, and laboratory based methods for calibrating PV cells. These workshops have also discussed measurement of multijunction cells and radiation damage testing and have also conducted measurement intercomparisons (round robins) of the different calibration methods.

International standards for space solar cells are supported by the International Organization for Standardization (ISO). ISO Technical Committee (TC) 20 is for Aircraft and Space Vehicles, while ISO TC20 Subcommittee (SC) 14 focuses on Space Systems and Operations. ISO TC20 SC14 Working Group (WG) 1 is for Space Systems Design and Engineering. The US members of the Space Solar Cell Calibration and Measurement Workshop also serve as a Technical Advisory Group (TAG) to support the US delegation to the ISO TC20 SC14 WG1.

The most recent workshop, the 7th International Workshop on Space Solar Cell Calibration and Measurement, was held in Girdwood, Alaska on September 25–27, 2000. Twenty-eight representatives from Argentina, Canada, Germany, Japan, The Netherlands, Spain, the UK, and the USA represented ESA, NASA, NASDA, and industry. This workshop had several working groups. One working group focused on single junction crystalline cell calibration, one on single crystal multijunction cell measurement, one on round robin measurement intercomparisons and one on irradiation damage testing.

SINGLE JUNCTION CELL STANDARD

A document entitled "Single Junction Space Solar Cell Measurement and Calibration Procedure" was first proposed by the Japanese in 1994 and became ISO working draft WD15387 in 1996 with inputs from the six workshops which were held every year from 1995 to 2000. The six calibration methods described in this document include the NASA JPL and French CNES high altitude balloon methods, the NASA GRC high altitude aircraft method, ground level sunlight methods including global and direct normal, and synthetic sunlight methods including solar simulator or differential spectral response. The calibration results will have a standard deviation of ±1 percent using an AM0 solar constant of 1367±7 W/m^2. Measurements are done at a cell temperature of 25±1 °C. Further work with the error analysis for each of the methods is in progress. The document is a Draft International Standard and will be voted upon by the member nations this year. The six methods are shown in the figure and are grouped by light source. The references in the bibliography give further information about the methods.

ROUND ROBIN INTERCOMPARISON

A series of round robin measurement intercomparisons has been performed for each of the single junction calibration methods. A standardized cell holder that is compatible with each of the methods was developed. The latest round robin with three silicon and three gallium arsenide single junction cells will be completed by the end of this year. Because of the seasonal nature of some of the calibration methods, it takes a year or longer to complete a round robin intercomparison calibration at the seven laboratories worldwide. Future round robins will include multijunction cell calibration.

OTHER STANDARDS

The 7th Workshop reviewed an outline of a standard proposed by the Japanese participants on Electron and Proton Irradiation Test Methods. This document has been proposed to ISO as an official New Work Item (NWI) for their standards creation process. A favorable vote for its adoption as a NWI is anticipated.

A NWI on multijunction solar cell calibration and measurement will be proposed to ISO this year.

The 8th International Workshop is being planned by ESA for October 17–19, 2001 in Freiburg, Germany.

A partial bibliography of selected publications for some of the calibration methods is shown below.

BIBLIOGRAPHY

1. Phillip Jenkins, David Brinker and David Scheiman, "Uncertainty Analysis of High Altitude Aircraft Air Mass Zero Solar Cell Calibration," 26th IEEE Photovoltaic Specialists Conference, Anaheim, CA, Sept. 29–Oct. 3, 1997.
2. B.E. Anspaugh and R.S. Weiss, "Results of the 1995 JPL Balloon Flight Solar Cell Calibration Program," JPL Publication 95–23, Jet Propulsion Laboratory, California Institute of Technology, Pasadena, California, USA, Dec. 1, 1995.
3. Brue Anspaugh, "Errors Estimates in the Calibration of Solar Cells on Jet Propulsion Laboratory High Altitude Balloon Flights," 5th International Workshop, Aldershot, UK, July 13–16, 1998.
4. A. Cuquel and M. Roussel, "Calibration of Solar Cells Outside the Atmosphere," RA/DP/EQ/QM/92–398, CNES, Toulouse, France, Dec. 22, 1992.
5. R. Shimokawa, F. Nagamine, Y. Miyake, K. Fijisawa and Y. Hamakawa, "Japanese Indoor Calibration Method for the Reference Solar Cell and Comparison with the Outdoor Calibration," Japanese Journal of Applied Physics, Vol. 26, No. 1, Jan. 1987, pp. 86–91.
6. O. Kawasaki, S. Matsuda, Y. Yamamoto, Y. Kiyota and Y. Uchida, "Study of Solar Simulator Calibration Method and Round Robin Calibration Plan of Primary Standard Solar Cell for Space Use," 1st World Conference and Exhibition on Photovoltaic Solar Energy Conversion (WCPEC), Hawaii, USA, Dec. 5–9, 1994.
7. Japanese Space Solar Cell Calibration Technical Committee, "Results of Uncertainty Analysis for AM0 Standard Solar Cells Calibration Methods," 4th International Workshop, Cleveland, USA, Oct. 6–9, 1997.
8. Klaus Bucher, "Calibration of Solar Cells for Space Applications," Progress in photovoltaics: Research and applications, Vol. 5, pp. 91–107, 1997
9. Trinidad J. Gomez, "Calibration of Space Solar Cells–Terrestrial Global Method," 4th International Workshop, Cleveland, USA, Oct. 6–9, 1997.
10. CAST Technical Report, "Uncertainty Estimation of CAST Direct Normal Calibration Facility," 4th International Workshop, Cleveland, USA, Oct. 6–9, 1997.
11. Yang Yiqiang, Zeng Lingru and Ma Zong Cheng, "Calibration of AM0 Reference Solar Cells Using Direct Normal Terrestrial Sunlight," 9th International Photovoltaic Science and Engineering Conference (PVSEC), Miyazaki, Japan, Nov. 11–15, 1996.
12. PTB Technical Report, "CD15387, Para.7.3.4, Differential spectral response calibration method," 5th International Workshop, Aldershot, UK, July 13–16, 1998.

IECEC2001-AT-36

THIN FILM PHOTOVOLTAICS FOR SPACE APPLICATIONS

Sheila G. Bailey and Aloysius F. Hepp
NASA Glenn Research Center

Ryne P. Raffaelle
Rochester Institute of Technology

ABSTRACT

Future NASA and Air Force space missions are incredibly diverse. Air Force missions encompass both hundreds of kilowatt arrays to tens of watt arrays in various earth orbits. While NASA missions also have small to very large power needs there are additional unique requirements to provide power for near sun missions, and planetary exploration including orbiters, landers and rovers both to the inner planets and the outer planets with a major emphasis on Mars. Missions requiring high power will fall into roughly three categories: electric propulsion, whether inter-orbit (LEO to GEO, e.g.) or from the earth to either the moon or Mars; surface power systems to sustain an outpost and eventually a permanent colony on the surface of either body; and finally, large earth orbiting power stations which can serve as central utilities for other orbiting spacecraft, or perhaps even for sending power to the earth itself. NASA, primarily to see if the technical capability exists to launch and operate such systems at a reasonable cost, has recently studied the latter application. This paper will discuss the current state-of-the-art, address some of the technology development issues required to make thin film photovoltaics a viable choice for future space power systems, and conclude with related technology that may be enabling for other space power missions.

INTRODUCTION

In general, deployment of a large earth orbiting space power systems will require major advances in the photovoltaic array weight, stability in the space environment, efficiency, and ultimately the cost of production and deployment of such arrays. It has become clear to NASA and others that the development of large space power systems, and a host of other proposed space missions, will require the development of viable thin film arrays.[1] Studies have shown that the specific power or power per mass that will be required (i.e., 1 kW/kg) cannot be achieved with single crystal technology.[2] The specific power required is almost 40 times what is presently available in commercial arrays.[3] While high efficiency ultra lightweight arrays are not likely to become commercially available anytime soon, advances in thin film photovoltaics may still impact other space technologies (i.e., thin film integrated power supplies) and thus support a broad range of missions in the next century.[4] Mission examples include micro-air vehicles, ultra-long duration balloons (e.g. Olympus), deep space solar electric propulsion (SEP) "Tug" Array, Mars SEP Array, and Mars surface power outpost (see Figure 1).

Lighter power generation will allow more mass to be allocated to the balance-of-spacecraft (e.g. more payload). In addtion, less expensive power generation will allow missions with smaller budgets and/or the allocation funds to the balance-of-spacecraft (e.g. more payload). This is an essential attribute in enabling such missions as the Mars Outpost SEP Tug. Example benefits for the now-cancelled ST4/Champollion indicate a $50 million launch cost savings and 30% mass margin increase when thin-film PV solar array power generation was combined with advanced electric propulsion. A parametric assessment showed similar advantages for other solar system missions (e.g., main belt asteroid tour, Mars solar electric propulsion vehicle, Jupiter orbiter, Venus orbiter, Lunar surface power system).

Figure 1. Proposed Mars solar electric propulsion vehicle.

Much of the original development of thin film photovoltaic arrays was performed with the terrestrial marketplace in mind. This has been a tremendous benefit to researchers hoping to develop such arrays for space. Features such as cell efficiency, material stability and compatibility, and low cost and scalable manufacturing techniques are important to both environments. However, many key array aspects necessary for space utilization are not important for terrestrial use and thus have not experienced a similar progress. Features such as radiation tolerance, air mass zero (AM0) performance, lightweight flexible substrates, stowability and lightweight space deployment mechanisms must be developed before a viable space array can become a reality. Unfortunately, the costs associated with developing these features along with the subsequent space qualification studies mitigate the savings of using a thin film array for space and thus has inhibited the development.

THIN FILM TECHNOLOGY

On-going efforts by NASA and the Air Force are now addressing these issues associated with the development of thin film arrays for space. NASA is currently supporting joint research efforts in thin film array development with researchers at the Florida Solar Energy Center (FSEC), Daystar, Inc., and Global Solar, Inc. Copper indium diselenide (CIS), cadmium telluride (CdTe), and amorphous silicon (α-Si) thin film materials are three materials that appear to have a good chance of meeting several proposed space power requirements.[5] Reasonably efficient (~ 8 % AM 1.5) large area triple junction blankets using α-Si are already being manufactured.[6] Large area CIS cells are now reaching as high as 10% AM 1.5.[7] Development of other wide bandgap thin film materials that can be used in conjunction CIS to produce a dual junction device are underway. As has already been demonstrated in III-V cells for space use a substantial increase in the single junction efficiency is possible with a dual junction device.[8] NASA has initiated a dual-junction CIS-based thin film device program with researchers at the University of South Florida. The use of Ga to widen the bangap of CIS and thus improve the efficiency is already well known.[9] The substitution of S for Se also appears to be an attractive top cell material.[10] AM0 cell efficiencies as high as 7% have been measured for $CuIn_{0.7}Ga_{0.3}S_2$ (E_g 1.55 eV) thin film devices on flexible substrates (see Figure 2).

Unfortunately most of the high efficiency thin film devices developed thus far have been on heavy substrates such as glass. However, progress is being made in reducing substrate mass through the use of thin metal foils and lightweight flexible polyimide or plastic substrates.[11] A major problem with the use of plastic substrates is the incompatibility with many of the deposition processes. The most efficient thin film cells to date are made by some combination of co-evaporation of the elements and subsequent annealing. The use of such plastic substrates as Upilex or Kapton puts an unacceptable restriction on the processing temperatures. This of course can be obviated by the use of metal foil if one is willing to accept the mass penalty.

Another approach that may allow for the use of plastic substrates is to develop low temperature deposition techniques. Efforts are being made by NASA and others to develop low temperature chemical vapor deposition and electrochemical deposition methods. Low temperature processing techniques (< 400 °C) that are compatible with plastic substrates must be developed if the goal of achieving 1000 W/kg (1 kW/kg) cell is to be achieved. Recent use of a single-source precusor for low temperature chemical vapor deposition of $CuInS_2$ and related compounds has demonstrated promise in this regard.[12]

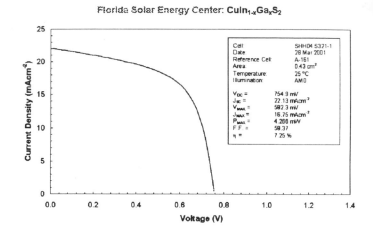

Figure 2. AM0 photoresponse for $CuIn_{0.7}GaS_2$ thin film solar cell deposited on Mo coated stainless steel foil.

ARRAY SPECIFIC POWER

To achieve an array specific power of 1kW/kg a cell specific power considerably higher than that will be necessary. Similarly, the blanket specific power (i.e., interconnects, diodes and wiring harnesses) must be over 1 kW/kg as well. The Advanced Space Photovoltaic Array (APSA) assessment determined that the mass of the deployment mechanism and structure is essentially equal to the blanket mass for a lightweight system. Therefore, a blanket specific power of approximately 2000W/kg, would be necessary to achieve a 1 kW/kg array.[13] NASA is currently sponsoring an effort by AEC Able Engineering to develop lightweight thin film array delployment systems.

Gains in specific power can addressed through an increase in the operating voltage. Higher array operating voltages can be used to reduce the conductor mass, especially at the high power levels that may be encountered. The APSA was designed for 28 volt operation at several kilowatts output, and the resulting wiring harness comprised about 10% of the total array mass, giving it a specific mass of about 0.7 kg/kW. Designing the array for operation at 300 V could easily allow a reduction of the harness specific mass by 50% or more, with a concurrent increase in the APSA specific power of 5% or more without any other modification. In any event, it is clear that future thin film, light weight solar arrays must be capable of high voltage operation in the space plasma environment, and it is likely that the required voltages will approach 1000v to 1500v to be compatible with direct drive electric propulsion spacecraft (i.e., no voltage step up is required to operate the thrusters). Such a requirement is completely compatible with the demand for high specific power. NASA has bench marked a thin film stand-alone array specific power that is 15 times the state-of-the-art (SOA) III-V arrays; area power density that is 1.5 times that of the SOA III-V arrays, and specific costs that are 15 times lower than the SOA II-V arrays.[14]

SPACE QUALIFICATION

The cells, blankets, and deployment mechanisms must all be able to withstand the rigors of space utilization. This will require that they are stable with respect to thermal cycling, vibration and mechanical stresses, and exposure to radiation (specifically high energy protons and electrons) and atomic oxygen. This poses tremendous challenges for substrate and cell stability and module electrical and physical interconnects. A benefit to the use of thin film CIS in this regard is its radiation tolerance. CIS has been shown to retain much more of its beginning of life (BOL) power than a comparable III-V cell.[15] In fact, little or no degradation has been measured for $CuInGaSe_2$ thin film cells due to 1 MeV electron irradition at a fluence of 10^{16} (see Figure 3).

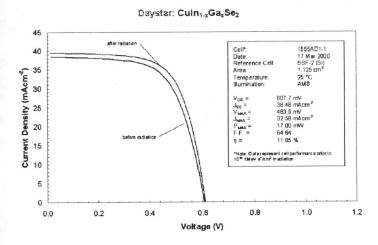

Figure 3. AM0 photoresponse of $CuInGaSe_2$ thin film cells on Mo coated Ti foil before and after electron irradiation.

INTEGRATED THIN-FILM POWER SYSTEMS

NASA has also been working to develop a lightweight, integrated space power system on a flexible substrates (see Figure 4).[16] The system will consist of a high efficiency thin film solar cell, a high energy density solid state Li ion battery, and the associated control electronics in a single monolithic package. This requires the development of suitable materials and low temperature chemical processing methods necessary to produce battery components that are compatible with thin film solar cells and other microelectronic components.

Thin film Li ion batteries have recently been directly integrated with monolithically interconnected photovoltaic modules and other electronic components for monolithic energy conversion and storage devices and distributed power nanosatellites.[17] These systems have the ability to produce constant power output throughout a varying or intermittent illumination schedule as would be experienced by a rotating satellite or "spinner" and by satellites in a low earth orbit (LEO).

An integrated thin film power system has the potential to provide a low mass and cost alternative to the current SOA systems for small spacecraft. Integrated thin film power supplies simplify spacecraft bus design and reduce losses incurred through energy transfer to and from conversion and storage devices. It is hoped that this simplification will also result in improved reliability. NASA is looking for a 5-fold improvement in specific power over SOA for earth-orbiting systems.[14]

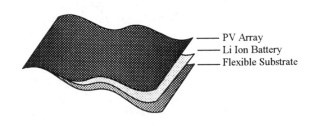

Figure 4. Thin film integrated power supply.

CONCLUSIONS

NASA has identified several areas in which large area thin film photovoltaic arrays can be of tremendous benefit for space exploration and development. Space qualified blankets (i.e., large area arrays on lightweight flexible substrates) with efficiencies as low as 10% (AMO) will serve as an enabling technology for a host of proposed missions. NASA plans to utilize the demonstrated radiation tolerance and possible efficiencies offered by materials such as $CuInSe_2$ by developing a low temperature deposition process that is compatible with plastic substrates. It is also hoped that this process may help to enable a dual junction CIS-based thin film device. Finally, it is anticipated that large area thin film arrays can be combined with other thin film devices (i.e., Li ion batteries) to produce completely autonomous and lightweight power supplies.

ACKNOWLEDGEMENTS

The authors wish to thank John Tuttle of Daystar Technologies, Inc. and Neelkanth Dhere of the Florida Solar Energy Center for their contributions to this work. We would also like to acknowledge the NASA Cooperative Agreements NCC3-710 and NCC3-563 with the Rochester Institute of Technology.

REFERENCES

[1] G. Landis and A.F. Hepp, "Applications of Thin Film PV for Space," 26th IECEC, 2, 256 (1991).

[2] D.J. Hoffman, T.W. Kerslake, A.F. Hepp, M.K. Jacobs, and D. Ponnusamy, "Thin film Photovoltaic Array Parametric Assessment," 35th *IECEC*, 1, 670 (2000).

[3] D. Flood, "Thin Film Photovoltaics: Apllications to Space Power," 34th *National Heat Transfer Conference*, Pittsburg, PA, August 20-22, (2000).

[4] R.P. Raffaelle, J. Underwood, D. Scheiman, J. Cowen, P. Jenkins, A.F. Hepp, J.D. Harris, and D.M. Wilt, "Integrated Solar Power Systems," 28th *IEEE PVSC*, Anchorage, AK September 17-22 (2000).

[5] J. Tringe, J. Merrill, and K. Reinhardt, "Developments in Thin-Film Photovoltaics for Space," 28th *IEEE PVSC*, Anchorage, AK September 17-22 (2000).

[6] W. Fuhs and R. Klenk, "Thin-Film Solar Cells Overview," 2nd *World Conf. On Solar Energy Conversion*, Vienna, Austria (1998).

[7] R.W. Birkmire, "Compound Polycrystalline Solar Cells: Recent Progress and Y2K Perspective," *Photovoltaics Science and Engineering Conf.*, Sapporo, Japan (1999).

[8] K. Barnham and G. Duggan, *J. Appl. Phys.* 67, 3490 (1990).

[9] M. Contreras, B. Egaas, K. Ramanathan, J. Hiltner, F. Hasoon, and R. Noufi, "Progress Toward 20% Efficiency $Cu(In,Ga)Se_2$ Polycrystalline Thin-Film Solar Cell," *Progress in Photovolt.*, 7, 311 (1999).

[10] C. Dzionk, H. Metzner, S. Hessler, and H.E. Mahnke, *Thin Solid Films* 299, 38 (1997).

[11] K. Tabuchi, S. Fujikake, H. Sato, S. Sito, A. Takano, T. Wada, T. Yoshida, Y. Ichikawa, H. Sakai, and F. natsume, "Improvements of Large-Area SCAF Structure a-Si Solar Cells with Plastic Thin Film Substrate," 26th *IEEE PVSC*, Anaheim, CA (1997).

[12] J.D. Harris, D.G. Hehemann, J. Cowen, A.F. Hepp, R.P. Raffaelle, and J.A. Hollingsworth "Using Single Source Precursors and Spray Chemical Vapor Deposition to Grown Thin Film $CuInS_2$," 28th *IEEE PVSC*, Anchorage, AK September 17-22 (2000).

[13] P. Stella and R.M. Kurland, *The Advanced Photovoltaic Solar Array (APSA) Technology Status and Performance*, 10th Space Photovoltaic Research and Technology Conference, NASA CP-3107, 421-432, (1986).

[14] A.F. Hepp, NASA GRC Thin Films Technology Product Agreeement (2000).

[15] H.W. Schock and K. Bogus, "Development of CIS Solar Cells for Space Applications," 2nd *World Conf. On Solar Energy Conversion*, Vienna, Austria (1998).

[16] D.J. Hoffman, G.A. Landis, R.P. Raffaelle, and A.F. Hepp, "Mission Applicability and Benefits of Thin-Film Integrated Power Generation and Energy Storage," 36th *IECEC*, Savannah, GA, July 29-Aug. 2 (2001).

[17] R.P. Raffaelle, J. Underwood, J. Maranchi, P. Kumta, O.P. Khan, J.D. Harris, C.R. Clark, D. Scheiman, P. Jenkins, M.A. Smith, D.M. Wilt, R.M. Button, and A.F. Hepp, "Integrated Microelectronic Power Supply (IMPS)," 36th *IECEC*, Savannah, GA, July 29-Aug. 2 (2001).

Proceedings of IECEC'01
36th Intersociety Energy Conversion Engineering Conference
July 29–August 2, 2001, Savannah, Georgia

IECEC2001-AT-66

INTEGRATED MICROELECTRONIC POWER SUPPLY (IMPS)

R.P. Raffaelle and J. Underwood
Rochester Institute of Technology

J. Maranchi and P. Kumta
Carnegie Mellon University

O.P. Khan and J. Harris
Cleveland State University

C. R. Clark
ITN Energy Systems

D. Scheiman, P. Jenkins and M.A. Smith
Ohio Aerospace Institute

D.M. Wilt, R.M. Button, and A.F. Hepp
NASA Glenn Research Center

W.F. Maurer
AKIMA

ABSTRACT

We have developed a microelectronic power supply for a space flight experiment in conjunction with the Project Starshine atmospheric research satellite. This device integrates a 7 junction small-area GaAs monolithically integrated photovoltaic module (MIM) with an all-polymer $LiNi_{0.8}Co_{0.2}O_2$ lithium-ion thin film battery. The array output is matched to provide the necessary 4.2 V charging voltage and minimized the associated control electronic components. The use of the matched MIM and thin film Li-ion battery storage maximizes the specific power and minimizes the necessary area and thickness of this microelectronic device. This power supply was designed to be surface mounted to the Starshine 3 satellite, which will be ejected into a low-earth orbit (LEO) with a fixed rotational velocity of 5 degrees per second. The supply is designed to provide continuous power even with the intermittent illumination due to the satellite rotation and LEO. An overview of the Starshine project and the IMPS experiment is given. Results on the ground-based characterization of this device under simulated space illumination and temperature conditions are presented.

INTRODUCTION

The development of small micro and nano-satellites have generated a need for smaller lightweight power systems. Thin film batteries and solar cells are ideally suited to such applications. The necessity for both generation and storage of power for microelectronic applications can be achieved by combining a thin film photovoltaic array with a thin film lithium ion battery into what is called an integrated microelectronic power supply (IMPS). These supplies can be combined with individual satellite components and are capable of providing continuous power in a variety of illumination schemes.

As a demonstration of the utility of an IMPS we have developed several such supplies for an atmospheric research satellite (i.e., Project Starshine).[1] Project Starshine is a micro-satellite that is designed to measured the drag in the upper earth atmosphere due to changes in solar activity (see Figure 1).

Figure 1. Starshine 1 (Photo from Reference 1).

The Starshine 3 satellite is ~1.0 m in diameter and has a mass of 88 kg. Its surface is covered with 1500 student polished mirrors, 31 laser retroreflectors, 48 - 2 cm x 2 cm triple junction solar cells manufactured by Emcore, and our 5 integrated power supplies. It is scheduled to be launched from a Lockheed Martin Athena I rocket from Kodiak, Alaska on August 31, 2001. It will be inserted into a 67 ° inclination with a fixed rotational velocity of 5 ° per second in a low earth orbit (LEO) with a period of 92 minutes. The data will be downloaded using a transmitter operating at a frequency of 145.825 MHz (2 M HAM, AX 25).

The IMPS we have developed combines a 7 junction - 1 cm^2 monolithically interconnected GaAs module (MIM) (see Figures 2 and 3) with a lithium ion battery. Ideally, the output of the high-voltage small area MIM would be designed to match the open circuit voltage of the lithium ion battery.[2] The MIM we used in this case has more than enough voltage and current to both charge a Li ion battery with a CoO_2-based cathode (e.g., $V_{oc} = 4.2$ V)[3] and power an equivalent load.

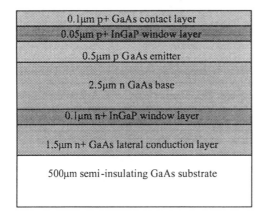

| 0.1μm p+ GaAs contact layer |
| 0.05μm p+ InGaP window layer |
| 0.5μm p GaAs emitter |
| 2.5μm n GaAs base |
| 0.1μm n+ InGaP window layer |
| 1.5μm n+ GaAs lateral conduction layer |
| 500μm semi-insulating GaAs substrate |

Figure 2. GaAs cell schematic.

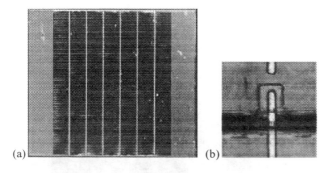

(a)　　　　　　　　　　(b)

Figure 3. a) 7-Junction 1 cm x 1 cm monolithically inter-connected module (MIM); b) Scanning electron microscope image of an interconnect.

The matching of the MIM to the battery is performed in order to minimize the control electronics. In an ordinary satellite power system, the charging of the battery back-up power is performed using a charge controller. Unfortunately, for a small power system the parasitic power losses of these controllers can make them prohibitive. When the PV array and battery are ideally matched all that is required is a blocking diode to prevent the battery from back discharging through the array when in eclipse. Since the 7-junction MIM is more than what is required for our Li batteries a low-power voltage regulator is added to prevent overcharging. Also, a MOSFET is used to cut off the battery from the load in the event that its voltage drops below a threshold value to prevent deep discharge. The circuit board schematic for our IMPS is shown in Figure 4.

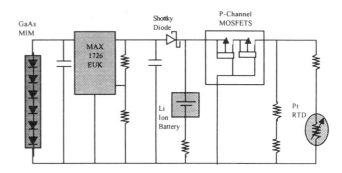

Figure 4. IMPS circuit schematic.

The loads for this demonstration are small temperature sensors (i.e., ~ 100 μW). These sensors are strategically placed about the spacecraft to monitor the temperatures of the critical components (e.g., batteries, solar cells, electronic boards), both inside and outside of the satellite.

Three prototype IMPS were developed. The first incorporated a commercial Panasonic ML2020 rechargeable manganese dioxide Li ion battery (see Figure 5).[4] This 3.0 V This 3.0 V "coin cell" has a diameter of 2.0 cm, thickness of 2.0 mm, mass of 2.2 g, and a nominal capacity of 45.0 mAh.

(a)　　　　　　　　　　(b)

Figure 5. a) Starshine 3 IMPS prototype with Panasonic battery; b) IMPS with Litestar battery on mounting stub.

The second was a solid state thin film Litestar battery developed by Infinite Power Solutions.[5] This 4.2 V had a $LiNi_xCo_{1-x}O_2$ cathode, a LiPON electrolyte, and Li metal anode. It had an active area of 2.69 cm^2, capacity of 800 μAh, and approximate mass of 1.23 g, including its substrate, contacts and sealants. The third battery was a 4 cm^2 and 1.38 g thin film polymer battery developed at the NASA Glenn Research Center. The electrolyte for this 4.2 V battery consisted of lithium (bis) trifluoromethane-sulfonimide, ethylene and propylene carbonate, and the polymer polyacrylonitrile (PAN). The cathode was this same polymer impregnated with $LiNi_{0.8}Co_{0.2}O_2$. The anode was also the same polymer impregnated with graphite.

RESULTS

The IV photoresponse of the MIM was measured at the Glenn Research Center using a simulated air mass zero (AM0) spectrum. The array had a short circuit current of 3.16 mA, an open circuit voltage of 6.93 V, and an 80% fill factor.

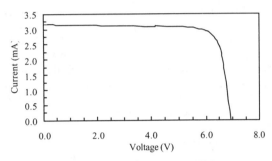

Figure 6. AM0 photoresponse of the GaAs MIM.

The capacity of the various batteries used in the IMPS were measured. A typical charge/discharge curve for the Glenn polymer battery is shown in Figure 7. The coin cell has the largest capacity but is also the largest and most massive. Conversely the Litestar has less capacity but is substantially lighter and thinner. It has the highest specific energy at 200 Wh/kg.

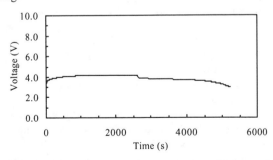

Figure 7. Charge/discharge behavior of 4.2 V lithium polymer battery with a 100 μA discharge rate.

The prototype IMPS were cycled under a simulated AM0 illumination scheme designed to mimic the on-orbit conditions. The IMPS will experience a cyclic charging due to the fixed 5 ° per second rotation which will result in 24 s of charge followed by 48 s of discharge for each 72 s rotation. In addition, this cycling will occur for 56 minutes and then be followed by 36 minutes of eclipse or continuous discharging due to the 92 minute LEO.[6] The IPS was tested under a illumination scheme based on the above times using a simulated AM0 source (see Figure 8). Examining a single LEO shows the charge discharge behavior due to the rotation of the satellite (see Figure 9).

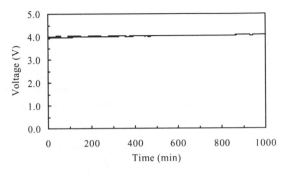

Figure 8. IMPS (with Litestar battery) voltage under simulated illumination consistent with Starshine 3 orbit (i.e., 5° /sec rotation and 92 min LEO).

In addition to the variation in illumination the IMPS will also see a thermal variation associated with the LEO orbit. It is anticipated that the IMPS will experience temperatures from 44.0 °C, while in the sun, to − 1.0 °C, when in eclipse. Figures 10, 11, and 12 show the voltage on the various batteries during a simulated LEO with an even more rigorous thermal cycle than anticipated (e.g., plus-or-minus 40 °C). Figure 13 shows a single cycle of the IPS with the panasonic battery initited at a partially discharged state. The overshoot in the voltage upon charging is due to the thermal behavior of the circuit components.

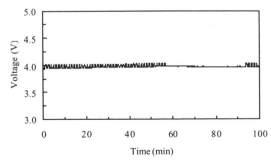

Figure 9. IMPS (with Litestar battery) voltage under simulated illumination consistent with a single Starshine 3 orbit (i.e., 5°/sec rotation and 92 min LEO orbit).

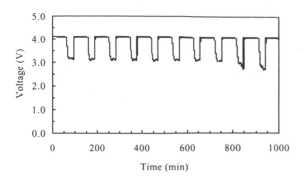

Figure 10. IMPS (with Litestar battery) voltage under simulated illumination consistent with a Starshine orbit and thermal fluctuation (i.e., $5°$ /sec rotation, 92 min LEO, and −40 to 40 ° C temperature variations).

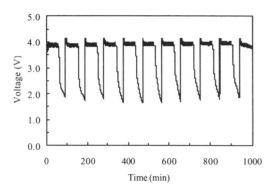

Figure 11. IMPS (with GRC polymer battery) voltage under simulated illumination consistent with a Starshine orbit and thermal fluctuation (i.e., $5°$ /sec rotation, 92 min LEO, and −40 to 40 ° C temperature variations).

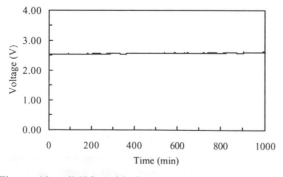

Figure 12. IMPS (with Panasonic battery) voltage under simulated illumination consistent with a Starshine orbit and thermal fluctuation (i.e., $5°$ /sec rotation, 92 min LEO, and −10 to 40 ° C temperature variations).

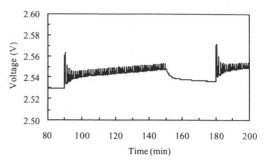

Figure 13. IMPS (with Panasonic battery) voltage under simulated illumination showing one Starshine orbit with thermal fluctuation (i.e., $5°$ /sec rotation, 92 min LEO, and −10 to 40 ° C temperature variations).

SUMMARY/CONCLUSION

We have developed an integrated microelectronic power supply which combines monolithically interconnected small-area GaAs photovoltaic modules with thin film lithium ion rechargeable batteries. This supply is capable of providing continuous power output even under varying illumination and temperature consistent with a spinning satellite in low earth orbit. These devices should be an ideal power supply for future nano and pico-satellite microelectronic applications.

ACKNOWLEDGMENTS

The authors would like to thank Kurt Kelty of Panasonic for providing the ML2020 batteries. The author would also like to acknowledge the support of the Glenn Research Center Directors Discretionary Fund, and NASA Cooperative Agreements NCC3-710 and NCC3-563 with Rochester Institute of Technology.

REFERENCES

1. http://www.azinet.com/starshine/
2. R.P. Raffaelle, J.D. Harris, D. Hehmann, D. Scheiman, G. Rybicki, A.F. Hepp, "Integrated Thin-Film Solar Power System,"IECEC paper #AIAA-200-2808.
3. R. Koksbang, J. Barker, H. Shi, and M.Y. Saidi, *Solid State Ionics*, **84** (1996)1.
4. Lithium Handbook, Panasonic Batteries, Matsushita Electric Corp. (1998).
5. C. Clark, Summers, B. Armstrong, "Innovative Flexible Lightweight Thin-Film Power Generation and Storage for Space Applications", IECEC paper #AIAA-2000-2922.
6. R.P. Raffaelle, J. Underwood, D. Scheiman, J. Cowen, P. Jenkins, A.F. Hepp, J.D. Harris, D.M. Wilt, "Integrated Solar Power Systems," 28th *Photovoltaics Specialists Conference*, Anchorage, AK (2000).

IECEC2001-AT-54

MULTI-MEGAWATT SPACE POWER TECHNOLOGY COMPARISON

Douglas M. Allen
Schafer Corporation
4027 Colonel Glenn Highway, Suite 409
Dayton, Ohio 45431
Phone: (937) 427-4279; Fax: (937) 427-1242
E-mail: dallen@schafercorp.com

ABSTRACT

The technology options and the feasibility of power systems for future electric powered reusable weapons in space that require multi-megawatts (MMW) of power in a short time frame are assessed. The study considered the entire power system including power generation, power conditioning, the recharging system, spacecraft baseload power, and special hardware requirements. Multiple technologies were considered in each of these categories to select the optimum approach in each. The best technology options were used to optimize the best approach to a rechargeable MMW power generation system for further development. The selection was accomplished in a system-engineering environment where the thermal management system and weapon system interactions were included to truly select the best overall approaches. A system using a regenerative solid oxide fuel cell with hydrogen and oxygen reactants was selected as the best approach.

INTRODUCTION

Electrically powered directed energy weapons are potentially an attractive option for space based systems because they can be recharged following their use and then used again. Schafer recently completed a study for the Air Force Research Laboratory (AFRL) that looked at the feasibility of such a system, including the power and thermal management technologies that would need to be developed to make the system viable. We defined feasibility as a system that could meet the general mission requirements and be able to be launched on a single launch vehicle. The study included assessments of power and thermal management technologies, along with the payload and the other subsystems required on the spacecraft. This paper summarizes the results of the power part of the study.

We assessed the spacecraft subsystems by developing a spreadsheet model of the entire spacecraft, including multiple options for the power subsystem. The options were compared at the spacecraft level on the basis of mass and volume to identify the best technologies. A 20% contingency was included in all the mass calculations due to the conceptual nature of the design points used in the study.

POWER REQUIREMENTS

The power requirements, simply stated, are to generate the power needed by the electric weapon system, then to recharge the power source for reuse of the weapon. For the class of system that can be placed into low earth orbit, this represents >10 MW of electric power.

Other significant requirements for the power subsystem include minimizing the mass and volume of the system. The system must be designed to deliver 270V DC to the payload. The payload drives the DC power requirement, and the voltage was selected to minimize the cabling mass while keeping the voltage low enough to avoid significant arcing problems in a space vacuum environment.

Other items of importance include lifetime/reliability, minimum impact on the spacecraft (vibration, torque, momentum, etc.), ability to operate at reduced power levels, quick start-up times, ability to tolerate long standby times (months to years) and then operate reliably, and minimum development and acquisition costs.

POWER SUBSYSTEM OPTIONS

Most studies of rechargeable space based electric weapons have either used batteries for the power source or considered both batteries and fuel cells. Our study started with those two options and added two more. In addition to batteries and flywheels, we included regenerative fuel cells and heat engines as power subsystem options.

Batteries are the simplest option to integrate into the spacecraft, as they have no moving parts. However, they do require power conditioning to provide a constant voltage into the payload. Our study included assessments of lithium ion batteries and lithium polymer batteries, plus very high specific power batteries [1] that use the double layer effect. The double layer batteries were quickly dropped because the run time capability is typically on the order of microseconds to milliseconds; while our requirement was for run times of hundreds of seconds. The required switching network to use a large number of batteries in a time sequence would have been prohibitively heavy. We assessed both high specific power and high specific energy versions of the lithium ion system, looking at specific power of up to 20 kW/kg and specific energy of up to 100 W-hr/kg. The lithium polymer battery was configured for a high specific energy of 300 W-hr/kg. The battery was recharged directly by the spacecraft solar array following weapon operation.

For the flywheel option, we used projected data from the Compact High Power System (CHPS) program that SAIC is performing under sponsorship from DARPA for a terrestrial electric weapon application [2]. Performance estimates were for a 25 W/kg specific energy and 10 kW/kg specific power. This system also requires power conditioning to convert the power output from the AC produced by the generator (which converts the flywheel shaft power to electrical power) to DC. The flywheel was recharged directly by the spacecraft solar array following weapon operation.

Regenerative fuel cell options considered included alkaline, polymer exchange membrane (PEM), and solid oxide fuel cells (SOFC). All the options were based on high purity hydrogen and oxygen reactants. The alkaline fuel cells were dropped early due to a lack of active development of the system and also the need for asbestos for the cell separator, which is no longer available. PEM fuel cells were optimized for high specific power, with a projected stack specific power of 5,500 W/kg. A separate PEM stack, operating at 1,000 W/kg, was used to electrolyze the exhaust water back into hydrogen and oxygen during recharge. The SOFC option, with an operating temperature of 1,000°C, was also optimized for high specific power at 5,000 W/kg. In this case, the same stack was also used for water electrolysis during recharge. Previous papers and studies have projected stack specific powers of up to 20,000 W/kg [1], but based on our discussions with developers, we believe the numbers we used represent the best performance that is realistically achievable for each respective technology. A fuel cell was the only one of the power system options that did not require power conditioning, as they are capable of delivering the required 270V DC continuously during operation. A block diagram of the SOFC power subsystem option is shown in Figure 1.

Turbine engines are usually the power system of choice for high power applications due to their compact size, but only when an open cycle system is an option. For rechargeable or regenerative needs however, they usually are not considered. Because of their potential for very high specific power, we decided to include heat engines in this study. We used a power

system design that included a hydrogen-oxygen combustor with the engine configurations, resulting in steam exhaust. We then used a PEM fuel cell electrolyzer to convert the steam back into hydrogen and oxygen during recharge. We included both Brayton cycle turbine engines and Stirling cycle engines in our technology assessments. The Brayton engine provides very high specific power at 10,000 W/kg [3]. We further pushed the turbine by assuming operation at 118% of design performance, consistent with our specified operating times based on over-design operations for turbine engines scaled from data published by GE for one of their engines [4]. While the engine itself is much heavier, Stirling engines offer the advantage of higher efficiency, at 41% (using a combustor temperature of 2,000K), resulting in reduced requirements for both reactants and cooling fluid. Like the flywheel option, heat engines require power conditioning to convert the power output from the AC produced by the generator (which converts the engine shaft power to electrical power) to DC. A block diagram of the heat engine power subsystem option is shown in Figure 2.

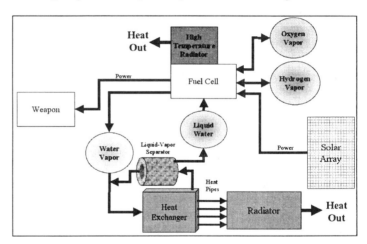

Figure 1: SOFC Power Subsystem Block Diagram

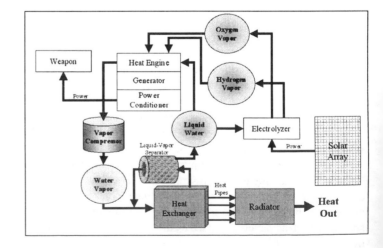

Figure2: Heat Engine Power Subsystem Block Diagram

For the options that required power conditioning, we used American Superconductor's cooled converter technology [5]. This was for either DC to DC converters or AC to DC converters depending on the power source. This technology was selected because of its very high conversion efficiency (typically 99%) coupled with very high specific power. We used 5,500 W/kg for the specific power of the AC to DC converters and 1,100 W/kg for the DC to DC converters. This compares to 650 W/kg projected for future improvements to traditional spacecraft DC to DC converters, which also typically have lower efficiency. The ASC cooled converters require operating temperatures of –30 to –40°C. Since we already were planning to use an ammonia cooling loop for the weapon subsystem operating at –33°C, this was easily accommodated in the system designs.

The cooling for the battery and flywheel options (plus the power conditioner for the heat engine option) was accomplished using the ammonia cooling loop already on the spacecraft for the weapon subsystem. The heat for the battery and flywheel was transferred through a cold plate incorporating enough thermal resistance for the power source to operate at its optimum temperature (usually 10 to 20°C). The fuel cell and heat engine options first used their reactants/steam exhaust products for heat removal and storage. Extra cooling that was required beyond the heat capacity of the exhaust was accomplished by using cooling water, with the resulting steam added to the exhaust steam storage. The SOFC was an exception, as the high operating temperature made it better to use direct radiation during weapon operation to remove excess heat, instead of using cooling water. The spacecraft thermal management portion of this study is described in another paper at this conference [6].

The solar array power required for baseload power and power source recharging was a variable in our study, changing with recharge time. Power requirements varied from less than 10 kW to more than 100 kW. The designs used an advanced solar array with a specific power of 150 W/kg and cell efficiency of 27%. Others have projected higher specific power and efficiency, but we believe that these are realistically achievable goals for power levels between 10 and 100 kW. A solar thermionic power system is also a realistic option in this range, based on a new program being sponsored by the Air Force Research Laboratory, with similar performance possible. The spacecraft used a 100 W-hr/kg lithium ion battery at 60% depth of discharge, to meet baseload power requirements estimated at 5 kW.

RESULTS

The mass of the four options discussed above is shown in Figure 3. For all variations of the spacecraft that we analyzed, the mass of the SOFC power subsystem was the lowest. The biggest reason for this is that the fuel cell did not require power conditioning to deliver the required DC power to the payload. The SOFC option was a little lighter than the PEM fuel cell. This was driven largely by the mass required for extra cooling

fluid and the associated storage tank for the PEM fuel cell, which was significantly higher than the mass of the high temperature radiator used with the SOFC.

Figure 4 shows how the spacecraft mass scales with the operating time for the weapon payload. As mentioned previously, the SOFC is the lowest mass option for the full range considered. The figure only shows the best mass for each power subsystem option. The fuel cell curve represents a SOFC. If the PEM fuel cell had been included, its mass would fall between the SOFC curve and the heat engine curve.

The flywheel's high specific power makes it good for very short run times, but it's low specific energy results in a heavy power subsystem for the realistic range of operating parameters. The Brayton cycle heat engine option is the second choice to fuel cells over most of the operating time range that we studied, but the mass differential with the fuel cell grows with operating time due to its lower efficiency.

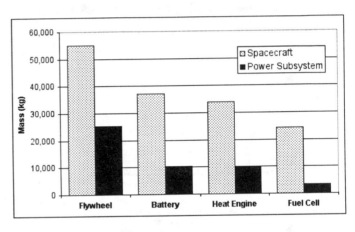

Figure 3: Mass of Power Subsystem Options

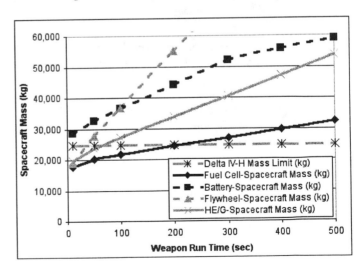

Figure 4: Spacecraft Mass as a Function of Weapon Run Time for Each of the Power Subsystem Options

We recommended a one day recharge time for the system based on an analysis of the total spacecraft mass variation with recharge time. The spacecraft mass associated with each of the power subsystem options is plotted vs. recharge time in Figure 5. This shows that the knee in the curve is below a one-day recharge time, which is the basis for our recommendation. This allows a reasonable availability of the payload without excess mass penalty.

However, even at one day, the size of both the solar array and the radiators are very large and longer recharge times may be desired to minimize the complexity of deploying these very large structures in orbit. Figure 6 shows how solar array power and area vary with recharge time.

For a one-day recharge, the solar array area is 213 m^2 for a 70 kW array. Extending the recharge to 5 days instead of 1 day reduces this total solar array area by almost than a factor of 4, to a much more manageable 55 m^2 (18kW). However, this limits the availability of the payload to be used in condensed periods of conflict.

believe that decreasing the mass of the power conditioning system beyond our projections is realistic.

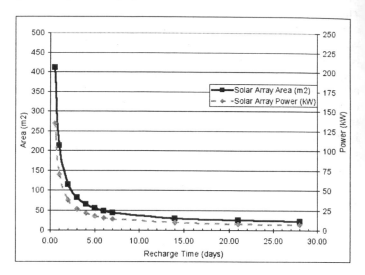

Figure 6: Solar Array Area and Power as a Function of Recharge Time

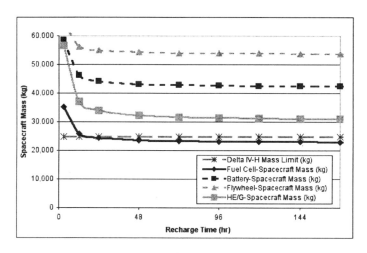

Figure 5: Spacecraft Mass as a Function of Recharge Time for Each of the Power Subsystem Options

POWER SUBSYSTEM OPTION COMPARISONS

Since we were projecting performance for all of the power subsystem options, we felt it was important to assess the sensitivity of our projections to the selection of the best power source. Because the flywheel subsystem was so heavy, we did not do any further analysis with that option.

For the battery, we studied the impact of specific energy on spacecraft mass and compared that to the mass of a spacecraft with the SOFC power subsystem. The results are shown in Figure 7. The graph shows that even as the battery itself nears zero mass (using the Y-axis on the right side of the graph), the spacecraft is still heavier than the one using a fuel cell. This is due to the mass of the power conditioning, and we do not

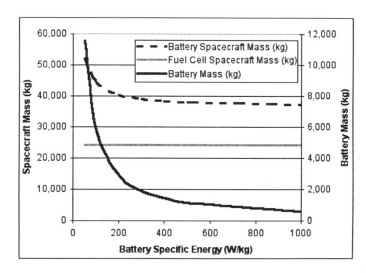

Figure 7: Impact of Battery Specific Energy on the Comparison with a Fuel Cell Power Source

As noted previously, the Brayton cycle system is preferred over Stirling for our MMW power levels due to lower engine mass. Figure 8 illustrates this, showing that even though the balance of plant components are over 1,000 kg lighter for the more efficient Stirling engine, the low engine mass of the Brayton cycle system drives the selection. Weapon run time would have to extend to 20 minutes before the balance of plant mass difference would offset the engine mass difference. Table 1 shows a mass breakdown comparison for the heat engine and fuel cell power systems.

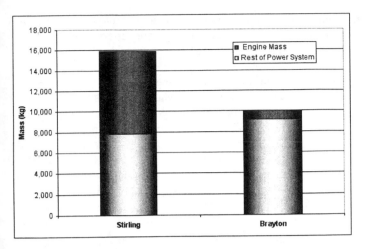

Figure8: Comparison of Stirling and Brayton Power Subsystem Mass

This illustrates that the higher mass of the heat engine power system is due primarily to the mass of the power conditioner plus the difference in the mass of the cooling systems required for each option. These include the water cooling mass plus tank mass for the heat engine and the high temperature radiator mass for the SOFC. Of course, the cooling water and tank mass is a function of engine efficiency. We also looked at the sensitivity of this comparison to efficiency and found that even at 90% turbine efficiency, the spacecraft with the fuel cell is still over 2,000 kg lighter. For the fuel cell power subsystem, we looked at the sensitivity of system mass to both stack specific power and stack efficiency. The results are shown in Figures 9 and 10.

Figure 9 shows how the spacecraft mass varies with stack specific power, with curves for both the SOFC and the PEM fuel cells. The mass of a comparable spacecraft using a Brayton heat engine power system is shown as a horizontal line on the graph for comparison. This shows that the stack specific power must be nearly a factor of five worse than our projections (5,000 W/kg for SOFC and 5,500 W/kg for PEM) before the heat engine becomes a better option. It also shows that if development results in stack specific power below 3,000-4,000 W/kg, there will be a significant system mass penalty that could impact mission feasibility.

Figure 10 shows how the spacecraft mass varies with stack efficiency, again with curves for both the SOFC and the PEM fuel cells. As with Figure 9, the mass of a comparable spacecraft using a Brayton heat engine power system is shown as a horizontal line on the graph for comparison. This shows that the spacecraft with the SOFC power subsystem is not very sensitive to stack efficiency, while spacecraft is impacted more with the PEM fuel cell subsystem. This is driven by the different cooling approaches, with the cooling water mass and pressurized steam tank mass driving the higher sensitivity for the PEM subsystem. The high temperature radiator mass used by the 1000°C SOFC has a lower sensitivity to efficiency.

Another aspect of this plot is that it shows that end result the trade between heat engines and fuel cells is not effected by stack efficiency, within a reasonable range.

Brayton Power System			SOFC Power System	
Turbine Engine Power System Components	Mass (kg)	Mass (kg)	SOFC Power System Components	
Engine	871	2058	Fuel Cell Stack	
Combustor	26			
Starter	309			
Controls and Ancillaries	82	111	Ancillaries and Insulation	
Generator	278			
Generator Cooling System	209			
AC to DC Converter	1868			
Electrolyzer	93			
Electrolyzer Ancillaries	19			
Hydrogen Tank	312	145	Hydrogen Tank	
Oxygen Tank	157	73	Oxygen Tank	
Steam Tank	1425	144	Steam Tank	
Steam (Reactants)	502	233	Steam (Reactants)	
Extra Water for Cooling	1801	121	Radiator	
Residual Hydrogen	14	6	Residual Hydrogen	
Residual Oxygen	111	52	Residual Oxygen	
Water/Steam Separator	92	9	Water/Steam Separator	
Plumbing (pumps, valves, pipes)	691	70	Plumbing (pumps, valves, pipes)	
Water Vapor Compressor	92	9	Water Vapor Compressor	
Liquid Water Storage Tank	46	5	Liquid Water Storage Tank	
Solar Array	962	465	Solar Array	
Spacecraft Battery	42	42	Spacecraft Battery	
Total Power System	10,001	3,542	Total Power System	

Table 1: Heat Engine – Fuel Cell Component Mass Comparison

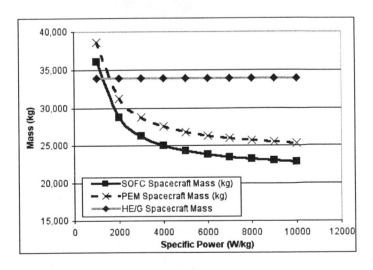

Figure 9: Spacecraft Mass as a Function of Fuel Cell Stack Specific Power

247

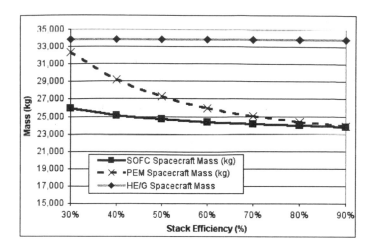

*Figure 10: Spacecraft Mass as a
Function of Fuel Cell Stack Efficiency*

Figure 11 illustrates the importance of stack specific power to the SOFC power subsystem, as the stack accounts for well over half of the total subsystem mass.

We also considered the volume of each of the power subsystem options. The heat engine based spacecraft was nearly twice as big as the other options, but the spacecraft was still driven by mass, when considering launch vehicle constraints.

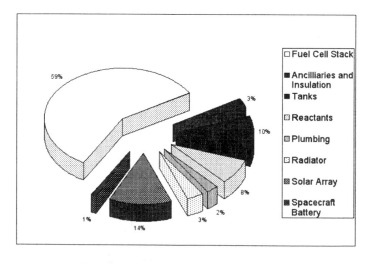

Figure 11: Mass Breakout for SOFC Power Subsystem

QUALITATIVE DISCUSSION OF POWER OPTIONS

Of course, such a system can not be designed on the basis of only mass and volume considerations. Many other aspects must be considered. This section addresses potential issues with each of the power subsystem options.

As mentioned previously, batteries are the easiest of the options to integrate into the spacecraft with no moving parts and

no fluid flows. The only issues here are the mass, including the requirement for DC to DC power converters, and possibly difficulties in scaling up to MMW power levels.

Flywheels are not so simple to integrate with the spacecraft because of the rotational energy in the wheel. First of all, at one or more pairs of counter-rotating wheels must be used to avoid spinning the spacecraft. The biggest issue here, other than mass, is that the rotational momentum of each wheel in the pair must be exactly matched over all operating conditions and depths of discharge. The energy in the wheels for this application is so high that a mismatch on the order of 0.001 to 0.01% might be enough to spin the spacecraft out of control.

While the Brayton cycle engines are well suited to MMW power levels, momentum and vibration concerns must be considered. Like the flywheels, counter-rotating turbines must be used to eliminate spacecraft attitude control impacts. The concern is much lower here than with flywheels, however, since the rotational energy is several orders of magnitude lower. Fluid loops for the reactants must also be designed in reverse directions due to the fluid flow momentum, due to total flow rates on the order of 11-12 kg/sec during payload operation. In addition to the turbine, pumps and compressors are required that could be significant sources of mechanical noise that could effect payload operation. System start-up is another possible area of concern that would have to be addressed.

While the fuel cell stacks don't have moving parts, the system still requires fluid flow, so some of the same concerns listed above for the heat engine power subsystem are valid here too. This includes pump and compressor vibrations and fluid momentum, although flow rates are lower (about 0.1 kg/sec) due to eliminating the cooling water with the SOFC. The SOFC also results in concerns related to temperature. While the 1000°C operating temperature helps significantly in the mass required to cool the stack, it also requires thermal isolation from the rest of the spacecraft. Additionally, heaters will be required to keep the stack at temperature during idle periods on orbit once the system is fully charged and waiting to be used. Probably the biggest issue is scaling up the fuel cell stacks to the MMW power levels, as this will require large stack footprints (area), with the requirement for fairly constant current density operation across the entire cell area.

For both the heat engine and the fuel cell power subsystem options, we compared the mass of storing the hydrogen and oxygen reactants in high-pressure (3500 psi) tanks or cooling and storing them at liquid temperatures. Pressurized storage was the lowest mass of these options. The mass of the cryo-cooler more than offset the extra mass of the pressurized storage tanks until recharge times reached about 40 days. This was primarily due to the hydrogen cryo-cooler, which was sized at 70 kW for a one day recharge.

TECHNOLOGY INVESTMENT RECOMMENDATIONS

We recommend development of the fuel cell power system. Given the importance of stack specific power, we suggest an initial parallel development of both SOFC and PEM stacks, with a later down select. The stack programs should focus on performance at high current densities for single cells operating with hydrogen and oxygen reactants. Operation with pure reactants allows increased current density and better specific power. All ongoing programs in fuel cell technology use air instead of oxygen and most use a hydrocarbon fuel instead of hydrogen. The cell work should include an investigation of alternate or modified composition cell materials to emphasize specific power of the stack instead of optimizing on low cost.

Once progress has been demonstrated with single cells, multi-cell stacks of 3-5 cells should be developed. At this point, a down selection can be accomplished between the SOFC and PEM fuel cell technologies, followed by a technology program to scale-up cell area while maintaining high current density.

While the cell area is being scaled up, the other components of the fuel cell power system should be developed, including the compressor, pumps, liquid-vapor separator, insulation, high-pressure tanks, high temperature radiator, and required ancillary components. Then an integrated fuel cell system can be developed and demonstrated with the payload.

CONCLUSIONS

Generating MMW for short periods of time in space is an achievable goal. And it can be accomplished with a regenerative system that allows the payload to be reused many times during the life of the spacecraft.

The key approach to minimize the mass of the power subsystem on the spacecraft is to use regenerative fuel cells. Solid oxide fuel cells appear to be the best approach, but PEM fuel cells are also an option.

ACKNOWLEDGMENTS

The Aerospace Power Division, Propulsion Directorate, Air Force Research Laboratory (AFRL/PRP) sponsored this work under a subcontract to Schafer Corp. from UES, Inc., as a part of Air Force contract F33615-99-C-2921.

REFERENCES

1. Fellner, Joseph P., "Electrochemical Power/Energy Storage for Diode Lasers," Proceedings of the Diode Pumped Solid State Laser Technology Review, Air Force Research Laboratory, Albuquerque, NM, 2000

2. McNab, Ian R., and Frazier, George, "Recent Developments in High Power from Pulsed Alternators, Flywheels, and Batteries," Proceedings of the Diode Pumped Solid State Laser Technology Review, Air Force Research Laboratory, Albuquerque, NM, 2000

3. Avco Lycoming T52/T55 web site, http://www.gas-turbines.com/aircraft/t53.htm, March 2001

4. GE Aircraft Engines web site, http://www.geae.com/military/t700_ge_t6e1.html , March 2001

5. Ramalingam, Mysore L., Donovan, Brian D., and Beam, Jerry E., "Thermal Systems Analysis And Benefits Of A Cooled Radar Power System," Proceedings of the 33rd Intersociety Energy Conversion Engineering Conference, American Nuclear Society, La Grange Park, IL, 1998.

6. Allen, Douglas M. and Bautch, James R., "Space Thermal Management Technology Comparison for a Gigajoule Class Mission," Proceedings of the 36th Intersociety Energy Conversion Engineering Conference, American Society of Mechanical Engineers, New York, New York 2001.

Proceedings of IECEC'01
36th Intersociety Energy Conversion Engineering Conference
July 29–August 2, 2001, Savannah, Georgia

IECEC2001-AP-15

Near Earth Asteroid Rendezvous – Shoemaker Spacecraft Power System Flight Performance

Jason E. Jenkins and George Dakermanji
The Johns Hopkins University Applied Physics Laboratory

ABSTRACT

The Near Earth Asteroid Rendezvous (NEAR) – Shoemaker spacecraft launched in February 1996 and successfully completed its mission in February of 2001, having spent one year in orbit about the asteroid 433-Eros. The NEAR-Shoemaker spacecraft is the first to launch of the NASA Discovery programs. Its power subsystem is a direct energy transfer architecture with a Super™ nickel cadmium battery and four fixed solar panels populated with single junction gallium arsenide on germanium solar cells. The distance from the Sun varied significantly during the mission including a record-setting 2.2-AU aphelion distance resulting in large variations in solar array I-V characteristics. The power subsystem has performed as designed and has demonstrated robust performance in-flight though launch, mission aphelion, boost charging, an attitude anomaly, and landing on the asteroid surface.

INTRODUCTION

The Near Earth Asteroid Rendezvous (NEAR) – Shoemaker spacecraft was developed by The Johns Hopkins University Applied Physics Laboratory. It was launched in February 1996; the first launch of the NASA Discovery missions. Its primary mission was to achieve orbit about the asteroid 433-Eros and perform the most comprehensive study of asteroid composition and morphology to date, with an instrument suite consisting of a multispectral imager, a x/γ-ray spectrometer, a near-IR spectrograph, a magnetometer, and a laser altimeter[1]. This mission is the first to orbit an asteroid and though not originally designed to do so, is the first to successfully land on an asteroid. In an end-of-the-mission attempt to capture high resolution images of the asteroid surface, the NEAR-Shoemaker team commanded a controlled descent to a successful landing.

En route to Eros, the spacecraft illumination varies from 1.0 Sun to a low of 0.207 Sun at a distance of 2.2 astronomical units (AU). With the exception of an eclipse in early operations and one attitude perturbation, the panels were in continuous sun throughout the mission.

The power subsystem for this spacecraft was designed using low-risk, mass-effective technology to operate in the highly variable interplanetary environment. This article briefly

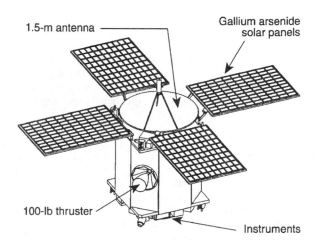

Figure 1 The NEAR spacecraft configuration employs fixed solar arrays for simplicity and reliability.

describes the power subsystem and reports flight performance from launch to landing as it compares to predicted performance.

POWER SYSTEM DESIGN

TOPOLOGY

The power system[2] is a Direct Energy Transfer (DET) system illustrated by Figure 2. The solar array is divided into 20 segments each with a digital shunt transistor. The "coarse" solar array power control and bus voltage regulation are performed by sequentially turning ON/OFF the digital shunts. "Fine" control is maintained by a six-stage linear sequential full dissipative shunt. The bus is regulated at 33.5V ± 1% when the solar array power is adequate to supply the load and battery charge power. The bus follows the battery voltage during the launch phase and whenever the solar array power capabilities are exceeded. Software algorithms perform heater load management to avoid solar array lockup conditions following peak load activity. The battery is isolated from the main power bus by redundant battery chargers and discharge diodes.

The selection of the power subsystem topology for NEAR was strongly influenced by the need to use heritage circuits to minimize power system electronics hardware development time and risk while optimizing the utilization of power from the array to minimize the size and weight of the power subsystem. The solar array size for NEAR was minimized by selecting the number of cells in the array string such that the operation can be on either side of the I-V curve. This optimization was tailored to variation of solar illumination due to the mission trajectory. The maximum array temperature, and minimum voltage occurred at 0.985 AU during an Earth fly-by, determining the required number of cells in series. The 2.2 AU aphelion determined the paralleled cell area. The operation on the Isc side during this minimum power and coldest temperature case was about 8% of the maximum power point. This compares favorably with power utilization of a maximum power tracking power system topology after taking into account the inefficiency of the maximum power point tracker electronics and the inaccuracy of the tracking operation.

SOLAR ARRAY

The solar array is the primary power source for the NEAR mission, and was sized to provide power throughout the mission without reliance upon the battery, except during launch and emergency conditions. Spectrolab, Inc. performed fabrication and laydown of the cells.

The four coplanar fixed solar panels total 8.919 m^2 in area and are populated with 960 7.48 × 2.89 cm solar cells. The cells are 190-micron single-junction gallium arsenide on germanium. The cell has an antireflective coating and is protected by a 152-micron ceria-doped microsheet coverglass with an ultraviolet reflective coating. The minimum average efficiency of the CIC flight lot is greater than 19.1%. All cells were screened to a 1.1-Isc reverse bias current to protect the array from shadowing during a tumble.

Each panel is populated with 24 40-cell circuits that are electrically and physically grouped in parallel into five digital

shunt groups of four or five circuits each. The isolation diodes for each string to the main power buses and digital shunt group busses are located on the backside of each panel. Each panel also has three analog shunt elements mounted to the backside.

BATTERY

The battery provided power in the launch phase of the mission and was to be a backup source of energy in the event of an excessive load excursion or temporary decrease in solar array power. The 9-Ah Super™ nickel cadmium (S-NiCd) battery cells were fabricated by Eagle Picher Industries. Hughes Aircraft Company performed the 22-cell battery assembly and testing. The S-NiCd design is described by Pickett, et al.[3] The manufacturer recommended a storage condition of 100% state-of-charge sustained by a trickle charge, which made the S-NiCd well suited to the NEAR-Shoemaker application. During the mission the battery was maintained in its optimum storage regime, with no need for reconditioning during the mission. The flight spare battery was maintained in a flight-like, refrigerated, VT-limited trickle storage regime

The battery operating temperature was maintained between -2°C and 5°C during flight. It is thermally isolated from the spacecraft deck using fiberglass standoffs. The battery is radiatively decoupled from the spacecraft by a thermal blanket that completely surrounds the assembly with the exception of the radiator surface. The thermal fins between the cells conduct heat toward the radiative baseplate located opposite the mounting surface. Redundant heater elements are located on this surface with thermostats located on an edge member.

POWER SYSTEM ELECTRONICS

The PSE houses two Bus Voltage Regulators (BVRs). The primary BVR is set to regulate the bus at 33.33 V and the redundant on-line unit regulates to 33.67 V, providing on-line protection from an over-voltage failure of the primary unit. A hard 40-V voltage clamp is also implemented to protect

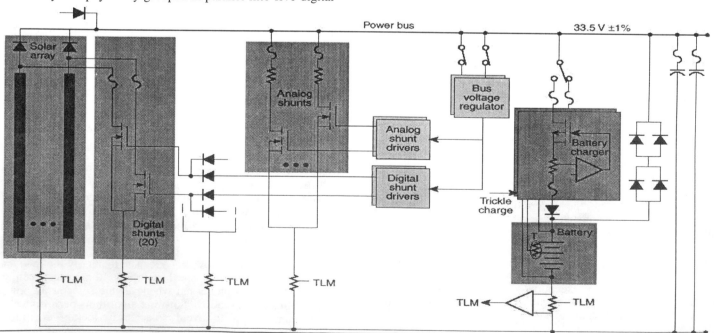

Figure2 The NEAR-Shoemaker power system achieves robustness with redundancy, fault tolerance, and graceful degradation.

against voltage transients that are faster than the BVR control loop. By spreading the power removal capability of the analog shunts over six elements, the total dissipation in the analog shunt FETs is reduced to facilitate thermal management of the PSE box.

The PSE also houses two identical battery linear chargers providing off-line redundancy. The primary charge control regulates the battery voltage to a temperature compensated voltage (VT) limit. A current limited control loop is run concurrently with the VT controller to limit the maximum charge rate to C/20 (0.45-A) to recharge the battery after discharge or to top-off the battery in preparation for possible discharge. A trickle charge current limit is selectable by relay command, limiting the current to C/75 (0.12-A).

MISSION PERFORMANCE

LAUNCH

The battery was sized to a 40% depth of discharge from launch to first contact plus ten minutes without reliance upon solar array power. However, power input from the stowed solar array was apparent following fairing release five minutes after launch. Sufficient power was generated to stop battery discharge and bring the bus to regulation for 11.5 minutes prior to entering a 30-minute eclipse. Integration of the discharge current yields an approximate discharge depth of 15%.

Upon eclipse exit the battery was recharged at a C/20 rate, followed by C/75 trickle charge with a battery voltage below the VT-5 limit.

Upon exit from the launch eclipse, the telemetered solar array power was 2 kW at 53°C. The telemetered BOL performance translates into an array level specific energy of 36.8 W/kg including panels and array hinges The power system level specific energy was 27.6 W/kg including the solar panels, hinges, PSE, and battery, but excluding spacecraft harness.

RECORD-SETTING APHELION IN CRUISE

On July 22, 1996 the NEAR-Shoemaker solar array became the first photovoltaic array to be used as primary power source beyond the orbit of Mars. In February 1997, the new photovoltaic solar distance record was set as NEAR-Shoemaker flew through mission aphelion at 2.2 AU. Figure 3 shows a plot of the modeled and telemetered total solar array current throughout the mission. The average modeling error in the first three-year pre-solar maximum period was $0\pm1\%$ (1-σ). Array model comparative results are obtained using actual flight solar distances, sun angles, temperatures, and digital shunt configurations. These parameters are input into the original array analysis software, which models the I-V response to radiation, temperature, illumination (including LILT), and voltage of the array circuits.

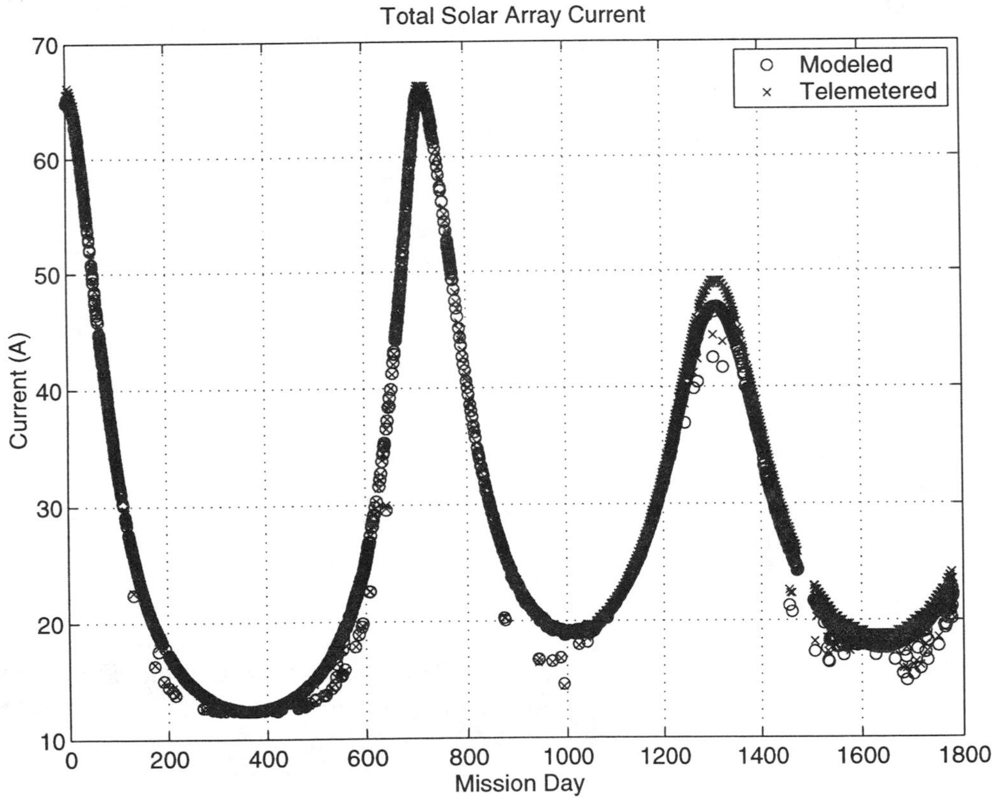

A unique problem of interplanetary missions away from the sun is the effect of low intensity low temperature (LILT) on array operation. A NEAR-Shoemaker program study included a review of industry effort to quantify the phenomena as well as LILT testing of single junction GaAs cell assemblies. The results are conservatively consistent with those of industry, revealing a correlation between the beginning-of-life cell electrical performance and the degree of LILT voltage reduction. Evaluating only single junction GaAs cells with efficiencies greater than 18% under mission aphelion conditions yields 3.5% degradation in open-circuit voltage (Voc) and 4.8% degradation in the voltage at maximum power (Vmp). However, considering the inclusion of lower performance GaAs cells

Total Solar Array Current

Figure 3 *The solar array model shows a good correlation to telemetered current for the first 3 years of the mission, but becomes increasingly conservative with application of the solar maximum radiation model in the last two years.*

in the flight distribution, 3.9% Voc degradation and 7.6% Vmp degradation are evident.

MATHILDE FLYBY

En route to Eros, the spacecraft trajectory came within 750 km of the asteroid Mathilde, offering a rare opportunity for additional fly-by science of a Class-C asteroid. To attain images of Mathilde, the spacecraft was required to direct its non-articulated solar arrays 52° away from sun-pointing while at a solar distance of 2.0 AU. To perform the Mathilde observation with this diminished solar input, aggressive load management was implemented, and in preparation for contingency operation, the battery was prepared to maximize capacity in the event of discharge during the mode. The battery was, however, not required to discharge during the observation.

After 14 months of in-flight trickle charge storage, a C/20-VT7 boost charge (the highest available current and voltage limits) was performed for 4.3 hours in preparation for the Mathilde observation to maximize the available capacity in the unlikely event of discharge during the solar off-pointing. This boost charge event was performed a total of four times within a month of the Mathilde encounter, three times during dress rehearsals of the Mathilde event and once in preparation for the observation itself.

Upon application of the high-rate charge, the battery voltage immediately reached the VT-7 limit. As shown by Figure 4, the current spiked and quickly decreased to about 0.15A. With the voltage held at VT-7, the charge current gradually rose over the 4.3-h duration of the boost charge event. Concurrent review of the battery temperature shows that the current increase doesn't appear to be a simple response to battery temperature as is usually seen in VT limited charging, but is more indicative of a gradual decrease in effective battery impedance.

Following the boost charge event, the battery charge rate and voltage limits were commanded back to the C/75 trickle charge with a VT-6 voltage limit (formerly VT-5).

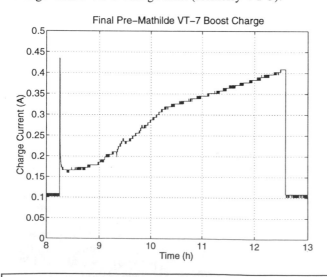

Figure 4 The fourth in-flight VT-clamped boost charge shows a capacitive response and active material conversion.

The observed current-voltage behavior was not consistent with initially expected behavior, which would predict a gradual voltage rise with a current taper if the VT limit were reached. The boost charge events all resulted in similar behavior. After review of the flight data of the first Mathilde dress rehearsal, a test was performed on 8 cells that had been in lifetest at Eagle-Picher Industries, Inc. in Colorado Springs. The 8 cells were tested in two 4-cell packs to simulate the boost charge experienced in flight. Similar voltage and current characteristics were observed.

The immediate current spike is attributed to capacitive charging which soon dissipates due to electrolytic conduction reducing the current draw. The subsequent gradual increase of charge results from oxygen gassing and active material conversion. Upon return to trickle charge, the battery trickle charge voltage had reduced due to removal of polarization during the boost charge.[4]

ASTEROID RENDEZVOUS BURN

On December 20, 1998, the NEAR-Shoemaker spacecraft attempted to perform a burn to slow the spacecraft for its scheduled rendezvous with Eros in February of 1999. Shortly after the burn sequence began, it was aborted and the spacecraft lost attitude control. During this period, the battery sustained spacecraft health while the guidance and control subsystem attempted to regain control. Limited safing telemetry indicates that the battery reached a minimum voltage of 25.4V sustaining a load of approximately 200 to 250W with varied solar array input over 6 hours. Maximum discharge current was as high as 1.06C. Following the discharge, the battery was recharged at a C/20, VT-6 limited rate finishing in C/75 trickle charge after 48 hours.

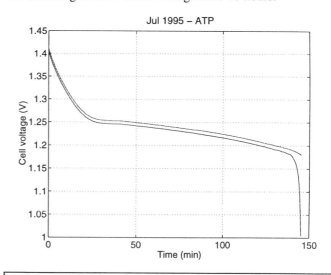

Figure 5 Baseline discharge of the spare battery in incoming acceptance test (Highest & lowest cells shown)

The spare battery had been in flight-like ground storage since launch. Figure 5 illustrates a 10°C capacity discharge performed in incoming acceptance test, 7 months before storage began. The ampere-hour capacity in this test was 10.94 Ah (299 Wh). Voltage performance was good with a mid-discharge cell voltage spread of 1.5 (1-σ). The end of discharge spread was 40.8 mV (1-σ) with a battery voltage of 25.3V.

Following the rendezvous burn anomaly, the spare was discharged at a C/2 rate at approximately 10°C until the first cell reached 1 V (Figure 6). Cell voltages were a little depressed but fairly well matched at the mid-point of the discharge with a 1-σ spread of ±2.8 mV. The battery delivered 11.29 Ah (298 Wh). During the last 25 minutes of discharge a secondary plateau of approximately 1.05 to 1.1 V/cell was apparent. End of discharge voltage was 22.8V with a 1-σ spread between cells of ±15mV.

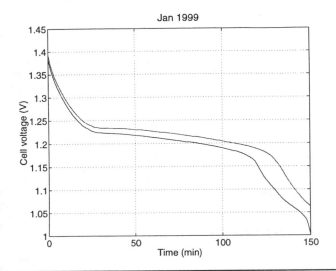

Figure 6 Discharge of the spare battery after 3 years of trickle charge shows the occurrence of a secondary plateau. (Highest & lowest cells shown)

BATTERY AGING

Following launch recharge, the battery was maintained below a VT-5 voltage limit with a current-limited C/75 trickle charge. At 3 months from launch, the battery charge started a slight VT-5 taper during the cold portions of the thermostat cycles. After one year, the charge rate was constantly in a VT-5 taper as low as C/100 while cold. Just prior to the first Mathilde top-off the trickle charge was varying from C/80 to C/110. After the first top off, trickle charge current returned to the C/75 level. The VT limit was then increased to VT-6. Charge rate tapering began again 10 months after the final Mathilde top-off. 17 months later, just prior to the aborted burn discharge, the trickle charge had tapered to as low as C/85. By the end of the mission, 14 months later, the charge had tapered to C/90.

The spare battery in trickle charge storage also experienced charge tapering with age. However, the VT limit remained set at VT-5. By the end of the mission, the spare battery trickle charge had tapered as low as C/150.

2 YEARS OF SOLAR MAXIMUM

The original mission was to have experienced one year of solar maximum. However, the aborted burn resulted in an additional year of cruise phase prior to rendezvous, and with it, an additional year of solar maximum. Solar array telemetry did not contain any voltage data. Solar array current however shows no step changes correlate to solar flare activity detected elsewhere on the space craft.

The telemetered and modeled solar array currents illustrated by Figure 2 show the two years of solar maximum at the end of the mission. The larger variations in the latter years are due to spacecraft attitude variations during orbital observations. As the power model is applied to this solar maximum period, the time-average modeling error increased from 0±1% in the year before solar maximum to 3±1% in the first year of solar maximum to 5±1% in the second year. This apparent margin increase is attributed to the modeled radiation environment being greater than the actual fluence. Since the modeled degradation of current was about 5%, it may be inferred that the actual radiation damage was insignificant.

PREPARATION FOR END OF MISSION

As the scheduled end of the mission approached, the NEAR-Shoemaker program elected to attempt a controlled descent of the spacecraft toward the surface of the asteroid to obtain the "bonus" science of high resolution images. In preparation for this activity, the spare battery was again pulled from flight simulated storage to determine its end of mission capacity in the event that the battery would be needed in descent.

The discharge was conducted in the same manner as the previous test after the aborted burn. Cell voltages were again well matched with a mid-discharge 1-σ spread of 2.0 mV. A secondary plateau was again apparent in the last 25 minutes of discharge. The battery delivered 11.17 Ah (296 Wh) with an end-of-discharge voltage of 22.5 V and 1-σ spread between cell voltages of 12.4 mV.

Following the discharge, the battery was recharged at C/10 for 20h at 10°C, and the capacity discharge was repeated. The mid-discharge voltage was slightly elevated with a spread of 1.7 mV (1-σ). The capacity was 10.7 Ah (291 Wh). The end-of-discharge cell voltages had a 37mV spread (1-σ) and a virtual absence of a second plateau. The battery was recharged at C/20. The charge rate was then reduced to a C/75 limited VT5 trickle and the battery returned to storage. Within 8 hours of beginning trickle charge, the VT-5 taper was down to C/150 and down to C/175 a few days later. After 3 additional months the trickle charge appears stable at C/175.

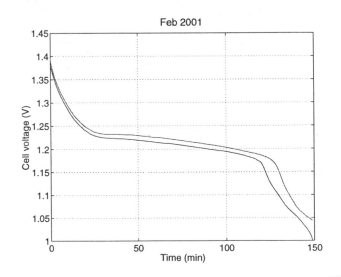

Figure 7 A discharge after 5 years of storage again shows good capacity and the second voltage plateau. (Highest and lowest cells shown).

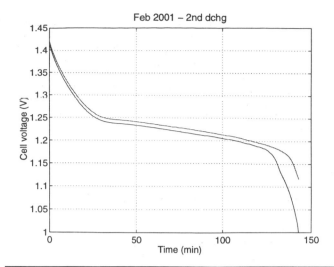

Feb 2001 – 2nd dchg

Figure 8 A second 5-year discharge after a resistive letdown and recharge shows a diminished second plateau. (Highest and lowest cells shown)

LANDING

The controlled descent culminated in a successful landing with the spacecraft operating and in earth contact. The landing occurred near the asteroid's spin axis, which permitted non-eclipsed solar illumination with a variable sun angle in the range of 10 to 27 degrees from array normal. Since the high-gain antenna is not gimbaled, RF contact was via the low-gain antenna at 10 bits per second. In the remaining 16 days of mission funding, a total of 56 housekeeping telemetry frames were received.

The power system electronics appeared to survive the landing remaining on the primary BVR maintaining steady bus voltage regulation at 33.33V. Since the load did not exceed the array output, the battery was not discharged in descent or landing. While on the surface of Eros, the battery temperature rose from 5°C to 25°C. The charge rate was current limited at C/75 below the VT-6 limit.

The solar array continued to produce current at about 97% of the predicted level (corrected to actual pre-landing performance and landing attitude). Digital and analog shunts were shunting 20 to 23 A of excess current. Panel temperatures ranged from 30°C to 55°C.

There are several possible explanations for the 3% array power deficit including:

- panel deflection from the coplanar orientation

- minor panel damage upon impact

- occultation by dust or nearby surface features

An ongoing review of data from the attitude subsystem, temperature telemetry, and magnetometer measurements in concert with the asteroid shape model, spacecraft geometry, and panel structural properties may eventually settle on some likely scenarios.

CONCLUSION

The power subsystem of the NEAR spacecraft is a simple, effective design that provides power in an interplanetary environment. The direct energy transfer topology, GaAs solar array, and S-NiCd battery have been appropriately selected to provide a low-risk system over a highly variable range of operating conditions. Performance during a spacecraft tumble, an additional year of solar maximum, and an impromptu asteroid landing demonstrates a robust power system design. The NEAR power subsystem represents a design that is candidate for use as a baseline in future small-size, high-performance interplanetary missions.

ACKNOWLEDGMENTS

The authors of this article wish to acknowledge NASA for the sponsorship of the NEAR-Shoemaker program and vendors Spectrolab, Eagle-Picher, and Hughes Aircraft Company for their efforts in the development of the solar array, battery cells, and battery assembly.

REFERENCES

1. Santo, A.G, Lee, S.C, and Gold R.E., "NEAR Spacecraft and Instrumentation, "Journal of the Astronautical Sciences, Vol. 43. No. 4, October-December 1995, pp. 373-398

2. J. Jenkins, G. Dakermanji, M. Butler, and U. Carlsson, "The Near Earth Asteroid Rendezvous Spacecraft Power Subsystem", Proc. 4th European Space Power Conference, SP-369, Vol.1, pp.277-282 (1995)

3. David F. Pickett, Jr., Hong S. Lim, Stanley J. Krause and Scott A. Verzwyvelt, *J. Power Sources*, 22(1988), pp. 243-259.

4. Jason E. Jenkins, Jeff W. Hayden and David F. Picket, "Near Earth Asteroid Rendezvous Flight Battery Performance: Boost Charge Response of Super-NiCd Following Extended Trickle Charge Flight Storage", Proc. The Thirteenth Annual Battery Conference on Applications and Advances, IEEE 98TH8299, pp. 259-263 (1998)

Proceedings of IECEC'01
36th Intersociety Energy Conversion Engineering Conference
July 29–August 2, 2001, Savannah, Georgia

IECEC2001-AT-21

MISSION APPLICABILITY AND BENEFITS OF THIN-FILM INTEGRATED POWER GENERATION AND ENERGY STORAGE

David J. Hoffman
NASA Glenn Research Center

Geoffrey A. Landis
NASA Glenn Research Center

Ryne P. Raffaelle
Rochester Institute of Technology

Aloysius F. Hepp
NASA Glenn Research Center

ABSTRACT

This paper discusses the space mission applicability and benefits of a thin-film integrated power generation and energy storage device, i.e. an "Integrated Power Source" or IPS. The characteristics of an IPS that combines thin-film photovoltaic power generation with thin-film energy storage are described. Mission concepts for a thin-film IPS as a spacecraft main electrical power system, as a decentralized or distributed power source and as an un-interruptible power supply are discussed. For two specific missions, preliminary sizing of an IPS as a main power system is performed and benefits are assessed. IPS developmental challenges that need to be overcome in order to realize the benefits of an IPS are examined. Based on this preliminary assessment, it is concluded that the most likely and beneficial application of an IPS will be as the main power system on a very small "nanosatellite", or in specialized applications serving as a decentralized or distributed power source or uninterruptible power supply.

INTRODUCTION

Thin-film photovoltaic (TFPV) power generation has been under development for some time. TFPV sample cells and panels have flown in space, but a full TFPV solar array has not yet been built. The principle benefits of TFPV arrays include very high mass specific power (W/kg), radiation tolerance and good stowability. The mission benefits of TFPV solar arrays have been identified [refs. 1 and 2], and may soon be realized once full scale TFPV arrays are constructed and space qualified.

In comparison to TFPV power generation, thin-film energy storage (TFES) is a relatively recent development. Very small thin-film lithium-ion batteries have been developed and tested in the lab for use in multi-chip modules (MCMs) [ref. 3]. With a

typical operating range between 3.0 V and 4.2 V, the useable capacity of these initial TFES batteries is very small, ranging from 0.2 to 10 mAh/cm^2.

Because of the similarity in the materials and processes that go into TFPV and TFES devices, it is practical to consider combination of the two. It is feasible to combine a TFPV cell on a Kapton™ substrate with a Li-ion thin-film battery sandwiched in Kapton™. With the further addition of very small power conditioning and control electronics, an Integrated Power Source (IPS) is possible, as depicted in figure 1.

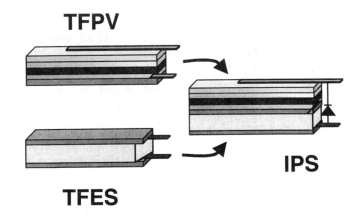

TFPV

TFES

IPS

Figure 1 – Conceptual diagram depicting IPS construction.

In fact, a number of devices of this type have been built and tested in the laboratory [refs. 4 and 5]. The first in-space demonstration of an IPS, although with a GaAs monolithically integrated module (MIM) solar cell and a Li-ion thick "coin" battery (figure 2), should occur on launch of the Starshine-3 satellite in late summer 2001 [ref. 6].

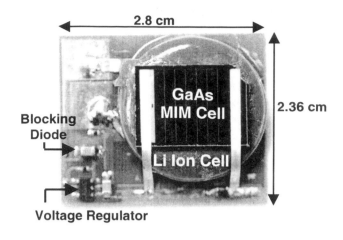

Figure 2 – Starshine-3 satellite flight unit IPS.

In the remainder of this paper, the characteristics of a thin-film IPS under development at NASA's Glenn Research Center will be described. The potential benefits and mission applicability of an IPS is assessed. In addition, the challenges that must be overcome in order to realize the benefits are mentioned.

NOMENCLATURE

A	Amperes
mA	milli-amperes
Ah	Ampere-hours
a-Si	amorphous Silicon; solar cell
BOL	Beginning Of Life
CIS	Copper-Indium-diSelenide; solar cell
CPU	Central Processor Unit
DOD	Depth-Of-Discharge
DRACO	**D**ynamics, **R**econnection, **A**nd **C**onfiguration **O**bservatory
EOL	End-Of-Life
GaAs	Gallium Aresenide; solar cell
IPS	Integrated Power Source
Kg	Kilograms
LEO	Low Earth Orbit
Li	Lithium
MCM	Multi-Chip Modules
MEM	Micro-Electrical-Mechanical
MIM	Monolithically Integrated Module
NASA	National Aeronautics and Space Admin.
PV	Photovoltaic
TF	Thin-Film
TFES	Thin-Film Energy Storage
TFPV	Thin-Film Photovoltaic
UPS	Uninterruptible Power Supply
V	Volts
W	Watts

IPS CHARACTERISTICS/DESCRIPTION

The physical characteristics of an IPS can differ dramatically, and to a large extent will be governed by the specific application. Regardless of the configuration, every IPS will include devices for power generation, energy storage and power conditioning. So far, TFES and IPS systems created at NASA Glenn have been developed to meet the needs of microelectronic devices in space, such as in Multi-Chip Modules (MCM) like the one in figure 3 [rcf. 3].

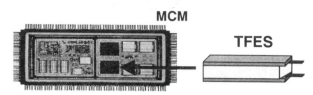

Figure 3 – Integration of TFES into an MCM.

NASA Glenn IPS systems have focused on the use of photovoltaic (PV) power generation and Li-ion battery energy storage. However, one can easily envision the use of other energy generation sources (e.g., alpha or beta voltaics, micro fuel cells, etc.) and other storage devices (e.g., super-capacitors, MEM flywheels, etc.) based upon energy needs and mission requirements.

The power requirements placed on an IPS will play a large role in determining the ultimate size of the device. The voltage of the PV portion of the device is determined by the nature of the p-n junction, or, in other words, the materials used. In the case of a GaAs homo-junction device this will be around 1.0 V. For thin-film a-Si or CuInSe₂ (CIS) PV, the voltage generated will be somewhat less (0.4-0.8 V). However, through the use of monolithically interconnected modules (MIM), many junctions can be put together in series to increase the voltage. Unfortunately, the available current will always be a function of the active surface area of the device. The current density presently available from a thin-film CIS cell is rather small due to its low photovoltaic conversion efficiency, although the goal of NASA Glenn's in-house TFPV program is >20% efficiency via a dual junction thin-film PV cell like the one illustrated in figure 4.

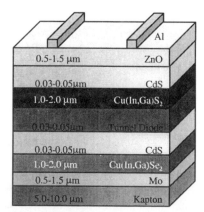

Figure 4 – Dual junction thin-film cell concept.

The voltage of a Li-ion battery is based on its chemistry and is primarily determined by the material used in its cathode. A vanadium pentoxide or manganese oxide battery will have and open circuit voltage of 3.0 V, whereas a nickel cobalt cell will be 4.2 V [ref. 7].

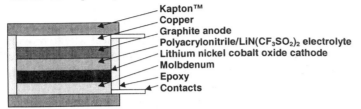

- Kapton™
- Copper
- Graphite anode
- Polyacrylonitrile/LiN(CF$_3$SO$_2$)$_2$ electrolyte
- Lithium nickel cobalt oxide cathode
- Molbdenum
- Epoxy
- Contacts

Figure 5 – Li-ion thin-film energy storage (TFES)

In a way similar to PV cells, Li battery cells can be connected in series configurations to produce different voltages. However, the amount of energy that can be stored in a cell, its capacity, is determined primarily by its volume. Thus for a thin-film Li-ion battery, the capacity will be determined in the same way the current capability of the PV cell is determined - by the area of the device. The size also impacts the rate at which a battery can be charged and discharged (i.e., the smaller the battery the smaller the charging and discharging currents it can handle).

Ideally, in order to minimize the control electronics associated with an IPS, the photovoltaic array is designed such that its output voltage matches the voltage needs of the battery and its current output is sufficient to charge the battery while simultaneously providing power to the load. The precise sizing of the array and battery will also be dependent on the anticipated illumination scheme. For example, in a typical 90-minute low-earth orbit (LEO) period, the battery will have to support the electrical load for 35 minutes of eclipse. During the 55 minute insolation (daylight) period, the solar array has to provide load power while fully re-charging the battery.

The matching of the solar array and batteries for these small power systems is essential as the parasitic power loss in a conventional charge controller normally used in a larger power system actually exceeds the output of a small IPS. Once the PV and battery are matched, the only additional components required are a blocking diode to prevent the battery from discharging through the PV array during eclipse.

The Li-ion batteries play a large role in determining the temperature regime in which these systems are suitable. Li-ion cells will deliver a sizeable fraction (i.e. 80%) of their capacity at temperatures as low as –20 °C [ref 8]. Below such a temperature they do not perform well. However, they do not exhibit permanent damage if they are cycled between larger temperatures regimes (i.e., plus or minus 80 °C) [ref. 4]. The high temperature performance is much less of an issue with thin-film Li-ion batteries as they have been shown to operate well at temperatures up to 60 °C [ref 9]. Thermal control issues associated with IPS applications are discussed later.

POTENTIAL BENEFITS

As one might anticipate, the primary benefit resulting from the combination of two extremely light weight devices providing distinct functions is a less complex, reduced volume, light weight system providing an integrated function. A thin-film IPS could serve as the main power system on a spacecraft or satellite. Scaling up the manufacturing methods should allow an IPS to deliver the highest specific power and energy for the lowest cost. Reducing power system mass, which is typically 20% to 30% spacecraft dry mass, will help reduce launch mass, perhaps enough to enable a mission concept previously too heavy to fly, or allow the use of a smaller, cheaper launch vehicle. Incorporating energy storage with power generation reduces volume formerly required by traditionally separately located chemical batteries, freeing up valuable space for other systems or an increased payload.

Almost completely opposite in approach to a fully integrated single power source is a decentralized or distributed power bus. In this instance, numerous IPSs are used to provide continuous power to loads, either spacecraft bus components or payload instruments, *in situ*, wherever the component is located. Of course this would require components to be located such that they have view of the sun for at least some portion of the orbit. The main benefit of this approach is a reduction in spacecraft complexity, especially with respect to power distribution wiring, simplifying spacecraft integration.

Related to this concept is the use of IPSs as power sources for MCM sensors that may be placed wherever they are needed in a "postage stamp" fashion.

Finally, a hybrid concept would be to use IPSs as un-interruptible power sources, providing "stay-alive" power for select spacecraft components. The main power system may or may not be an IPS in this case.

CANDIDATE MISSION CONCEPTS

In keeping with the potential benefits outlined in the previous section, specific mission/application concepts for an IPS as a main power system, a decentralized power bus and an uninterruptible power source will now be examined. For two concepts employing an IPS as the main power system, a novel nanosatellite mission and a more traditional LEO spacecraft mission, specific requirements are identified enabling a preliminary IPS applicability and benefits assessment to be performed.

Main Power System

The requirements for use of an IPS as the main power system will be considered for novel micro/nano satellite missions and a typical LEO earth or space science mission.

An IPS will most likely find its first application on one of the many emerging micro/nano spacecraft mission concepts [ref. 10]. The bus voltage for many of these missions will be 3.3 V, which is ideally suited to a thin-film Li-ion battery. Some of the more interesting micro/nanosat mission concepts entail the use of many satellites in constellations, working together to make simultaneous measurements from various, widely separated locations. NASA's Magnetotail Constellation **D**ynamics, **R**econnection, **A**nd **C**onfiguration **O**bservatory (DRACO) Mission is one proposal among the Solar Terrestrial Probe mission class of NASA's Office of Space Science Sun-Earth Connections theme. The DRACO mission calls for 50 to 100 nanosatellites, defined here as having a mass equal to or less than 10 kg, in nested, near-equatorial orbits sharing the same perigee, with varying apogees out to 40 earth radii. The nominal two-year mission is proposed for launch in 2010/2011. The total load power requirement is 4.5 W [ref. 11].

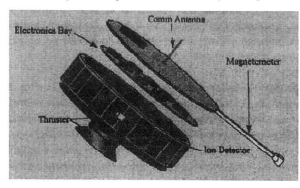

Figure 6 – DRACO Nanosatellite Concept.

Similar to the DRACO nanosatellite application, IPSs can be also used to make even smaller, extremely tiny satellite systems consisting of little more than a single chip and a power system. These IPSs would be useful in applications where a constellation of many remote sensors monitors a single, low bit-rate variable. A good example is swarms of nanosatellite weather monitors on Venus or Mars, such as the "Pascal" microprobe network proposed by Haberle and Catling [ref. 12].

However, the low Mars nighttime temperature can be an issue. In this case, the IPS could benefit from the use of an alpha-voltaic cell with an integrated lithium polymer battery; the alpha-voltaic cell would trickle charge the lithium battery, allowing short weather measurement sessions (e.g., two seconds once per hour). In the Venus application, thousands of micro-balloon weather stations, each one with a mass under one gram, can be sprinkled in the atmosphere to compile a global climate map.

On a more traditional earth science mission application, a typical low earth orbit (LEO) Earth or space science spacecraft, similar to the one depicted in figure 7, requires about 500 to 1000 W at 28 V. Mission durations are on the order of two to five years. Accounting for system losses and including a 60% depth-of-discharge (DOD) limit, about 22 Ah of energy storage is required (based on a 500 W load and a 35 minute eclipse).

Figure 7 – NASA's SeaWinds QuickSCAT LEO spacecraft.

Decentralized/Distributed Power Bus

An IPS is unique in that it combines three formerly separate functions of an electrical power system into an integrated package. Taking advantage of this feature, and applying it locally, so to speak, results in a decentralized or distributed power bus.

Use of a local IPS could allow distant portions of a satellite or space probe to be operated without a physical wire to provide power. Wiring can be 10% of the mass of a spacecraft in some cases. Removal of the physical wires would considerably simplify design. The command and control could use infrared or microwave remote control (similar to "wireless internet" control). Potential applications include:

- Actuators for deformable mirrors for large (15-25 meter) space telescopes
- Interferometric sensors
- Wireless remote actuators for spacecraft attitude control
- Gossamer spacecraft controls
- GPS attitude sensors
- Dipole array antenna elements

Uninterruptible Power Source

A decentralized power bus concept applied to discrete components, leads to the notion of an IPS as an uninterruptible power source, or UPS, to increase the reliability of essential spacecraft functions. Two specific functions that could benefit from this are computer memory and spacecraft communications.

CMOS ("volatile") memory is faster than non-volatile memory and has higher density and lower mass. However, if power is not maintained on the memory, it is erased. The amount of power required for this is extremely small, and a tiny IPS could be incorporated to make certain that even in a low-power condition, the memory remains charged.

Loss of attitude control on many satellites is a fatal error. This can occur when solar arrays lose pointing and batteries discharge. When battery voltage drops so low that the spacecraft central processing unit (CPU) and radio receiver lose power, there is no way to regain control of the satellite. An IPS could be used as a back-up power system, designed to provide enough power to run a low bit-rate omni-directional receiver and the spacecraft CPU only when the main power system failed.

IPS APPLICABILITY AND BENEFITS

The preliminary results of sizing an IPS for the DRACO nanosatellite and a representative LEO spacecraft are given below. A first-order benefits assessment is also performed.

DRACO Nanosatellite

Given the 3.3 V bus voltage and low power requirement of only 4.5 W, the applicability of an IPS for this mission as a main power system should be quite promising.

Preliminary studies indicate the DRACO nanosatellites will be 10 kg cylindrical disks 30 cm in diameter and 10 cm in height. The DRACO reference power system has triple-junction GaAs-based crystalline solar cells mounted on eight 11.5 cm x 10 cm panels with a beginning-of-life (BOL) capability of 7 W. Energy storage is to be provided by two lithium-ion batteries sized to support the load during a 1.17 hour eclipse at 60% depth-of-discharge. (All requirements and preliminary design options are from reference 11.)

Since only half of the DRACO nanosatellite cylindrical disk will see the sun at any time, and accounting for the incidence angle on the illuminated panels, the equivalent normally illuminated panel area will be about 0.03 m^2. Using a yearly average AM0 solar flux of 1350 W/m^2 and a solar cell packing factor of 0.85 while accounting for array integration, power system losses, and including power margins, the minimum required end-of-life (EOL) bare-cell efficiency at operating temperature will be about 15%. Allowing for minor degradation, a TFPV solar cell BOL efficiency will need to be about 18% to 20% at 28 °C. While this TPFV cell efficiency is challenging, it should be attainable before the 2010/2011 timeframe of this mission given development efforts at NASA Glenn and elsewhere. Alternatively, an IPS using a crystalline solar cell would suffice, although with a mass penalty.

As for the energy storage portion of the IPS, using a 60% maximum DOD and accounting for conversion system losses, 3.3 Ah total battery capacity will be required. Since there is 920 cm^2 available on the eight 115 cm^2 panels, the required specific capacity is 3.6 mAh/cm^2. Given that thin-film Li-ion specific capacity is presently at about 2.4 mAh/cm^2, improvements are necessary. As alternatives, either the size of each panel could be increased 50% to provide the area required by thin-film Li-ion TFES or alternate Li-ion battery technology could be used. In any case, since the TFES portion of the IPS prefers a warm environment, its location on the inside of the eight cylindrical panels is ideal from a thermal perspective.

The discussion thus far has indicated that a thin-film IPS is feasible for this mission. In terms of benefits, there are both mass and systems integration advantages of using an IPS in this case. First, the TFPV cells will be about two to three times lighter than the baseline crystalline high-efficiency cells. Also, the lithium-ion TFES is lighter than alternative lithium-ion batteries. While these relative benefits are significant, the absolute benefit will be small given the original mass of the baseline components are small. However, with a total mass goal of only 10 kg, any mass reduction can have a significant effect.

While mass is typically a prime discriminator in space missions, the primary benefit of an IPS in this case will be in the spacecraft layout and system integration. With the power generation and energy storage functions integrated and included on the cylindrical structural walls of the spacecraft, three typically separate spacecraft subsystems/functions are combined into one. Also, since the energy storage system is removed from the interior of the spacecraft, there is more room for other system's components. So, not only are there synergistic mass savings, but also the integration and assembly of the spacecraft is simplified. Given TFPV and TFES advancements, or a slightly larger spacecraft bus, nanosatellite applications such as DRACO could be ideally suited to an IPS.

LEO Spacecraft

Because of the higher voltage and power requirements and the thermal environment experienced by a deployed, sun-tracking solar array wing (as opposed to the small, body-mounted IPS in the previous section), this will be a challenging application for an IPS.

Sized to provide 500 W to the spacecraft loads at EOL and accounting for solar array integration and degradation losses, battery charging during insolation, and other electrical system losses, the total TFPV array area with 20% efficient cells will need to be about 8.5 m^2. In comparison, the total array area using 30% efficient multi-junction GaAs-based cells is estimated at less than half of this, or about 4 m^2. However, the mass of the flexible TFPV array should be about one-third of the typically rigid crystalline cell array.

To provide the required 22 Ah energy storage, a total TFES area of about 1 m^2 will be needed based on an area capacity of 2.4 mAh/cm^2. Assuming that the TFES operates near 4.0 V for this application, eight TFES layers will need to be placed in series to meet the 28.5 V bus requirement.

Comparing the required areas, it is seen that the TFES area is only a small fraction of the TFPV. The IPS in this case would consist of a flexible 8.5 m^2 thin-film solar array (or pair of 4.25 m^2 arrays) with the TFES located on the lower portion of the array along with its associated power condition and charge control electronics.

The benefits of an IPS in this LEO mission application include even better spacecraft volume advantages along with the mass and systems layout/integration advantages identified in the nanosatellite application. The use of flexible TFPV arrays for power generation could reduce array mass by about 67% compared with rigid arrays with crystalline solar cells, although at the expense of increased array deployed area and consequent spacecraft-level impacts (possible field-of-view obstructions and increased moments-of-inertia). In addition, the volume associated with more traditional chemical battery energy storage, especially NiH_2 batteries at about 10 Wh/liter, is significantly reduced with placement of the energy storage on the array wing within an IPS. If replacing NiH_2 batteries, about 64000 cm^3 would be available for other systems and payloads.

However, these benefits may not be fully realizable unless challenges associated with the thermal environment on the solar array wing and series connectivity of the TFES are overcome. A flexible solar array in LEO typically experiences a temperature variation from +80 °C during insolation to –100 °C during eclipse when the energy storage system is active. Passive thermal control techniques such as coatings, or baffles covering the back surface of the IPS where the TFES is located, may be sufficient to maintain the minimum required TFES operating temperature. If not, resistance heaters powered by additional TFES will be necessary.

Based on this preliminary assessment, an IPS as a LEO spacecraft main power system seems feasible, although maybe not the most directly applicable or beneficial application of this device.

CHALLENGES

There are several developmental challenges that need to be overcome in order to expand the mission applicability of the current IPS technology. With respect to TFPV, attaining 20% cell efficiencies using low temperature processes required for polyimide substrates is a significant challenge. With respect to TFES, as previously mentioned, a major drawback of using current Li-ion batteries is their inability to function at low temperatures as would be experienced on a solar array wing during eclipse, or on the surface of Mars. Another challenge is the limited cycle lifetime associated with the larger capacity polymer batteries. Although, the solid-state thin film batteries do show good cyclability, it is unclear as to whether this can be maintained when they are scaled-up to provide larger capacities. Also, the efficient use of parallel and series combination of Li thin-film batteries has yet to be demonstrated. Finally, one of the largest challenges today is in finding the appropriate sealing technologies for the Li batteries that will inhibit their degradation while also being able to withstand the rigors of space flight.

CONCLUSION

Combining three traditionally separate power system functions into a single, integrated device is a unique concept made possible by advances in thin-film technology. Assuming further improvement in both thin-film power generation and energy storage performance, applications most likely to first use and benefit from a thin-film IPS as a main power system will be those with low power and low voltage requirements. Upcoming missions with these characteristics are those using constellations of very small spacecraft, or nanosatellites. IPSs may also enjoy nearer-term applicability in specialized instances where they can serve as de-centralized or distributed power sources or un-interruptible power supplies for discrete components.

Given the early stage of development and their inherent performance limits, it remains unclear as to whether or not thin-film IPSs will find widespread applicability. However, since thin-film power generation and energy storage will be developed independently for other reasons, the techniques for combining each of these functions into an IPS, along with any required power conditioning, will be further refined. As better performing devices are built, any applications and associated benefits that can possibly be imagined will undoubtedly be further explored.

ACKNOWLEDGMENTS

The authors would like to acknowledge the work performed under NASA Cooperative Agreements NCC3-710 and NCC3-563 with the Rochester Institute of Technology.

The authors would also like to acknowledge and thank their NASA Glenn colleagues who reviewed drafts of this paper and provided valuable comments: Thomas Kerslake, Jeffrey Hojnicki and David Kankam.

REFERENCES

1. Hoffman, D.J., Kerslake, T.W., Hepp, A.F., Jacobs, M.K., Ponnusamy, D., "Thin-Film Photovoltaic Solar Array Parametric Assessment", AIAA-2000-2919, *35th IECEC*, Vol.1, pp. 670-680, July 2000 (also NASA/TM-2000-210342).

2. Landis, G., Hepp, A.F., "Applications of Thin-Film PV for Space", *Proceedings of the 26th Intersociety Energy Conversion Engineering Conference*, Vol. 2, pp. 256-261, Aug. 1991.

3. Raffaelle, R.P., Harris, J.D., Hehemann, D., Scheiman, D., Rybicki, G., Hepp, A.F., "A Facile Route to Thin Film Solid State Lithium Microelectronic Batteries", *J. of Space Power Sources*, **89** (2000) pp. 52-55.

4. Raffaelle, R.P., Harris, J.D., Hehemann, D., Scheiman, D., Rybicki, G., Hepp, A.F., "Integrated Thin-Film Solar Power System", IAIAA-2000-2808, *35th IECEC*, Vol.1, pp. 58-62, July 2000.

5. Clark, C., Summers, J., Armstrong, J., "Innovative Flexible Lightweight Thin-Film Power Generation and Storage for Space Applications", AIAA-2000-2922, *35th IECEC*, Vol.1, pp. 692-698, July 2000.

6. Raffaelle, R.P., Underwood, J., Jenkins, P., Scheiman, D., Maranchi, J., Khan, O.P., Harris, J., Smith, M.A., Wilt, D.M., Button, R.M., Maurer, W.F., and Hepp, A.F., "Integrated Microelectronic Power Supply (IMPS)", *36th IECEC*, July 2001.

7. Koksbang, R., Barker, J., Shi, H., Saidi, M.Y., *Solid State Ionics* **84** (1996) 1.

8. Lithium Handbook, Panasonic, Sept. 2000.

9. Takada, K., Kondo, S., *Ionics* 4 (1998) p. 42.

10. Panetta, et al, "NASA-GSFC NanoSatellite Technology Development", SSC98-VI-5, *12th AIAA/USU Conference on Small Satellites*.

11. Report of the NASA Magnetospheric Constellation Science and Technology Definition Team, November, 1999, available on the Internet at the following URL http://sec.gsfc.nasa.gov/magcon.htm.

12. Haberle, R. M., and Catling, D. C., "A Micro-Meteorological Mission for Global Network Science on Mars: Rationale and measurement requirements," *Planetary and Space Science, Vol 44*, No 11, pp. 1361-1393, 1996.

Proceedings of IECEC '01
36th Intersociety Energy Conversion Engineering Conference
July 29-August 2, 2001, Savannah, Georgia

IECEC2001-AT-06

FLYWHEEL ENERGY STORAGE SYSTEM FOR THE INTERNATIONAL SPACE STATION

Julie A. Grantier
NASA Glenn Research Center
Bruce M. Wiegmann
NASA Glenn Research Center
Rob Wagner
Ohio Aerospace Institute

ABSTRACT

A Flywheel Energy Storage System (FESS) is under development for the International Space Station (ISS) Electric Power System (EPS). This technology will replace the existing Nickel Hydrogen (NiH_2) batteries in performing the functions of providing power on the dark, or eclipse, portion of the orbit as well as contingency power in the event of a total EPS failure. The advantages of the FESS technology over batteries include higher reliability, longer life, higher energy density, greater energy storage capability and the ability to operate over a wider temperature range on orbit. It is estimated that by replacing the entire NiH_2 batteries, a life cycle cost savings of over $150 million will be realized, not including the cost of reduced upmass and Extravehicular Activities (EVAs) for maintenance. The current plan is to produce a flight-qualified FESS for launch in 2006, directly replacing the Battery Charge/Discharge Unit (BCDU)/Battery string. Subsequent production FESS units will replace the remainder of the BCDU/Battery string as a matter of routine maintenance. The FESS will be installed in the existing "footprint"; there will no modifications to the ISS to accommodate the FESS. The projected life of the FESS is 15 years as compared to a 5-year life of the battery.

The FESS consists of a Flywheel Module (FM), developed by The University of Texas at Austin-Center for Electromechanics (UT-CEM) and Flywheel Energy Storage Electronics (FESE), developed jointly by the National Aeronautics and Space Administration (NASA) Glenn Research Center (GRC) and TRW Space Electronics Group, Redondo Beach, California, under a Cooperative Agreement. The work focuses on the design, development and manufacture of a flight-qualified FESS, which will utilize a high-speed rotating composite rotor, to minimize weight and maximize the energy stored, given the weight, volume and thermal constraints of the existing BCDU/Battery Orbital Replacement Unit (ORU) "footprint." This project is an extension of the earlier Attitude Control and Energy Storage System (ACESES) experiment and has the potential offering proving the technology for other satellite applications.

This paper will describe the FESS design for the ISS, advantages over the current battery system, constraints as well as a discussion on flight certification activities to ensure the 15-year life requirements for the FESS.

INTRODUCTION

The International Space Station (ISS) is currently under construction. After completion in 2006, it will be operational for a minimum of fifteen years in low earth orbit. The Electric Power System (EPS) relies upon photovoltaic solar arrays during the 57-minute insolation (light side) portion of the orbit and Nickel Hydrogen (NiH_2) batteries to supply continuous power for the 35-minute eclipse (dark side) portion. These batteries have several drawbacks in Low Earth Orbit (LEO) driven by limited life and operational capabilities. The life of battery is estimated to be 5 years, which can be substantially reduced should high depth of discharges (DOD) be required, for example in contingency power scenarios. Operationally, batteries require a rigid charging profile in order to maintain health. As an alternative, Flywheel Energy Storage Systems (FESS) are being developed to replace the batteries. They are expected to last throughout the life of the ISS without replacement. These flywheels are being designed to be a direct replacement of the batteries and will be located on each of the Integrated Equipment Assemblies (IEAs) as shown in Figure 1. There will be no changes to the charge/discharge characteristics as the flywheels will operate seamlessly as batteries.

FESS offers more operational flexibility: They are able to take full solar array current for recharging as well as provide frequent, high power pulses. Repeated deep DODs do not have an effect on flywheel life. Although not applicable for the ISS program, flywheels have an added advantage by integrating with satellite stability and attitude control systems, replacing the momentum wheels. The comparison of the FESS to NiH_2 batteries are given in Table 1.

FESS DESCRIPTION

A flywheel is an electromechanical devise that stores energy in the form of rotating kinetic energy and delivers electrical energy upon demand. The equation for storing rotational energy is $E=I\omega^2$, where:

E is the rotational energy
I is the moment of inertia
ω is the tip speed

Flywheels are not new. It is with the recent advances in high strength carbon fiber technologies and miniaturization of electronics for accurate power and magnetic bearing control that make flywheels feasible for space applications. High strength carbon fibers allow for high cyclic rotational speeds without the fatigue associated with metal rotors. This drastically increases the overall energy density as compared to batteries. For example, the FESS as designed for the ISS, operates between 53,000 and 41,500 rpm which translates to a tip speed of 916 m/sec. It has an energy density of 12 Whr/lb, for NiH_2, the energy density is 4.5 Whr/lb. To minimize frictional losses, the rotor is levitated on radial and axial suspension magnetic bearings in a vacuum of 10^{-6} torr. Mechanical rotating element touchdown bearings are provided in order to minimize collateral damage due to a failure in the magnetic bearings.

Each FESS will replace a Battery Charge/Discharge (BCDU)/Battery string as shown in Figure 2.

It will utilize two of the three Orbital Replacement Unit (ORU) slots currently occupied by the BCDU/Battery string as well as the ammonia thermal control system via Radiant Fin Heat Exchangers, located on the bottom of the ORUs. Each FESS is composed of two Flywheel Energy Storage Units (FESUs) configured similarly to the Battery ORU boxes. These meet the same Robotics and on-orbit maintainability requirements. The rotors will be counterotating to ensure zero net torque on the ISS structure during operation. The subsystems in each FESU are the Flywheel Module (FM) and Flywheel Energy Storage Electronics (FESE) and are depicted in Figure 3.

Each FESS ORU measures 36"W x 34"L x 25" H and weighs approximately 450 lb.

FLYWHEEL MODULE

The FM is being developed by the University of Texas at Austin-Center for Electromechanics. It acts as the repository for the stored energy for conversion to electrical power. The components making up the FM are:

a) Composite carbon fiber rotor
b) Titanium Shaft
c) Radial and axial magnetic bearings
d) Touchdown Bearings
e) Motor/Generator
f) Flywheel Housing
g) Caging Mechanism

A cutaway view of the FM is shown in Figure 4.

The composite carbon fiber rotor is a series of twelve nested tapered rings, each of which are under radial preload compression. This preload compression ensures that the rotor rings are not under positive radial stress during operation. The service life of the rotor is designed for 15 years. This translates to 92,000 normal duty cycles (between 53,000 and 41,500 rpm, which is 39% DoD) and 1,100 startup and shutdown cycles. For contingency orbits (where the FESS provides the total ISS power), the design is 30 cycles at 89% DoD (53,000 to 17,800 rpm). The rotor will be safe-life tested to 60 years, 4 times the nominal life as required for manned space flight fracture critical components. The ratio of polar moment of inertia to the transverse moment of inertia, I_p/I_t is 0.43, with a margin of maximum speed to the first bending mode of 59%. The rotor temperature range for the normal operation is between 72 and 180^0F. The rotor is mounted on a monolithic titanium shaft with laminations at each end. These laminations ensure the proper magnetic properties at high speed operation for the magnetic bearings.

The magnetic bearings provide axial and radial suspension. They are permanent magnet, homopolar bearings which minimize losses and power requirements. Located on tombstone bearing supports, they have a load capacity of 165 lb per axis, which meets a 2 g load requirement. A combination magnetic bearing consists of x- and y-radial actuators and a z- axial actuator and is located on the end opposite of the motor generator. A radial bearing for x- and y-axis suspension only is collocated with the motor generator. Control for the magnetic bearings is provided in the FESE.

Touchdown bearings at each end of the shaft are provided to prevent damage due to failures in the magnetic bearing actuators. These bearings are designed to survive on touchdown event, which is defined as when the rotor intermittently contacts the bearing surfaces during operation with exception of a shutdown event. The bearings are rolling element type, consisting of ceramic balls and dampers for stiffness.

Energy into and out of the FM is provided by a motor generator. It is also mounted on a tombstone support along with the Radial Magnetic Bearing. It is a high-speed, high-efficiency permanent magnet motor, capable of providing a maximum of 6.6 kW of power. The motor windings are stationary. Control for the motor generator is provided in the FESE.

The flywheel housing provides structural support to the FM as well as a secondary barrier for micrometeoroid/orbital debris (MM/OD). Since the ORU box that the FM will reside is vented to the environment, this housing will not be hermetically sealed to provide a vacuum for the rotor.

During launch and abort landing, the FM will need to prevent damage to the FESS from vibration and shock. Current studies are focusing on a series of options that will either load the bearing with an inflatable bladder or suspend the rotor via clamps and/or bladder. It is also possible that no caging mechanism is required, the free movement of the rotor will not cause damage. These trades are being evaluated to minimize weight, length and complexity to the FESS.

FLYWHEEL ENERGY STORAGE ELECTRONICS

The FESE is under development by TRW Space Group. Its purpose is to interface with the primary power and data buses for the ISS as well as provide control within the FESS. It consists of:

a) Input/Output Manger
b) Secondary Power Control
c) Drive Control
d) Inverter

The Input/Output Manager (IOM) interfaces with the ISS Photovoltaic Control Applications (PVCU) via 1553 bus. It provides the data on the state of the health of the FESS to the ISS as well as the control logic for commanding the FESS into various operating modes.

The secondary power control provides +15, -5 volt power for sensors and general housekeeping.

Two drive controls, each support the Magnetic Bearings and Motor Generator. Control algorithms reside in the drive control modules.

The inverter converts the incoming DC power to 3Φ AC from the primary bus during charging cycles and vice versa during discharge.

STATUS OF THE FESS PROJECT

Detailed design of the FESS is scheduled to be completed by the summer of 2001. It will take approximately 10 months to build and test the FESS subsystems. Therefore, testing of the FESS at GRC is planned for the summer of 2002 and FESS operation in the Space Power Electronics Lab at Boeing Canoga Park is scheduled for 2003. The first flight-qualified FESS is to be launched in 2006.

Currently, the rotor certification program which is required to demonstrate safe life is planned and will be implemented upon funding. The rotor certification program is a comprehensive test program designed to ensure the rotor will operate for 15 years on the ISS.

SUMMARY

Using state of the art flywheel technology, the FESS is being designed to develop a replacement for the NiH_2 batteries on the ISS. The advantages of FESS over batteries include higher reliability, longer life, higher energy density, greater energy storage capability, higher efficiency and greater operational flexibility. The FESS will be a direct replacement for the batteries. Each FESS will be located on the same footprint as the BCDU/Battery string. In addition, there will be no changes to the charge/discharge characteristics as the FESS will operate seamlessly as batteries. No modifications to the ISS are required to accommodate the FESS. The FESS has a projected life of 15 years as compared to a 5 year life of the batteries. Therefore, FESS will not need to be replaced for the life of ISS resulting in a substantial savings in cost, up mass, and maintenance.

The current plan is to build a FESS to be tested at the SPEL in 2003. Flight-qualified FESS units will be launched in 2006 after completion of the ISS. The FESS units will replace the existing BCDU/Battery strings at the end of their life.

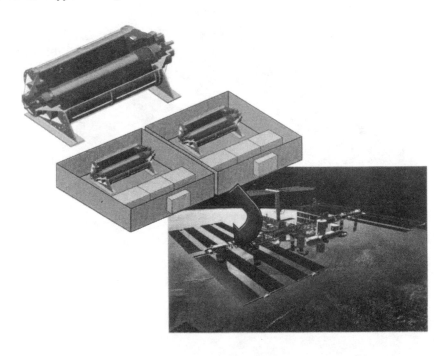

Figure 1- Flywheel Location On the IEA

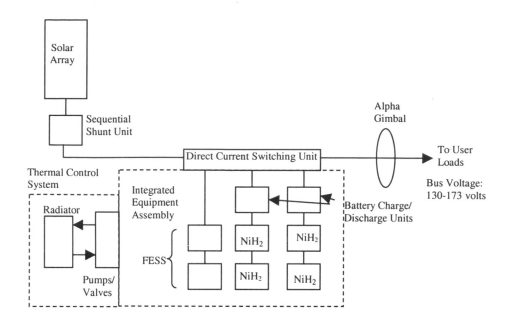

Figure 2 ISS Electric Power System Schematic

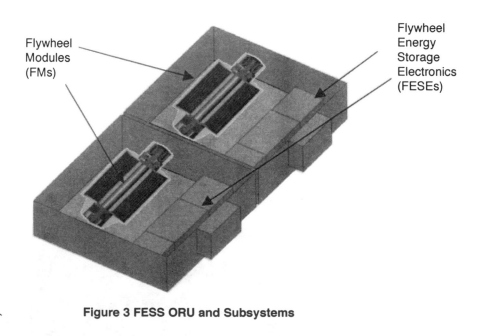

Figure 3 FESS ORU and Subsystems

266

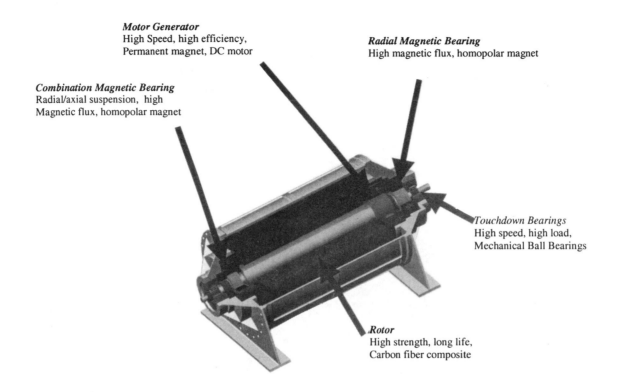

Motor Generator
High Speed, high efficiency,
Permanent magnet, DC motor

Radial Magnetic Bearing
High magnetic flux, homopolar magnet

Combination Magnetic Bearing
Radial/axial suspension, high
Magnetic flux, homopolar magnet

Touchdown Bearings
High speed, high load,
Mechanical Ball Bearings

Rotor
High strength, long life,
Carbon fiber composite

Figure 4 FESS FM

Table 1 FESS vs. NiH$_2$ Comparison

	FESS	NiH$_2$
Power	4.1 kW	4.1 kW
Peak Power	5.5 kW	5.5 kW
Energy Storage	5.5 kWhr	4.6 kWhr
Contingency Power	2 Orbits	1 Orbit
Life Expectancy	15 years	5 years
RoundTrip Efficiency*	83%	65%

* Measured at the Power Bus

Proceedings of IECEC '01
36th Intersociety Energy Conversion Engineering Conference
July 29–August 2, 2001, Savannah, Georgia

IECEC2001-AT-10

INTERNATIONAL SPACE STATION BUS REGULATION WITH NASA GLENN RESEARCH CENTER FLYWHEEL ENERGY STORAGE SYSTEM DEVELOPMENT UNIT

Peter E. Kascak[1]

Barbara H. Kenny[2]

Timothy P. Dever[3]

Walter Santiago[2]

Ralph H. Jansen[1]

1. Ohio Aerospace Institute
NASA Glenn Research Center
21000 Brookpark Road
Cleveland, Ohio 44135

2. NASA Glenn Research Center
21000 Brookpark Road
Cleveland, Ohio 44135

3. QSS Group Inc.
NASA Glenn Research Center MS 23-1
21000 Brookpark Road
Cleveland, Ohio 44135

ABSTRACT

An experimental flywheel energy storage system is described. This system is being used to develop a flywheel based replacement for the batteries on the International Space Station (ISS). Motor control algorithms which allow the flywheel to interface with a simplified model of the ISS power bus, and function similarly to the existing ISS battery system, are described. Results of controller experimental verification on a 300 W·hr flywheel are presented. Keywords: flywheel, motor control, battery, charge, discharge, charge reduction.

INTRODUCTION

A developmental flywheel test facility has been built at NASA Glenn Research Center (GRC), in Cleveland, Ohio. This system includes a carbon composite high-speed flywheel, which is coupled to a motor/generator, and is suspended by active magnetic bearings. This test facility allows development and testing of control algorithms for the flywheel motor/generator and magnetic bearings. Flywheel based energy storage systems are being considered as a possible replacement for the battery based system currently in use on the ISS because flywheel systems feature longer life, higher efficiency and greater depth of discharge than battery based systems. In order to allow direct replacement, the flywheel system must be made to operate in the same fashion as the battery systems currently in place on the ISS. This paper describes motor/generator control algorithms which allow the flywheel system to interface with a simplified version of the ISS electrical bus, enabling the flywheel to be "charged" by storing energy mechanically in the wheel, and "discharged" by converting mechanical energy to current on the bus, while regulating bus voltage.

NOMENCLATURE

I_{charge}^* – charge current setpoint

I_{DC}^* – DC current regulator setpoint

i_{ds}^r – d-axis current in the rotor reference frame

i_{qs}^r – q-axis current in the rotor reference frame

I_{CRD}^* – current set point for charge reduction/discharge modes

$I_{flywheel}$ – DC Current into the Flywheel

I_{load} – load current

$I_{s/a}$ – solar array current

ISS – International Space Station

BCDU/BS – Battery Charge Discharge Unit/Battery System

L_{ds} – d-axis stator inductance, H

p – time derivative operator (d/dt)

P – number of poles

$P_{flywheel}$ – Power on the AC side of the inverter

R_S – Stator resistance per phase, Ω

SAS – Solar Array System

SAW – Solar Array Wings

SSU – Sequential Shunt Unit

V_{dc} – DC bus Voltage

v_{ds}^r – d-axis voltage in the rotor reference frame

v_{qs}^r – q-axis voltage in the rotor reference frame

$V_{s/a}^*$ – Sequential Shunt Unit voltage set point

$V^*_{discharge}$ – flywheel (or BCDU) discharge voltage set point

λ_{af} – flux due to the rotor magnets

ISS POWER BUS CONFIGURATION

Power for the ISS is generated by the station solar array system (SAS), which includes the solar array wings (SAW) and the sequential shunt unit (SSU). Since the SAS cannot generate energy within the 35 minute eclipse period of the 92 minute ISS Earth orbit, some energy needs to be stored aboard the ISS. Presently, this energy is stored and regulated in a battery-based system called a Battery Charge Discharge Unit/Battery System (BCDU/BS). In order for a flywheel system to replace the BCDU/BS on the ISS, it must mimic the electrical performance of the BCDU/BS. In the following text, present BCDU/BS operation is described, and required flywheel operation is notated in parentheses. Figure 1 is a schematic of the present BCDU/BS (proposed flywheel) system.

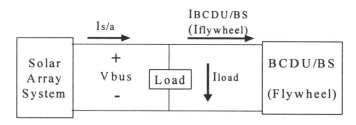

Figure 1: Block Diagram of ISS BCDU/BS (Flywheel) Power System

The BCDU/BS (flywheel) has three operational modes - the "charge" mode, the "discharge" mode, and the "charge reduction" mode. These modes of operation are summarized in Table 1.

Mode	BCDU/BS (Flywheel) DC Current	Regulated Bus Voltage
Charge	$I_{BCDU/BS} = I^*_{charge}$ $(I_{Flywheel} = I^*_{charge})$	$V_{bus} = V^*_{s/a}$
Discharge	$I_{BCDU/BS} < 0$ $(I_{Flywheel} < 0)$	$V_{bus} = V^*_{discharge}$
Charge Reduction	$I^*_{charge} > I_{BCDU/BS} > 0$ $(I^*_{charge} > I_{Flywheel} > 0)$	$V_{bus} = V^*_{discharge}$

Table 1: BCDU/BS (Flywheel) Modes of Operation

Charge mode on the energy storage system occurs when the SAS is generating enough current to supply the ISS user (designated by the "load" in Figure 1), and to charge the batteries (accelerate the flywheel) at its charge mode current setpoint. In this mode, ISS DC bus regulation is provided by the SAS. This charge mode typically takes place when the station is in full sun.

Discharge mode on the energy storage system occurs when the batteries are discharging (flywheel is decelerating) and providing power to the load. In this mode, the BCDU (flywheel) regulates the DC bus voltage at $V^*_{discharge}$. This discharge mode typically takes place when the station is in full eclipse.

Charge reduction mode on the energy storage system occurs when the SAS provides some current, but not enough to both supply the load and charge the batteries (accelerate the flywheel) at the charge mode current setpoint. In this mode, the BCDU (flywheel) provides regulation of the DC bus at $V^*_{discharge}$. This charge reduction mode typically takes place when the station is in partial sun.

FLYWHEEL TEST CONFIGURATION

A simplified schematic of the flywheel test configuration is presented in Figure 2. In this system, the ISS bus is modeled using a commercially available DC power supply as the source, and a resistor as the load. For charge mode operations, the supply was set to 60V and 30A current limit; for discharge mode, the current limit of the supply was brought down to 0A. The load resistance value was changed to simulate different loading conditions during discharge mode. Although the flywheel is rated at 60,000 RPM, speed was limited to under 10,000 RPM during controller testing. The input filter to the inverter was a 13.2 mF capacitor.

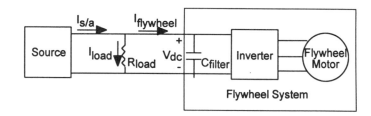

Figure 2: Simplified Flywheel Test Configuration

The entire motor/generator portion of the flywheel development system is shown schematically in Figure 3; this system includes the hardware presented in Figure 2, along with details of the motor control system and the DC bus controls. A commercially available computer was used for the controller (DC bus control system design and operation is described in detail in the following sections). The motor control signals are converted via a PWM board, and passed on to a commercially available inverter, which drives the flywheel motor. Inverter power was supplied by the simplified ISS bus (see Figure 2). Feedback signals for the controller include two motor phase currents (I_a and I_b), the motor once around (OAR) signal, and the DC bus current and voltage (I_{dc} and V_{dc}).

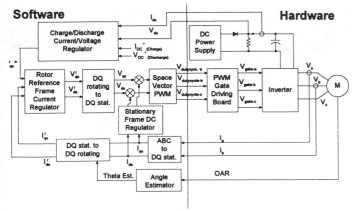

Software **Hardware**

Figure 3: Flywheel System Control Software/Hardware Configuration

DC BUS CONTROL OVERVIEW

The DC bus controller must control the flywheel motor such that the dc bus current, $I_{flywheel}$, (in charge mode) and the dc bus voltage, V_{dc}, (in discharge mode) are regulated. This is done in the block labeled "Charge/Discharge Current/Voltage Regulator" (CDCVR) in Figure 3. The input variables to this block are the commanded and measured values of the dc bus voltage and current. The output variable is the current command to the inner loop motor control algorithm (i_{qs}^r). This motor control algorithm is based on the field orientation technique that is described in [1].

The CDCVR block for the DC bus control consists of three main functions:

1. Regulate the dc charge current to the flywheel system, $I_{flywheel}$, to the commanded value, I_{charge}^*, set by a higher-level ISS control during charge mode.
2. Regulate the station DC bus voltage, V_{dc}, to commanded value, $V_{discharge}^*$, set by the higher-level ISS control during charge reduction and discharge modes.
3. Transition smoothly between charge, charge reduction and discharge modes.

Two types of CDCVR controls were investigated: proportional-integral (PI) control, and PI plus feedforward control.

CDCVR PROPORTIONAL INTEGRAL (PI) CONTROL

The PI version of the CDCVR control configuration is shown in Figure 4. This PI control configuration consists of two control loops, the DC voltage loop (for discharge mode and charge reduction mode) and the DC current loop (for charge mode), and also two transition conditions.

When the flywheel system is in charge mode, the $I_{flywheel}^*$ switch is in position 1 making it equal to I_{charge}^*. At this stage, the output of the PI regulator sets the inner loop motor current command, i_{qs}^{r*}, based on the error between the

commanded flywheel current (I_{charge}^*) and the measured flywheel current ($I_{flywheel}$, as defined in Figure 2). In the charge mode, $I_{flywheel}$ is positive and the solar array system (SAS) regulates the dc bus voltage.

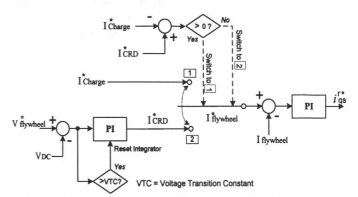

Figure 4: PI CDCVR Control Algorithm

When the flywheel system is in discharge mode or charge reduction mode, the $I_{flywheel}^*$ switch is in position 2 (see Figure 4) making it equal to I_{CRD}^*. In these modes, the flywheel system regulates the dc bus voltage by using the two PI loops connected in series. The first PI loop, going from left to right on Figure 4, processes the voltage error between the commanded dc bus voltage, $V_{flywheel}^*$ and the measured dc bus voltage, V_{dc}. The output of this 1st PI loop generates I_{CRD}^*, which is the charge reduction and discharge mode current set point that feeds the 2nd PI loop. In discharge mode, the flywheel system is providing power to the station loads thus $I_{flywheel}$ as defined in Figure 2 is negative. In charge reduction mode, the flywheel is charging, and thus $I_{flywheel}$ is positive. In either case, the flywheel system provides the dc bus voltage regulation.

From the perspective of the flywheel control system, there is really only one transition point. The flywheel control system is either set to regulate the flywheel current (charge mode) or the dc bus voltage (charge reduction and discharge modes).

In current regulation (charge mode), the solar array provides enough current to supply all of the load demand plus the charge current set point, I_{charge}^*. As the solar array moves into eclipse, the current that the array can provide drops off. The dc bus voltage will begin to fall because the solar array can't provide enough current to meet both the load demand and the charge current set point to the flywheel system. Once the dc bus voltage falls below a certain level, the flywheel system must transition from current regulation to voltage regulation and begin to regulate the dc bus voltage.

In voltage regulation (charge reduction and discharge modes) the flywheel system is regulating the dc bus. As the solar array moves out of eclipse, it can provide more and more current. Once $I_{flywheel}$ reaches the charge current set point, I_{charge}^*, it is an indication that the solar array is now producing enough current to provide all of the station loads plus I_{charge}^*. At this point, the flywheel system must transition from voltage regulation to current regulation and the solar array system will begin to regulate the dc bus voltage.

Current and voltage regulation are detected in the PI CDCVR Control in the following way and will be described using an example. Starting in charge mode, the dc bus voltage, V_{dc}, is at 130 volts, controlled by the solar array. The flywheel voltage set point, $V_{flywheel}^*$, is at 120 volts. The voltage transition constant (VTC) is −5 volts (see Figure 4). Thus in charge mode, $V_{flywheel}^* - V_{dc}$ is less than the VTC (-10 < -5) and the integrator is not reset. These conditions produce a I_{CRD}^* value larger than I_{charge}^*, which causes the $I_{flywheel}^*$ switch to move to position 1 (charge mode).

As the solar arrays move into eclipse, the dc bus voltage falls. As the bus voltage falls below 125 volts, $V_{flywheel}^* - V_{dc}$ falls below the VTC, and the integrator resets to 0. This sets I_{CRD}^* to a small positive number. Since I_{CRD}^* is now less than I_{charge}^*, the $I_{flywheel}^*$ switch is moved to position 2 and the system transitions into voltage regulation (charge reduction/discharge mode).

As long as the flywheel current, $I_{flywheel}$, is less than the charge set point current, I_{charge}^*, the system remains in voltage regulation because the solar arrays are not providing enough current to meet the load demands and the current charge set point. As the station moves out of eclipse, the SAS begins to contribute current, increasing the bus voltage (V_{dc}) until $V_{flywheel}^* - V_{dc}$ drops below VTC. At this point, voltage PI loop integrator will no longer be reset, I_{CRD}^* will exceed I_{charge}^*, and the $I_{flywheel}^*$ switch will move to position 1, transitioning the system to current regulation (charge mode).

CDCVR PI PLUS FEEDFORWARD CONTROL

The PI control described in the previous section and in Figure 4 is structured such that the inner loop motor command current, i_{qs}^{r*}, is derived from a PI controller operating on a $I_{flywheel}^*$ command. In charge mode, $I_{flywheel}^*$ is equal to the charge current set point, I_{charge}^*. In discharge mode, $I_{flywheel}^*$ is equal to I_{CRD}^*, the charge reduction/discharge current set point. In either case, i_{qs}^{r*} is the output of a PI control operating on the error between $I_{flywheel}^*$ and $I_{flywheel}$.

There is another approach to producing the motor current command based on the desired $I_{flywheel}$ value that is shown in Figure 5. In this approach, the open loop steady state relationship between i_{qs}^r and $I_{flywheel}$ is used to determine i_{qs}^{r*}

from $I_{flywheel}^*$. This approach allows the use of a feedforward command for both voltage and current regulation, which results in a fast response with lower gains on the PIs. The lower PI gains result in less noise in the system as will be shown in the experimental results section.

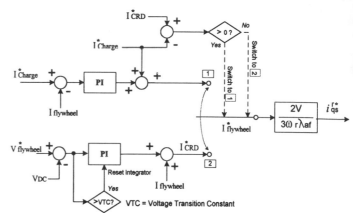

Figure 5: PI Plus Feedforward CDCVR Control Algorithm

The relationship between i_{qs}^r and $I_{flywheel}$ is based on the steady state power balance between the dc power going into the inverter and the ac power going into the flywheel motor. The ac power used by the flywheel motor can be expressed as shown in (2)[3].

$$P_{flywheel} = \frac{3}{2}(v_{qs}^r i_{qs}^r + v_{ds}^r i_{ds}^r) \qquad (2)$$

Neglecting minor losses in the inverter, the dc power supplied will equal the ac power used by the flywheel motor.

$$V_{DC}I_{DC} = \frac{3}{2}(v_{qs}^r i_{qs}^r + v_{ds}^r i_{ds}^r) \qquad (3)$$

In the motor control algorithm used in this system, i_{ds}^r is regulated to 0 [1]. Therefore the power balance in (3) is reduced to

$$V_{DC}I_{DC} = \frac{3}{2}(v_{qs}^r i_{qs}^r) \qquad (4)$$

The q-axis voltage equation is [2]

$$v_{qs}^r = i_{qs}^r R_S + L_{qs}p i_{qs}^r + i_{ds}^r \omega_r L_{ds} + \omega_r \lambda_{af} \qquad (5)$$

In steady state, the derivative term is zero. Furthermore the d-axis current is regulated to zero. Therefore (5) reduces to

$$v_{qs}^r = i_{qs}^r R_s + \omega_r \lambda_{af} \qquad (6)$$

Substituting (6) into (4), the power balance equation results in the following relation

$$V_{DC}I_{DC} = \frac{3}{2}((i_{qs}^r R_s + \omega_r \lambda_{af})i_{qs}^r) \qquad (7)$$

R_S is small compared to the back emf term, $\omega_r \lambda_{af}$, especially at high speeds, so the power balance can be approximated as

$$V_{DC}I_{DC} \cong \frac{3}{2}(\omega_r \lambda_{af} i_{qs}^r) \qquad (8)$$

This can be used to relate a current command on the dc side of the inverter to a current command on the ac side of the inverter:

$$i_{qs}^r = \frac{2V_{DC}I_{DC}^*}{3\omega_r \lambda_{af}} \qquad (9)$$

From (9) it can be seen that during current regulation (charge mode) the feedforward command to the motor current is

$$i_{qs}^{r*} = \frac{2V_{DC}I_{charge}^*}{3\omega_r \lambda_{af}} \qquad (10)$$

Figure 5 shows that the charge current set point, I_{charge}^*, sums with the output of the PI regulator that operates on the error between I_{charge}^* and $I_{flywheel}$. Under current regulation (charge mode), $I_{flywheel}$ is equal to this sum. If the feedforward relationship of (10) is exactly accurate, then the contribution of the PI portion to $I_{flywheel}^*$ will be zero. However, approximations were made in the derivation of (10) plus the back emf constant, λ_{af}, may not be known exactly. So the PI is used to make up for any errors in the feedforward calculation. However, the feedforward portion is contributing most of the $I_{flywheel}^*$ command so the gains on the PI can be set lower than in the previous case (PI only algorithm).

Similarly for voltage regulation (discharge and charge reduction modes) the feedforward command to the motor current is

$$i_{qs}^{r*} = \frac{2V_{flywheel}^* I_{flywheel}}{3\omega_r \lambda_{af}} \qquad (11)$$

Figure 5 also shows that $I_{flywheel}$ sums with the output of the PI regulator that process the error between $V_{flywheel}^*$ and V_{dc}. Under voltage regulation (discharge and charge reduction modes), $I_{flywheel}^*$ is equal to this sum. Similar to the current regulation (charge mode) control loop, the feedforward portion of the voltage regulation loop is contributing most of the $I_{flywheel}^*$ command so the PI gains can be set to lower values.

Furthermore, Figure 5 shows the $I_{flywheel}^*$ command converted to the i_{qs}^{r*} command through the power balance relationship of (9). For current regulation, V in this block is equal to the measured value V_{dc}. For voltage regulation, V in this block is equal to the commanded value, $V_{flywheel}^*$. In both cases, the speed, ω_r, is measured and used as an input to this block.

The transition characteristics between current regulation and voltage regulation for this PI plus feedforward control are the same as described in the previous section for the PI only control.

EXPERIMENTAL RESULTS

To verify the proper operation of the two described control loops and prove that the flywheel system can perform like a BCDU/BS, two types of tests were performed. The first test was a step change on the charge set point and the second test was a load step while the flywheel was in discharge mode. These tests were performed on both the PI and the PI plus feedforward controllers.

The two controllers were tuned to have similar responses. This was achieved by using approximately 1 order of magnitude higher gains for the PI only control than the PI plus feedforward control. There was very little difference between the PI only and the PI with feedforward performance in the response of the DC bus variables, $I_{flywheel}$ and V_{dc}. Thus for the DC bus variables, only the results for the PI plus feedforward control are presented.

Figure 6 shows the response to a step change in commanded current during charge mode. The transient response time was about 12 msec. As mentioned previously, the combination of the SAW and SSU is experimentally modeled using a DC power supply. Because the measurement of DC current is taken on the power supply side of the capacitor, most of the current transient is caused by the reaction of the power supply to the change in current taken by the flywheel system.

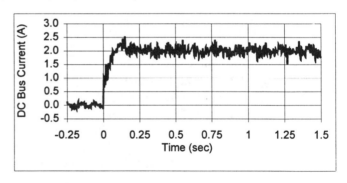

Figure 6: I $_{flywheel}$ Response for Charge Mode Current Step Change Using PI Plus Feedforward Control Algorithm

Figure 7 shows the system operating in discharge mode (flywheel system regulating the dc bus voltage) with a step change in load. Discharge mode is experimentally modeled with the power supply turned off, therefore the step response is entirely due to the flywheel system. The discharge voltage set point, $V_{flywheel}^*$, is set to 60 volts. Figure 7 clearly shows that the bus voltage regulation is maintained for this load step.

Figure 8 shows the flywheel speed for the same load step shown in Figure 7. It can be seen that when the load is applied the slope of the speed trace becomes more negative. This is because more power is taken from the flywheel system and delivered to the load.

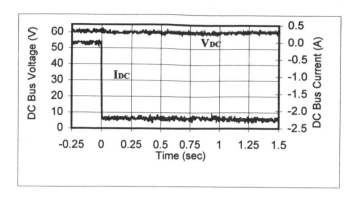

Figure 7: Discharge Mode Load Step Change – Voltage and Current Response (PI Plus Feedforward Control Algorithm)

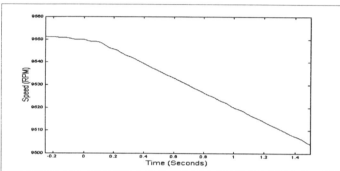

Figure 8: Discharge Mode Load Step Change – Speed Response (PI Plus Feedforward Control Algorithm)

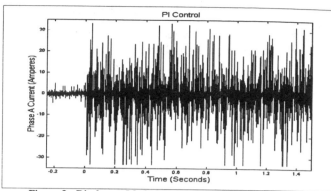

Figure 9: Discharge Mode Load Step Change – Phase Current Response (PI Control Algorithm

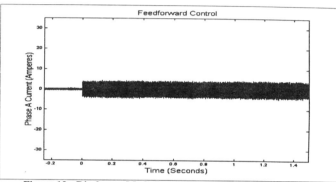

Figure 10: Discharge Mode Load Step Change – Phase Current Response (PI Plus Feedforward Control Algorithm)

Figures 9 and 10 show the flywheel system motor phase current response for the same load step shown in Figure 7 for pure PI and PI plus feed foward controllers respectively. It is clear that the phase currents for the pure PI controller are much noisier than with the PI plus feedforward controller. This is because the correcting PI loop on the PI plus feedforward controller can have very small gains due to the action of the feedforward portion of the control. The pure PI controller requires larger gains to achieve similar performance; these large gains increase the control bandwidth, allowing the controller to act on noisy feedback signals (i.e. the DC bus voltage).

CONCLUSIONS

DC bus regulation during changes in charge current set point and discharge load value was demonstrated on the flywheel development unit. Additionally, CDCVR regulator mode transition algorithms were implemented and discussed. These results demonstrate that a flywheel system can successfully mimic the operating modes of the ISS battery system. This is an important milestone because it is a first step in demonstrating the feasibility of replacing the ISS battery system with a flywheel system.

CDCVR regulators using PI and PI plus feedforward compensation were implemented and tested, and the results were presented. Although the pure PI controller is simpler, the resulting motor phase currents are noisier. This phase current noise is irrelevant to the DC bus control; however, it does cause energy loss and unnecessary heating of the flywheel motor.

REFERENCES

[1] Kenny, B., P. Kascak, H. Hofmann, M. Mackin, W. Santiago, R. Jansen, "Advanced Motor Control Test Facility For NASA GRC Flywheel Energy Storage System Technology Development Unit", *to be published at 2001 IECEC*, Savannah, Georgia.

[2] Krishnan, Ramu, *Permanent Magnet Synchronous and Brushless DC Motor Drives: Theory, Operation, Performance, Modeling, Simulation, Analysis and Design*, Virginia Tech., Blacksburg, Virginia, 1999.

[3] Novotny, D.W. and T.A. Lipo, "*Vector Control and Dynamics of AC Drives*", Oxford Science Publications, Oxford, 1996.

IECEC2001-AT-30

ELECTRICAL POWER DISTRIBUTION SYSTEM FOR THE EUROPEAN 'COLUMBUS' ATTACHED PRESSURISED MODULE (APM) OF THE INTERNATIONAL SPACE STATION

Giuliano Canovai,
European Space Agency,
European Space Technology Centre,
Keplerlaan 1,
P. O. Box 299,
2200 AG,
NOORDWIJK,
The Netherlands.
E-Mail: giuliano.canovai@esa.int

Antonio Ciccolella,
European Space Agency,
European Space Technology Centre,
Keplerlaan 1,
P. O. Box 299,
2200 AG,
NOORDWIJK,
The Netherlands.
E-Mail: antonio.ciccolella@esa.int

Jim Haines,
European Space Agency,
European Space Technology Centre,
Keplerlaan 1,
P. O. Box 299,
2200 AG,
NOORDWIJK,
The Netherlands.
E-Mail: jim.haines@esa.int

ABSTRACT

The COLUMBUS Attached Pressurised Module (APM) represents the European contribution to the infrastructure of the International Space Station (ISS). The paper describes the electrical power distribution system of the COLUMBUS module and highlights some detailed design aspects of the main equipment involved namely the Power Distribution Unit (PDU).

INTRODUCTION

The pressurised COLUMBUS Laboratory Module an illustration of which is shown in figure 1, represents the European contribution to the infrastructure of the International Space Station (ISS). This module is being built in Europe by an industrial team led by Astrium GmbH of Bremen, Germany who is also responsible for the related on-board avionics and the final module system integration. Alenia of Turin Italy is a major co-contractor within the COLUMBUS programme, being responsible for the module structure, its thermal control and environmental control and life support aspects.

Figure 1. COLUMBUS Attached Pressurised Module (APM)

MISSION REQUIREMENTS

As an attached element of the International Space Station, detailed below are the mission requirements and system parameters that are defined for the COLUMBUS Attached Pressurised Module (APM).

Orbit	51.6° inclination
Altitude	nominal range between 335 km and 500 km
Mission Duration	10 years with on-orbit maintenance
Payload Accommodation	10 active International Standard Payload Racks (ISPRs)

SYSTEM DESCRIPTION

Configuration	4 rack lengths Pressurised Module permanently ISS attached
Dimensions	6.4 m x 4.5 m diameter
Launch Mass	12,400 kg (including 2,500 kg of Payload)
On-orbit Payload Mass	9,000 kg
Resources	Provided by the ISS except for local Data Processing
Electrical Power	Electrical Power Distribution System (EPDS), sized for maximum capability of 20 kW (120 VDC) (13.5 kW for Payload Operations)
Comms Infrastructure	33Mbps down via ISS TDRSS 32 Mbps down via JEM/Artemis 10 kbps up via ISS S-Band
Environmental Control	Sized for 3 crew members : heat rejection up to 22 kW
Data Management	Layered Multi Computer Architecture Control via Military Standard 1553 B busses Data Communications via Ethernet LAN
Video	Display, recording, compression and routing

The Electrical Power Distribution System (EPDS) is responsible for the receipt of power from Space Station provided Main Busses and its distribution within the COLUMBUS Module to individual payload elements and subsystem/support items. It also provides capabilities for ON/OFF switching of supply lines to both payload and subsystem units and associated protection against line overcurrents and ground fault leakages within portable equipment.

Additionally the EPDS is responsible for the provisioning of illumination within the COLUMBUS Module.

The EPDS receives 120 volt DC power from two independent Space Station Main Busses (SSMB), each bus being sourced by two parallel connected 6.25 kW, NASA DC-DC Converter Units (DDCUs) housed within Node 2 of the ISS. Dependent upon the operational mode existing within Node 2, each of the two feeders is thus capable of providing up to 12.5 kilowatts of power to the COLUMBUS Module. In addition the EPDS is also responsible for furnishing several 300 watt, 28 volt DC power lines, for specific low voltage users.

ELECTRICAL OPERATIONAL MODES

The COLUMBUS EPDS itself operates in one of four different mode configurations, these being :-

Launch Configuration	All equipment de-activated, this mode being used during APM Passive Mode and APM Unberthed Survival Mode
Survival Configuration	With the Power Distribution Units (PDUs) in passive cooling mode and all other EPDS equipment in OFF configuration. This mode is associated with Berthed Survival Mode.
Reduced Configuration	With the Power Distribution Units (PDUs) in active cooling mode and with the possibility of activated lighting for module internal, video surveillance. This configuration is associated to the (unmanned) Routine Operations Mode and the Housekeeping Mode of the APM.
Nominal Configuration	With the Power Distribution Units (PDUs) in active cooling mode and the lighting system active. This configuration is associated with all manned modes of the APM, such as Support Mode and Routine Operations Mode.

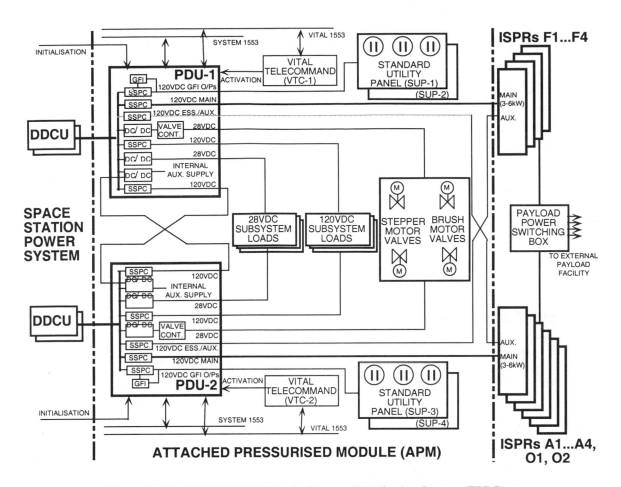

Figure 2. The COLUMBUS Electrical Power Distribution System (EPDS)

ELECTRICAL POWER DISTRIBUTION SYSTEM

Under the technical leadership of Astrium GmbH of Bremen, Germany, the COLUMBUS EPDS as depicted in figure 2, is comprised of the following equipments :

Two Power Distribution Units and associated power distribution harnesses

One External Payload Power Switching Box (PPSB)

Eight Module Lighting Units (NASA General Lighting Assemblies)

One Emergency Light Strip which in conjunction with the Emergency Light Power Supply forms the Emergency Egress Lighting

The two Power Distribution Units which are designed and manufactured by Alcatel Space, France, provide several 120 volt DC to 28 volt DC buck converters and make extensive use of current limiting Solid State Power Controllers (SSPCs) within their design. Inside each PDU these SSPCs are arranged

to provide the following output switching and protection functions :

For fixed Payload Racks at 120 volts DC

3 x 6 kW outputs and 2 x 3 kW outputs and 5 x 1.2 kW Auxiliary Power outputs

For Portable Payload Equipment at 120 volts DC

6 x 1.2 kW, Ground Fault Interrupt (GFI) protected outputs provided at the level of four Standard Utility Panels (SUPs)

For Subsystem functions at 120 volts DC

6 x 1.2 kW outputs and 2 x 200 W down-rated outputs

For Subsystem functions at 28 volts DC

9 x 300 W outputs and 6 x 60 W down-rated outputs

The electrical parameters of the different outlet types within each PDU are detailed in Table 1 below.

No.	OUTLET TYPE	VOLTAGE	POWER	No. OF OUTLETS	NOMINAL CURRENT	PEAK CURRENT
1	ISPR Main Power	120 V	6 kW	3	50 A	72 A
2	ISPR MAIN Power	120 V	3/3.6 kW	2	25 A	36 A
3	ISPR Aux. Power Lateral	120 V	1.2(3) kW	4	10 A	36 A
4	ISPR Aux. Power Overhead	120 V	1.2(1.8) kW	1	10 A	18 A
5	Standard Utility Panel (SUP)	120 V	0.6/1.2 kW	6	5/10 A	12.2 A
6	External Payload Facility (EPF)	120 V	1.25(6) kW	4	10.4 A	(72 A)*
7	System, Nominal	120 V	1.2 kW	6	10 A	12A
8	System, Downrated	120 V	0.2 kW	2	1.7 A	1.8 A
9	System, Nominal	28 V	0.22 kW	9	10.7 A	12.9 A
10	System, Downrated	28 V	0.05 kW	6	2.1 A	3.0 A

)* EPF Power is shared with the 6 kW ISPR F3 and A3 outlets

Table 1. Electrical Parameters of the PDU Power Outlets

120 VOLT DC POWER DISTRIBUTION OF THE PDUs

In order to satisfy these switching and protection requirements two basic types of Solid State Power Controllers (SSPCs) are utilised within the Power Distribution Units, these being the specific designs for the 120 volt DC and 28 volt DC control functions.

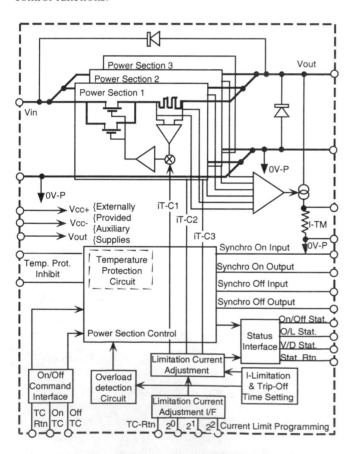

Figure 3. 120 V –15 A SSPC Block Diagram

For the 120 volt DC application as is shown in figure 3, all of the different outputs are designed around one single type SSPC, which however can be used in one of two derivatives, selectable by an external strap. These are :

A 15 A version with maximum current limitation of 18.9 A and trip-off time value higher than 1.5 milliseconds.

A 10 A version with maximum current limitation of 12.9 A and trip-off time value higher than 4.5 milliseconds.

Depending on the current rating of the specific output, up to four parallel connected, individual SSPCs, are used to satisfy the overall switching function.

Each 120 volt DC SSPC is comprised of six Rad-Hard, die size 6, Harris FRK 260 N-Channel MOSFETs and the use of discrete transistors for the three current control loops. Programmability of the output current limitation of the SSPC is possible in 8 steps using 3 numeric control inputs. An external power supply unit (PSU) is provided for the low level circuits of the SSPC, this PSU providing individual galvanically isolated supplies for up to five SSPCs. The basic SSPC also provides an analog telemetry output of its output current value.

28 VOLT DC POWER DISTRIBUTION OF THE PDUs

The 28 volt power supply system of the Power Distribution Unit is derived from the output of nine 300 watt, 120 volt to 28 volt buck type, DC to DC converters. The outputs of seven of these converters are directly utilised by the essential loads of the COLUMBUS APM. The power outputs from the additional two converters are further distributed to 28 volt Solid State Power Controllers (SSPCs), in order to obtain six 60 watt power feeds as shown in figure 4.

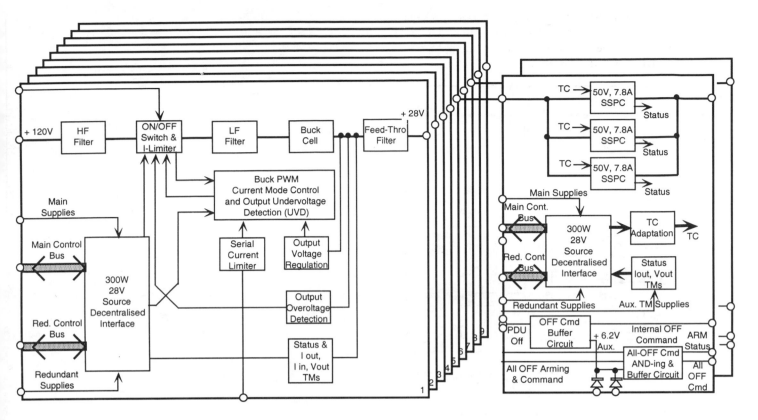

Figure 4. 28V Power Distribution System

The principle architecture of the 300 watt, 120 volt to 28 volt buck type, DC to DC converters is closely based on previous Alcatel Space equipment designs. The buck control circuits are powered by an external auxiliary supply generated by one of two 11 volt power supplies common to the nine power sources. These buck control circuits are referenced to the output of the 28 volt power source and transformer isolated commands are used to PWM control the buck switch.

Each 300 watt power source is basically comprised of :

- A high frequency input filter circuit

- An ON/OFF switch including I/P current limitation

- A low frequency input filter circuit

- A PWM buck switching circuit and associated inductor

- An O/P overvoltage detection circuit

- Analogue and status telemetry conditioning circuits

Additionally three specific protection features are applied to the converter design, these being :

- Current limitation which is provided by the inherent current control mode of the buck cell and is presettable in the range 12.9A to 16.7A. Whenever the output current exceeds a level of 11.3A ± 0.6A, the converter is switched off after a delay lasting between 34 to 64 milliseconds.

- Output overvoltage protection which is detected by appropriate circuitry in order to rapidly switch off the buck input current limiter and thus avoiding that a permanent overvoltage is imposed.

- Input current overload which is protected by the input current limiter of the converter circuit and is pressettable in the range 8.7A to 14.2A. Whenever the input current exceeds a level of 7.15A ± 0.45A, the converter is switched off after a delay lasting between 430 to 500 microseconds.

For the 28 volt DC, 60 watt 'down-rated' power outputs, the specific power control utilises one specific low voltage type of Solid State Power Controller (SSPC), this being based on an existing hybridised 50 volt/7.8 amp design.

As depicted in figure 5, each of these 28 volt DC SSPCs is comprised of a single P-Channel MOSFET, one current sensing resistor and control amplifier. Programmability of the output current limitation of the SSPC is possible by connection

279

of an appropriate resistor external to the hybrid package. Similarly programmability of the output current limit duration of the SSPC is possible by connection of an appropriate capacitor external to the hybrid package.

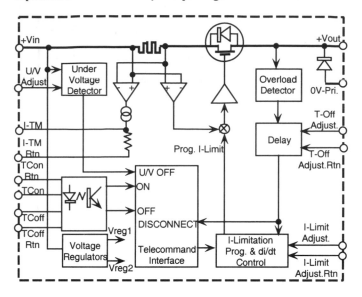

Figure 5. 50 V –7.8 A SSPC Block Diagram

CONCLUSIONS

The COLUMBUS Attached Pressurised Module and its Electrical Power Distribution System was subjected to a Critical Design Review at the end of 2000. Based on the development status as of that time and the few design recommendations proposed, the flight standard EPDS is on track to be integrated into the COLUMBUS APM Laboratory, which is scheduled to be attached to the International Space Station in 2004.

REFERENCES

Various COLUMBUS EPDS Project Documentation

Proceedings of IECEC'01
36th Intersociety Energy Conversion Engineering Conference
July 29 – August 2, 2001, Savannah, Georgia

IECEC2001-AT-48

DEVELOPMENT AND IMPLEMENTATION OF STABILITY REQUIREMENTS FOR THE INTERNATIONAL SPACE STATION ELECTRIC POWER SYSTEM

H. David Fassburg / Boeing Co.   Edward Gholdston / Boeing Co.   Alvin Mong / Boeing Co.

ABSTRACT

The International Space Station (ISS) primarily distributes 120-Volt dc power to individual loads. This paper describes the development and implementation of system- and box-level requirements that are used to specify small signal, large signal, and local stability. The principles used are applicable to the development of payloads for the ISS, and the development of distributed dc power systems in general. For the small signal stability criterion, a minimum gain and phase margin is based on the complex load and source impedances at the system interfaces. The concept of gain or phase separation is also described and related to gain and phase margin, providing means to specify stability with load and source impedance requirements using scalar values.

1.0 INTRODUCTION

A crucial factor in the design and implementation of any dc power network using switching converters is the stability of the system under all expected conditions of loading and transient perturbations. The power system of the ISS has several characteristics that drove the development of a set of impedance criteria for both the source converters, as well as the load converters. These criteria ensure a robust and stable system that accommodates significant variations in load connectivity over the life of the station. A description of the system and its characteristics follows.

2.0 DESCRIPTION OF THE POWER SYSTEM

2.1 Solar Arrays

The ultimate power source for the station is the sunlight, which is incident on the silicon solar arrays deployed as paired sets mounted on specialized solar power modules. The arrays are the largest ever flown in space, and have a combined surface area of over an acre. Each array wing consists of two thin blankets held under tension on each side of a central collapsible mast. The entire assembly turns on a "beta gimbal", which provides one axis of rotation for solar pointing. A second orthogonal axis of rotation is provided at the "alpha gimbal", where the entire solar power module connects to the rest of the truss structure of the station (see Fig. 1). These arrays each provide a total of 25 kW of power during the sunlit portion of the 90-minute, low-earth orbit of the station. Due to solar cell degradation over time, and different pointing angles, the optimum output voltage will vary between 160 and 140 Volts over the life of the station [1].

Figure 1. International Space Station

2.2 Battery Charge/Discharge Units (BCDUs) and Batteries

The first switching converters in the power string are the BCDUs, which regulate charging of the nickel-hydrogen (NiH2) batteries during insolation, and serve the dual role of discharge control and bus regulation during eclipse. The units are bi-directional, stepping the voltage down to the battery level of approximately 72 Volts when charging, and stepping the battery voltage up to the main bus voltage when discharging. Switching between charge and discharge is automatically triggered by the bus voltage setpoint, even during sunlight. If load demand exceeds the ability of the solar arrays to provide full power, the batteries will switch in to pick up the slack.

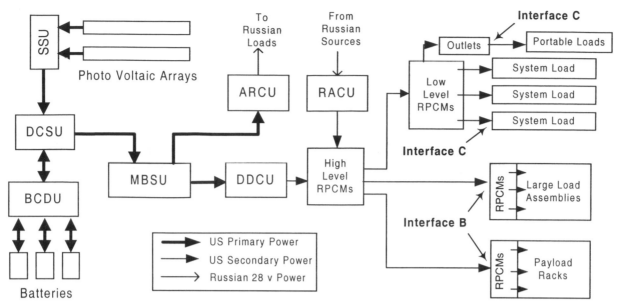

Figure 2. International Space Station Electric Power System

(This latter condition has to be only short-term, or the batteries will not be fully recharged at the end of the sunlit portion of the orbit.) The Battery Orbital Replacement Units (ORUs) are each rated at 81 Amp-hours and each contains 79 pressure cells connected in series.

2.3 Direct Current Switching Units (DCSUs) and Main Bus Switching Units (MBSUs)

These are remotely controlled switching boxes, used to route power between redundant channels, and to direct power to Dc-to-Dc Converter Units (DDCUs) and local loads.

2.4 Dc-to-Dc Converter Units (DDCUs)

The DDCUs provide the interface between the primary 160-Volt system and the more tightly regulated 120-Volt secondary system, which provides power directly to core system loads and to all payloads. The DDCUs provide excellent isolation between input and output ports, ensuring that perturbations on one branch of the primary system will have minimal impact on other branches. Downstream from each DDCU are Remote Power Controller Modules (RPCMs), which are computer-controlled, solid-state circuit breakers that branch out the power to the loads.

2.5 Russian-to-American Converter Units (RACUs) and American-to-Russian Converter Units (ARCUs)

These ORUs allow interconnection between the US-built 160/120-Volt system and the Russian-built 28-Volt system. Several of the Russian modules have their own independent solar arrays, providing important levels of redundancy for power availability. When power is being transferred between systems, these converters play a role in the overall bus impedance of both systems.

The top-level architecture of a single US power channel is shown in Fig. 2. (There is currently one Photovoltaic Module (PVM) on orbit with two such channels. When the station is fully assembled, there will be eight channels.) It shows the US-built primary and secondary systems, as well as the interface points, for transferring power to and from the Russian 28-Volt system. Because the entire station, and consequently the power system, is assembled in stages, a key criterion in the design is the flexibility to accommodate a wide variety of cable lengths and changing architecture. Virtually all payloads will be brought onboard the station long after it is in orbit, so it is not possible to test the entire system on the ground in advance, as it is with satellites and other spacecraft.

3.0 NOMENCLATURE

ARCU – American-to-Russian Converter Unit
BCDU -- Battery Charge/Discharge Unit
DDCU – Dc-to-Dc Converter Unit
DCSU -- Dc Switching Unit
Flight Element – Major module of the ISS
Insolation – Solar illuminated portion of an orbit
ISS – International Space Station
MBSU -- Main Bus Switching Unit
ORU – Orbital Replacement Unit
PV – Photovoltaic
RACU – Russian-to-American Converter Unit
RPCM -- Remote Power Controller Module
SSU – Sequential Shunt Unit

4.0 POWER SYSTEM STABILITY

To ensure a stable power system at each assembly stage of the station, an approach was developed that imposed requirements on the source and load impedances of the converter elements within the system [1]. The details of the approach are given in [1], but are summarized here as a preface to explain how specific impedance values (both magnitude and phase) were derived for interface points throughout the system. Any network can be divided into two series-connected sub-networks as shown in Fig. 3, where the source block and load blocks may consist of multiple converters connected in parallel. The stability and interface performance is governed by the ratio of source and load impedance, which can be considered the loop gain of the integrated system, $Z_S Z_L = T_m$.

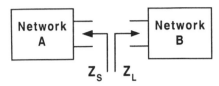

Figure 3. Two Series-Connected Networks

If $|Z_S| < |Z_L|$ for all frequencies, then a stable source and load ensures the stability of the entire connected system. For the ISS (and indeed most large distributed systems), it is not practical to require that $|Z_S| < |Z_L|$ for all interfaces and frequencies, so the Nyquist criterion was applied to deal with the typical case where the load and source impedances overlap. For a system to be stable, the plot of T_m must not encircle the (-1,0) point in the complex plane. To ensure that this does not occur over a reasonable range of system parameters, a level of margin was calculated that was developed into specific criteria on sources and loads. The bus impedance is defined as the parallel combination of Z_S and Z_L, as shown in Eq. 1, and was used to set the gain and phase margins for the ISS [1].

$$\vec{Z}_{bus} = \vec{Z}_S \parallel \vec{Z}_L = \frac{\vec{Z}_S}{1 + \vec{Z}_S / \vec{Z}_L} = \vec{Z}_S \left[\frac{1}{1 + \vec{T}_m} \right] \quad (1)$$

A plot of Z_{bus} showed that the resonant peaking at the cross-over points for Z_S and Z_L was acceptably low (i.e., ringing damped out quickly) when the gain margin was better than 3 dB and the phase margin better than 30 degrees. On a Nyquist plot, the phase margin is specified only at the cross-over points where T_m intersects the unit circle. The gain margin is specified only where T_m intersects the -Re axis. These three points define a T_m "forbidden zone", shown as the hatched area in Fig. 4. This then can be mapped onto a Bode plot of magnitude and phase versus frequency, as shown in Fig. 6, which can be used by independent designers to achieve stability at the interface points of sources and loads through control of the input or output impedance of their converters.

Using this concept and approach, the next step was to determine realistic boundaries based on the analysis of multiple loads with a known source (the DDCUs) to levy requirements on load and subsystem designers that would not be overly conservative or impractical to implement.

5.0 APPLICATION TO DISTRIBUTED SYSTEMS

The T_m "forbidden zone" becomes a constraint on loads when the complex source impedance has been defined at the load interface. In establishing this impedance, the effects of parallel loads must be included, as shown in simplified form in Fig. 5.

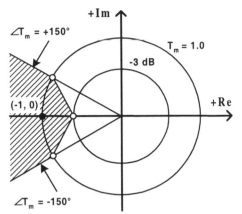

Figure 4. Gain and Phase Margin "Forbidden Zone" of Z_S / Z_L

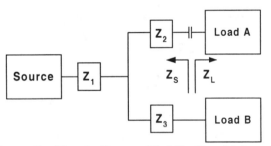

Figure 5. Simple Power Distribution Network

Figs. 6a and 6b show plots of the source impedance magnitude and phase that might be seen by a typical Load A on a 12-Ampere line 100 ft from the source due to two different Load B, and with no load at the Load B location. The variations in Z_S shown illustrate one reason why it is generally not possible to predict the load stability margins from the margins at the source. Thus, interaction between two or more loads can, in fact, result in mutual instability while the source remains stable. For this reason, it is helpful to think of stability across a complex power distribution system as being regional. To ensure stability in all regions of the ISS secondary power system, the 3-dB and 30-degree (or equivalent) criterion was applied at the output of the DDCUs, the interfaces between flight elements (major ISS modules), and at the load interfaces.

6.0 CHALLENGES IN APPLYING THE CRITERIA

In the above example, Load B itself may be time varying, or a utility outlet for portable loads or payloads. It becomes clear that the source impedance at the Load A interface can only be bounded. The criterion requires that the input impedance of Load A provide the minimum gain and phase margins with all possible (physically realizable) complex source impedance functions within the prescribed bounds on Z_S. At first, this appears to be a formidable task since the load characteristics are normally not known when the system source impedance boundaries need to be defined. However, the boundaries can, in fact, be determined if only the minimum gain and phase requirement is known. This point is illustrated in Section 8.0.

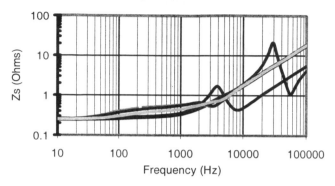

Figure 6a. Source Impedance Magnitude at Outlet A Due to Two Different Loads or No Load at Outlet B

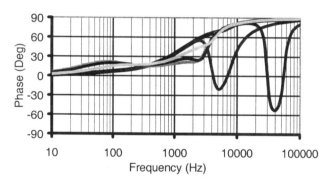

Figure 6b. Source Impedance Phase at Outlet A Due to Two Different Loads or No Load at Outlet B

Evaluation of the gain and phase margin between the complex impedance of specific loads and the boundaries of the source impedance would present yet another source of difficulty. In this case, the actual source impedance may only be known in terms of its boundaries or limits. The required margins on T_m must be between Z_L and the worst-case possible functions Z_S within the bounds, not the bounds on Z_S. Carrying this out would be a formidable task. It is much easier to instead evaluate stability of the interface using the comparable gain or phase (G/P) separation criterion.

7.0 GAIN OR PHASE SEPARATION

Fig. 4 shows the "forbidden zone" on the Nyquist diagram for a 3-dB and 30-degree minimum gain and phase (G-P) margin criterion. As noted, this zone can be approximated by another "forbidden zone" that is defined by the separations between the Bode magnitude and phase plots of the source and the load impedances, over all significant frequencies [2]. The gain separation (Gs) is defined as:

$$G_S = 20 \cdot Log|Z_L| - 20 \cdot |Z_S|$$ (2)

The positive and negative phase separations (Ps) are defined on the upper and lower load phase boundaries from the farthest source phase boundaries as:

$$P_S = \phi_L - \phi_S$$ (3)

Where:
$|Z_S|$ and $|Z_L|$ are the magnitude component of the source and load impedance functions, or limits/boundaries, and:
ϕ_S and ϕ_L are the phase components of the source and load impedance functions, or farthest limits/boundaries.

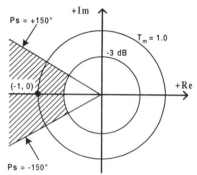

Figure 7. Gain or Phase Separation "Forbidden Zone" of Z_S / Z_L

Fig. 7 shows the (hatched) "forbidden zone" on the complex plane of T_m with boundaries at Gs = 3 dB ($|T_m|$ = 0.707) and Ps - ±150 deg (phase margin = 30 deg). The "forbidden zone" of Fig. 7 cannot be entered at any frequency unless both Gs < 3 dB AND |Ps| > 150 degrees are true. This is equivalent to requiring that EITHER the magnitude separation satisfies the inequality Gs ≥ 3 dB OR the phase separations satisfy the inequality |Ps| ≤ 150 degrees. Hence, this is called the gain or phase separation criterion. Note that the "forbidden zone" of Fig. 4 is contained within the forbidden zone of Fig. 7. So the (G-P) margin criterion is satisfied if the (G/P) separation criterion is satisfied at all significant frequencies, as discussed later in Section 15.0.

When the source impedance can, at best, be defined by the boundaries on its actual impedance function, the task of verifying stability is simplified if the load complex impedance can be shown to satisfy the gain or separation criterion with respect to the source impedance boundaries.

8.0 DEFINING SOURCE IMPEDANCE BOUNDARIES

The source impedance boundaries were determined by a series of system simulations using commercially available circuit simulators, comprising over 1,200 simulation runs. The primary objectives were to define the envelope of the source impedances that broad categories of loads would see wherever they might be located onboard the ISS. A secondary objective was to allow for more narrow categories that could minimize potential for impacting loads that had already been designed.

Loads were first categorized as complex (load assemblies and equipment racks), and simple (individual loads). The input to load assemblies and equipment racks is called Interface B. Interface B loads are required to have their own internal power distribution and protection for individual loads deriving power from their input. Power provided to individual loads is called Interface C power, and is defined at the input terminals of fixed loads and at the outlets where portable loads are connected. The Interface B category was subdivided on the basis of the system feeder ratings of 50-60, 25-30 and 10-12 Amps. The Interface C category was subdivided on the basis of the system branch line ratings of 25-30, 10-12, and 1.5–3.5 Amps. These six categories are now the basis for the standard source impedance limits for the ISS load interfaces. The limits define the boundaries for source impedances that will occur at all but the most remote loads from the DDCUs.

Overall cable lengths of 12 to 220 ft are encompassed by the above categories, and include virtually all ISS secondary power distribution. For the exceptions beyond 220 ft, impedance compensation is used to bring the source impedance at the load within the standard limit. By designing for stability with the standard limit, the load is expected to be stable at any location within its category. The standard limits are used for payloads, portable equipment, and general use loads such as lighting and system control computers.

Non-standard limits were developed for location-specific loads. These limits were developed to minimize risk of impact to loads intended for a few specific locations, but already had been designed. These categories were created by dividing the six standard categories into subcategories based on the length of the cables between the DDCU and the load. The four ranges of 14-25, 25-50, 50-100, and 100-220 ft in overall length have more narrow boundaries, compared to the standard limits, which provided significant benefit to existing fixed loads. The system source impedance was analyzed for each of the defined categories using circuit models synthesized to accurately represent the significant characteristics of the sources, cables, and switchgear.

It often seems desirable to perform this type of analysis using source models that attempt to approximate the actual output impedance magnitude and phase limits. However, this is frequently not possible with arbitrarily drawn limits. In general,

the output impedance of actual sources will often be within these limits by a wide margin at most frequencies. System source impedance boundaries generated from limit-based models would tend to exaggerate the predicted Z_S boundaries at load interfaces, with potential for undue impact on pre-existing loads or the evaluation of their stability. To avoid this problem, the simulation used source models that were in good agreement with actual DDCU test data at power levels from the 6.25-kW power rating down to as low as 70 Watts. Considerable attention was given to the way in which the effects of parallel loads were included in the analysis. Many factors could influence these effects, including the point along the line where the parallel loads are connected, the number and kinds of loads connected, as well as the characteristics of the individual loads. These effects were extensively analyzed.

One series of simulations investigated the crucial characteristics of the load that needed to be included in the final models. These simulations confirmed that the dominant eigenvalue created with the rest of the system was most significant to the overall analysis, and allowed simplifications of the load models used, as shown in Fig. 8.

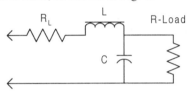

Figure 8. Typical Model for Load B

Next, a series of simulations that investigated the effects of parallel loads, assuming both similar (multiples of the same) and dissimilar loads. As expected, increasing the number of dissimilar loads reduced the effect of the individual loads. Increasing the number of similar loads did not significantly increase the effect when individual circuit component values were allowed to have expected tolerance variations around the nominal values. Overall, the simulations showed that a single parallel load adequately represents the worst-case effect upon the source impedance seen by another load. This result allowed simplification of the architecture used to represent the system in the analysis to determine the Z_S boundaries at the load interfaces. It should be noted that the worst-case parallel loads discussed here are significant because they can occur during system start-up or with light system loading.

The source impedance boundaries must also account for the worst-case effects of the system architecture itself. It was recognized that parallel loads would have a greater effect upon the source impedance seen by another load when connected through short branch lines to a common long feed line. Conversely, the impedance of the source itself would be much more dominant when the overall line lengths are short. The worst-case analysis had to account for these effects in the actual system. Fortunately, the locations of the DDCUs and RPCMs had already been established, so it was possible to estimate the cable lengths of the feeders and branch lines fairly accurately. This also allowed the identification of the worst cases in terms of where parallel loads needed to be located along a given overall cable length being analyzed in each category.

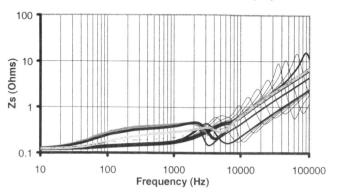

Figure 9a. Source Impedance Magnitude at Outlet A Due to Different Loads at Outlet B

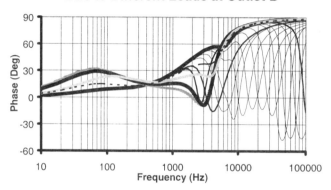

Figure 9b. Source Impedance Phase at Outlet A Due to Different Loads at Outlet B

To be realistic, load interface source boundaries must strike a balance between potential impacts to both the power distribution system and the loads. This meant that the boundaries defined by the analysis needed to have a very modest amount of pad around the expected variations in the source impedances due to loads that satisfy the 3-dB and 30-degree criterion. So, marginally acceptable parallel loads were assumed. The analysis to define the source impedance boundaries varied the synthesized parallel load parametrically in a way that maintained the phase margin at its interface six degrees outside the specified 30-degree phase margin limit. This constraint was imposed while varying the frequency of the created system eigenvalue in approximately half-octave intervals up to 100 kHz. The lower end of the eigenvalue sweep stopped at frequencies short of load circuit components that would be unreasonably large for a given size load. This analysis was repeated for cable lengths incremented in half-octave increments from 14 to 200 ft in each category. The effects of varying DDCU power levels were also included in simulations for each category and cable length. Figs. 9a and 9b illustrate the composite results of the simulations that were used

to define the boundaries for 12-Amp branch lines at individual loads 70 feet from the source.

9.0 SETTING THE SOURCE IMPEDANCE LIMITS

The process of drawing the source impedance location-specific limits was straightforward. Within each category, the impedance envelopes defined for each cable length were grouped by cable length range.

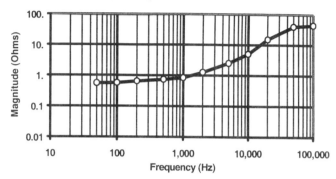

Figure 10a. Standard Source Impedance Upper Magnitude Limit for Loads on a 10- to 12-Amp Line

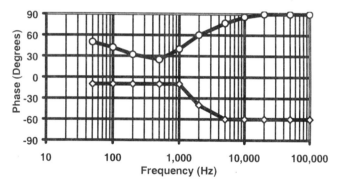

Figure 10b. Standard Source Impedance Outer Phase Limits for Loads on a 10- to 12-Amp Line

For each range, the impedance magnitude and phase boundaries are the maximum corresponding value produced by the simulations for the longest, shortest, and the geometric mean cable lengths within the cable length range. The source impedance magnitude (specification) limits were then drawn approximately 2 dB above the resulting values. The source impedance phase (specification) limits were drawn approximately 5 degrees outside the resulting phase boundaries. The standard (station-wide) limits are the envelope of these location-specific limits that apply to the Interface Type (B or C) and rating. Figs. 10a and 10b show the standard source impedance magnitude and phase limits that resulted for loads connected to branch lines rated for 10-12 Amp.

10.0 MINIMIZING IMPACT ON EXISTING LOADS

In applying these principles to the ISS, a number of issues were immediately apparent. Many loads had already been designed, and may not satisfy the gain or phase (G/P) separation criterion relative to the standard source impedance limits.

To avoid undue impact, the standard and location-specific impedance limits were incorporated into a system of four alternative stability criteria that could be used depending upon the circumstances. The source impedance at the standardized Interface B and C outlets are verified to be within the standard impedance limits. Hence, loads that operate from these outlets must show adequate G/P separations relative to the standard source impedance limits. These include payloads and portable equipment. However, applications of fixed loads used in a limited number of locations are allowed to satisfy any one of three additional alternative criteria. These are: a) demonstrate required G/P separations with the appropriate location-specific limit, b) demonstrate required gain and phase (G-P) margin with the actual system source impedances and variations thereof, and c) demonstrate that the load damps the interface response to injected transients to less than 10% of maximum within 1 millisecond, using a specified test method.

The alternative criteria have avoided impacts to several ISS system loads. At the same time, no instances of instability have been detected during integration tests or on orbit to date.

11.0 EVALUATING INTERFACE STABILITY

As noted earlier, the load impedances reflected to the interface must maintain a 3-dB and 30-degree minimum gain and phase margin with all complex impedance functions that are possible within the defined limits. However, the task of verifying these margins between yet-to-be-defined loads and time-varying source impedances can be formidable. Because of this, it is often not practical to evaluate the actual system gain and phase margins when verifying system stability. The task of verifying the system is simplified significantly when the impedances on either, or both, side(s) of the interfaces are shown, or can be assumed to be within defined boundaries. The G/P separation criterion, or similar criteria [2], allows evaluation of stability margins when one or both sides of the interface are defined only by the scalar boundary values for the impedance magnitude and phase.

Using this criterion, load stability verification no longer needs to consider the actual complex impedance functions possible within the specified source impedance boundary. Instead, it is sufficient to verify that the G/P separation between the load complex impedance (or the boundaries on its magnitude and phase) satisfy the criterion with respect to the boundary values on the source impedance Bode plots at all specified frequencies. As noted in Section 7.0, the G/P "forbidden zone" is a little larger than the "forbidden zone" defined by the comparable gain and phase margin criterion. Hence, applicability of the alternative criteria discussed in Section 10.0 should always be considered if the interface does not satisfy the G/P separation criterion.

12.0 DEVELOPMENT OF LOAD IMPEDANCE LIMITS

As noted in Section 7.0, a 3-dB and 30-degree minimum gain and phase margin is ensured when the impedance of the loads satisfy the 3-dB (minimum) or 150-degree (maximum) gain or phase (G/P) separation.

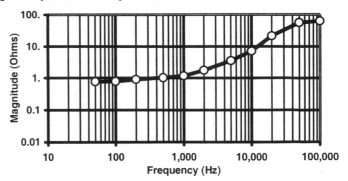

Figure 11a. Standard Load Impedance Lower Magnitude Limit for Loads on a 10- to 12-Amp Line

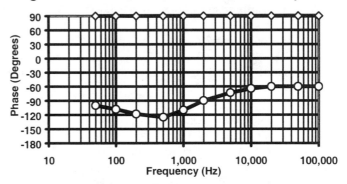

Figure 11b. Standard Load Impedance Outer Phase Limits for Loads on a 10-to 12-Amp Line

With these criteria, the boundaries for the load impedance can be defined once the boundaries on the source impedance have been established. The lower load magnitude limit is set at 3 dB above the upper magnitude limit of the source impedance. The upper limit for positive load impedance phase is set 150 degrees above the lower limit for negative source impedance phase. Similarly, the lower limit for negative load impedance phase is set at 150 degrees below the upper limit for positive source impedance phase.

For example, Figs. 11a and 11b show the load impedance requirement for portable loads connected to the ISS 120-Volt 12-Amp outlets. The G/P separation criterion requires that the load impedance phase be between the upper and lower phase limits at any frequency where the impedance magnitude is below its limit. This method is different from, but equivalent to, the method described by Wildrick, et. al [2].

13.0 LOCAL STABILITY CRITERIA

As noted in Section 4.0, the interface stability criteria can ensure that otherwise stable sources and loads remain stable when integrated into a larger system. These methods are not a substitute or criteria for good design for local (stand-alone) stability of sources or loads. At the same time, it was recognized that loads must be highly stable locally to meet the ISS electromagnetic compatibility (EMC) conducted emissions and susceptibility requirements (which are similar to those of MIL-STD-461). All ISS loads must have sufficient local stability to maintain performance during the EMC conducted susceptibility tests, while meeting the EMC conducted emission requirements. The principles described by R. D. Middlebrook [3] are important in meeting these requirements.

14.0 LARGE SIGNAL STABILITY

It was recognized that the small signal criteria discussed so far have limited effect when significant non-linearities are encountered. Large signal stability is addressed in two ways in the ISS secondary power system.

The first is a test requirement involving the injection of ± 15-Volt, 0.5-millisecond pulses on the positive input power line allied to loads. The source impedance is that of a 100-ft line, or an equivalent Line Impedance Simulation Network (LISN). The load is required to maintain function, and resulting transients at the interface are required to damp to 10% of the maximum value within 1 millisecond.

The second large signal stability criterion requires that loads power up correctly when energized by the 1- to 10-millisecond RPCM soft-start function. During tests, several loads needed modifications to start and initialize correctly when powered up by this soft-start function.

The two criteria cover the largest input variations that can occur under normal system operations. Larger variations are associated with abnormal system conditions, so there was no need to include these in the large signal stability requirement. These conditions are associated with action of the protective switchgear, such as RPCMs, which were extensively tested for a wide range of abnormal/fault conditions. In general, the ISS switchgear will remove power from loads when faults, current overload, undervoltage, or unstable load conditions are detected.

15.0 FREQUENCY RANGE OF SMALL SIGNAL STABILITY REQUIREMENT

On the ISS, the gain/phase criteria are specified from 50 Hz to 100 kHz. The lower end is sufficiently low to detect impedance conditions with potential for a stability problem, but avoids instrumentation problems that are common below 50 Hz. Most loads have some kind of regulation and appear as a constant power load to the bus. In these cases, the phase component of their input impedance will be between -180 and -90 degrees at frequencies that lie within their inner loop unity gain bandwidth. The gain and phase margin ensures that stable loads will remain stable with the system over this band of

frequencies where the load is able to contribute negative real impedance to the interface.

The unity gain bandwidth of the regulator within loads is dependent upon its design, and normally cannot be predicted by the system integrator. The gain and phase margin interface criterion is applied up to 100 kHz to ensure both that the inner loop bandwidth of all loads is covered, and to control bus impedance resonances. It should be noted that bus impedance resonances can give rise to excessive distribution system ripple and noise when they coincide with the frequencies of significant electromagnetic interference (EMI) emissions. Above 100 kHz, bus ripple and noise are essentially controlled by the loads, which must satisfy the EMC conducted emission requirements.

16.0 SUMMARY AND CONCLUSIONS

A stability approach has been developed and implemented, which is applicable for the ISS and other large distributed dc power systems. A system of alternative stability criteria was developed to avoid undue impact on limited-use loads, while ensuring overall stability of the system. A system of load impedance requirements were developed that, when satisfied, ensure stability of the loads with ISS. The requirements have worked well with minimal impact to program costs and schedules. Ripple, noise, and transient responses of power distributed from the US sources has been a fraction of the specified limits, and no instances of instability have been seen to date during integration testing, or in the on-orbit EPS.

17.0 REFERENCES

[1] E. W. Gholdston, K. Karimi, F. C. Lee, B. Manners, "Stability of Large Dc Power Systems Using Switching Converters, With Application to the International Space Station", IECEC, August 11-16, 1996.

[2] C. M. Wildrick, F. C. Lee, B. H. Cho, B. Choi, "A Method of Defining Load Impedance Specification for a Stable Distributed Power System", IEEE Transactions on Power Electronics, May 1995.

[3] R. D. Middlebrook, "Input Filter Considerations in Design and Application of Switching Regulators", Proceedings of the IEEE Industry Applications Society Annual Meeting, 1976 Record, pp. 366-382, Chicago, IL, October 11-14, 1976.

Proceedings of IECEC'01
36th Intersociety Energy Conversion Engineering Conference
July 29 – August 2, 2001, Savannah, Georgia

IECEC2001-AT-51

INTERNATIONAL SPACE STATION ELECTRIC POWER SYSTEM STABILITY VERIFICATION AND ANALYSIS USING COMPUTER MODELS AND TEST DATA

Shen Wang / Boeing Co.   Hrair Aintablian / Boeing Co.   Erich Soendker / Boeing Co.

ABSTRACT

The International Space Station (ISS) Electric Power System (EPS) is the world's largest orbiting direct-current (DC) power system. ISS electric power is generated by solar arrays, and distributed to utility loads and payloads in two different voltage ratings -- 120 and 28 Volts. To ensure power quality of the distribution system, it is very important to maintain the EPS stability. Technical requirements are specified at interfaces between adjacent modules and/or within each module. Element-level and system-level tests are performed for stability verification prior to each assembly sequence. Since not all operational scenarios can be tested on the ground, computer simulation is necessary in analyzing the EPS stability. This paper presents a framework for EPS computer model construction, validation, and system stability analyses. Several examples from the EPS verification are included to show how a model is validated and system stability is analyzed. Loads impact the stability of the EPS in different ways depending on their types. Constant power loads have a destabilizing effect, while constant resistance loads improve stability margins. A theoretical explanation of small signal stability of a DC power distribution is presented. The impacts of constant power loads, constant resistance loads, and passive filter components on the system stability are assessed. Furthermore, the theoretical evaluation of stability is verified by simulation.

1.0 INTRODUCTION

The ISS EPS is the world's largest orbiting DC power system. ISS electric power is generated by solar arrays, regulated by Sequential Shunt Units (SSUs), stored in batteries, converted to secondary systems in 120 and 28 Volts, and distributed through Remote Power Controller Modules (RPCMs) to utility loads and payloads [1].

To ensure power quality of the distribution system, it is very important to maintain the EPS stability. Technical requirements are specified at interfaces between adjacent modules and/or within each module. Element-level and system-level tests are performed for stability verification prior to each assembly sequence. Since not all operational scenarios can be

tested on the ground, computer simulation is necessary in analyzing the EPS stability.

This paper presents a framework for ISS EPS computer model construction, model validation, and system stability analyses. Several examples from the EPS verification are included to show how a model is validated and system stability is analyzed. The impacts of constant power loads, constant resistance loads, and passive filter components on the system stability are also assessed through theoretical analysis and computer simulation.

2.0 COMPUTER MODEL CONSTRUCTION

The ISS EPS computer models are constructed at different levels. All systems are broken down into source Orbital Replacement Units (ORUs), load ORUs, switchgear, and cables. For example, a typical system consists of many ORUs such as Dc-to-Dc Converter Units (DDCUs) and RPCMs to provide switching and fault protection. The RPCM is broken down further into sub-units such as individual Remote Power Controllers (RPCs), power supplies, and cables. The computer models are constructed on the sub-unit, ORU, and system levels. The unit/sub-unit models are integrated into a system-level model according to its schematic or architecture. Although the models are categorized at different levels, they follow the same modeling principle described below:

1. Models are based on schematics or architecture.
2. Some models are based on test data when schematics are not available to protect manufacturer's proprietary information.
3. Since load filters dominate the system response to the load, complex loads may be represented, depending on the application, as passive input filters connected to power converters and output filters, or as input filters and constant power loads.
4. While most heaters are modeled as resistors, heaters with thermal controllers are modeled to include their input filters and controllers.

3.0 MODEL VALIDATION

Before stability analyses are performed, the ORU and system models must be anchored to test data. In the ISS program, the following tests have been conducted:

1. Electrical Power Consuming Equipment (EPCE) Characterization Test. The purpose of this test is to check the electrical characteristics of each element manufactured by Boeing or its subcontractors. The characteristics include load impedance, inrush current, step response, etc. This test is conducted either at Boeing or at the manufacturer's facility by engineers from Boeing and the manufacturer.

2. Space Station Module Power Quality Test. This is a system-level test used to verify the electrical characteristics of each ISS module. It is conducted either at Boeing or NASA facilities by Boeing and NASA engineers.

3. Multiple Element Integration Test (MEIT). The purpose of this test is to ensure functionality and compatibility among different ISS modules before orbit assembly. The test is conducted at NASA Kennedy Space Center (KSC) by Boeing and NASA engineers.

These tests involve the measurement of impedance magnitudes and phases, voltage and current transients due to load changes, system recovery from simulated faults, electromagnetic emission, and susceptibility checkout [2]. The test data have been used for the validation of element and system computer models [3]. The computer-simulated results must be in agreement with the test results before the models can be used to analyze system stability for more complicated operational scenarios that can not be tested on the ground.

4.0 SYSTEM STABILITY ANALYSES

The system stability analyses consist of three types of simulations: small signal analysis, large signal analysis (or transient analysis), and fault analysis.

4.1 Small Signal Analysis

During small signal analysis, the system stability is checked under small perturbations to the system. The perturbing signals have small magnitudes while sweeping the frequencies from 10 Hz to 100 kHz. This analysis checks the source and load impedances and computes stability margins at different interfaces [1]. The minimum requirements for stability margins are defined by ISS as 3-dB gain margin and 30° phase margin. There are three major steps in performing the small signal analysis:

1. Document requirements for source impedance and stability margins.
2. Validate the system models with impedance test data.
3. Generate source and load impedances, compare them with the impedance requirement envelopes, and calculate gain and phase margins.

Fig. 1 shows a sample of the ISS Destiny Module model validation for small signal analysis. The computer results have

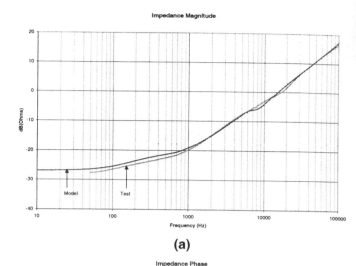

(a)

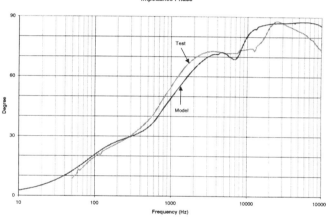

(b)

Figure 1. Destiny Module Model Validation for Impedance Magnitude and Phase

a strong correlation to the test data, therefore, anchoring the model.

Fig. 2 shows part of the small signal analysis results for the Destiny Module. Fig. 2(a) displays the magnitudes of the source impedance (ZS) and load impedance (ZL) at one of the Destiny Module interfaces. Fig. 2(b) is the corresponding Nyquist plot, showing a 62° phase margin.

4.2 Large Signal Analysis

The purpose of large signal analysis (or transient analysis) is to verify the system stability during a load step change (i.e., when a load is turned on or off) to ensure that transients damp out or decay within a specified time. During the load step change, the load inrush current should stay within the RPC trip threshold rating, and voltages at different interfaces must remain within specified envelopes to guarantee the system power quality. Three major steps are involved in performing the large signal analysis:

1. Document requirements for step load changes (increase and decrease).

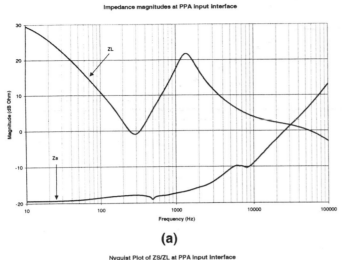

Impedance magnitudes at PPA input interface

(a)

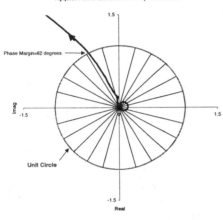

Nyquist Plot of ZS/ZL at PPA input interface

(b)
Figure 2. Destiny Module Small Signal Analysis

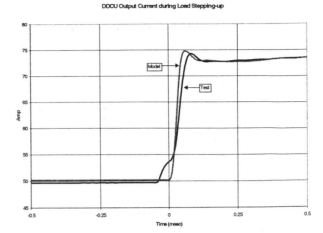

DDCU Output Current during Load Stepping-up

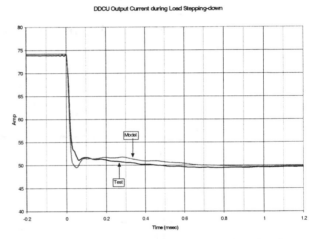

DDCU Output Current during Load Stepping-down

Figure 3. Destiny Module Model Validation for Large Signal Analysis

2. Validate system models with transient current and voltage test data.
3. Display voltage and current transients, and compare them with the requirement envelopes.

Fig. 3 shows a sample of the ISS Destiny Module model validation for large signal analysis. Fig. 4 shows part of the Destiny Module large signal analysis results, demonstrating that voltage transients fall within the requirement envelopes.

4.3 Fault Analysis

The purpose of the fault analysis is to check the system response when the system is faulted by hard or soft faults on a system bus or at a load (such as an electrical short at load input terminal). The system must demonstrate the capability of a DDCU or RPCM to isolate the fault and show that all bus voltages can recover to the specified level within a specified time. There are three major steps to follow in performing the fault analysis:

1. Document requirements for fault clearing and protection.
2. Validate models with current and voltage test data.

3. Apply faults at different interfaces, display voltage, and current transients, and compare them with the requirement envelopes

Fig. 5 shows a sample of the Z1 Module fault analysis results. The DDCU output voltage recovers to its normal range after its downstream RPC isolates the fault from the system.

5.0 STABILITY ASSESSMENT OF A DC DISTRIBUTION SYSTEM

In DC power systems, switching power converters, which are constant-power loads, play important roles. These loads exhibit negative resistance characteristics that tend to destabilize the system [4]. Constant-resistance loads (heaters) improve system stability. From system design and analysis perspectives, it is important to evaluate the stability of the system.

5.1 Theoretical Analysis

The equivalent circuit of a DC system with source and load subsystems is shown in Fig. 6. The parameters L_S, R_S, and C_S

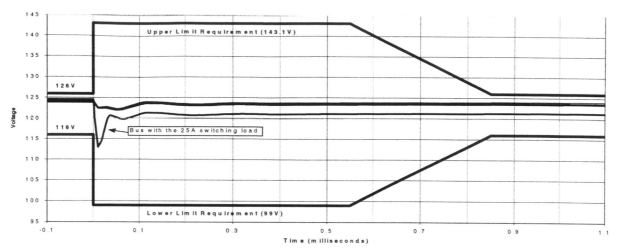

Figure 4. Destiny Module Large Signal Analysis

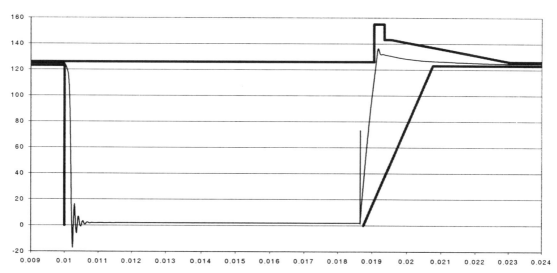

Figure 5. Z1 Module Fault Analysis

represent the source output filter and the parameters L_F, R_F, and C_F represent the input filter of the load. P_{CP} represents the constant power of the load. To simplify the theoretical analysis, a simplified circuit as shown in Fig. 7 is analyzed.

Applying Kirchoff's laws to Fig. 7 yields:

$$i = C_{eq}\frac{dv_O}{dt} + \frac{P_{cp}}{v_O} + \frac{v_O}{R_L} \tag{1}$$

$$v_S = L_{eq}\frac{di}{dt} + R_{eq}i + v_O \tag{2}$$

Performing Laplace Transformation with approximation leads to:

$$\frac{v_O}{v_S} = \frac{\dfrac{1}{L_{eq}C_{eq}}}{S^2 + \left[\dfrac{R_{eq}}{L_{eq}} + \dfrac{1}{C_{eq}}\left(\dfrac{1}{R_L} - \dfrac{P_{cp}}{V_O^2}\right)\right]S + \dfrac{1}{L_{eq}C_{eq}}\left(1 + \dfrac{R_{eq}}{R_L} + \dfrac{R_{eq}P_{cp}}{V_O^2}\right)} \tag{3}$$

For the system to be stable, the poles of Eq. (3) must be in the left-half plane, i.e., the coefficient of S and the constant in the denominator must be positive. The stability condition leads to:

$$\frac{R_{eq}}{L_{eq}}C_{eq}V_O^2 + \frac{V_O^2}{R_L} - P_{cp} > 0 \tag{4}$$

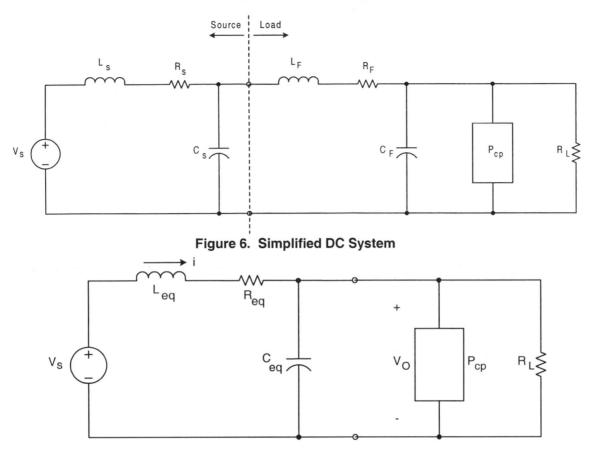

Figure 6. Simplified DC System

Figure 7. Equivalent Circuit

There are six parameters in Eq. (4) that affect system stability. Increasing C_{eq}, R_{eq}, or V_O and decreasing L_{eq}, P_{CP}, or R_L will make the system more stable.

5.2 Verification by Simulation

A hypothetical system is simulated in SABER (circuit simulation tool) to confirm the theoretical assessment for stability. The parameters of this system are as follows: R_S=100 mohms, L_S=2 uH, C_S= 5 uF, R_F=100 mohms, L_F=50 uH, C_F=100 uF, P_{CP}=1000 Watts, and R_L= ∞. Fig. 8 compares the source and load impedance magnitudes of the circuit of Fig. 6 for the hypothetical system. Fig. 9 displays the source and load impedance phases.

Since the magnitudes of the source and load impedances cross, the Nyquist criterion is applied. Fig. 10 shows the Nyquist plots for different values of R_F. Increasing R_F results in greater phase margins. Figs. 11, 12, 13, and 14 show the Nyquist plots when C_F, L_F, P_{CP}, and V_S are varied (Vo is changed by varying Vs). These results confirm the theoretical assessment of a representative DC system.

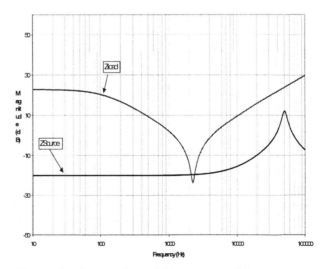

Figure 8: Source/Load Impedance Magnitudes

293

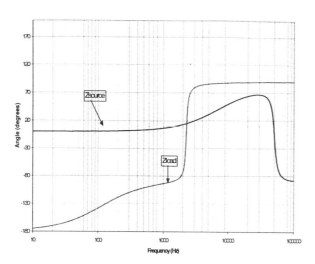

Figure 9: Source/Load Impedance Phases

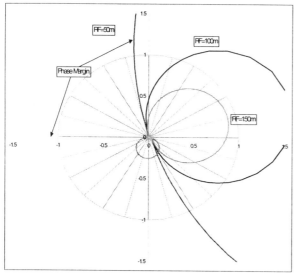

Figure 10: Nyquist Plots for Varying R_F

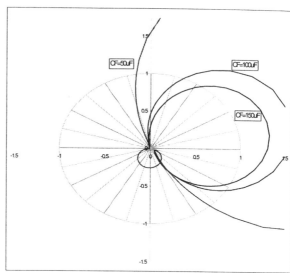

Figure 11: Nyquist Plots for Varying C_F

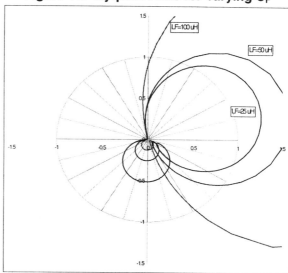

Figure 12: Nyquist Plots for Varying L_F

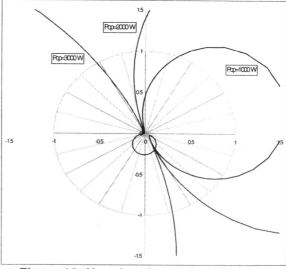

Figure 13: Nyquist plots for varying P_CP

6.0 CONCLUSIONS

The modeling and analysis processes described in this paper are used by Boeing for the ISS EPS verification. Simulations of several orbital assemblies have been completed successfully. The results of these analyzes have allowed violations of requirements and other areas of risk to be highlighted. Recommendations have been made to correct the violations and reduce the risk of instability.

The ISS EPS is considered to be large and complex. The system is partitioned into subsystems, which are analyzed for stability. The design of the system includes ample margin for stability. The source output filters and the load input filters improve system stability. Heater loads and high bus voltage provide added stability margin.

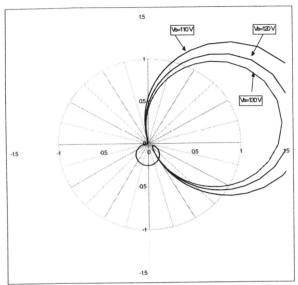

Figure 14: Nyquist Plots for Varying V_S

7.0 REFERENCES

[1] Gholdston, E. W., Karimi, K., Lee, F. C., Manners, B., "Stability of Large DC Power Systems Using Switching Converters, with Applications to the International Space Station," IECEC 1996, pp. 166-171.

[2] Wilde, R. K., Aintablian, H. O., Gholdston, E. W., "International Space Station U. S. Major Element Laboratory Power Quality Test," IECEC 1999, Paper No. 1999-0102433.

[3] Aintablian, H. O., Wang, S., Silva, C. D., "Simulation and Analysis of the International Space Station U. S. Laboratory Electric Power System," IECEC 2000 Paper No. AIAA-2000-2952.

[4] Emadi, A., Ehsani, M., "Negative Impedance Stabilizing Controls For PWM DC/DC Converters Using Feedback Linearization Techniques," IECEC 2000, Paper No. AIAA-2000-2912.

IECEC2001-AT-52

INTERNATIONAL SPACE STATION NICKEL-HYDROGEN BATTERY START-UP AND INITIAL PERFORMANCE

Fred Cohen/Boeing Co.   Penni J. Dalton/NASA Glenn Research Center

ABSTRACT

International Space Station (ISS) Electric Power System (EPS) utilizes Nickel-Hydrogen (Ni-H$_2$) batteries as part of its power system to store electrical energy. The batteries are charged during insolation and discharged during eclipse. The batteries are designed to operate at a 35% depth of discharge (DOD) maximum during normal operation.

Thirty eight individual pressure vessel (IPV) Ni-H$_2$ battery cells are series-connected and packaged in an Orbital Replacement Unit (ORU). Two ORUs are series-connected utilizing a total of 76 cells, to form one battery. The ISS is the first application for low earth orbit (LEO) cycling of this quantity of series-connected cells.

The P6 Integrated Equipment Assembly (IEA) containing the initial ISS high-power components was successfully launched on November 30, 2000. The IEA contains 12 Battery Subassembly ORUs (6 batteries) that provide station power during eclipse periods. This paper will describe the battery hardware configuration, operation, and role in providing power to the main power system of the ISS. We will also discuss initial battery start-up and performance data.

1.0 INTRODUCTION

At Assembly Complete, the ISS EPS will be powered by 24 batteries during eclipse and extended operation periods. The battery (see Fig. 1) is designed to operate for 6.5 years with a mean-time-between-failure (MTBF) of 5 years when run in the reference design 35% DOD LEO regime. Typical expected discharge currents can range from <25 Amps in a low-demand orbit to as high as ~75 Amps to meet short peaking load requirements at a battery operating voltage range of 76 to 123 Vdc. The ORUs are individually fused to protect the ISS EPS from fault propagation that could result from a cell-to-EPS ground event. Primary charge control is accomplished by a pressure temperature algorithm that incorporates acceptance test data to initialize basic reference parameters.

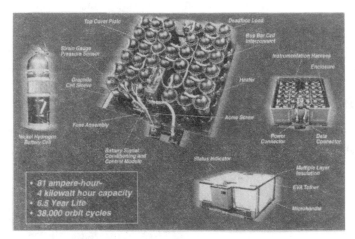

Figure 1. ISS Battery Subassembly ORU

Table 1. Reference Orbit Design Parameters Per Battery Subassembly ORU

Condition	Time (min) Start	End	Energy (Watt-hrs)	Power (Watts)
CONTINUOUS POWER REQUIREMENTS				
Constant Power Charge	0.0	43.9		1995*
Taper Charge	43.9	57.0		
Total Charge			1677*	
Constant Power Discharge	57.0	92.0	1342	2300
PEAKING POWER REQUIREMENTS				
Constant Power Charge	0.0	7.5		1554*
Constant Power Charge	7.5	43.9		2072*
Taper Charge	43.9	57.0		
Total Charge			1677*	
Constant Power Discharge	57.0	84.5	967	2110
Constant Power Discharge	84.5	92	375	3000
Total Discharge			1342	
CONTINGENCY POWER REQUIREMENTS				
Constant Power Discharge	0.0	92.0	997*	650
*Designates a maximum value				

The ISS power system is the first on-orbit use of such a large quantity of series-connected IPV Ni-H$_2$ battery cells (38/76), in an ORU/Battery configuration. Previous ground testing had been performed on 22 IPV NiH_2 cells in series [1]. Therefore, during the ISS program development stage, it was important to demonstrate that the "as-designed" battery could be successfully run (see Table 1). This was accomplished at the Power Systems Facility (PSF) Laboratory at NASA Glenn (then NASA Lewis) Research Center in Cleveland, Ohio in 1992 [2].

2.0 BACKGROUND: INITIAL BATTERY PERFORMANCE TEST SUMMARY

Two Space Station Engineering Model (EM) ORUs were initially tested using an orbital rate capacity (ORC) test, as well as individually LEO cycled at the 35% DOD reference orbit to provide baseline characteristics. After completion of the baseline testing, the hardware was configured as a "battery" by connecting them in series and subsequently running them for 3,000 simulated ISS reference design cycles at a recharge ratio (RR) of 1.043 (as described in Table 2). The ISS design power requirements are specified in units of Watts and, therefore, the cycle regime is power based. The 3,000 "peaking" cycles (see Tables 1 and 2) were performed using the maximum discharge power delivery requirement and a recharge regime that incorporates a taper charge that reduces charging stress at high states of charge (SOCs). The test was performed while maintaining the cell sleeve temperatures at $5 \pm 5°C$.

Table 2. ISS Simulated Peaking Reference Design Orbit, ≤35% DOD, 1.043 RR

Charge	57.0 Minutes (total)
3,108 Watts	7.5 minutes
3,746 Watts	36.4 minutes
3,746 taper to 700 Watts	13.1 minutes
Discharge	35.0 Minutes (total)
4,220 Watts	27.5 minutes
6,000 Watts	7.5 minutes

Following completion of 3,000 cycles, the ORUs were subjected to individual orbital rate capacity tests to determine any degradation in performance.

The result is that the ORUs exceeded the ISS design requirements for electrical performance, heat generation, thermal uniformity, and charge management.

3.0 ORU DESIGN CONSIDERATIONS

Remembering that the original ISS battery design effort began in 1988, a long-life, high-performance battery was needed. Therefore, state-of-the-art $Ni-H_2$ IPV chemistry was chosen at that time, and designed to meet the following ORU requirements:

- 6.5-year design life
- 81-Amp-hr nameplate capacity to limit the maximum reference DOD to less than 35%
- Contingency orbit capability consisting of one additional orbit at reduced power after a 35% DOD without recharge
- 5-year MTBF
- Easy on-orbit replacement using the ISS robotic interface

The cells selected for use in the Battery ORUs are manufactured by Eagle Picher Industries. The cells are RNH-81-5 EPI IPV NiH_2, and utilize a back-to-back plate configuration. They are activated with 31% potassium hydroxide (KOH) electrolytes. The ORUs are assembled and acceptance tested by Space Systems/Loral.

4.0 ISS BATTERY CONFIGURATION

The Battery Subassembly ORU, as designed and built, is pictured below in Figs. 2 and 3.

The NiH_2 cells for the current 12 ISS Battery ORUs were manufactured 3.5 to 4.4 years before the November 30, 2000 launch date. The flight ORUs were used for IEA systems ground testing and final checkout, but were stored open-circuit, discharged, and at –10 °C when they were not in use.

The 12 Battery ORUs were integrated onto the P6 IEA in July 2000 at the Kennedy Space Center (KSC). Two ORUs in series form one battery, for a total of 76 cells in series. These 12 ORUs form six separate batteries, with three batteries on each of two power channels. For the P6, these power channels are designated as 2B and 4B. During insolation, power is supplied to the source bus by solar arrays that meet the demand for user loads, as well as battery recharging. The batteries, through a Battery Charge/Discharge Unit (BCDU), provide the power to the source bus for the ISS during eclipse periods.

Each ORU contains a Battery Signal Conditioning and Control Module (BSCCM). The BSCCM provides conditioned battery monitoring signals from the ORU to the Local Data Interface (LDI) located within the BCDU. Available data includes 38 cell voltages, four pressure (strain gauge) readings, six cell and three baseplate temperatures and are provided as an analog multiplexed voltage. A separate signal provides ORU total voltage output. The BSCCM also accepts and executes commands from the BCDU/LDI to control ORU cell heater and letdown functions.

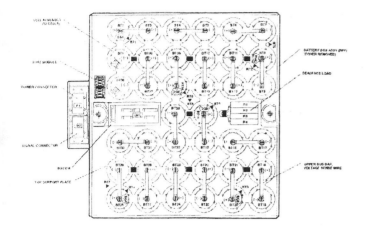

Figure 2. Baseplate Layout - ISS Battery Subassembly ORU

For battery charging, the BCDU conditions power from the source bus and charges the battery at charge setpoints as calculated from the charge algorithm (reference paragraph 6.0). During periods of eclipse, the BCDU extracts power from the battery, conditions this power, and supplies power to the source bus.

Figure 3. ISS Flight Model Battery Subassembly ORU with Cover Removed

The batteries are actively cooled using the ISS Thermal Control System (TCS). The battery cells are assembled in an ORU box, using a unique finned radiant heat exchanger baseplate. The baseplate is then mounted on the IEA using ACME screws and mated to the TCS. The TCS was designed to maintain the Battery ORUs at a nominal operating temperature range of 5 ± 5°C (41 ± 9°F) with minimum heater operation when run at a 35% DOD LEO regime.

5.0 ISS ON-ORBIT START-UP

The ISS batteries are launched in a discharged state. As a result a multi-orbit start-up was necessary to begin orbital operation. Battery charging was not begun until after solar array deployment and thermal conditioning. System control and operational power was supplied by the National Space Transportation System (NSTS) Auxiliary Power Control Unit (APCU). As a result of the limited capability of this power source and the desire to quickly charge the batteries to 100% SOC, heater operation and battery discharge were inhibited during eclipse.

After thermal conditioning, which consisted of warming the ORUs using their internal heaters to nominal operating temperature (between 0 and 10°C), battery charging was initiated using an initial low-rate charge of ~10 Amps. This continued until they reached a voltage of 76 Volts (1 Volt per cell average), and was followed by three consecutive insolation periods of charging at 30 Amps. Charging was completed during the 4th insolation period using a programmed taper charge. This start-up regime charged the batteries to 100% SOC with a total input of 103 Amp-hrs. Nominal operations were subsequently initiated and battery charge control was provided by the temperature-pressure algorithm.

At beginning of life (BOL), total capacity of the ISS P6 batteries was measured at KSC during IEA final electrical checkout. The battery total capacities during final IEA checkout ranged from 83.0 to 89.9 Amp-hrs when charged using the ISS charge algorithm.

6.0 ISS CHARGE ALGORITHM

The temperature-pressure charge algorithm provides a low-stress charge profile that allows the initial charge current to reach a pre-set maximum and then "tapers" (reduces current) at a rate that is SOC dependent. This profile is designed to maximize the use of available array power, reduce charging stress, and minimize ORU heat generation.

Charge control of this type is necessary in order to ensure orbit-to-orbit energy balance, since power to recharge the batteries varies due to a combination of seasonal orbit conditions:

- User loads
- Extravehicular activity (EVA) operations
- ISS operational scenario (i.e., locked, or non-sun-tracking array mode)

BOL battery 100% SOC is user set at nameplate capacity (81 Amp-hrs). The charge algorithm calculates SOC using a VanDerWaal's equation and a pressure vs. SOC relationship. Basic or initial parameters taken from battery acceptance data are used to initialize the system before flight. These parameters include strain gauge calibration, initial moles of H2, and pounds per square inch (PSI) per Amp-hr. During LEO operation, the point of recharge where charge efficiency begins to noticeably fall off is 94%. It is at this point where charge current reduction ("taper") begins.

7.0 ISS ON-ORBIT OPERATION

The ISS main power system charge algorithm has pre-set parameters. Maximum charge rate is determined and set based on the on-orbit operation need. Currently, a 50-Amp maximum charge rate setpoint is employed due to operating scenarios that feather arrays to save fuel and/or reduce the possibility of charge build-up on the ISS structure during EVA activity. As such, it is necessary to replenish the battery energy used during eclipse as quickly as possible when it is available from the solar arrays. The taper charge profile is pre-programmed in a look-up table with the following parameters:

SOC%	20	85	90	94	96	98	1.00	1.01	>1.05
Chg Rate (Amps)	50	50	50	50	40	27	10	5	1

The above table is on-orbit programmable and can be revised to allow optimal charge rates for changing operational scenarios, as well as for compensation of changing battery performance characteristics caused by aging.

8.0 ISS ON-ORBIT DATA

The ISS on-orbit data is telemetered to the ground, and is available real time through data screens on console at the Engineering Supports Rooms (ESRs) and the Mission Control Center. Stored, long-term data can be accessed from the Orbiter Data Reduction Complex (ODRC) through the consoles. Representative on-orbit data is shown below in Figs 4, 5, and 6. This data is for Flight Day #101 (April 11 2001). As of this date, the batteries had completed approximately 1,600 LEO cycles. The data depicts the three Channel 2B batteries (6 ORUs). Spaces in the data are caused by data drop-out and are not intentional omissions. The data clearly shows operational ranges of:

- Battery voltage (76 cells) 95 to 115 Vdc
- Maximum charge rate 50 Amps (note that due to ISS EPS conventions, charging current is shown as negative)
- SOC ~85 to ~103% (average DOD 15%)
- ORU temperature range ~1.0 to 2.5°C (Note heater cycling due to ISS operation at less than ORU power design loads)
- Pressure ~580 to ~730 psi
- Cell voltages ~1.26 to ~1.5 Vdc
-

9.0 CONCLUSIONS

The ISS EPS is successfully maintaining power for all on-board loads. This power is currently supplied by six NiH_2 batteries (three per channel) during eclipse. The batteries are designed for a LEO 35% DOD cycle, however, due to the low power demands at this point in the ISS assembly phase, they have been operating at 15% DOD. The batteries are operating nominally and have exceeded all ISS requirements.

10.0 REFERENCES

1. Lowery, J. E., Lanier, J.R., Hall, C.I., and Whitt, T.H., "Ongoing Nickel-Hydrogen Energy Storage Device Testing at George C. Marshall Space Flight Center," Proceedings of the 25th Intersociety Energy Conversion Engineering Conference, Reno, NV, August 1990

2. Cohen, F., and Dalton, P. J., Space Station Nickel-Hydrogen Battery Orbital Replacement Unit Test," Proceedings of the 29th Intersociety Energy Conversion Engineering Conference, Monterey, CA., August 1994

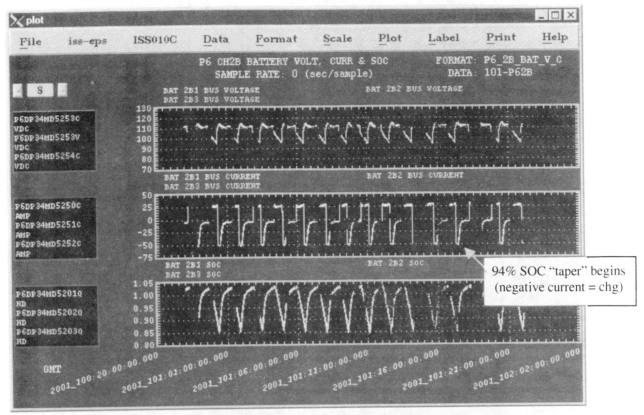

Figure 4: On-Orbit Data Battery Voltage, Current, and SOC

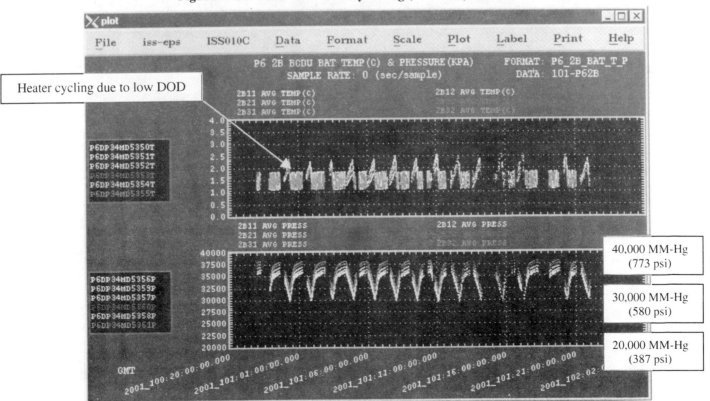

Figure 5. On-Orbit Data Battery Temperature and Pressure

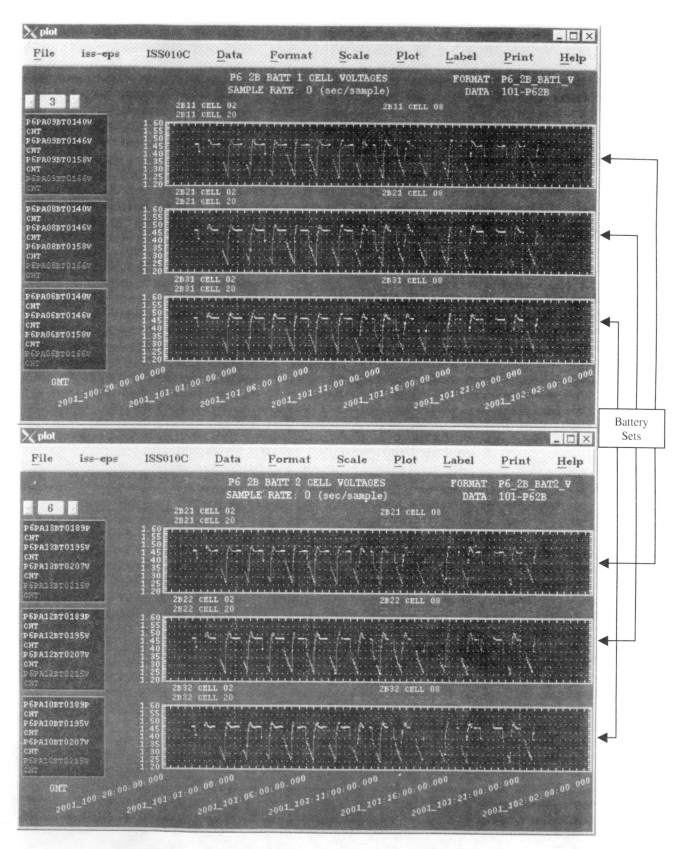

Figure 6: On-Orbit Data, Battery ORU Monitored Cell Voltages
(4 cells per ORU 02, 08, 20, and 28)

302

Proceedings of IECEC'01
36th Intersociety Energy Conversion Engineering Conference
July 29–August 2, 2001, Savannah, Georgia

IECEC2001–ES–2001–AT–42

TURBOMACHINERY DESIGN SENSITIVITY

J. Mark Janus

Computational Simulation and Design Center
Mississippi State University
P.O. Box 9627
MSU, MS 39762
e–mail: mark@erc.msstate.edu

Luca Massa

University of Illinois at Urbana Champaign
College of Engineering
256 Engineering Sciences Building, MC–278
Urbana, IL 61801
e–mail: luca1@galileo.cse.uiuc.edu

ABSTRACT

A numerical algorithm for the sensitivity analysis of unsteady turbine stages is briefly described and exercised for a number of test cases. The analysis software has been in use by aircraft engine manufacturers but has not crossed over into ground–based applications. Aerodynamic and thermal sensitivity derivatives are computed using the complex Taylor series expansion (CTSE) technique. Test cases involve both a 2D linear cascade and a 3D stator–rotor stage geometry. Design sensitivity is examined for several design variables including; surface shape and other geometric properties in two dimensions, and axial gap, rotation speed, and blade indexing (clocking) in three dimensions. Results indicate that selective complexification produces a CTSE algorithm with the same accuracy as black–box (full) complexification at a great reduction in computational cost. Although current costs still show the sensitivity data to cost approximately the same as the initial analysis, continued improvements in implementation are indicating that these cost will soon be only a fraction more than the initial cost of analysis.

Keywords: Turbomachinery; Sensitivity Derivatives; Complex Taylor Series

INTRODUCTION

Power plants of the future are envisioned to operate (efficiently and reliably) on various feedstocks of fuel which have different heating values and thus will require system operation at various points on a performance map. In this context, the concept of a "design point" must be broadened to encompass a "design range". To achieve the efficiency and reliability goals currently set forth by the DOE, US industry will be called upon to design equipment which far supersedes present technology. In order to accomplish this goal, the design of tomorrows power generation equipment can not rely on current practice. Regarding turbomachinery components of ground–based power generation, present day empirical correlations, time–steady calcula-tions and experimental data which is based on conventional fuel turbines are not readily applicable to the design of nonconventional fuel (e.g. synfuel, biomass gas, H_2, etc.) and/or fuel–flexible turbines. Given the current risk/reward scenario, development costs for new turbine designs suitable for highly–efficient fuel–flexible operation is viewed as a very risky investment and thus may not be pursued by industry. Research is currently underway toward developing design tools to enable US industry a cost–effective means of designing highly–efficient fuel–flexible turbomachinery components for the power plant of the future. Central to this initiative is the use of computational simulation technology as a foundation to obtaining system sensitivity to critical design parameters.

The three–dimensional design sensitivity effort presented in this paper, begins with the software tool *MSU_TURBO* as the base unsteady analysis tool. NASA Glenn sanctioned the development of *MSU_TURBO* and has pledged continued support of the software tool. The software is in use by all aircraft engine manufacturers as well as many research facilities. *MSU_TURBO* utilizes a structured multiblock domain decomposition comprised of structured body–fitted moving meshes. The flow model solves the unsteady three–dimensional Reynolds–Averaged Navier–Stokes equations using an unsteady K–ε turbulence model. Design optimization based on computational simulation requires a numerical algorithm with the ability to accurately provide information about the system response (system sensitivity) to perturbations in the "so–called" design variables. The direct differentiation of a three–dimensional Navier–Stokes flow solver produces software capable of accurately evaluating the sensitivity derivatives, but introduces additional complexity and increases computing resource requirements.

This design effort exploits the advantageous features of a new technique to compute derivatives of real functions using complex variables. The approach, referred to as complexification, requires little modification to the software package. Projections indicate the expense of obtaining a design analysis and its sensitivity to any design variable will likely be only fractionally more than the analysis alone. An effort is also planned to implement this novel approach of obtaining sensitivity derivatives in a multidisciplinary environment to explore the cross–disciplinary system response to design variable perturbations. For example, in the power generation industry, it is common to question how a particular design would perform when operating on alternate fuels. Variations in fuel feedstocks result in different turbine inlet conditions, subsequently causing operation of the equipment at various points on the performance map. The software being developed under this project can be used to explore the impact of these variations on the system equipment.

The importance of unsteady phenomena in turbomachinery analysis has recently been recognized. Adamczyk [1] estimated that for a compressor blade the total loss could be reduced by 10.5% recovering the energy lost in the wakes by means of reduction of the wake depth. Huber, et al. [2] obtained an adiabatic efficiency improvement of 0.4% through appropriately aligning the airfoils of a two–stage turbine. The maximum efficiency of the two–stage turbine was achieved when the wakes from the first stator impinged on the leading edge of the second stator airfoils. The research reported on in these papers provides evidence of the significant impact of the wake–blade interaction on turbomachine performance and justifies the effort in developing a code for the time–dependent sensitivity analysis of turbine stages. A full three–dimensional time–dependent Navier–Stokes simulation of a multiple blade row turbomachine is very expensive but necessary to accurately resolve the blade row interaction. The unsteady aerodynamics is mainly due to the time evolution of the wakes as they are convected by the main stream. Including the unsteady effects in the sensitivity analysis further complicates the derivative evaluation process but is critical to the effectiveness of the design process.

Some researchers contend that simplified models work well enough and there is no reason to implement a design module which utilizes a high–fidelity unsteady solver. The authors contend that using a reduced equation set, in effect, limits the design space. There may be a more optimal design which results from unsteady physics that one would not be driven to, due to assumptions made regarding the mathematical model. Conversely, a reduced equation set may indicate a design is optimal yet when analyzed with higher–fidelity physics, unsteady effects may lead to the contrary. Moore's Law postulates that computer power doubles every eighteen months. Currently, computational simulations using high–fidelity physics are being performed on affordable workstations, and are producing scientific breakthroughs that were not possible with the previous generation of computers. In addition, the efficiency of solution of *MSU_TURBO* is continuing to be addressed with techniques such as using *APNASA* (average passage equation solver) as a starting solution, using time–shift boundary conditions to enable modeling only one passage in a stage with uneven blade count, and a parallel implementation.

Some preliminary design sensitivity studies have been performed in both two– and three–dimensions regarding the efficient calculation of sensitivity derivatives using software com-

plexification. Comparisons between sensitivity derivatives obtained via complexification and those obtained via finite difference have been made, emphasizing the efficiency and accuracy of calculation. This computational design effort, through its development of design sensitivity and analysis capabilities, could greatly enhance the performance of ground–based turbomachinery by affording US industrial manufacturers the ability to determine design sensitivity to various design parameters and establish tradeoffs in the design process, e.g. between efficiency and reliability, availability, and maintainability (RAM). This could all come about while simultaneously incorporating high–fidelity (multidisciplinary) unsteady physics into future designs.

COMPUTATIONAL DESIGN

Flow Analysis Software

This design effort begins with the software tool *MSU_TURBO* as the base unsteady analysis tool. For technical details regarding the software implementation, see Janus [3], and Chen [4]. *MSU_TURBO* utilizes a structured multiblock domain decomposition comprised of structured body–fitted moving meshes. This domain decomposition strategy is used to partition the field axially as well as azimuthally into a ordered arrangement of blocks which exhibit varying degrees of similarity. Block–block relative motion is achieved in the axial direction using a sliding mesh. A sliding interface technique is used to pass information axially across blocks with different rotating speeds, Barter, et al. [5]. Block–block interfaces, including dynamic interfaces, are treated such as to mimic cell communication interior to a block (i.e. high–order spatial resolution). A general high–order numerical scheme is applied to satisfy the geometric conservation law (space conservation law) for dynamic grid capability. The flow model uses the unsteady three–dimensional Reynolds–Averaged Navier Stokes equations, discretized as a finite–volume method, utilizing a high–resolution approximate Riemann solver for cell interface flux definitions. The numerical scheme is an implicit, symmetric Gauss–Seidel, Newton iterative–refinement method. Turbulence is modeled using an unsteady K–ε turbulence model developed for turbomachinery at CMOTT (NASA Glenn). Hot gas effects are simulated using a variable ratio of specific heats (assuming a linear relationship with temperature following Turner [6]). In addition, to better enable turbine geometry simulation, film cooling is introduced via source terms which are received as input.

Sensitivity Derivative Calculations

The central element for design sensitivity analysis is the evaluation of sensitivity derivatives. The use of state–of–the–art design optimization techniques such as the evaluation of sensitivity derivatives has the potential to provide the design engineer with the insight toward which design variable(s) (e.g. blade count, stage row spacing, cooling methods, etc.) have the greatest impact on system efficacy. This is true whether efficiency or durability is the primary goal.

For a central finite–difference approximation to the derivative, one may expand the function in a Taylor series about a given point using a forward step and a backward step, and then subtracting to yield an approximation for the gradient. This expression for the derivative has a truncation error of $O(h^2)$. The advantage of the finite–difference approximation to obtain sensitivity derivatives is that any existing code may be used without modification. The disadvantages of this method are the computational time required and the possible inaccuracy of the derivatives. The former is due to the fact that for every derivative, two well–converged solutions for the function evaluations are required. In the case of nonlinear aerodynamics, for example, these solutions may become extremely expensive. The latter is attributed to the sensitivity of the derivatives to the choice of the step size. To minimize the truncation error one selects a smaller step size, however, an exceedingly small step size may produce significant subtractive cancellation errors. The optimal choice for the step size is not known a priori, and may vary from one function to another, and from one design variable to the next.

The flow solver(s) described herein were differentiated using the complex Taylor series expansion technique (CTSE). This technique was first introduced and applied by Newman, et al. [7]. The advantages of the CTSE are the easy implementation (i.e., automated, black–box implementation) and the high accuracy of the results. This technique, since its inception, has been abundantly applied to transform complex CFD flow solvers. Martins, et al., [8] favorably compare the accuracy and the efficiency of the sensitivity codes obtained using the CTSE to that of codes obtained using other available tools such as: finite differences (FD) and ADIFOR [9]. Both ADIFOR and CTSE transform the code by directly differentiating the solution algorithm (discrete approach). One of the advantages of this approach is the ease of implementation: the differentiation tools can be applied as a black–box, in the sense that the transformation of a code into a sensitivity version does not require any knowledge of the problem being solved.

Black–box differentiation means full differentiation of the numerical code without regard for what is actually computed. When the algorithm is an iterative solver, black–box differentiation can produce a poor performing sensitivity algorithm and does not provide control on the convergence of the derivative. When the *entire* code is converted to complex, a significant time penalty is observed. This is due to the fact that all the arithmetic is now performed with complex arguments. Furthermore, at every iteration of the solution process, terms that are not a function of the design variables and thus have zero contribution, or are constant throughout the computation, are recomputed. These are a major contribution to the time penalty.

Griewank, et al. [10], investigated the treatment of the preconditioner matrix (in the case of the pure Newton method the preconditioner matrix is the inverse of the Jacobian) in a differentiated algorithm that uses a Newton–like update to converge the solution. The authors concluded that a sensitivity code in which the preconditioner matrix was left undifferentiated required only a few iterations more to converge than a fully differentiated algorithm, but was significantly better performing when the run time was compared. In [11], Massa describes a technique referred to as selective differentiation which takes into consideration the convergence of the derivative and what needs to be differentiated to this end. Selective differentiation aims at producing an algorithm with the same accuracy but reduced run time

and memory allocation. Massa [11] gives a detailed description of selective differentiation along with a comparison of computer resource requirements for various CTSE techniques.

From the CTSE standpoint, a partially differentiated Newton method leads to an algorithm in which the Jacobian matrix is not complexified. The ease of implementation of a partially differentiated algorithm is one of the advantages of CTSE when compared to other differentiation techniques. The complexified sensitivity algorithm is formally identical to the original one and the only modification to the black–box version is the use of real variables in the subroutines that compute the Jacobian matrix. The advantages of selective differentiation are memory and run–time reduction. The run–time is reduced because the Jacobian matrix is computed with real numbers and the inversion of two uncoupled real linear systems of equations is substituted for the inversion of a complex linear system of equations. Notice that the inversion of the linear system is a very time–consuming task in implicit algorithms and the inversion of two uncoupled real systems is considerably cheaper than the inversion of a complex system of equations. The memory requirement is drastically reduced because the Jacobian matrix accounts for approximately 70% of the overall memory allocation, i.e. 175 scalar entries per each computational point. Therefore, in light of the memory allocation for the Jacobian matrix being reduced by 50%, the total memory allocated in the partially differentiated algorithm is 35% lower than in the fully differentiated algorithm.

RESULTS

A two–dimensional test project [12] was initiated to test the feasibility of some of the techniques slated for incorporation in the full three–dimensional package. Among the project tasks were; the development of techniques to accurately obtain sensitivity derivatives, methods to parameterize blade (airfoil) shape, simulation of a conjugate heat transfer problem, and design cycle optimization tests utilizing this information. In addition, some three–dimensional tests on stage sensitivity to unsteady aerodynamics (for example, axial gap, wheel speed and clocking) have been performed.

2D Cascade Study

For the 2D portion of this study, a cascade code was constructed which utilizes the same basic algorithm as *MSU_TURBO*. The aerodynamic solver has allowances for adiabatic and constant wall temperature boundary conditions. The heat transfer within the turbomachinery blading is considered to be governed by the transient heat conduction equation. Two–dimensional conjugate heat transfer studies have been performed regarding aerothermal interaction in turbine geometries. The research software has options to solve the heat conduction equation via a finite–element or finite–volume discretization where both approaches utilize an unstructured (i.e., triangular) grid. The time integration was performed using an explicit m–stage Runge–Kutta scheme. High–order time accurate solutions could be obtained if desired. Boundary condition options were of Dirichlet, Neumann, and mixed–type.

A von Karman Institute turbine airfoil cascade [13], with it's accompanying suite of test cases was selected to examine the performance of the two–dimensional design sensitivity software in this study. The method adopted to represent the design surface is referred to as a Bezier–Bernstein surface parameterization. The Bezier control points are a logical choice for shape design variables, and were used in the two–dimensional effort. The parameterization of the pressure surface (PS) utilizing 9 Bezier control points is shown in Fig. 1. Parameterized separately, the suction surface (SS) and trailing edge of the turbine blade are shown in Fig. 2. The SS Bezier curve contains 13 control points. Constraints were placed at both ends of this curve to ensure continuity between the pressure and suction surfaces.

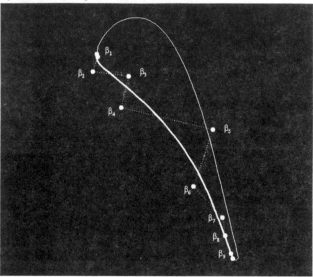

Figure 1. Bezier–curve parameterization for the pressure surface

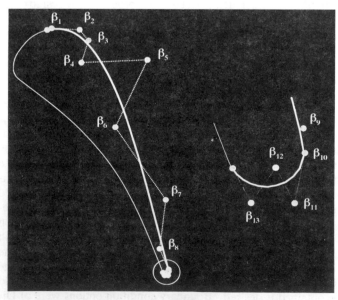

Figure 2. Bezier–curve parameterization for the suction surface and trailing edge

In the 2D effort, a structured grid was used for the aerodynamic analysis, and an unstructured grid was used for the thermal analysis. The viscous mesh about the turbine blade was gener-

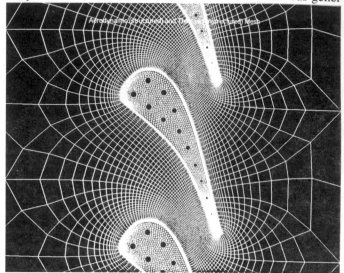

Figure 3. Discretizations for the aerodynamic (structured 169x46) and thermal (unstructured 4275 cells, 2361 nodes) grids

ated with *GRAPE* [14], and is shown in Fig. 3. The unstructured grid used for the thermal analysis of the turbine blade, also shown in Fig. 3, was generated using *SolidMesh* [15]. To ensure mesh quality throughout the design process a re–generation strategy for the structured, aerodynamic grid was utilized. The *GRAPE* software was used to re–generate the aerodynamic mesh after each design surface change. For the unstructured grid, a grid adaptation approach, previously developed by Batina [16] which considers the unstructured mesh as a system of interconnecting springs was adopted. This system is constructed by representing each edge of each triangle by a tension spring. The stiffness of this spring is usually assumed inversely proportional to the length of its edge. Then for each interior node, the force equilibrium equations are solved for the displacements. In this 2D effort, the complex variable technique was applied to obtain the grid sensitivity derivatives.

Test case MUR46 [13] with an inlet angle of 0.0 degrees, an isentropic exit Mach number of 0.875, and an exit Reynolds number of 1×10^6 was used to assess the accuracy of the aerodynamic analysis. The computed and experimental blade Mach number distributions were shown in [12] to be in excellent agreement. In addition, the laminar heat transfer coefficient for test case (MUR132) [13] which was instrumented for heat transfer characteristics also demonstrated good agreement as was shown in [12]. The conditions for that case were an inlet total temperature of 408.5K, a cooled wall at 299.75K, and an exit Mach number of 0.72.

The accuracy of the 2D aerodynamic sensitivity derivatives using the complex variable approach was established and also reported on in [12]. The complex variable approach demonstrated true second–order accuracy. As a qualitative illustration of the influence of shape changes on the flowfield solution corresponding to test case MUR46, Figs. 4, 5, and 6 depict the computed local loss coefficient and the sensitivity of the loss coefficient with respect to shape design variables SS β_{13} and PS β_1, respectively.

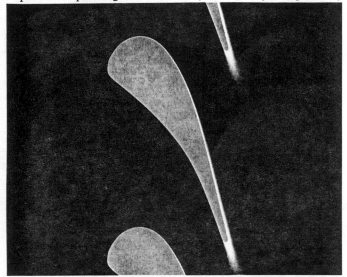

Figure 4.   Shaded local loss coefficient

Figure 5.   Shaded local loss coefficient
sensitivity with respect to SS β_{13}

It is important to note the magnitude of the sensitivity derivatives for these two design variables as seen in the legend. From the figures, it is evident that design variable SS β_{13} has a much more prounced impact (orders of magnitude greater) on the loss coefficient than does design variable PS β_1. This is important in that a designer can use this information to concentrate the effort of shape redesign in the vicinity of SS β_{13}. Small changes in this region will realise comparatively greater payoff in reduction of

Figure 6.   Shaded local loss coefficient
sensitivity with respect to PS β_1

loss coefficient. In the interest of developing tools for cross–disciplinary studies, the thermal sensitivity analysis of the turbine blade was performed and is shown in Fig. 7. Properties of a typical turbine blade material were assumed. Illustrated in Fig. 7 are the shaded contours of the temperature sensitivity with respect to the radius of the first cooling passage. Other possible design variables could be the shape and location of the cooling passages. A conjugate heat transfer solution was also obtained and shown in [12] but will not be reported on here.

3D Turbine Stage Study

Preliminary stage sensitivity tests were performed on a very low aspect ratio (essentially 2D) one and one–half–stage turbine geometry to better understand the unsteady aerodynamics associated with clocking (or indexing). Figure 8 demonstrates the effect indexing can have on the flowfield within the turbine. Shown in the figure is the entropy (indicating the viscous losses due to the boundary layers on the turbine blades). In Fig. 8a, the wake of the first stator is "out–of–alignment" with the second stator and hence two stator wakes contribute to the integrated loss at the exit of the turbine. On the other hand, in Fig. 8b, the second stator is aligned with the wake of the upstream stator and thus less loss is observed with this configuration. Design sensitivity derivatives can be used to determine the precise alignment to maximize this loss reduction.

In 3D, the sensitivity algorithm was tested on a conservatively designed stator–rotor turbine stage. The geometry, experimental measurements, and test conditions are reported in [17]. The computational grid used was a two–block elliptically–smoothed H–type mesh. The size of the stator grid block was 151x31x41 (191,921 computational cells) while the size of the rotor grid block was 171x31x41 (217,341 computational cells). The first number refers to the axial computational coordinate, the second to the radial coordinate (hub–to–casing) and the third to

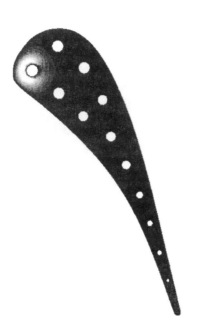

Figure 7. Temperature sensitivity contours
with respect to passage radius for
the VKI turbine with internal cooling

a. non–aligned stators

b. aligned stators

Figure 8.   Gray–shaded entropy

the circumferential coordinate (blade–to–blade). Two design parameters were chosen to test the sensitivity algorithm: the stator–rotor axial gap and the shaft speed. The time dependency of the sensitivity derivatives was a primary issue in this research. The first design parameter determines the depth of the wakes at the rotor inlet and, consequently, the importance of the unsteady effects in the solution and its derivative. The system sensitivity to the second design parameter provides knowledge about the response of the turbomachine to partialization of the load. The sensitivity analysis to the shaft speed will address the issue of the periodicity of the derivative approximations.

The experimental report provides information only about the performance and not about the derivatives of the performance of the turbomachine. The accuracy of the sensitivity analysis is demonstrated using three criteria: accuracy of the discretization/model, accuracy of the differentiation technique and accuracy of the derivatives when compared to qualitative results from empirical and theoretical turbine knowledge. The CTSE technique evaluates the sensitivity derivatives by differentiating the discrete flow solution rather than by solving discretized equations for the derivatives. Therefore, the accuracy of the sensitivity analysis depends on the accuracy of the discretization/model, which was tested by comparing the computed performance with available experimental data and on the accuracy of the differentiation of the code which was tested by comparing the results with FD approximations. However, the comparison of computed performance with experiments does not guarantee that the code is able to predict the variation of the performance with the parameter. Hence, it is important to check the derivatives with qualitative information based upon theory and empirical data.

Performance Analysis.

The comparison of computed performance with experiment was shown in [11] and demonstrated very good agreement. In addition, the accuracy of the sensitivity derivatives shown in this section was established in [11] by comparing the CTSE derivatives with derivatives obtained from finite difference methods. For unsteady analyses such as presented here, sensitivity derivatives obtained using central finite differences proved to be are far less robust than those obtained from complexification.

The time variation of the total efficiency is shown in Fig. 9. The non–dimensional time in the abscissa, $\bar{t}$, is defined in the following formula:

$$\bar{t} = \frac{\omega t N_b}{2\pi},$$

where N_b is the number of blades in the stator–rotor system and ω is the wheel speed. The unsteadiness of the flow in a turbine stage is due to two contributions: periodic potential interaction and periodic wake interaction. For a typical design value of the axial gap, the second is largely dominant over the first. The periodic wake interaction is caused by the stator wakes as they are convected downstream by the main flow in the gap region. The blade passing frequency (BPF) is the frequency with which the stator wakes impinge on the rotor blades. For an even blade count stator–rotor system, the BPF is the frequency of the largest Fourier component of the time variation of the performance, as the plot in Fig. 9 displays. In the next two subsections the calculated derivatives are compared to qualitative results.

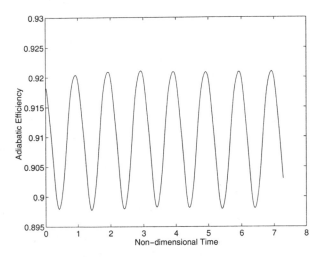

Figure 9. Turbine Stage Efficiency Time Variation

Sensitivity to the Axial Gap.

A test was performed to evaluate the design sensitivity to axial gap between the rotor and the stator of this configuration. Sensitivity derivatives were obtained using complexification and central finite differences of varying step size. The results presented in [11] indicate that subtractive cancellation errors begin to corrupt the derivatives with only moderate step sizes. This creates a situation where the appropriate stepsize for accurate sensitivity derivatives using finite difference methods is ambiguous. Complexification does not suffer from this problem due to inherent robustness regarding step size as it relates to subtractive cancellation errors.

Qualitatively, the derivatives have been validated against two criteria: time dependency and accuracy of the period–averaged values when compared to experimental/theoretical result. The time dependent derivative of the adiabatic efficiency with respect to the axial gap is plotted versus the non–dimensional time

in Fig. 10. The contribution (harmonic magnitude) of each discrete frequency can be isolated by performing a discrete Fourier transform (DFT) of the derivative as shown in Fig. 11. The abscissa in Fig. 11 is the frequency expressed as a multiple of the BPF, while the ordinate is the magnitude of the corresponding harmonic. The harmonic associated with the BPF is largely dominant in magnitude.

The computed value of the adiabatic efficiency derivative averaged over the time period is $\frac{\Delta\eta}{\Delta gap(mm)} = \frac{0.0174}{25.4}$, which represents an efficiency improvement of 1.91% for every inch the axial gap is increased. Note, the original gap width (of the experimental apparatus) was only about a half and inch. Notice that, if the period does not depend on the parameter with respect to which the differentiation is carried out, the average value of the evaluated derivative is equal to the derivative of the average value. Therefore, the numerical experiments show that the most significant component in the time variation of the derivative is the BPF and that the period–averaged adiabatic efficiency increases slightly as the axial gap increases.

Experimental results, Wu [18], and theoretical models, Smith [19], both predict a slight increase of the adiabatic efficiency with the axial gap in turbine stages. Wu's experiments show that the magnitude of the derivative is dependent on the value of the gap and on the total pressure ratio. The derivative is expected to be small for typical design values of the axial gap, and to increase as the axial gap is reduced. Thus the small positive computed derivative agrees with Wu's experiments. Smith's wake recovery theory links a decrease in efficiency to the increase in the magnitude of the stator wake depth when it passes through the rotor passage of a turbine. The wake recovery mechanism has a less negative effect if a weaker disturbance reaches the rotor inlet plane. The viscous forces in the gap region act to decrease the magnitude of the wake disturbances before they enter the rotor passages, therefore a larger axial gap results in weaker disturbances at the rotor inlet plane. This theory explains the positive sign of the period–averaged sensitivity derivative. The wake recovery theory also validates the time variation of the computed derivative. In fact, an increased value of the axial gap leads to weaker wakes at the inlet of the rotor blade row and therefore a weaker interaction, which is the cause of the BPF. As a result of this, the BPF component of the performance is the most affected by the increased value of the gap and the component associated with this frequency is expected to be the largest in the derivative.

Sensitivity to the Shaft Speed.

The sensitivity analysis to the shaft speed, ω, presents an interesting computational problem to the unsteady analysis. The performance of a stator–rotor turbine stage is time dependent and periodic. The derivatives are expected to be time dependent but not necessarily periodic. Since the overall effects of the design parameter on the performance is judged by looking at the value of the period average value of the derivative, it is important in most applications to obtain a periodic sensitivity derivative.

The computed performance is a function of the parameter and the time as well. The sensitivity derivative can be defined

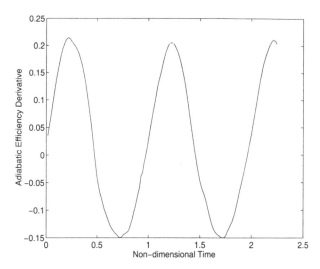

Figure 10. Sensitivity to Axial Gap (Efficiency Derivative)

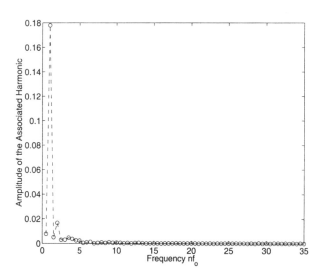

Figure 11. DFT of the Efficiency Derivative (Harmonic Magnitude)

and computed as the partial derivative with respect to the parameter at constant time. It is not difficult to see that when the shaft speed is the parameter, the derivative evaluated at constant time is not time periodic. The performance function can be also considered as a function of the parameter and of the blade row rotation angle, θ. Thus, the time dependent sensitivity derivative, SD_ω, can be defined and evaluated as the partial derivative with respect to the shaft speed evaluated at constant θ, thus,

$$SD_\omega = \left[\frac{\partial F}{\partial \omega}\right]_\theta \neq \left[\frac{\partial F}{\partial \omega}\right]_t .$$

The SD_ω defined above, is time periodic and the period–averaged value of the derivative is equal to the derivative of the period–averaged value of the performance, which represents the overall sensitivity of the performance to change in the parameter. From the computational stand–point, keeping θ constant in the differentiation means that the derivative of time discretization step, Δt, with respect to the parameter is different from zero. That is, the sensitivity derivatives are defined and evaluated at non–constant time discretization. The dependency of Δt on the parameter can be the cause of the non–periodic numerical contributions independent of BPF appearing in the derivative plots presented later on. Moreover, the fact that the sensitivity derivatives are not evaluated at constant time can be regarded as inconsistent with the requirement that the derivatives have to be time accurate.

The wheel speed was reduced from the design value of $13,043$ RPM to $10,000$ RPM, in order to enhance the response of the system to change in wheel speed. The derivative of the adiabatic efficiency and the torque with respect to ω are shown in Fig. 12 (a) and (b), respectively. The period–averaged adiabatic efficiency derivative is $\frac{\Delta\eta}{\Delta\omega(RPM)} = \frac{0.014}{1000}$, which indicates an efficiency improvement of 1.6% for an increase of $1,000$ RPM in wheel speed. The period–averaged torque derivative is $\frac{\Delta T(N \cdot m)}{\Delta\omega(RPM)} = -\frac{38.91}{1000}$, which indicates that an increase in wheel speed of $1,000$ RPM yields a decrease of 8.2% in torque. Both the torque and the efficiency derivatives are in agreement with experimental measurements and theoretical analysis. Change in wheel speed affects the total–to–total adiabatic efficiency mainly because of variation of incidence angle on the rotor blades and consequent variation of the profile losses. Experimental tests run on a similar geometry, Ainley, et al. [20], show that the total loss function is a weak function of the incidence angle for values lower than the stalling incidence. This means that the derivative of the total–to–total adiabatic efficiency is small away from stall. A slightly positive efficiency derivative is expected, because the incidence on the blade decreases as the wheel speed is increased and the design value is approached.

Torque is more affected by change in ω than adiabatic efficiency. The computed derivative can be qualitatively validated using the turbine mean section analytical theory. A change in wheel speed leads to a change in the exit swirl component which causes the torque value to decrease as the shaft speed is increased (negative derivative). This result can be proven by correlating the torque to the change in exit swirl using the Euler theorem and considering the influence of ω on each term, see [11] for details. Therefore the sensitivity derivatives are in agreement with what is expected, based upon experiments and analytical theory.

ACKNOWLEDGEMENTS

The authors would like to acknowledge Dr. David Whitfield, Director of the SimCenter who provided financial support to the second author for this investigation. In addition, thanks go out to Dr. James C. Newman, III for providing some of the 2D solutions contained within and for the many discussions regarding sensitivity derivatives based on complexification. The Mississippi Center for Supercomputer Research must also be ac-

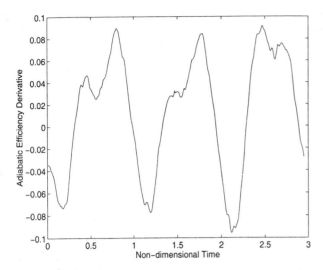

Figure 12.a   Sensitivity to Shaft Speed
Efficiency Derivative

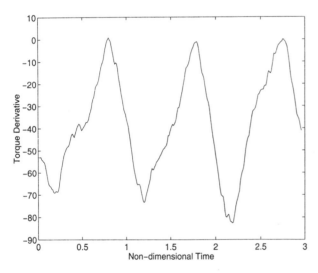

Figure 12.b   Sensitivity to Shaft Speed
Torque Derivative

knowledged for providing the runtime necessary for the 3D simulations.

CONCLUSIONS

Ultimately the goal of this effort is to further develop a computational tool such that a design engineer can use it and glean some insight into choices/tradeoffs which will be encountered throughout the design process. Bringing to bear the computational capabilities described herein toward a specific design or optimization problem has not been the intent of this project. Computational costs are currently thought to be prohibitive regarding a complete design optimization cycle. The authors are well aware of the vast expense represented by cross–disciplinary analysis for turbomachinery, let alone design optimization cycles. On the other hand, it is their belief that the utility of the

tool can be realized, near term, simply by having the added knowledge of sensitivity derivative information. Armed with this additional knowledge the design engineer can make more informed decisions regarding tradeoffs in design. This project seeks to put into place tools which can be routinely used by design engineers and used in a design optimization cycle when, not if, computational means become available (based on the observed validity of Moore's Law regarding computer power).

An algorithm for the sensitivity analysis of unsteady turbine stages has been described. The CTSE technique has been used to differentiate the original field solver(s). The sensitivity derivatives have been investigated for a stator–rotor turbine stage test case for a couple of pertinent design variables. The sensitivity derivatives agree with the qualitative result provided by experiments and theory.

The sensitivity analysis requires the convergence of both the derivative and the solution. Obtaining a time–periodic derivative is considerably more expensive than obtaining a periodic solution, both because the sensitivity algorithm is slower than the original flow solver and because the number of time iterations required is higher. Black–box differentiation of the algorithm produces an accurate sensitivity code but with poor computational efficiency. The CTSE technique lends itself well to selective differentiation because it maintains the same algorithm structure as in the original code.

REFERENCES

1. Adamczyk, J. J., "Aerodynamic Analysis of Multistage Turbomachinery Flows in Support of Aerodynamic Design", 1999 IGTI Scholar Lecture, 1999.

2. Huber, F. W., Johnson P. D., Sharma O. P., Staubach J. B. and Gaddle S. W., "Performance Improvement through Indexing of Turbine Airfoils, Part 1 – Experimental Investigation", ASME Paper 95–GT–27, 1995.

3. Janus, J. M., *Advanced 3–D CFD Algorithm for Turbomachinery*, Ph.D. Dissertation, Mississippi State University, Mississippi, May 1989.

4. Chen, J. P., *Unsteady Three–Dimensional Thin–Layer Navier–Stokes Solutions for Turbomachinery in Transonic Flow*, Ph.D. Dissertation, Mississippi State University, Mississippi, December 1991.

5. Barter, J. W. and Chen J. P., "Comparison of Time–Accurate Calculations for Unsteady Interaction in Turbomachinery Stage", AIAA Paper 98–3292, 1998.

6. Turner, M.G., "Multistage Turbine Simulations with Vortex–Blade Interaction", Journal of Turbomachinery, Vol. 118, Oct. 1996.

7. Newman III, J.C., Anderson, W.K., and Whitfield, D.L., "Multidisciplinary Sensitivity Derivatives Using Complex Variables," MSSU–EIRS–ERC–98–08, July 1998. (*AIAA J.* to appear)

8. Martins, J. R. R. A., Kroo I. M. and Alonso J. J. "An Automated Method for Sensitivity Analysis using Complex Variables", AIAA Paper 2000–0689, January 2000.

9. Bischof, C., and Griewank, A., "ADIFOR: A Fortran System for Portable Automatic Differentiation," AIAA Paper No. 92–4744, Sept. 1992.

10. Griewank, A., Bischof C., Carle A., Corliss G. and Williamson K., "Derivative Convergence for Iterative Solvers", *Optimization Methods and Software*, vol. 2, pages 321–355, 1993.

11. Massa, L., *Computational Simulation and Aerodynamic Sensitivity Analysis of Film–Cooled Turbines*, Ph.D. Dissertation, Mississippi State University, Mississippi, December 2000.

12. Janus, J.M. and Newman, III, J.C., "Aerodynamic and Thermal Design Optimization for Turbine Airfoils", AIAA–2000–0840, January, 2000.

13. Arts, T., Lambert de Rouvroit, M., and Rutherford, A. W., "Aero–thermal Investigation of a Highly Loaded Transonic Linear Turbine Guide Vane Cascade", VKI–TN 174, von Karman Institute for Fluid Dynamics, Sept. 1990.

14. Sorenson, R.L., "A Computer Program to Generate Two–Dimensional Grids About Airfoils and Other Shapes by the Use of Poisson's Equation," NASA TM–81198, May 1980.

15. http://www.erc.msstate.edu/thrusts/grid/solid_mesh

16. Batina, J.T., "Unsteady Euler Airfoil Solutions Using Unstructured Dynamic Meshes," AIAA Paper 89–0115, 1989.

17. Heller, J. A., Whitney R. L. and Cavicchi R. H., "Experimental Investigation of a conservatively Designed Turbine at Four Rotor–Blade Solidities", NACA RM E52C17, 1952.

18. Wu, Chung–Hua, "Survey of Available Information on Internal Flow Losses through Axial Turbomachines", NACA RM E50J13, 1951.

19. Smith, L. H., Jr., "Wake Dispersion in Turbomachines", *Journal of Basic Engineering*, pp. 688–690, September 1966.

20. Ainley, D. G. and Mathieson G. C., "An Examination of the Flow Pressure Losses in Blade Rows of Axial–Flow Turbines", Rep. R&M 28 Aeronautical Research Council, Gt. Britain, 1955.

Proceedings of IECEC'01
36th Intersociety Energy Conversion Engineering Conference
July 29–August 2, 2001, Savannah, Georgia

IECEC2001-ES-2001-AT-75

GLOBAL CHARACTERISTICS OF HYDROGEN-HYDROCARBON COMPOSITE FUEL TURBULENT JET FLAMES

Ahsan R. Choudhuri
Combustion and Propulsion Engineering Laboratory
Department of Mechanical and Industrial Engineering
University of Texas, El Paso
e-mail: ahsan@utep.edu

S. R. Gollahalli
Combustion and Flame Dynamics Laboratory
School of Aerospace and Mechanical Engineering
University of Oklahoma, Norman
e-mail: gollahal@ou.edu

ABSTRACT

The global flame characteristics (flame length, pollutant emission, radiative heat loss fraction, and volumetric soot concentration) of hydrogen-hydrocarbon composite fuel turbulent jet diffusion flames are presented. The correlation of flame length and hydrogen concentration in the fuel mixture is shown. It is found that the flame length decreases with the increase of hydrogen concentration in the mixture. The reactivity of fuel mixture increases with the increase of hydrogen concentration, which ultimately shortens the combustion time, and thereby reduces the overall flame length. Convective time scale decreases with the increase of hydrogen content in the mixture. The measured and predicted flame lengths show a similar trend, however, the predicted values are 1.4 times higher than the measured values. Axial soot concentration decreases with the increase of hydrogen content in the fuel mixture because of lower carbon input and higher soot oxidation due to higher OH concentration. The CO emsission index decreases with the increase of hydrogen concentration in the mixture. On the other hand, NO and NO_X emission indices increase because of the higher flame temperature at higher hydrogen concentration in the mixture.
Keywords: Composite Fuel; Turbulent Flame; Hydrogen; Combustion Characteristics

NOMENCLATURE

ρ	Density
d_j	Burner diameter
f	Mixture fraction
F	Radiative heat loss factor
Fr	Froude number
Fr_f	Global flame Froude number
Fr_j	Jet exit Froude number
H_c	Enthalpy of combustion
L^*	Nondimensional flame length
L_f	Visible flame length
L_H	Normalized lift-off height
L_n	Normalized blowout velocity
m	Mass
M	Mixture molecular weight
q_g	Gas Radiation
Q_R	Total flame radiation
q_R	Radiation flux
q_s	Soot radiation
Q_T	Total Radiation
R	Distance between radiaometer and flame
Re	Reynolds number
Re_j	Jet exit Reynolds number
r_j	Burner radius
T	Local flame temperature
t	Time
T_∞	Surrounding temperature
T_{ad}	Adiabatic flame temperature
T_f	Flame temperature
U_j	Jet Exit velocity
V_f	Flame volume
W_f	Flame width
X_i	Species mole fraction

Y	Mass fraction
Y_s	Stochiometric mixture fraction
μ	Molecular (laminar) dynamic viscosity
ϵ	Emissivity
ξ	Fuel and Air density ratio (ρ_F/ρ_A)
ρ_A	Air Density
ρ_F	Fuel Density
ρ_f	Flame gas density
μ_F	Dynamic viscosity of fuel
ν	Kinematic viscosity
τ_G	Global flame residence time

INTRODUCTION

Recently, hydrogen-hydrocarbon blends have received increased attention as alternative fuels for terrestrial and aerospace power generation applications. Lean burning of hydrocarbons yields exceptionally low pollutant emissions and superior combustion characteristics. However, the lean flammability limit of most hydrocarbon fuels makes achieving stable combustion conditions in the lean burning regime extremely difficult. In contrast, due to the extremely low lean flammability limit, burning of hydrogen is quite attractive. However, severe flash back and difficult storage problems due to its low volumetric density make the use of hydrogen in most practical combustion applications difficult. In this context, a hydrogen-hydrocarbon hybrid fuel is an attractive solution. The relatively slow reaction rate of a typical hydrocarbon fuel can be accelerated by mixing it with hydrogen, which results in an improved ignitability and flame holding.

In order to use the hydrogen-hydrocarbon composite fuel effectively in a practical combustion system, the global characteristics such as flame length, flame radiation etc. of the composite fuel flame need to be properly determined. On the other hand, in many cases pollution emissions of the flame is one of the limiting factors for combustion chamber design. In this paper, the effects of hydrogen addition on flame length, flame radiation and pollutant emissions are presented. Also, the scaling issues of these global characteristics with jet exit Reynolds number and Froude numbers are also discussed.

As mentioned earlier, information regarding flame length, flame radiation and pollutant emission are highly important for combustion chamber design. Estimation of flame length is required for combustion chamber sizing procedure. Depending upon the fuel used, turbulent diffusion flames can radiate significantly. For certain practical applications such as industrial furnaces, radiant heaters etc., flame radiation is a desired attribute to contribute in heating loads. On the other hand, in some other applications, radiation losses can contribute to loss of efficiency (diesel engine) or to safety hazards (e.g., industrial flaring operations). Also, in aero-propulsion and terrestrial power generation gas turbine engines excessive flame radiation losses impose a difficult challenge on combustion

chamber linings and turbine blade inlet vanes. And, on top of everything, due to present stringent environmental regulations, emission of pollutants such as NO_x and CO is one of the most important considerations that have to be evaluated prior to combustion system design. In a series of earlier papers Choudhuri and Gollahalli (1997, 2000a, 2000b, 2000c), have presented the thermochemical characteristics of *laminar* diffusion flames of hydrogen-hydrocarbon mixtures. However, it is not clear how those results can be interpreted for *turbulent* flames. Hence, this study deals with *turbulent* diffusion flames of hydrogen-hydrocarbon mixtures This paper describes the relation of flame length, radiation heat loss, and NO_x and CO emission indices with hydrogen content in the mixture. A scaling technique is also presented for general representation of the data.

ANALYSIS
TURBULENT FLAME LENGTH

The primary factors which affect the turbulent flame lengths are (i) ratio of jet exit momentum force and active buoyant force on flame, Fr_f, (ii) flame stochiometry, $Y_s = [1/(\phi_s+1)]$ (iii) ratio of air and fuel density, ξ, and obviously (iv) exit diameter of the burner, d_j. A global flame Froude number can be defined based on the analysis presented by Delichatsios, 1993 and Bahador, 1993

$$Fr_f = \frac{U_j Y_s^{1.5}}{\left(\dfrac{T_{ad} - T_\infty}{T_\infty} gd_j\right)^{0.5} \xi^{0.25}} \qquad [1]$$

where, ξ is the density ratio of fuel to air. A condition $Fr_f > 5$ represents the regime of pure momentum-dominated flame. Delichatsios, 1993 proposed a general expression for flame length

$$L^* = \frac{L_f Y_s}{d_j \xi^{0.5}} \qquad [2]$$

Also, it was shown that for $Fr_f < 5$ i.e., for buoyancy-dominated region the relation of equivalent flame length and Flame Froude number is

$$L^* = \frac{13.5 Fr_f^{\frac{2}{5}}}{\left(1 + 0.07 Fr_f^2\right)^{\frac{1}{5}}} \qquad [3]$$

and for momentum dominated regime i.e. $Fr_f > 5$ the equivalent flame length L^* is constant and has value of 23. With the visible flame length, a global flame residence time (τ_G) can be defined based on the argument presented in Turns and Myhr, 1991.

$$\tau_G = \frac{\rho_f W_f^2 L_f Y_s}{\rho_F d_j^2 U_j} \qquad [4]$$

Where ρ_f is the density of the flame gases and it can be approximated as the density at flame equilibrium temperature.

SCALING PARAMETERS FOR RADIATION HEAT LOSSES

Radiant heat losses from hydrocarbon flames primarily occur through two sources: First, due to the molecular sources i.e., infrared spectrum broadband radiation from CO_2 and H_2O, and second, continuum radiation from soot particles at short wavelengths. For a methane diffusion flame the wavelength ranges associated with molecular radiation and continuum radition are 2.5-4.5μm and 1.2-1.5μm respectively. So, the total radiation from the flame can be written as the summation of radiation from soot particles and radiation from gas molecules, i.e.,

$$Q_R = q_s + q_g \qquad [5]$$

where q_s and q_g are radiation from soot particles and gas molecules respectively. Although, the radiation from soot particles can be modeled with a more complex method, for scaling purposes a simple model can be conceived by considering the whole flame as a total source.

$$Q_R = \in \sigma V_f (T_f^4 - T_\infty^4) \qquad [6]$$

where V_f is flame volume (approximated as $0.0027\ L_f^3$), T_f is the apparent flame temperature based on flame radiation, $\in$ is the flame emissivity, and σ is the Boltzman constant (5.67051×10^{-8} W/m^2-K^4). The primary problem of the simplified analysis is to assign a proper value of ε. For the present investigation a constant value of 0.55 was used (Choudhuri, 2000). Although this approach neglects the wavelength and temperature dependence of ε, the approximation is fairly reasonable because it is an average value determined from an actual flame measurement. Now if a wide-angle radiometer located at distance R (half of the visible flame length) from the flame axis and measuring a radiant heat flux of q_R is measured than the radiant heat loss can be alternatively modeled as

$$\dot{Q}_R = q_R'' (4\pi R^2) \qquad [7]$$

The total heat release by combustion of fuel can be expressed as

$$\dot{Q}_T = \dot{m}_F \Delta H_c \qquad [8]$$

where m_F is the fuel mass flow rate and ΔH_c is the enthalpy of combustion of this fuel. For hydrogen-hydrocarbon composite fuel the above expression can be rewritten as

$$\dot{Q}_T = (\dot{m}_{HC} \Delta H_{c_{HC}}) + (\dot{m}_{H_2} \Delta H_{c_{H_2}}) \qquad [9]$$

Hence, the radiative heat release factor can be defined as the ratio of radiative heat loss rate to the total heat released rate by the fuels.

$$F = \frac{\dot{Q}_R}{\dot{Q}_T} \qquad [10]$$

If the flame volume is assumed to be proportional to the cube of the burner diameter i.e., d_j^3 (Turns and Myhr, 1991) and the

fuel flow rate to $d_j^2 U_j$, then the radiative heat loss fraction can be scaled as

$$F \approx \in T_f^4 \frac{d_j}{U_j} \qquad [11]$$

Hence, the scaling analysis shows that the radiation heat loss fraction decreases and mean flame temperature increases as the convective timescale d_j/U_j decreases. Furthermore, the radiation heat loss fraction can be related to the jet exit Reynolds number ($Re_j = \rho U_j d_j/\mu$) and the jet exit Froude Number ($Fr_j = U_j^2 g/d_j$) in the following fashion (Turns and Myhr, 1991)

$$F \approx \in \sigma T_f^4 (\nu Re_j)^{0.33} (gFr_j)^{-0.75} \qquad [12]$$

This expression can be further written as a function of densimetric Froude number (Choudhuri, 2000)

$$F \approx \in \sigma T_f^4 (\nu Re_j)^{0.33} (\xi gFr_d - gFr_d)^{-0.75} \qquad [13]$$

where ξ is the ratio of fuel and air density.

EXPERIMENTAL PROCEDURE

The circular fuel burners used in these experiments consisted of stainless steel tubes of 1 mm, 2.3 mm, 3.8 mm, and 4.5 mm ID, through which fuel was injected into an atmosphere of air. These flames were located in a vertical steel combustion chamber of 76cm × 76cm cross-section and 163cm height. The chamber was fitted with rectangular windows of dimensions 20cm × 20cm × 145cm on all its four side-walls. Three of the windows were fitted with high temperature air-cooled Pyrex glass and the fourth window was fitted with a slotted metal sheet for introducing probes. Air was admitted into the test chamber through a 20cm diameter circular opening at the base plate due to natural convection. Three layers of fine-wire-mesh screens were used to provide a uniform flow. The top of the test chamber was open to atmosphere through an exhaust duct. The ambient pressure of the lab was maintained slightly above the atmospheric pressure to ensure positive draft inside the test chamber (Fig.1). Nominal operating conditions and measurement uncertainties are listed in Table 1 and 2.

Mixtures of commercially available Natural gas (NG) (95.8% CH_4, 1.19% C_3H_8, 1.15% C_2H_6, 0.50% CO_2, 0.10% N_2 and remains are pentane, hexane and water) and hydrogen (98%+) were used. The two fuels were mixed inside an annular mixing device. Natural gas (NG) was injected into the stream of NG through a concentrically located injector. The length of the mixing device was sufficiently large (>150 hydraulic dia.) to ensure a homogeneous mixture of the two fuels. The mixed fuels were then supplied to the fuel nozzle through Teflon tubing. For a fixed jet exit Reynolds number, the volume flow rate of the hybrid fuel was calculated at different mixture conditions. The volume flow rate of NG and hydrogen into the mixing device was regulated with calibrated rotameters. For measuring exhaust emissions a quartz flue-gas collector was mounted over the visible flame, and was axially aligned with the burners. A sample profile across the collector diameter showed a variation of less than 1.5% in the species concentration, and hence, center point data were treated as the average values. Gas samples were collected from the combustion products through an uncooled quartz probe of tip diameter 1mm and were treated to remove particlulates and moisture with a series of filters and an ice-chilled moisture trap. The sampling flow rate was adjusted such that suction and local free stream velocities in the flow-field were close enough to ensure *quasi-isokinetic* sampling. A chemiluminescence analyzer (Thermoenvironmental Inc. Model 42H) was used to measure NO and NO_x. Two non-dispersive infrared (NDIR) analyzers (Rosemount Analytical model 880A) were used to measure CO and CO_2. Since the temperature of the exhaust stream at sampling location was less than 1000 K, the measurements of CO mole fraction were not significantly affected by the lack of quenching inside the sampling probe. The mole fractions of O_2 were measured by using a polarographic (Catalytic research model Miniox-II) analyzer. To determine the visible flame height, a high-speed video camera was used. Strobe recording technique (1/6 seconds interval progression, 1/2000 S.S.) along with back-light illuminating and DEIS (digital electronic image stabilization system) method were used to visualize the flame image in a dark background. The back-light illuminating method has the advantage of sharpening the image boundary of a very bright image in contrast with its dark background; also it helps to remove the pixel distortion due to imbalance in the pixel illumination of the CCD (charge couple device). A 2-second recording time was used to measure the visible flame height. The technique proposed by Yagi and Iino (1966) was used to measure the soot concentration. A He-Ne laser (Spectra Physics) beam was passed through the flame. Due to the presence of soot the beam intensity was attenuated. The amount of attenuation was measured using a pyro-electric laser power meter placed on the other side of the beam. Yagi and Iino (1966) proposed a correlation between the soot concentration and the beam attenuation. The process is slightly modified by using a narrow-band laser line-filter to block the luminous flame background and the *Anti-Stoke* component of the Raman spectra due to light scattering from the soot particle. This method is fast and simple compared with the direct probing method. However, it results in the line-of-sight average concentration.

RESULTS AND DISCUSSIONS
Turbulent Flame Length
The variation of normalized visible flame length (L_f/d_j) with hydrogen content in the mixture is shown in Fig. 2. It can be seen that flame length decreases as hydrogen content increases in the mixture. With the increase of hydrogen in the mixture, the chain carrying radicals/atom pool such as of H, O and OH increases, which enhances the combustion rate of natural gas (Choudhuri and Gollahalli, 2000a-c). As a global effect of this, the composite fuels with higher hydrogen content burn faster and thereby reduce the residence time and consequently, the

overall convective time scale. Hence, the fuel travels a short length before combusting, which ultimately shortens the visible flame length. The global convective time scale, (T_G), defined by Eq. 4 also reveals the same. The variation of global convective time scale with hydrogen content in the mixture is shown in Fig. 3. The convective time scale, which is a representation of global flame residence time, decreases as hydrogen concentration increases in the mixture. This also supports the argument mentioned earlier i.e., the increase in hydrogen content increases the overall combustion performance. Also, on a global scale, the percent hydrogen content in the mixture shows a linear correlation [*-1.4356(%H₂) +142.51; R²=0.99*] with the convective time scale. The variation of nondimensional flame length (L*; defined by Eq. 2) with hydrogen content in the fuel mixture is shown in Fig.4. The L*, which primarily shows the effect of flame stochiometry and density, increases as the hydrogen concentration increases in the mixture. This trend indicates that the decrease in visible flame length is slower than the decrease in flame stochiometry (f_s) and density as the hydrogen content increases in the mixture. The nondimensional flame length (L*) correlates fairly well ($R^2=0.94$) with hydrogen content in the mixture and yields a linear relation [0.0233(%H₂)+16.755]. The effects of Flame Froude number on flame length are shown in Fig. 5. Clearly, the effects of buoyancy on the composite fuel flames are still important for the range of Flame Froude Number encountered in the present investigation. Although the trend predicted by Eq. 3 is similar to that measured, the values are higher by a factor of approximately 1.4. The non-dimensional flame length increases with Flame Froude number and follows a power correlation of $L^* = 17.16 Fr_f^{0.16}$ with a R^2 value of 0.93.

Radiative Heat Loss Fraction
Fig. 6 shows the variation of heat release factor with the increase of hydrogen content in the mixture. Radiative heat loss decreases as hydrogen content increases in the mixture. Since an increase in hydrogen content decreases soot and carbon dioxide formation, the radiative heat losses from both of them decrease. Initially, the rate of decrease is high, however the rate decreases as more hydrogen is added to the mixture. Most likely, a comparative increase in radiation from H_2O near 2.5μm at higher hydrogen contents slows down the decrease rate of radiant heat release factor at higher hydrogen concentrations. The variation of apparent flame temperature, T_f, derived[*] from radiation measurement is shown in Fig. 7. The apparent flame temperature shows an indication of how hot or cold the flame was felt to the surroundings. This information is essential to design the combustion chamber linings. The variation of computed adiabatic equilibrium flame temperature with hydrogen content in the mixture is also shown in the same

---

[*]From Eq [6] $T_f = [T_\infty^4 + \dfrac{Q_R}{84.2 \times 10^{-12} L_f^3}]^{1/4}$

figure. The apparent fame temperature decreases with hydrogen content in the mixture despite the slight increase in adiabatic flame temperature. The adiabatic flame temperature increases since hydrogen has a higher adiabatic flame temperature, 2382 K, than that of natural gas 2225K (Choudhuri, 1997). Hence, despite an increase in the adiabatic flame temperature the surroundings feel a much cooler flame as the hydrogen content is increased in the fuel mixture. The relation between convective time scale, T_G, and radiative heat loss fraction is shown in Fig. 8. Although the data are not perfectly correlated [$T_G=9.71F+3.64$; $R^2=0.84$], in general an increase in convective time scale increases flame radiation.

Emission Indices

The variation of NO emission indices with hydrogen content in the mixture is shown in Fig. 9. The NO emission increases as hydrogen content increases in the mixture. As discussed earlier, the overall flame temperature increases as hydrogen is added to the system because of the higher energy input and lower flame radiation. In a methane diffusion flame *thermal* or *Zeldovich Mechanism* is a dominant route of NO production. An increase in flame temperature and radical pool (H, OH, O) (Choudhuri, 2000) concentration enhances the NO formation through this route. On the other hand, the lower CH formation (Choudhuri and Gollahalli, 2000a-c) at higher hydrogen concentration decreases the NO destruction through *Fenimore mechanism*. As a combined effect the NO emssion increases as hydrogen concentration increases in the fuel mixture. For the 100% NG the EI_{NO} is 0.23 g/kg, of fuel which increases by 8.6% as 5% hydrogen is added to the mixture. The variation of NO_X emssision with hydrogen content in the mixture is shown in Fig. 10. The NO_X emission also increases with the increase of hydrogen concentration in the mixture. For the 100% NG the EI_{NOX} is 0.58 g/kg of fuel which increases by 3% as 5% hydrogen is added to the fuel mixture. The NO_2 portion decreases as the hydrogen concentration increases in the mixture. This can be attributed to an increase in H concentration, consequently N_2O destruction through N_2O intermediate mechanism.

The variation of CO emssion index with hydrogen concentration in the mixture is shown in Fig. 11. As expected, the CO emission index decreases with the increase of hydrogen concentration in the mixture. Primarily there are two reasons behind it: First, for the same Reynolds number, an increase in hydrogen concentration results in a lower carbon input to the flame, and thereby reduces CO formation; Second, an increase in hydrogen concentration prompts more OH formation, and since OH is a dominant radical for CO oxidation, the overall CO formation in the flame drops. This phenomenon is evident in the decreasing trend of CO emission index.

The scaling of Emission indices with Flame Froude Number is shown in Figs. 12, 13 and 14. It can be seen that an increase of Flame Froude Number results in a higher NO and NO_X

formation. The NO and NO_X emission indices exhibits exponential correlations with the Flame Froude number, whereas the CO emission index yields power law relation with the Flame Froude Number. The correlations are shown in the figures with the respective R^2 values. The scaling of emission indices with the heat release rate is shown in Figs. 15-17. With the increase of heat input to the flame, the emssion indices of NO and NO_X increase. Since a higher heat input in a low radiating flame results in a higher flame temperature, this trend is justifiable. The exponential correlations of NO and NO_X with the heat release rates are [$EI_{NO}=1.19e^{0.35QT}$] and [$EI_{NOx}=0.42e^{0.15QT}$] with correlation coefficients of 0.9 and 0.97. The CO emission index decreases with the flame heat release rate. The higher flame temperature with lower carbon input and higher availability of OH radicals is responsible for this trend. The CO emission index shows a power relation [$EI_{CO}=2.67Q_T^{-0.48}$; $R^2=0.9$] with the total heat release rate.

Volumetric Soot Concentration

The axial distribution of volumetric soot concentration, as a line-of-sight average normal to the flame axis, with the axial distance from the burner is shown in Fig. 18. For all mixture conditions soot concentration increases from the near-burner to mid-flame regions attains a peak value in the mid-flame regions and decreases in the far-burner region. This trend indicates a three-regime soot history inside the flame. The region with increasing concentration is the *soot inception* region followed by the *soot growth* and *burning region* in the mid-flame and far-burner regions. With the increase in hydrogen content in the mixture, the soot concentration decreases. This is expected since a lower carbon input at higher hydrogen concentration results in lower PAH (Polycyclic Aromatic Hydrocarbons) formation, which ultimately decrease the soot inception process. For the 100% NG flame the maximum measured volumetric soot concentration is 16.9×10^{-7} g/cc which decreases to 10.7×10^{-7} g/cc as 10% hydrogen is added to fuel mixture. The maximum soot concentrations in the 80-20% and 65-35% NG-H_2 mixtures are 7.9×10^{-7} g/cc and 4.9×10^{-7} g/cc respectively.

CONCLUSIONS

The present study of the global characteristics of the turbulent diffusion flames of hydrogen-natural gas fuel mixtures lead to the following results:
With the increase of hydrogen concentration in the mixture:
i. visible flame length decreases. A correlation [$t_G=-1.4(\%H_2)+142.51$; $R^2=0.99$, $0 \leq \%H_2 \leq 35$] between convective time scale (derived from visible flame length) and hydrogen concentration is formulated. The general correlation between the nondimensional flame length, L^*, and the volumetric percentage of hydrogen in the mixture is $L^*=0.023(\%H_2)+16.75$; $R^2=0.95$; $0 \leq \%H_2 \leq 35$.
ii. radiative heat loss fraction decreases. The convective time scale (t_G) exhibits a linear relation [$t_G = $

9.71(%F)+3.64; R^2=0.84; 0≤%H$_2$≤35] with the radiative heat release factor of the flame.

iii. the CO emission index decreases. The CO emission index shows a power relation [EI$_{CO}$=1.5Fr$_f^{-1.34}$, R^2=0.92, 0≤%H$_2$≤35] with the flame Froude number. The correlation between CO emission index and flame heat release rate (Q$_T$) is EI$_{CO}$=2.667Q$_T^{-0.4823}$, R^2=0.90, 0≤%H$_2$≤35.

iv. the NO$_X$ emission index increases and shows an exponential relation [EI$_{NOx}$=0.183e$^{1.338Frf}$, R^2=0.96, 0≤%H$_2$≤35] with the flame Froude number. The correlation between NO$_X$ emission index and flame heat release rate (Q$_T$) is EI$_{NOx}$=0.4168e$^{0.145QT}$, R^2=0.978, 0≤%H$_2$≤35.

v. the axial volumetric soot concentration decreases. The peak soot concentration decreases from a value of 17 x10^{-7} g/ml to 10.2 x 10^{-7} g/ml as 10% hydrogen is added to the 100% NG flame. the axial volumetric soot concentration decreases. The peak soot concentration decreases from a value of 17 x10^{-7} g/ml to 10.2 x 10^{-7} g/ml as 10% hydrogen is added to the 100% NG flame.

Where, %F= percent radiative heat loss fraction, %H$_2$= volumetric percentage of hydrogen in the mixture, H= characteristic flame height, EI =emission indices, Fr$_f$= flame Froude number, L*= nondiemsional flame length, L$_b$= maximum lift-off height at blowout condition, Q$_T$= Total heat release rate by the fuel, R^2= correlation coefficient, r$_j$= burner radius, S$_{L, max}$= maximum laminar flame speed, t$_G$= convective time scale, U$_b$= blowout velocity, Y = fuel Mass Fraction, Y$_s$= stochiometric fuel mass fraction, ρ$_F$= fuel density at burner exit condition, ρ$_A$=ambient air density, μ$_F$ = kinenatic viscosity of the fuel at burner exit condition.

REFERENCES

Bahador, M.Y., Stocker, D .P., Vaughan, D.F., Zhou, L. and Edelman, R. B., 1993, "Effects of Buoyancy on Laminar, Transitional and Turbulent Gas Jet Diffusion Flames," Modern Developments in Energy, Combustion and Spectroscopy. Pergamon Press, NY. pp. 49-66.

Choudhuri, A. R and Gollahalli, S. R., 2000a, "Combustion Characteristics of Hydrogen-Hydrocarbon Hybrid Fuels," International Journal of Hydrogen Energy, Vol. 25, pp. 451-462.

Choudhuri, A. R and Gollahalli, S. R., 2000b, "An Experimental And Numerical Study of Intermediate Radical Concentrations In Hydrogen-Hydrocarbon Hybrid Fuel Jet Flames" ASME ETCE'2000 Meeting, New Orleans.

Choudhuri, A. R. and Gollahalli, S. R., 2000c, "Laser Induced Fluorescence Measurements of Radical Concentration In Hybrid Gas Fuel Flames," International Journal of Hydrogen Energy, Vol. 25, pp. 1119-1127.

Choudhuri, A. R., 2000, An Experimental and Numerical Investigation on Hydrogen-Hydrocarbon Composite Fuel Combustion, Ph.D. Dissertation, University of Oklahoma, Norman, Oklahoma

Choudhuri, A. R., 1997, Experimental Studies in Hybrid Fuel Combustion, Published M.S. Thesis, University of Oklahoma, Norman, Oklahoma.

Delichatsios, M. A., 1993, "Transition from Momentum to Buoyancy-Controlled Turbulent Diffusion Jet Flames and Flame Height Relationship," Combustion and Flame. Vol. 92. pp. 349-364.

Turns, S. R. and Myhr, F. H., 1991, "Oxides of Nitrogen Emissions from Turbulent Jet Flames: Part I- Fuel Effect and Flame Radiation," Combustion and Flame. Vol. 87, pp. 319-335.

Yagi, S. and Iino, H., 1966, "Radiation from soot particles in luminous flame," Eighth Symposium (International) on Combustion, The Combustion Institute, Pittsburgh, PA, pp. 288-293.

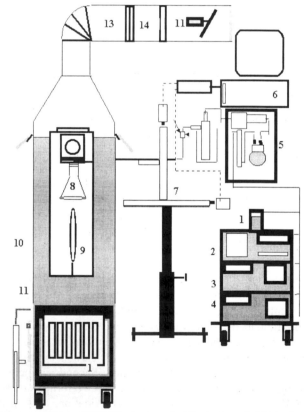

1. Oxygen Analyzer 2.NO-NO$_x$ Analyzer 3. CO$_2$ Analyzer 4. CO Analyzer
5. Exhaust Treatment System: Particulate Filter, Ice-Bath, Vacum Pump
6. Data Acquisition System 7. Stepper Motor Driven 2-D Traverse
8. Exhaust Collection Cone 9. Flame with Burner 10. Combustion Chamber
11. Mixing Device 12. Rotameters 13. Damper 14. Filter 15. Relay Valve

Fig.1 Experimental Setup

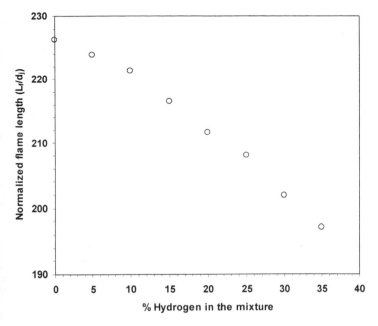

Fig. 2 Variation of Normalized Flame Length with % Hydrogen in the Mixture

Fig. 3Variation of Convective Time Scale with % Hydrogen in the Mixture

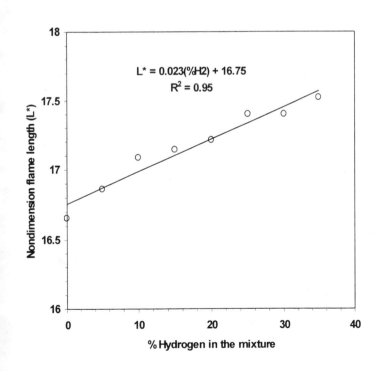

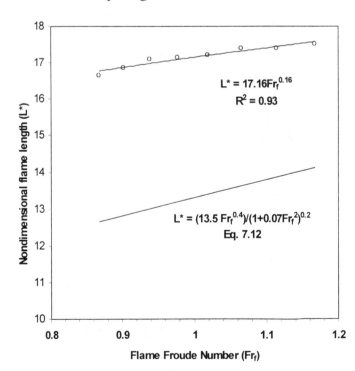

Fig. 4 Variation of Nondimensional Flame Length with % Hydrogen in the Mixture

Fig. 5 Variation of Nondimensional Flame Length with Flame Froude Number

319

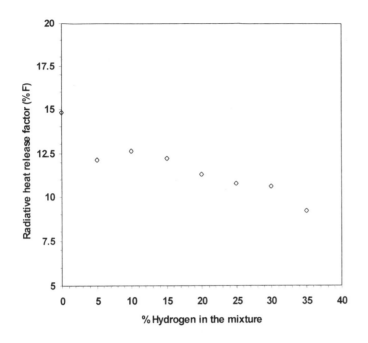

Fig. 6 Variation of Radiative Heat Release Factor with % Hydrogen in the Mixture

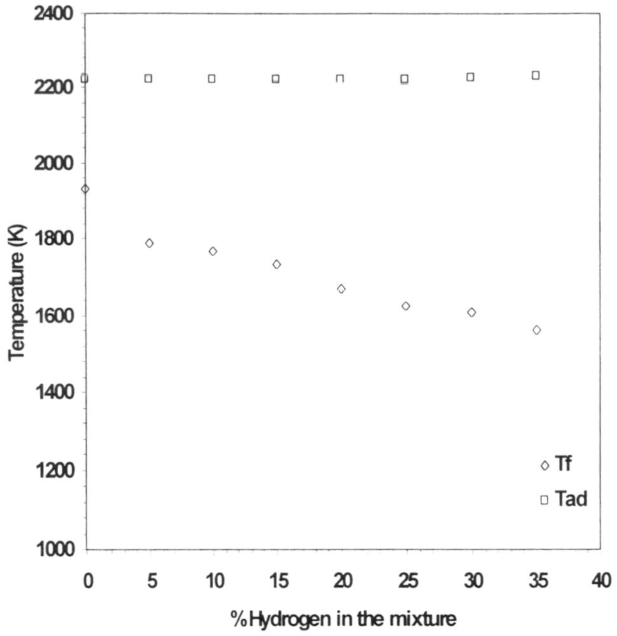

Fig. 7 Variation of Apparent (T_f) and Adiabatic Flame Temperature (T_{ad}) with % Hydrogen in the Mixture

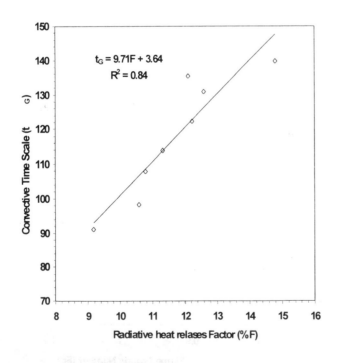

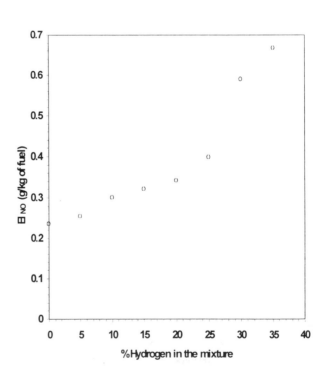

Fig. 8 Variation of Convective Time Scale (t_G) with Radiative Heat Loss Factor

Fig. 9 Variation of NO Emission Index % Hydrogen in the Mixture

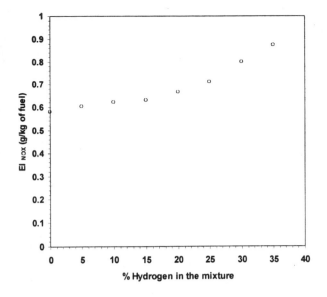

**Fig. 10 Variation of NO$_X$ Emission Index %
Hydrogen in the Mixture**

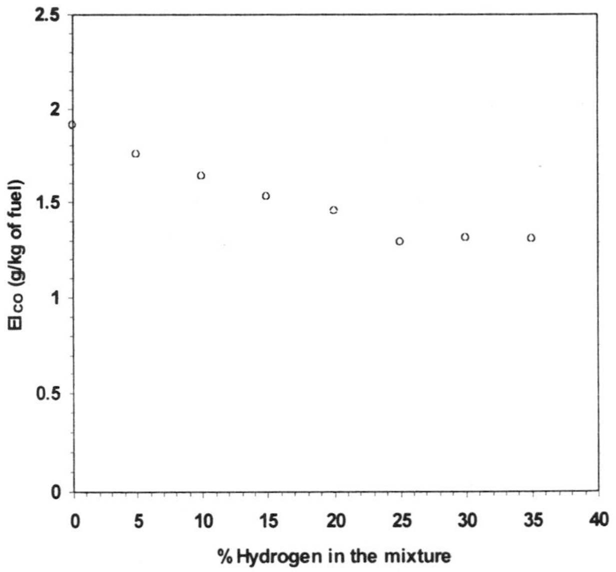

**Fig. 11 Variation of CO Emission Index % Hydrogen
in the Mixture**

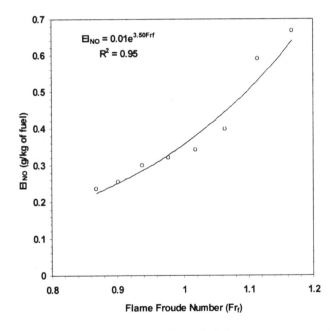

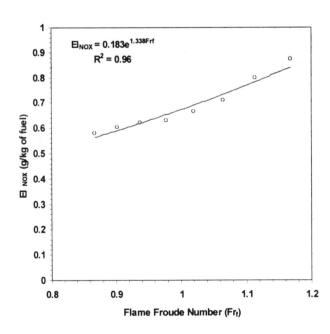

**Fig. 12 Variation of NO Emission Index with Flame
Froude Number**

**Fig. 13 Variation of NO$_X$ Emission Index with
Flame Froude Number**

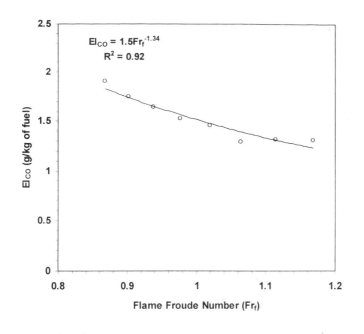

Fig. 14 Variation of CO Emission Index with Flame
Froude Number

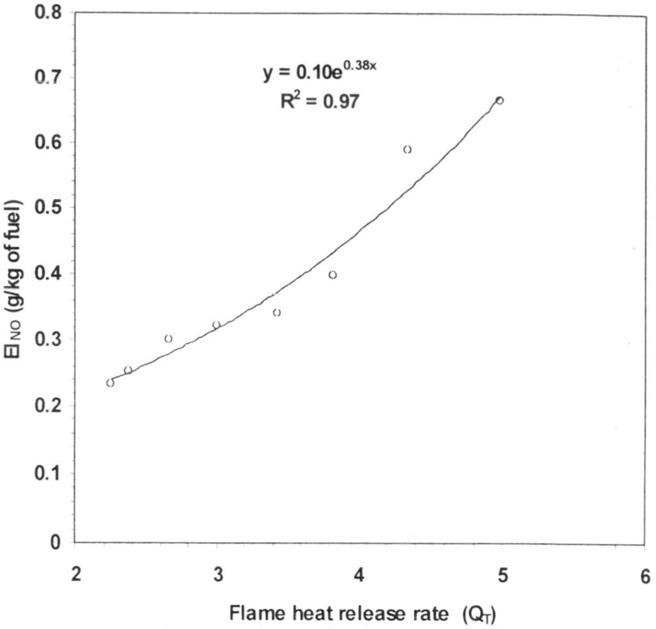

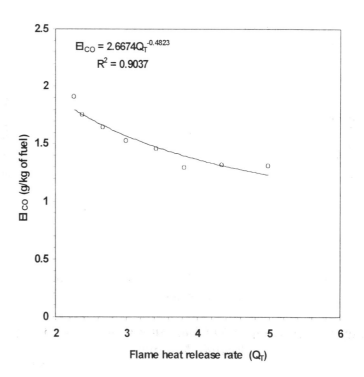

Fig. 15 Variation of NO Emission Index with
Total Heat Release Rate

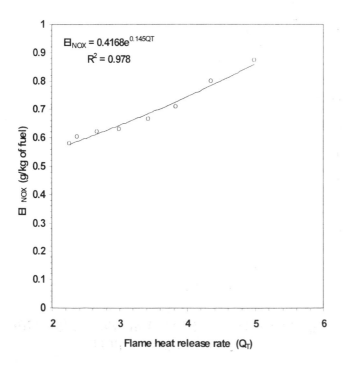

Fig. 16 Variation of NO_X Emission Index with Total Heat
Release Rate

Fig. 17 Variation of CO Emission Index with
Total Heat Release Rate

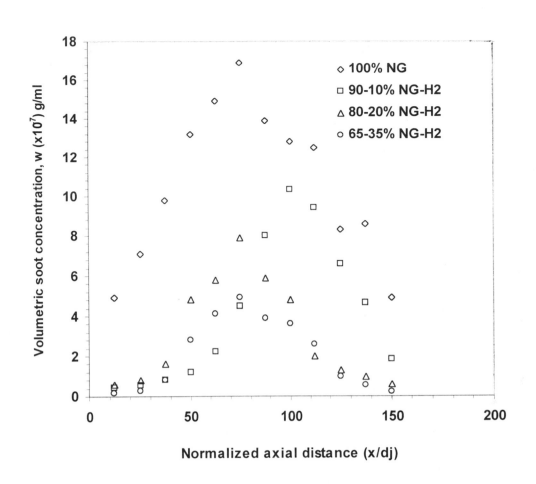

Fig. 18 Axial Distribution of Volumetric Soot Concentration

Table 1: Nominal operating

Burner diameter (ID)	1mm, 2.3 mm, 3.8mm, and 4.5mm
Fuel	H_2 (98%+):0-100%* NG (95.8% CH_4, 1.19% C_3H_8, 1.15% C_2H_6, 0.50% CO_2, 0.10% N_2 and remains are pentane, hexane and water)): 0-100%*
Ambient Temperature	295 K
Ambient Pressure	102 KPa
Jet Reynolds Number	8700
Jet Exit Froude Number	39000-85000
Flame Froude Number	0.8-1.6

Table 2: Estimated Uncertainties

Measurements	% of Mean Value
Emission Index	1.7
Radiative Heat Loss	2
Visible Flame Length	1.8
Volumetric Fuel Flow rate	2.1
Concentration of NO	7.9
Concentration of NO_x	8.2
Concentration of CO_2	8.6
Concentration of CO	8.8
Concentration of O_2	4
Temperature	1.4
Soot Concentration	6.4

Proceedings of IECEC'01
36th Intersociety Energy Conversion Engineering Conference
July 29–August 2, 2001, Savannah, Georgia

2001-AT-77

A CFD STUDY OF THE EFFECTS OF INLET DROPLET VARIABLES ON WATER MIST FIRE SUPPRESSION EFFICIENCY

K.C.Adiga
ApTech Research Center
151 Osigian Blvd., Suite 199
Warner Robins, GA 31088
Email: kcadiga@aptechrc.com

ABSTRACT

A computational study was conducted to understand the role of mist droplet size on the efficiency of fire suppression by a coflowing water mist injection system. The initial velocity and the location of injection were chosen to ensure a complete entrainment of the mist mass flow into the firebase. Fluent, a commercial CFD program was used to simulate the turbulent fluid flow with heat and mass transfer, along with a coupled treatment for stochastic droplet tracking and drying.

CFD simulation results show that within the range of fire plume velocities of this fire, water mist droplets in submicron or nano ranges vaporize quickly near the firebase and provide an instantaneous cooling behavior. However, as the droplet size increases, the vaporization zone moves relatively away from the firebase and the extent of fire cooling becomes low. For relatively coarse mists, droplets do not exhibit appreciable vaporization due to their shorter residence time as compared to the droplet vaporization times.

At fire plume velocities above droplet terminal velocities, residence times for various droplet diameters are not much different. However, the vaporization times increase sharply with increasing droplet diameters. Therefore, the fire cooling process not only depends on the droplet size but also on the fire flow field and the scale of fire through droplet residence time.

INTRODUCTION

Water mist fire suppression systems are being considered seriously as replacements for halon fire suppression agents in a wide range of applications. Some of these applications may include Class B pool fires, shipboard machinery, aircraft cabins, computers and electronic equipment, turbine hoods, incinerator installations, emergency generator rooms, switch gear rooms, and engine rooms [1]. Several conferences and workshops have been conducted in recent years as ongoing efforts to demonstrate the water mist fire suppression technology [2]. Basic studies have also been reported on some important variables affecting the suppression efficiency of water mist such as the mist mass flow, droplet size and size distribution, injection velocity, the location of injection, and spray orientations [3-5]. These studies have been conducted on relatively small-scale laboratory flames.

The key to the success of water mist fire suppression systems is the ability to generate a mist or fog of fine water droplets. Smaller droplets exhibit higher rates of mist entrainment into the fire plume and produce faster droplet vaporization rates, yielding higher suppression efficiency. Since heat extraction from the flame and oxygen displacement dominates in water mist systems, mechanisms that deliver very fine droplets capable of quick evaporation and instantaneous cooling effects are highly desirable. New generation NanoMist fire suppression technologies [6] may present opportunities to produce ultrafine droplets using simple and cost-effective mist generators.

The interaction of mist with the fire is a coupled process between dynamics of fires and the mist inlet properties. For example, the varying plume velocities of different scales of fires influence the droplet trajectories and hence the residence times of droplets. The residence time in turn influences the vaporization and heat absorption processes responsible for the fire-cooling behavior.

The objective of this study was to investigate the influence of mist droplet size on the fire suppression efficiency of coflowing water mist fire suppression systems.

SIMULATION

Description of Fire-Mist Model

A commercial CFD package, Fluent, was used to simulate the dynamics of fire-mist interaction. In the present study, Navier-Stokes equations along with energy and species equations are solved with suitable boundary and initial conditions. A thermal plume representing a medium scale fire is generated using a volumetric heat generation source within a specified volume at the center of the room. Since the primary objective of this study is to investigate the droplet size effect on the overall cooling behavior of a hot gas stream, combustion chemistry was not included. A standard κ-ε model was used to treat the turbulent flow. The radiation submodel was not used in this study. The mist was considered as discrete phase particles vaporizing and drying in the presence of a hot gas environment. The Lagrangian discrete phase model solves the particle trajectory at each time step and updates the droplet path. Stochastic modeling is activated for the turbulent flow. The conservation equations are solved by a finite volume method. Description of discretization, solution method, and the discrete phase droplet model are described in a series of Fluent User's Guide volumes [7]. Many of these model elements were described earlier [8,9].

Geometry and Grid Topology

Figure 1 shows the room and fire source geometry used in this study. The room geometry has a 3.6 x 3.6 x 3.6 m³ volume. The fire source is located at the center of the room.

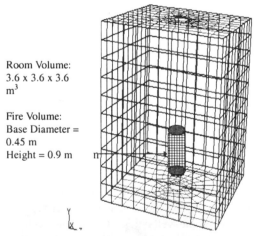

Room Volume:
3.6 x 3.6 x 3.6
m³

Fire Volume:
Base Diameter =
0.45 m
Height = 0.9 m

Figure 1: The geometry and grid topology of fire in a room

Fire source has a cylindrical shape with a diameter of 0.45 m and a height of 0.9 m and is located 0.3 m above the ground level. The total number of unstructured cells within the room and the fire source volume was in the range 50,000 –75,000.

Boundary Conditions

The sidewalls of the room were treated as pressure inlets and the top wall was considered as a pressure outlet boundary. The bottom of the room included a wall boundary. The volumetric heat release rate inside the fire volume was adjusted to provide a maximum plume temperature of approximately 1900 K. Boundary conditions for standard κ-ε turbulence model included intensity and hydraulic diameter approach at the boundary [7].

Mist Inlet Properties

Inlet variables for the water mist droplet includes the mist mass flow rate, initial droplet diameter, injection velocity, and injection location. The mist inlet location was fixed just below the firebase. The injection velocity was 3 m/s, which is fairly above the terminal velocity for all the droplets. This approach was used to eliminate uncertainties of droplets not being entrained into the fire or droplets falling back from the firebase. The mass flow rate was kept constant at 0.01 kg/s. The mono-disperse droplet spray sizes consisted of 0.5, 150 and 500 μm.

RESULTS AND DISCUSSION

Steady State Fire Plume

Numerical simulations were carried out to generate a steady state turbulent fire plume located inside a room. Figure 2 shows the steady state velocity vectors both inside and around the fire plume region. As expected, ambient air is drawn into the fire due to the lateral pressure gradient driven by the buoyancy force generated by the upward flow of hot gases. The total mass flow of air drawn from all sides into this non-combusting fire is 17 kg/s. The maximum fire temperature is 1900 K and the plume velocity is 8 m/s.

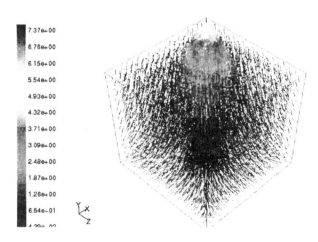

7.37e+00
6.76e+00
6.15e+00
5.54e+00
4.93e+00
4.32e+00
3.71e+00
3.09e+00
2.48e+00
1.87e+00
1.26e+00
6.54e-01

Figure 2: Velocity vectors (m/s) inside the room containing the fire.

Fire temperature contours are shown in Figure 3 below at two vertical panes. The maximum temperature is close to 1900 K and is located along the centerline closer to the firebase.

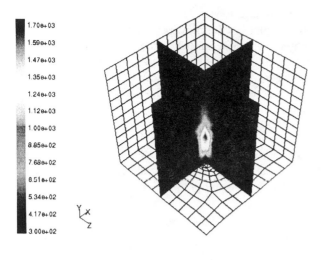

Figure 3: Temperature contours (K) of fire plume located inside the room.

Fire-Mist Interaction

Figure 4 shows the centerline temperatures 1.0 second after the application of various mono-disperse mists containing droplets of 0.5, 150 and 500 μm. The lowering of the peak temperature due to vaporization and heat extraction by the mist is substantial for submicron-size droplets and is about 800 K. Larger droplets also contribute to cooling, but to a lesser degree. Model calculations also indicated that 0.5 and 150 μm droplets vaporized completely within the fire zone. However, 500 μm droplets showed only partial vaporization. This was due to the shorter residence time compared to the required droplet vaporization time (d^2 –law).

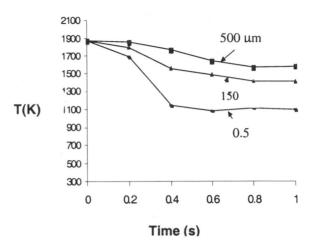

curves, all attaining their maximum effect within a second. This because of the direct injection of droplets at the firebase.

Next, rates of vaporization and enthalpy extraction along the centerline of the fire are shown in Figures 6 and 7 respectively. These profiles support the fact that smaller droplets vaporize and extract heat quickly near the firebase due to the very short droplet vaporization time, once they are

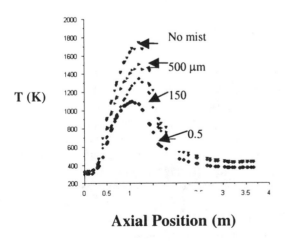

Figure 4: The effect of mist droplet size on the centerline temperatures of the fire plume.

The role of droplet diameter on the cooling histories of mists is summarized in Figure 5. The droplets of submicron size cool the flame within a second. Although to a lesser degree, larger droplets also show relatively sharp cooling

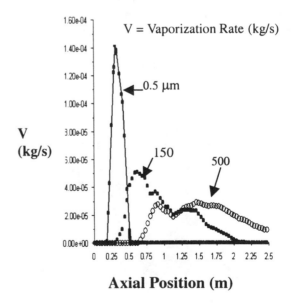

Figure 6: Vaporization rate of droplets along the centerline.

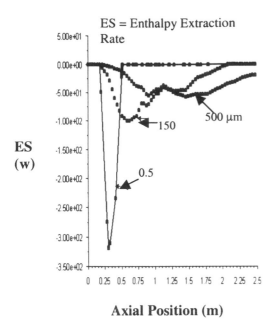

ES = Enthalpy Extraction Rate

ES (w)

Axial Position (m)

Figure 7: Rate of enthalpy extraction by water mist droplets along the centerline.

inside the firebase. As the diameter increases, the vaporization rate and hence the enthalpy extraction rate decreases. For large droplets such as 500 μm, the vaporization time is relatively longer compared to the droplet residence time. Therefore, most of them escape at the plume exit without vaporization. However, it is easy to show that if the flame height was doubled or the droplet path-length was increased, the droplets would continue to vaporize, contributing significantly to the cooling process. This observation suggests the implication of the scale of the fire in the cooling behavior brought about by the mist.

Figures 8A and 8 B show profiles of mass fractions of water

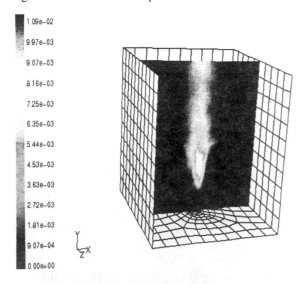

Figure 8A: Mass fraction of water vapor released by water mist droplets of diameter 150 μm.

within the fire plume for a relatively large droplet of 150 μm and a small droplet of 50 μm respectively. Results show that smaller droplets vaporize quickly close to the flame base due to the shorter vaporization time needed for submicron droplets. However, larger drops vaporize slowly and release vapors downstream depending on the droplet diameter and residence time of the droplets within the fire. This observation is important in the fire suppression technology since the specific location of cooling of the real flame is a crucial factor in reducing the rates chemical reactions. Cooling the flame downstream near the tip of the plume will not have a significant effect on the ability to put out the fire.

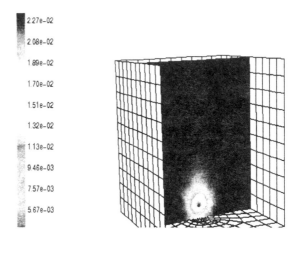

Figure 8B: Mass fraction of water vapor released by water mist droplets of diameter 0.5 μm.

The Effect of Fire Velocity Field on the Mist Cooling Behavior

Fire plume velocity is an important parameter in quantifying the ability of the mist to cool the fire. At low velocities typical of laboratory scale laminar flames, larger droplets such as 150μm show considerably different residence times compared to submicron particles such as 0.5 μm, as shown in Figure 9. Because the drag force acting on larger particles cannot counter balance the weight of particles, larger particles show longer residence time at low velocities. At velocities larger than the terminal velocity of most of the droplets, residence times become shorter and they are not much different for different size droplets. However, the vaporization times for large droplets increase considerably with increasing droplet diameter by the d^2-law. Hence at practical fire plume velocities such as 5-10 m/s, the cooling behavior of large particles differs considerably from small droplets since the vaporization times for these droplets become considerably longer compared to the short residence time. The fire scale effect thus influences the cooling through residence times of droplets within the fire.

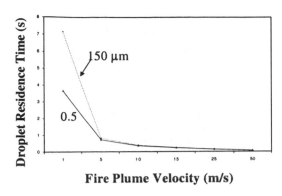

Figure 9: Variation of residence time of water mist droplets with fire plume velocity.

CONCLUSIONS

The interaction of water mist with a turbulent fire was investigated with a special emphasis on the effects of droplet size on the cooling behavior of water mist fire suppression system. In a coflow mist configuration, smaller droplets vaporized and extracted heat quickly from the fire very close to the flame base. As the droplet size was increased, the vaporization rate decreased and the heat extraction process moved farther away from the firebase. The cooling behavior became relatively inefficient. For a sufficiently coarse mist, the droplet vaporization time become longer than the residence time and droplets escaped the fire zone without significant vaporization. The fire cooling process thus depends not only on the droplet size but also on the fire flow field and the fire scale through droplet residence time.

REFERENCES:

1. Mawhinney, J.R, and Soloman, R., "Water Mist Fire Suppression Systems," in The Fire Protection Handbook, 18th ed., edited by A.E.Cote, Boston, and National Fire Protection Association. 1977, pp 216-248 (1997).

2. Committee on Assessment of Fire Suppression Substitutes and Alternatives to Halon, "Fire Suppression Substitute and Alternatives to Halon for U.S. Navy Applications, National academy Press, Washington, D.C (1997).

3. Prasad, K., Li, C., Kailasanath, K. "Simulation of Water Mist Suppression of Small Scale Methanol Liquid Pool Fires," *Fire Safety J.*, 33:185-212 (1999).

4. Prasad, K., Li, C., Kailasanath, K., Ndubizu C, Gopal, R., Tatem, P.A., "Numerical Modeling of Water Mist Suppression of Methane-air Diffusion Flame," *Combust. Sci. Technol*, 132:325 (1998).

5. Prasad, K., Li, C., Kailasanath, K. "Optimizing Water-mist Injection Characteristics for Suppression Methane-air Diffusion Flames," Twenty-Seventh Symposium (International) on Combustion, The Combustion Institute, (1998).

6. Adiga, K.C., "NanoMist Fire Suppression Technologies: Concepts, Droplet Generation, Application and Merits," Patent Pending (2001).

7. Fluent User's Guide Volumes 1-4, Fluent Incorporated, Centra Resource Park, 10 Cavendish Court, Lebannon, NH 03766 (1998)

8. Adiga, K. C., Ramaker, D. E., Tatem, P. A., and Williams, F. A., " Modeling Pool-like Gas Flames of Propane," *Fire Safety J.*, 2141:250(1989).

9. Adiga, K. C., Ramaker, D. E., Tatem, P. A. and Williams, F. A., " Numerical Simulation for a Simulated Methane Fire," *Fire Safety J .*, 443:458 (1990).

Proceedings of IECEC'01
36th Intersociety Energy Conversion Engineering Conference
July 29–August 2, 2001, Savannah, Georgia

IECEC2001-ES-2001-AT-78

USE OF CFD MODELING IN ASSESSING EXPLOSION CONSEQUENCES IN PETROCHEM AND POWER PRODUCTION FACILITIES

J. Keith Clutter
University of Texas at San Antonio
jclutter@utsa.edu

ABSTRACT

A computational fluid dynamic (CFD) model for industrial vapor cloud explosions (VCEs) is presented. The utility of the model in determining explosion consequences and performing accident investigations is discussed. The computational model is described with attention on a reduced combustion model used for efficient solutions. Example simulation results for realistic petrochemical facilities are presented. A methodology for using the model to determine the cause of accident explosions is also presented. **Keywords:** CFD; Explosions; Combustion; Accidents.

NOMENCLATURE

Cp_i	Specific heat of species i
E	Total Energy
e	Specific energy
g_i	Gravity force in i
H	Total Enthalpy
h_i	Enthalpy of species i
$h^\circ f_i$	Heat of formation of species i
$J_{h,j}$	Diffusion of enthalpy in direction j
$J_{i,j}$	Diffusion of species i in direction j
P	Pressure
T	Temperature
t	Time
u_j	Velocity component in direction j
x_j	Spatial dimension j
α_i	Mass fraction of species i
ρ	Density
τ_{ij}	Shear stress in i on surface normal to j
ω_i	Reaction source term for species i

INTRODUCTION

Industrial explosions resulting from vapor clouds are of interest to petrochemical and power production companies. Recently, computational fluid dynamic (CFD) models have been routinely used to access the potential consequences from such explosions. These same models have proven very useful in helping to determine the cause and origin of explosion accidents.

The analysis of the consequences from VCEs typically occurs as part of a broad risk assessment for a facility. The number of explosion scenarios to be considered is large and efficiency in simulations is a necessity. To this end, the author has developed a CFD code tailored for the VCE scenario. To ensure efficiency, a reduced combustion model has been developed for the simulation code. This model is detailed by Clutter & Luckritz (1999) and a brief review is provided here.

After a description of the computational model, example results for two VCE scenarios are presented. The first is a VCE initiated in the simplified configuration of Figure 1 for which experimental data is available (Mercx, Popat & Minga). Next, prediction for a VCE inside a full-scale off-shore module are presented. This data has been used by many investigators to evaluate complex and simplified explosion analysis tools (Selby & Burgan). Even though the scenario is an off-shore module, the environmental factors are similar to on-shore facilities as well. These explosion scenarios have been used to evaluate the accuracy of the CFD model using the reduced combustion model.

This same model used to access potential VCE consequences for risk assessment studies has been used very successfully in investigating VCE accidents that have caused both losses in life and property. Here, a brief outline of the methodology using the CFD model for accident investigations is presented. The CFD model used here is called the Computational Explosions and Blast Analysis Model (CEBAM). CEBAM has been used over the last couple of years to determine the cause and origin of several explosion incidences. Many times, information from such cases is sensitive and release of data is not possible. Therefore, here a fictitious incident is used to demonstrate the analysis process.

COMPUTATIONAL MODEL

The chemically reacting flow found in the VCE problem is governed by the Navier-Stokes equations with additional equations for the reaction process and multispecies composition involved. The series of governing equations include one for continuity,

$$\frac{\partial}{\partial t}(\rho) + \frac{\partial}{\partial x_j}(\rho u_j) = 0, \tag{1}$$

and momentum,

$$\frac{\partial}{\partial t}(\rho u_i) + \frac{\partial}{\partial x_j}(\rho u_i u_j) + \frac{\partial P}{\partial x_i} = +\frac{\partial}{\partial x_j}(\tau_{ij}) + \rho g_i \tag{2}$$

where the shear stress in the ith direction on a surface normal to the jth direction is represented by τ_{ij}.

The energy equation, in terms of enthalpy, is

$$\frac{\partial}{\partial t}(\rho E) + \frac{\partial}{\partial x_j}(\rho u_j H) = -\frac{\partial}{\partial x_j}(J_{h,j}) \qquad . \tag{3}$$

The diffusion of enthalpy in the jth direction is represented by the term $J_{h,j}$. The internal energy is related to the individual species enthalpy (h_i) through the relationship,

$$e = E - \frac{1}{2}\vec{u}^2 = \sum_{i=1}^{NS}\alpha_i h_i - \frac{P}{\rho} \tag{4}$$

where

$$h_i = h_{f_i}^o + \int_{T_R}^T C_{p_i}\, dT \tag{5}$$

with C_{pi} and $h_{f_i}^o$ being the specific heat at constant pressure and the heat of formation of species i, respectively. A continuity equation of the form

$$\frac{\partial}{\partial t}(\rho \alpha_i) + \frac{\partial}{\partial x_j}(\rho \alpha_i u_j) = -\frac{\partial}{\partial x_j}(J_{i,j}) + \omega_i \tag{6}$$

is needed for each species involved in the combustion process. The term $J_{n,j}$ represents the mass diffusion of species n and ω the chemical reaction source term for the species.

In reacting flows, viscous effects are manifested through dissipation of species and heat as well as the dynamics of the flame involved in the combustion. For the VCE application, the explosion is driven by the ignition of a preformed cloud. Therefore, the combustion process is not one dominated by reactant-oxidizer mixing and neglection of the viscous stress terms is a reasonable reduction.

The viscous and turbulence effects on the combustion process are incorporated through a reduced model based on flame speeds. This is detailed in Clutter & Luckritz (1999) along with comparisons to other VCE analysis tools. The model is briefly reviewed next.

Reduced Combustion Model

The VCE problem is represented as composed of three potential species; (1) a reactant mixture (Rx), (2) a product mixture (Pd), and (3) ambient air. The reaction source term of Equation 6 can be represented in a variety of ways. For the current model, a single-step reaction of the form

$$Rx \rightarrow Pd \tag{7}$$

is used. The source term for the reactant mixture takes the form

$$\omega_{Rx} = -k_e \rho \alpha_{Rx} \tag{8}$$

where α_{Rx} is the local concentration of the reactant mixture and k_e is an effective reaction rate.

To derive the reduced model, first consider the relationship between the laminar flame speed of a mixture (u_L), its diffusivity characteristics (υ) and the reaction rate is of the form (Warnatz, Maas & Dibble, 1995)

$$k \propto u_L^2/\upsilon \qquad . \tag{9}$$

Viscous effects such as turbulence enhances the diffusivity of a mixture. This in-turn will effect the reaction rate and can be quantified on a global scale by measuring the speed of the flame since the velocity of the flame is a result of the balance between the reaction and diffusion processes (Echekki & Chen, 1999). This is the general idea behind several simplified models used by the petrochemical industry for VCE analysis and is incorporated in the current model.

The current implementation adds more details, beyond simpler VCE models, in terms of representing the explosion environment (Clutter & Luckritz 1999). For a given area of congestion, which will generate a particular turbulence field, empirical data has been collected and analyzed to correlate the level of congestion to a characteristic flame speed. Using this approach, the effective reaction rate takes the form $k_e = 1/\tau_c$ where τ_c is a chemical time scale related to the flame speed in the congested area and the reaction rate varies accordingly.

The implementation of the flame speed based model requires the definition of congested zones when a simulation is set up. Consider the scenario shown in Figure 1. The volume containing the array of pipes would be defined as a congested zone and have assigned to it a characteristic flame speed based on empirical data. This will define the reaction rate within this volume. Outside this volume, where there is no congestion, the flame is assumed to decelerate.

SIMULATION OF EXPERIMENTAL CONFIGURATIONS

Two sets of experimental data have been used to evaluate the reduced combustion model developed in the current CFD codes. The first set of data was generated through the Extended Modeling and Experimental Research into Gas Explosions (EMERGE) Project. In this project, explosions in rectangular configurations of pipes such as Figure 1 were initiated and overpressures were measured at various distances. Experiments were conducted for small, medium, and large configurations and a variety of fuels were used. A review is provided in Mercx, Popat & Linga.

Clutter and Luckritz (2000) show comparisons between output from CEBAM and some of the experimental data from the small rig. Clutter (2001) presents a complete comparison using CEBAM for small and medium size tests. Some of the data is presented here in Figures 2 and 3. The purpose of these earlier studies was to determine the accuracy of the reduced combustion model. The comparison between measured and predicted overpressures suggests the reduced model captures the relevant combustion aspects of the VCE.

The second set of data used here was collected through the Blast and Fire Engineering for Topside Structures Joint Industry Project, Phase 2 carried out in Europe (Selby & Burgan). This project collected data from explosions inside structures representing full-scale, off-shore modules found on processing platforms. Tests in rigs of various size were performed using a range of fuel equivalence ratios. The fuel used was natural gas.

A variety of configurations (i.e., objects in the rig and boundary confinements) were tested. The case simulated here is for a rig measuring 25.6 m x 8 m x 8 m and the representation of the rig within CEBAM is shown in Figure 4. In addition to the roof and the north wall, there was a south wall in place on the rig. It is not shown in Figure 4 to facilitate the presentation of the interior of the rig.

The scenario simulated with CEBAM was tested twice and referred to as Test 1 and Test 3 in Mercx, Popat & Linga. Figures 5 through 7 show comparisons between the measured and predicted pressure time histories at locations inside and outside the rig. The data from the tests was provided in the form of idealized triangular pulses which is evident in the figures. Figure 5 shows some of the spread seen in the experimental measurements inside the rig. Outside the rig, the pressure time histories measured in the appeared more repeatable.

For this configuration and level of congestion, the characteristic flame speed, based on empirical data, was on the order of M=.2. This value used with the reduced combustion model produces a very good comparison between the measured and predicted overpressures. This further increases the confidence in the flame-speed based combustion model.

Using the reduced combustion model approach produces a very efficient computational tool. The use of defined congestion regions such as process units, instead of defining each item of the unit, as some simulation tools require (Clutter and Luckritz, 2000), dramatically reduces preprocessing time. The flame-speed based combustion model requires the solution of fewer equations which reduces calculation times.

ACCIDENT INVESTIGATION

CEBAM has also been used to investigate various accidental explosions. The methodology is reviewed here using a fictitious facility. The facility is depicted in Figure 8 and has been created to include items found in typical petrochemical and power production facilities. There are four buildings depicted along with a circular storage tank and an elevated deck. On three of the buildings, doors are included. The transparent boxes denote process units consisting of various levels of congestion.

An explosion event is composed of a multitude of physical and chemical processes. The chemical phenomenon involved is associated with the combustion process discussed earlier. This is the avenue through which flame-speed and other issues enter the problem. The physical phenomenon includes the propagation of the blast from the explosion and the response of the structures such as the building walls and doors.

To simulate an explosion incident for a consequence analysis or accident investigation, a representation of the facility is constructed. Structural items such as buildings and storage tanks are explicitly defined as well as areas of congestion such as process units. CEBAM allows for the definition of buildings as a collection of components such as wall and roof sections, windows, doors, etc. In the example facility four doors have been defined. When using models to investigate VCE incidents, it is essential to include items that respond to blast loadings. Here, the doors are used to show how items such as metal panels, windows, and other structural members are used in the analysis process.

To identify the most probable explosion source, various scenarios are evaluated within the virtual framework of the CFD model. Figure 9 shows an example initial setting for the CEBAM analysis. In this scenario, at the time of ignition a vapor cloud is located in the congestion zone near the circular storage tank. Also visible in the figure is numerous probes which record information during the analysis process.

The use of the CFD based model ensures phenomena important in the explosion process are captured. For example, vapor cloud explosions typically occur in congested regions and it is the turbulence field generated in these regions that sets the severity of the explosion. During the VCE, the pressure build-up can push the explosive cloud out of these regions. For some cases, only a portion of the cloud initially situated in areas such as process units contribute. Another possibility is that during the explosion, explosive mixture can be pushed into congested areas and generate a more severe explosion than expected if just the location of the initial cloud is considered. To capture such occurrences, the explosion model must include cloud migration and interaction with facility components as well as an accurate accounting of the flame dynamics. The CFD based analysis model can correctly represent this aspect of VCEs.

The propagation of blast from the VCE is highly dependent on the layout of the facility and directly affects the damage created during a VCE incident. Examples are present in the scenario used here and are visible in Figure 8. In this case, the presence of the elevated deck will focus the blast towards some buildings and the blast will not decay with distance in the same manner as seen in free field conditions. The relative locations of the facility components to each other have varying effects. The corner and alley created by some of the buildings will elevate the pressure and impulse in the region of one door beyond that seen in simple reflections.

The building nearest the circular tank will generate a shielding effect and depending on the VCE location could reduce the loads on the building behind it below that expected from free field data. The dynamics of blast propagation in an environment including focusing and shielding is a highly nonlinear process that can not be accurately represented with simplified models currently available. This is another advantage of using a CFD based model for VCE accident investigations.

The process of determining the most probably source for a VCE incident is an effort to reproduce an actual physical event. Any shortfalls in the employed explosion and blast model in representing all aspects of real VCEs hinder the investigation effort. Given the variety of physical and chemical phenomena involved and their highly nonlinear nature, use of a model that does not accurately represent these phenomena could result in inaccurate conclusions. This is a major reason the CFD based model has been used and has shown good results in recent accident investigations.

The investigation of a VCE accident can be considered to be a reverse engineering effort. There are specific parameters that can vary over defined ranges. Example parameters are cloud size, cloud location, fuel involved, equivalence ratio in the cloud, and ignition location. There are also variables dependent on these parameters which can be measured. The two key variables are pressure and impulse at various locations. These variables produced in the damaged items found after an accident. The parameters essentially establish a box within which lies the true solution to the VCE incident. The variables function together as a compass to point to the location where the solution lies within the box.

For the example analysis here, the simulated doors are used as damage indicators. In practice, the damage to these doors is analyzed, as described by Clutter and Whitney to determine the blast load they experienced during the VCE incident. The load criteria may not be a single value for pressure and impulse but may be represented by a curve in P-i space that denotes a collection of loads that would have created the damage. An example data set is shown in Figure 10. Also, if the damage indicator experiences total failure (i.e. a window breaking) then it can only be concluded that the experienced load lies somewhere in the region of P-i space above one of the damage curves.

Here, the accident has been created with the computational model. Therefore we know the point in P-i space that each indicator, doors d1-d4, experienced. This also streamlines the presentation of the analysis process. However, the same principles and practices are valid for cases where the criteria for some indicators are in the form of a curve or space in P-i space. For this example VCE incident, the experienced loads are shown in Figure 11 for the four doors.

Various parameters affect the level of severity in a VCE accident. The unknown parameters might be the size of the vapor cloud and the fuel involved, if the source is located in a unit with multiple reactive compounds. In this example, only the cloud location and size are assumed to be unknown to demonstrate the analysis process.

After determining the details of the damaged items, various candidate solutions to the question of what source was involved are evaluated. Here, a set of six simulations is carried out in which cloud location and size are varied. For all cases, the cloud was assumed to be a stoichiometric methane-air mixture, ignited at the center of the cloud, at the ground plane. Figure 12 shows the footprints of the 3 assumed locations. The size of the smaller cloud was 80 ft x 50 ft in plane and 20 ft in height with the larger cloud having the same footprint but a height of 50 ft.

To measure the loads, virtual pressure probes are placed in front of the four doors. Example load histories are shown in Figure 13. The simulated loads on the 4 damage indicators are shown in Figure 11. Consider the output from the three cases with the smaller cloud. The first noticeable fact is that even though the cloud at location 1 comes close to reproducing the experienced loads on doors 2 and 4, there is a wide difference on doors 1 and 3. It is also clear that changing the cloud location generates a different trend in the resulting load for each door. This is due to the highly nonlinear nature of the governing equations and the focusing and shielding effects discussed earlier.

With the larger cloud, the load trend with cloud location depends on the particular door. For doors 1, 3, and 4, there is some similarity in the behavior of the load for a given cloud size as the location is moved. However, in the case of door 2, the trend for the larger cloud is quite different. It is clear that out of all six candidates, the case of the larger cloud at location 3 seems to come the closest at matching the experienced load on all four doors. This conclusion is based on visual inspection of the P-i space results and a more rigorous approach is desired.

The VCE accident investigation is analogous to a reverse engineering problem. In such problems, it is advantageous to identify trends between adjustable parameters and dependent variables. Optimization techniques can be employed to automate the analysis process and help select other candidates to be evaluated. In this example, a candidate denotes a particula cloud size and location combination. Optimization techniques have been developed primarily for use in identifying the ideal settings of specific parameters to achieve a certain design value in the variables of interest (Gill, Murray, Wright). The design variables are dependent on the design parameters and certain quality factors are established in order to identify trends between the parameter settings and the variable output. Here our "design" parameters are the size and location of the vapor cloud and our "design" variables are the blast magnitude and impulse at the doors.

To use an optimization approach, functions are needed that provide measures of quality for various candidate solutions. In practice, a finite set of candidates will be initially selected and evaluated. In this example, the six cases shown in Figure 11 would serve as an initial set. The quality of this set would be used to identify trends as variables such as cloud size and location are changed. These trends would be used to select subsequent candidates.

There can be weight functions imposed if certain criteria is more important, but in this example case, it is equally important to match the experienced loadings on all doors. For the VCE accident problem, the form of the quality function most incorporate the difference in pressure magnitude and impulse from that experienced and that generated by the candidate at all doors.

Figure 14 shows a collection of contours denoting the magnitude of deviation from the experienced load and the predicted load, as a function of cloud location and volume. There is a plot for variation in pressure (dP) and impulse (di) at the four doors, denoted by the subscript. The contours are based on calculations at a finite set of candidates shown in Figure 14 as solid black circles. The plotted variable is the L2-norm of the percent of deviation. For example, $dP = \log[abs(P_{exp}-P_{cand})/P_{exp}]$ where P_{exp} is the experienced load determined by the damage analysis and P_{cand} is the load generated by the candidate solution. The same formula is used for impulse. The more negative the value, the closer the candidate is to the actual explosion source.

Even though the data for each door shows an obvious minimum, the trend across the remainder of the design space varies differently for each door. Early in the candidate selection and evaluation process, these differences will be key and the problem of local minimum can appear. This is why a conclusion should not be based on a single damage indicator and a quality factor must be used that incorporates all indicators. Here, a quality factor (Q) is defined based on the sum of all the deviations in pressure and impulse. Figure 15 shows the variation of Q with cloud location and volume. This combination of data from all the damage indicators processes a better-behaved function across the parameter space. The minimum appearing at a cloud location of 200 ft and a volume of 100,000 ft^3 is in fact the explosive source used to produce the simulated damage on the doors.

Clutter and Whitney (2000) discuss in detail the complete VCE accident investigation methodology. Here, an example of how candidate explosive sources are selected and evaluated using CEBAM has been provided. The goal is to determine a source that produces a best match to actual structural damage. The parameter space can be of very high order and the use of an optimization technique can improve the investigation process.

CONCLUSION

Industrial vapor cloud explosions can produce tremendous losses in both petrochemical and power production facilities. CFD models have proven valuable in assessing potential consequences. This class of analysis tool incorporates many important factors that simpler models, typically used for VCE analysis ignore. The reduced combustion model used in CEBAM has proven accurate while improving efficiency.

The use of CFD models has also proven advantageous in investigating VCE accidents. These models include more of the realistic aspects of the explosion than simplified models. This results in simulations much closer to the actual event. Using the CFD model with an optimization process helps to efficiently select and evaluate numerous potential explosion sources till a best match is identified.

REFERENCES

Baker, Q.A., Tang, M.J., Scheier, E.A., & Silva, G.J. (1996) Vapor Cloud Explosion Analysis. *Process Safety Progress.* (Vol. 15, No. 2, pp. 106-109).

Clutter, J.K., (2001). A Reduced Combustion Model for Vapor Cloud Explosions Validated Against Full-Scale Data, *Journal of Loss Prevention in the Process Industry*, (Vol. 14, pp. 181-192).

Clutter, J.K., & Luckritz, R.T. (1999). Comparison of Blast Curve and a Reduced Explosion Model to Experimental Results. *International Conference and Workshop on Modeling the Consequences of Accidental Releases of Hazardous Materials*, ISBN 0-8169-0781-1(pp. 515-540).

Clutter, J.K., & Luckritz, R.T. (2000). Comparison of a Reduced Explosion Model to Blast Curve and Experimental Data, *Journal of Hazardous Materials*, (Vol. 79, No. 1/2, pp. 41-61).

Clutter, J.K., and Whitney, M. (2000). Use of Computational Modeling to Identify the Cause of Vapor Cloud Explosion Incidents. *CCPS International Conference and Workshop on Process Industry Incidents*, ISBN 0-8169-0821-4 (pp. 497-517).

Echekki, T., & Chen, J.H. (1999) Analysis of the Contribution of Curvature to Premixed Flame Propagation. *Combustion and Flame.* (Vol. 118, No. 1/2, pp. 308-311).

Gill, P.E., Murray, W. and Wright, M.H., (1981) Practical Optimization, Academic Press, Inc.

Mercx, W.P.M., Popat, N.R., & Linga, H. Experiments to Investigate the Influence of an Initial Turbulence Field on the Explosion Process. *Final Summary Report for EMERGE, Commission of the European Communities.* Contract EV5V-CT93-0274.

Selby, C.A., & Burgan, B.A. Blast and Fire Engineering for Topside Structures - Phase 2: *Final Summary Report U.K. Steel Construction Institute* (SCI Pub. No. 253). Ascot.

Warnatz, J., Maas, U., & Dibble, R.W. (1995) Combustion, Physical and Chemical Fundamentals, Modeling and Simulation, Experiments, Pollutant Formation, *Springer-Verlag.* New York.

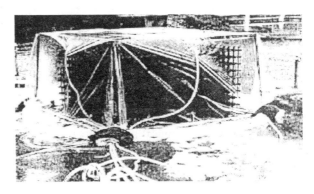

Figure 1. Test rig used in EMERGE tests.

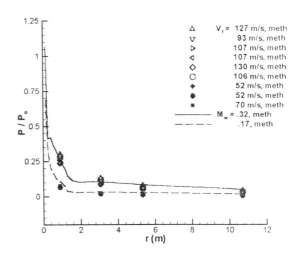

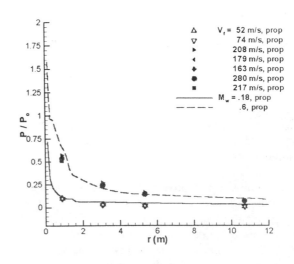

Figure 2. CEBAM predictions (lines) compared to test data (symbols) for small-scale EMERGE tests.

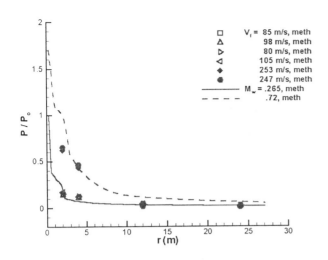

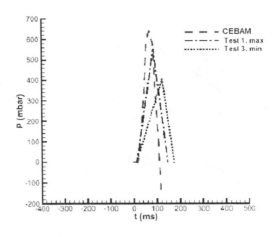

Figure 5. Comparison of measured and predicted blast loads inside the BFETS rig.

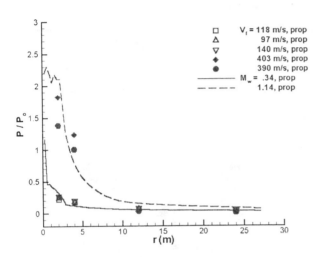

Figure 3. CEBAM predictions (lines) compared to test data (symbols) for medium-scale EMERGE tests.

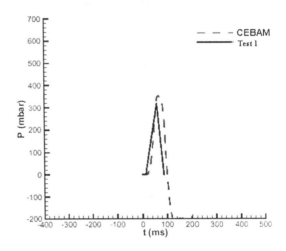

Figure 6. Comparison of measured and predicted blast loads at a second location in the BFETS rig.

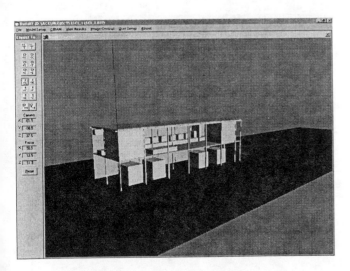

Figure 4. CEBAM representation of an off-shore rig tested in the BFETS Project.

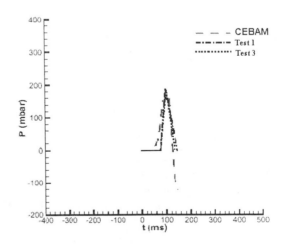

Figure 7. Comparison of measured and predicted blast loads outside the BFETS rig.

337

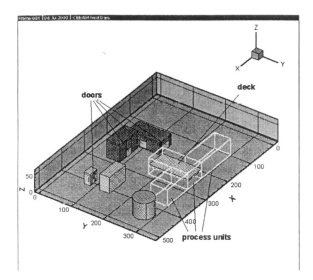

Figure 8. Representation of fictitious facility used in the accident example.

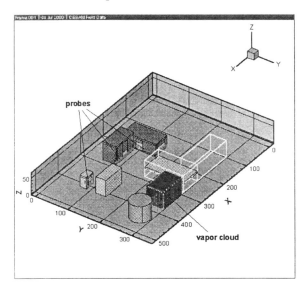

Figure 9. Location of probes and vapor cloud for one of the simulations.

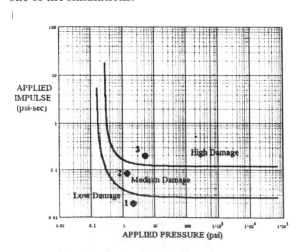

Figure 10. Example structural response data shown in P-i space.

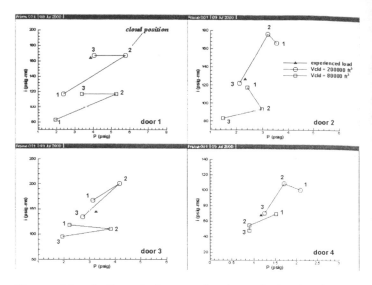

Figure 11. Indicated damage levels and simulated blast loads on the doors for various cloud locations and size.

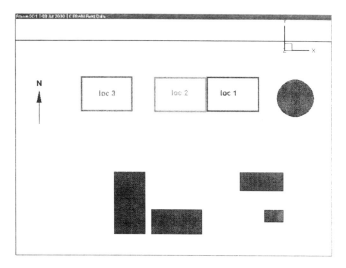

Figure 12. Location of simulated candidate explosion sources.

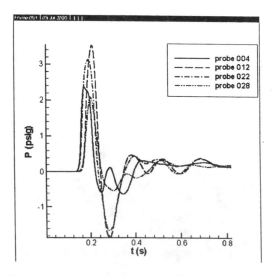

Figure 13. Example load histories at the center of the four doors taken from one of the

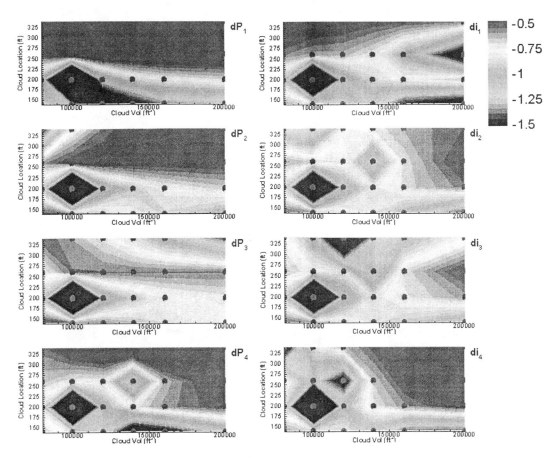

Figure 14. Deviation in pressure and impulse as a function of cloud volume and location.

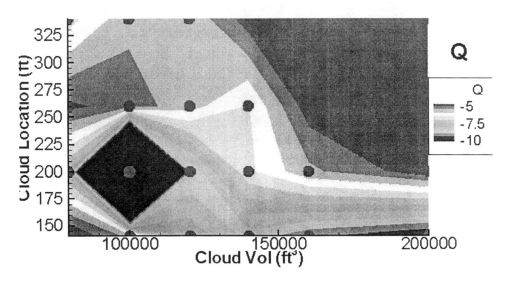

Figure 15. Quality factor as a function of cloud volume and location.

339

IECEC2001-ES-20

Control of Kerosene Spray Flame Characteristics from a Swirl Bu

B. Habibzadeh and A. K. Gupta
The Combustion Laboratory
University of Maryland
Dept. of Mechanical Engineering
College Park, MD 20742

Abstract

Results are presented on the effect of radial distribution of swirl and distribution of combustion airflow in a burner on spray flame characteristics. Direct flame photography and droplet characteristics in terms of size, velocity and number density have been obtained for different combination of swirl and combustion airflow in the double concentric swirl burner using an air-assist fuel nozzle. The burner allowed independent variation of swirl and combustion airflow in the inner and outer annulus of the burner. Maintaining either the radial distribution of swirl or airflow distribution in the burner constant and varying the other parameter have examined the flame features. Results have been obtained for a swirl combination of 30°/30°, 30°/50 ° and 50°/30 ° in the inner/outer annulus of the burner. The distribution of air was changed as 25/75, 50/50 and 75/25 percent in the inner/outer annulus, respectively.

The results show significantly different distribution of droplet size in the resulting spray as well as the flame plume configuration. The noise levels were also very different. The results elucidate the role provided by swirl and airflow distribution in a burner to control the flame structure and combustion instability.

Introduction

The control of flame structure is one of the most important parameter since this can provide means to control flame instability, flame signatures and emission of trace pollutants. The size, velocity and distribution of droplets in the spray flame significantly affect the flame structure. Smaller droplet size in spra
more desirable since the smalle
instrumental for enhancing flame stab
Smaller size also reduces the residen
droplets in the hot combustion
vaporization and reduced emission leve

Our objective has been to use
combination of burner input
parameters for reducing the size of d
relocating the droplets within the f
relocation of droplets at specific reg
flame is important from the point c
combustion instability. If the local h
distribution in a flame can be changed
be possible to eliminate the combustior
at a specific operational condition w
need for external energy, such as s
control.

The usual approach adapted to
combustion instability in flames is by s
or passive techniques. Passive control
as useful and reliable than the acti
without using any external force or
Furthermore, active control is in ger
expensive, complicated and requires inc
of some external energy to control the
in harsh combustion environments. T
requirements as well as the operation of
under harsh combustion conditions ca
challenging. Therefore, finding means
the flame features using some passive a
attractive, reliable and economical
requires less components and control de

Control techniques using passi
allow one to obtain a stable flame
changing or controlling the flame char

If the spatial distribution of local chemical energy in the flame can be changed then the driving mechanism for the instability can be removed or relocated to some other power setting. So one can anticipate it should be possible to control the flame structure through changes in the distribution of local heat release rate in order to alleviate the combustion instability in the flame. In the case of spray combustion this change in local heat release rate can be accomplished via changing the droplet size, velocity and number density distribution of the droplets.

Smaller size of droplets from a fuel spray can be obtained by changing the nozzle operational parameters. The operating fuel pressure is often a fixed quantity for a given nozzle. Changing to a different type of nozzle may translate to increased costs and hardware matching difficulties with the combustor. In most cases air-assist or air-blast atomizer is used for the atomization fuels.

Effervescent atomizers, in which the atomization gas and liquid fuel are mixed inside a channel or large cavity prior to the discharge of fluid from the nozzle exit, provide smaller size of droplets. In most cases air is used as the atomization fluid. However, our studies using atomization gases of different dielectric properties have shown that it is possible to achieve better atomization with other gases.

The studies carried out using CO_2, steam, oxygen, Ar, He, N_2 have shown that the atomization quality, combustion and emission characteristics and flame plume configuration are significantly affected by the dielectric properties of the atomization gas[1-7]. As an example the use of oxygen enriched atomization air provided a very intense spray flame[2]. The flame with oxygen enriched atomization gas was short, wider and of bluish-white color. Steam assisted atomization provided smaller size droplets[3]. It is therefore possible to select an atomization gas or gas mixture that provides good atomization characteristics, desired flame stability, and also results in reduced emission of pollutants. Achieving some physical insights into the spray characteristics is therefore very important.

The increase in noise levels from the flame at a single tone may be an indication of possible propensity for combustion instability in a combustor. Less tonal noise may mean less

instability. It is therefore desirable to achieve uniform and finer spray as much as possible. A control of such features should allow one to change flame shape or size as well as other important physical characteristics. A non-fluctuating flame showed less noisy flame and high stability.

There are several variables that affect the spray flame characteristics. These include fuel atomization nozzle type, atomization gas property, fuel nozzle location in the combustor, and swirl and flow distribution in the burner. A series of experiments have been conducted using the double concentric swirl burner having a air-assist fuel nozzle located on the central (longitudinal) axis of the burner.

Experimental facility

A double concentric swirl burner used here has a centrally located fuel nozzle surrounding which there are two annuli. The inner and outer annuli of the burner allow independent control on the variation of swirl and flow distribution through the burner. The fuel nozzle type and location in the burner can also be changed which permits variation to the premixing distance in the burner. The facility allows examination of the effect of atomization gas and radial distribution of swirl and combustion airflow in the burner on the spray flame characteristics. A swirl blade cascade of any known blade angle setting can be placed in the two annuli of the burner to provide the desired combination of co- or counter-swirl distribution in the burner.

A schematic diagram of the experimental burner is shown in Fig. 1. An important feature of this facility is the capability to vary radial distribution of swirl and airflow distribution in the burner and the combustor dome geometry. A stainless steel air-assist fuel nozzle, nominally rated for 0.5 GPH, is used as the fuel nozzle. Any fuel nozzle type can be easily adapted and examined in the facility. The burner is mounted to a computer controlled traversing mechanism assembly so that spatially resolved measurements in the flame can be made at any position in the spray and spray flame downstream from the nozzle exit. In the present study kerosene has been used as the fuel.

The effect of swirl and combustion airflow distribution in the burner on spray flame

characteristics has been examined using an air-assist spray nozzle. The objective here is to control the spray flame features. This is achieved by determining the global features of flame as well as size, velocity and number density distribution of droplets produced with different distribution of swirl and combustion airflow in the burner. The goal is to determine if one can control the distribution of droplets in the sprays and subsequently controls the flame plume configuration. This control should allow one to transport droplets from one location in the flame to some other desired location in the flame.

Our approach is to use different combination of swirl in inner and outer annulus of the burner, airflow distribution in the two annuli, and atomization gas. These parameters significantly affect the flame stability and control the flame features. Air is used as the atomization gas. We also plan to use other atomization gases such as methane, hydrogen, CO_2, N_2, O_2, steam or some combination of these gases. These gases provide the role of chemical and physical property of the atomization gas on fuel droplet atomization. In each case the momentum of the gas is maintained constant. The atomization gas provides important role in achieving smaller size droplets while the swirl and flow distribution provides control for spray droplet distribution.

Results and Discussion

The effect of swirl and combustion airflow distribution in the burner, using air as the atomization gas, on global flame features is shown in Fig. 2. In all cases the equivalence ratio and momentum of the atomization gas was maintained constant. The parameters varied include radial distribution of swirl and flow into the burner. The results showed significant effect of radial distribution of swirl and airflow distribution in the burner on flame plume characteristics.

Stronger swirl in the inner or outer annulus at low flow rates in the inner annulus resulted in a thin long flame while the stronger swirl both annuli resulted in a compact flame. Introduction of larger proportion of airflow via the inner annulus resulted in a more compact flame in comparison to the case with more air flowing to the outer annulus. The flame standoff distance from the burner exit could be changed with changes in input parameters to the burner. This

revealed that the distribution of droplets in the burner must be significantly different under these conditions. In order to determine the distribution of droplets in these flames measurements have been obtained from selected flames.

Droplet size characteristics have been obtained for the effect of swirl distribution and flow distribution in the burner. The swirl distribution in the annulus 1/annulus 2 of the burner was varied as 30°/30°, 30°/50° and 50°/30°. The effect of flow distribution in the burner was varied by changing the airflow distribution in the burner in the ratio of 50%/50% or 25%/75% in the inner/outer annulus, respectively, so that the overall equivalence ratio of the flame remained unchanged. The results on the droplet size, velocity and number density and droplet time of arrival in the control volume are presented in Figs. 3-7. Data has been obtained at various spatial positions in the flame.

The results presented in Fig. 3 show the effect of flow distribution in the burner on droplet SMD. The results are reported at 5, 10, 20, 30, and 40 mm downstream from the fuel nozzle exit. Much smaller droplet sizes were found near the spray boundary with equal distribution of flow in the two annuli. Similarly the arithmetic mean diameter is much smaller with equal flow distribution in the two annuli, see Fig. 4. Near to the nozzle exit, at the spray centerline, larger droplet sizes are found with higher flow through the inner annulus, see Figs. 3 and 4.

Mean axial velocity in the center of the spray is higher with smaller flow in the inner annulus (see Fig. 5). However, at the spray boundary higher velocity is seen with smaller flow in the outer annulus. This suggests that less proportion of flow in the inner annulus results in increased number of smaller droplets enter the center region of the spray which assists combustion and these droplets then undergo lack of deceleration. Similar trend can be seen for the outer annulus with low and high proportion of flow. The effect of swirl distribution on the mean axial velocity at a fixed flow distribution in the burner is shown in Fig. 6. The results show higher mean axial flow velocity on the central axis of the spray with stronger swirl in inner annulus 1. Similarly the droplet number density and volume flux distribution from the burner was affected. Information was obtained on the time of arrival of droplets so as to determine if the droplets are transported from one region of the spray to some other location.

The results on the time of arrival of droplets at 5 (top row) and 40 mm (bottom row) downstream from the fuel nozzle exit are shown in Fig. 7. Results shown in Fig. 7(a) are at the spray centerline while those presented in Fig 7 (b) is at the spray boundary. Indeed it is possible to transport the droplets from a location in the spray to some other location with a change in flow distribution in the burner.

Summary

The results show that the radial distribution of swirl has an effect on the spatial distribution of droplet size, droplet number density and axial velocity. An increase in swirl strength to the inner annulus increased both the axial velocity and SMD at the longitudinal axis through the spray centerline. This effect is more pronounced with increase in axial distance downstream of the burner exit. An increase in swirl strength in the outer annulus provided an increase in SMD and negligible effect on axial velocity at the spray longitudinal (central) axis. However, at radial positions outside the flame boundary the axial velocity for the $30^\circ/30^\circ$ swirl distribution case was found to be more than for the $50^\circ/30^\circ$ swirl distribution cases. The effect of combustion

Air-flow distribution showed significant effect on the size and velocity distribution of droplets in the spray flames. The flame was much shorter with 50%/50% (inner/outer annulus) flow distribution as compared to 25%/75% distribution. A capability to change the distribution of droplet size, velocity and number density distribution in spray flames means that one can control the flame structure, and hence combustion instability, at any thermal loading of the combustor.

Acknowledgements

This research was supported by ONR under contract N00014-99-1-0491, Program Manager Dr. Gabriel D. Roy. This support is gratefully acknowledged.

References

1. Aftel, R., A. K. Gupta, C. Presser and C. Cook.: Gas property Effects on Droplet Atomization and Combustion in an Air-Assist Atomizer, Proc. 26th Symposium (International) on Combustion, pp. 1645-1651, 1996.
2. Gupta, A. K., T. Damm, C. Cook, S. R. Charagundla and C. Presser.: Effect of Oxygen-Enriched Atomization Air on the Characteristics of Spray Flames, 35th Aerospace Sciences Meeting, January 6-10, 1197, Reno, NV, 1197, Paper No. 97-0268.
3. Gupta, A. K. and Megerle, M.: Effect of Steam Assisted Atomization on Spray Flame Characteristics, 33rd AIAA/ASME/SAE/ASEE Joint Propulsion Conference, July 6-9,1997, Seattle, WA, 1997, Paper No. 97-2839.
4. Gupta, A.K. Megerle, M, Charagundla, S.R. and Presser, C.: Spray Flame Characteristics for High Temperature Gas-Assisted Atomization, ILASS-98, Sacramento, CA, May 1998, pp.354-358.
5. Presser, C., Gupta, A.K., and Semerjian, H.G.: Aerodynamic Characteristics of Swirling Spray Flames, Combustion and Flame, Combustion and Flame, vol. 92, 1993, pp.25-44.
6. Presser, C., Gupta, A.K., Semerjian, H.G., and Avedisian, C.T.: Journal of Propulsion and Power, vol. 10, No. 5, 1994, pp. 631-638.
7. Presser, C., Gupta, A.K., Avedisian, C.T., and Semerjian, H.G.: Atomization and Sprays, vol. 4, 1994, pp.207-222.

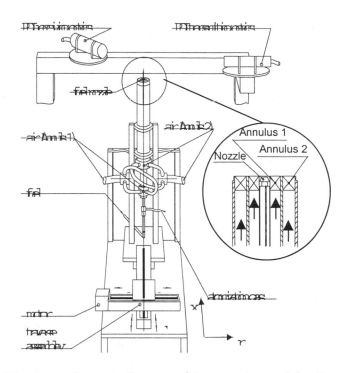

Fig. 1. A schematic diagram of the experimental facility

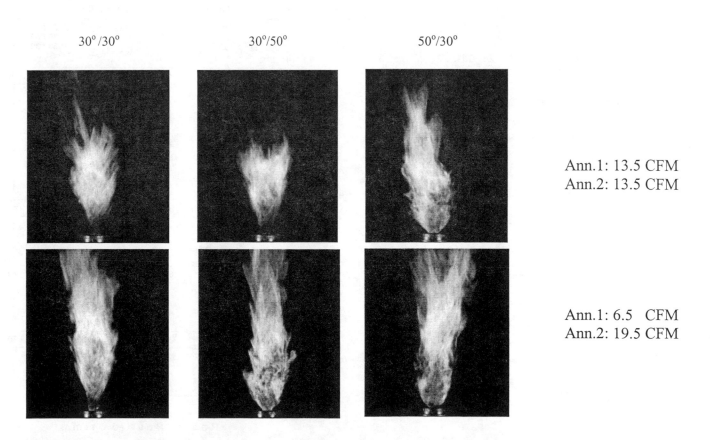

30°/30°	30°/50°	50°/30°	
			Ann.1: 13.5 CFM Ann.2: 13.5 CFM
			Ann.1: 6.5 CFM Ann.2: 19.5 CFM

Fig. 2. Effect of radial distribution of swirl and combustion airflow in the burner on global features of flames.

345

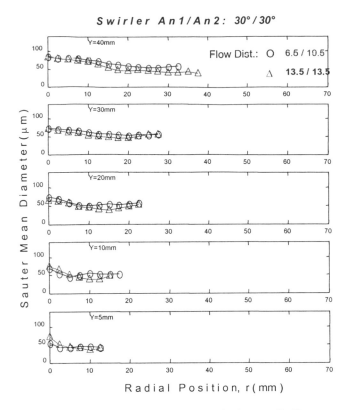

Fig. 3. Effect of airflow distribution on SMD.

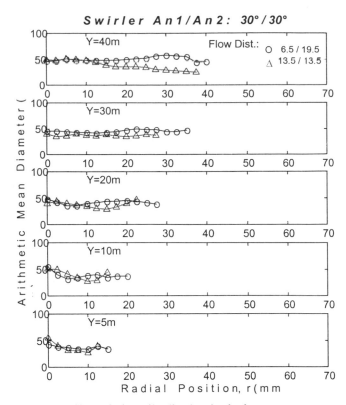

Fig. 4. Effect of Flow distribution in the burner on mean Arithmetic diameter.

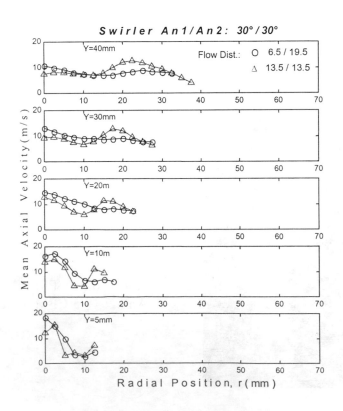

Fig. 5. Effect of airflow distribution on droplet mean axial velocity.

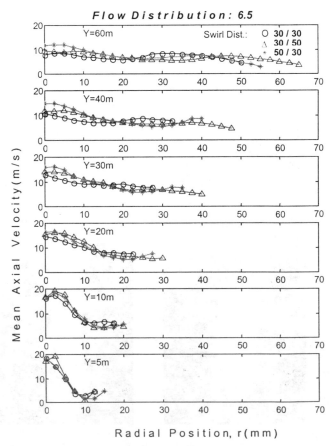

Fig. 6. Effect of swirl distribution in the burner on droplet mean axial velocity.

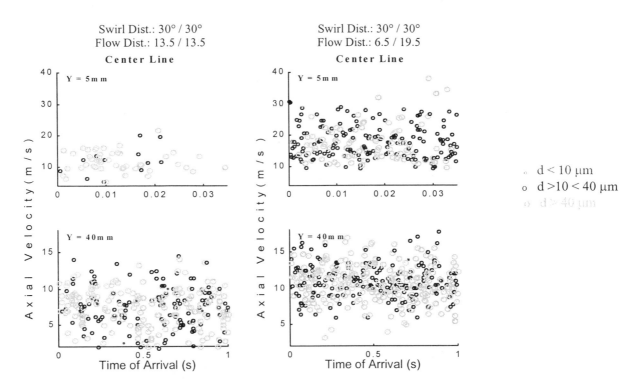

Fig. 7(a). Droplet time of arrival on the spray centerline with 30°/30° swirl and equal or unequal flow distribution.

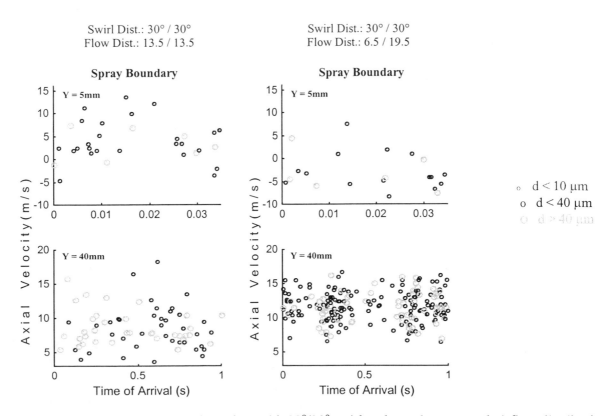

Fig. 7(b). Droplet time of arrival on the spray boundary with 30°/30° swirl and equal or unequal airflow distribution.

IECEC2001-AT-05

COMPARISON OF SPACE SOLAR POWER SYSTEMS AND TERRESTRIAL SOLAR ARRAYS

Finley R. Shapiro
Finley Shapiro Consulting, Inc.
2021 Rodman Street
Philadelphia, PA 19146
USA

ABSTRACT

Comparisons are made between three recently proposed space solar power systems and matching terrestrial photovoltaic arrays. The space solar power systems involve a satellite with a large photovoltaic array that transmits power to the earth by a microwave beam, a large terrestrial receiving antenna, and equipment to connect the system to the terrestrial electric power grid. Each space system is compared to a grid connected terrestrial photovoltaic array of the same area as the receiving antenna of the space system.

For two of the three systems, the space solar array and the terrestrial array would produce about the same average daily output power, and for the third the terrestrial array would produce significantly more power. Comparisons of other features of the systems are also made.

INTRODUCTION

The concept of a space based solar power system to supply electricity to the terrestrial electric power grid has received renewed interest in recent years. [1-4] Various designs have been discussed, but all such systems are based on one or more satellites where solar cells convert sunlight to electric power. This power is then converted to microwave power and transmitted to earth. The terrestrial part of the system includes one or more large receiving antenna arrays, converters to low frequency electric power, and transmission lines connecting the system to the electric power grid.

The comparison of a space based solar power system to a possible large scale terrestrial photovoltaic system is interesting because both systems ultimately derive their power from sunlight, use solar cells to convert the sunlight to electricity, and require a large receiving area on earth.

The use of large scale terrestrial photovoltaic arrays as a major source of electricity for the electric power grid has several difficulties which have prevented its realization. These include the limited hours per day that sunlight is available at a given location, variations in sunlight due to weather conditions, the large amounts of land necessary to produce electricity from sunlight on a large scale, the limited efficiency of solar cells, and, most importantly, the cost of solar cells and other system components.

However, the technology for grid connected terrestrial solar electricity generation already exists. Small scale systems supply electricity to public electric grids every day. Performance information on many systems is publicly available. Existing infrastructure includes factories that produce solar cells, modules, and other system components, technicians experienced in installing them, regularly scheduled technical conferences, and a growing knowledge base in all aspects of system design, installation, and management. In addition, the output of terrestrial solar power systems correlates somewhat with the variation of electricity demand in warmer parts of the United States, and weather forecasting can be used to determine the short term scheduling of other generation resources based on the expected availability of solar electric power.

RESULTS

Specific information on three recently proposed space solar power systems is shown in Table 1. These systems were chosen because their power and sizes were readily available. For the Abacus/Reflector [5] and SunTower [6] systems, the overall conversion efficiency of microwave power to electric power is assumed to be 84%, although this is an overestimate. For the

Table 1. Comparisons of three space solar power systems, the Abacus/Reflector [5], the SunTower [6], and the GEO Symmetrical Concentrator [7] with a terrestrial photovoltaic system of the same area as each space solar power system's receiving antenna.

System		Abacus/Reflector	SunTower	GEO Integrated Symmetrical Concentrator
Receiving antenna	shape	circle	circle	oval
	diameter or axes (km)	7.45	4	6.5 x 8.6
Receiving antenna area (km^2)		43.6	12.6	44
Microwave beam power (GW)		1.2	0.20	not stated
Output electrical power (GW)		1.0	0.18	1.2
Output electricity per day (GW-h)		24	4.3	27
Average terrestrial PV system output per day (GW-h)		25	7.2	25

GEO Integrated Symmetrical Concentrator [7] the electrical output power from the reference is used. The output energy per day of the system is based on full power for 24 hours a day, which is also an upper limit or overestimate.

Each space solar power system is compared to a terrestrial photovoltaic array with the same area as the space system's receiving antenna. The power output of the terrestrial array is calculated using an average insolation of 5.7 kW-h/m^2/day. This is the calculated year-round average insolation striking a flat panel pointing straight up in Phoenix, Arizona based on 30 years of collected data [8]. The efficiency for the conversion of sunlight to AC output power is assumed to be 10%, which is within the capabilities of today's technology. Solar panels with 12% panel conversion efficiencies to DC output are available on the commercial market, and some sites in the United States report overall conversion efficiencies to AC power that are greater than 10% [9].

It can be seen in Table 1 that the output power of both the Abacus/Reflector and GEO Integrated Symmetrical Concentrator systems are about the same as an equally sized terrestrial system. The output of the SunTower system is significantly less than that of an equally sized terrestrial system. This is not only true for an array in Phoenix. An average insolation of 5.7 kW-h/m^2/day is also reported for Las Vegas, Nevada and Tucson Arizona, and 5.6 kW-h/m^2/day for Albuquerque, New Mexico. Values of 5.0 kW-h/m^2/day are reported in the southwestern United States from southern California into Colorado [8]. Even the value of 4.1 for Billings, Montana would only require an overall conversion efficiency of 14% for the same result.

We can also compare other features of a space solar power system and a corresponding terrestrial photovoltaic array as follows.

Technological Risk

The terrestrial system can be built entirely with equipment that is presently available on the commercial market. The space system will require favorable outcomes from many years of research and development. It is difficult to predict how long this effort will require.

Time Delay Until Operation

The space based system cannot begin working until the satellite part is developed, built, and functional. Also, due to concerns about worker exposure, part of the receiving antenna cannot be used while other parts are still under construction. By contrast, the terrestrial system can begin to generate power as soon as a small part of the array is completed and at least some of the power transmission lines. This can begin to generate electricity and provide cash flow and operating experience while the remainder of the system is still under construction.

Environmental Concerns

Both the space and terrestrial systems block sunlight from large areas of land. However, the space system also exposes plants, birds, and terrestrial animals to microwave radiation in both the antenna area and the larger safety zone. The terrestrial system can readily be spread out over a larger area to leave gaps through which some sunlight can pass and to avoid environmentally sensitive sites.

Land Use

Land under the terrestrial photovoltaic system can be used simultaneously for a variety of other purposes, including factories, shopping malls, and parking lots. People would hesitate to make similar use of the land under the space system's microwave receiving antenna, so other land would need to be used for these purposes. Concerns about microwave exposure would probably limit the choice of sites for the receiving antenna to very remote areas. Also, unlike the space system, the terrestrial system can be split up into separated segments. This expands the choices of usable sites by making it possible to build around areas that should not or cannot be used, and facilitating the dual use of the land. The additional location options may also lead to shorter power transmission lines and reduced labor costs. In addition, the terrestrial system does not require a safety zone around it.

Maintainability

Maintenance work can be done on one part of the terrestrial system while nearby parts continue to operate. Maintenance on the microwave receiving antenna will require directing the microwave beam away from that part of the antenna, and possibly away from the entire antenna, leaving the entire facility out of operation during the work. Maintenance of the satellite part of the space system will have numerous difficulties.

Reliability

As the terrestrial system is based on established technology, its overall reliability will be higher that that of a space based system for some time. Twenty year warranties on solar panels are now common. If necessary, day-to-day fluctuations in output can be accommodated by scheduling other power generation facilities based on weather forecasts.

Public Acceptance

Substantial public opposition to a space solar power system can be expected, due to fears of human and wildlife microwave exposure and displeasure at the cost. The history of nuclear power shows the delays and cost increases that can be caused by this opposition. The public has generally reacted favorably to solar power. As substantial public money and public land will probably be used for either system, the public's preferences need to be given strong consideration.

Research and Development Costs

The terrestrial photovoltaic system can be built with equipment that is already commercially available. The space solar power system will require substantial research and development even before a decision is made whether or not to proceed.

Implementation Costs

Labor costs of building the terrestrial photovoltaic system or the space solar power system's receiving antenna at the same place should be similar. Both will require a large trained construction force to cover the same area with similar support structures. However, the increased flexibility in site selection for the terrestrial system could allow the use of a less remote site, which could reduce the costs of recruiting and providing for the workers.

Equipment costs are likely to be dominated by the cost of building factories for the components and the labor to operate them. The terrestrial system has the advantage that there is already a substantial knowledge base on building and operating factories for photovoltaic systems, some existing factories can be used, and there is already a market for the output of the factories after the system is built. The reduced cost uncertainty from experience in building factories for components of the terrestrial system should make it less expensive to raise the capital for new factories.

If the system is not built entirely on public land, the increased flexibility in a choice of a site for the terrestrial photovoltaic system will probably keep land acquisition costs lower for this system. The ability to locate parts of the system in disjoint locations may also reduce land acquisition costs.

The terrestrial photovoltaic system is likely to require more expensive power transmission lines because the peak power is higher for the same average power as a space solar power system. However, the cost increase should be less than linear with peak power. The added flexibility in site selection may also lead to shorter power lines, and may make it possible to share the lines with other facilities, including those that make up for power that the terrestrial system is not generating at night or in bad weather.

The other major cost components of the space solar power system are for the space parts of the system, most notably launch and construction costs. These cost components are absent from the terrestrial photovoltaic system.

DISCUSSION

All of the comparisons here are critically dependent on the intensity of the microwave beam in each proposed space solar power system. If the beam intensity of the Abacus/Reflector or GEO Integrated Symmetrical Concentrator is doubled while keeping the receiving antenna area the same, then the output from the space solar power system will be about twice the power of the corresponding terrestrial photovoltaic system.

However, such a doubling of the beam intensity would require a doubling of the area of the satellite solar collector, and the use of microwave transmitting components that can transmit the higher intensity from space. Focusing the beam will become more critical, and a larger safety zone may be necessary. Environmental concerns and public opposition will probably also increase. The likely results are longer development and implementation times, and higher costs.

An alternative way to double the intensity is to focus the same power to a smaller receiving antenna. While this does not increase the size of the satellite solar collector, the transmitting components would still need to produce a more focused beam, and concern for the exposure of humans and wildlife would still increase even though a smaller land area is affected.

Thus, if the beam intensities of the systems considered here have been optimized for satellite size, the output power and focusing ability of future microwave transmitting components, the ability to place in orbit and operate the space part of the system, concerns about human and wildlife exposure, and other possible environmental risks, then such systems do not show clear advantages over terrestrial photovoltaic systems of the same area. The risks involved in developing the new technology for a space solar power system are further reasons to favor the terrestrial photovoltaic system.

ACKNOWLEDGMENTS

The author gratefully acknowledges the encouragement from and discussions with Richard A. Wakefield, Thomas R. Schneider, and other members of the IEEE-USA Energy Policy Committee.

REFERENCES

1. Mankins, J., 1997, "A Fresh Look at Space Solar Power: New Architectures, Concepts and Technologies," www.sunsat-energy.org/Ver.1/freshlook2.htm.

2. Mankins, J. C., 1998, "The Space Solar Power Option," Ad Astra, January/February 1998, pp. 25-29/

3. Macauley, M. K., et al., 2000, "Can Power from Space Compete?" Resources for the Future, Discussion Paper 00-16, www.rff.org/nat_resources/space.htm.

4. Mankins, J. C. 2000, Statement to the Subcommittee on Space and Aeronautics, Committee on Science, U.S. House of Representatives, September 7, 2000.

5. Carrington, C., et al., 35th Intersociety Energy Conversion Engineering Conference, 2000, AIAA-2000-3067.

6. Mankins, J. C., 1997, 48th International Astronautical Conference, IAF-97-R.2.03

7. Mankins, J. C., 2000, copies of transparencies for an oral presentation, April 14, 2000.

8. National Renewable Energy Laboratory, Solar Radiation Data Manual for Flat-Plate and Concentrating Collectors.

9. For example, go to the web site of the Solar Electric Power Association, http://www.ttcorp.com/upvg/sindex, click on "Go to Site Data" and then on Michigan, then click beneath "Goto site" for "Detroit Edison, Southfield f, Southfield Substation-Fixed Array," and then click on "Tables – Month."

2001-AT-56

EFFECTS OF HYPERVELOCITY IMPACT ON SOLAR CELL MODULES AT HIGH VOLTAGE

Henry W. Brandhorst, Jr.
Space Power Institute
231 Leach Center
Auburn University, AL 36849-5320
Tel: 334-844-5894
brandhh@auburn.edu

Stevie R. Best
Space Power Institute
231 Leach Center
Auburn University, AL 36849-5320
Tel: 334-844-5894
bestste@auburn.edu

ABSTRACT

As satellite power levels of advanced spacecraft climb above 20 kW, higher solar array operating voltages become attractive. For Solar Power Satellites, voltages of 1000 V have been considered. However, micrometeoroid impacts on such high voltage arrays may have catastrophic results. To assess these effects, contemporary GaAs modules were exposed to hypervelocity impact with 100 micron soda lime glass spheres at velocities up to 12 km/sec. The two strings were held at differential voltages above 60 V at absolute voltages above 400 V. A plasma environment typical of geosynchronous orbit (GEO) was also present.

INTRODUCTION

As the requirements for increased power and operating voltage levels for satellites have been identified, serious problems have been encountered that have resulted in the loss of power, science data and the loss of mission. These newly discovered high voltage plasma effects have resulted in significant schedule and cost "hits" to current satellites as they attempt to accommodate for these interactions with the space plasma. Tests on space station hardware disclosed arcing at 50 V on anodized Al structures that were struck with hypervelocity particles in a Low Earth Orbit (LEO) plasma. Thus an understanding of these effects is necessary to design reliable high voltage solar arrays of the future, especially for Space Solar Power applications. For these very high power systems, low array and cable masses are imperative, and these can only be practically achieved by using high voltages. Yet, we don't know how high a voltage is practical on a realistic solar array in GEO. Most solar arrays arc into the space plasma (sometimes catastrophically) at voltages of 200 - 300 V. Thus there is an absence of reliable information on which to base high power satellite designs. The greatest unknown is the effect of high velocity micrometeoroids on future high voltage arrays. Existing NASCAP-GEO models can provide guidance about solar cell string design, encapsulation approaches and field

control for the expected environment, but it cannot include the effects of micrometeoroid penetration in that environment.

Therefore, he objective of this work is to study the effect of hypervelocity impacts on the design of high voltage solar arrays for use in GEO environments.

DESCRIPTION OF THE HYPERVELOCITY IMPACT FACILITY

A unique hypervelocity impact facility (HYPER) shown in figure 1, has been used extensively to determine the effects of small particles on spacecraft surfaces. Complete solar cell assemblies exposed to determine damage resulting from exposure to the debris flux. Many particles can be accelerated simultaneously, with a velocity distribution roughly typical of that of the man-made debris spectrum. One exposure in the HYPER facility corresponds to the number of impacts received in a 7-year exposure in space for the same particle size distribution.

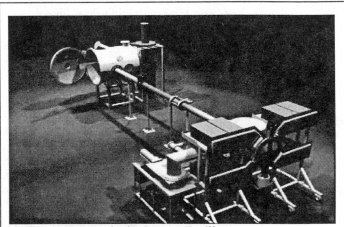

Figure 1: Hypervelocity Impact Facility

The Hypervelocity Impact Facility's gun consists of an arc discharge gun, a 5-meter flight tube and a 1-m diameter target chamber. Discharging eight capacitors that are charged to a potential of 40,000 V fires the gun. This discharge vaporizes an aluminum foil placed behind the impact particle charge. In these tests, soda lime glass particles nominally 40-120 μm in diameter were used. A "skimmer cone" is used to select only those particles that will pass directly into the flight tube. Approximately 40 – 70 of these particles will impact the target. A magneforming shutter placed in the middle of the 5-meter flight tube eliminates the low speed (<5 km/s) particles. In general, peak velocities in the 10-11 km/sec range were achieved for all these tests.

Because the primary acceleration process is "plasma drag", there is always a gradient in the velocity of the particle stream. The maximum velocity obtainable and the velocity distribution are a complex function of the particle shape, particle density, absolute particle dimensions, and the gun parameters. Just as in space, there is always a wide range of particle velocities in a given experiment. From the point of view of simulation, this is desirable since it probably is a more accurate representation of the actual conditions in space.

The target chamber is equipped with a streak camera that records particle x-y coordinates and velocities as well as observing any plasma discharges that may occur. A photomultiplier tube also observes the sample to provide impact information as well as discharge information. A Mylar film in front of the sample allows determination of particle size and LEDs surrounding the sample define sample edges for the streak camera images.

Thus, impacting particle coordinates, size and velocity can usually be obtained. However, in these tests, the presence of arcing during impact produced additional light that tended to obscure subsequent impact events. This was shown in the first impact test where the streak camera film was nearly completely saturated due to the arcing events. We used extensive image processing to try to discern impact locations and were partially successful. On the next shot, the streak camera aperture was reduced by one stop. This was better, but saturation still occurred. However, in this case, more impact locations were defined. The presence of additional arcing thus impacted our ability to accurately determine impact locations.

A photomultiplier tube (PMT) that imaged the test plane also observed the arcing. Although impact locations could not be determined with the tube, the signal provided timing and shape of both the impact and the arc emissions.

Two approaches were used to provide a plasma environment for the impact tests. The primary mode was to use a hollow cathode device provided by MSFC. The cathode was ignited and power reduced to the lowest level to maintain steady operation. A Langmuir probe placed within a few centimeters of the sample was used to establish plasma densities and energies. A second approach was also developed. A Tesla coil was used to excite an argon gas stream and provide low temperature plasma. This approach was also successful and in fact, provided an excellent simulation of low temperature plasma, such as is found in GEO, as confirmed by the Langmuir probe measurements.

The HYPER facility has evolved into a sophisticated method of analyzing the effects of space debris on any material contemplated for use in space. The debris spectrum and the choice of particles that can be used, allow the experimenter to more closely duplicate the natural conditions, which will be encountered in space. The optical diagnostics are unique and allow adequate characterization of both the impacting particle stream and the damage inflicted at impact. Inherent in the facility is the ability to look at long term effects. By choosing the particle size in correlation with the expected number of impacts based on known space flux, it is possible to estimate end-of-life characteristics due to debris in addition to local damage.

SOLAR CELL MODULES

Solar cell modules were obtained from a major supplier of space solar cells. These modules were composed of four SOA 2x4 cm GaAs solar cells with 6 mil cover glasses connected in two-cell series strings on a Kapton substrate that was attached to a plexiglass substrate. A picture of one module is shown in figure 2. The series strings are on the left and right sides of the picture, with a gap between them. The supplier measured the efficiency of all 15 modules. The amount of coverglass overhang and the spacing between the series strings were measured at Auburn. These data were used to select modules for testing.

Figure 2: Typical GaAs Solar Cell Module

A plot of efficiencies of the series strings is shown in figure 3. Efficiencies were above18% except for two modules that had low efficiency strings caused by poor fill factors. Open circuit voltages were tightly clustered (2.010 +/-0.007V). Similarly the short circuit currents were also well grouped (0.252 +/- 0.002 A). Fill factors were all near 80% except for the two modules mentioned previously. These were certainly typical of state-of-the-art GaAs cells and modules.

When inter-string spacing and coverglass overhang were measured, a wide range of results were apparent. Interstring spacing varied from a low of 707 μm to a maximum of 1235 μm, however within any one sample, the maximum spread was generally between 100 and 200 μm. Of most interest was the cover glass overhang on the cell corners in the region between the series strings. It is in this region that the differential bias between the strings will be the greatest. When the samples were measured, to our surprise, all samples had at least one cover

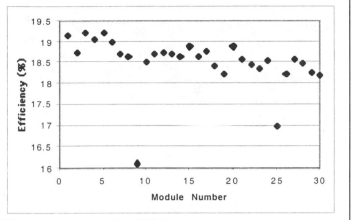

Figure 3: GaAs Module Efficiencies

glass with zero overhang. Of the 15 samples, eight had only one interstring cover glass with zero overhang, five had three zero overhangs, one had two with zero overhang and one sample had six of the eight corners with zero overhangs. In fact, one sample had a –37 μm "underhang" along with two zero overhang corners.

Because of the zero overhang conditions, the likelihood of biasing the samples to high voltages was felt to be minimal. To test that assumption, the module with six zero overhang corners was selected for the first shot.

TEST RESULTS

Shot SSP-1: The initial test was conducted on a sample with six areas of zero coverglass overhang in the region between the series strings. Because of the large number of "zero-overhang" areas, it was difficult to obtain a high voltage on the sample. The two cells in each series string were shorted together and a bias applied across them. Each string could be biased independently of the other so both a maximum bias voltage and a differential voltage between the two strings could be maintained. In order to determine the maximum voltage possible in the plasma, one string, called the "red string" (color of wire) was biased to -300V. Arcing was observed at that point and the voltage was then gradually reduced to a level where no arcs were observed over about a ten-minute period. This stable point occurred at -200V. The second series string ("blue") was then biased at –140V to establish a 60V differential between the strings and the shot initiated. The 60V differential bias was selected to be what is generally believed to be the arcing threshold. The hollow cathode source provided the plasma environment.

Based on post-shot inspection, approximately 35 particles impacted the target. An arc was observed on the "red" string after impact of the 14th particle at 1273 μsec into the shot. After a delay of 56μsec, the "blue" string discharged. Figure 4 shows the voltage traces from the two strings and figure 5 shows the PMT output. Because the current traces saturated there are no data are available for current behavior. The lowest

trace in figure 4 is the "red' string has a major discharge at 1273 μsec and the voltage drops from –200V to –8.5V in 20 μsec. The impact is also seen on the PMT trace (labeled 13th impact). However, the arc appears to extinguish and the voltage begins to recover on its own. At 56 μsec after that first arc, a second arc begins at a voltage differential of –126.6V.

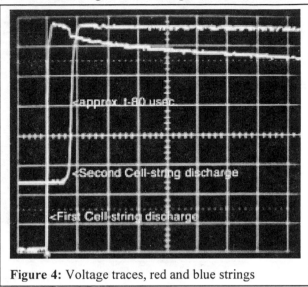

Figure 4: Voltage traces, red and blue strings

This does not seem to be associated with any particle impact (note the particle impact about 32 μsec later). It appears that the recharging seen on the "red" curve takes a sudden increase just at the onset of the second arc, so the second arc appears to be feeding the first. The voltage on that string then continues a gradual rise back to the –200V level. The shape of that discharge pulse is also different from the one at 1273 μsec.

The PMT trace shows that the maximum intensity of the second arc occurs precisely at the point of maximum dV/dt. Interestingly, the voltage of the second string drops to near –12V, but in this case, no recharging is seen at the time scales that were recorded. However, several minutes after the test was complete, all string voltages had returned to their pre-test values (-200 and –140 V). It is interesting to note that the recharging of the red string as assisted by the blue string terminated when the voltage of the red string had increased to –21V and the voltage of the blue string had decreased to –87V – or at a voltage differential of 66V.

Because of the saturation of the streak camera image and

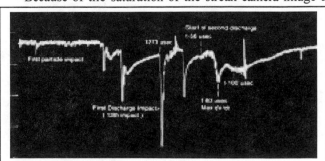

Figure 5: Photomultiplier traces for shot 1

alignment of the LEDs, only the x-coordinate was recorded. However, it was possible to determine the streak coordinate line along which the plasma flashes were observed. In the first arc event (1273 μs), several possible impact sites were near the

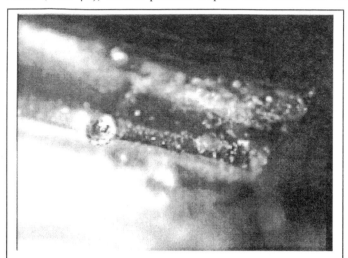

Figure 6: Impact site on the solder connection (SSP-1)

line, but only one fell on the line. That was on top of the red wire solder connection as shown in figure 6. We did not observe signs of an electrical discharge at this site. Some of the sites near the line appeared to nearly penetrate the cover glass, but no arcing evidence was apparent there either. Also observed on the streak camera record was another plasma flash event that appeared to start and stop at nearly identical times as the discharge event on string "blue". When the streak coordinate line was determined, it passed over the second string of cells. There were no craters along this line, but the line passed through the solder connection point of the blue wire. Hence it is possible, but not confirmed, that the second discharge may have been initiated from the end of this wire.

In this first test, a <u>new phenomena</u> was observed wherein a second arc, not associated with an impact event, served to partially recharge the recovery of the voltage that occurred in the first impact-related arc event. Shapes and light emission patterns of the two events are noticeably different.

<u>Shot SSP-2:</u> The sample for this test had two "zero overhang" areas directly across from one another. Test conditions were nominally the same as Shot 1. The hollow cathode was operated at its minimum stable operating voltage and current. The velocity of the first particle impacting the target was 10.2 km/s. By putting appropriate attenuators in the current line for both strings, good current traces were observed. In addition, the shutter for the streak camera was reduced by one stop. This helped us identify impacts, but arc emission still made exact impact location less accurate. In the following paragraphs, a detailed description of the PMT, voltage and current traces is provided along with less detailed oscilloscope traces.

<u>PMT Diagnostics:</u> As can be seen in figure 7, about 2 μs after what is believed to be the 8th particle (853.4 μs), the PMT

saw a slow noisy increase in light level peaking at 855.2 μs and decreasing to a minimum by 857 μs. Then, after 858 μs, the PMT saw the start of a noisy increase to the light level,

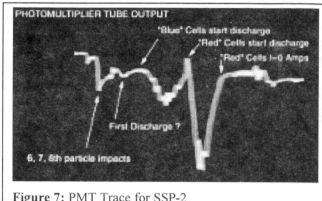

Figure 7: PMT Trace for SSP-2

peaking at approximately 864.6 μs, then decaying to a minimum by 871 μs. Immediately after 871 μs, the optical level increased rapidly with a rise-time that is characteristic of a hypervelocity particle impact plasma flash. However, when nearing the first peak intensity at 871.6 μs, a noisy signal pattern emerges within the waveform indicative of electrical discharges. After approx. 8 μs, the signal decays rapidly to a minimum at 879.4 μs where the waveform remains noisy until 890.0 μs. There is a very high correlation between the PMT trace shown here and the current trace shown below in figure 9.

The voltage and current traces for shot SSP-2 are shown in figures 8 and 9 respectively. In the voltage trace shown in

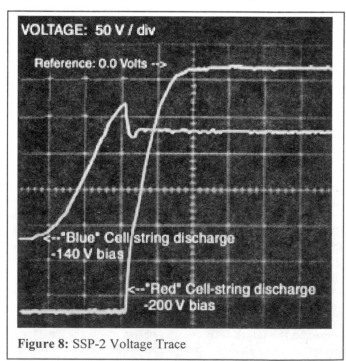

Figure 8: SSP-2 Voltage Trace

Figure 8, the "blue" (−140V) string began a gradual discharge from −139.5V starting at 857.8 μs. When it reached a minimum voltage of −31.7V at 870.8 μs, there was a sudden increase back to –58.3V at 871.8 μs. After that the voltage takes a small decrease to –52V and remains constant. The "red" (-200V) string begins to discharge just after 870.8 μs, at essentially the same time the "blue" string begins it's recharging. Next, the voltage reaches –126V at 872 μs, just when the recharging of the "blue" string is at an end. Thereafter it continues its discharging, reaching 0V at 886 μs.

The current trace for this shot is shown in figure 9. The

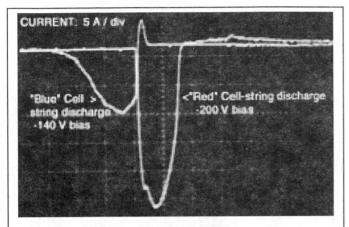

Figure 9: Current trace for shot SSP-2

results The "blue" current measured 0.0 A at 854 μs, started to change at ~856.8 μs, reaching -10.0 A at 865.6 μs, peaking negatively at –11.25 A at 868.1 μs, decaying to –8.45 A at 870.8 μs, zero-crossing (0.0 A) at 871.15 μs. From then on the positive current rebound, reaches +4.7 A at 871.9 μs, decaying to +0.30 A at 873.0 μs then holding steady at approximately +0.30 A until 886.6 μs. Thereafter it gradually reduces to 0A over time.

For the "red" (-200V) channel, At 870.8 μs, the current starts a rapid linear change from 0.0 A to –24.6 A at 872.0 μs where it hesitates, decreasing only slightly to –24.3 A at 872.275 μs. Then, at a slower rate, the current continues to increase from –24.6 A at 872.5 μs to a negative maximum of –28 A at 873 μs. From here it decays to zero-crossing (0.0 A) at 878.7 μs moving positively to approx. +1.35 A at 880 μs and decreasing slowly to +0.6 A at 906 μs to approx. zero by 1250 μs as limited by the scope's resolution.

A third test was run using a concentrator solar cell module supplied by ENTECH, Inc. This module consists of a series string of Spectrolab-supplied concentrator multijunction solar cells. These cells were completely covered by a coverglass and had been hi-pot tested at ENTECH to a bias of 2250 V as shown in figure 10. The overhang extended well beyond the cell boundaries and was filled with silicone. This sample was placed in the Hypervelocity Impact Facility and exposed to two shots. Because of the damage to the hollow cathode, a Tesla coil was used to provide the background plasma. This plasma was an excellent simulation of low temperature plasma. In the first shot, particle velocities were only 9.4 km/sec while in the second test a maximum of 11.6 km/sec was achieved. In the first test the sample was biased to –400 V and in the second, to –438 V. In both cases NO arcing was observed. It is

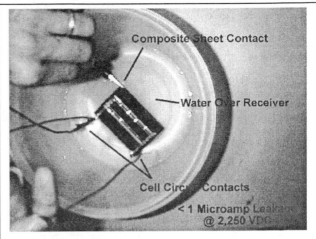

Figure 10: Hi-Pot testing of concentrator cell module

important to note that in this sample, the electrical contacts are fully insulated from the space plasma environment.

DISCUSSION

The results for the SSP-2 bear special discussion. The immediate event of a particle impact does not appear to have directly initiated the first discharge seen in the "Blue" (–140 V) bias channel. However, it is possible that charge carriers contained in the plasma were produced by the prior three impact events. Those events included a bright impact at 848.8 μs, followed by two smaller impacts at 849.8 μs and 851.6 μs. These three events could have initiated the discharge event that started at 853 μs that was optically detected by the PMT. This optical-discharge event does not appear to have immediately affected the solar cell strings because it was not detected by the solar cell electrical diagnostics except as a 1/2bit ADC noise shift. It is possible that this could have been a discharge between the plasma in the chamber provided by the hollow cathode plasma source and some other charge accumulation; perhaps a capacitive build-up of charge at the solar cell surface. This discharge may have in turn provided the charge carriers triggering the "Blue" string discharge.

The discharge of the "Blue" string caused the voltage on that string to decay at rate and waveform shape comparable to that of the "Blue" string discharge of test SSP-1 also biased at –140 V. Despite this being the first cell string to discharge, it did so at a moderate rate unlike that of the SSP-1 test "Red" string with –200 V bias being the first discharge of that test. This may suggest that the dynamics of discharge may perhaps be potential related; i.e. moderate potentials cause gradual discharges whereas higher potentials lead to abrupt discharge rates. However, this is pure speculation at this time.

The discharge of the "Red" string in this test was originally thought to have been initiated partially by an impacting particle at 871 μs. However, on further examination of the PMT waveform, the optical intensity did not decay immediately as is characteristic of impact flashes. Rather, it had a noisy sustained peak for approximately 8 μs followed by decaying to a noisy baseline for about another 10 μs.

It is possible that another particle impacting at this time could have initiated the discharge that allowed the optical

357

signature to last longer than usual. Due to the degree of optical saturation of the streak camera film, it could not be absolutely determined if there were a particle impacting at this time.

What may perhaps be more interesting is to examine the potential difference between the two strings during the discharge of the "Red" string. At approximately 870.8 µs, the "Blue" string potential was -31.7 V and "Red" string was –200 V, or a 168.3 V differential ("Red" over "Blue") when the discharge of the second string, the "Red" string starts. Compare this with the onset of the second discharge event of test SSP-1 of a 126.6 V differential ("Blue" over "Red"). There is indication that the discharge of current from "Red" to "Blue" string occurred that partially recharged the "Blue" string. Its voltage rapidly goes from –31.7 V to -58.3 V before this current "conduit" closes. The potential difference at the moment when this connection stops is approximately 68.3 V (126.6 to 58.3). This is very close to that of 66 V from test SSP-1 when that conduit also apparently closed.

The second discharge in this test of the "Red" string had a rapid rise-time comparable to that of the "Red" string of SSP-1 (see comment above about voltage magnitude). Also, note that the "Blue" string again, as in test SSP-1, holds a nearly constant voltage while the "Red" string recharges during the time the waveform data is recorded. Several minutes after the test, both strings had recovered their full initial bias potentials.

SUMMARY

From the series of tests performed here, several conclusions may be drawn. Based on the testing done to date, it appears as if solar arrays with unprotected contacts are susceptible to arcing upon hypervelocity particle impacts. Although coverglasses were penetrated and other cell contacts damaged, no arcing occurred at those sites to the best of our detection ability. The GaAs samples had numerous areas where there was zero coverglass overhang, hence it was impossible to obtain bias voltages above –200 V. With larger coverglass overhang and insulation of bare interconnects it may be possible to achieve voltages near 1000V. The ENTECH samples had both cells and contact strips that were fully insulated. These samples showed no arcing upon hypervelocity particle impact at velocities as high as 11.6 km/sec and bias voltages up to –438 V. Thus it appears that these preliminary tests have uncovered basic design approaches that can lead to high voltage (>400 V and perhaps 1000 V) solar arrays

ACKNOWLEDGMENTS

The authors gratefully acknowledge the support of Dr. Dale Ferguson of the Glenn Research Center for this work performed under Contract NAS3-99200.

Proceedings of IECEC'01
36th Intersociety Energy Conversion Engineering Conference
July 29-August 2, 2001, Savannah, Georgia

IECEC2001-CT-01

RECENT INNOVATIVE SYSTEMS - ANALOGIES AND DIFFERENCES FROM A THERMODYNAMIC POINT OF VIEW

Giacomo Bisio/University of Genoa   Giuseppe Rubatto/University of Genoa

ABSTRACT

In these last years, one has proposed several innovating systems, which are suitable to accomplish operations already carried out by other classical systems, usually with advantages of reliability and simplicity, but with limitations in power and exergy efficiency. Pulse-tube systems and standing wave and traveling wave systems are among these. Recently, new acoustic resonators (indicated by the acronym RMS) have been developed. They allow us to reach so high pressures that they can be utilized as usual compressors of air and other gases or for reversed cycle plants. In this paper, one firstly outlines some innovative systems and deals with acoustical resonators RMS in a more particular way. Then, in the light of the generalized thermodynamics, one carries out an analysis of analogies and differences of these systems among them and in comparison with other systems that seem similar from a purely external examination.

NOMENCLATURE

p pressure [bar]
p_a amplitude of acoustic pressure [bar]
p_m mean pressure [bar]
T temperature [K, °C]
ε performance [dimensionless]

1. INTRODUCTION

Reversibility hypothesis is frequently assumed since it allows the attainment of relations of exchanged work and of other parameters easily. This hypothesis, however, is questionable, because the irreversibility of phenomena is one of the essential implications of the second law of thermodynamics. Nevertheless, it is evident that the consequences of the above-said hypothesis are very unlike in the different processes and that in some cases reversibility hypothesis can lead to results qualitatively (and not only quantitatively) different from reality. It seems interesting to look for a criterion to judge the suitability of the hypothesis of partial or total reversibility in a given process. With reference also to previous papers, some typical processes are analyzed, while an element, which distinguishes them basically in reference to the entropy production, is outlined.

(i) Systems with irreversibilities, which are remarkable but not «essential» in order that a phenomenon can actually take place. Within these systems the following are mentioned: (a) galvanic cell which presents limited irreversibilities if electric current intensity is low (Epstein, 1947); (b) ideal Curzon-Ahlborn machine in which only the irreversibilities due to heat transfers from heat source and to heat sink are taken into consideration (Curzon and Ahlborn, 1975; Rubin, 1979; De Vos, 1987; Bisio, 1996); (c) traveling wave thermoacoustic devices in which there is not a valuable mass transfer and irreversibilities are not unavoidable at least in principle (Ceperley, 1979, 1982, and 1985); (d) Stirling engines which are thermodynamically similar to the above thermoacoustic devices, but present valuable mass transfers (e.g., Walker, 1980; Urieli and Berchowitz, 1984).

(ii) Systems with "constructive" or "essential" irreversibilities according to the meaning of the "generalized thermodynamics"; this terminology has been used by Prigogine (1980), Paty (1982), Haken (1983 and 1988). Within these systems the following are mentioned: (a) the standing wave thermoacoustic devices in which there is not a valuable mass transfer and irreversibilities are unavoidable, also in limit conditions (Wheatley et al., 1983a, 1983b, and 1985; Swift et al., 1985; Wheatley, 1986; Bisio, 1996, and 1997; Swift, 1988; Benvenuto and Bisio, 1989); (b) the Gifford-Longsworth pulse-tube refrigerators, which are thermodynamically similar to the above thermoacoustic devices, but present valuable mass transfers (Gifford and Longsworth, 1964, 1965 and 1966; Radebaugh et al., 1986; Richardson, 1986; Thirumaleshwar and Subramanyam, 1986; Longsworth, 1988; Radebaugh, 1990).

(iii) Systems with very limited irreversibilities, when a suitable parameter is lower than a given value. Acoustic wave propagation is particularly mentioned. In this case, the hypothesis of an adiabatic-isentropic process seems appropriate (Shapiro, 1953).

2. PULSE-TUBE REFRIGERATORS

Pulse-tube refrigerators were firstly described by Gifford and Longsworth. Until recently, however, they remained something of a curiosity, whereas the majority of researches in the field of cryocoolers were directed towards Stirling and Gifford-McMahon machines (Gifford and Longsworth, 1964, 1965 and 1966; Mikulin et al., 1984; Richardson and Evans, 1997). The great advantage of the pulse tube (operating at frequencies of 1-60 Hz) is that it has no moving parts at low temperature.

Basic pulse tube refrigerators (BPTR) operate at a pulse rate of a few Hz and can reach temperatures of ≈ 120 K in a single stage, whereas lower values are possible by multistaging. A layout is shown in Fig. 1. The basic elements are heat exchangers (A1, A2) and regenerator (B). In addition to these components, apparatus is required to provide a pulsed high-pressure gas supply. This may be either a reciprocating compressor (C) directly linked to the pulse tube or a free-running compressor system (D) with the pulse produced by using control valves (E).

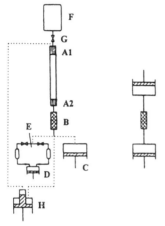

Fig. 1. Layout for a comparison of basic, orifice and double inlet pulse tubes with a Stirling cooler: A1) high-temperature (environment) heat exchanger; A2) low-temperature heat exchanger; B) regenerator; C) reciprocating compressor; D) continuous-flow compressor; E) control valves; F) buffer volume; G) orifice; H) stepped piston compressor.

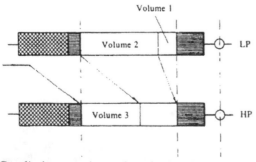

Fig. 2. Gas displacement in a pulse tube (LP: low pressure; HP: high pressure).

A significant advance was made by Mikulin et al. (1984) who added a buffer volume (F) with an orifice (G) as a gas distributor and receiver at the warm end of the pulse tube. This new device was named "orifice pulse tube refrigerator" (OPTR) and allows one to reach temperatures below 70 K in a single stage and temperatures as low as 4 K in a multi-staging. OPTR operate at pulse rates that are usually many times the optimum for a BPTR. Typically, an OPTR using direct compressor drive

will operate at 50-60 Hz. If valves are used to produce the pressure pulses, the operating pressure is usually less than 10 Hz. A variation of the OPTR, the double inlet design, was introduced by Zhu et al. (1990) and achieved 42 K in a single stage. Double inlet design does not necessarily need an orifice and buffer volume. The same result, a variation in the phase relationship between pressure change, gas displacement and heat exchange, can be reached by means of a double piston arrangement with appropriate displaced volume ratio and phasing (H in Fig. 1).

The potential disadvantage of these multi-stage, orifice and double inlet types is that they are, in general, still not as efficient as a Stirling cooler of comparable capacity and, almost without exception, are both unwieldy and complex, detracting from the initial concept of the pulse tube.

The geometry of all the types of pulse tubes is such as to promote temperature stratification and to minimize bulk mixing of the gas, although local mixing is unavoidable.

In all pulse tubes, thermal energy pumping from the cold end heat exchanger (A2) to the hot end depends on the movement of gas that accompanies cyclic pressurization. Gas entering the pulse tube via the regenerator compresses gas already present, thus increasing its temperature and displacing it towards the closed end (or orifice, or secondary inlet) of the tube. With reference to Fig. 2, gas in volume (1) comes into direct contact with the hot end heat exchanger at some time during the cycle, gas in volume (2) moves up and down the tube as the pressure fluctuates, and gas in volume (3) flows between the pulse tube and regenerator. The temperature change experienced by a particular element of gas will depend on its initial position in the case of gas in volumes (1) and (2) and the point in cycle when it first enters the pulse tube (from the regenerator) in the case of gas in volume (3).

The fundamental difficulty of thermodynamic analysis is that the compression and expansion processes are not well defined and each element of gas undergoes a slightly different series of processes, which in no case approximate reversibility, as sometimes supposed for simplicity's sake.

In the pulse tube, work is not done directly on a solid boundary but rather by one element of gas on another.

3. RESONANT PULSE-TUBE REFRIGERATORS (OR THERMOACOUSTIC REFRIGERATORS)

The resonant pulse-tube (or thermoacoustic) refrigerators are partially similar to BPTR and OPTR systems. However, thermoacoustic refrigerators operate at higher resonant frequencies (from 100 to 1000 Hz) and are driven by an oscillating diaphragm. Thermoacoustical phenomena in pipes with a closed end and with both open ends were firstly examined by Sondhauss and Rijke, respectively, and after by other authors (Feldman, 1968a and 1968b; Bisio and Rubatto, 1999).

Wheatley et al. (1983a) presented the results of their experimental and theoretical researches on thermoacoustical effects in acoustically resonant tubes. One of these systems is represented in Fig. 3. Near the closed end of a tube, there is a stack of fiberglass plates of thickness 0.125 mm, length 100 mm (much less than the acoustic wavelength), and spaced by an amount which is substantially larger than the thermal penetration depth. The stack is fitted with five thermocouples as indicated in Fig. 3 and the tube is well insulated from the environment. At the other end of the tube an acoustic power at the frequency of 411 Hz is applied while the entire structure is initially at uniform temperature. By using helium at pressure (p_m = 4.9 bar) and with (p_a/p_m = 0.04), where p_a is the amplitude of acoustic pressure, a minute or so later, the temperature difference between points 1 and 5 is even >100 K with $T_1<T_o$ and $T_5>T_o$, where T_o is the environment temperature. In points 2, 3, 4 there is only a small temperature increase.

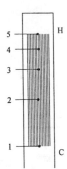

Fig. 3. Layout of a standing wave thermoacoustic system; H: closed end; C: open end where thermoacoustic power is applied.

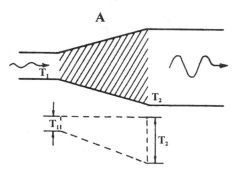

Fig. 4. Layout of a traveling wave thermoacoustic system: A) regenerator.

Afterwards, Wheatley and others examined several thermoacoustic systems both as heat engines and as refrigerators (Wheatley et al., 1983b and 1985; Swift et al., 1985; Wheatley et al., 1985; Wheatley, 1986; Swift, 1988; Benvenuto and Bisio, 1989; Swift and Keolian, 1993; Olson and Swift, 1994; Cao et al., 1996; Wetzel and Herman, 1997; Mozurkcwich, 1998; Gopinath et al., 1998; Adeff et al., 1998). The phenomena of this section and those of section 2. are intrinsically irreversible, since, in order to operate, they mandate thermal inertia and heat transmission with finite temperature differences. This remark led Wheatley and others to examine a particular class of idealized thermal engines that are intrinsically irreversible. The idealization consists in the fact that the only irreversibilities taken into account are those due to heat transfer with finite temperature differences, owing to thermal inertia of system parts. For a system operating in a reversed cycle with monatomic ideal gases having (c_p/c_v = const.), one has found a performance ε depending on the system geometry and on the gas properties, but not, in a direct way, on the extreme temperatures. Only in the case of no irreversibility (fact which is contradictory with the hypotheses) ε would be a function of the extreme temperatures.

In addition, a fluid-acoustic coupling mechanism was put forward by Eisinger and Sullivan (1993) as the probable cause of the intense acoustic vibration experienced in heat exchanger

and steam generator tube banks. Subsequently, Eisinger (1994) analyzed low-frequency acoustic vibrations of a vertical gas turbine recuperator during cold start-up. The vibrations were identified as fluid-thermoacoustic instabilities driven by a modified Sondhauss tube-like thermoacoustic phenomenon. Recently, many studies on the topic have been developing and several have been presented at ECOS 2000; we quote, e.g., two of them, one regarding combustion driven oscillations in a gas turbine combustor (Abbott, 2000) and the other relative to pulsations in hot blast stove burners (Boonacker and van dem Bemt, 2000).

4. TRAVELING WAVE THERMOACOUSTIC DEVICES

Ceperley (1979, 1982 and 1985) explored a different, but similar, class of acoustical heat engines, based on traveling waves. These traveling-wave heat engines use a Stirling thermodynamic cycle, which is reversible at the limit, allowing the engines to serve also as heat pumps.

The basic process involves an acoustical traveling wave propagating through a differentially heated regenerator, as is shown in Fig. 4. The regenerator consists of a casing packed with metal or ceramic parts, small enough to insure that gas in any part of the regenerator is essentially at the temperature of the packing at that point, but not so fine as to cause excessive attenuation of acoustic waves. A continuous temperature gradient is set up along the length of the regenerator by external sources that heat one end and cool the other.

Let us compare standing and traveling waves. The pressure and velocity peaks of standing waves are phased to occur one after the other, as is shown in Fig. 5. Thus, compression and heating would occur simultaneously and similarly expansion and cooling, were it not for the thermal delay in the heating and cooling processes which allow some of the heating to occur after the compression and some of the cooling after the expansion. This renders the process essentially irreversible.

On the contrary, the pressure and velocity that a traveling wave imparts to the gas volume, in which it is propagating, is shown in Fig. 6. For a wave traveling from cold to hot through a differentially heated regenerator, there are four phases: compression, heating, expansion, cooling. Since this is the same type of cycle a gas volume would undergo in a Stirling engine, one would expect a similar conversion of thermal into mechanical energy: this energy will amplify the traveling wave. In the case of acoustical wave traveling in the reverse direction, from hot to cold through the regenerator, acoustical energy is converted into thermal energy; the device is functioning as a heat pump.

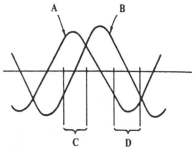

Fig. 5. Pressure and velocity of a gas volume in a standing wave thermoacoustic system: A) velocity; B) pressure; C) compression and heating phase; D) decompression and cooling phase.

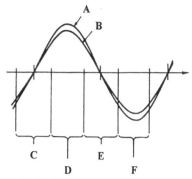

Fig. 6. Pressure and velocity of a gas volume in a traveling wave thermoacoustic system: A) velocity; B) pressure; C) compression phase; D) heating phase; E) decompression phase; F) cooling phase.

5. ACOUSTIC RESONATORS OF LUCAS EMPLOYING NON-LINEAR STEADY STATE WAVES

Some researchers devised a new technology to use sound waves with a power higher than those previously realized. In this manner, they opened the way for simple acoustic compressors, speedy chemical-process reactors, and clean electric power generators (Lawrenson et al., 1998; Ilinskii et al., 1998).

Ever since electricity became a familiar part of everyday life, people have grown accustomed to the idea of getting the power for various mechanical tasks from unseen electromagnetic waves traveling through metal wires. Few, however, have witnessed the acoustic analogue of electromagnetism-sound waves, or pressure waves propagating through gas-filled chamber, doing useful work with a power acceptable for industrial applications. The situation may change as Lucas has developed a technique by which standing sound waves resonating in specially shaped closed cavities can be loaded with thousands of time more power than it was previously possible. Lucas's wave-shaping technology is known as "resonant macrosonic synthesis" (RMS). Lucas states (and his statement seems acceptable) that the inherent simplicity of this technology could result in increased reliability and durability, and oilless operation, which is critical for the semiconductor and pharmaceutical industries. We, however, do not agree with the statements of Lucas on the possibility of obtaining high values of exergy efficiency. In these devices, indeed, there is the presence of "constructive irreversibilities" that can not be reduced. This fact is analogous to what already seen for other phenomena with such kinds of irreversibilities (Benvenuto and Bisio, 1989).

Finite-amplitude acoustic phenomena in resonant cavities have been of practical interest since the 1930s, when German researchers studied them in connection with the development of mufflers for tanks. Historically, researchers have believed that there is an intrinsic limit for sound waves in gases that would never allow high-power level to exist. Previous experimental work had shown that sound waves in a resonator would build up power to a certain level and no more. This acoustic saturation point occurs when shock waves start to form. Once a shock wave exists, any power added to the wave is wasted as thermal power, without any increase in the dynamic pressure of the wave.

For resonant sound waves, Lucas discovered that the geometry of the resonator cavity through which the sound waves travel

is the most important factor in determining the shape of the wave. In fact, many researchers in the past had used cylindrical resonators. This configuration, however, is the most likely to produce shock waves. In 1990, Lucas found that he could create relatively large-amplitude, or macrosonic, sound waves up to a pressure of 4 bar by properly shaping the resonator. Figure 7 shows four interior geometries of resonators: (a) cylinder, (b) horn-cone, (c) cone, and (d) bulb. Only the three last geometries are suitable to obtain technically meaningful pressures.

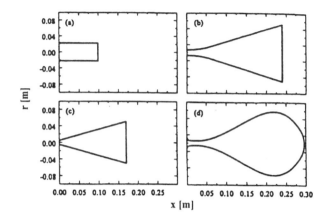

Fig. 7. Interior geometries of resonators: (a) cylinder; (b) horn-cone; (c) cone; (d) bulb.

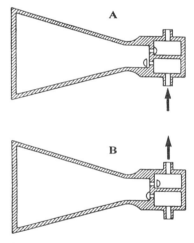

Fig. 8. Layout of an acoustic compressor: A) suction phase; B) discharge phase.

The next technical hurdle that researchers addressed was to figure out how to transfer a lot of power into a cavity. Lucas decided to shake the entire cavity with a linear motor, a vibrator that is nothing more than a glorified electromagnet. In that way, the whole inner surface of the resonator transferred power to the enclosed gas. Thus, the cavity acted like one large piston.

Lucas demonstrated his RMS technology using a resonator cavity shaped like an elongated pear. When this cavity is vibrated with a linear motor so that its walls move back and forth a distance of about 100 mμ, it resonates with a smooth, shockless wave of high power. Acoustic particle velocity reaches Mach numbers higher than 0.5. The first major application of RMS technology was an acoustic compressor. Inside the cavity, dynamic gas-molecule displacement was about 5-6 cm, about one-third of the resonator length.

Let us consider an acoustic compressor. The standing wave in the resonator causes the pressure to oscillate during one acoustic cycle of about 2 ms. When the pressure is low at the thinner end, the suction valve opens and low pressure gas flows in and enters the plenum (Fig. 8A). When the pressure is high at the thinner end, the discharge valve opens and high pressure gas flows from the resonator into the discharge plenum (Fig. 8B). High-pressure gas in the resonator holds the suction valve shut.

Lucas with his team has also designed a compressor that does not need valves. It is based on creating a static pressure distribution in which a high-pressure area is located at a port on one end of the resonator and a low-pressure area is placed near another port at the other end. This design generates a lower pressure head, but probably it can allow higher values of efficiency owing to lower "essential" irreversibilities.

According to Lucas, the new RMS concepts will make previously unattainable physical effects possible. Such effects could be used in a range of new industrial devices and processes with cavities shaped specifically for each application. For example, specialized non-contaminating acoustic compressors and pumps for commercial gases, ultrapure fluids, and hazardous fluids are a possibility. These devices are important for the pharmaceutical and semiconductor industries.

RMS technology could be used to drive and control thermal and kinetic chemical reactions by producing localized heating with rapid pressure changes.

Acoustic chambers could be used for separation, agglomeration, mixing and pulverization of materials. For example, improved acoustic agglomeration of particulate contaminants, causing them to stick together, could lead to improved gas scrubbers for power-plant flue gases by allowing smaller units to operate at higher flow rates. Electric-power generation using pulse combustion of hydrocarbon fuels (Bisio and Rubatto, 1999) could also be realized through this technology.

6. SIMILARITIES AND DIFFERENCES AMONG SEVERAL SYSTEMS

As done in previous papers (e.g., Benvenuto and Bisio, 1989), a table (Table 1) with similarities and differences among several systems, including the new RMS, is reported.

Table 1. - Similarities and differences among several systems

	Systems with "essential" irreversibilities	Systems without "essential" irreversibilities
Systems without a valuable mass transfer and (100 – 1000 Hz)	Standing wave thermoacoustic devices (Sondhauss and Rijke tubes within these)	Traveling wave thermoacoustic devices – Magneto-acoustic hydrodynamic transducers
Systems with a valuable mass transfer and (1-60 Hz)	Gifford-Longsworth pulse-tube refrigerators Arkharov et al. wave cryogenerator Stirling engines in which the velocity wave leads the pressure wave by a small phase angle	Classical Stirling engines
Systems with a valuable mass transfer and (500-600 Hz)	Acoustic resonators of Lucas with non-linear steady state waves (RMS)	

The following distinguishing criteria have been assumed: (i) the presence or absence of constructive irreversibilities, (ii) the presence or absence of a valuable mass transfer and then of a valuable power, (iii) the range of the employed frequencies.

The systems characterized by irreversibilities having a "constructive" role present low exergy efficiency, and for these the reference to reversibility is devoid of any meaning. In these cases, indeed, useful effect tends to zero as an infinitesimal of higher order than entropy production rate (Bisio and Rubatto, 1999). On the contrary, when irreversibilities have not a "constructive" role, even if they are always present, the efficiency values can be higher, but only with an expedient choice of the various parameters.

7. CONCLUSIONS

Within the common characteristics, one can distinguish the systems considered in this paper in five different classes.
(a) Basic pulse-tube refrigerators (BPTR). They operate at the frequency of 1-3 Hz and can reach temperatures of about 120 K in a single stage, whereas lower values are possible by multi-staging.
(b) Orifice pulse-tube refrigerators (OPTR). A significant advance was made by adding a gas distributor and receiver at the warm end of the pulse tube. Typically, an OPTR using direct stepped piston compressor drive will operate at 50-60 Hz. If valves are used to produce pressure pulses, the operating frequency is usually <10 Hz. They can reach 70 K in a single stage and temperatures as low as 4 K by multi-staging.
(c) Standing-wave thermoacoustic systems. These systems operate at resonant frequencies of 100-1000 Hz and are driven by an oscillating diaphragm. They allow us to obtain a minimum temperature of about 200 K. One can obtain, however, meaningful refrigerating powers with somewhat higher temperatures.
(d) Traveling wave thermoacoustic systems.
(e) Acoustic resonators with non-linear steady state waves.

In these last years, the possibility of synthesizing unshocked waveforms of very large amplitude by careful design of the shape of a resonator has been demonstrated. Pressure amplitudes more than an order of magnitude larger than those obtained in an acoustically saturated cylindrical resonator have been measured. A key technological component in the generation of these macrosonic standing waves is the entire resonator drive system, which provides a practical method of transferring hundreds of watts of power into the standing waves.

The comparison among systems with fundamental analogies and differences, in this paper shown in a table, seems useful to visualize better the different physical operation of the several systems.

ACKNOWLEDGMENT

The authors wish to thank Dr. Roberto Martini for his valuable help in the edition of the Figures of this paper.

REFERENCES

Abbott D.J., 2000, Combustion driven oscillations on a gas turbine combustor, *Proc. of ECOS 2000*, Enschede, The Netherlands, July 5-7, G.G. Hirs, Editor, **4**, 2033-2047, Febodruck BV, Enschede.

Adeff J.A., Hofler T.J., Atchley A.A., Moss W.C., 1998, Measurements with reticulated vitreous carbon stacks in thermoacoustic prime movers and refrigerators, *J. Acoust. Soc. Am.*, **104**, 32-38.

Benvenuto G., Bisio G., 1989, Thermoacoustic systems, Stirling engines and pulse-tube refrigerators - Analogies and differences in the light of generalized thermodynamics, *Proc. of 24th IECEC*, W.D. Jackson, D.A. Hull, Eds., **5**, 2413-2418, IEEE, New York.

Bisio G., 1996, The weight of heat transfer mode on heat engines operating at maximum power - the reference ideal cycle, *Proc. of ECOS '96*, Stockholm, June 25-27, P. Alvfors, L. Eidensten, G. Svedberg, J. Yan, Eds., 127-134.

Bisio G., 1997, On the suitability of reversibility hypothesis, *Proc. of FLOWERS '97*, Florence, July 30 - August 1, 1143-1155, SGE, Padova.

Bisio G., Rubatto G., 1999, Sondhauss and Rijke oscillations – Thermodynamic analysis; possible applications and analogies, *Energy - The International Journal*, **24**, 117-131.

Boonacker R., van den Bemt J., 2000, Pulsations in hot blast stove burners, *Proc. of ECOS 2000*, Enschede, The Netherlands, July 5-7, G.G. Hirs, Editor, **4**, 2079-2087, Febodruck BV, Enschede.

Cao N., Olson J.R., Swift G.W., Chen, S., 1996, Energy flux density in a thermocouple, *J. Acoust. Soc. Am.*, **99**, 3456-3464.

Ceperley P.H., 1979, A pistonless Stirling engine - The traveling wave heat engine, *J. Acoust. Soc. Am.*, **66**, 1508-1513.

Ceperley P.H., 1982, Gain and efficiency of a traveling wave heat engine, *J. Acoust. Soc. Am.*, **72**, 1688-1694.

Ceperley P.H., 1985, Gain and efficiency of a short traveling wave heat engine, *J. Acoust. Soc. Am.*, **77**, 1239-1244.

Curzon F.L., Ahlborn B., 1975, Efficiency of a Carnot engine at maximum power output, *Am. J. Phys.*, **43**, 22-24.

De Vos A., 1987, Reflections on the power delivered by endo-reversible engines, *J. Phys. D: Appl. Phys.*, **20**, 232-236.

Eisinger F.L., Sullivan R.E., 1993, Unusual acoustic vibration in heat exchanger and steam generator tube banks possibly caused by fluid-acoustic instability, *J. Engineering for Gas Turbines and Power*, **115**, 411-417.

Eisinger F.L., 1994, Fluid-thermoacoustic vibration of a gas turbine recuperator tubular heat exchanger system, *J. Engineering for Gas Turbines and Power*, **116**, 709-717.

Epstein P.S., 1947, *Textbook of Thermodynamics*, J. Wiley & Sons, New York.

Feldman K.T., Jr., 1968a, Review of the literature on Sondhauss thermoacoustic phenomena, *J. Sound Vib.*, **7**, 71-82.

Feldman K.T., Jr., 1968b, Review of the literature on the Rijke thermoacoustic phenomena, *J. Sound Vib.*, **7**, 83-89.

Gifford W.E., Longsworth R.C., 1964, Pulse tube refrigeration, *Trans. ASME, J. Eng. Ind.*, **86**, 264-268.

Gifford W.E., Longsworth R.C., 1965, Pulse tube refrigeration progress, *Int. Advances in Cryogenic Engineering*, K.D. Timmerhaus, Ed., **X**, sec. M-U, 69-79, Plenum Press, New York.

Gifford W.E., Longsworth R.C., 1966, Surface heat pumping, *Advances in Cryogenic Engineering*, K.D. Timmerhaus, Ed., **11**, 171-179, Plenum Press, New York.

Gopinath A., Tait N.L., Garret S.L., 1998, Thermoacoustic streaming in a resonant channel: The time-averaged temperature distribution, *J. Acoust. Soc. Am.*, **103**,1388-1405.

Haken H., 1983, *Synergetics*, Springer Verlag, Berlin.

Haken H., 1988, *Synergetics - From physics to biology. Order and Chaos in Nonlinear Physical Systems*, 167-192, Plenum Press, New York.

Ilinskii Y.A., Lipkens B., Lucas T.S., Van Doren T.W., Zabolotskaya E.A., 1998, Nonlinear standing waves in an acoustical resonator, *J. Acoust. Soc. Am.*, **104**, 2664-2674.

Lawrenson C.C., Lipkens B., Lucas T.S., Perkins D.K., Van Doren T.W., 1998, Measurements of macrosonic standing waves in oscillating closed cavities, *J. Acoust. Soc. Am.*, **104**, 623-636.

Longsworth, R.C., 1988, 4 K Gifford McMahon/Joule-Thomson cycle refrigerators, *Adv. Cryog. Eng.*, **33**, 689-698, Plenum Press, New York.

Mikulin E.I., Tarasov A.A., Shkrebyonock M.P., 1984, Low-temperature expansion pulse tubes, *Advances in Cryogenic Engineering*, R.W. Fast, Ed., **29**, 629-637, Plenum Press, New York.

Mozurkewich G., 1998, Time-average temperature distribution in a thermoacoustic stack, *J. Acoust. Soc. Am.*, **103**, 380-386.

Olson J.R., Swift G.W., 1994, *J. Acoust. Soc. Am.*, **95**, 1405-1412.

Paty M., 1982, Une philosophie de la science des métamorphoses, *Scientia*, Annus LXXVI, **117**, 29-41.

Prigogine I., 1980, *From Being to Becoming*, W.H. Freeman and Co., New York.

Radebaugh R., et al., 1986, A comparison of three types of pulse tube refrigerators: new methods for reaching 60 K, *Adv. Cryog. Eng.*, **31**, 779-789, Plenum Press, New York.

Radebaugh R., 1990, A review of pulse tube refrigeration, *Adv. Cryog. Eng.*, **35**, 1191-1205, Plenum Press, New York.

Richardson R.N., 1986, Pulse tube refrigerator - an alternative cryocooler, *Cryogenics*, **26**, 331-340.

Richardson R.N., Evans B.E., 1997, A review of pulse tube refrigeration, *Int. J. Refrig.*, **20**, 367-373.

Rubin M.H., 1979, Optimal configuration of a class of irreversible heat engines, I, *Phys. Rev. A*, **19**, 1272-1276.

Shapiro A.H., 1953, *Compressible Fluid Flow*, **I**, Roland Press Co., New York.

Swift G.W., Migliori A., Hofler T., Wheatley J., 1985, Theory and calculations for an intrinsically irreversible acoustic prime mover using liquid sodium as primary working fluid, *J. Acoust. Soc. Am.*, **78**, 767-781.

Swift G.W., 1988, Thermoacoustic engines, *J. Acoust. Soc. Am.*, **84**, 1145-1180.

Swift G.W, Keolian R.M., 1993, Thermoacoustics in pin-array stacks, *J. Acoust. Soc. Am.*, **94**, 941-943.

Thirumaleshwar M., Subramanyam S.V., 1986, "Two-stage Gifford-McMahon cycle cryorefrigerator operating at 20 K", *Cryogenics*, **26**, 547-555.

Urieli I., Berchowitz D.M., 1984, *Stirling Cycle Engine Analysis*, Adam Hilger Ltd., Bristol.

Walker G., 1980, *Stirling Engines*, Oxford University Press, Oxford.

Wetzel M., Herman C., 1997, Design optimization of Thermoacoustic refrigerators, *Int. J. Refrig.*, **20**, 3-21.

Wheatley J., Hofler T., Swift G.W., Migliori A., 1983a, An intrinsically irreversible thermoacoustic heat engine, *J. Acoust. Soc. Am.*, 1983, **74**, 153-170.

Wheatley J., Hofler T., Swift G.W., Migliori A., 1983b, Experiments with an intrinsically irreversible thermoacoustic heat engine, *Phys. Rev. Lett.*, **50**, 499-502.

Wheatley J., Hofler T., Swift G.W., Migliori A., 1985, Understanding some simple phenomena in thermoacoustics with applications to acoustical heat engines, *Am. J. Phys.*, **53**, 147-162.

Wheatley J., Buchanan D.S., Swift G.W., Migliori A., 1985, Nonlinear natural engine: Model for thermodynamic processes in mesoscale systems, *Proc. Natl. Acad. Sci. USA*, **82**, 7805-7809.

Wheatley J., 1986, Intrinsically irreversible or natural engines, *Frontiers in Physical Acoustics*, 395-475, North-Holland Phys. Publ. Amsterdam.

Zhu S., Wu P., Chen Z., 1990, Double inlet pulse tube refrigerator: an important improvement, *Cryogenics*, **30**, 514-520.

2001-CT-05

FIRST EXPERIMENTAL RESULTS ON HUMIDIFICATION OF PRESSURIZED AIR IN EVAPORATIVE POWER CYCLES

Farnosh Dalili , Mats Westermark

Department of Chemical Engineering and Technology
Energy Processes
Royal Institute of Technology
S-100 44 Stockholm, Sweden
Phone: +46 8 7906000 , Fax: +46 8 7230858
E-mail: farnosh@ket.kth.se

ABSTRACT

Humidification of compressed air before combustion is a key operation in evaporative power cycles. However, little work has been done to study this operation at high pressures and temperatures. A tube humidifier pilot plant was designed and constructed to fill this void.

Experiments at different pressures and flow conditions have been carried out. The results show that the theoretical design methodology developed parallel with the experiments can be considered reliable. The tube dimensions and design gave satisfactory efficiencies regarding the flue gas cooling and the humidification of compressed air. Furthermore the results show a new behavior of the working line, which may have an impact on designing such equipment.

This paper mainly describes the pilot humidifier facility and its components and the first results obtained.

NOMENCLATURE

C	Molar concentration [kmol/m^3]
h	Individual heat transfer coefficient [W/m^2 °C]
h_D	Mass transfer coefficient [m/s]
h_{lat}	Latent heat transfer coefficient [W/m^2 °C]
i	Enthalpy of humid air per unit mass of dry air [kJ/kg]
$i*$	Enthalpy of saturated air per unit mass of dry air [kJ/kg]
T	Temperature [°C]
ω	Humidity [g water vapor/kg dry air]
M	Mass flow rate [g/s]
M_W	Molar weight of water [kg/kmol]
N'	Overall rate of mass transfer [kmol/s, m^2]
P	Total pressure [bar]

Subscripts

a	Compressed air
d	wet-bulb
i	Water-air interface
v	Water vapor
w	Water

INTRODUCTION

The gas turbine market has expanded rapidly during the last decade. Only between 1999 and 2000, the number of gas turbine units ordered increased by 37 percent (McNeely 2000). The technical development of gas turbines is towards higher inlet temperatures (TIT) and more complex systems by introduction of inter-cooling, multi-stage combustion, and exhaust heat recovery implementations. These implementations include recuperation, steam generation, humidification and condensation. The exhaust gas from the turbine usually has a temperature above 500°C. Hence, exhaust heat recovery is the key to high electrical and overall efficiencies. The combined cycle is an approved approach to gain high efficiencies. However, steam turbines are expensive and not available for small gas turbines. Other alternatives with similar performance are advanced gas turbine cycles e.g. the steam injected gas turbine (STIG) and the evaporative gas turbine (EvGT). The STIG cycle as well as the combined cycle recovers high-quality heat, i.e. down to temperatures above the boiling point, in the heat recovery steam generator (HRSG). The EvGT cycle even recovers low-quality heat, i.e. to temperatures far below the boiling point, in a humidifier.

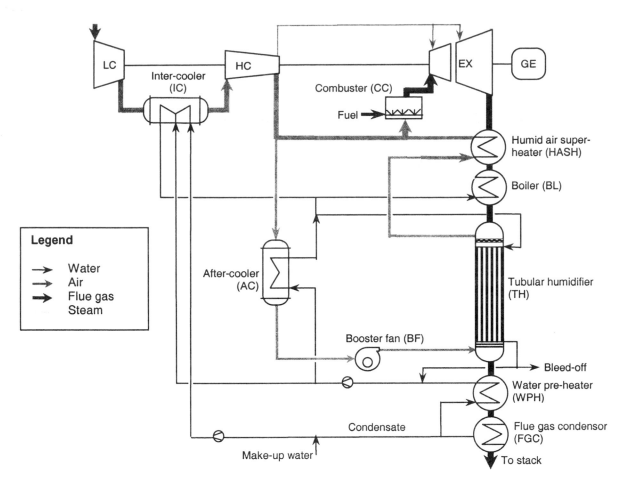

Figure 1. Proposed evaporative gas turbine cycle with tubular humidifier

The first evaporative gas turbine pilot plant (600 kW), located at the Lund Institute of Technology in Sweden, has been operating successfully since 1998. Invaluable experience is obtained from that plant, (Lindquist 1999) and (Ågren 2000). One key component in EvGT cycle is the humidification apparatus. Though the humidification process at atmospheric conditions is well known and practiced, e.g. in cooling towers, only little experimental data are available at high pressures and temperatures. These data are necessary for designing the humidification equipment. Therefore a separate project was initiated at the Royal Institute of Technology in Stockholm, to build up a firm understanding of the humidification process and to provide reliable procedure for designing such equipment.

This paper presents first experimental results from a humidifier pilot plant. A comparison is made between these experimental results and simulation results to validate the design procedure developed by the authors.

THE EVAPORATIVE POWER CYCLE

The EvGT power cycle, also referred to as the HAT cycle, was first presented in the mid 80s, (Nakamura et al., 1985, 1987) and (Rao 1989). The EvGT power cycle is basically characterized by high efficiency, low NOx emissions and low investment cost. Other advantages are quick start-up times, ready availability and compact size. Its efficiency is comparable to the combined cycle, but the absence of a bulky and expensive steam turbine makes the EvGT cycle favorable, especially in the small sizes (1-20 MW).

The EvGT concept involves the addition of water vapor to high-pressure air by humidification. Introduction of water vapor increases the mass flow rate through the expander, resulting in a higher power output and a high exhaust heat recovery potential. The heat required for humidification is mainly taken from the exhaust gas, as mentioned above, augmenting the overall cycle efficiency. Water consumption is extremely low, since sufficient humidification water is provided by exhaust gas condensation. However a small bleed-off is necessary to avoid salt enrichment.

Figure 1 shows an EvGT configuration with a tubular humidifier (TH). Only a part of the total compressed air flow, approximately 30 percent is suggested for humidification. This was first presented by Westermark (1996) and later used by Dalili & Westermark (1998) and Ågren (2000). The exhaust gas heat at high temperatures is recovered in the humid air superheater (HASH) and the boiler (BL).

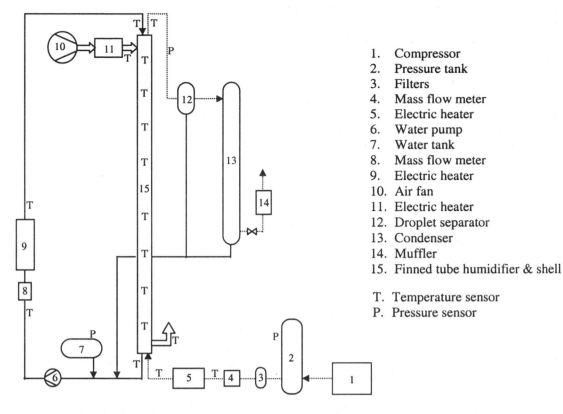

Figure 2. Tube humidifier pilot flow sheet

1. Compressor
2. Pressure tank
3. Filters
4. Mass flow meter
5. Electric heater
6. Water pump
7. Water tank
8. Mass flow meter
9. Electric heater
10. Air fan
11. Electric heater
12. Droplet separator
13. Condenser
14. Muffler
15. Finned tube humidifier & shell

T. Temperature sensor
P. Pressure sensor

The remaining exhaust heat is recovered in the TH. A comparison between TH and HRSG displays that in the HRSG the exhaust gas is cooled on the shell side causing the water to boil inside the tubes. In the TH however, the heat source is the same, although at lower temperatures, while there are two phases inside the tubes, water and high-pressure air.

Humidification is subsequently carried out by bringing the high-pressure air into countercurrent contact with water, causing the water to evaporate due to the concentration gradient of water vapor in the gas phase. This process thereby comprises simultaneous heat and mass transfer across the interface between the two phases.

The humidification process permits water to evaporate on a large scale at temperatures below the boiling point, because of the lower partial pressure of water in the air-water vapor mixture This lower partial pressure is a result of the diluting effect of the air. Thus the exhaust gas heat can be recovered down to significantly lower temperatures compared to steam generation.

The moist air (10-40% water vapor on mass dry air basis) from the humidifier is reunited with the rest of the compressed air before combustion. The steam generated in the boiler can be injected to increase the mass flow rate further. This is optional since steam may be favored elsewhere, e.g. in a pulp processing plant etc.

THE TUBULAR HUMIDIFIER CHARACTERISTICS

Tubular humidifiers have a compact design and are considered to be effective, since heat exchanging is carried out directly between the heat source (exhaust gas) and the heat sink (compressed air). Furthermore, economical considerations make the tubular humidifier favorable to other alternatives such as a packed bed and an economizer, especially in small size gas turbines (Wahlberg 2001).

Some specific features of humidifiers in EvGT systems are summerized below:

- Humidifiers operate optimally, considering the whole EvGT system, at relatively low exhaust gas temperatures.
- EvGT humidifiers operate at high pressures and high water temperatures compared to cooling towers.
- Because of elevated water temperatures, high water vapor pressures are feasible leading to high humidity levels in the air.
- Since humidification is boosted by direct exhaust gas cooling, the process is diabatic.
- Part-flow humidification is advantageous, since the required heat exchanging surface is less (Westermark 1996).
- The investment costs of tubular humidifiers are rather low compared to other components (Wahlberg 2001).

In designing humidifiers, the following factors should be taken into consideration:

- Water is sub-cooled by 10-15 °C to prevent any boiling.

367

- The moist air exiting the humidifier may be assumed to be saturated and has a temperature below the inlet temperature of water.
- The exhaust gas exiting the humidifier shell has a temperature well above its dew-point temperature.
- The flow rate of compressor air and the water available in the system decides the number of tubes in the humidifier.
- Flooding must be avoided. Proper wetting of the inner tube walls is essential (Dalili & Westermark 1998).
- Minimum entrainment of water droplets in the passing compressed air is desirable. Effective droplet separation should be implemented.

Flooding in a wetted wall tube can occur at high gas rates. This condition is exhibited by a sudden large increase in the pressure drop, considerable entrainment of water droplets or surging of the liquid in the tube. In this work, the flooding velocity correlation by Alekseev (McQuillan & Whally, 1984) is applied.

The minimum liquid rate for proper wetting of a vertical plane surface, according to Perry & Green (1997), is 0.03-0.3 kg/m s water at room temperature. Since the tube diameter in this case is relatively large, the wetting rate limits above can be applied.

THE TUBULAR HUMIDIFIER PILOT PLANT

Figure 2 shows the pilot humidifier schematically. The heart of the humidifier is a vertical extended-surface tube of stainless steel with an outer and an inter diameter of 60.3 and 51.3 mm, respectively. The tube has a total length of 9.2 m, of which 8.6 m is available for heat exchanging. The remaining length is used for inlet and outlet. Its exterior surface is equipped with carbon steel rods enhancing its total heat-exchanging surface by a factor of about 7. The rods have a diameter of 6 mm and there are 332 rods with a length of 82 mm and 664 rods with a length of 68 mm per meter tube forming a quadratic cross-section. The rods are angled downwards, to decrease the pressure drop. A thin stainless steel shell (17×17 cm) with insulation surrounds the tube (figure 3).

Figure 3. Extended-surface tube and shell

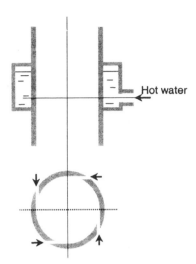

Figure 4. Humidification water inlet

A two-step compressor with inter-cooler produces pressurized air for humidification. It has a maximum flow capacity and pressure of 100 g/s and 40 bars, respectively. Two mechanical and one active carbon filter, in that order, clean the compressed air, before its pressure is reduced to desired level (5-35 bars) for each specific experiment of run. The flow rate is set by a needle valve and is measured by a mass flow meter. The mass flow meter gives also the temperature and the density of the compressed air.

The pressurized air coming from the compressor has a temperature slightly above ambient temperature. After cleaning and pressure reduction, it is heated in an electric heater to 60-150°C, to simulate an aftercooled gas turbine process. Water is fed from the top of the tube through four tangentially drilled holes, each with a diameter of 2.3 mm (figure 4). The water film formed falls continuously down the inner tube wall (figure 5). Boiling in the water film should be avoided because of the associated increased risk for entrainment. Hence the temperature of the entering water is held a few degrees below the boiling point.

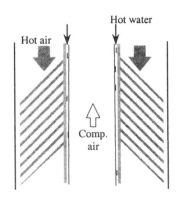

Figure 5. Flow directions in the tube and the shell

Since the falling water film evaporates into the pressurized air, its temperature falls and its flow rate decreases gradually. Cooled falling water (80-120°C) is collected at the bottom of the tube and recirculated after addition of fresh distilled water. An electric heater (0-60 kW, max. 250°C) heats the water before entering at the top again. The water flow rate is controlled by a speed regulated gear pump and is measured by a mass flow meter (same type as above).

A temperature-regulated electric heater (0-30 kW, max 350 °C), generates hot air, simulating the exhaust gas from a gas turbine. A wing-wheeled mass flow meter measures the hot air flow rate. The hot air flows down along the tube on the shell side.

Heat is transferred from the exhaust gas to the tube wall on the shell side. The heat transfer speeds up the evaporation of the falling water film into the countercurrent pressurized air stream. Thus the humidity of compressed air increases rapidly on the tube side.

The humid air exiting the tube may contain entrainment in the shape of mist and small droplets. Generally in EvGT cycles, separation of entrainment is necessary since water contains small amount of salts that will attack the hot turbine blades, resulting in corrosion. Munters Euroform GmbH provided the droplet separator. Inside the separator humid air passes first through a wire mesh pad. Mist and small droplets colliding with the wires form bigger droplets. The humid air passes then through waved plates, where the droplets are trapped and separated.

After droplet separation, the pressurized humid air is led to a condenser, with 6-m² of heat exchange surface. Most of the water is condensed and recirculated back to the plant. The recirculation is optional since collecting the condensate reveals the amount of water being evaporated. The air is then expanded to atmospheric pressure by a needle valve and released to the surroundings. The mass flow rate of pressurized air is set by the same needle valve.

Thermocouples (type K) measure the temperature of water, pressurized air and hot air streams at the inlets and outlets and along the tube. The temperatures of the tube wall and in the hot air stream on the shell side are measured in one-meter intervals. The measured figures are collected in data storage for further computer processing.

Two pressure sensors at the tube inlet and outlet measure the system pressure and also indicate the pressure drop of the compressed air in the tube.

The power consumed by water heater, pressurized air heater and air heater are separately displayed. The total power consumption, after reduction of losses, is included in the energy balance relations.

This equipment is designed to operate at different gas flow rates, temperature and pressure conditions up to 35 bars.

The experimental results considering the following design parameters will be reported: heat and mass transfer coefficients; entrainment; wetting limits and flooding boundaries of the system, entrainment of water droplets and elimination of them. The results will be used for designing humidifiers for evaporative gas turbine cycles.

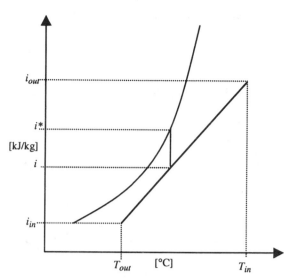

Figure 6. Humidification diagram for a packed bed humidifier

Due to limited space of this paper only a selected results at a pressure of 10 bars are reported. More extensive results will be published in the near future.

HEAT AND MASS TRANSFER MODEL

A detailed description of the modeling of simultaneous heat and mass transfer is complicated and beyond the scope of this paper. As mentioned before, humidification is a process of enhancing the water vapor content in pressurized air by bringing the two phases, compressed air and water, into countercurrent contact.

The required heat is mainly provided by transferring sensible heat from the exhaust gas on the shell side to the falling film. Latent heat (mass transfer) is conveyed into the air by evaporation, while cooling of the falling water film (additional sensible heat) also contributes to the process. It is reasonable to assume that the system is gas-film-controlled.

The equilibrium curve/working line concept is a convenient tool to describe the humidification process. Enthalpy difference is employed as the driving force for combined heat and mass transfer, first suggested by Mickley (1949). Figure 6 shows a humidification diagram for a packed bed humidifier. The difference in this case compared to a tubular humidifier is that the cooling of the exhaust gas with water occurs separately in an economizer. The second step is the humidification of compressor air by heated water in a packed bed. The working line is only slightly curved since there is no interaction from the exhaust gas. As will be shown later, the working line for tubular humidifier is significantly curved. The X- and Y-axis present temperature of water and humid air enthalpy respectively. The equilibrium curve describes the enthalpy of the saturated air in thermal equilibrium with the liquid water as a function of the equilibrium temperature. The enthalpy of humid air in this work is calculated using the Hyland and Wexler model for real-gas mixtures (Dalili et. al. 2001).

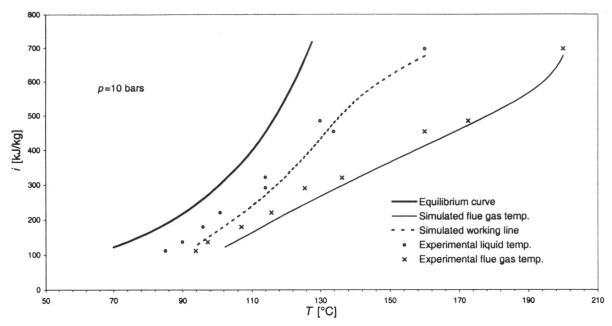

Figure 7. A comparison between the experimental and the simulated results

The working line describes the enthalpy of the humid air at any point along the humidifier as a function of the water temperature at that point.

One way to express the driving force is to implement the number of the transfer units methodology. The number of transfer units (NTU) is described by the following expression:

$$NTU = \int_{i_{in}}^{i_{out}} \frac{di}{(i^* - i)}$$

where, (i^*-i) is equivalent of the enthalpy driving force.

The rate of transfer of water vapor is a function of the mass transfer coefficient and the concentration driving force, as shown in the following expression:

$$N_v^{'} = h_D \left(C_{vi} - C_v \right)$$

The rate of transfer may also be expressed as a function of the latent heat transfer coefficient on the tube side with the following equation:

$$N_v^{'} = h_{lat} \frac{(T_{iw} - T_{da})}{M_w \lambda}$$

where T_{iw} and T_{da} are the temperature of the water surface and the wet-bulb temperature of pressurized air, respectively; thus presenting the temperature driving force for the evaporation of water. Heat transfer coefficients on the tube and the shell side are compared to investigate the resistance to heat transfer

The solution to mass and energy balances and the NTU can be obtained by dividing the tube into 50 height segments (figure 8). The same amount of energy is assumed to be transferred to the compressed air in each segment. First, mass and energy balance equations for the total system and then for each segment, starting from top, are established. The solution to these equations gives the temperatures of all three flows at each segment. The results are then plotted in a humidification diagram similar to figure 6. This is done in a spreadsheet simulation program, developed at the department.

RESULTS AND DISCUSSIONS

The first experiments carried out on the tubular humidifier were to validate the design procedure and the results provided by the simulation program. Hence, the same input data were used for the experiments. The design procedure applied in this work was first presented by the authors in 1998 (Dalili & Westermark 1998) and later by Wahlberg (2001).

As mentioned before, the temperatures and flow rates of the streams are measured frequently in the humidifier. Measuring the humidity of the moist air on-line proved to be difficult, since the state of the moist air is very close to saturation. Thus water vapor will instantly condensate when brought into contact with a slightly colder object. One simple and accurate way to estimate the humidity is through a water mass balance.

The data collected from the experiments are stored and processed. The enthalpy of the humid air, which is a function of temperature, pressure and humidity, is readily provided since all the properties are known.

The experimental and simulated results are then plotted in the same diagram for comparison. The results from the simulation program are apparently accurate if they coincide with the experimental results.

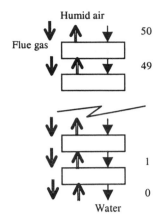

Figure 8. Humidifier Calculation segments

Experiments were carried out for a pressure of 10 bars. The mass flow rate and the inlet temperature of the hot air on the shell side were kept constant at 90 g/s and 200 °C respectively. The inlet temperature of water was also kept constant at 160 °C. The mass flow rates of water and compressed air were varied in each session to verify the results.

The results from one of these experiment sessions are presented in figure 7. All the other results exhibited similar trends and features. The hot air temperatures (also called the flue gas temperature in this work) are also included in figure 7 for two purposes: firstly, to show the total temperature driving force; and secondly, to compare the two results. A selection of important operation data is shown in Table 1.

The heat transfer coefficient on the shell side is estimated to be approximately 300 W/m² °C at a velocity of 4 m/s. The flue gas velocity should be boosted to about 10 m/s for effective heat transfer. The latent heat transfer on the tube side is strongly dependent on the temperature. Its value at the bottom and the top of the tube is estimated to 300 and 1400 W/m² °C, respectively. Hence, most resistance to heat transfer is on the shell side.

The experimental and simulated working lines are clearly in accordance, although the experimental points are slightly shifted to the left. In other words the temperatures of the water and the hot air are lower than expected. This can be explained by the heat loss to the surrounding, since the insulation is somewhat deficient. This heat loss is not included in the simulation calculations.

The working line deviates significantly from the straight line employed with a packed bed humidifier, to describe the mass transfer. The explanation for the total process is complicated, since there are three different streams that interact. The curvature of the working line becomes significant at low driving forces (small distance between the equilibrium curve and the working line).

In the particular case described above the curvature enables the evaporation and the cooling of the falling water film to proceed to relatively low water outlet temperatures at the bottom without crossing the equilibrium curve. The crossing point presents zero driving force.

Table 1. Experimental operation data

	T [°C]		M [g/s]		ω [g/kg dry air]	
	Inlet	Outlet	Inlet	Outlet	Inlet	Outlet
Flue gas	200	94	90	90	6	6
Water	160	85	30	23	-	-
Comp. air	67	127	37	44	17	203

A tube length of 8.0 m was estimated from the simulation, which differs from the actual tube length, 8.6 m, by 7 percent. This is probably due to the heat loss to the surrounding.

The humid air at the top has a humidity of 0.2 kg water vapor per kg dry air. This is relatively high considering the low hot gas inlet temperature and the low water flow rate. The humid air velocity at the top was set to 2 m/s, which is approximately 65 percent of the theoretical flooding velocity.

CONCLUSIONS

The tubular humidifier is an effective and cost efficient option for simultaneous exhaust gas heat recovery and the humidification of compressed air. The first experimental results show a satisfying agreement with the design methodology developed by the authors. Hence the methodology can be considered reliable.

The working line deviates significantly from a straight line drawn between the points representing the conditions at inlet and outlet. The curvature of the working line has an impact in designing such equipment to maintain sufficient driving forces for the desired performance.

The tube dimensions chosen and the construction of the pilot plant proved to be efficient. No complication was detected in the falling water film formation. This may become a challenge at higher pressures (above 20 bars). The lower limit for proper wetting is estimated to 30 g water/s.

Additionally, no evidence of flooding was detected. The flooding correlation may be reliable for the conditions in this work, but should be tested further at higher pressures and air flow rates.

FUTURE WORK

Experiments are being carried out for higher pressures (10-35 bars). The parameters being studied include: heat and mass transfer coefficients; wetting; flooding velocity; and entrainment of water droplets. The results will be presented in future publications.

ACKNOWLEDGMENTS

Financial support from the Swedish National Energy Administration, Vattenfall and Munters is gratefully acknowledged. Also, special thanks are directed to Aalborg for data on tubes and to M.Sc. Jens Wolf for practical assistance during experiments.

REFERENCES

Dalili, F., and Westermark, M., 1998, "Design of Tubular Humidifiers for Evaporative Gas Turbine cycles", ASME Paper No. 98-GT-203, International Gas Turbine & Aeroengine Congress & Exhibition, Stockholm, June 2-5.

Dalili, F., Andrén, M., Jinyue Y., Westermark, M., 2001, "The Impact of Thermodynamic Properties of Air-Water Vapor Mixtures on Design of Evaporative Gas turbine Cycles", ASME Paper No. 2001-GT-0098, International Gas Turbine & Aeroengine Congress & Exhibition, New Orleans, June 4-7.

Lindquist, T., 1999, "Theoretical and Experimental Evaluation of the EvGT-Process", Licentiate Treatise, Lund Inst. of Tech., Dept. of Heat & power Eng., Lund, Sweden.

McQuillan, K. W., and Whally, P. B., 1984, "A comparison between Flooding Correlations and Experimental Flooding Data for Gas-Liquid Flow in Vertical Circular Tubes", Chem. Eng. Sci., 40, No. 8, pp. 1425-1440.

McNeely, M., 2000, "Power Generation Survey Results-2000", Diesel and Gas Turbine World Wide, Waukesha, Wisconsin.

Nakamura, H., Takahashi, T., Narazaki, N., Yamamoto, F., Sayama, N., 1985, "Regenerative Gas Turbine Cycle", US Patent 4,537,023, August 27.

Nakamura, H., Takahashi, T., Narazaki, N., Yamamoto, F., Sayama, N., 1987, "Regenerative Gas Turbine Cycle", US Patent 4,653,268, March 31.

Perry, R. H., Green, D. W., 1997, "Perry's Chemical Engineers' Handbook" 7th Ed., McGraw-Hill, USA.

Rao, A.D., 1989, "Processes for Producing Power", US Patent No. 4829763, May 16.

Wahlberg, P. E., 2001, "Design and Comparison between a Finned Tube Humidifier and a Packed Bed Humidifier Concerning Performance and Costs", M.Sc. Thesis, Dept. of Chem. Eng. & Tech., Royal Inst. of Tech., Stockholm.

Westermark, M., 1996, "Method and Device for Generation Of Mechanical Work and, if Desired, Heat in an Evaporative Gas Turbine Process", International Patent Application No. PCT/SE96/00936.

Ågren, N., 2000, "Advanced Gas Turbine Cycles with Water-Air Mixtures as Working Fluid", Doctoral Thesis, Dept. of Chem. Eng. & Tech., Royal Inst. of Tech., Stockholm.

IECEC2001-CT-28

ADVANCED ENERGY CONVERSION TECHNOLOGIES BASED SMALL POWER SYSTEMS FOR REMOTE SITE APPLICATIONS

Mysore Ramalingam
UES, Inc., 4401 Dayton-Xenia Road, Dayton, OH 45432-1894
Ph.:(937) 253-3986; Fax:(937)253-9386
E-Mail: mysore.ramalingam@wpafb.af.mil

Edward Doyle, Kailash Shukla
Thermo Power Corporation, 45 First Avenue, Waltham, Massachusetts 02454-9046
Ph: (781)622-1053; Fax: (781)622-1025
E-Mail: edoyle@tecogen.com

Brian Donovan
Power Division, AFRL/PRPS,1950 Fifth St., Bldg. 18, Wright Patterson Air Force Base, OH 45433-7251
Ph.:(937) 255-6241; Fax:(937) 656-4781
E-Mail: brian.donovan@wpafb.af.mil

ABSTRACT

Remote site power generation is of primary importance for military applications such as off-grid monitoring stations, ground operations, air combat test range instrumentation and aircrew training area communications. The Air Force is responsible for providing electricity for seismic sensing and other data acquisition activities at remote locations in Alaska. Typically, low power levels in the range of 100 Watts to 500 Watts are provided to many separate remotely located areas. Though Radioisotope-powered Thermoelectric Generators (RTGs) have been the power sources of choice for several years, primarily due to their reliable operation over extended periods of time without fuel or maintenance, during recent years there has been a growing resistance to the deployment of devices whose operation relies on radioactive materials.

The current investigation deals with advancements associated with the development of remote site power systems based on advanced energy conversion technologies such as Thermionic Energy Conversion (TEC), Free Piston Stirling Engines (FPSE), Alkali Metal Thermal Electric Conversion (AMTEC) and Thermo Photo Voltaic energy conversion (TPV). While several government agencies and their contractors are involved in this evaluation and development of improved remote site power systems, the Power Division of the Propulsion Directorate at US Air Force Research Laboratories (AFRL) is primarily focussed on AMTEC and TPV related technologies for the present. The power levels associated with these power systems (100 - 500 watts) do not pose the logistics and costs associated with the alternate technologies mentioned above, which are key factors used in these evaluations. These factors were weighed in with system efficiencies to arrive at feasible configurations for the power system even if it entailed the use of hybrid energy conversion technologies.

NOMENCLATURE

AEPS	Auxiliary Electric Power Supply
AFRL	Air Force Research Laboratory
AMPS	Advanced Modular Power Systems
AMTEC	Alkali Metal Thermal Electric Converter
BASE	Beta Alumina Solid Electrolyte
FPSE	Free Piston Stirling Engine
RTG	Radioisotope Thermoelectric Generator
TBE	Teledyne Brown Engineering
TEC	Thermionic Energy Conversion
TPV	Thermophotovoltaic Energy Conversion

INTRODUCTION

Remote site power systems must operate continuously with minimal noise and vibration and without frequent maintenance or refueling. Currently, power for remote sites is supplied by operating radioisotope-powered thermoelectric generators. Thermoelectric converters are low efficiency devices, and clearly there are more efficient energy conversion devices. If a new power generation system can meet the operational requirements with a higher efficiency, there would be significant cost savings.

One candidate energy conversion technology is TPV conversion. Filters are used in the TPV generators to reflect unwanted out-of-band radiation back to the emitter and transmit near-monochromatic radiation to the photovoltaic cell greatly improving the conversion efficiency in this process. To further enhance system efficiency, the bandgap of the photovoltaic cell is matched to the emission characteristics of the filtered radiation. As a result of this there is an ongoing evaluation of alternate power sources such as engine-driven generators, solar and wind generators, propane thermoelectric generators, batteries, fuel cells and power systems based on advanced energy conversion technologies such as AMTEC, TPV and TEC have been considered.

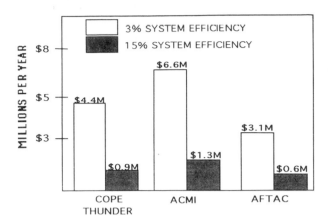

Figure 1. Fuel Transportation Cost Estimates

Economical Implications of RTG Replacement

Cost and safety are the primary considerations in the economical implications of replacing low efficiency RTG power sources with those based on high efficiency energy converters. The most demanding aspects of providing power at Arctic and Antarctic remote site locations are the unique problems of supply and maintenance for any potential power system depending on consumable fuels. Due to virtually non-existent surface transportations, all supplies and personnel must be flown into the sites by transport aircrafts. In order to obtain a measure of the costs of fuel transport to the high Antarctic polar plateau as a function of the efficiency of the power system, three different USAF operations for fuel transport in Alaska were considered. A propane fired modular power system with 3% system efficiency and one with 15% system efficiency were chosen for comparison of propane transportation costs using a C-130 transport aircraft with snow skis. The choice was based on data available for similar logistics support for geophysical observatories in the Antarctic polar plateau. Figure 1 clearly illustrates the influence that power system efficiency has on the logistics expenses associated with the three different operations in Alaska [Lamp, 1994]. The savings in all three instances were about 80% for the propane fired remote site power system with a system efficiency of 15%. Higher efficiency power systems require less fuel for the same duration and this translates to a lower level of operation. Lower level of operations reduces the risk of serious accidents considering that the helicopter accident rates in Alaska are among the highest recorded for helicopter operations. Thus the development of higher efficiency power systems is justified both from a cost and safety perspective as lower efficiency drives the need for more frequent refueling.

Power Sources With Advanced Energy Conversion

In view of attempting to replace the RTG remote site power sources, several emerging advanced energy conversion technologies are being considered for remote site applications. The technologies that show significant promise in the far-term, from a system efficiency point of view, are Thermionics (TEC), ThermoPhotovoltaics (TPV), Free piston Stirling Engines (FPSE) and Alkali metal conversion (AMTEC). While thermionic energy conversion technologies [Ramalingam, et al., 1996] are being

pursued primarily for space related applications, the US Air Force is actively pursuing AMTEC-based [Sievers, et al.,1995] and TPV-based [Becker, et al., 1997 and Doyle, et al., 1999] power systems for 50 - 200 W remote site power systems. Analytical and experimental verification activities are currently underway at the Propulsion Directorate of Air Force Research Laboratory (AFRL), small businesses such as Amps, Inc., ThermopowerCorp. and a larger business in General Atomics in California, USA. The results and status of developmental activities in these advanced energy conversion technological areas are summarized in subsequent sections.

Table 1 provides a comparison of current and far-term efficiencies for the various advanced energy conversion technologies being considered [Lamp, 1994]. Though an increase in system efficiency is predicted for thermoelectric energy conversion in the far-term, the system efficiencies of the other technologies are expected to make larger strides in performance improvement. Significant measures are being incorporated in an effort to raise system efficiencies and subsequently reduce refueling costs and safety by reducing the total number of refueling trips required for continuous operation of these power systems.

Alkali Metal Thermoelectric Conversion

The US Air Force and NASA Glenn Research Center have sponsored developmental work in AMTEC technology. An Auxiliary Electric Power Supply (AEPS) developmental program with AMPS to develop a 60 Watt AMTEC-based power system is scheduled to be completed this year even though life testing on one of the converters will be continued. An AMTEC converter on test for the US Air Force now has over 1400 hours of operation in a propane burning system without significant power degradation. This continuous performance without deterioration, coupled with advances in the design of the original AMTEC converter have essentially provided a springboard for complete systems readiness for field testing at an Air Force remote site. The testing system developed at AMPS was equipped to handle 12 converters and provide the 60-Watts of electrical power from a series-parallel electrical array of these converters. Table 2 provides a list of the AEPS operating specifications, which were used as a guideline for the development of the converters [Giglio, et al., 2001]. The 60-Watt system design arranges 12 AMTEC converters in 3 rows in a radial arrangement. Each row of converters is offset 45 degrees for maximum packing densities and minimal bulk insulation parasitic heat losses. The design uses a counterflow heat exchanger to maximize system

Table 1. Efficiencies for Various Technologies

TECHNOLOGY	PRESENT EFFICIENCY (%)	FAR-TERM EFFICIENCY (%)
Thermoelectrics	5-8	10-15
Thermionics	10-15	20-25
Thermophotovoltaics	15-20	25
Stirling Converters	20	30
AMTEC	20	30

Table 2. Operating Specifications for the AEPS.

Operating Environment	233-300 K 0-100% Relative Humidity 0-10,000 Ft. Elevation
Power	60 Watts
Voltage (nominal)	12 VDC
Voltage Range	12 VDC nom: 11-16 VDC 24 VDC nom: 20-28 VDC
Voltage Ripple	0.3 VDC Maximum
Load Regulation	0.1 VDC Maximum
Output Isolation	100 k
Lifetime	10 years
Maintenance Interval	6 Months
Fuel	Propane

efficiency and locates the combustor at the center of the system housing. The burner/recuperator was designed and built by Teledyne Brown Engineering (TBE) while AMPS supplied the housing/insulation package for the converters. The converter itself which was originally designed for space application had to undergo a series of design changes to facilitate terrestrial, fossil fuel applications. The original converter design that AEPS started with was constructed using a stainless steel housing, had no Molybdenum BASE tube sleeve and used no getter materials to trap hydrogen and other non-condensable gases. Significant improvements in structural integrity and performance were observed after incorporating these changes.

The actual power output of one of the converters AEPS6-1 tested with a hot end temperature of 1123 K and a cold end temperature of 573 K is depicted in Figure 2. Peak power of 4.8 Watts was observed at about 2.1 Amps. Improved converter

interconnects are expected to minimize power losses and raise the peak power level of the converter. One of the major issues addressed by the AEPS program was hydrogen diffusion through the converter walls into the AMTEC converter. Increase in cathode pressure due to non-condensable gases can potentially depress performance and eventually lead to failure as shown in Figure 3. This figure represents performance deterioration as a result of non-condensable gas diffusion over 1200 hours of testing. This situation was eliminated by using non-condensable gas absorbing getters along with Haynes 230 for the housing material to make it impervious to hydrogen. AMPS is currently working on series-connected multiple converter systems to bring the output level to 60 Watts with the intention of delivering a small fossil fuel AMTEC power system to be tested in a remote location in Alaska or the Antarctic region.

Thermophotovoltaic Energy Conversion

Thermophotovoltaic energy conversion (TPV) is a technology that may be ideal for the development of highly efficient, compact and reliable sources of electricity. Thermo Technologies and its subcontractor, JX Crystals, are developing a logistic-fueled 130 W TPV power source for an Air Force remote ground power application. The major components of a TPV system are illustrated in Figure 4. These are the burner/emitter, optical filter, photovoltaic cell array, and the recuperator. Fuel and air are combusted in the burner and the heat from combustion is transferred to an emitter by convection and radiation. The emitter, in turn, radiates energy to the PV cells through an optical filter, which is used to reflect non-convertible (off-band) radiation back to the emitter. The PV cells convert a portion of the convertible (in-band) radiation into electric power. The recuperator is used to transfer the thermal energy in the combustion products leaving the emitter cavity to the incoming combustion air.

The net fuel-to-electric efficiency, $\eta_{fe,}$ for a TPV power source can be defined as the product of four factors, as follows:

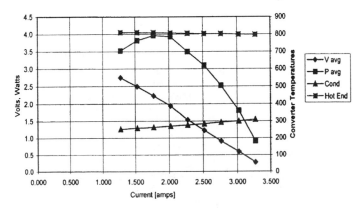

Figure 2. AEPS6-1 Step Test Data @ 1073 K (800°C) Hot End [Giglio, et. al., 2001]

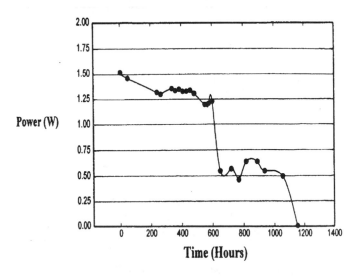

Figure 3. AEPS 1 Performance Deterioration Due to Gas Diffusion [Giglio, et al, 2001]

375

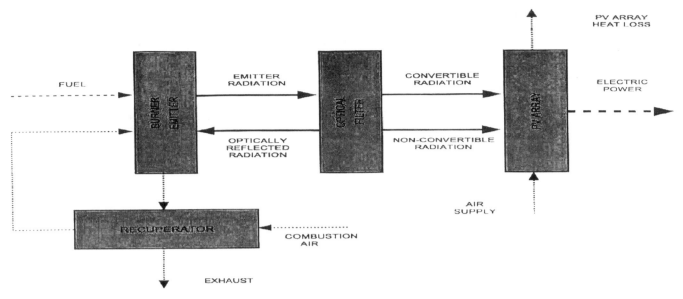

Figure 4. TPV Power Source Components

$$\eta_{fe} = \eta_{fr} \times \eta_{sp} \times \eta_{array} \times \eta_{aux}$$

η_{fr} = fuel to net radiation conversion efficiency

η_{sp} = spectral efficiency (in-band radiation @ PV array / net radiation from emitter)

η_{array} = PV array conversion efficiency for in-band radiation

η_{aux} = efficiency factor for auxiliary components (1 - auxiliary power/array power)

The specific objectives of this program are to develop and demonstrate a 130-W TPV power source that:

- Operates on logistic fuels (DF2 and JP8) over a wide range of ambient conditions

- Has a minimum fuel to electric power efficiency of 5% with a long-term potential to achieve 15% or higher
- Has a weight of no more than 12 kg with a long term potential of 8 kg

The values for each of the efficiency factors needed to achieve the fuel-to-electric efficiencies of 5% (near-term) and 15% (long-term) are listed in Table 3.

The design of the logistic-fueled 130 W TPV power source prototype is presented in Figure 5. The power conversion module, which contains a cylindrical burner/emitter/recuperator surrounded by a nine-sided gallium antimonide PV array and heat sink, is packaged in a cylindrical configuration and mounted in a rectangular frame. The overall dimensions of the package, which also includes the auxiliaries, are 11"x11"x26". The main auxiliaries are the cooling air fan, the combustion air blower, the atomizing air pump, the fuel pump, the ignition system, and the start-up battery.

The spectral characteristics of the emitting surface are the most critical factor for obtaining high efficiency in a TPV system.

The emitter approach selected for this project consists of a base Kanthal metal alloy tube surrounded by a platinum (near-term) or tungsten (long-term) foil re-radiator with an anti-reflective (AR) coating for in-band emissivity enhancement. AR-coated platinum (AR/Pt) foil was selected as the near-term approach, since it can be operated in air, though it has a lower in-band fraction than AR-coated Tungsten (AR/W). Use of AR/W will require a hermetically sealed enclosure filled with an inert gas, since tungsten cannot be operated in an oxidizing atmosphere at high temperatures. If an inert gas with low thermal conductivity, such as xenon, is used, it will also reduce the heat sink load and improve efficiency. The AR coated metal foil emitters are being developed and supplied by JX Crystals.

Table 3. Efficiency Factors for the 130 W TPV Power Source

	Near Term	Long Term
η_{fr}	65%	78%
η_{sp}	40%	75%
η_{array}	30%	30%
η_{aux}	66%	85%
η_{fe}	5%	15%

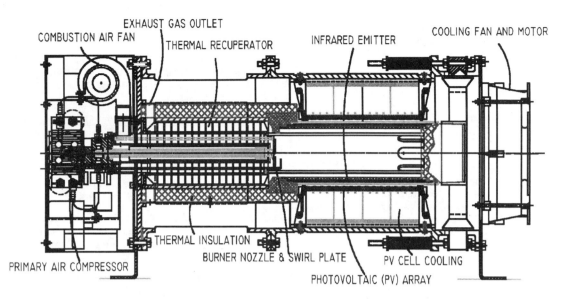

Figure 5. Layout Drawing of the 130-W TPV Power Source

Another major contributor to the system efficiency is the burner/recuperator system. The logistic-fueled burner design, shown schematically in Figure 6, is an air assist, siphon-type, atomizing burner. It makes use of a small portion (5 to 10%) of cold combustion air, called primary air, supplied by a miniature air pump, for aspirating and atomizing the fuel, as well as for nozzle cooling to prevent thermal degradation of the fuel. The use of cold ambient air in the air-aspirated nozzle helps to prevent fuel-cooking problems associated with logistic fueled burners operating with high temperature preheated air. The remaining (90 to 95%) combustion air, called secondary air, is preheated in the recuperator and rapidly mixed with the atomized fuel and primary air mixture in the combustion zone. A secondary air swirl plate is used to facilitate rapid mixing, to produce a short flame, and for flame stabilization. The fabrication of components for the first prototype is nearly complete and the assembly is underway. The completed components, including the housing, are shown in Figure 7.

CONCLUSIONS

Significant advances have been made in the attempts to develop 100W-200W power sources based on advanced energy conversion technologies in support of the US Air Force remote site power requirements. Demonstration of reliability has reached a point where power systems based on these technologies can begin transition to terrestrial commercial and government applications. The key to the development of these advanced power systems appears to be tied with the performances of the components where any significant breakthroughs in materials and processing translate into higher efficiencies and subsequently lower costs. A significant amount of the developmental activities have focused on the coupling between the heat source and the converter in the cases of AMTEC and TPV converter development. Especially in military applications, the emphasis is on multi-fuel machines that are capable of reliable, efficient operation on diesel, propane or any

of the JP-fuels. The prototypical AMTEC, TPV and Thermionic converters being developed for the remote site applications, will be performance mapped and life tested at the propulsion directorate of AFRL to assess performance deterioration, if any, over extended periods of time.

Some typical applications that are often cited for these technologies are base cogeneration, space power, remote power, self-powered appliances, and hybrid electric vehicles.

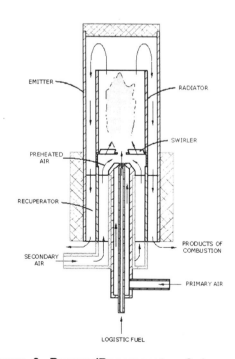

Figure 6. Burner/Recuperator Schematic

377

Two additional USAF applications for advanced conversion technologies include ground support power for unmanned aerial vehicles, and high performance power systems for combat aircraft deployment. Mobility and rapid deployment capabilities, which are strongly emphasized by today's military planners, are enhanced by lighter, smaller, more efficient power systems.

ACKNOWLEDGMENTS

Research activities were sponsored through UES contract F33615-99-C-2921 and AMPS contract F33615-96-C-2688 for the Power Division of USAF Air Force Research Laboratory. The Authors gratefully acknowledge the assistance of Mr. Joe Giglio, Mr. Robert Sievers/Dr Rahul Mital of Advanced Modular Power Systems, Dr. Fred Becker of Thermopower Corp. in the preparation of this manuscript.

Figure 7. Components of the 130-W TPV Prototype

REFERENCES

Lamp, T.R., "Power System Assessment for the Burnt Mountain Seismic Observatory," Final Report Number WL-TR-94-2026, Wright-Patterson AFB, OH., 1994.

Ramalingam, M. L., Lamp, T. R., Jacox, M., and Kennedy, F., "The Integrated Solar Upper Stage (ISUS) With High Temperature Thermionic Energy Conversion," 34th Aerospace Sciences Meeting and Exhibit, Reno, NV., 1996

Sievers, R. K and Ivanenok, J. F., , "Innovative AMTEC Auxiliary Electric Power Supply," Final Report WL-TR-96-2027, Wright-Patterson AFB, OH 45433., 1995.

Rose, M.F., Adair, P., Schroeder, K, "High Temperature Emitters for Thermophotovoltaic Power Systems for Aerospace Applications," 33rd Aerospace Sciences Meeting, Reno, NV. 1995

Fatemi, N.S., "A Combustion-Based Thermophotovoltaic Electrical Power Generator," Final Report WL-TR-97-2017, Wright-Patterson AFB, OH 45433. 1996

Becker, F. E., Doyle, E. F., Shukla, K., " Development of a Portable Thermophotovoltaic Power Generator ", Third NREL conference on the Thermophotovoltaic Generation of Electricity, Colorado Springs, CO., 1997

Doyle, E. F., Shukla, K. C., Becker, F. E., and Fraas, L. M., " Design and Fabrication of a 20W Propane Fueled Thermophotovoltaic Battery Substitute ", SAE Aerospace Power Systems Conference, Paper # 99-01-1397, Mesa, AZ., 1999

Giglio, J., Sievers, R. K., and Hunt, T., " Auxiliary Electric Power Supply (AEPS) Final Report ", Final Report, AMPS TR-01-001, USAF Contract # F33615-96-C-2688, AMPS, Inc., 2001.

Proceedings of IECEC'01
36th Intersociety Energy Conversion Engineering Conference
July 29–August 2, 2001, Savannah, Georgia

IECEC2001-CT-42

DEVELOPMENT OF CHEMICALLY RECUPERATED MICRO GAS TURBINE

Takao Nakagaki, Takashi Ogawa,
Haruhiko Hirata, Koichi Kawamoto
and Yukio Ohashi
Toshiba Corporation
Power and Industrial Systems Research and
Development Center
1, Toshiba-cho, Fuchu-shi, Tokyo, 1838511, Japan
Phone +81-42-333-2564
Fax +81-42-340-8060
E-mail takao.nakagaki@toshiba.co.jp

Kotaro Tanaka
Toshiba Corporation
Public and Industrial Systems Div.
Electrical and Control Systems Engineering Dept. 1
1-1, Shibaura 1-chome, Minato-ku, Tokyo,
1058001, Japan
Phone +81-3-3457-8501
Fax +81-3-5444-9299
E-mail kotaro3.tanaka@toshiba.co.jp

ABSTRACT

Micro gas turbines (MGTs) are subject to certain problems, notably low thermal efficiency of the system and high emission including NOx. The chemically recuperated gas turbine (CRGT) system introduced in this paper is one of the most promising solutions to these problems. The CRGT system we propose uses an endothermic reaction of methane steam reforming for heat recovery. It is usually thought that the reaction of methane steam reforming does not occur sufficiently to recover heat at the temperature of turbine exhaust, but we confirmed sufficient reaction occurred at such low temperature and that applications of the chemical recuperation system to some commercial MGTs are effective for increasing the efficiency.

INTRODUCTION

As a consequence of deregulation of energy markets in recent years, distributed power sources have become widespread in some parts of the world. MGTs are small-scale power sources which have attracted attention due to their good cost performance and easy operation. With a view to competing in markets, many manufacturers are developing MGTs, which have certain common specifications; (1) the power output is 30-300kW, (2) radial compressors and turbines with comparatively low pressure ratio are used, (3) natural gas, LP gas or kerosene are available as gas turbine fuel. However, certain problems have been identified under practical operating conditions, namely machine reliability, low thermal efficiency of the system and high emission including NOx.

Methane, the principal component of natural gas, is converted into hydrogen-rich gas with steam. This steam-reforming reaction of methane is well known in the case of hydrogen refineries, which are generally operated at about 800°C with Ni-based catalyst, but 20-50% conversion could be realized under the equilibrium conditions using turbine exhaust at 600°C. The chemical heat recovery from turbine exhaust using an endothermic reaction of steam reforming increases generating power and efficiency beyond what is possible with direct combustion. Also, NOx emission is reduced because the good combustibility of hydrogen-rich gas and the existence of much steam in the gas decrease adiabatic combustion temperature.

Concerning CRGT systems, there are some reports on the use of methanol, but few on the use of methane. Carapellucci et al. (1994, 1996) have reported feasibility studies of a CRGT system using methanol as fuel. Kuroda et al. (1989) and Nagaya et al. (1993) have reported on the practical tests that were conducted at 1.5MPa. Methanol is almost completely converted to hydrogen-rich gas at 300°C using Cu-Zn catalyst, which is a lower conversion temperature than is attainable for methane. Therefore, it is easier to design a methanol reformer than it is to design a methane one. However, methanol is a less widely used fuel than methane. Also, change in enthalpy in the case of methanol steam-reforming reaction is 49.5kJ/mol, which is smaller than that for methane (165.1kJ/mol), and so improvement of generating efficiency is inferior to that in the case of methane.

A number of studies on methane steam reforming have been reported. For example, Xu and Froment have presented a useful formula of reaction rate in their reports (1989).

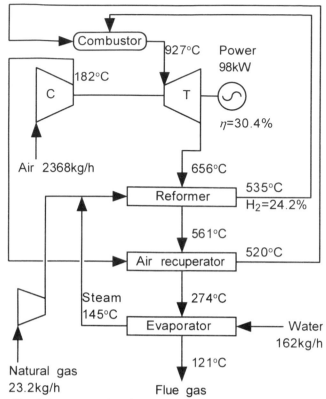

Fig.1 An example of chemically recuperated micro gas turbine system

Simulations of a methane reformer have been an object of study for several decades (Hyman, 1968). Murray et al. (1985) have estimated methane conversion, considering energy balance only when configuration of the reformer is changed. Alatiqi et al. (1989) have presented a design of methane steam reformer, for which purpose they used a reaction rate equation from the literature. There are many papers that report on heat and mass transfer in a methane reformer. Fukusako et al. (1997) have described a reformer in a fuel cell system, and Altman and Wise (1956) and Wijngaarden and Westerterp (1989) have studied the effect of chemical reactions on heat transfer.

In this report, we describe the feasibility of applying chemical recuperation to a commercial MGT system. At first, the mass and heat balance of the CRGT system was estimated in order to design the heat recovery reformer, and the basic performance data of commercial catalysts was evaluated to select suitable catalysts for the reformer. Next, the performances of heat transfer and conversion rate to hydrogen-rich gas were examined using a tube 300mm in length packed with catalyst. Based on the results of this evaluation, a simulation code calculating reformer performance was developed in order to design the heat recovery reformer. Using this code, a shell-and-tube reformer unit was designed and the performance of this reformer unit was evaluated.

OUTLINE OF CRGT SYSTEM

Fig.1 shows a block diagram of a CRGT system. In an MGT system, turbine exhaust temperature is about 600°C, and the generating efficiency is increased by heat recovery of air recuperator, but in the CRGT system, the reformer recovers the turbine exhaust heat first. As shown in Fig.1, the equipment after the reformer is the air recuperator, and next is the evaporator, but the system works if the sequence of equipment after the reformer is arranged in accordance with the layout of the engine system of micro gas turbines. The amount of heat recovered by the reformer is determined by enthalpy change of chemical reaction which converts fuel and steam into hydrogen-rich gas. The reactions in a reformer are expressed as follows.

$$CH_4 + H_2O \Leftrightarrow CO + 3H_2 \qquad \text{(Steam-reforming)}$$
$$CO + H_2O \Leftrightarrow CO_2 + H_2 \qquad \text{(Water gas shift)}$$

The reaction rate of water gas shift is larger than that of steam reforming, and it is regarded as equilibrium. Thus, the molar stoichiometric ratio between methane and steam (i.e. steam carbon ratio, S/C) in the above reactions is 2.0, but S/C is operated from 3 to 4 in order to avoid deposition of carbon by Boudouard's reaction which is expressed as follows.

$$2CO \Leftrightarrow CO_2 + C \qquad \text{(Boudouard's)}$$

The upper limit of methane conversion into hydrogen-rich gas is determined thermodynamically, which is influenced by physical conditions: temperature, total pressure and S/C. Fig.2 shows the relation between equilibrium conversion and S/C or temperature. In the CRGT system, reaction pressure P in the reformer is higher than that in the combustor. In the MGT system, it ranges between 0.3 and 0.5MPa; $P=0.4$MPa in Fig.2. As shown in Fig.2, equilibrium conversion becomes higher when temperature or S/C becomes higher, and equilibrium conversion is over 35% under the conditions of T > 500°C and

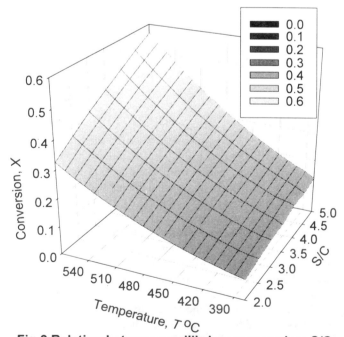

Fig.2 Relation between equilibrium conversion, S/C and temperature

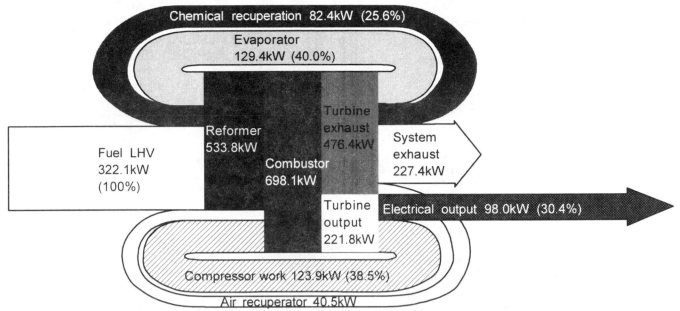

Fig.3 Heat balance for chemically recuperated micro gas turbine system
*(   ) is expressed as percentage of natural gas LHV

S/C > 4.0. Therefore, it is considered that temperature and S/C in the reformer should be high to enlarge heat recovery. However, characteristics of the gas turbine and the evaporator restrict the amount of steam, and S/C has an upper limit. In the gas turbine, surge limit of the compressor determines the maximum flow rate including steam through the turbine. In the evaporator, since pinch point temperature difference becomes small, upper limit of steam generation is determined. Hence, the maximum S/C is approximately 7. Besides, temperature in the reformer is influenced by the configuration of the reformer because the reaction occurs in the reformer exchanging heat with turbine exhaust, and design of the reformer is very important.

Fig.1 shows the conditions of the principal stream of a chemically recuperated MGT based on commercial 75kW MGT system using natural gas as fuel. Fig.3 shows a schematic illustration of heat balance in the system shown in Fig.1. Fig.3 also shows each energy percentage of natural gas LHV (Lower Heating Value). In the system shown in Fig.1, in order to enlarge heat recovery by increase of conversion, steam is generated as much as possible and S/C=6.3. In the reformer, heat duty is estimated considering reactions as equilibrium, and the reformer uses exhaust heat from 656°C to 561°C. Mole fraction of hydrogen in the reformer output is 24%, and the conversion reaches 51%. The heat recovery at the reformer is 82.4kW, which recuperates 25.6% of fuel LHV. The air recuperator is placed after reformer and it heats compressed air until 520°C using exhaust heat from 561°C to 274°C, which recuperates 40.5kW. The evaporator uses bottom of exhaust heat from 274°C to 121°C. According to this system analysis, the output and efficiency were expected to be improved to 98kW and 30.4%, respectively, compared with 75kW and 28% for the original MGT.

Another advantage of the CRGT system is reduction of NOx emission. NOx is created in combustion, and it is roughly classified as prompt NOx and thermal NOx according to reaction mechanisms. In the CRGT system, addition of steam lowers combustion temperature and decreases thermal NOx. Fig.4 shows the effect of S/C on adiabatic combustion temperature. In the figure, horizontal axis expresses equivalence ratio (air fuel ratio / stoichiometric air fuel ratio). In evaluation of adiabatic combustion temperature, including results of system analysis, combustion air temperature was applied as the outlet temperature of the air recuperator. The fuel temperature was 535°C and outlet temperature of the fuel compressor in the CRGT system and the original one, respectively. It is clearly shown in Fig.4 that the CRGT system lowers adiabatic

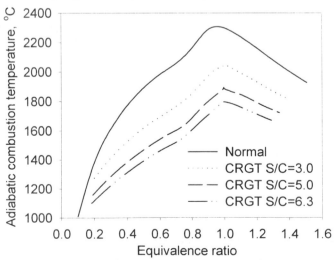

Fig.4 Effect of S/C on adiabatic combustion temperature

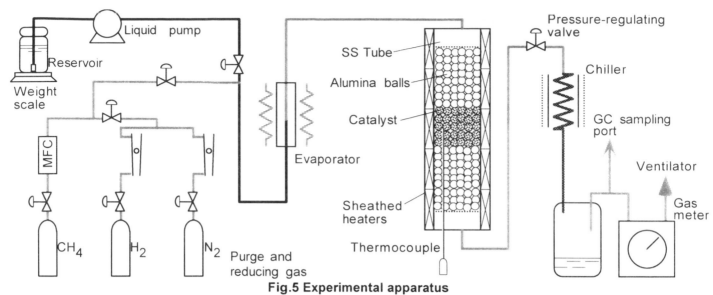

Fig.5 Experimental apparatus

combustion temperature, and the lowering becomes more marked with larger S/C.

APPROACH TO DEVELOPMENT

Catalyst Examination

In the first step of development of a chemical recuperation system, selection of catalysts for steam reforming is necessary and their performances should be examined. For the catalyst, following specifications are required: difficulty of carbon deposit or easy desorption by steaming, high activity at low temperature below 500°C, chemical and mechanical robustness for DSS (Daily Start and Stop), long life, no necessity of reduction treatment, ability of preservation purging with steam or air at cool down, large thermal conductivity and small pressure drop. We sought to identify commercially available catalysts whose basic performances largely satisfy the above specifications.

The experimental apparatus is shown in Fig.5. The material of steam reforming was methane controlled by mass flow controller (MFC) and steam generated by vaporization of demineralized water sent by constant flow pump, and they were mixed in the required S/C.

The mixture gas was heated and adjusted to the required temperature and passed to the packed bed of catalyst with a small volume kept at the required pressure by the regulating valve. Components of output gas were analyzed by TCD-type gas chromatograph (GC) and flow rate was measured by gas meter.

Fig.6 shows an example of results with some kinds of catalysts, which are expressed as relation between reaction rate and temperature. Ruthenium-based catalyst shows higher activity at low temperature than does nickel-based catalyst. It was confirmed by an additional experiment that Ru-based catalyst satisfied almost all of the above required specifications. These basic characteristic data were arranged in reaction rate

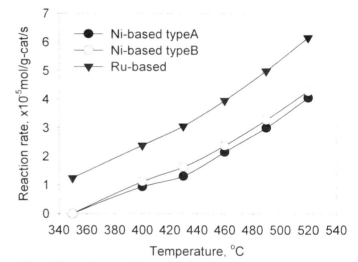

Fig.6 Experimental results for catalyst performance

formula with Arrhenius's law to design and analyze the reformer.

In order to examine the adequacy of reaction rate formula, a reforming test was conducted, which used a tubular reformer shown in Fig.5 with 300mm-long packed bed of the Ni-based type A catalyst. The tube wall of the reformer was kept at constant temperature. Fig.7 shows the temperature profile in the packed bed and composition of outlet gas. Total pressure P is 1.0MPa, inlet gas temperature T_{in} is 330°C, T_{u} is 400°C and mass flux G of material is 0.67kg/m²s. In this figure, an analytical result is also shown, which includes the reaction rate formula in the simulation code mentioned below. Both temperature profile and gas composition are estimated within an error of 2% for experimental data. According to this result, it is confirmed that the reaction rate formula is applicable to design and analysis of the reformer.

Design of Reformer

382

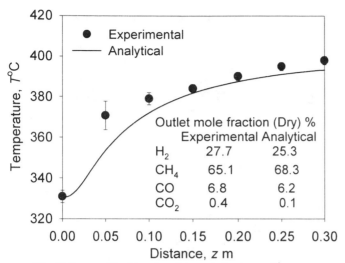

Fig.7 A result of test using a tubular reformer
(P=1.0MPa, T_{in}=330°C, T_w=400°C, S/C=4.5,
G=0.67kg/m^{2}s)

Outlet mole fraction (Dry) %

	Experimental	Analytical
H_2	27.7	25.3
CH_4	65.1	68.3
CO	6.8	6.2
CO_2	0.4	0.1

For the CRGT system shown in Fig.1, a shell-and-tube type reformer advantageous in cost was designed. Fig.8 shows an illustration of the reformer. The reformer has two bundles of staggered tube banks, each bank consisting of 48 tubes whose inner diameter is 22mm and length is 430mm. The mixture of fuel gas and steam is sent into the inlet manifold and distributed into the 48 tubes of the first bundle. A catalyst is packed in each tube, and the mixture gas takes a U-turn into the second bundle at the lower manifold, is converted into hydrogen-rich gas and reaches the outlet manifold. The first and second bundles are placed so that the mixture gas may flow counter from downstream of turbine exhaust to upstream. Turbine exhaust flows through the bundle and exchanges heat with internal catalyst or hydrogen-rich gas through the tube walls. In order to design the reformer shown in Fig.8, the following configuration parameters are considered: pitch of tube bank, diameter and

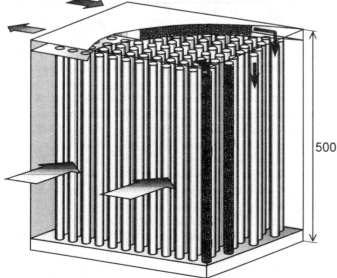

500

Fig.8 An illustration of the reformer for 75kW MGT

length of tube, number of tubes in a bundle and number of bundles. In order to determine these parameters, influence on pressure drop and heat transfer inside and outside of the tube should be surveyed, which is similar to the case of designing heat exchangers. For instance, since tightening pitch of tube bank makes pressure drop of turbine exhaust increase, backpressure of turbine increases and generation of electricity decreases. However, simultaneously, since it makes velocity of turbine exhaust though the tube bank increase, heat transfer coefficient of external tube wall and heat recovery enlarge. Thus, configuration parameters affect pressure drops and heat transfers positively or negatively, and they affect system efficiency, too. The reformer shown in Fig.8 is one of the optimized solutions, which balances pressure drops and heat transfers so as to maximize system efficiency.

Numerical Analysis

In the side view of the reformer shown in Fig.8, 8 tubes are visible. In order to investigate the state inside and outside these 8 tubes, a numerical analysis was conducted. The analysis solved continuous, momentum (modified Darcian), energy, each chemical component (forced convection and diffusion) and gas state (ideal) equations simultaneously and 2-dimensionally (Nakagaki, 1999). The reaction rate formula included the results shown in Fig.4, temperature drop of duct gas was also solved to estimate heat flux through the tube wall, and all equations were made to converge until the approaching temperature of the duct gas was equal to the required value. In the energy equation, Zukauskas's formula was used as heat transfer coefficient of forced convection in the tube banks, and Yagi-Kunii's fomula was used as heat transfer in packed bed. The analyzed area was surrounded by a given inlet a boundary, symmetric boundary along the center axis of tube, a free outlet boundary and a known boundary of velocity and heat flux at internal tube wall. An example of analysis is presented in Fig.9. The figure shows contour at intervals of 10°C in 8 tubes under the conditions of T_g=655°C, P=0.4MPa and S/C=6.3, which are required for the system shown in Fig.1. The mixture gas of methane and steam flows from the right side of the figure to the left in a U-shape, and the duct gas flows inversely from left to right. Temperature profile of duct gas is also shown in the figure. In the first bundle, intervals of contour are narrow and an ascent of the temperature is large. In the second bundle, contour is vertical and radial gradient of temperature is larger than in the first bundle. It is considered that since reaction rate of reforming gas at nearly 500°C increases exponentially, temperature gradient enlarges to increase heat flux. Besides, pressure drop of this tube bank is under 1kPa and it is equal to or less than pressure drop of a conventional air recuperator, but a further decrease is expected to be achieved.

CONCLUSION

The results of this study indicate the practicability of the CRGT system. CRGT systems can be attached to existing systems and are applicable to small and medium-size GTs

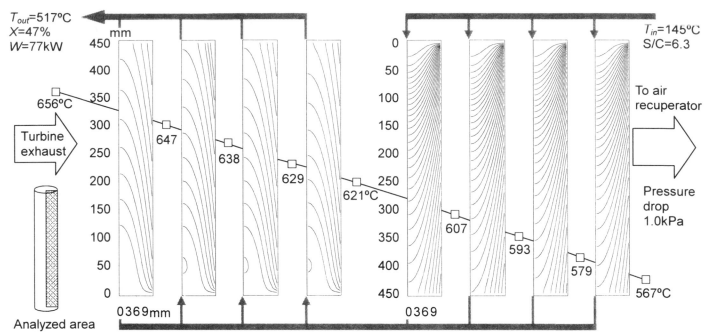

Fig.9 Analytical result of temperature distribution in reformer (Contour at intervals of 10°C)

ranging from micro GT to MW class GT. In the next step of this work, we intend to perform a combination test using an actual GT engine.

NOMENCLATURE

G	Mass flux, kg/m²s
P	Total pressure, Pa
T	Temperature, °C
W	Heat recovery, W
X	Methane conversion
z	Distance from inlet of packed bed, mm

Subscript

C	Catalyst
D	Duct
g	Turbine exhaust
in	Inlet
out	Outlet
w	wall

REFERENCES

Carapellucci, R., Risalvato, V., Bruno C., and Cau, G., 1996, "Performance and Emissions of CRGT Power Generation Systems with Reformed Methanol", *Proc., IECEC 96*, Vol.2, pp. 707-12

Carapellucci, R., Risalvato, and Cau, G., 1994, "Characteristics of the Heat Recovery Methanol Reforming in CRGT Power generation Systems", *Proc., ASME COGEN-TURBO*, IGTI-Vol.9, pp. 257-64

Kuroda, K., Imai, T., and Yanagi, M., 1989, "Development of Methanol Reformer", *Mitsubishi Jukou Gihou*, Vol. 26, No. 5, pp. 408-12

Nagaya, K., Oshiro, H., Ito, H., and Yoshino, N., 1993, "Demonstration Test of Reformed Methanol Gas-Turbine Power Generation", *Hitachi Zousen Gihou*, Vol. 54, No. 3, pp. 180-4

Xu, J. and Froment, G.F., 1989, "Methane Steam Reforming, Methanation and Water-Gas Shift: I. Intrinsic Kinetics", *AIChE. J.*, Vol. 35, No. 1, pp.88-96

Hyman, M.H., 1968, "Simulate Methane Reformer Reactions", *Hydrocarbon Processing*, Vol. 47, No. 7, pp.131-7

Murray, A.P., and Snyder, T.S., 1985, "Steam-Methane Reformer Kinetic Computer Model with Heat Transfer and Geometry Options", *Ind. Eng. Chem. Process Des. Dev.*, Vol. 24, pp. 286-94

Alatiqui, I.M., Meziou, A.M., and Gasmelseed, G.A., 1989, "Modeling, Simulation and Sensitivity Analysis of Steam-Methane Reformers", *Int. J. Hydrogen Energy*, pp. 241-56

Fukusako, S., Yamada, M. and Usami, Y., 1997, "Heat and Mass Transfer Characteristics in a Reforming Catalyst Bed", *J. HTSJ*, Vol. 36, No. 140, pp.18-26

Altman, D. and Wise, H., 1956, "Effect of Chemical Reactions in the Boundary Layer on Convective Heat Transfer", *Jet Propulsion*, No. 4, pp.256-69

Wijngaarden, R.J. and Westerterp, K.R., 1989, "Do the Effective Heat Conductivity and the Heat Transfer Coefficient at the Wall Inside a Packed Bed Depend on a Chemical Reaction? Weakness and Applicability of Current Models", *Chem. Eng. Sci.*, Vol. 44, No. 8, pp. 1653-63

Nakagaki, T., Ogawa, T., Murata, K. and Nakata, Y., 1999, "Development of Methanol Steam Reformer for Chemical Recuperation", *Proc., IJPGC'99*, Vol. 1, pp. 569-76

2001-CT-11

PREDICTION OF PERFORMANCE OF SIMPLE COMBINED GAS/STEAM CYCLE AND CO-GENERATION PLANTS WITH DIFFERENT MEANS OF COOLING

Sanjay Yadav
Assistant Professor
Department of Mechanical Engineering
R.I.T. Jamshedpur
Pin-831014 (India)
sanjay_y@altavista.com

Onkar Singh,
Reader , Department of Mechanical Engineering,
H.B.T.I , Kanpur, (INDIA)

B.N. Prasad ,
Professor & Head, Department of Mechanical
Engineering, R.I.T, Jamshedpur, (INDIA)

ABSTRACT

This paper deals with comparative study of the influence of different means of cooling on the performance of simple combined gas/steam cycle and co-generation plants. All possible open and closed loop cooling with air, water and steam as the cooling medium have been considered. The prediction is based on the modeling of various elements of simple combined gas/steam cycle considering the real situation. The study shows that open loop water-cooling and closed loop steam-cooling are superior as compared to air-cooling. Closed loop steam cooling is the promising option as its technology is now available.

INTRODUCTION

Combined cycle and co-generation power plants are gaining increasing acceptance as alternatives to conventional or nuclear steam systems. This leads to the development of gas turbines dedicated to combined cycle and co-generation applications such as the AGT J-100 "Moonlight" project in Japan and ATS program in the US. The increase in the electrical efficiency of combined cycle power plant in the last few years has mainly been caused by gas turbine design improvements and to some extent due to better configuration of steam cycles. The reduction of fuel consumption in the case of gas turbine can be achieved not only by improvement in component efficiencies but most importantly by increasing both turbine inlet temperature and compressor pressure ratio. These improvements require active cooling of hot turbine components in order to avoid a reduction of operating life due to an unfavorable combination of oxidation, creep, thermal stresses and by adopting advanced aerodynamic compressor design. However, the coolant mass flow rate can become so large that excess compressor power for cooling can offset the power gains

associated with incremental increase in turbine inlet condition. So better cooling means and methods are paramount factors in turbine cooling technologies.

Pioneering work in this field have been done by Louis et al [1983], El.Masri M.A [1985,86,88], Bolland [1991], Bolland and Stadass [1995], etc.

In this work, following cooling means have been attempted to have a comparative study of the influence of different means of cooling on the performance of simple combined cycle and co-generation plants.

1. Open loop cooling
   a. Internal convection and impingement air cooling
   b. Internal and film air-cooling.
   c. Transpiration air-cooling.
   d. Water cooling
   e. Internal convection steam cooling.
   f. Internal convection film steam cooling.
2. Closed loop cooling.
   a. Steam cooling.   b.   Water-cooling.

NOMENCLATURE

c_p = specific heat at constant pressure
F= factor
h = specific stagnation enthalpy
LCV = lower calorific value
m = mass flow rate
N= number of stages ,
 p= pressure ,
 q=heat supplied ,
rp= pressure ratio,
S= blade perimeter ,
s= entropy

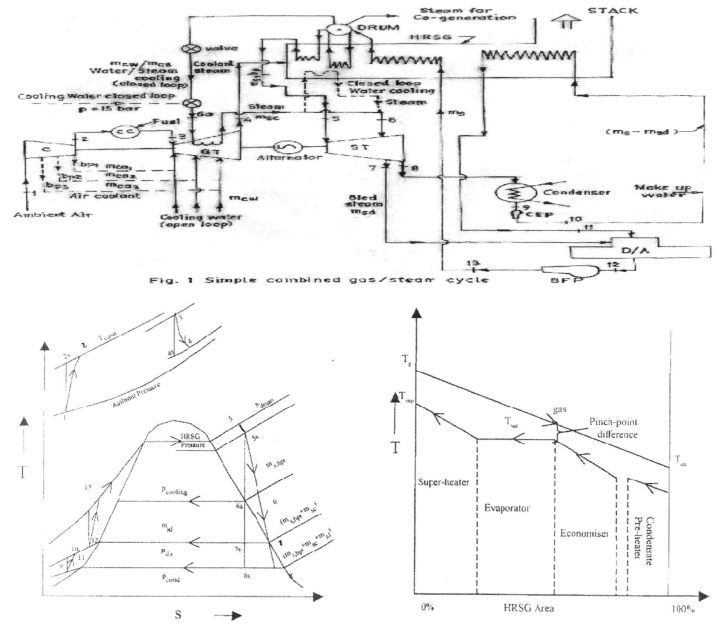

Fig. 1 Simple combined gas/steam cycle

Figure 2. T-s representation of cycle and T-HRSG area for HRSG.

St= average Stanton number
T= absolute temperature
t= blade circumferential spacing at pitch line
TIT= gas turbine inlet temperature
W= specific work
η = Efficiency
η_{iso} =average isothermal (film / transpiration) effectiveness
α = inlet gas flow angle
ε = effectiveness
Ω = exergy
ω = specific exergy

Subscripts

a= air , mass flow rate ratio
b=blade
c=compressor , coolant
cc = combined cycle , combustion chamber
f = fuel
gt = gas turbine
g = gas
HRSG = Heat recovery steam generator
in = inlet
o= total, stagnation values

386

out = outlet
p = pump
s = steam
sa = surface area
t = turbine , transpiration

Modeling and Governing Equations.

A simple combined gas /steam cycle is shown in figure 1. and the corresponding T-s representation is shown in figure 2. The modeling of various elements of combined cycle is given below:

Gas Model:

For accounting the realistic situation in gas model the specific heats of air and combustion products at constant pressure have been chosen as a function of temperature in the form of polynomials.

From the specific heat polynomials, entropy, enthalpy and exergy of air and gases will be calculated by

$$s = \phi - R \ln (p/p_o) \qquad ----- (1)$$
$$\text{Where } \phi = \int c_p (dT/T)$$
$$h = \int c_p(T) dT \qquad ------ (2)$$
$$\omega = h - T_o \cdot s = h - T_o \phi + T_o \cdot R \cdot \ln \cdot (p/p_o) \qquad ------ (3)$$

The atmospheric air is assumed to have zero enthalpy , entropy and exergy.

Compressor-:

The compression in axial flow compressor has been assumed polytropic . Equal pressure ratio in each stage has been assumed and is calculated as
$$(rp)_{stage} = (rp_c)^{1/N}$$
where N = number of stages
For a compressor with a specified inlet state and polytropic efficiency, the temperature is expressed in terms of local pressure by integrating
$$dT/T = [R /(\eta_{pc} \cdot c_{pa}] \cdot dp/p \qquad ------ (4)$$
The mass , enthalpy balance and exergy loss in compressor for open loop air cooling are
$$m_{in} = m_{out} + \Sigma m_c \qquad ------- (5)$$
$$W_c = [m_{out} \cdot h_{out} + \Sigma m_c \cdot h_c - m_{in} \cdot h_{in}] \qquad ------- (6)$$
For open loop cooling by steam and closed loop cooling by water and steam no air is bled from compressor, so
$$W_{c,in} = W_{c,out} \qquad ------- (7)$$
$$W_c = [m_{in} \cdot (h_{out} - h_{in})] \qquad ------- (8)$$
$$(\Omega_c)_L = W_c + m_{in} \cdot (\omega_{in} - \omega_{out}) \qquad ------- (9)$$

Combustor

The losses which occur inside the combustion chamber due to incomplete combustion and frictional losses are taken care by introducing the concept of combustion efficiency and percentage pressure drop respectively. For a specified combustion efficiency, fuel type, pressure drop, the state of inlet air and fuel , the discharge pressure and temperature, the fuel

flow rate is calculated from the traditional mass and energy balance.

$$m_{out} = m_{in} + m_f \qquad ------ (10)$$

$$\eta_{cc} \cdot m_f (LCV)_f = m_{out} \cdot h_{out} - m_{in} \cdot h_{in} \qquad ------ (11)$$

Losses of availability in combustion chamber can be represented in terms of constituents availability loss, which are of the four types :- [El-Masri, 1988] namely incomplete combustion loss (Ω_{cc})$_{L,IC}$, pressure drop loss (Ω_{cc})$_{L, po}$, throttling loss (Ω_{cc})$_{L,FT}$, thermal degradation loss $(\Omega_{cc})_{L, TD}$ as given below

$$(\Omega_{cc})_{L,IC} = m_f \cdot (1 - \eta_{cc}) \cdot (LCV)_f \qquad ------- (12)$$
$$(\Omega_{cc})_{L, PD} = m_{out} \cdot R_g \cdot T_o \cdot \ln[p_{in} / p_{out}] \qquad ------- (13)$$
$$(\Omega_{cc})_{L,FT} = m_f \cdot R_f \cdot T_o \cdot \ln (p_f / p_{cc}) \qquad ------ (14)$$
$$(\Omega_{cc})_{L, TD} = m_f \cdot \eta_{cc} \cdot (G_r - LCV)_f + T_o[m_{out} \cdot \phi_{out} + m_{in} \cdot \phi_{in}] \qquad -- (15)$$

The total availability loss in combustion chamber is the sum of the above.
$$(\Omega_{cc})_L = (\Omega_{cc})_{L,IC} + (\Omega_{cc})_{L, PD} + (\Omega_{cc})_{L,FT} + (\Omega_{cc})_{L, TD} \qquad ----(16)$$

Cooled Gas Turbine

Following assumptions are made for cooled turbine model which is a refined version of Louis and Chuan [1984].

1. Blades internally cooled by convection either in open or closed loop are treated as heat exchanger operating at constant temperature and the coolant exit temperature is simply expressed as a function of heat exchanger effectivness, ε.

2. In the case of film or transpiration cooling , a concept of isothermal effectiveness (η_{iso}) is introduced.

3. The blade surface area consideration is based on the pitch line condition . In order to account for the actual blade surface area , a factor , F_{sa} is introduced whose value may be taken as more than unity , around 1.05 .

For open or closed loop internal cooling by air or steam , the mass flow rate of coolant to main gases is expressed as

$$a_c = \frac{m_c}{m_g} = \left(\frac{St_{in} \cdot c_{pg}}{\varepsilon \cdot c_{pc}} \right) \cdot \left(\frac{S_g}{t \cdot \cos \alpha} \right) \left[\frac{T_{og,in} - T_{ob}}{T_{ob} - T_{oc,in}} \right] \cdot F_{sa} \qquad -- (17)$$

For internal and film cooling or transpiration cooling by air or steam the mass flow rate of coolant to main gases is given by

$$a_c = \frac{m_c}{m_g} = [1 - (\eta_{iso})_{f/t}] \cdot \left(\frac{St_{in} \cdot c_{pg}}{\varepsilon \cdot c_{pc}} \right) \cdot \left(\frac{S_g}{t \cdot \cos\alpha} \right) \left[\frac{T_{ogin} - T_{ob}}{T_{ob} - T_{ocin}} \right] \cdot F_{sa} \qquad --- (18)$$

For open and closed loop water cooling , the mass of coolant to mass of gas gas is given by

$$a_c = m_c / m_g = R_c \cdot St_{in} \cdot (S_g / t \cdot \cos \alpha) \cdot F_{sa} \qquad --------(19)$$

where R_c = cooling factor = $\left[\left(\frac{h_{ogin} - h_{ob}}{(c_{pw} \cdot (T_{sat} - T_w)) + x \cdot h_{fg}} \right) \right]$

For calculating the turbine work following assumptions are made:

1. The expansion work is polytropic and it is taken care by suitable polytropic efficiency, η_{pt}.

2. The irreversible loss of stagnation pressure caused by mixing of coolant air / steam / water has been assumed continuous throughout the turbine surface and the loss of work done is taken care by a factor , F_m which is less than unity (0.97)

3. There is an equal pressure ratio in each stage and the number of turbine stages will be determined by

$$N_t = (\ln(p_z/_{pz+1})) / \ln(rp_{stage})_{max} \quad \text{------ (20)}$$

$$\text{where } (rp_{stage})_{max} = 2.4$$

N_t is rounded off to the next higher integer number and maximum N_t is set to 5

For open loop cooling, the mass and energy balance are :-

$$(m_{gt})_{ex} = (m_{gt})_{in} + \Sigma m_c \quad \text{------ (21)}$$

$$W_{gt} = [(m_{gt})_{in}.(h_{gt})_{in} + \Sigma m_{c,in}.h_{c,in} - (m_{gt})_{ex}.(h_{gt})_{ex}]F_m \quad \text{------ (22)}$$

The availability loss for open loop is given by

$$(\Omega_{gt})_L = [(m_{gt})_{in}.(\omega_{gt})_{in} + \Sigma m_c\omega_c + m_{gt}(\omega_{gt})_{out}] - W_{gt} \quad \text{------ (23)}$$

For closed loop cooling , $F_m = 1$ and so

$$W_{gt} = [(m_{gt})_{in}.(h_{gt})_{in} - \Sigma m_c (h_{c,out} - h_{c,in}) + (m_{gt})_{ex}.(h_{gt})_{ex}] \quad \text{------(24)}$$

The loss of availability for closed loop is expressed as

$$(\Omega_{gt})_{lost} = [(m_{gt})_{in}.(\omega_{gt})_{in} - \Sigma m_c\omega_c + (m_{gt}).(h_{gt})_{out}] - W_{gt} \quad \text{----- (25)}$$

Heat Recovery Steam Generator (HRSG)

The mass of steam generated from the exhaust of gas turbine is established from the energy balance inside the HRSG. The loss in HRSG is accounted by effectiveness .

$$\varepsilon_{HRSG} . m_g (h_g - h_{stack}) = m_s.q_{HRSG}$$

The loss of availability in HRSG is expressed as

$$(\Omega_{HRSG})_L = m_g[(\omega_g)_{in} - (\omega_g)_{out}] + [\Sigma(m_w.\omega_w) - \Sigma(m_s.\omega_s)] \quad \text{-----(26)}$$

Steam Turbine

The loss in steam turbine is taken care by the concept of isentropic turbine efficiency . The steam turbine work and availability are expressed by

$$W_{st} = \Sigma [m_s.(h_{st,in} - h_{st,out})] \quad \text{------ (27)}$$

$$(\Omega_{st})_L = \Sigma m_s[(\omega_{st})_{in} - (\omega_{st})_{out}] - W_{st} \quad \text{------ (28)}$$

Condenser

The loss of availability and cooling water requirement are in condenser are given by

$$(\Omega_{cond})_L = m_s.[(\omega_{cond,s})_{in} - (\omega_{cond,s})_{out}] - m_w(\omega_{w,out} - \omega_{w,in}) \quad \text{----(29)}$$

$$\varepsilon_{cond}.m_{cond,s}[(h_{cond,s})_{in} - (h_{cond,w})_{out}] = m_w.c_{pw}(T_{w,out} - T_o) \quad \text{----(30)}$$

Pumps:

The losses in the pump is taken care by the concept of overall efficiency . The pump work and exergy loss are given by

$$W_p = \Sigma [m_w.v_w(p_{out} - p_{in})] \quad \text{-------- (31)}$$

$$(\Omega_p)_L = W_p + m_w[(\omega_p)_{in} - (\omega_p)_{out}] \quad \text{-------- (32)}$$

Overall Performance

The combined cycle plant specific work and combined cycle plant efficiency are given by

$$W_{gt,net} = [(W_{gt}.\eta_m) - (W_c/\eta_m)] \quad \text{------(33)}$$

$$W_{st,net} = [(W_{st}.\eta_m) - (W_p/\eta_p)] \quad \text{------(34)}$$

$$W_{plant} = (W_{gt,net} + W_{s,net})\eta_{alternator} \quad \text{------(35)}$$

$$\eta_{plant} = (W_{plant}) / (m_f.(LCV)_f) \quad \text{---- (36)}$$

Power to heat ratio (PHR) for co-generation plant is expressed as

$$PHR = W_{plant} / q_{HRSG} \quad \text{------- (37)}$$

Fuel Utilization efficiency (η_f) for co-generation plant is given by

$$\eta_f = (W_{plant} + q_{HRSG}) / m_f.(LCV)_f \quad \text{-------- (38)}$$

For the prediction of performance the computer code has been written in C++.

Results and Discussions

Results have been plotted for throttle steam pressure p_s = 50 bar and maximum throttle temperature 650°C. The cooling water and steam temperature are taken as 288K and 488K respectively and at a pressure of 15bar. The condenser pressure is 0.05bar and de-aerator pressure is 3 bar.

Figure 3 shows the effect of cooling means on the mass flow rate of coolant required for various cooling means at different

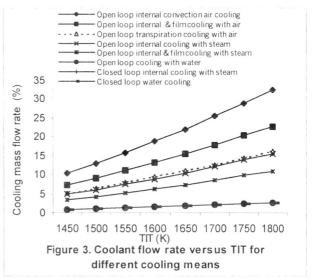

Figure 3. Coolant flow rate versus TIT for different cooling means

TIT for rp_c=14. It is observed from the results that for all means of cooling , the cooling requirement increases with TIT as expected. The results show that the minimum cooling flow rate is required in the case of open and closed loop water cooling (curves overlapped on each other) followed by open loop film steam cooling , open & closed loop internal steam cooling and transpiration air cooling. The maximum coolant is required by open loop internal air cooling.

The effect of cooling means on plant efficiency for various TIT is depicted in figure 4. It is obvious from the results that the plant efficiency achieved the maximum value in the case of closed loop water cooling at any TIT followed by closed loop steam cooling. The open loop film steam cooling exhibits plant efficiency lower than closed loop steam cooling. This is because of the fact that in water cooling the vaporization of water takes partly or fully in the cooling passage and the specific heat of water/steam is very high which results in high cooling capacity . But still the open loop or closed loop water cooling is not commercialized due to corrosion of blade channels and other engineering problems. The next option is the closed loop steam cooling which has now been commercialized but the purity of cooling steam has to be maintained in both closed loop steam cooling and open loop steam cooling. There is an optimum TIT around 1600K for open loop air-cooling.

Figure. 5 shows the effect of cooling means on plant specific work for various TIT. Except for two cases – open loop

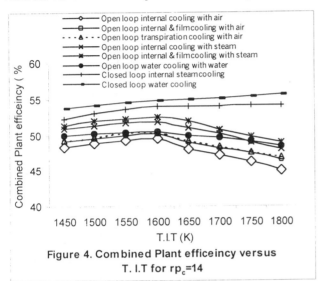

Figure 4. Combined Plant efficiency versus T.I.T for rp_c=14

internal air , impingement cooling and open loop internal air , film cooling the specific work increases with TIT in all other means of cooling.

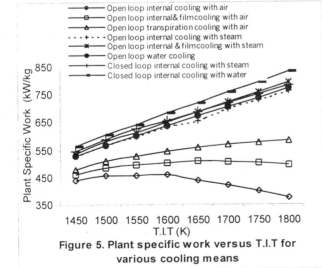

Figure 5. Plant specific work versus T.I.T for various cooling means

Figure 6 shows the variation of plant efficiency with compressor pressure ratio for various TIT for closed loop internal steam cooling. The plant efficiency in general decreases

with compressor pressure ratio , however there is an optimum value around rp_c=14.

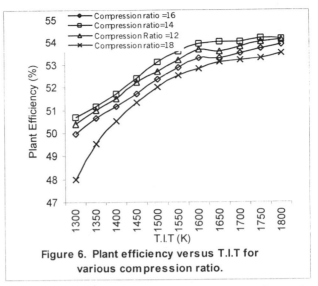

Figure 6. Plant efficiency versus T.I.T for various compression ratio.

The work and exergy loss for the case of transpiration air-cooling is shown in figure 7. The maximum exergy loss is found in combustion chamber due to high temperature followed by steam cycle. This graph quantifies the losses in every sector (component) and being a designer one has to reduce the losses in the high profile sector.

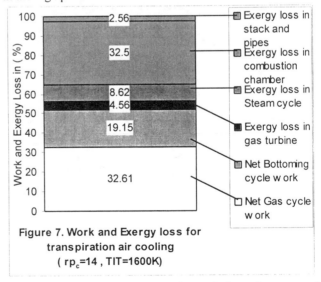

Figure 7. Work and Exergy loss for transpiration air cooling (rp_c=14 , TIT=1600K)

Figure 8 and 9 show the variation of power to heat ratio (PHR) and fuel utilization efficiency (η_f) as a function of compressor pressure ratio. PHR increases slowly with compressor pressure ratio while fuel utilization efficiency decreases . This suggests that the selection of the plant for co-generation work has to be carefully dealt so far as the selection of compressor pressure ratio and TIT are concerned.

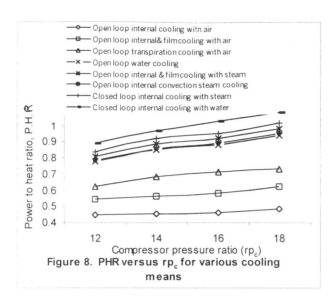

Figure 8. PHR versus rp_c for various cooling means

Conclusion

Following conclusions have been drawn from this study:

1. A systematic methodology has been developed by using modeling, for visualizing the effect of cooling means on performance and to select the better cooling means for future development.

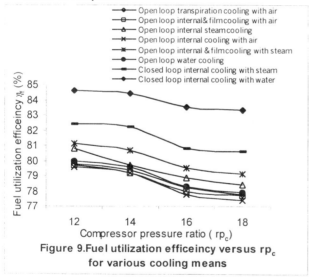

Figure 9. Fuel utilization efficeincy versus rp_c for various cooling means

2. The closed loop water cooling yields highest plant efficiency followed by closed loop steam cooling. There exists an optimum turbine inlet temperature for all open loop air cooling means which is around 1600 K . However in the case of closed loop water and steam cooling , higher TIT gives better performance but in the closed loop steam cooling the increase in plant efficiency beyond 1600K is slow. The optimum compressor pressure ratio is around 14.

3. The performance of open loop film steam cooling is slightly superior to open loop internal steam cooling.

4. The power heat ratio and fuel utilization efficiency increases and decreases respectively with compressor pressure ratio and for selecting the co-generation plant, the careful selection of compressor pressure ratio and TIT is a must.

REFERENCES

1. Wu.Chuan Shao and Louis J.F, 1984, "A Comparative Study of the influence of Different Means Cooling on the Performance of Combined Cycle", Trans. Of ASME Journal of Engg. for Gas Turbines and Power, **106**, p 750.

2. Huang F.F and Wang Ling , 1987, "Thermodynamic Study of An Indirect Fired Air Turbine Co-generation System with Reheat", Trans. of ASME, Journal of Engg. for of Gas Turbines and Power", **109**, pg. 16.

3. Bolland .O , 1991, " A Comparative Evaluation of Advanced Combined Cycle Alternatives ", Trans. Engineering for of ASME , Journal for Gas Turbines and Power ,**113**, pg. 190.

4. Bolland.O and Stadaas. J.F. 1995, " Comparative Evalution of Combined Cycle and Gas Turbine Systems with Injection , Steam Injected and Recuperation", Trans. Of ASME , Journal of Engineering for Gas Turbines and Power ,**107**, pg. 138.

5. El-Masri. M. A, 1988, " GASCAN- An Interactive code for Thermal Analysis of Gas Turbine Systems, Trans. Of ASME , Journal of Engineering for Gas Turbines and Power , **110**,pg. 20.

6. El-Masri. M.A, !986, " A modified High efficiency, recuperated Gas Turbine Cycle", Trans. Of ASME , Journal of Engineering for Gas Turbines and Power ,**110**,pg. 223.

7. El-Masri.M.A, !986, " On Thermodynamic of Gas Turbines Cycle – Part-2 – A Model for Expansion in Cooled Turbines", Trans. Of ASME , Journal of Engineering for Gas Turbines and Power ,**108**,pg. 15.

8. El-Masri. M .A, !985, " On Thermodynamics of Gas Turbines Cycles , Part-1 Second Law Analysis of Combined Cycles", Trans. Of ASME , Journal of Engineering for Gas Turbines and Power ,**107**,pg. 880.

Proceedings of IECEC'01
36th Intersociety Energy Conversion Engineering Conference
July 29–August 2, 2001, Savannah, Georgia

IECEC2001-CT-13

Test and Demonstration Results of a New Magnetically Coupled Adjustable Speed Drive

Kenneth J. Anderson, PE, CEM, ASME
Northwest Energy Efficiency Alliance
522 SW Fifth Ave.
Portland, Oregon 97214
USA
(503) 827-8416 Ext. 249

Alan Wallace, Fellow IEEE
Oregon State University
Electrical and Computer Engineering Dept.
Corvallis, Oregon 97331-3211
USA
(541) 737-2995

Ron Woodard, President
MagnaDrive Corp.
1177 Fairview Avenue North
Seattle, Washington 98109
USA
(206) 694-4711

ABSTRACT

Motor energy savings are demonstrated in laboratory tests and field demonstration of a new magnetically coupled variable speed drive called-- MagnaDrive. Units rated at 50, 100 and 200 HP (37.3, 74.5 and 149.2 kW) reduced fan energy by 30% compared to a baffle and pump energy by 44% compared to a throttling valve. This is an average of 65% of the fan savings and 62% of pump savings of a variable frequency drive (VFD).

Industrial process and building management engineers most often cite non-energy benefits such as reduced motor vibration leading to lower maintenance costs, less noise, and shorter duration power sags during motor start-up eliminating problems with electronic devices. In a wastewater treatment facility, speed control of the load allowed downsizing of the pump motor. In an area where a VFD is unsuitable due to harsh environmental factors (poor power quality, corrosive or dusty ambient air, etc.) the magnetic coupling is an excellent replacement technology.

INTRODUCTION

In 1999 the Northwest Energy Efficiency Alliance (Alliance) funded the Magna-Drive Corporation to develop, demonstrate and market its magnetically coupled adjustable speed motor coupling in the Pacific Northwest (PNW) (Oregon, Washington, Idaho and Montana). To demonstrate energy savings in a controlled laboratory environment and to improve the design, the Motor Systems Resource Facility (MSRF) at Oregon State University located in Corvallis, Oregon was added to the team. This paper summarizes the test results of the MSRF at Oregon State University [1] but it focuses on the results of four specific field applications— a cement mining operation [2], a 55-story office tower [3], a wastewater treatment plant [4] and a pulp and paper plant [5].

NOMENCLATURE

Cm/s, Centimeters per second
GJ, Gigajoules
HP, Horsepower
kW, kiloWatt
kWh, kiloWatt-hour
MJ, Megajoules
MN/m2, MegaNewtons per square meter
pf, Power Factor
psi, pounds per square inch
RPM. Revolutions per Minute
RPS, Revolutions per Second
TJ, Terajoules

Product Description -- Installed primarily on pump and fan loads, the MagnaDrive coupling is a mechanical alternative to a VFD for controlling speed.

Figure 1. MagnaDrive Adjustable Speed Coupling

Fig. 1 shows the assembled MagnaDrive coupling which consists of a series of discs with sets of alternatively polarized, neodymium/iron/boron (NeFeB) permanent magnets on the disc attached to the load.

Figure 2. MagnaDrive Conductor and Rotor

In Fig. 2 on the left side, the conductor assembly with steel-backed copper rings forms a rotating cage that is connected to the motor shaft. On the right side of Fig. 2 is the rotor with high-energy magnets mounted to steel plates. The rotor assembly is connected to the load shaft. When the motor rotates the conductor assembly, its slip relative to the magnets induces currents in the conducting discs and enables torque transmission without physical contact. Slip and torque are controlled by mechanically moving the magnetic rotor along the axis of rotation to open or close the gap between the magnets and the copper discs. Units up to 250 HP (186.5 kW) have been tested and installed. Units over 550 HP (410.3 kW) are under development. The average installed unit cost is just under $100 per HP ($74.6/kW). However, energy savings are just over 60%

of the electrical energy saved by an equivalent VFD. For a wide range of variable load applications in the Pacific Northwest, the composite average energy savings are estimated at 854 kWh per horsepower per year.

MagnaDrive unit have numerous non-energy benefits—soft loading and unloading, isolating load vibration from the motor, eliminating complex shaft alignment, protecting the motor from load jams, isolating bearing currents, reducing the duration of high motor start-up current, and tolerating poor power quality and harsh environments.

Alliance Project – The total Alliance investment was $2.3 million and about half was used to conduct laboratory tests to prove savings and to complete field tests to demonstrate the technology to prospective consumers. After the first two years of the three-year project, thirty (30) field installations have been completed in the PNW on motors ranging from 20 to 250 HP (14.9 to 186.5 kW) for a total installed capacity of about 3,000 HP (2,238 kW). The Alliance is trying to get the MagnaDrive coupling and VFDs to replace throttling valves and dampers in a majority of the variable load applications in the PNW by 2010. In 1999 the variable load market size was estimated at 0.67 million horsepower, but it is estimated to grow to 1.14 million horsepower by 2010. MagnaDrive is assumed to capture 30% of that market by 2010 for a total savings of over 290 GWh (1,044 TJ) of electricity per year with a benefit-to-cost ratio of 1.9, i.e. the region's consumers will gain $1.90 back for every dollar invested in a MagnaDrive application.

Laboratory Test Results – In early 2000, a team at the Motor Systems Resource Facility (MSRF) at Oregon State University located in Corvallis, Oregon conducted a series of laboratory tests on MagnaDrive couplings rated at 50, 100, and 200 HP (37.3, 74.5 and 149.2 kW) using fan and pump variable load profiles [1]. The magnetic coupling reduced fan energy by about 30% compared to a baffle, and reduced pump energy by about 44% compared to a throttling valve. The MagnaDrive captured on average about 65% of the savings that a VFD would obtain for a variable fan load. For a pump load the magnetic coupling reduced energy use by an average of 62% when compared with a VFD. The 200 HP (149.2 kW) magnetic coupling operating a fan load experienced a temperature rise of up to 260 degrees C (500 degrees F) in the copper disc due to heat dissipation. The basic design was changed to add finned heat sinks to the exterior surface of the disc. Under the same load, the new disc experienced a temperature rise of approximately 100 degrees C (212 degrees F). Cooling fins have now been integrated into the standard design of the MagnaDrive coupling. In addition to saving energy compared to throttling systems, the magnetic coupling reduced the need for precise alignment of the load and motor shafts and it isolated load vibrations from the motor. For the 200 HP (149.2 kW) coupling on a fan load the tests showed a

50% reduction in the duration of the motor inrush current from 20 milliseconds to 10 milliseconds.

One potentially negative impact was the MagnaDrive's rapidly decreasing power factor (pf) as the motor is off-loaded. The VFD tested at 0.7 pf at full load dropping to 0.4 pf at 20% load. The MagnaDrive full load power factor (pf) was higher, 0.9 pf but dropped rapidly to 0.25 pf at 20% load.

Demonstration Site Results – The following discussion addresses results from four specific demonstration sites— a cement mining operation [2], a 55-story office tower [3], a wastewater treatment plant [4] and a pulp and paper plant [5].

In late 1999 the **Ash Grove Cement Company** [2] plant in Durkee, Oregon was selected as the first field test for the MagnaDrive coupling. It was installed between a 125 HP (93.3 kW), 1800 RPM (30 RPS) motor and a centrifugal fan moving air across a bag house to remove airborne dust.

Figure 3. Horizontal Axis MagnaDrive

The coupling replaced the existing belts and pulleys. The existing control system was disconnected from the fan inlet dampers and attached to the MagnaDrive coupling. Immediately the maintenance crew noticed a vibration reduction from 0.5 inches/second (1.27 cm/second) to 0.05 inches/second (0.13 cm/second). Mike Henningan, maintenance planner for Ashgrove Cement reported, "I can see (MagnaDrive) eliminating a lot of our day to day and year to year maintenance on these machines, and extending the life of other equipment." He estimated a maintenance cost savings of $10,000. An electrical meter recorded motor start-up current of over 800 Amps causing a 10-second brownout in the bag house. After the MagnaDrive coupling was installed, start-up current peaked at 750 AMP and lasted only 1.5 seconds. Running current dropped from 142 to 49 Amps. Mike noted, "We see a tremendous drop in energy consumption, especially on start-up, compared to before when the motor was running, at times, actually in an overload condition. We're pretty impressed with this thing." Annual energy savings were about 25% or 155,000 kWh/year (0.56 TJ/year).

Washington Mutual Tower [3] is a 55-story, 1 million square foot premier office building in Seattle, Washington. Two MagnaDrive couplings were installed, one on a 125 HP (93.3 kW) 1800 RPM (30 RPS) vertical solid shaft motor running a centrifugal chilled water pump and the other on a 75 HP (56 kW) 1800 RPM (30 RPS) vertical solid shaft motor driving a centrifugal condenser water pump. The pumps were experiencing cavitation and vibration problems; and voltage sags during motor start-up sometimes disrupted the tenant's sensitive electronic equipment.

Figure 4. Vertical Axis MagnaDrive

Jeff Kasowaski, Chief Engineer noted "MagnaDrive came at a perfect time. We were looking into a Variable Frequency Drive, but we had a problem finding room for it. We were worried about the harmonics and filtering devices we would have to add, not to mention powering up a VFD that large and the inevitable extra heat loads. Then came the MagnaDrive. The large footprint isn't needed, harmonics aren't an issue, no power is required, and the control is better than imagined." After the MagnaDrive couplings were installed Jeff noticed that, "We don't have to replace couplings annually and deal with aligning the pumps and motors. The cavitation noise we lived with for ten years disappeared as soon as we started the MagnaDrive Coupling." The 75 HP (56 kW) condenser pump motor dropped its peak demand by 66%, from 39 kW to 13 kW. The 125 HP (93.3 kW) chilled water pump reduced its peak demand from 65 kW to 45 kW. Because Seattle's climate is relatively cool, Washington Mutual's cooling system operates only about 100 days per year; however, the MagnaDrive coupling still reduced electrical energy use by 394,000 kWh (1.4 TJ) per year.

Unified Sewerage Agency Rock Creek Advanced Wastewater Treatment Facility [4] in Hillsboro, Oregon serves over 150,000 residents and industries. A parallel system of three 60 HP (44.8 kW) 1800 RPM (30 RPS) single vane centrifugal pumps supplies digester bio-solids to belt presses to be de-watered. Two of the three pumps operate continuously with the third being a

393

backup. They were controlled by two six-year old VFDs, and the operators could no longer get replacement parts. The single vane impellers in the pumps were creating torque pulses in the downstream piping at speeds in excess of 1100 RPM (18.3 RPS). Even though one pump at full speed could supply water for the four belt presses, the system could not tolerate the vibration and shaking and the system piping was completely replaced. Then the plant used the VFDs to operate two pumps at slower speed trying to reduce the vibration but the piping still required constant repair. Bob Kennedy, Electrician II of Rock Creek Wastewater Treatment said, "The MagnaDrive Coupling has performed above and beyond our expectations. It has smoothed out our process and stopped vibrations and shaking in our pipes. We're now running one pump instead of two, which reduces energy consumption." A test of the system showed that only 10 HP was needed to supply flow to the four presses. As a result, the motors and couplings were downsized from 60 HP (44.8 kW) 1800 RPM (30 RPS) to 20 HP (14.9 kW) 1200 RPM (20 RPS). Electrical demand was reduced from 9.6 kW for two 60 HP (44.8 kW) with VFDs to 5.5 kW for one 20 HP (14.9 kW) motor and a MagnaDrive coupling.

Ponderay Newsprint [5] in Usk, Washington was searching for a way to improve operational efficiency. A single 250 HP, 1800 RPM motor operating at 2300 Volts drove a centrifugal pump to move thermal-mechanical pulping (TMP) whitewater to both the pulping process, 24 hours a day, and to the de-inking system, 12 minutes each hour. For 48 minutes per hour the system re-circulated TMP water through a bypass valve to maintain 50 psi (0.435 MN/m2) on the line when downstream valves were throttling. During the by-pass period, the piping system experienced excessive vibration, up to 0.3 inches per second (0.12 cm/s). During the 12 minutes that the de-inking system drew water the by-pass was closed and all of the water was used. The motor's electrical demand was 173 kW 24 hours a day even though it was not needed for 80% of the time.

A VFD was considered but rejected because of the high cost for a 2300-Volt system. So a MagnaDrive coupling was installed. During the 48 minute per hour period when flow could be reduced, the peak demand dropped from 173 kW to 65 kW, a peak reduction of 108 kW. Assuming 5,800 hours operation per year, the overall energy savings was about 633,000 kWh/year (2.28 TJ). Pump cavitation was eliminated and vibration greatly reduced. Don Guenther, Energy Manager for Ponderay Newsprint, commented, " We are excited about the potential for this equipment and have budgeted for future applications throughout the plant."

ACKNOWLEDGEMENTS
The authors acknowledge the work of Amy Cortese and Susan Hermenet, Alliance Project Coordinators.

REFERENCES
[1] Wallace, A., von Jouanne, A., Jeffryes, R., Matheson, E., and Zhou, X., June 2000, "Comparison Testing of an Adjustable-Speed Permanent-Magnet Eddy-Current Coupling," IEEE Pulp and Paper Conference, Atlanta, GA, Conference Record on CD-ROM.

[2] Woodard, R., 2000, "Ash Grove Cement Company, Durkee, Oregon, Case Study," MagnaDrive and Northwest Energy Efficiency Alliance.

[3] Woodard, R., 2000, "Washington Mutual Tower, Seattle, Washington, Case Study," MagnaDrive and Northwest Energy Efficiency Alliance.

[4] Woodard, R., 2000, "Unified Sewerage Agency Rock Creek Advanced Wastewater Treatment Facility, Hillsboro, Oregon, Case Study," MagnaDrive and Northwest Energy Efficiency Alliance.

[5] Woodard, R., 2000, "Ponderay Newsprint, Usk, Washington, Case Study," MagnaDrive and Northwest Energy Efficiency Alliance.

Proceedings of IECEC'01
36th Intersociety Energy Conversion Engineering Conference
July 29–August 2, 2001, Savannah, Georgia

IECEC2001-CT-02

I/V Characteristic of ZnSe Schottky Diodes

by

Dr. Dirk M. CHIU
Moon Laboratory, 71 Long Valley Way,
Doncaster East, Victoria 3109, Australia.
E-mail: dmchiu@hotmail.com

Abstract

The ZnSe sample used to make the Schottky diode was grown using conventional thermal diffusion technique with iodine as the transport agent, and indium was subsequently diffused into. While gold was used for the rectifying contact, indium was used for the ohmic contact. From the analysis of the I/V characteristic of the diodes, it had been concluded that the transport mechanism of the diodes was thermal emission with an ideality factor of 1.4 that shows the important influence of the interfacial layer on the performance of the diode. Limited tunneling nature of the diode was also observed due to the thin depletion layer by the heavy doping. From the general diode equation of $I = I_s.\exp(V/V_0)$, we found a relationship of $\ln(I_s) \propto (V_0)^{1/4}$ existed for the diodes.

Introduction

The rectifying properties of a metal-semiconductor contact arise due to the presence of an electrostatic barrier between the metal and the semiconductor. The existence of this barrier is due to the difference in the work functions of the two materials.

There is not much literature describing the nature of Schottky diode using compound semiconductors, probably due to its limited application. However, in the II-VI materials, p-n junction fabrication is often difficult to accomplish. The investigation of the junction properties of such materials is best on the Schottky diodes.

In the process of producing p-n junction using ZnSe, we observed some special nature of the sample. Therefore, Schottky diodes were made using this material. The analysis of the nature and performance of the diodes produced many unexpected results. The following describes the findings.

I/V Characteristic of Schottky Diodes

Consider that in an metal-semiconductor diode the current flowing from the metal to the semiconductor is I_s, and that most of the voltage drop, V, occurs across the space charge region, then the effective height of the barrier will be $e(\varphi_B - V)$ where φ_B is the surface potential, while e is the electron charge.

The electron current, $-I_e$, from the semiconductor to the metal is given by

$$-I_e = A.exp[-e(\varphi_B-V)/kT] \quad \ldots\ldots \quad (1)$$

where A is the Richardson constant.

When V=0, we have $I_e=I_s$, so that

$$-I_s = A.exp(-e\varphi_B/kT) \quad \ldots\ldots\ldots \quad (2)$$

The total current, I, is thus given by

$$I = -(I_s + I_e) = I_s[exp(eV/kT)-1] \quad \ldots \quad (3)$$

which is practically the same as for p-n junction.

To evaluate the saturation current, I_s, the value of A should be calculated. For thin layer, A is given by [1]

$$A = n.e. (kT/2\pi m_e)^{1/2} \quad \ldots\ldots\ldots \quad (4)$$

where m_e is the electron effective mass.

By combining Schottky's diffusion theory and Bethe's thermionic emission theory, it was shown that the Schottky barrier resembles the metal-oxide-semiconductor (MOS) structure. Therefore, for a Schottky diode, the Interface State plays the major role on the current/voltage characteristic [2].

The interfacial layer (oxide) is transparent to current flow but it can sustain a voltage drop which thus gives rise to a dimensionless parameter, n (called the ideality factor) that larger than unity. Hence, the modified current transport expression becomes [3]

$$I = I_s[exp(eV/nkT)-1] \quad \ldots\ldots \quad (5)$$

Pressure contacts experiments, however, show that the I/V technique can used, under certain conditions, successfully for carrier profiling when the diode is reverse-biased, instead of impurity profiling [4].

By using modified Richardson constant, A^*, the saturation current, I_s, can be obtained from [5]

$$I_s = A^*ST^2.exp(-e\varphi_B/kT) \quad \ldots\ldots \quad (6)$$

where S is the area of the diode. The value of I_s is determined from the extrapolation to zero voltage of the ln(I)/V characteristic, so that φ_B can be obtained from Equation (6).

Since the barrier height obtained using Equation (2) or Equation (6) often differs from that obtained using the C/V plot technique, a modified version of Equation (6) was suggested when the ideality factor, n, was included in the expression for the saturation current [6], so that

$$I = A^*ST^2.exp(-e\varphi_B/nkT)[exp(eV/nkT)-1]$$
$$\ldots\ldots\ldots\ldots\ldots\ldots\ldots\ldots\ldots\ldots \quad (7)$$

I/V Measurements and Results

The basic ZnSe crystal was grown using closed-tube conventional thermal diffusion technique with iodine as the grown agent. The resulting crystal was made to go through the zinc-extraction treatment so that its resistivity was reduced to about 0.1 Ω–cm. The low-resistivity sample was then diffused with Zn-In compound powder. The eventual propose of the experiment was to make p-n junction. By such heavy doping with indium, the blue-emission spectrum had been quenched off [7-10].

The Schottky diodes were made using gold (by evaporation) as the rectifying contact, and the ohmic contact was made by alloying indium pellet at about 300^oC in an argon atmosphere for about one minute (the timing is crucial).

The variation of the forward biased I/V curves with temperature is shown in Figure 1. Diodes made from the zinc treated only crystals and those from subsequently doped with Zn-In show similar characteristic. Since all the crystals have the same emission peak (at about 2.01 eV) [10] as revealed by the cathodoluminescent spectra, the diffusion region of them must be similar. The slightly larger breakdown voltage of the subsequently Zn-In doped diodes (less than 0.1 eV compared with the diodes made from the zinc treated only crystals) is probably due to the higher concentration of impurities which reduce the width of the space-charge layer.

From Figure 1, we can see that four regions are present, the space-charge region at low fields, the diffusion region where light emits, the current modulation region, and the high current region.

From the linear region of ln(I)/V characteristic, i.e. the diffusion region, it can be approximated satisfactorily by the Sah-Noyce-Shockley formula [11] for eV>>nkT :-

$$I = I_s.exp(eV/nkT) \qquad\text{............. (8)}$$

which takes into account the carrier recombination in the junction.

At room temperature, the diffusion region starts at about 1.4V and ends at about 1.6V. The high current region starts at about 2.6V. The ideality factor, n, is about 1.40±0.05 which shows very clearly the influence of the interfacial layer upon the performance of the diode.

Analysis of the I/V Characteristic

In the ln(I)/V characteristic of Figure 1, the small space-charge recombination current is due to carriers that diffuse into the depletion region, and do not emit lights [11].

The space-charge recombination component continues with unchanged slope up to a voltage near 1.6V at room temperature; and that it remains a small part of the total current above this voltage, a further current component can be isolated by subtraction. This component initially follows an exponential relationship as shown in Equation (8) with n=1.40±0.05 at room temperature. At high field density, where this further current component becomes the dominant current, the exponential relationship breaks down and the characteristic becomes curved.

The beginning of the linear region of the ln(I)/V curves corresponds to the electroluminescence voltage threshold, V_T. When the diode forward bias voltage reaches V_T, the electron can tunnel from the indium contact towards the ZnSe substrate where they recombine and generate lights. Therefore, for $V \geq V_T$ we have a double-injection mechanism whereas for $V < V_T$ the current is dominated by the electron flux injected from the n-type substrate throughout the space-charge region.

The double-injection mechanism is always present whatever the temperature as shown in Figure 1. The resistivity of the substrate is not significantly affected by the variation of temperature, so the injection conditions at the boundaries of the interfacial (oxide) region (x_i) remains constant. In here, we have considered that the present Schottky diodes resemble the MOS structure as discussed previously, such as that shown in Figure 2. The slight shift of V_T in Figure 1 at different temperatures is due to the variation of the carrier capture cross sectional areas which have a temperature dependency [12].

The double-injection mechanism of the present diodes is made more evident when we plot the forward voltage at a constant current (e.g. 5 mA) as a function of temperature, as shown in Figure 3. The voltage decreases monotonically as the temperature is increased. No step variation in the V/T characteristic appears.

From the result of Figure 1, we obtain the general empirical diode equation

$$I = I_s.\exp(V/V_0) \qquad \ldots\ldots\ldots \quad (9)$$

The value of I_s can be obtained by the extrapolation of the linear region of the curve until it intersects the current axis. At room temperature, I_s had been found to be in the order of 10^{-19} ampére.

Using temperature dependence of the values of V_0 from Figure 1, we obtain the V_0/kT graph of Figure 4 which gives the value of the ideality factor, $n=1.40\pm0.05$. If all the minority carriers recombine in the junction, the factor n has its maximum value of 2; if there is no recombination, then n equals to unity (the Shockley formula for $eV \gg kT$). The value of $n=1.40\pm0.05$ in our diodes indicates that although recombination in the junction region happens, it is not particularly strong. Verification of this assumption had been carried out by making two diodes with contacts made on freshly cleaved surfaces, and the ideality factor of them was found to be $n=1.31\pm0.05$, thus confirms our suggestion.

At room temperature, the dependence of $\ln(I)$ upon the applied voltage is linear up to about 1.6V. At voltage above 2.2V, the rise of current slows down very appreciably, which is probably due to the influence of the bulk resistance. Therefore, it follows that the junction barrier is in the region of $2.2V \geq \varphi_B \geq 1.6V$, and so the barrier height is also within this range.

The strong temperature dependency of the I/V characteristic can be considered as an indication that the current transport mechanism in our diodes is of the thermal emission nature. According to the thermionic emission theory, the barrier height $e\varphi_B$ of the diodes is much larger than kT, the electron collisions within the depletion region are neglected, and the effect of the image force is neglected. Based upon Equation (6), a plot of $\ln(I_s T^{-2})$ via $(kT)^{-1}$ is obtained as shown in Figure 5. A straight line is obtained and the thermionic emission transport mechanism for our diodes is thus confirmed. From the slope of this plot, we get 2.0 eV$\geq e\varphi_B \geq 1.7$ eV that confirms our finding previously for the barrier height result.

Since the samples are heavily doped, the depletion layer is expected to be reasonably thin so that the carriers can tunnel through the edge of the depletion region [1]. If this is so, then V_0 will be weakly dependent on temperature; and if transport of carriers is completely dominated by tunneling, then V_0 should be temperature independent. Our present investigation reveals that a temperature dependence of V_0 is observed. Therefore, we can assume that tunneling does occur in our diodes due to the heavy doping, but it is not the predominant transport mechanism, i.e. our samples are not strongly degenerated.

In the empirical diode expression of Equation (9), no linear relationship between $\ln(I_s)$ and V_0 is observed. Instead, we have a somewhat less rigid relationship of the form

$$\ln(I_s) \propto (V_0)^{\frac{1}{4}} \qquad \ldots\ldots\ldots \quad (10)$$

as shown in Figure 6 by extrapolation, assuming linear relationship for low current value.

Conclusion

The analysis of the ZnSe Schottky diode revealed that the transport mechanism of such diodes is of the thermal emission nature. The ideality factor was found to be about 1.4 that indicated the important influence of the interfacial layer upon the performance of the diode. Limited tunneling capacity of the carriers was also observed. By using the extrapolation technique, assuming linear characteristic, we discovered that the current constant and the voltage constant bore such relationship as $\ln(I_s) \propto (V_o)^{1/4}$.

References

1. Crowell, C.R., Solid-State Elec. **12**, 55 (1969).

2. Crowell, C.R., and Sze, S.M., Solid-State Elect. **9**, 1035 (1966).

3. Goodman, A.M., J. Appl. Phys. **34**, 329 (1963).

4. Ladbrooke, P.H. and Carroll, J.E., Int. J. Elect. **31**, 149 (1971).

5. Sze, S.M., "Physics of Semiconductor Devices", Chapter **8**, J. Wiley, NY, 1969.

6. Hackam, R. and Harrop, P., Solid-State Elect. **15**, 1031 (1972).

7. Chiu, D.M., J. Phys. D:Appl. Phys. **16**, 2281 (1983).

8. Chiu, D.M., Proc. IECEC91, SNE-3, 190 (1991).

9. Chiu, D.M., Proc. IECEC94, AIAA-2, 1053 (1994).

10. Chiu, D.M., Proc. IECEC2000 AIAA-2, 1222 (2000).

11. Sah, C.T, Noyce, R.N., and Shockley, W., Proc. IRE **45**, 1228 (1957).

12. Lambert, M.A. and Mark, P., "Current Injection in Solids", Academic Press, NY, 1970.

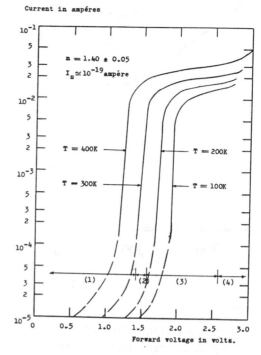

Fig. 1 Temperature dependency of I/V curves of ZnSe Schottky diodes
(1) space-charge region,
(2) diffusion region,
(3) current modulation region,
(4) high current region.

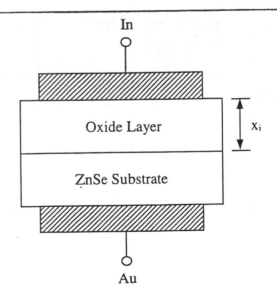

Fig. 2 Schematic representation of ZnSe Schottky diodes.

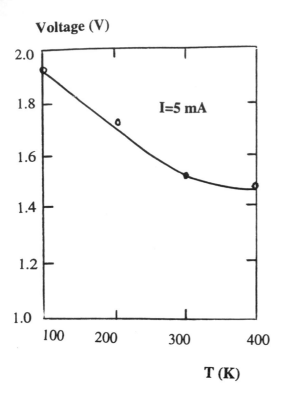

Fig. 3 Forward voltage variation with temperature at constant current.

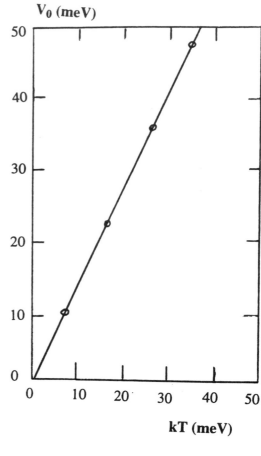

Fig. 4 Temperature dependency of the voltage constant of I/V equation.

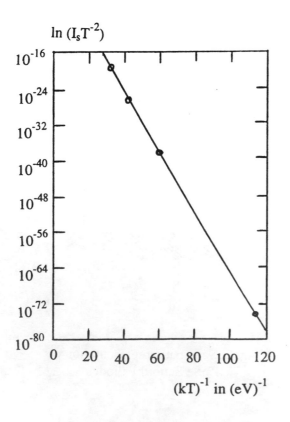

Fig. 5. Variation of T with $I_s T^{-2}$.

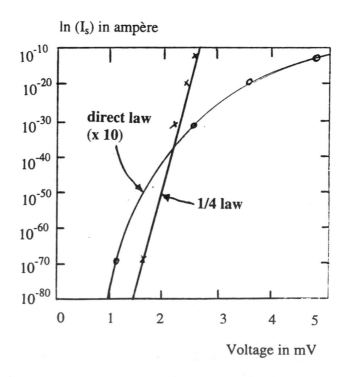

Fig. 6 Proposed saturation current variation law.

Proceedings of IECEC'01
36th Intersociety Energy Conversion Engineering Conference
July 29 – August 2, 2001, Savannah, Georgia

IECEC2001-CT-45

THE ELECTRIC AIR ARC IS AN MHD GENERATOR

Neal Graneau, Oxford University, Dept. of
Engineering Science, Parks Rd., Oxford,
OX1 3PJ, UK, neal.graneau@eng.ox.ac.uk

Peter Graneau, 205 Holden Wood Rd., Concord,
Massachusetts, 01742, USA

George Hathaway, Hathaway
Consulting Services, 39 Kendal
Av., Toronto, Ontario, M5R 1L5,
Canada

ABSTRACT

MHD power generation in air arcs is a new discovery and area of active interest. As these arcs are always known to explode, charged ions must be driven radially outward and travel across the lines of encircling magnetic flux. This motion induces axial e.m.f.'s which augment the arc current. When the increased current flows through a resistor, it is found to produce more heat than the energy that was supplied by the capacitors to initially create the arc. It is argued that ions are propelled by the release of the stored potential energy of repulsion between the atomic nuclei in molecular N_2 and O_2. Electricity generation from the energy stored in the atmosphere is of great interest because air is freely available anywhere on earth.

Keywords: MHD generators, arcs, plasma, bond energy

INTRODUCTION

Despite a century of research on lightning and exploding air arcs, it has only recently been discovered that pulsed air arcs generate MHD (magneto-hydrodynamic) electric power in a manner similar to commercial MHD generators. However one major difference between the two technologies is that commercial units rely on a separate magnet while an exploding arc utilizes the encircling magnetic field of the arc current itself without requiring an external magnet. This was first discovered at Oxford University in 1999 [1].

The forces which drive the ions across the magnetic field lines of a modern commercial generator are of thermal origin. For example, the ions may be propelled by the flame jet of an oil burner. However in 1989 it was found by one of the authors (PG) that pulsed air arcs of short duration (<1 ms) are not hot and the expansion that is always observed by the presence of a

shockwave is not the result of thermal forces [2]. It is therefore probable that the arc ions are accelerated by the release of stored potential energy during the dissociation of N_2 and O_2 molecules. The fact that arcs produce large quantities of nitrous oxides (NO_x) and ozone (O_3) demonstrates that many bonds must be broken during the arc. The energy was first stored in the nitrogen and oxygen molecules when the atoms became bonded upon collision with each other. As the particles were brought to rest relative to each other, their kinetic energy was partly converted into stored potential energy. This energy of repulsion between the atomic nuclei is referred to as the bond energy. It is not to be confused with the thermal dissociation energy which is the quantity of thermal energy that needs to be supplied in order to break the bond. The kinetic energy of the pre-collision nitrogen and oxygen atoms, some of which is converted into the stored potential energy, is derived ultimately from incident solar radiation.

EXPERIMENTAL PROOF OF ENERGY GENERATION

Experimental proof of MHD electric energy generation in air arcs was obtained with the simple discharge circuit of Fig.1. This relied on a capacitor bank of $0.64 < C < 3.75$ μF charged to a voltage, V_0 in the 10-30 kV range. The initially stored energy supplied by the capacitor bank was

$$E_S = \text{stored capacitor energy} = \tfrac{1}{2} C V_0^2 \text{ (J)} \quad . \qquad (1)$$

The capacitor voltage was applied across a load resistor, R_L, in series with a triggered air spark gap of 4.5 mm length between domed electrodes with tungsten arcing inserts. The trigger energy was less than one Joule while the capacitor bank could store up to 1.69 kJ. The load resistor was a stack of 15 cm. diameter, 2.5 cm. thick, carbon ceramic disks capable of withstanding large current pulses up to 80 kA peak current. The resistance of the stack used in the experiments described here came to $R_L = 2.38 \, \Omega$.

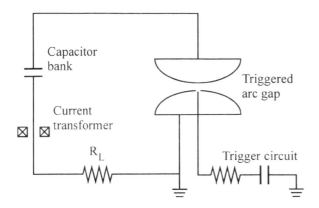

Figure 1 : Air arc discharge circuit with load resistor, R_L

The instantaneous discharge current, i, was measured with a current transformer and recorded on a digital oscilloscope. The signal was then processed on a computer to yield the action integral ($\int i^2 \, dt$) where t represents time. The electrical energy converted to Joule heat in the load resistor is

$$E_L = \text{Joule heat in } R_L = R_L \int i^2 \, dt \quad \text{(J)} \quad . \qquad (2)$$

For appropriate values of V_0, C and R_L, it was found that

$$R_L \int i^2 \, dt > \tfrac{1}{2} C V_0^2 \quad \Rightarrow \quad E_L / E_S > 1 \quad . \qquad (3)$$

In fifty shots of input energies $30 < E_S < 500$ (J), the Joule heat dissipated in the load resistance, E_L, was found to be greater than the energy supplied from the capacitor bank, E_S. The results for the measured (E_L/E_S) ratio are plotted in Fig.2 as a function of the supplied energy, E_S. The scatter of the points is primarily due to the statistical nature of the breakdown process which causes the electron emission site location and efficiency to vary from shot to shot.

It must also be appreciated that the capacitors supplied the energy losses in the rest of the discharge circuit as well, including the electromagnetic and acoustic energy radiated into the environment. The major components of the circuit losses were (a) the ionization energy which is then stored in the arc plasma and later returns as heat, (b) the losses in the capacitor itself and (c) all ohmic losses in the conductors of the circuit and in the plasma. Hence we know that the total MHD energy

generated by the arc explosion was greater than indicated in Fig.2.

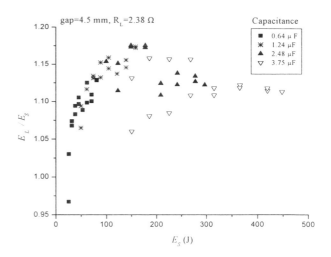

Figure 2 : MHD energy gain (E_L /E_S) vs. energy drained from the capacitor bank, (E_S)

THE CAUSE OF LOUD AIR ARC EXPLOSIONS

To fully appreciate the implications of the inequality in Eq.(3), confirmed by the results shown in Fig.2, the cause of the arc explosion has to be determined. It has been demonstrated that electrodynamic forces are incapable of causing the explosion of the arc with the magnitude and direction of the measured forces [2,3]. Until 1972, it had been assumed that thunder and the loud explosion of laboratory air arcs were caused by a thermal shockwave. Then Viemeister [4] published his findings with regard to hot and cold lightning which he summarized as follows:

"Cold lightning is a lightning flash whose main return stroke is of intense current but of short duration. Hot lightning involves lesser currents but of longer duration. Hot lightning is apt to start fires while cold lightning generally has mechanical or explosive effects."

This was confirmed by direct measurements of the force and current dependence of confined arc explosions [2]. Further, a pulsed arc does not cause steam when striking a pool of water, and while it will tear paper placed across its path, it will not ignite it.

Faced with the fact that neither electrodynamic nor thermodynamic forces are responsible for the arc explosions, it became clear that a second and independent source of energy, in addition to the capacitor energy, was involved in the phenomenon. Since chemical changes occur in the gaseous arc medium, we have been forced to examine whether stored chemical energy can cause a rapid expansion of the arc column before the ionic motions become collision dominated and are slowed down. In order to produce the known arc byproducts, the diatomic bonds in N_2 and O_2 molecules must first be broken. This leads to the possibility that the explosion energy is in fact the liberation of stored diatomic bond energy.

MHD E.M.F. GENERATION MECHANISM

Without pursuing the details of the chemical energy conversion processes, there can be no doubt, on account of the loud explosion, that in our experiments nitrogen and oxygen ions in varying states of ionization are driven radially outward from the arc column with a velocity v_r in the direction indicated in Fig.3.

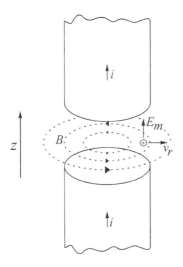

Figure 3 : Induction of the MHD electric field, E_m, in an arc gap

In crossing the encircling lines of magnetic flux, these ions generate an axial electric field strength, E_m, in the direction of current flow given by

$$E_m = v_r \times B \quad , \qquad (4)$$

where B is the magnetic flux density. Integrating the E_m field along the arc gap leads to the motionally induced e.m.f, e_m, between the electrodes in the z direction,

$$e_m = \int v_r\, B\, dz \quad . \qquad (5)$$

This represents the e.m.f which should drive current in the direction of the discharge current, i. It may be described as a forward e.m.f resulting from the conversion of mechanical (kinetic) energy into electrical energy. Conversely, back e.m.f's are produced by the conversion of electrical to mechanical energy as in a motor. The explosive nature of pulsed air arcs is

well documented in the arc physics literature [5-7]. Allen and Craggs [6] found radial expansion velocities of the optical arc channel in atmospheric air up to 2.5×10^3 m/s.

Unfortunately, it is not possible to accurately calculate the MHD energy generated by the pulsed arc current. It would require unavailable knowledge regarding the number of ions involved and their charge state as well as their velocity distribution. The current distribution is also unknown and it determines the strength and variation of the encircling magnetic flux density, B. Faced with these unknown quantities, the only means of determining the magnitude of the MHD effect is to measure it by experiment.

CHEMICAL ENERGY LIBERATION

To satisfy the inequality in Eq.(3), two requirements must be met. First of all, in addition to the capacitor energy, another stream of energy must flow into the discharge circuit. Secondly, the additional energy must be converted to electrical energy so that it can be absorbed in the load resistor. These two requirements are satisfied by the proposed MHD generator principle. No other mechanism has been found or suggested.

The additional energy source must be in the form of ion kinetic energy in order to achieve MHD generator action. As suggested in the Introduction, this second stream of energy is assumed to be liberated chemical bond energy derived from the dissociation of N_2 and O_2 molecules.

The treatment of bond energy in the chemistry literature is confusing. Many authors treat the bond energy as the amount of thermal energy required to break the bond ignoring the energy that is stored in the bond itself. However the bond is also discussed as a balance of attractive and repulsive forces acting like a compressed spring. In the same way that a spring and its compressing mechanism store potential energy, so must a chemical bond. This is the energy that can be liberated to produce high ion kinetic energy.

The diatomic or covalent bond is caused by a strong force of attraction between the two atoms. This attraction is sometimes described as a Van der Waals force and may be due to spin coupling of the two valence electrons in common orbitals. Whatever the precise microscopic mechanism of atomic attraction, it has to be counteracted by an equal Coulomb repulsion between the two nuclei in order to prevent fusion. The repulsion becomes greater as the nuclear distance shrinks, and as a result the stored potential energy in an N_2 molecule is particularly large because of the strong attraction between two atoms with 3 unpaired electrons each. This is why nitrogen is very stable and therefore abundant in the atmosphere and also a large source of stored energy.

The only way of liberating this potential energy is to, somehow, remove the electronic attraction. The thermal bond dissociation process relies for this purpose on the collision of two molecules with sufficient mutual kinetic energy. The quoted dissociation energy for N_2 molecules is 225.5 kcal/mole. The pulsed arc plasma is quite cool and as a result thermal dissociation is not expected to occur to any great extent. More importantly, the final outcome of the explosion is to heat the surroundings. If heat were also the cause of the explosion, then by the principle of conservation of energy embedded in the first law of thermodynamics, we could not expect that this process could lead to a gain in energy. As a consequence, thermal dissociation cannot be the cause of the second energy source.

The mechanism of the attractive force in a covalent bond relies on the highly complex electromagnetic interactions between electrons with magnetic spin components. It is possible that these interactions are severely affected by the high current density of the arc, leading to a reduction or even elimination of the attractive force in the chemical bond while the electrostatic repulsion force between the atomic nuclei remains unaffected. It is quite likely that this electromagnetic decoupling may require less energy than thermal collision dissociation pointing to the possibility of an energy gain.

If the arc is fired in a closed vessel containing pure nitrogen gas, then the individual atoms created during the explosion will eventually recombine back into N_2 when the two atoms collide. As a consequence of the principle of chemical irreversibility as contained in the second law of thermodynamics, this recombination can only occur at the expense of heat drawn from the surroundings. The energy that is drawn from the surroundings plus some of the existing kinetic energy of the free atoms is then stored in the newly formed covalent bonds ready to be exploited by a later arc. This demonstrates that such a device would ultimately be running on atmospheric heat which is caused by solar heating. The MHD generator and load produce heat which compensates for the energy drawn out of the environment and thus the device could produce electrical output without a net warming or cooling of the atmosphere.

While it may seem surprising that we are proposing an energy generator that runs on atmospheric heat, we can draw a comparison with hydroelectric power which also uses atmospheric heat to evaporate low altitude water and return it to the top of the mountain in order to use it to drive the turbine. To continue the analogy, N_2 is like the water at the top of the mountain and atomic nitrogen acts as the water after it has passed through the generator. The MHD generator behaves like the turbine and extracts energy while the working substance changes state.

REFERENCES

[1] Graneau N., 1999, "MHD Activity in High Current Arcs", IEE Pulsed Power '99, Oxford, IEE Digest No:99/030, pp5/1-5/4

[2] Graneau P., 1989, "The Cause of Thunder", J.Phys.D:Appl.Phys, **22**, pp1083-1094

[3] Graneau P., Graneau N.,1996, *Newtonian Electrodynamics*, World Scientific, Singapore, pp192-227

[4] Viemeister P.E., 1972, *The Lightning Book*, MIT Press, Cambridge, MA, p.123

[5] Flowers J.W., 1943, "The Channel of the Spark Discharge", Phys.Rev, **64**, pp225-235

[6] Allen J.E., Craggs J.D., 1954, "High Current Spark Channels", Brit.J.Appl.Phys, **5**, pp446-453

[7] Uman M.A., Cookson A.H., Moreland J.B., 1970, "Shock Waves from a Four-Meter Spark", J.Appl.Phys, **41**, pp3148-3155

**Proceedings of IECEC'01
36th Intersociety Energy Conversion Engineering Conference
July 29–August 2, 2001, Savannah, Georgia**

IECEC2001-CT-03

SOME RESULTS OF PRELIMINARY BENCH TESTS ON A 3-kW$_e$ STIRLING ENGINE BASED CO-GENERATION UNIT

**K. Makhkamov, V. Trukhov, I. Tursunbaev, E. Orda,
A. Lejebokov, B. Orunov, A. Korobkov, and I. Makhkamova**
Physical-Technical Institute
Uzbek Academy of Sciences
700048, 2-b Moavlyanova, Tashkent, Uzbekistan

D. B. Ingham
School of Mathematics
University of Leeds
LS2 9JT Leeds, UK

ABSTRACT

A 3-kW$_e$ Stirling Engine based co-generation unit and a test bench for the validation of the performance of the unit have been developed and manufactured at the Laboratory for Stirling Engines at the Physical-Technical Institute of the Uzbek Academy of Sciences. Pictures of some elements of the installation and a scheme of the system for the hot water supply and for heating, which is incorporated in the micro co-generation unit, are presented and described in this paper. A "V" type Stirling Engine is used as a thermodynamical converter in this Micro Combined Heat and Power (MCHP) system. The Engine is fuelled by natural gas and has a sealed dry crank-case with a built-in three-phase induction alternator. A Combustion Chamber of the Engine is made to work from the pressure of the natural gas in the in-house home pipelines and it has neither an air fan or an air pre-heater. Exhaust gases from the Combustion Chamber heat up the water in the boiler. Heat rejected in the cooling circuit of the engine and in the cooling jacket of the alternator is also used for providing the hot tap water and for heating purposes. This paper focuses on the results from the first preliminary tests which were run on the co-generation unit just after finishing its manufacture. During these experimental investigations, the Engine with the Combustion Chamber has been tested with different values for the charge pressure of the working fluid. The Stirling Engine has been instrumented with pressure and temperature sensors to provide information on the parameters of the working process. Additionally, several thermocouples and flow meters have been installed in different locations of the heat utilization system and a data acquisition system has been used to obtain an overall heat balance of the installation.

INTRODUCTION

In [1,2] the activities of the Laboratory for Stirling Engines on the development of small energetic units on the basis of different types of kinematical and fluid piston Stirling Engines are described. These energetic units have been intended for solar, radio-isotope and fossil fuel applications.

One of the present research projects of the Laboratory is concerned with the development of a laboratory prototype for a 3-kW$_e$ MCHP unit for application in small business market. Recently, a 3-kW$_e$ Stirling Engine with a Combustion Chamber for natural gas have been manufactured at the Physical-Technical Institute and the MCHP installation has been assembled at the Laboratory for bench tests.

DESCRIPTION OF THE MCHP INSTALLATION

Figure 1 presents a scheme of the prototype for the MCHP system and a sketch of the Stirling Engine which is undergoing preliminary tests. The Stirling Engine is a "V" type or α-configuration machine and consists of the cold (1) and hot (2) cylinders which are installed on a pressurised crank-case (3). The heater's cage (4) of the Engine consists of two collectors and two sets of semicircular tubular rings which are made from stainless steel tubes (37 tubes in one set; the outer diameter of tubes is 5 mm). Each collector is rectangular in cross-section and they form a frame for the conically shaped heater cage and carry sets of the semicircular tube rings. Semicircular tube rings are connected to the collectors and are distributed uniformly along the heights of the collectors. The semicircular tube rings are placed symmetrically relative to the vertical plane, in which the collectors are installed, and in such a way that the tubes form circles in a plane. The Combustion Chamber (5),

405

with the gas collector (6) and ejection pipes (7), surrounds the heater. The meshes, which are made from stainless steel type material, surrounds the heater tubes and form a porous media in which the combustion of the natural gas occurs. The working temperature of the heater is 923K. The Combustion Chamber is developed for operation on natural gas pressure for domestic gas appliances and no air fans and air pre-heaters are used in this design. A boiler (8) with the heat exchanger (9) are installed on the top of the Combustion Chamber for utilization of the heat of the flue gases. The Engine has a typical cylindrical regenerator (10) with mesh gauzes which are also made from a stainless steel type of material. The cooler (11) consists of a package of tubes which are washed with water. In order to reduce the heat losses, the heater, the regenerator and the cylinder of the expansion space are surrounded by thermoinsulation (12). The pistons (13) have two guide and sealing rings which are made from a fluoroplastic material. The stroke of the pistons is 40 mm and the speed of the machine is 25 Hz. A three-phase induction alternator (14) is built into the special shell (15) which is connected to the crankcase. Rolling bearings are used throughout in the design of the Engine and Helium is the working fluid and the maximum pressures in the cycle is 7.5 MPa.

The hot tap water is stored in the hot water vessel (17) which has a thermocouple unit (18) to control the temperature of the water. The level regulator (19) provides the constant level of the water in this vessel. When there is a decrease in the temperature of the water, due to filling up with a fresh portion of water or in the case when the building's heating system is switched on, then a water pump (20) starts to circulate the water in the system. This pump also starts to work when the temperature of the water in the vessel (17) reduces in time because of heat losses through its walls. When water circulates in the system then natural gas is supplied to the Combustion Chamber through a valve (21). This valve also provides a continuous supply of the gas to the pilot light (22) in the Combustion Chamber. The water in the system runs from the water pump into two separate circuits. In the first circuit the water firstly goes to the cooler of the engine and then to the water jackets of the Engine's cylinders. In the second circuit the water flows initially into the water jacket of the alternator and then into the water jacket of the Combustion Chamber. These two separate flows join each other before the water enters the channels of the heat exchange which is placed in the boiler. Then the water, which is heated in the boiler, is pumped into the water jacket of the water vessel to provide the hot tap water. If the building's heating system is switched on by the heating regulator (23) then the water flow is re-directed by the valve (24) and water runs though the radiators (25) in the building.

DESCRIPTION OF THE METHODOLOGY FOR CONDUCTING EXPERIMENTS

After the completion of the manufacture of the parts and units for the Stirling Engine and the Combustion Chamber then the Engine, Combustion Chamber and the MCHP installation have been assembled on the test rig at the Laboratory and preliminary tests have been started.

Figure 2 shows an appearance of the manufactured Engine with the Combustion Chamber and the boiler being removed. Figure 3 presents the Combustion Chamber which was fabricated for this Engine. Finally, the Engine which is assembled with the Combustion Chamber and the boiler is shown in Figure 4. Other elements of the MCHP installation were assembled using commercially available hardware.

The data acquisition system which was used during the first preliminary tests allows us to register the speed of the Engine; the electrical power consumed or produced by the alternator; measure the average temperature of the working fluid in the hot and cold cylinders and the average pressure of the working fluid in the cycle. This system also measures the average temperatures on the surface of the heater's tubes and of the flue gases in the Combustion Chamber and in the boiler in several locations. Finally, the system measures the flow rate of the natural gas, flow rates of the water in the separate circuits and the difference in the temperature of the water entering and leaving the various water jackets and the boiler.

In the first stage of the preliminary tests, the Engine has been vacuumed and driven by the alternator of the engine which was used as an electrical motor in order to evaluate the mechanical losses in the driven mechanism of the machine. In the second stage, the Engine was driven for cases when the charge pressure of the helium was 0.3, 0.5 and 0.8 MPa (the Combustion Chamber was not operating). In the third stage of the preliminary experiments, the Engine run with the Combustion Chamber being switched on and with the charge pressure of the helium being at the level of 1.5 MPa. The cold tap water has been supplied separately to all the water jackets in the system for the cooling on this stage of experiments. Some results registered during the above tests are presented in Table 1 and discussed in the following section.

DISCUSSION OF THE RESULTS OBTAINED

From the experimental results obtained in the first stage of the investigations it can be assumed that mechanical losses in the drive mechanism of the unpressurased Engine are approximately 490-500 W.

When the engine was driven by the alternator in the second stage of investigations with the charge pressure of the helium being 0.3, 05 and 0.8 MPa then the value of the electrical power N_{el} consumed by the alternator was registered being 1640W (at the speed 1493 revolutions per minute), 2400W (at the speed 1487 revolutions per minute) and 3550W (at the speed 1480 revolutions per minute), respectively. These results indirectly reflect the very high level of frictional losses in the sealing and guiding rings of the pistons.

In the third stage the Combustion Chamber operated at its full capacity (Q_{cc} = 20.5 kW) and the Engine worked steadily when the charge pressure of the helium was 1.5 MPa and produced 1000 W of electrical power with the speed at 1560 revolutions per minute. The average temperature on the tubes ranged from 690 to 715 °C in different locations in the Combustion Chamber. At the same time the average temperature in the expansion space was registered at a rather low value, 390 °C.

The temperature of the cold tap water entering all the water jackets and the boiler was 18 °C. The measured value of the heat Q_{cooler} rejected in the cooler of the machine is very high and equal to 5792 W and the reason for this might be that the present design of the regenerator for the Engine does not have a sufficiently high heat capacity. The value of the heat Q_{hcyl} rejected in the water jacket of the hot cylinder is also relatively high and this may indicate that heat losses in this zone also are very intensive. Another large source of heat losses is the water jacket of the Combustion Chamber: $Q_{cc-wj} = 2114$ W.

The measured rate of the heat of flue gases utilized in the boiler is $Q_b = 4564$ W. A second heat utiliser has been installed in this regime of the experiments and the value of the heat rejected at this second stage was measured to be 1960 W. The calculated value of Q_{nr}, the immeasurable heat losses, was also 1960 W.

The total heat balance for the second set of experimental tests is presented in Figure 5. The fuel/electricity efficiency of the installation in this regime is 5% at this stage and this is as a direct result of the high level of mechanical and heat losses in the design of the Engine and the Combustion Chamber.

OUTLOOK

In the near future it is planned to perform preliminary experimental tests on the MCHP installation with the Engine running at different values of the charge pressure of the working fluid which will be increased gradually from 2.5 up to 6.5 MPa and the performance of the co-generation system will be experimentally evaluated for the case when all the cooling water jackets and the boiler are connected in a single enclosed loop of the MCHP installation. After these preliminary tests have been performed, the data obtained will be analyzed and improvements will be introduced into the design of the Stirling Engine in order to increase its efficiency and its overall performance.

ACKNOWLEDGMENTS

This research work was partially supported by a NATO Science for Peace Grant (SfP- 972296) and an EU INCO-Copernicus Grant (Contract ERB IC15-CT98-0501).

REFERENCES

[1]. K. Makhkamov, D.B. Ingham, et al. Development of Solar and Micro Co-generation Power Installations on the Basis of Stirling Engines. Proceeding of the 35th Intersociety Energy Conversion Engineering Conference "Energy and Power in Transition", 24-28 July 2000, Las Vegas, Nevada, V.2, pp. 723-733.

[2]. K. Makhkamov and D.B. Ingham. Development of Experimental Micro Co-generation Systems with Stirling Engines. Proceedings of the European Stirling Engine Forum 2000, 22-24 February, 2000, Osnabruck, Germany.

Table 1. Some results from preliminary experimental tests of the Engine and the MCHP installation.

The average pressure of the working fluid in the cycle, MPa	Vacuum	1.5
The engine's speed, rpm	1497	1560
N_{el} - the electrical power produced by the alternator, W	------	1000
The electrical power consumed by the alternator, W	680	-----
Q_{cc} -the power of the Combustion Chamber (assuming full combustion of the fuel), W	-------	20484
Q_{cooler} –the heat rejected in the cooler, W	46,8	5792
Q_{ccyl} - the heat rejected in the cold cylinder, W	96,7	681
Q_{hcyl} – the heat rejected in the hot cylinder, W	243	1590
Q_{alter} – the heat rejected in the water jacket of the alternator, W	190	353
Q_{cc-wj} – the heat rejected in the water jacket of the Combustion Chamber, W	------	2114
Q_b – the heat utilized in the boiler, W	------	4564
Q_{fg} – the heat of flue gases utilized in the second heat utilizer , W	------	1960
Q_{nr} – the immeasurable heat losses, W	-------	1960
The average temperature of the helium in the cold cylinder, K/°C	271/ - 2	348/75
The average temperature of the helium in the hot cylinder, K/°C	0	663/390
The average temperatures of the surface of the heater's tubes, K/°C	-----	963-988 / 690-715
The average temperature of the flue gases in the entrance into the boiler, K/°C	-----	825/552
The average temperature of the flue gas in the exit from the boiler, K/°C	----	453/180
Ambient temperature, K/°C	295.8/22.8	293.2/20.2

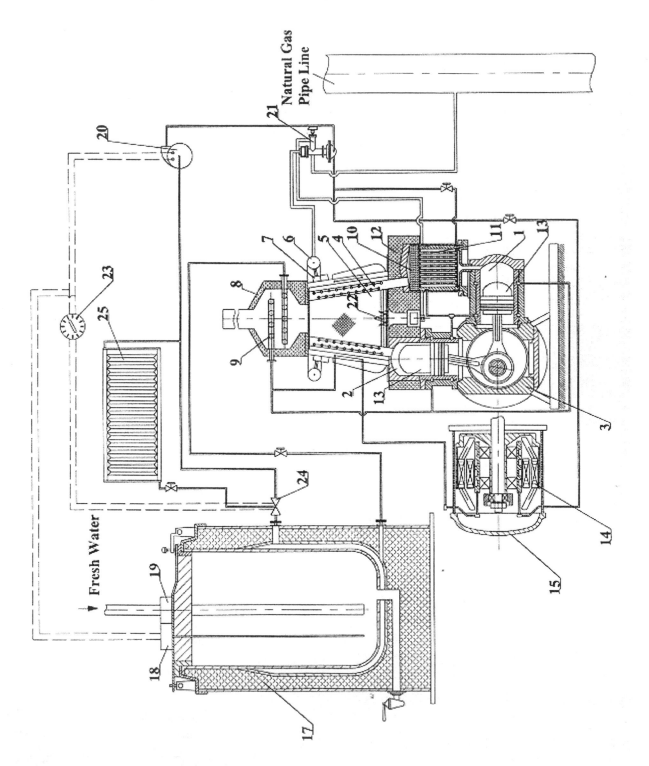

Figure 1. A schematic diagram of the MCHP installation.

Figure 2. An appearance of the 3-kW Stirling Engine.

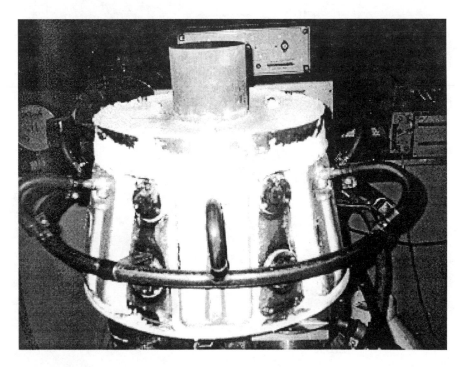

Figure 3. An appearance of the Combustion Chamber for the 3-kW Stirling Engine.

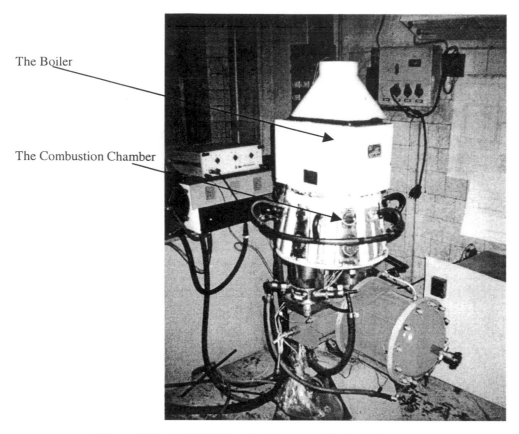

The Boiler

The Combustion Chamber

Figure 4. The 3-kW Stirling Engine assembled with the Combustion Chamber and the boiler.

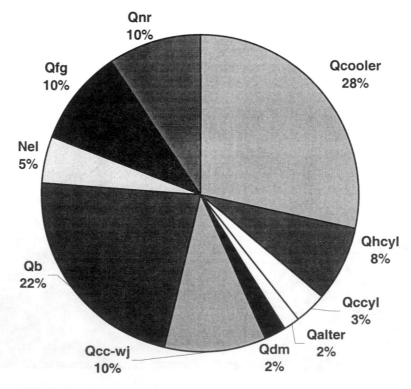

Figure 5. The heat balance for the MCHP installation obtained from the preliminary tests.

Proceedings of IECEC'01
36th Intersociety Energy Conversion Engineering Conference
July 29–August 2, 2001, Savannah, Georgia

IECEC2001-CT-04

Determination of the Temperature Distribution in a Combustion Chamber for a Stirling Engine

K. Makhkamov	**D.B. Ingham**	**M. Pourkashanian**
Physical –Technical Institute,	School of Mathematics,	Department of Fuel and Energy
Uzbek Academy of Sciences	Centre for CFD	Centre for CFD
700048, 2-b Mavlyanova, Tashkent,	The University of Leeds	The University of Leeds
Uzbekistan	LS2 9JT Leeds, UK	LS2 9JT Leeds, UK
Email: khamid@amsta.leeds.ac.uk	Email: amt6dbi@amsta.leeds.ac.uk	Email: fue5lib@sun.leeds.ac.uk
Fax: +998 712 354291	Fax: +44 113 233 5090	Fax: +44 113 244 0572

ABSTRACT

In this paper some of results obtained from a mathematical simulation of the working process of the Combustion Chamber (CC) or burner are described. The CC has been designed as a component of the micro co-generation unit based on the Stirling Engine (SE) and is fuelled by natural gas. CFD modeling has been used to describe the combustion of methane in the axi-symmetric CC. Furthermore, the tubes of the engine's heater, which are in the CC, have been assumed to be at a constant temperature. The numerical investigation enables the distribution of the temperature in the internal space of the burner to be obtained. These results can be used at the design stage for choosing the rational configurations of both the CC and the heater in order to provide the maximum heat input into the internal circuit of the SE.

INTRODUCTION

In [1] CFD modeling has been employed for the determination of the heat losses due to the free convection of air in the cavity of a solar heat receiver of a Dish/SE installation. In [2] a simple axi-symmetric CFD model for the simulation of the working processes of a SE has been described and computational results presented. These CFD models for the theoretical investigation of SEs were developed in collaboration between the Laboratory for Stirling Engines (LSE) at the Physical-Technical Institute of the Uzbek Academy of Sciences in Tashkent and the Centre for Computational Fluid Dynamics at the University of Leeds. At present these models are intensively used in the activities on the development of different types of SE at the Physical-Technical Institute.

The next stage of the continuing joint research is concerned with the development of CFD models for the description of the working process of a CC for SEs.

DESCRIPTION OF PROBLEM

Currently, the LSE is developing Micro Combined and Heat Power installations on the basis of a SE fuelled by natural gas. Clearly, both the power of the installation and its overall efficiency are directly influenced by the performance of the CC and therefore it is very important to correctly design the burner with the heater inside the CC in order to maximise the heat input into the internal circuit of the machine.

At the design stage it is useful to have visual information of the flow of combustion gases inside the burner and around the tubes of the heater, together with the temperature distribution in the CC. Additionally, it is very important to obtain data on the influence of the shape and dimensions of the burner, locations of air and fuel inlets, temperature and flow rates of the fuel and oxidizer on the performance of the burner. "Zero"- and "one"-dimensional mathematical models, which are used for the calculations of the combustion processes and the determination of the physical dimensions of the burner, assume the uniform distribution of the parameters of the working process in the internal space of the CC, and hence they are unable to provide the detailed information required for a successfully completion of this project. At present, although CFD modeling is in widespread use for the simulation of boilers and combustion chambers in numerous applications, there have not been any serious attempts to introduce CFD techniques into the design of the CC in Stirling Engines.

Therefore collaborative work has recently commenced between the LSE (Tashkent) and the CFD Centre (Leeds) on

411

the development of simple two-dimensional CFD models for the CC of Stirling Engines which could assist engineers to gain a much greater insight of the working process of the engine's burner. Although this work is in its initial stage some of the preliminary computational results, which have been obtained so far for simplistic cases, are illustrated.

GEOMETRY OF THE COMBUSTION CHAMBER AND THE CFD MODEL

In the first phase of the investigation it was decided to apply CFD modeling to the analysis of simplistic cases in which the designs of the CC have separate inlets for fuel and air. These separate inlets provided parallel flows of fuel and air in the entrance region of the burner. In addition, the reacting species were not premixed and there was no special swirl to mix the methane and air. Several different designs of the CC were investigated numerically and the results for two of these are presented in this paper. Figure 1 shows the physical dimensions of the first version of the burner's design in which the CC is of a conical shape with a central fuel inlet and annular air inlets, and the tubes of the engine heater are fitted in the inner space of the burner. Figure 2 presents the second version of this design with three annular fuel and air inlets. In this modification the fuel inlets are located in such a way that the combustion products are directed towards the tubes of the heater and all the walls of the CC are assumed to be perfectly insulated. The inlet velocities of the natural gas and air for both the design versions were 40 m/s and 0.5 m/s, respectively and the inlet temperatures of the gases were set at 300 K compared with a surface temperature of the tubes of 923 K.

The k-ε turbulent model for the incompressible fluid flow has been used to describe the combustion of methane in the axi-symmetric burner. The CFD modeling of the combustion of the methane is well established and is described in numerous scientific publications. In this model the governing equations describing the working process of the CC have been solved with the use of the control volume method. Whilst solving the governing equations it has been assumed that all the properties of the fuel and air mixture remain unchanged. Triangular elements have been employed for the discretization of the internal space of the burner and a typical computational grid used in performing the calculations is presented in Figure 3.

DISCUSSION OF RESULTS

Figures 4 and 5 show the velocity and temperature fields inside the burner, respectively, for the first design of the CC. Calculations were repeated over a wide range of values of the methane's inlet velocity (4-40 m/s) and there appeared no significant changes in the character of the flow in the CC. The tubes of the engine's heater remain virtually untouched by the flu gases and therefore there is negligible heat input in the SE. The temperature of the gases leaving the CC varies from 301 K

to 2520 K, with an average exit temperature of 768 K. Since the mixing of fuel and air in this design is not intensified there is a considerable quantity of unburnt fuel which leaves the CC, see Figure 6.

It can been seen from Figures 7 and 8 that in the second version of the design of the burner the heater's tubes are washed by the combustion products more intensively, although there is still not a complete motion of flu gases between the tubes. The reason being that the distance between the two rows of the tubes is too small and consequently, the local hydraulic resistance is large. The combustion gases flow mainly around the two rows of tubes and as such the character of the flow significantly reduces the magnitude of the heat transfer rate into the internal circuit of the machine. Figure 9 shows the temperature distribution of the gases for this design of burner and it can been observed that the majority of the heater's tubes remain within the low temperature zone. Computational results show that in this design also a considerable amount of unburnt fuel leaves the burner.

In general, the results obtained from CFD modeling show that the designs of a burner with parallel separate inlets for fuel and air (fuel not pre-mixed with air) will not provide the necessary rate of the heat input in SE and leads to substantial fuel losses.

FUTURE WORK

The development of the CFD modeling technique for the CC of a SE fuelled by natural gas is underway at the moment. In the near future it is planned to perform CFD analysis for different designs of CC fuelled by natural gas which is pre-mixed with air. In additional, several configurations of the CC with the different geometries of the heaters will be studied.

ACKNOWLEDGMENTS

This research work was partially supported by a NATO Science for Peace Grant (SfP- 972296) and an EU INCO-Copernicus Grant (Contract ERB IC15-CT98-0501).

REFERENCES

[1] K. Makhkamov and D.B. Ingham. A Two-Dimensional Model of the Heat Transfer and Gas Dynamics in a Cavity-Type Heat Receiver for a Stirling Engine as Used in a Solar Power Unit. ASME Journal of Solar Energy Engineering, Vol.121, pp. 210-216, November, 1999.

[2] K. Makhkamov and D.B. Ingham. Theoretical Investigations on the Stirling Engine Working Process. Proceeding of the 35th Intersociety Energy Conversion Engineering Conference "Energy and Power in Transition", Vol.1, pp. 101-110, 24-28 July 2000, Las Vegas, Nevada.

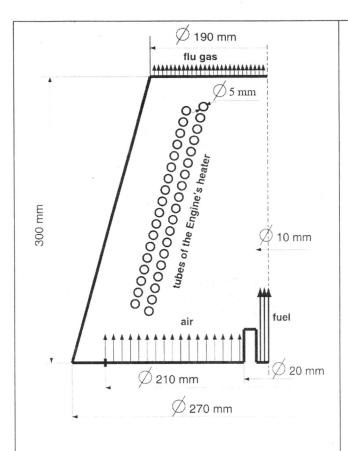

Figure 1. A schematic sketch of a Combustion Chamber with the fitted tubes of the Stirling Engine heater and with the central fuel inlet.

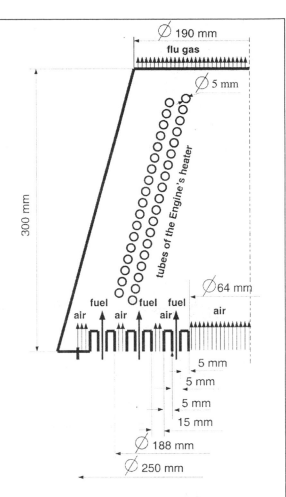

Figure 2. A schematic sketch of a Combustion Chamber with the fitted tubes of the heater and the annular fuel and air inlets.

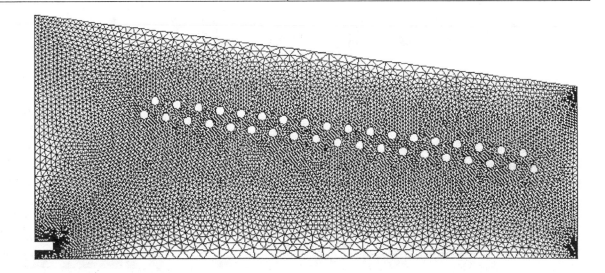

Figure 3. A typical computational grid for the internal space of a Combustion Chamber with the fitted tubes of the Stirling Engine heater and with the central fuel inlet.

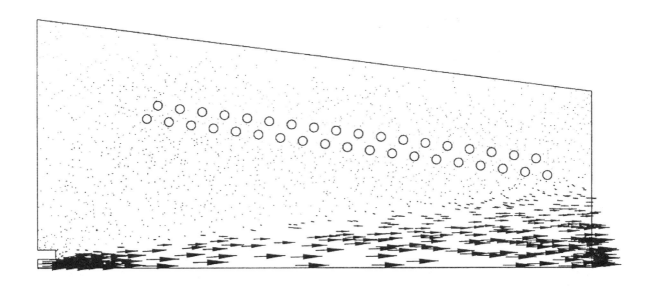

Scale: ⟶ 50.78 m/s

Figure 4. The velocity field in the Combustion Chamber (the first design version).

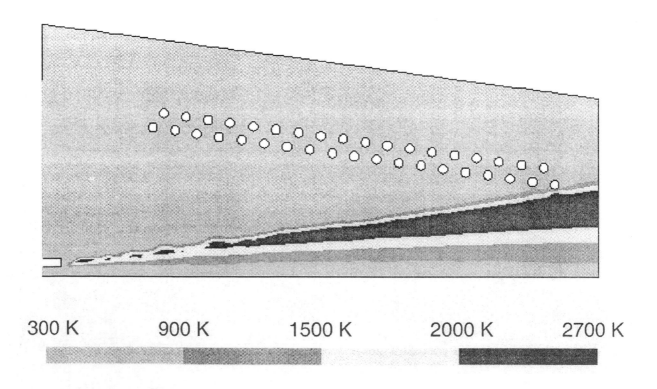

300 K   900 K   1500 K   2000 K   2700 K

Figure 5. The temperature distribution in the Combustion Chamber (the first design version).

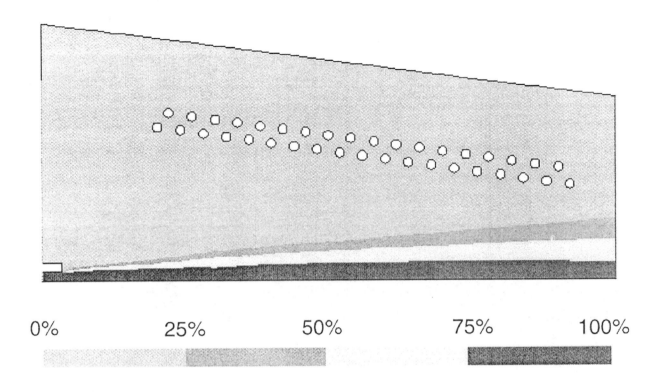

0%   25%   50%   75%   100%

Figure 6. The distribution of the methane mass fraction in the Combustion Chamber (the first design version).

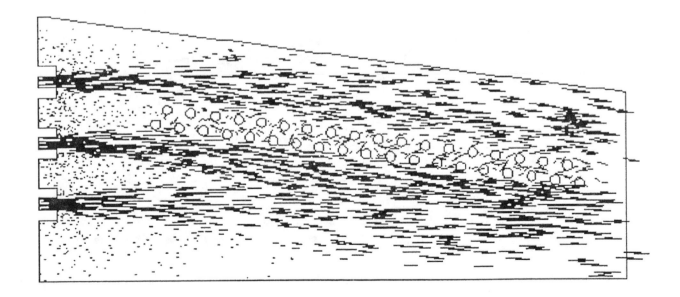

Figure 7. The velocity field in the Combustion Chamber (the second design version).

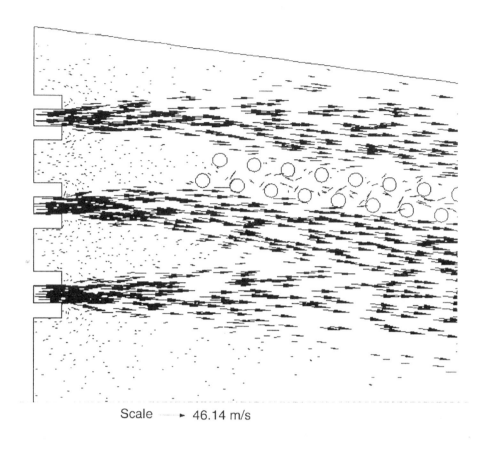

Scale ———► 46.14 m/s

Figure 8. The velocity field in the Combustion Chamber (the second design version).

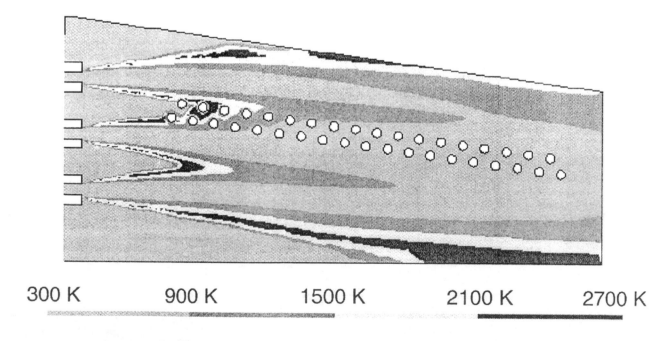

300 K   900 K   1500 K   2100 K   2700 K

Figure 9. The temperature distribution in the Combustion Chamber (the second design version).

Proceedings of IECEC'01
36th Intersociety Energy Conversion Engineering Conference
July 29-August 2, 2001, Savannah, Georgia

IECEC2001-CT-07

A STUDY OF STIRLING ENGINE WITH CIRCULAR DISK
HEAT EXCHANGER CONSISTED OF PIN-FIN ARRAYS

Seita Isshiki, Shuya Kamei, Akira Takahashi
Fukushima National College of Technology
Aza-Nagao 30, Taira Kamiarakawa, Iwaki-city, Fukushimaken, 970-8034, JAPAN
Naotsugu Isshiki
Professor Emeritus of Tokyo Institute of Technology
2-29-6 Kyodo Setagaya-ku, Tokyo 156-0052
JAPAN

ABSTRACT

In this study, we present the structure and numerically predicted results of a new Stirling engine with a circular disk heat exchanger consisting of pin-fin arrays. The heater structure comprises the circular disk, an annular shroud plate, and a connecting ring. Square pin-fin arrays are carved into the circular disk's face. An annular shroud plate upper is attached to these tips. A cooler of similar shape is located under the displacer cylinder. The cooler structure consists of an annular shape disk, a shroud, and a connecting ring. Pin-fin arrays are carved into the inner face of the disk, as in the heater mentioned above. The shroud plate is attached to the tips of the pin-fin. The type of proposed experimental engine is called a "β configuration." The bore diameter of the displacer and power pistons is 100 [mm] with a 60 [mm] stroke. The operating pressure is 500 [kPa]. We carried out simple numerical predictions of the engine using first-order approximation methods. Adjusted one-way constant velocity was used as the representative velocity of oscillating flow. We obtained the indicated power by an ellipsoidal approximation method. We also calculated the fluid resistance of each part but disregarded all mechanical friction losses. The output power is obtained by the aforementioned indicated power and fluid-resistance losses. Our calculations show the output power of the engine becomes 1.0 [kW] when using helium, with the temperature of 773 [K] on the heating side and 298 [K] on the cooling side. In conclusion, we believe the Stirling engine described in this paper is appropriate for use with solar and biomass fuel. Also, the engine does not need any heat exchanger pipes to be connected at all, unlike the conventional Stirling engine.

1. INTRODUCTION

Recently from the viewpoint of sustainable energy systems and preserving the environment of the earth, studies on the Stirling engine have become popular in Japan. In the present study, an improvement of the heater and cooler, which are important components of the Stirling engine, is proposed. We previously proposed a Stirling engine whose heater and cooler have pin-fin arrays. [1] Square pin-fin arrays are carved into the circular disk's face. The disk is situated on the upper part of the displacer cylinder. An annular shroud disk is attached to the tips of pin-fin arrays of the heater circular disk. A cooler of similar shape is located under the displacer cylinder. The cooler structure consists of an annular shape disk, a shroud, and a connecting ring. Pin-fin arrays are carved into the inner face of the disk, as in the heater mentioned above. The shroud plate is attached to the tip of the pin-fin. When the displacer moves downward in a β-type engine, the working gas from the cooling side of the displacer expands in the cooler disk through its pin-fin arrays to the outer edge. The working gas then passes up to the regenerator situated annularly around the displacer cylinder. After passing the regenerator, the working gas proceeds through the rear of the heater's shroud plate to its edge. Finally, the working gas contracts in the heater disk through its pin-fin arrays into the heating side of the displacer cylinder. A Stirling engine having fewer pipes and components with a compact structure can thus be realized. The higher heat transfer performance of flow in the pin-fin arrays enables the proposed Stirling engine to deliver almost the same performance as the conventional style Stirling engine.

2. Pin-fin Stirling engine
2.1 Shape of pin-fin

Figure 1 shows a projected view of the pin-fin arrays. By carving many rows of grooves in both the lengthwise and crosswise directions of a circular disk, pin-fin arrays whose cross section is square as shown in Fig.1 can be made. As for manufacturing the pin-fin arrays, a wire electrical discharge cutter can be utilized in addition to a milling machine. In the engine described here, square pin-fin arrays with thickness of 2 [mm], height of 10 [mm] and pitch of 4 [mm] are assumed. Stainless steel SUS304 is used for the heater

to withstand the temperature of 800 [K] and pressure of 1 [MPa]. Super duralumin A2024-T6 is used for the cooler because of its high material strength and high thermal conductivity at room temperature.

2.2 Structure of engine

Figure 2 shows the cross sectional view of the "β-configuration" Stirling engine proposed in this paper. The engine consists of heater, cooler, regenerator, displacer piston, power piston and crankcase. The crankcase situated in the lower space has a crank mechanism, which drives the displacer piston in advance of the power piston by a phase angle of 90 degrees. Table 1 shows the specifications of this engine. The structure of the heater consists of a circular disk having square pin-fin arrays, an annular shroud plate attached to the tips of the pin-fin, an annular flange, a connecting ring, and so forth. The connecting ring is welded to both the heater and annular flange pin-fin on the entire circumference. Adequate air-tightness and pressure-proofness up to 1 [MPa] can be obtained. As the shroud is connected to a displacer cylinder with a small fit allowance, internal gas leakage is very small even if the pin-fin disk expands due to heat and pressure. The structure of the cooler consists of a pin-fin annular disk made of super duralumin, an annular shroud, an annular flange and a connecting ring, which are all welded along their entire circumference. Thus the cooler has good air-tightness and is pressure-proof to 1 [MPa]. Next, we explain the mechanical motion of the engine. The power piston has guide rollers that move smoothly on the inside surface of the power cylinder so as to cancel the side thrust of the power piston crank. Also, the power piston has a linear bearing in its center in order to keep the motion of the displacer rod in a straight line, up and down exactly. The main shaft has a crankshaft, which is connected vertically to the power piston and horizontally to one side of a bell crank. The other side of the bell crank is connected to the displacer cylinder by a "C" shape rod. The main shaft drives a generator situated inside the pressurized crankcase.

2.3 Flow of working gas

When the displacer moves downward, the working gas from the cooling side of the displacer expands in the cooler disk through its pin-fin arrays to its outer edge. Here, the working gas cannot escape to outside of the pin-fin arrays as the cooler's circular shroud plate is attached to the tips of the pin-fin. Therefore, the working gas contracts in the rear side of the cooler's shroud plate to its inner edge. The working gas then passes up to the regenerator situated annularly around the displacer cylinder. The temperature of the working gas rises as the gas receives heat energy from the regenerator. After passing the regenerator, the working gas expands in the rear of the heater's shroud plate to its edge. Finally, the working gas contracts in the heater's disk through its pin-fin arrays into the heating side of the displacer cylinder. Here, the temperature of

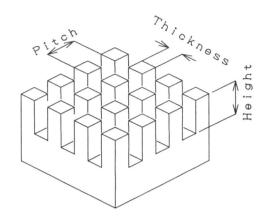

Figure 1 Projected view of pin-fin arrays

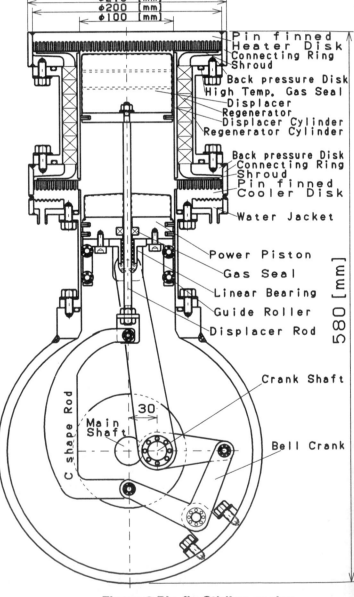

Figure 2 Pin-fin Stirling engine

the working gas rises as the gas receives heat energy from the heater's pin-fin arrays. In the case that the displacer moves upwards, the flow pattern of working gas is simply the reverse.

3. Numerical analysis

We carried out numerical analysis using the first-order approximation method. We considered indicated output power, fluid resistance losses, heat transfer of heater, cooler, regenerator and direct heat conduction losses, but disregarded all mechanical friction losses.

3.1 Effective velocity of oscillating flow

We used steady one-way flow speed as a representative flow velocity of oscillating flow phenomena. Assuming the phenomena are proportional to the n-th power of flow speed, the ratio of effective velocity to the maximum velocity during oscillation can be described by Eq.(1). Here, the flow is assumed to be exactly sinusoidal oscillatory flow.

$$\frac{u_{eff}}{u_{max}} = \sqrt[n]{\frac{1}{\pi}\int_0^\pi \sin^n x\, dx} \qquad (1)$$

Figure 3 shows the value of this ratio and its relation to power number n. In the calculation of flow resistance, n=2 as the dynamic pressure is proportional to the square of flow speed. In this case, the ratio becomes 0.707 from Fig.3. In the calculation of convective heat transfer ratio, the value of 0.60 was used as n=0.6 to 0.8 because the Nusselt number of flow inside the pin-fin arrays is proportional to the 0.63th power of Reynolds number from the study described later, and the Nusselt number of flow through the regenerator mesh is proportional to the 0.63th power of Reynolds number from the previous study [2].

3.2 Fluid friction factor and Nusselt number of flow through pin-fin arrays

Figure 4 shows the previous data for the fluid friction factor of flow inside pin-fin arrays and crossflow over the bank of circular cylinders. The horizontal axis denotes the Reynolds number whose length scale is a diameter (the perimeter in the square case) and velocity scale is a maximum velocity u_{max} in the smallest flow spacing area of pin-fin arrays. The vertical axis denotes fluid friction factor f defined as Eq. (2).

$$f = \frac{\Delta p}{2\rho u_{max}^2 N} \qquad (2)$$

where fluid friction loss is non-dimensionalized by number of rows N, fluid velocity u_{max} and fluid density ρ. In Fig.3, the solid line shows data of crossflow over the bulk of circular cylinders arranged in line by Bergelin et al. [3]. Here, stream-wise and span-wise pitches are 1.5 times of cylinder's diameter. The dotted line shows the empirical equation of Jacob [4] for the same arrangement of cylinders at the Reynolds number range of 4,000<Re<32,000. □ and ■ marks

Table 1 Specifications

List	Value
Heater disk temperature	773 [K]
Cooler disk temperature	298 [K]
Displacement volume	471 [cc]
Expansion Volume	471 [cc]
Bore diameter	100 [mm]
Heater disk diameter	200 [mm]
Stroke	60 [mm]
Regenerator matrix	#30 mesh
mesh number, sheets	360 sheets
number, wire diameter	ϕ 0.14 [mm]
Dead volume ratio	a=1.413
Overlap volume ratio	a'=0.345
Mean pressure	500 [kPa]
Working gas	Air, Helium

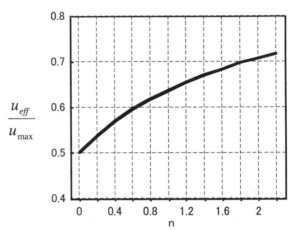

Figure 3 Ratio of effective velocity to maximum velocity

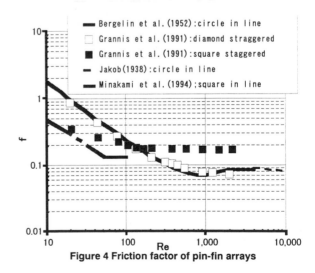

Figure 4 Friction factor of pin-fin arrays

show data of rhombus shape pillars in a staggered arrangement by Grannis et al. [5], denoting diamond shape pillars and square shape pillars, respectively. The dotted dashed line shows the empirical equation of square pin-fin arrays arranged in line whose stream-wise and span-wise pitches are both 3 times the perimeter, of Minakami et al. [6]. From Fig.4, it is clear that the fluid friction factor of circular cylinder's arrays decreases as Reynolds number increases and reaches a constant value over Re>500. On the other hand, in the case of square shape pin-fin arrays, the friction factor seems to be constant over Re>50 except the diamond shape case. Thus we determined the fluid friction factor of square pin-fin arrays irrespective of Reynolds number as f=0.17.

Figure 5 shows Nusselt number of pin-fin arrays. Here, Prandtl number is arranged as Pr=0.71. In Fig.5, the solid line shows the data of Bergelin et al. [3], and the dotted line shows the empirical equation of Zukauskas [7] in the case of crossflow over circular cylinders arranged in line, in the range of 10<Re<100 and 10^3<Re<2·10^5. The dashed dotted line shows data of square pin-fin arrays' by Minakami et al. [6]. Nusselt number of pin-fin arrays seems to increase gradually as Reynolds number increases. Also, Nusselt number of square pin-fin arrays agrees well with that of circular pin-fin arrays in the region of Re<100. However, there have been no any previous studies concerning Nusselt number of square pin-fin arrays in the region of Re>100 from our survey of the literature [8]. Thus, in the present study, we used data of Bergelin et al. [3] in the range of 10<Re<4,000. and also used the experimental equation of Zukauskas [7] in the range of Re>4,000. Also from the previous study, Nusselt number of pin-fin arrays is proportional to 0.33 to 36th of Prandtl number. Thus we determined that Nusselt number is proportional to one-third of Prandtl number.

3.3 Method of calculation

First, we calculated the indicated power of this engine by the elliptical approximation method of Isshiki [9]. Second, we calculated fluid friction losses of the heater, cooler, regenerator and other gas passages. We defined engine output power as the difference between the indicated power and the fluid friction losses. We calculated the heat balance of this engine, considering engine efficiency, heat loss of regenerator, direct heat conduction loss of cylinders' wall and heat transfer efficiency of pin-fin. We obtained the efficiency of the engine by dividing the engine output power by the sum of the aforementioned heat losses and input heat energy.

Fluid friction losses in the heater and cooler were calculated by the following equation.

$$\Delta p = \int_{r0}^{r1} 2\rho \cdot f \cdot u_0^2 \frac{r_0^2}{r^2} dr \qquad (3)$$

where, r is the radius position from the center of the circular disk, r_0 is the inner radius of the disk, r_1 is the outer radius of the disk, and u_0 is effective gas velocity in the smallest spacing of pin-fin arrays at the position of $r=r_0$, respectively. The value of f was determined as f=0.17. The heat resistance of the circular disk shape heater and cooler was obtained as the sum of the heat resistance of the base

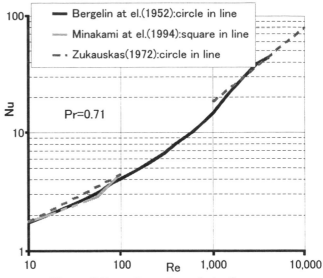

Figure 5 Nusselt number of pin-fin arrays

plate and heat resistance between pin-fin arrays and working gas. The heat resistance of the base plate was obtained by dividing the thickness of the base plate by the heat conductivity ratio of the plate. The heat resistance between pin-fin arrays and working gas was obtained by repeating the calculation of the following Eqs. (4) to (7) until all values converged. In the calculation, the heat transfer ratio of pin-fin arrays and the heat transfer efficiency of pin-fin were considered. The heat resistance between pin-fin arrays and working gas, R_2, is given by Eq. (4):

$$\frac{1}{R_2} = \int_{r0}^{r1} \eta_{fin} Nu(\text{Re}) \frac{\lambda}{d} \sigma 2\pi r dr \qquad (4)$$

where, d is the perimeter of square pin-fin, λ is the heat conductivity of working gas, Nu(Re) is Nusselt number corresponding to Reynolds number at the position r of pin-fin arrays, η_{fin} is heat transfer fin efficiency, σ is the surface area ratio of pin-fin arrays and Reynolds number is given by Eq. (5):

$$\text{Re} = \frac{u_0 d}{\nu} \frac{r_0}{r} \qquad (5)$$

where ν is the dynamic viscosity coefficient of the working gas. The index of fin efficiency is given by Eq. (6):

$$u_b = 2\frac{l}{d}\sqrt{\frac{\lambda}{\lambda_{fin}} Nu(\text{Re})} \qquad (6)$$

where 1 and λ_{fin} are length of pin-fin and heat conductivity coefficient of pin-fin, respectively. Fin efficiency is given by Eq. (7):

$$\eta_{fin} = \frac{\tanh(u_b)}{u_b} \qquad (7)$$

4. Consideration of the results of calculation

Figure 6 shows the calculated results of engine output power. □ and ○ marks show the results for air and helium, respectively. The maximum output power becomes 486 [W] in the case of air at the engine speed of 1300 [rpm]. In the case of helium, it becomes 1013 [W] at the engine speed of 2700 [rpm]. This maximum output power agrees well with the empirical cubic power law of Kolin [10] at the temperature difference of 440 [K]. Also, the free rotational engine speed becomes 2200 [rpm] in the case of air, and 4900 [rpm] in the case of helium. Both the maximum output power and free rotational speed in the case of helium are about twice that of the case of air. This result supports the view that the performance of the Stirling engine is generally inversely proportional to the square root of the molecular quantity of working gas proposed by Isshiki [9]. Figure 7 shows the calculated results of thermal efficiency of this engine. □ and ○ marks denote the results for air and helium, respectively. The maximum thermal efficiency becomes 35[%] at the rotational speed of 700 [rpm] in the case of air, and becomes 43[%] at the rotational speed of 900 [rpm] in the case of helium.

The heater and cooler in the conventional Stirling engine usually have a large number of very complicated tubes. We compared the fluid friction and heat transfer of the proposed pin-fin heat exchanger with the conventional heat exchanger having nearly the same surface area. Both the compared heater and cooler consist of one hundred circular pipes made of stainless steel (SUS304) with 100 [mm] length of 3 [mm] inner diameter and 4.5 [mm] outer diameter.

Figure 8 shows the results of fluid friction. Here, ● and ■ marks show the case of pin-fin arrays, and ○ and □ marks show the case of circular cylinders, respectively. The horizontal axis denotes the rotational speed of the engine and the vertical axis denotes the pressure drop. The pressure drop due to the pin-fin arrays increases linearly. The pressure drop due to the circular cylinders is higher than that due to the pin-fin arrays at the rotational speeds of under 800 [rpm]. The reason seems to be because flow inside the tubes is laminar flow at lower engine speed, resulting in higher friction factor than flow through the pin-fin arrays. The pressure drop of the heater due to the circular cylinders seems to be similar to that due to the pin-fin arrays at rotational speeds of over 1000 [rpm]. This reason seems to be because flow inside the tubes becomes turbulent, resulting in higher friction factor. The pressure drop of the cooler due to the circular cylinders is nearly the same as that due to the pin-fin arrays.

Figure 9 shows the calculated results of heat transfer resistance. Here, ● and ■ marks show the case of pin-fin arrays, and ○ and □ marks show the case of circular cylinders, respectively. The heat transfer resistance of both the heater and cooler due to the pin-fin arrays decreases linearly as the rotational speed increases. In the case of the pin-fin arrays, the heat transfer resistance of the cooler is clearly smaller than that of the heater. This is considered to be because the cooler in the pin-fin arrays is made of super duralumin which has a very high heat conductivity. The heat transfer resistance of the heater due to the circular cylinders is higher than that of the pin-fin arrays at rotational speeds of under 800 [rpm] because the flow inside the circular cylinders becomes laminar flow. When the rotational speed exceeds 1000 [rpm], the heat transfer resistance of the heater due to the circular cylinders is lower than that due to pin-fin arrays because the flow inside the circular cylinders becomes

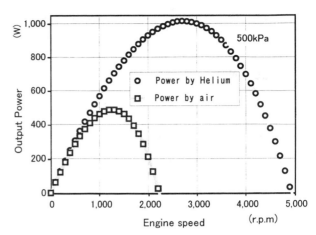

Figure 6 Output power of pin-fin Stirling engine

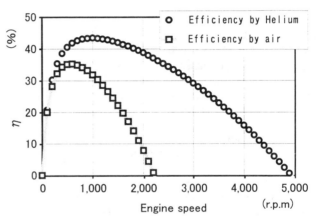

Figure 7 Thermal efficiency of pin fin Stirling Engine

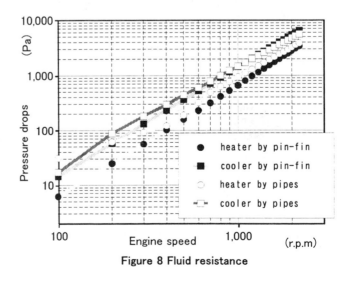

Figure 8 Fluid resistance

turbulent, resulting in higher heat transfer ratio. The heat transfer resistance of the pin-fin arrays of the cooler is clearly lower than that of the circular cylinders.

5. CONCLUSION

(1) A Stirling engine with circular disk heat exchanger consisting of pin-fin arrays is easy to manufacture at low cost as the structure of the engine is simple.

(2) Our calculations show that the output power of the present engine becomes 1.0 [kW] when using helium with the temperature of 773 [K] on the heating side and 298 [K] on the cooling side.

(3) A comparison with the conventional heat exchanger shows that the fluid friction losses of the proposed heat exchanger consisting of pin-fin arrays are smaller than those of the conventional heat exchanger consisting of circular cylinders. Also, the heat transfer performance of the present heat exchanger is superior to that of the conventional heat exchanger consisting of circular cylinders.

(4) We believe the Stirling engine described in this paper is appropriate for use with solar and biomass fuel as the shape of the heater is a circular disk.

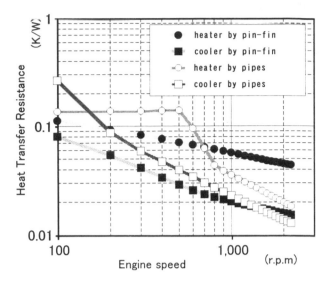

Figure 9 Heat transfer resistance

REFERENCES

1. Isshiki, S., and Isshiki, N., 2000, "Stirling cycle machine," Japanese patent application, No. 2000-205453., Japan

2. Tanaka, M., Yamashita, I., and Chisaka, F., 1989, "Flow and heat transfer characteristics of Stirling engine regenerator in oscillating flow," (in Japanese), Trans. JSME, series B, Vol. 55, No. 516, pp. 2478- 2485., Japan

3. Bergelin, O. P., Brown, G. A., and Doberstein, S. C., 1952, "Heat transfer and fluid friction during flow across banks of tubes," ASME, Vol. 74, pp. 953- 960.

4. Jacob, M., 1938, "Heat transfer and flow resistance in cross flow of gases over tube banks," ASME, Vol. 60, pp. 384- 386.

5. Grannis, V. B., and Sparrow, E. M., 1991, "Numerical simulation of fluid flow through an array of diamond-shaped pin fins," Numerical Heat Transfer, series A, Vol. 19, pp. 381- 403.

6. Minakami, K., Mochizuki, S., Murata, A., Yagi, Y., and Iwasaki, H., 1994, "Heat- transfer characteristics of pin-fin arrays," Trans. JSME, series B, Vol. 60, No. 569, pp. 255- 262.

7. Zukauskas, A., 1972, "Heat transfer from tubes in cross flow," Advances in Heat Transfer, Vol.8, pp. 93- 160.

8. Survey in 2000 year by Japan Science and Technology Corporation, Japan

9. Isshiki, N., 1982, "Development on Stirling engine," Kogyo Chosakai Publishing Co., p.48, 70. (in Japanese), Tokyo, Japan

10. Kolin, I., 1995, "Thermodynamic theory for Stirling cycle machine designs," Proceedings on 7th International Conference on Stirling Cycle Machines, sponsored by JSME, Tokyo, Japan, pp.1- 6.

Proceedings of IECEC'01
36th Intersociety Energy Conversion Engineering Conference
July 29-August 2, 2001, Savannah, Georgia

IECEC2001-CT-19

A DISCUSSION FOR THE MSR – STIRLING POWER SYSTEM

L. Berrin Erbay

Osmangazi University
School of Engineering And Architecture, Eskisehir, Turkey
Phone: 90-222-2303972, Fax: 90-222-2213918, e- mail: lberbay@ogu.edu.tr

ABSTRACT

In this paper, the analysis of the alternative combinations for conceptual design of a power plant consisting of a Molten – Salt Nuclear Reactor (MSR) and Stirling Engine has been presented. The combination of these two systems, each having different characteristics has been supplied by a heat exchanger. The design constraints and particular problems of the present power system have been evaluated.

INTRODUCTION

MSRs are thermal fission reactors with fluid thorium based nuclear fuel. Small Molten Salt Reactors (MSR) have been proposed by Furukawa [1990] in the THORIMS - NES (Thorium Molten Salt Nuclear Energy Synergetics) system and studied by Furukawa et all [1990], Furukawa and Lecocq [1990]. THORIMS - NES is a new philosophy depending on the principles: Thorium breeding cycle, molten - fluoride technology and separation of fissile and power producing plants. In the standard designs of MSRs proceeded in THORIMS - NES, fuel has self - sustaining characteristics without continuous fuel processing and the classical steam cycle with a steam turbine is considered for the heat removal system. Considering the fact that when a small MSR is substituted as a continuous high temperature heat source, the possibility of power generation by a combination of MSR and Stirling Engine has been introduced and preexamined by Erbay[1999]. In her study, the steam - turbine has been replaced with Stirling heat engine and the hot molten - salt fuel transfers heat through a heat exchanger which has constituted the heater (hot - end head) of the Stirling engine. She has assumed that the flexibility in size of MSRs can make the use of a Stirling engine possible instead of the salt / steam cycle and hence the combined system can yield a prosperous closed power producing system.

In this study, the alternative combinations of the MSR with Stirling engine have been presented and examined. By considering the liquid – coupled indirect – type exchanger system has been preferred and a parallel connection of the Stirling engines has been introduced. The particular problems of the power system have been shown considering the requirements of a conceptual design.

THE MSR – STIRLING POWER SYSTEM

In the MSR – Stirling power system, the molten – salt ^{7}LiF-BeF$_2$ (Flibe) based fuel which is molten ^{7}LiF-BeF$_2$-^{232}ThF$_4$-^{233}UF$_4$ exits the reactor at around 1000 K and is pumped to the heat exchanger. The inlet temperature of the fluid fuel to the reactor is around 850 K. Molten ^{7}LiF-BeF$_2$-^{232}ThF$_4$-^{233}UF$_4$ is chemically inert, has low vapor pressure, moderate viscosity and thermal conductivity, and high heat capacity in fairly high working temperature (750 - 1100 K) and promises high safety and economy assurance. The use of a circulating fluid fuel in the form of a molten – salt may simplify the heat transfer problem within the reactor itself but has its own particular problems both in the reactor and the flowing system. Because of the intense radioactivity of the fluid fuel the design of the heat transfer system attracts more attention. The determination of the temperature distribution in the flowing system including heat exchangers and at the walls of the pipes has unusual aspects of heat transfer and fluid flow analysis.

LIQUID – COUPLED INDIRECT – TYPE EXCHANGER SYSTEM

Due to the intense radioactivity of fission products in the circulating fluid fuel, a liquid – coupled indirect – transfer – type exchanger system becomes necessary. Besides the disadvantages such as greater total heat transfer area requirement and some complications of the coupling – liquid circuit, this system has important advantages. The principle advantages, stated by Kays and London [1984], are : 1) Since the hot – fluid flow area is not coupled directly with the cold - fluid flow area a less awkward heat exchanger shape may result, particularly if there is as much as a 6:1 disparity in flow densities, and 2) the liquid coupling will allow for a better and more compact machinary arrengement for gasses. In the present case, the disparity in flow densities at the reactor side is

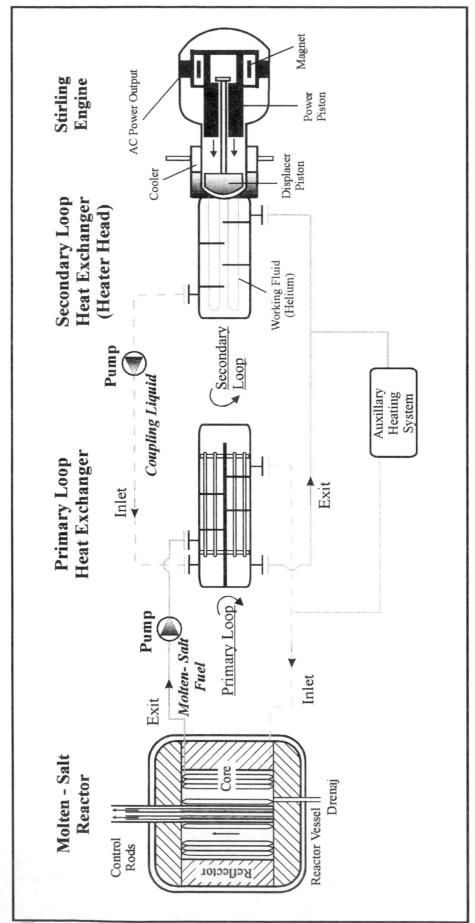

Figure 1. A Schematic fo the power plant of MSR - Stirling combination

about 1.6 but is worse at the Stirling side due to the density of gas working fluid.

The liquid – coupled indirect – transfer – type exchanger system has been preferred in this study. The system consists of two direct type exchangers coupled together by the circulation of a satisfactory heat transfer medium. As it is seen in the Figure 1, two loops are obtained. In the primary loop there are a nuclear reactor, a primary loop heat exchanger, a circulation pump and an auxiliary heating system for preventing the solidification of fuel at an accidental case. Beside the primary loop heat exchanger, there is another direct - type heat exchanger in the secondary loop which is the heater - head of the Stirling engine and a liquid – gas type exchanger. There are also a circulation pump and an auxiliary heating system in the secondary loop.

THE COUPLING LIQUID AND COMPACT HEAT EXCHANGERS

The coupling liquid flowing in the secondary loop between the heat exchangers of primary and secondary loops has been chosen as Flibe. The coupling liquid in the secondary loop is pumped for circulation between the hot – molten - salt fuel exchanger and the working fluid of Stirling heater – head type exchanger. The volume of the fluid fuel heat exchanger in the primary loop must be small in order to prevent the critical mass hold – up, where as the heat transfer area of the exchanger must be large. On the other hand, the heater - head exchanger for the Stirling engine has a gas flow, therefore there are effectiveness problems. The gas – side controls the heat transfer. In gas flow heat exchangers low mass velocities are arranged due to the friction problem. Additionally, low thermal conductivities of gasses result in low heat transfer rates per unit area which in turn, requires large surface area for the Stirling heater – head type exchanger. Therefore, Compact Heat Exchangers (CHXs) characterized by large heat transfer area per unit volume of exchanger have been designed for both of the exchangers.

As explained by Shah [1983]; various techniques are emploid to make heat transfer surfaces compact. Such compact surfaces have fins between plates, finned circular tubes, or densely packed continious or interrupted cylindrical flow passages of various shapes. The two imcomparable characteristics of compact heat exchanger can be given as the availability of many surfaces having different orders of magnitude of surface area density and flexibility in distributing the area on the hot and the cold sides.

The reactor – side CHX presents a sizing problem considering the fuel properties and hold – up. The Stirling side exchanger constitues a very important design problem in comparison with the other one since the operating conditions of Stirling engines are found as laminar oscillating flow. The laminar oscillating pipe flow field and the heat transfer calculations necessary in the heat exchangers of Stirling engines have been executed by Walter et all.[1998] and also theoretically analized by Lee et all.[1998]. Organ [1995] applied the Method of Characteristics to the pulsatile flow of the gas circuit of the Stirling engine. The effect of oscillation on the design and the steady operation of the the primary loop has to be studied in detail.

MATERIAL PROBLEM

The materials of the flowing system, pipes and the heat exchangers are chosen by considering nuclear properties as well as heat transfer and mechanical ones. Due to the radioactivity of the fluid fuel and the fission products, the heat exchanger must be made out of corrosion resistant materials and be leakproof. Therefore, the heat exchangers are constructed from modified Hastelloy-N [Ni(15-18%) - Mo(6-8%) - Nb(1%) - Cr] which has been found to a satisfactory material and is compatible with molten - fluorides, easy manufacturable, weldable, high temperature resistant (up to 850 C) Furukawa et all [1992]. The inside of heat exchangers can be cladded by stainless - steel to prevent any salt leakage.

THE STIRLING ENGINE TYPE

The Stirling engine staying in the present power system is considered as a free – piston type working at 950 K. Stirling engines are known as the most efficient devices for converting heat into mechanical work or electric current as preferred in this study. They operate quietly, works on the principle of closed operating chambers, necessities long life designs with minimum maintenance and requires high temperatures for high efficiencies. Further improvement of the Stirling engines are currently being under taken for achieving less weight, more compactness, longer life, higher power level and efficiency. Beck [1999] has given a report over viewing of the state of development, present and future applications and noted that the future prospects for Stirling engines are better than ever before and the engine can be used efficienctly as refrigerator and cogeneration system, as well as a driving unit or even for the direct conversion of solar energy into electrical energy. In the literature, Stirling engines from the micro levels [Fukui et all,1999] to the 150 kWe power designs for lunar base applications [Mason et all,1989] have been found, which strongly supports the possibility of a new design completely applicable for the present system. The operation temperature of the Stirling engine bases on the heat rejection system. The thermal - to - electric conversion efficiency depends on the temperature ratio between the temperature of the hot source to the heat rejection temperature. The design and operation of the heater head require serious developments which completely effects both sides of the proposed combined system.

THE PARALLEL CONNECTION OF THE STIRLING ENGINES

Beside the expectations about small scale nuclear power production by MSR and more powerful Stirling engines in the future, the power levels of both sides in the present case are very different. The power inconsistency can be solved by coupling the primary loop to a number of Stirling engines to supply sufficient heat removal from the nuclear reactor through the primary loop heat exchanger. The case is explained as a second alternative including the parallel connection of the Stirling engines which is shown in Figure 2, schematically. Due to the parallel instrumentation, the effects of oscillation of the gas flow in the heater - head exchangers can be compensated and a steady non - oscillatory flow is obtained. A high mass flow rate is obtained by using parallel connection although a more complex installation system is necessary.

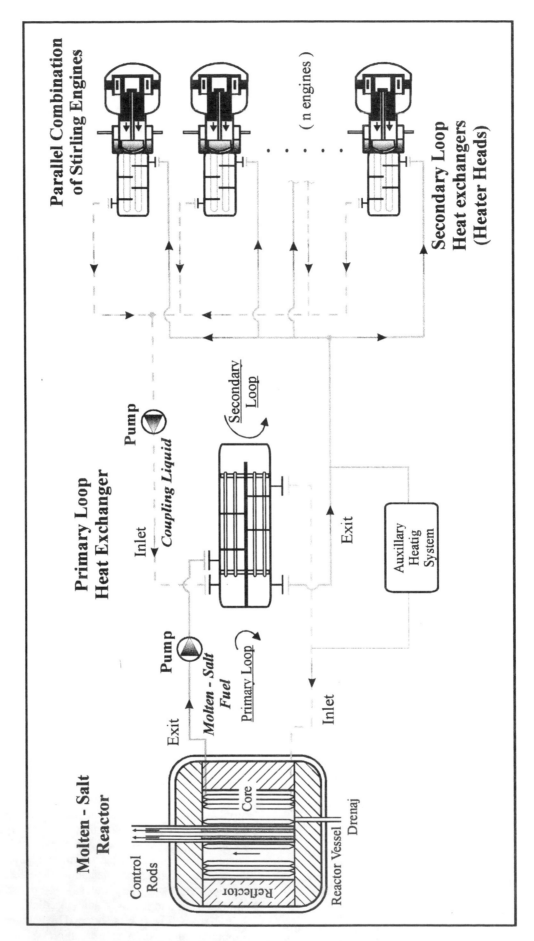

Figure 2. A Schematic of the present system with parallel Stirling Engines.

426

CONCLUSION

The liquid – coupled Indirect - type of heat exchanger is considered and evaluated to put forward of a new construction answering the requirements for the conceptual proposal. It is obvious that the present power system not only carries the excellent characteristics of the thorium molten - salt breeding cycle but also directly imports the advantages of the Stirling engine and improvements in the huge Stirling technology. As a result of the conceptual design study, it is obtained that the system has particular problems to be studied in detail. These are: 1) The oscillatory nature of the flow in the heater - head effecting the flow of the coupling liquid in the secondary loop; 2) Radioactivity of the fluid fuel flowing in the primary loop; 3) Heat exchanger sizing for preventing the hold - up problem; 4) Heater - head design as a heat exchanger for the flows of a molten - salt (Flibe) and a gas working in the engine; 5) The present power limit of Stirling engines; 6) Uranium - 233 production as a fuel for the Molten - Salt Nuclear Reactor.

REFERENCES

Beck P, 1999, The latest information about Stirling engines – The present satate of development and changes for the future, *Brennstoff – Warme – Kraft*, 51: (9) 38

Furukawa, K., 1990, Rational Restart of Safer Fission Energy Utilization Depending on the Small Thorium Molten - Salt Power Stations, *Proc. Of Technical committee Meetings and Workshops Organized by International Atomic Energy Agency*, Vienna, IAEA - TECDOC - 541, pp. 287 - 298.

Furukawa, K , Lecocq A, Kato Y and Mitachi K, 1990, Thorium Molten - Salt Nuclear Energy Synergetics, *Journal of Nuclear Science and Technology*, Vol.27, No. 12, pp. 1157 - 1178.

Furukawa, K. and Lecocq, A., 1990, New Safe Nuclear Energy for the Next Century - Thorium Molten - Salt Nuclear Energy Synergetics-, *Proc. Of the Florence World Energy Research Symposium: A Future for Energy FLOWERS'90*, 28 May - 1 June, Italy, Edited by Sergio S. Stecco and Michael J. Moran, Pergamon Press, pp.89 - 102.

Furukawa, K., Mitachi, K and Kato, Y., 1992, Small Molten - Salt Reactors with a Rational Thorium Fuel - Cycle, *Nuclear Engineering and Design*, 136, pp.157 - 165.

Fukui T, Shiraishi T, Murakami T and Nakajima N, 1999, Study on high specific power micro – Stirling engine, *JSMA Int. J. Series B- Fluids and Thermal Eng.*, 42: (4) 776-782.

Kays W.M. and London A.L., *Compact Heat Exchangers*, 3th ed., McGraw-Hill Book comp.,1984.

Lee DY, Park SJ, Ro ST, Heat transfer in the thermally developing region of a laminar oscillating pipe flow, *Cryogenics*, 38: (6) 585 – 594, 1998.

Mason LS, Bloomfield HS and Hainley DC, 1989, SP-100 power system conceptual design for lunar base applications, *NASA Lewis Research Center Group, NASA TM 102090*.

Organ A.J, 1995, Gas dynamics and Stirling engines, *IECEC paper no.SC – 256*, pp 465-474, ASME

Shah R.K., Classification of Heat Exchangers, published in *Low Reynolds Number Flow Heat Exchangers* edited by S.Kakaç, R.K. Shah and A.E. Bergles, Hemisphere Pub.Corp., pp.9-14,1983.

Walter C, Kuhl HD, Pfeffer T, Schultz S, Influence of developing flow on the heat transfer in laminar oscillating pipe flow, *Forschung im ingenieurwesen – Engineering Research*, 64: (3) 55 – 56, 1998.

Proceedings of IECE '01:
Intersociety Energy Conversion Engineering Conference
July29-August 2, 2001, Savannah, Georgia

IECEC2001-CT-26

UPDATE ON THE STIRLING CONVERTOR TESTING AND TECHNOLOGY DEVELOPMENT AT NASA GRC

Jeffrey G. Schreiber
Thermo-Mechanical Systems Branch
NASA Glenn Research Center
21000 Brookpark Rd.
Cleveland, OH 44135
USA
216 433-6144 jschreiber@grc.nasa.gov

Lanny G. Thieme
Thermo-Mechanical Systems Branch
NASA Glenn Research Center
21000 Brookpark Rd.
Cleveland, OH 44135
USA
216 433-6119 lanny.g.thieme@grc.nasa.gov

ABSTRACT

The Department of Energy (DOE), the Stirling Technology Company (STC), and the NASA Glenn Research Center (GRC) are developing a free-piston Stirling power convertor for an advanced radioisotope power system to provide spacecraft electric power for NASA deep space missions. A Stirling Radioisotope Power System has recently been identified as a candidate for potential use on a variety of fly-by or orbiting missions, and also for some surface power applications. The efficiency of the Stirling system is well in excess of 20% and will reduce the isotope inventory required by a factor of three or more when compared to the Radioisotope Thermoelectric Generators (RTGs) that have been used previously for deep space missions.

NASA GRC is conducting an in-house technology project to assist in developing the free-piston Stirling power convertor for space qualification and mission implementation. The technology project has many facets; this paper will highlight some of the key areas that have produced some of the more recent accomplishments. These areas include performance mapping of the Stirling convertor over a wide range of operating conditions, and a series of tests in the GRC Structural Dynamics Laboratory (SDL) to study operation of the power convertor when subject to launch environments. An additional test performed in the SDL was intended to study the dynamic signature of a pair of operating convertors. This paper will provide an update of the status, the plans for the future, and some results from these efforts.

INTRODUCTION

The candidate Stirling Radioisotope Power System (SRPS) is based on the use of the Technology Demonstration Convertor (TDC) that is being developed by STC under contract to DOE. The primary reason for interest in the SRPS is the substantially higher conversion efficiency when compared to the Radioisotope Thermoelectric Generators (RTG) that have been used on many of the past deep space missions. Compared to the RTG, the SRPS offers modest mass savings; the great benefit is in the cost savings provided in the reduced inventory of plutonium in the heat source, which is in limited supply. At this time, the TDC has been proven to be a high efficiency and rugged power conversion device and now must be transitioned to flight.

The TDC was originally designed as a laboratory device and is now being developed for flight [1]. To support the transition to flight, GRC has initiated a multi-faceted technology development project. Some of the areas included in this project are efforts to study structural life, material joining issues, aging of the permanent magnets, Electromagnetic Interference (EMI) and Electromagnetic Compatibility (EMC), linear alternator magnetic analysis, radiation tolerance, and reliability. In addition to these areas are Independent Validation and Verification (IV&V) of the performance, and a study of the ability of the convertor and power system to survive the dynamic launch environment. Many of these areas have been presented in earlier publications [2].

To support the IV&V effort, a Stirling Research Laboratory has been established at GRC. Figure 1 shows two of the free-piston Stirling test stands. Two pairs of TDCs have been received and operated to date. The first two convertors are known as Serial Numbers (S/N) 5 and 6. These convertors have undergone performance testing and have also been used in the vibration tests that will be highlighted later in this report. A second pair, known as Serial Numbers 7 and 8 was received later. These two convertors have been operated, however full performance testing has not been initiated.

Figure 1. – The Stirling Research Laboratory at GRC.

Prior to 1999, it was unclear if an operating free-piston Stirling power convertor would be able to survive the dynamic loads imposed during launch. Due to the nature of the heat source, a free-piston Stirling power convertor in a radioisotope power system would likely need to be operated at or near full power level during launch. As part of the DOE/NASA Stirling Technology Readiness Assessment performed in 1999 [3], GRC tested a single TDC operating at full power while being subjected to vibration levels beyond those used for any previous radioisotope power system. The test results showed that a Stirling convertor could survive the launch loads with only a temporary reduction in power output, and that the power output would return to normal following the vibration test. A post-vibration test inspection of the TDC components showed no signs of any damage incurred during the vibration test. This breakthrough vibration test was the first of a series of vibration tests in the technology project. Since that initial test, there have been three other tests in the GRC SDL. The first of these tests was intended to measure the dynamic signature generated by a pair of operating TDCs. The second test was the Modal Test, which used a series of accelerometers and optical probes to study the responses and deflections of components of the linear alternator when operated and subject to vibration. The third test was a vibration test with a pair of Stirling convertors mounted on an Engineering Test Structure. The purpose of this test was to study the interaction between the operating Stirling convertors and the random vibration when coupled with relatively soft and stiff mounting structures.

PERFORMANCE TESTS

To assist DOE as the TDCs are transitioned from laboratory devices to flight units, GRC is performing an IV&V effort. Two test stands have been assembled at GRC for this purpose. Each test stand consists of a rack with data system and other equipment associated with the operation of the Stirling convertors, a circulator/chiller used to remove the waste heat generated by the convertors, and a test stand to

which the convertors are mounted. Figure 2 shows TDCs S/N 7 and 8 mounted on the test stand. The rack and circulator/chiller have been designed to be portable to support testing at other facilities such as the SDL and the EMI laboratory.

Figure 2. – TDCs Number 7 and 8 in GRC test stand.

Acceptance Test

Following delivery of TDCs S/N 5 and 6 in August of 2000, the convertors were subject to an acceptance test. The purpose of this test was to verify that the convertors operated at the GRC Stirling laboratory with performance equivalent to that measured at STC. The acceptance test is conducted with a standard acceptance test procedure that was developed at GRC. The design and tuned operating point for TDCs S/N 5 and 6 was 650°C heater temperature and 120°C cold end temperature. The GRC test used the average of 4 thermocouples for the heater head temperature and the average of the inlet and outlet temperatures of the coolant to represent the cold end temperature. The performance reported by the manufacturer prior to delivery was 55.7 watts AC at 25.2% efficiency for TDC S/N 5, and 55.9 watts AC at 25.1% efficiency for TDC S/N 6. The performance was later measured at GRC. The data showed that the TDCs were performing as indicated by the manufacturer and therefore passed the acceptance test.

Thermal losses to ambient through the insulation on the heater head must be characterized to calculate the convertor efficiency. A subsequent test was performed to characterize the insulation thermal loss and the GRC test data was augmented to incorporate the efficiency. The efficiencies calculated for the GRC data was comparable to that reported by the manufacturer. The insulation loss test was conducted by evacuating the Stirling convertor with a vacuum pump and heating the heater head to various temperatures, including the design temperature of 650°C. The total heat input required to maintain the desired temperature was measured. The calculated conduction and internal radiation losses from the hot end of the head to the cold end were subtracted from the total heat input to indicate the amount of heat lost through the insulation. The gas conduction loss through the residual gas was included because there was a concern that the heat exchangers and displacer were not able to be evacuated sufficiently to eliminate conduction through the residual gas. The inclusion of these losses decreases the indicated insulation loss and therefore indicates lower convertor efficiency, thus giving a conservative result.

Performance Mapping

The SRPS configuration into which the TDCs will be integrated has not been selected and similarly, the operating points over the life of the mission are to be determined. To assist the potential system integrators in designing the SRPS, the performance of the TDC was mapped over a wide range of operating conditions. The heater head was varied from 450°C to 650°C in increments of 25°C. The coolant temperature was varied from 80°C to 120°C in increments of 10°C. Operation was conducted at temperature combinations with the temperature ratio less than or equal to the design temperature ratio of 2.3. At each combination of temperature, the stroke of the power piston was varied from 5.4 mm to the maximum amplitude permitted by the controller supplied by STC. The plutonium heat source that will be used in the SRPS has an 88 year half-life. The small variation in piston stroke for these tests resulted in the heat accepted by the heater head varying by more than would be experienced over a 15 year mission due to the decay of the heat source.

The results of this test have been posted on the GRC Stirling web site and are summarized in Figs. 3 and 4. There are several strategies that can be used in an SRPS as the heat generated by the plutonium heat source decays. One strategy would be to operate the Stirling convertors at full stroke and allow the heater head temperature to drop [4], and thus reduce conversion efficiency. Another strategy would be to slowly shorten the piston stroke and maintain the heater head temperature. Other strategies exist that would allow the temperature to drop until near the end of the mission at which time the stroke would be shortened and the temperature would thus increase to regain the high efficiency. The benefit of this option is that the creep of the heater head will be negligible during the bulk of the mission, and then full power would be provided to the spacecraft near the end of the mission as the temperature is increased.

Digital Controller Tests

The TDCs were delivered to GRC with a laboratory controller known as the zener diode controller. This controller uses a control logic based on a zener diode to prevent overstroke; however, it introduces significant higher harmonics to the output current and voltage waveforms. To obtain the EMI/EMC levels necessary for most any deep space mission, the controller must produce little harmonic content. A controller was developed at GRC that uses digital logic to apply the proper load on the linear alternator, and do so with a purely resistive load directly on the AC output of the alternator. Initial tests indicated that the harmonic content was less than 5%.

TDCs S/N 5 and S/N 6 have been operated with this controller, but full performance mapping has not yet been performed.

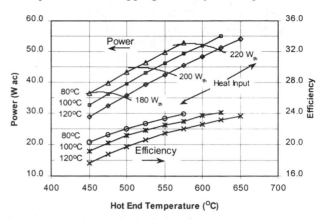

Figure 3. – Full stroke performance map for TDC 6.

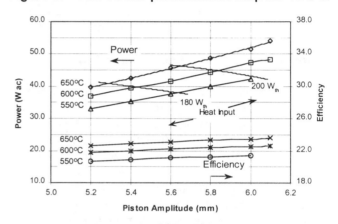

Figure 4. – Variable stroke performance map for TDC 6 at 120°C cold end temperature.

Plans for TDCs S/N 7 and 8

An additional set of convertors, known as TDCs S/N 7 and 8 have been delivered to GRC and are awaiting acceptance tests and performance mapping. These convertors will then be put on extended operation to accumulate approximately 1,000 hours of operation. Following the 1,000 hours of operation, the convertors will be operated in a vibration test at the GRC SDL to simulate a launch. The convertors will then be operated on extended operation to accumulate approximately 2,000 hours. This sequence will roughly simulate operation on earth prior to launch of a mission, the launch environment, and steady operation following a launch. The convertors will then be disassembled and carefully inspected to check for any evidence of wear or possible degradation.

LAUNCH ENVIRONMENTS TESTS

The testing and characterization of the free-piston Stirling TDCs in the GRC SDL has become a key part of the technology project. The SDL has two distinct capabilities that are utilized. One is a facility known as the Microgravity Emissions Laboratory (MEL), which is used to measure the vibration produced by operating devices. The SDL also has a variety of shake tables to perform more typical vibration tests. During calendar 2001, the TDCs have been used in one MEL test and two vibration tests. Each of these tests and a future test will be described briefly in the following sections.

MEL Tests

The MEL tests were performed on two opposing Stirling TDCs in January of 2001. The purpose of the test was to determine the effect of manufacturing tolerances on structure-borne disturbances generated during the operation of an opposed pair of convertors. To this end, the MEL fixture was designed such that the convertors could be misaligned by offsets and by angular misalignment. The maximum offset was 2.5 mm (0.10 in), and the angular misalignment was 4 degrees. The effect of rigidly mounted convertors was contrasted against a mounting structure resonant frequency of 40% below and 40% above the TDC operating frequency. The convertors were operated at the full piston stroke during all MEL testing and thus were generating their maximum signature.

The TDCs used in this test were S/N 5 and 6. When these convertors were manufactured, there was no attempt to make them identical nor to match parts. Going into this test, it was also known that the two convertors operated at slightly different piston strokes. It was therefore anticipated that a forcing function would be generated and measured. The test was viewed not as a demonstration of low vibration, but an initial attempt to exercise the test rig and to study the sensitivity to manufacturing tolerances and system alignment.

The primary data gathered from the MEL tests were the force and moment forcing functions. They were generated at the center of gravity of the dual TDC assembly. Acceleration time response data generated from the operating TDCs and acquired by the accelerometers on the MEL platform were post processed into rigid body forces and moments at the center of gravity of each axis of the TDCs. The MEL facility is shown in Fig. 5 and the TDCs mounted to the MEL is shown in Fig. 6.

Figure 5. - The MEL facility at GRC.

The results of the test showed the emissions to be distinctly tonal and generally benign. This is shown in Fig. 7. The responses are dominated by the fundamental 82.5 Hz operating frequency forcing function of 3.7 lbf and its associated harmonics. It is believed that simply tuning future convertors to operate at identical strokes should significantly reduce the emissions. The variation in response between the mechanical alignment configurations tested was minor.

Figure 6. - The TDCs mounted to the MEL test fixture.

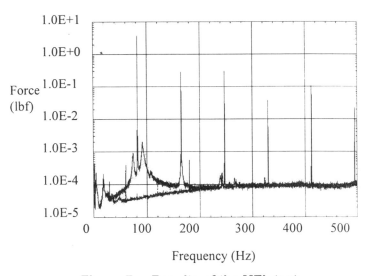

Figure 7. – Results of the MEL test.

Modal Tests

Following the results of the 1999 random vibration test, it was known that the TDC was able to operate at full power and suffer no damage when subjected to high levels of random vibration [5,6]. What was not known was the response of the internal components and what margin remained as the TDC passed the test. To begin to understand the internal structural dynamics of the TDC, a test was devised that used a single convertor mounted on the shake table with the pressure vessel removed. A suite of accelerometers, optical probes, and force sensing bolts was developed to measure the response of the key components, primarily the piston rod and linear alternator of the convertor. The convertor, TDC S/N 6, was motored at the resonant frequency of 73 Hz, which is lower frequency than the typical operation as a power convertor since it was now being operated at atmospheric pressure.

The test was performed in February of 2001. Test runs were performed with random vibration in the axial direction and then in the lateral direction, and also at full stroke and half stroke. The level of random vibration used in this test was 2.17 Grms with each run being 60 seconds in length. This level is 25% of the flight acceptance level since this would allow collection of sufficiently detailed data for a structural dynamicist to develop an understanding of the structural response and would not put the hardware at risk. A series of sine sweeps were also performed from 5 Hz to 2000 Hz at levels up to 0.5 G to provide more data to support the efforts to develop an understanding of the structural dynamics.

The tests were successfully completed and all the desired data were recorded. At this time, the data has not been fully analyzed and therefore no conclusions are presented in this paper.

Engineering Mounting Structure Tests

Random vibration tests were conducted with dual-opposed TDCs S/N 5 and 6 mounted in an engineering mounting structure. These tests were performed in April and May of 2001. The purpose of these tests was to study the interaction of the resonant operation of dual Stirling power convertors when subject to random vibration through a coupling that may be representative of the structure of the final SRPS. Mounting structures were devised with resonant frequencies of 11 Hz, 36 Hz, and 104 Hz. This range provides data with mounting structures having resonant frequencies lower than the operating frequency of the convertor, and also greater than the operating frequency. Similar to the Modal Test, random vibration was kept to 25% of flight acceptance to provide quality data without risking the hardware. The test fixture with the 11 Hz mounting structure is shown in Fig. 8.

The tests have been completed, and all the data have been recorded. The data have not been processed and therefore no definitive engineering conclusions are able to be presented in this paper. The TDCs did operate in a well behaved manner, even during sine sweeps when at the resonant frequency of the mounting structure. Following reduction and analysis of the data, the results will be reported and supplied to the selected SRPS system integrator.

Future Vibration Tests

A test has been proposed to characterize the response of the TDCs with a mounting structure that is dynamically similar to the one designed by the system integrator. If possible, this will also be the same test as previously discussed in the section of future testing of TDCs S/N 7 and 8. This test has been designated the Prototypical Mounting Structure test. Other tests may be run at the GRC SDL in support of DOE and STC

to validate modifications to the TDC as it nears the eventual flight configuration.

Figure 8. – TDCs mounted in the Engineering Mounting Structure test stand.

OTHER TECHNOLOGY AREAS

This paper has focused on two critical areas of the GRC Stirling technology project. There are also other areas that are active and are considered key to the ultimate successful deployment of the Stirling convertor for space power applications. These areas will be briefly discussed below.

The heater head is a high temperature thin walled pressure vessel fabricated from Inconel 718. Material tests for a heater head life assessment are continuing, and a facility for accelerated life tests on prototypical heater heads is being established. Results of these tests will be used to calibrate and validate a life prediction model. The analysis of the heater head will use probabilistic interpretation to allow trades in performance and life [7].

Even though most of the TDC is metallic, there are some limited applications of organic materials. These are primarily used as adhesives and insulators in the linear alternator and are also used as surfaces of wear couples. The four TDCs at GRC are the first TDCs built with radiation-hard organics and will serve as demonstration test beds to assist in proving the functionality of these organics. A plan to verify the acceptability of the organic materials used in the alternator and piston bearing coatings is now being drafted. Sample applications of the wear coatings have been prepared and will be irradiated for further characterization testing.

The linear alternator is one of the key elements for high performance. GRC linear alternator analysis is advancing the state-of-the-art in the understanding of this technology [8]. Short-term (200 hours) magnet aging tests were completed at 120°C and 150°C on various NdFeB magnet types to help in the selection of magnets for long-term aging tests. These long-term tests have been initiated and will characterize potential aging in the strength or demagnetization resistance of the magnets used in the alternator. The tests have now reached 2,000 hours. The long-term tests will be conducted for up to

12,000 hours at 120°C and with a DC demagnetizing field of 6 kOe.

Electromagnetic interference/electromagnetic compatibility (EMI/EMC) characterization testing of the TDCs will also be done to support evaluating concepts to reduce EMI/EMC. Plans are being developed to study the origin of EMI from the TDCs and to evaluate methods to reduce the effects. This may potentially lead to system level trades that should involve the System Integrator.

For the first deployment of a dynamic power system in space to be successful, high reliability is paramount. DOE has Westinghouse Government Services under contract to oversee the reliability issues prior to the selection of a System Integrator. NASA GRC is a part of the reliability committee and is an actively participant, drawing on the many years of experience in developing highly reliable systems that have been deployed in space.

A grant has recently been awarded to Cleveland State University (CSU) to develop a multi-dimensional Stirling computational fluid dynamics (CFD) code to significantly improve Stirling loss predictions and assist in identifying areas of the convertor for improvement. CSU and GRC collaborated on related loss understanding and initial two-dimensional CFD code investigations under GRC Stirling efforts in support of the SP-100 program. CSU is now developing a multi-dimensional CFD code for Stirling convertors of the general configuration of the Stirling radioisotope power convertors. The code results will be compared with test data and other codes as possible.

SUMMARY

The Department of Energy, the NASA Glenn Research Center, and the Stirling Technology Company are developing a free-piston Stirling convertor for a Radioisotope Power System to provide electric power for NASA deep space missions. This power system has recently been identified as a candidate for potential use on a variety of long life missions, including Mars rovers. The SRPS is expected to provide modest mass savings compared to the previously used RTGs; however, the major benefit is in the multi-fold increase in efficiency. The increased efficiency will result in substantial cost savings due to the significantly reduced plutonium inventory.

NASA GRC is conducting an in-house project to assist in transitioning the TDC Stirling convertor for readiness for flight qualification and use in a mission. GRC provided key support to a joint DOE/NASA technology assessment that has led to the current mission possibilities for the Stirling system. Test facilities for Stirling convertor performance mapping, magnet aging, and structural benchmark testing of the heater head have been established or are nearing completion at GRC. Tests have recently been completed relative to launch environment characterizations. Efforts are ongoing for an independent performance verification, convertor life assessment, permanent magnet aging characterization, and reduction of electromagnetic interference (EMI)

ACKNOWLEDGMENTS
The authors would like to acknowledge the support of the Office of Space Science (Code S) and the Office of Aerospace Technology (Code R) at NASA Headquarters for providing funding for this project.

REFERENCES
[1] White, M.A., Qiu, S., Augenblick, J., Peterson, A., and Faultersack, F., "Status Update of a Free-Piston Stirling Convertor for Radioisotope Space Power Systems," Proceedings, 2001 Space Technology Applications International Forum, M. El-Genk editor, Albuquerque, NM.

[2] Thieme, L.G. and Schreiber, J.G., 2001, "Update on the NASA GRC Stirling Technology Development Project," 2000, NASA TM-2000-210592, Proceedings, 2001 Space Technology Applications International Forum, M.El-Genk editor, Albuquerque, NM.

[3] Furlong, R., and Shaltens, R.K., "Technology Assessment of DOE's 55-We Stirling Technology Demonstrator Convertor (TDC)," NASA TM-2000-210509, Proceedings, 35th Intersociety Energy Conversion Engineering Conference, Las Vegas, NV.

[4] Schreiber, J.G., "Power Characteristics of a Stirling Radioisotope Power System Over the Life of the Mission," Proceedings, 2001 Space Technology Applications International Forum, M. El-Genk editor, Albuquerque, NM.

[5] Hughes, W.O., McNelis, M.E., and Goodnight, T.W., "Vibration Testing Of An Operating Stirling Convertor," 2000, NASA TM-2000-210526, Proceedings, 7th International Congress on Sound and Vibration, Garnisch-Partenkirchen, Germany.

[6] Goodnight, T.W., Hughes, W.O., and McNelis, M.E., "Dynamic Capability Of An Operating Stirling Convertor," Proceedings, 35th Intersociety Energy Conversion Engineering Conference, Las Vegas, NV.

[8] Halford, G.R., Shah, A., Arya, V.K., Krause, D.L., and Bartolotta, P.A., "Structural Analysis of Stirling Power Convertor Heater Head for Long Term Reliability, Durability, and Performance," Proceedings, 36th Intersociety Energy Engineering Conference, Savannah, GA.

[7] Geng, S.M., Schwarze, G.E., Niedra, J.M., and Regan, T.F., "A 3-D Magnetic Analysis of a Stirling Convertor Linear Alternator Under Load," Proceedings, 36th Intersociety Energy Conversion Engineering Conference, Savannah, GA.

Proceedings of IECEC'01
36th Intersociety Energy Conversion Engineering Conference
July 29–August 2, 2001, Savannah, Georgia

IECEC2001-CT-33

LONG-TERM CREEP ASSESSMENT OF A THIN-WALLED INCONEL 718 STIRLING POWER-CONVERTOR HEATER HEAD

Randy R. Bowman
NASA Glenn Research Center, Cleveland, Ohio

ABSTRACT

Stirling power convertors are candidate power systems for long duration, deep space exploratory missions. Projected design life requirements for these missions are in excess of 100,000 hours. Design reviews have identified the heater head of the Stirling power convertor as a critical component for long-term durability. This paper will discuss long-term creep assessment of the thin-walled, Inconel 718 (IN-718) nickel base superalloy heater head as one part of an overall life prediction methodology.

INTRODUCTION

DOE and NASA have identified Stirling power convertors as candidate power supply systems for long-duration, deep space science missions. A key element for qualifying the flight hardware is long-term durability assessment for critical hot section components of the power convertor. One such critical component is the power convertor heater head. The heater head is a high-temperature pressure vessel that transfers heat to the working gas medium of the convertor, which is typically helium.

An efficient heater head design is the result of balancing the divergent requirements of, on one hand, thin walls for increased heat transfer, and on the other hand, thick walls to lower the wall stresses and thus improve creep resistance/durability. In the current design, the heater head is fabricated from the Ni-base superalloy Inconel 718 and has a 0.05-cm wall thickness at the hot end. The vessel walls are subjected to a stress of approximately 110 MPa (16 ksi) at a maximum temperature of 649°C (1200°F) for up to 100,000 hours.

The chosen life prediction methodology consists of generating a material-specific database for the Stirling power converter application, defining the appropriate definition of failure, developing a probabilistic design methodology, and verifying the critical flight hardware using benchmark tests. Final validation of the flight hardware design will be accomplished by calibrating/verifying the life models using benchmark tests on actual heater heads under prototypical operating conditions. Details of the modeling efforts are presented elsewhere in these proceedings (Halford, et al.).

Although IN-718 is a mature alloy system (patented in 1962), there is little long-term (>50,000 hours) creep data available for thin-specimen geometries. Therefore one facet of the overall durability assessment program involves generating relatively short-term creep data at the design temperature of 649°C (1200°F). This data will be collected using 0.05-cm thick creep samples machined from the same heat of material from which the actual flight hardware will eventually be fabricated. In addition, the wealth of information available in the literature on long-term creep properties of thick samples will be incorporated into this thin-specimen, heat-specific database. The resulting aggregate database will be used as part of the overall life prediction methodology to assess the design life of the heater head.

Careful control of the microstructure is the key to achieving the optimal material performance needed to attain the long lives required by the mission profile. In this paper, microstructural evaluations of IN-718 including heat treatments, grain size, and creep data results will be presented from several heats of materials along with preliminary long-term life predictions.

MATERIALS AND PROCEDURES

Inconel 718 is an age-hardenable, Ni-Fe based superalloy, 52.5Ni-19Cr-18.5Fe-5.1Nb-3Mo-1Ti-0.5Al wt% [1], initially developed and used as a turbine disk material. It was chosen for disk applications due to its high resistance to creep and stress rupture. It is still considered to be one of the best alloys for applications under 649°C (1200°F). Inconel 718 is a fairly inexpensive alloy that can be readily processed into wrought bar, forgings, sheet, and castings, In addition, Inconel 718 is more weldable than most other superalloys Inconel 718 is somewhat unique among Ni-base superalloys in that after solutioning and aging two phases coexist in the matrix, namely the γ' [$Ni_3(Al, Ti, Nb)$] and γ'' [$Ni_3(Nb, Al, Ti)$].

Fabrication of the heater head involves rough machining of the outside diameter (4.85-cm O.D. for the flight hardware configuration) directly from an IN-718 bar stock of the appropriate length and diameter. After which, finish machining of the heater head inside diameter is performed. Initially it was thought that the heater heads would also incorporate flanges at the cold end of the shell. The presence of these flanges would require that the starting bar-stock be oversized to accommodate them. Accounting for both the 4.85-cm heater head O.D. plus the flange width dictated a starting bar-stock diameter of roughly 8 cm. The bar-stock dimension was eventually recognized as an important variable controlling the microstructure and hence mechanical properties of the material.

Numerous studies have reported that rupture life and creep rate of nickel base superalloys are greatly influenced by the cross-sectional area of the test sample. For specimens of a constant gage length and grain size, reducing the specimen thickness results in progressively shorter rupture lives and higher steady-state creep rates [2]. The relative significance of this phenomenon to overall mechanical properties is dictated by the number of grains (N_g) that are present through the thickness of the sample. This effect is of most concern in thin specimens where the thickness of the sample is on the same order of magnitude as the grain size. In situations where there are too few grains through the thickness, smaller grains (and hence more grains in the cross section) are desired for improved creep properties. Conversely, once the critical N_g is achieved, further refinement in grain size leads to degradation in creep resistance. The conclusions from various studies [2,3,4] on the effect of grain size-to-specimen thickness for a number of nickel base superalloys have been fairly consistent. The consensus is that, in general, approximately 15 grains are necessary through the thickness to avoid the thickness-induced debit in creep properties. So in general, while large grains are desired for good intrinsic creep resistance, small grains are required in thin samples to insure an adequate number of grains. These opposing trends imply that, in thin specimens, some optimum grain size exists where a minimum in steady-state creep rate and a maximum rupture life are obtained.

With these constraints, two trial heats of IN-718 were ordered from different vendors with a grain size specification of ASTM 8 or finer, which corresponds to an average grain diameter of 0.0025 cm (25 μm). This grain size results in about 20 grains through the thickness of the 0.05-cm creep samples. Heat U-245 was worked into an 8-cm diameter bar, and heat A-1345 into a 7.2-cm diameter bar. These specific heats of material were not intended to be used in the fabrication of the flight hardware; rather they were to provided baseline information on certain key parameters such as heat-to-heat variability, uniformity of the bar stock microstructure, Cr depletion in vacuum, effect of modified heat treatments, and creep properties of thin specimens.

After microstructural evaluation and mechanical testing was completed on these trial heats of material, a third heat was ordered for use in fabricating the flight certified hardware. This master heat of material (heat #SM9-20034) was worked to a slightly smaller diameter and with a finer grain size than the trial heats. The rational for this and the details of this heat will be discussed later.

All three bars were fabricated and certified according to ASTM B637 standards. Melting was accomplished by means of vacuum induction melting followed by electro-slag remelting. The bars were then hot rolled, annealed, and polished. For ease of readability, the larger diameter bars will be referred to as 718-LD1 and 718-LD2 corresponding to the 8-cm and the 7.2-cm diameter bars respectively. The smaller diameter (5.7 cm), master heat flight hardware material will be identified as 718-SD.

Test samples for the creep rupture experiments were taken longitudinally from the IN-718 bar stock. The 0.05-cm (0.02") thick dogbone samples were electro-discharge machined with a 3.18-cm (1.25") gage length. The location through the thickness of the bar-stock where the samples were taken will be discussed in detail in a subsequent section.

The machined creep samples were then heat treated in flowing argon as follows [1]: 1) Solution anneal at 924-1010°C ± 14°C (1700-1850°F) for 1-2 hour, followed by air-cooling, 2) age at 718 ± 14°C (1325 ± 25°F) for 8 hours, 3) furnace-cool to 621 ± 14°C (1150 ± 25°F) at 55 ± 28 °C /hr (100 ± 50°F/hr), 4) age at 621°C (1150°F) until the total duplex heat treatment time reaches 18 hours, 5) air-cool to room temperature. An alternate heat treatment was considered in order to better accommodate the brazing cycle used during fabrication of the heater heads. The heater head design incorporates fins that are brazed onto the inner surface of the heater head. The brazing cycle time and temperature are the same as that used during the solution step (heat treatment step #1). To avoid re-solutioning the samples in a separate step as part of the heat-treatment procedure, and as a consequence perhaps weakening the braze joints, the possibility of using the brazing cycle itself as the solution step was considered. The only difficulty was that the brazing furnace was not capable of air-cooling. Therefore an alternate heat treatment was considered that used furnace-cooling after solutioning, rather than air-cooling as prescribed above. Samples heat-treated with this modified procedure were examined both by optical metallography and by electron microscopy, and were also subjected to creep rupture testing.

Tensile creep rupture tests were conducted under constant load conditions in lever-arm creep machines. The testing temperature of 649°C (1200°F) was achieved using a resistively heated furnace and was controlled to ±2°C. Specimen temperature was monitored using three thermocouples located along the gage section. Temperature variation along the gage length was also ± 2°C. Creep strain was measured during the test using an extensometer attached to the specimen shoulders. The extensometer displacement was recorded using a super linear variable capacitance (SLVC) device connected to a computer data acquisition system.

RESULTS AND DISCUSSION

Because of the strong influence of grain size on the mechanical properties of thin structures, all three heats of IN-718 were metallographically examined for grain size. The samples were mechanically ground and immersion etched in a waterless Kallings reagent (100 ml ethanol, 100 ml HCL, 5-10 g CuCl$_2$). For each heat of material, the average grain size was measured by the linear intercept method. Fig. 1 shows the results of these measurements where grain size is plotted as a function of distance from the outer surface of the bars.

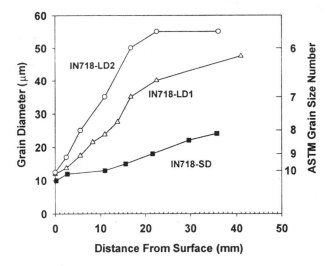

Figure 1 – Grain size as a function of distance from the surface of the IN718 bars.

It is immediately apparent from this plot that a great deal of variation in grain size was present through the thickness of all three heats of material. Heats 718-LD1 and 718-LD2 were specified to have an ASTM grain size of 8 (~25 μm grain diameter) or finer, but in actuality this grain size specification was met only in a 5 to 10-mm wide zone at the surface of the bar.

These observations illustrate the difficulty of obtaining uniform microstructures in hot worked structures. Hot working of even the simplest of shapes can produce widely differing microstructures throughout the workpiece. This occurs because gradients in the critical intrinsic variables (such as temperature, strain, and strain rate) are inevitably generated during hot working [5]. Since these variables strongly influence microstructural evolution, a gradient in their values results in a non-uniform microstructure in the finished part.

Even though the majority of the cross section of both large diameter bars failed to meet the grain size specification, the requested grain size was met in the near-surface regions. While certainly not ideal, this variation in grain size is not an insurmountable problem for the specific application of interest. The non-uniform grain size is manageable because fabrication of the heater heads involves machining away the interior of the bar in order to produce the thin-walled, hollow vessels. The relevant region of the bars cross section amounts to an approximately 0.25-cm wide zone at the surface of the bar. Within this zone, the outer 0.2-cm section is machined away, with the subsequent 0.05-cm thick section comprising the heater head walls.

Since grain size is not uniform throughout the cross-section of the IN718 bars, and because grain size is a primary factor controlling creep properties, it was important to machine the creep samples from the bars in a deliberate manner. Based on the consensus of the literature concerning the critical number of grains needed in the thickness, it was desirable to machine samples from those regions of the bar that would yield specimens with at least 15 grains through the thickness. Given the fact that the heater head wall thickness (and hence the specimen) thickness is 0.05 cm (500 μm), means that the grain size must be 33 μm or smaller to obtain the required 15 grains. To be well above critical N_g of 15, samples were taken from locations the 718-LD1 and LD2 bars corresponding to a grain size of 20 μm, thus giving 25 grains through the thickness of the test samples. It is with this 20-μm grain size that the majority of the creep testing was performed.

Although the variation in grain size across the bar thickness is not desirable in terms of uniformity of properties, it did offer the opportunity to document the effect of grain size on creep properties for these specific heats of material. The results can only approximate the true influence of grain size since other influential variables (such as precipitate size and distribution), in addition to grain size, may be also varying through the thickness. In spite of these limitations, an additional set of creep samples were machined from 718-LD1 at various locations through the thickness, so as to obtain samples with grain sizes ranging from about 12 to 50 μm. The resulting creep data should provide some indication of the optimum grain size for creep resistance, which as discussed earlier, is a consequence of the competing requirements of small grains to avoid the thickness effect versus large grains for improved intrinsic creep resistance.

Steady-State Creep Rate

The creep curves for IN-718, as for most materials, can be subdivided into distinct regions. The pertinent regions are easily distinguishable in the short-term creep test shown in Fig. 2, where strain versus time is plotted for a 718-L1 sample tested at 415 MPa (which is 3.5 times higher than the design stress). The period during which creep rate is relatively constant is referred to as the "steady-state" regime. After sufficient strain is accumulated, the material enters a stage where the strain rate increases rapidly, with final failure coming soon after. Since the material spends a large fraction of its life in the steady-state regime, the steady-state creep rate represents an important parameter describing creep behavior. The steady-state creep rate also correlates well with the rupture life.

In Fig 3, the steady-state creep rates for specimens with a 20-μm grain size taken from the two large-diameter IN-718 heats and tested at 649°C (1200°F) are presented as a function of applied stress. Also included in this figure is a band of data that represents the range of steady-state creep rates that have been reported in the literature [6-9] for large diameter specimens (where the thickness to grain size effect is not an issue).

437

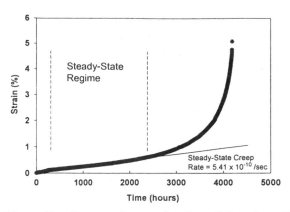

Figure 2 – Creep strain as a function of time for a 718-L1 sample tested at 649°C with an applied stress of 415 MPa (3.5 times the design stress).

Also included in Fig. 2 is a band of data that represents the range of steady-state creep rates that have been reported in the literature [6-9] for large diameter specimens (where the thickness to grain size effect is not an issue). Finally, data from another study [10] on thin sheet (0.0635-cm thick) IN-718 material is also included.

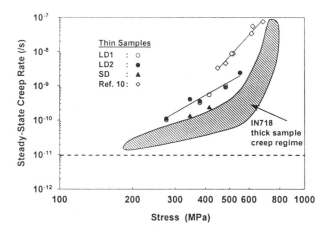

Figure 3 – Steady-state creep rates for thin (~0.05 cm) IN718 samples compared to typical values reported in the literature for thick specimens. Lower creep rates result in longer lives.

The effect of sample size on the creep properties, which has been reported in the literature, is clearly illustrated in Fig. 3. The steady-state creep rates of both the current study (IN718-LD1 and LD2) as well as those from Ref. 10 lie above the band of data measured on large diameter specimens. The creep rates from Ref. 10 are significantly higher than the current study even though the specimens were slightly thicker than those machined from IN718-LD1 and LD2. Since the grain size was not given in Ref. 10, it is impossible to reconcile this discrepancy in creep rates. Even though the two thin-sample studies shown in this figure are not directly comparable, they are consistent with the notion that thin samples tend to

have inferior properties to thicker specimens at all stress levels investigated. Documenting the behavior of thin samples is especially important because most literature and handbook creep data is generated using thick samples. Consequently, using these reference sources as the basis for predicting the lives of thin-walled structures leads to a non-conservative result.

To assess just how sensitive creep rate is to grain size in thin samples, the samples machined from bar LD1 at various positions through the thickness were tested at 649°C (1200°F) at an applied stress of 414 MPa (60 ksi). The samples had grain sizes of 12, 15, 20, 24, 39, and 46 μm. Figure 4 shows the results of these tests where the steady-state creep rate is plotted versus the number of grains through the sample thickness. It should be noted that the y-axis scale in Fig. 4 is linear and spans a relatively narrow range of strain rates. This is in contrast to Fig. 3 where the steady-state creep rates were plotted logarithmically over many orders of magnitude. So while the variation in strain rate as a function of grain size shown in Fig. 4 is significant, it is not as dramatic as might be suggested at first glance.

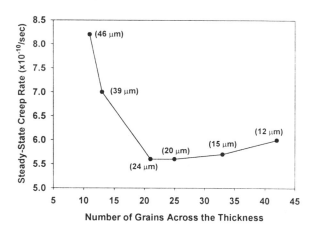

Figure 4 – Dependence of steady-state creep rate on grain size. Tested at 649°C/414 MPa using 0.05-cm thick samples machined from various locations through the thickness of bar IN718-LD1. Grain sizes are shown in parentheses.

The highest strain rate was measured on the sample with the largest grain size, and hence the fewest grains through the thickness. The average grain diameter in this sample was 46 μm, which corresponds to about 10 grains through the thickness. Refinement of this grain size resulted in a decreasing strain rate until about 20 grains through the thickness was achieved. After obtaining the critical number of grains through the thickness, further refinement resulted in an increase in creep rate. It is apparent that the increase in strain rate resulting from having smaller than the optimum grain size was not as severe as the increase due to having too few grains. This observation was important and guided the choice of grain size in the flight material.

Alternate Heat Treatment

Creep tests were performed on the samples machined from bar 718-LD1 and heat-treated with the alternate procedure, in which the samples were furnace-cooled from the solution temperature rather than air-cooled. As with the conventionally treated samples, the grain size was 20 μm. In all cases, the creep properties were found to be inferior to those obtained on samples subjected to the conventional heat treatment. The creep data is not included in Fig. 3 so as to avoid any confusion, but the creep rates at all stress levels were almost an order of magnitude higher than the IN718-LD1 and LD2 samples heat treated conventionally. Transmission electron microscopy found that the solution + air-cooling procedure resulted in a microstructure which was devoid of γ' and γ'', while the alternate slow cooling procedure allowed precipitation of both species. Subsequent aging of these two microstructures resulted in vastly different sizes and distributions of the γ' and γ''. Based on these observations, it is not surprising that the mechanical properties were dramatically different. Since the air-cooling step was so critical, the furnace-cooling solution treatment was abandoned and a new brazing furnace with air-cooling capabilities will be used to perform the brazing cycle.

Flight Material Specification

The wealth of information available in the literature regarding the microstructure and mechanical properties of IN718, combined with the experience gathered from the trial heats served as a guide for specifying the material that would be used in the actual hardware fabrication.

To begin with, it was decided that the flanges used on the ground-based Stirling heater heads were not required on the flight hardware units. Elimination of the flanges meant that the flight hardware bar stock could have a smaller diameter than the trial heats. Having the option to use a smaller diameter bar was beneficial because through-thickness microstructural variation should be minimized. The non-uniform grain size did not impact the creep data generated on the trial heats only because it was simply a matter of machining samples from the bars in locations that contained sufficiently fine grains to achieve the critical N_g. When fabricating the heater heads though, this positional flexibility will not be possible. It was critical therefore that the grain size requirements be met in the near-surface region of the bar corresponding to the eventual location of the heater head walls. Given this, the smallest diameter bar practical was chosen in order to achieve less grain size variation through the cross section of the bar, thereby maximizing the chances of achieving the required grain size in the relevant region of the bar. Based on these considerations the flight material (heat SM9-20034) was specified to be only 5.7 cm in diameter.

Since the heater head will be exposed to high temperatures for extended times, coarsening of the microstructure is a distinct concern. Even though Fig. 4 suggests a grain size of 24 μm would result in the lowest steady-state creep rate, any subsequent grain growth during service would lead to a significant increase in creep rate. Since the penalty in terms of creep rate that arises from having fine grains is not as severe as that due to having insufficient number of grains, it was decided to produce the small diameter bar of the flight material with a grain size specification of ASTM 10 (~12 μm grains) or finer. In this way, some grain growth could be tolerated during service without a significant degradation of creep resistance.

Flight Material Properties

The grain size distribution for the flight hardware material shown in Fig. 1 confirmed that the smaller diameter did result in less overall variation in grain size when compared with the larger diameter bars. In the smaller diameter bar (718-SD), even though much less overall variation in grain size is noted, again, as with the larger bars, the grain size specification was achieved only in a 15-mm region near the surface. Since the heater-head walls will be fabricated from within this surface region, the grain size variation in this bar was acceptable.

Creep samples were taken from bar IN718-SD at locations corresponding to the position of the heater head walls. The grain size in this region was about 12-μm, thus yielding about 40 grains in the 0.05-cm thick specimens. The creep samples are currently being tested at 649°C (1200°F), although only two tests have thus far accumulated enough strain to reliably measure the steady-state creep rate. Those creep rates are included in Fig. 3, and the initial results are encouraging. The creep rates for the IN718-SD specimens are lower than the other thin sheet data shown in this figure. In fact, the IN718-SD steady-state creep rates are just slightly higher than the values reported for thick samples.

The fact that the creep rates for the 718-SD material are lower than those for 718-LD1 and LD2 is probably due more to heat-to-heat variability rather than the refinement in grain size. In fact, based on the trends shown in Fig. 4, the 12-μm SD material should have had a slightly higher creep rate than the 20-μm LD specimens. This observation highlights the fact that the data in Fig. 4 is useful in describing the relative sensitivity of creep rate to small versus large grains, rather than predicting the precise rates. Heat-to-heat variability, as observed here, can overshadow the behavior attributable to grain size differences. Nevertheless, the data in Fig. 4 was useful in that it demonstrated that it is preferable to bias the microstructure toward fine grains rather than larger ones.

Rupture Lives Under Low Stress Conditions

Ultimately the goal of this study is to determine whether the IN718 heat head as designed can withstand the operating conditions for times of about 100,000 hrs. A full analytical life prediction model is under development and should provide meaningful life predictions as well as a statistical treatment of the confidence levels. At the present time, the model is still under development, and some open issues remain. One of these contested issues is the definition of failure. Certainly one criteria of failure would be when the heater head accumulates enough strain that catastrophic fracture occurs. Another definition might be to define failure as the point prior to fracture but when enough strain has accumulated that the gap clearance between the piston and heater head wall is such that the Stirling engine efficiency level falls below some critical level.

Although many of the details concerning the mission profile and failure criteria are currently under review, it is still instructive to make some first order approximations. One common predictive tool is based on the Monkman-Grant relationship. This relationship, $\dot{\varepsilon}_{ss} t_r = A$, says that the steady-state creep rate multiplied by the time to rupture equals a constant. There are several variations of this relationship, but for the present purposes this form is sufficient. Rewriting the equation, $t_r = A \dot{\varepsilon}_{ss}^{-1}$, and plotting t_r versus $\dot{\varepsilon}_{ss}$ on a logarithmic scale should yield a linear relationship.

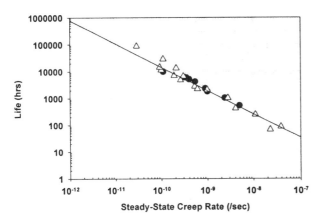

Figure 5 – Monkman-Grant representation of IN718 creep rupture lives at 649°C. Open symbols are data taken from Ref. 3 generated using thick samples. The filled symbols are data from the present study using thin samples.

In Fig. 5, the rupture life versus steady-state creep rate data from Ref. 6 (generated on thick samples) at 649°C is plotted in the Monkman-Grant form. Although this data was not generated on thin samples, it is invaluable because of the wide range of test duration, ranging from less than 100 hours to nearly 100,000 hours. As seen in this figure, a reasonably good regression fit is obtained. Also included in Fig. 5, is the thin-sample data generated in the present study from bars IN718-LD1 and LD2 (no samples from IN718-SD have as yet ruptured). The rupture data versus creep rate relationship generated on the thin specimen data agrees quite well with the thick sample data. Based solely on this relationship, rupture should not occur within 100,000 hours as long as the steady-state creep rate is 1×10^{-11}/sec or less. The dashed line in Fig. 3 indicates this critical creep rate. Unfortunately, no steady-state creep data for 718-SD is currently available for the expected

operation stress level of 110 MPa. A simple extrapolation of the thick-sample IN718 data in Fig. 3 suggests that at 110 MPa, the steady-state creep rate should be below this critical level. Of course this assumes that the low stress data continues to follow the trend shown in this figure. Even the thin sample data taken from IN718-LD1 and LD2, which have higher creep rates than the thick-sample data, extrapolate to creep rates of less than 10^{-11}/s at 110 MPa. Since the flight hardware material creep rates are similar to that of the thick-sample data it appears that outright failure of the heater head is not a concern. Tests are currently underway at the low stresses in order to measure the actual steady-state creep rates. Of course at the low stress levels, and hence low creep rates, it will be some time before sufficient strain is accumulated to accurately measure the creep rate and thus more accurately predict rupture life. It is for this reason that a more rigorous modeling program is underway to predict the lives of these components with greater accuracy and certainty based on the short-term data that is currently available and the heat-specific data that is being generated.

SUMMARY

Several heats of IN718 were purchased and analyzed for eventual use in the fabrication of a Stirling engine heater head. Particular attention was given to the grain size versus specimen thickness effect as it relates to creep properties. The knowledge gained from the literature and from trial heats of material resulted in a material specification that maximizes the likelihood of component survivability for mission durations in excess of 100,000 hours. Creep testing is ongoing on the actual flight hardware master heat material, with the initial results suggesting that the life goal is attainable within the currently envisioned operating conditions.

REFERENCES

1. ASTM B637-98, "Standard Specification for Precipitation-Hardened Nickel Alloy Bars, Forgings, and Forging Stock for High-Temperature Service".
2. Richards, E.G., 1968, Journal of the Institute of Metals, **96**, pp. 365-370.
3. Baldan, A., 1997, Journal of Material Science, **32**, pp.35-45.
4. Baldan, A., 1997, Journal of Material Science. Letters, **16**, pp. 780-783.
5. Mataya, M., 1999, JOM, **51**, pp 18-26.
6. Booker, M.K. and Booker, B.L.P, 1980, ORNL TM-7134.
7. Han, Y, Chaturvedi, M.C., and Cahoon, J.R., 1988, Scripta Metallurgica, **22**, pp. 255-260.
8. Hayes, R.W., 1991, Proceedings, <u>Superalloys 818, 625, and Various Derivatives</u>, E. Loria, ed., TMS, pp. 549-562.
9. Han, Y. and Chaturvedi, M.C., 1987, Material Science and Engineering, **89**, pp. 25-33.
10. Cullen, T.M. and Freeman, J.W., 1965, NASA CR-268.

Proceedings of IECEC'01
36th Intersociety Energy Conversion Engineering Conference
July 29 – August 2, 2001, Savannah, Georgia

IECEC2001-CT-34

A 3-D MAGNETIC ANALYSIS OF A STIRLING CONVERTOR LINEAR ALTERNATOR UNDER LOAD

Steven M. Geng and Gene E. Schwarze
Glenn Research Center, Cleveland, Ohio

Janis M. Niedra
QSS Group, Inc., Brook Park, Ohio

Timothy F. Regan
Sest, Inc., Middleburgh Hts., Ohio

ABSTRACT

The NASA Glenn Research Center (GRC), the Department of Energy (DOE), and the Stirling Technology Company (STC) are developing Stirling convertors for Stirling Radioisotope Power Systems (SRPS) to provide electrical power for future NASA deep space missions. STC is developing the 55-We Technology Demonstration Convertor (TDC) under contract to DOE. Of critical importance to the successful development of the Stirling convertor for space power applications is the development of a lightweight and highly efficient linear alternator.

This paper presents a 3-dimensional finite element method (FEM) approach for evaluating Stirling convertor linear alternators. The model extends a magnetostatic analysis previously reported at the 35th Intersociety Energy Conversion Engineering Conference (IECEC) to include the effects of the load current. STC's 55-We linear alternator design was selected to validate the model. Spatial plots of magnetic field strength (H) are presented in the region of the exciting permanent magnets. The margin for permanent magnet demagnetization is calculated at the expected magnet operating temperature for the near earth environment and for various average magnet temperatures. These thermal conditions were selected to represent a worst-case condition for the planned deep space missions. This paper presents plots that identify regions of high H where the potential to alter the magnetic moment of the magnets exists.

INTRODUCTION

NASA GRC, DOE, and STC are developing a Stirling convertor for an advanced radioisotope power system to provide spacecraft on-board electric power for NASA deep space missions. NASA GRC's role includes an in-house project to provide convertor, component, and materials testing and evaluation in support of the overall power system development. As a part of this work, NASA GRC has established an in-house Stirling Research Laboratory for testing, analyzing, and evaluating Stirling machines (Ref [1]). Four 55-We convertors (TDC 5, 6, 7, and 8) have been built by STC for NASA and are currently under test at GRC. A cross-sectional view of the 55-We TDC is shown in Figure 1. For more information on the TDC, see References [2,3,4]. As another part of this work, NASA GRC has been developing Finite Element Analysis (FEA) and Finite Element Method (FEM) tools for performing various linear alternator thermal and electromagnetic analyses, and evaluating design configurations.

A 3-dimensional (3-D) magnetostatic FEM model of STC's 55-We Technology Demonstration Convertor (TDC) linear alternator was developed by Geng et al (Ref [5]) to evaluate the demagnetization fields affecting the alternator magnets. The model, previously reported at the 35th IECEC, was an open-circuit model. The effect of load current flow through the alternator windings was not included in the preliminary analysis. The predicted alternator open-circuit voltage compared well with the experimentally measured value, which tended to verify the accuracy of the model. The model was then used to perform a preliminary evaluation of the demagnetization fields acting on the magnets. The preliminary evaluation showed that for the assumed magnet material, UGIMAX 37B, and magnet temperature of 75°C, the magnets margin to demagnetization was approximately 70 kA/m. The effect of current flow through the alternator windings was expected to further reduce this margin to demagnetization for the assumed magnet material.

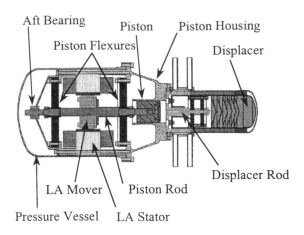

Figure 1 – Cross-section of STC's 55-We Technology Demonstration Convertor (TDC)

The 3-D magnetostatic FEM model of STC's 55-We TDC linear alternator has been enhanced to include the effect of current flow through the alternator coil windings. The current waveform is now a boundary condition for the magnetostatic model. The linear alternator current and terminal voltage was measured in NASA GRC's Stirling Research Laboratory for TDC 7 at full design stroke. The measured current was used as a boundary condition in the new linear alternator model. The magnetic properties of the magnets used in the TDCs built for NASA are now in the model and offer a better resistance to demagnetization than the UGIMAX 37B magnets assumed in the original model. The enhanced model was used to generate spatial plots of the magnetic field strength (H) in the region of the permanent magnets. The margin for magnet demagnetization was evaluated at various magnet temperatures.

OBJECTIVE

The objective of this work was to develop an analytical tool that could aid engineers in the design, development, evaluation and understanding of advanced linear alternator designs.

MAGNETOSTATIC/LOAD CURRENT MODEL

A 3-D magnetostatic model of STC's 55-We linear alternator design was developed using Ansoft's Maxwell 3-D Field Simulator software (Ref [6]). The magnetostatic model consisted of a quarter section of the linear alternator as described in Ref [5]. The components included the mover laminations, Neodymium-Iron-Boron magnets, stator laminations, and copper coils. Magnetostatic flux solutions were generated for a series of 13 different mover positions over a 1.2 cm mover stroke. For each mover position, the instantaneous current flow through the alternator coil windings was applied as a boundary condition. The linear alternator current was obtained experimentally by GRC for TDC 7. A plot of the experimental data is shown in Figure 2.

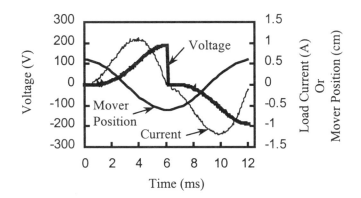

Figure 2 – Experimental Data of Terminal Voltage, Load Current, and Mover Position vs. Time for TDC 7 at Full Stroke

ELECTRO-MAGNETIC ANALYSIS AT ALTERNATOR OPERATING TEMPERATURE UNDER LOAD

A thermal model of STC's 55-We linear alternator was developed using the ANSYS™ finite element software to evaluate the alternator magnet temperatures for potential space missions. Many assumptions had to be made in preparing the thermal model. For instance, the Stirling convertor pressure vessel was assumed to have an unobstructed view to a radiation sink temperature of –40°C. This sink temperature is a conservative estimate for radiation from a body in space in near earth environment. These thermal conditions were selected to represent a worst-case condition for the planned deep space missions. The emissivity and absorptivity of the radiating surfaces were set to 0.91. Surface conductance between adjacent parts in contact with each other was set to 0.31 W/°C-cm². The efficiency of the alternator was assumed to be 85%. No convection heat transfer was modeled. Assuming a Stirling convertor cold-end temperature of 120°C, the thermal model predicted that the average magnet temperature would be approximately 75°C. At this temperature, the linear alternator magnets have a remanence B_r of 1.213 Tesla and a coercive force H_c of –900 kA/m.

Figure 3 shows a plot of the magnetic field strength (H) vs. distance along a pair of magnets attached to one stator leg for the mover positioned at the end of stroke. This is a mover position that leaves one of the magnets almost completely exposed. The model predicts that this is the most severe magnet loading over the entire stroke. The plot shows a maximum localized demagnetization field of roughly 725 kA/m on the inside surface of the uncovered magnet. At this magnet temperature, demagnetization can occur for demagnetization fields larger than 884 kA/m. The margin to demagnetization, defined as the difference between H at the knee of the normal induction curve and the maximum localized demagnetization field, is roughly 159 kA/m.

comparison with the magnet loading when the mover is at the end of stroke. The fringing fields in the magnet region near the advancing/receding edge of the mover due to the load current appears to be less severe than for the case where the magnet was almost completely exposed. The margin to demagnetization is roughly 224 kA/m.

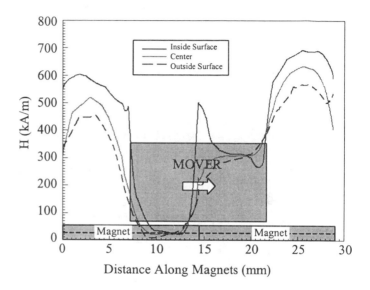

Figure 4 – Magnetic Field Strength $|\vec{H}|$ vs. Distance along Magnets at 75°C; Maximum Induced Voltage (Due to Magnet Flux)

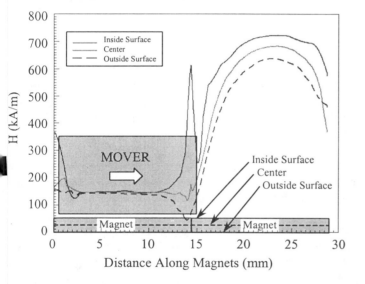

Figure 3 – Magnetic Field Strength $|\vec{H}|$ vs. Distance along Magnets at 75°C; End of Stroke

Figure 4 shows a plot of H vs. distance along the magnets for the mover positioned at mid-stroke. This mover position corresponds to the instant in the cycle of the peak-induced voltage in the coils due to magnet flux. The plot shows that the maximum demagnetization field is about 690 kA/m. The margin to demagnetization is 194 kA/m.

Figure 5 shows a plot of the magnetic field strength (H) vs. distance along the magnets for the mover positioned at maximum current flow. The current lags the induced voltage by approximately 40°. Therefore, the mover position shown in the figure is offset from its mean position. The plot shows a localized maximum demagnetization field of approximately 660 kA/m. The magnet loading is less severe in this case in

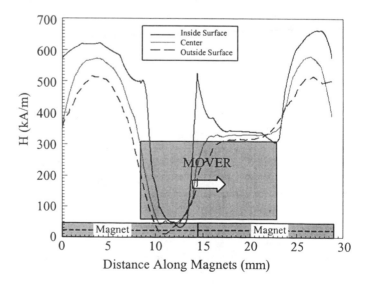

Figure 5 – Magnetic Field Strength $|\vec{H}|$ vs. Distance along Magnets at 75°C; Maximum Load Current

443

Both Figures 4 and 5 show an interesting characteristic of the linear alternator due to the load current. For the portions of the magnets covered by the mover, the effect of the load current tends to decrease the demagnetization field for one magnet while it tends to increase the demagnetization field for the other magnet over a half-cycle. The effect reverses itself during the second half of the cycle for each magnet. Even though the load current can boost the demagnetization field in the covered portion of the magnet, the demagnetization field for the uncovered portion of the magnet is substantially larger. Figures 4 and 5 also show a small kink in the demagnetization curve for the inside surface of the magnets adjacent to the trailing edge of the mover. This kink might represent a minor fringing effect at this location.

MAGNET LOADING AT VARIOUS TEMPERATURES

The thermal model of STC's 55-We linear alternator was used to evaluate alternator magnet temperatures for various Stirling convertor cold-end temperatures. For Stirling convertor cold-end temperatures of 100°C and 80°C, the average magnet temperatures were calculated to be approximately 62°C and 53°C, respectively. The linear portion of the magnet's normal induction curve at each temperature was used in the magnetostatic model of the linear alternator to evaluate the margin to magnet demagnetization as a function of temperature at the most severe magnet load condition (i.e. end of stroke). For a magnet temperature of 62°C, the maximum localized demagnetization field was calculated at 745 kA/m. At this magnet temperature, demagnetization can occur for demagnetization fields larger than 1018 kA/m. The margin to demagnetization is roughly 273 kA/m. For a magnet temperature of 53°C, the maximum localized demagnetization field was calculated at 760 kA/m. At this magnet temperature, demagnetization can occur for demagnetization fields larger than 1108 kA/m. The margin to demagnetization is roughly 348 kA/m.

Figure 6 shows a plot of three variables as a function of magnet temperature. The solid line represents the knee of the normal induction curve based on the magnet manufacturer's data. If the magnets are subjected to demagnetization fields above this line, some level of magnet demagnetization may occur. The long-dashed line in the figure represents the maximum localized demagnetization field affecting the magnets as predicted by the magnetostatic/load current model. The short-dashed line is the predicted margin to magnet demagnetization for TDCs 5, 6, 7, and 8. This plot shows that the maximum localized demagnetization field for the alternator magnets decreases as temperature increases but at a lower rate than the decrease in the magnet's ability to resist demagnetization. As a result, the margin to demagnetization decreases as temperature increases. The vertical lines shown

on the plot indicate the magnet temperatures calculated for the various Stirling convertor cold-end temperatures.

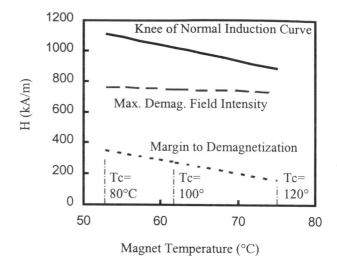

Figure 6 – Sensitivity of Resistance to Demagnetization, Maximum Localized Demagnetization Field Intensity and Margin to Demagnetization as a Function of Temperature

Table 1 summarizes the margin to demagnetization numbers presented in this paper and compares these margins on a percentage basis calculated in two different ways. The column in the Table labeled "Margin Based on Max Localized Demag. Field" computes the percentage by taking the difference between the demagnetization field limit and the maximum localized demagnetization field and dividing by the demagnetization field limit. The column labeled "Margin Based on Demag. Field Avg. Over Magnet Vol." computes the percentage by taking the difference between the demagnetization field limit and the demagnetization field averaged over the magnet volume and dividing by the demagnetization field limit. The demagnetization field limit represents the knee of the normal induction curve at the given magnet temperature.

Table 1 illustrates that the margin to demagnetization can vary widely depending upon the method chosen to perform the calculation. There is a large difference between the localized maximum demagnetization field and the volume averaged demagnetization field for a given magnet. A magnet that appears safe based on its volume averaged demagnetization field may, in fact, be in jeopardy.

Table 1 – Margin to Magnet Demagnetization

Avg. Magnet Temp. (°C)	Max. Localized Demag. Field (kA/m)	Demag. Field Avg. Over Magnet Volume (kA/m)	Demag. Field Limit (kA/m)	Margin Based On Max. Localized Demag. Field (%)	Margin Based On Demag. Field Avg. Over Magnet Vol. (%)
53	760	550	1108	31	50
62	745	539	1018	27	47
75	725	527	884	18	40

CONCLUSIONS

The results shown by this preliminary 3-D analysis indicate that the highest potential for demagnetization appears to be along the inside surface of the uncovered magnets. The demagnetization field for the uncovered magnet when the mover is at the end of stroke is higher than for when the mover is at the position of maximum induced voltage or maximum load current. The demagnetization field and the magnet's resistance to demagnetization decrease with increasing temperature; however, the magnet's resistance to demagnetization decreases at a higher rate. The magnet's margin to demagnetization decreases as magnet temperature increases as expected. Based on the results presented in this paper, it appears that the 55-We TDCs built for NASA can operate safely at magnet temperatures up to 75°C, provided that no abnormal load conditions occur.

ACKNOWLEDGMENTS

The authors wish to acknowledge the Stirling Technology Company for providing hardware information and data relative to the DOE/STC Technology Demonstration Convertor.

REFERENCES

[1] Thieme, L.G., Schreiber, J.G.: "NASA GRC Technology Development Project for a Stirling Radioisotope Power System," in Proceedings from 35th Intersociety Energy Conversion Engineering Conference, Las Vegas, Nevada, AIAA-2000-2840, NASA TM-2000-210246, 2000.

[2] Thieme, L.G., Qiu, S., White, M.A.: "Technology Development for a Stirling Radioisotope Power System," NASA TM-2000-209791, 2000.

[3] White, M.A., Qiu, S., Augenblick, J.E.: "Preliminary Test Results from a Free-Piston Stirling Engine Technology Demonstration Program to Support Advanced Radioisotope Space Power Applications," in Proceedings from STAIF-00, Space Technology and Applications International Forum-2000, Albuquerque, New Mexico, 2000.

[4] White, M.A., Qiu, S., Augenblick, J.E., Peterson, A., Faultersack, F.: "Status Update of a Free-Piston Stirling Convertor For Radioisotope Space Power Systems," in Proceedings from STAIF-01, Space Technology and Applications International Forum-2001, Albuquerque, New Mexico, 2001.

[5] Geng, S.M., Schwarze, G.E., Niedra, J.M.: "A 3-D Magnetic Analysis of a Linear Alternator for a Stirling Power System," in Proceedings from 35th Intersociety Energy Conversion Engineering Conference, Las Vegas, Nevada, AIAA-2000-2838, NASA TM-2000-210249, 2000.

[6] Ansoft's Maxwell 3-D Field Simulator, see World Wide Web page: www.ansoft.com

THE ELECTROMAGNETIC COMPATIBILITY (EMC) DESIGN CHALLENGE FOR SCIENTIFIC SPACECRAFT POWERED BY A STIRLING POWER CONVERTOR

Noel B. Sargent
Analex Corporation
3001 Aerospace Parkway
Brook Park, Ohio 44142

ABSTRACT

A 55 We free-piston Stirling Technology Demonstration Convertor (TDC) has been tested as part of an evaluation to determine its feasibility as a means for significantly reducing the amount of radioactive material required compared to Radioisotope Thermoelectric Generators (RTGs) to support long-term space science missions. Measurements were made to quantify the low frequency magnetic and electric fields radiated from the Stirling's 80 Hertz (Hz) linear alternator and control electronics in order to determine the magnitude of reduction that will be required to protect sensitive field sensors aboard some science missions. One identified "Solar Probe" mission requires a 100 dB reduction in the low frequency magnetic field over typical military standard design limits, to protect its plasma wave sensor. This paper discusses the electromagnetic interference (EMI) control options relative to the physical design impacts for this power system, composed of 3 basic electrical elements. They are: (1) the Stirling Power Convertor with its linear alternator, (2) the power switching and control electronics to convert the 90 V, 80 Hz alternator output to DC for the use of the spacecraft, and (3) the interconnecting wiring including any instrumentation to monitor and control items 1 and 2.

INTRODUCTION

An optimal solution to achieving EMC for the Stirling Radioisotope Power System (SRPS), (DC out for power distribution to the satellite), and protection of the electric and magnetic field instruments aboard the spacecraft requires the trade-off of many variables. This initial assessment is an attempt to quantify the magnitude of the EMI control issues, and to explain in broad terms the relationships that are at work in the determination of the design options. *Good EMI engineering enables the design process, and should not constrain it.*

The SRPS is assumed to be composed of 3 basic electrical elements: (1) the Stirling Power Convertor (SPC) with its heat source and linear alternator (LA) as shown in Fig. 1, (2) the power switching electronics (PSE) and associated control electronics to convert the nominal 90 V, 80 Hz SPC output to DC for the use of the spacecraft, and (3) the interconnecting wiring including any instrumentation to monitor or control items 1 and 2. More subtly involved are stray capacitance and inductive effects between these components and structure that cause currents to flow in paths that are not on the electrical schematic! There is significant stray capacitance between the stator coils and structure whether we like it or not. These leakage paths are also very significant for the power conversion and control electronics since they operate at higher frequencies. Furthermore these stray current paths take different routes (the one of least impedance) as a function of frequency.

THE METHODOLOGY OF EMI CONTROL

The first instinct is to solve all electromagnetic radiated field problems by shielding. Actually this is the last line of defense to be applied after proper electrical grounding (controlling current flow in structure and circuit loop area) and filtering (frequency-band limiting of unintentional currents). Indeed shielding will be required, particularly to control low frequency magnetic fields to protect the very sensitive science instruments. It is the overriding goal of this paper to sort out all the effects important to specifically

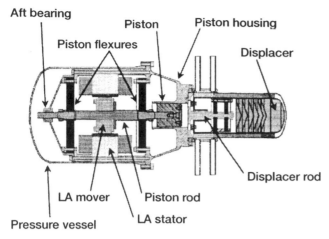

Figure 1.—Stirling power convertor.

solving the design issues of Stirling, so that informed decisions on weight, thermal and structural requirements can be made while providing minimally intrusive EMI design and shielding solutions. Project managers must provide for an EMI control plan that defines the EMI control methods in the very earliest stages of design if optimal solutions are to be applied. Design retrofits are rarely sufficient, or efficient!

THE LINEAR ALTERNATOR

The first area to attack (but perhaps not the most difficult) is that of the electrical machine, to ensure that its design results in the lowest residual magnetic field possible. It is not the purpose of this report to mandate specific design fixes but rather to suggest a list of options that will result in controlling the magnetic path to the maximum extent possible. These may include a copper band called a "shading ring" intimately contacting and completely covering the perimeter of the core laminations and stator coils at the outside boundary of the machine to control the eddy currents at the surface and hence the leakage flux. The shading ring, also known as the "shorted turn method," is a technique used on open-core transformers that may have merit for the proposed LA design. Then choose a material for the pressure vessel, or combination of materials, that gives improved magnetic shielding performance over 304 stainless steel alone (no real weight penalty here). Shielding materials will need to be an integral part of the electromechanical design for the entire SRPS. The mechanical aspects of the shielding design are equally as important as choosing the proper material. Close attention is required to holes, seams, sharp transitions, and wire penetrations that can destroy most of the shielding effectiveness of the chosen material. Additionally, care should be taken to minimize the total loop area of the interwinding connection path prior to exiting the machine. The fact that the flight SPC will be a

hermetically sealed unit is a big plus, however care must be taken to control all penetrations so as not to destroy its inherent shielding effectiveness.

THE SPC CONTROLLER

An electronic controller is necessary to regulate the LA output power of the SPC in a way that limits the travel distance of the LA mover. The method of control used on the TDC was a zener diode that clipped the LA voltage and hence disturbed the current waveform. This produces harmonic distortion that shows up both in the output current waveform, and in the radiated magnetic field, beyond the 40th harmonic, decreasing at the rate of 40 dB per decade of frequency, as shown in Figs. 2 and 3 respectively. The flight controller design should be one in which the electrical current output is sinusoidal so that the harmonic content is dramatically reduced. Such a digital controller design has been demonstrated at GRC with a reduction in harmonic content of 30 dB for the 2nd and 3rd harmonics and at least 60 dB beyond the 8th harmonic. Should the magnetic field shielding design prove too heavy to achieve the "solar probe" mission magnetic field requirement of -30 dBpT/$\sqrt{Hz}$ at 1 meter, it is far more acceptable to relax the limit at a single frequency that can be accounted for as a single line in the scientific data spectra, than throughout the entire 150 kHz passband of interest.

EFFECTS OF INTERCONNECTING WIRING

Maxwell's equations have been reduced to the absolute simplest algebraic form possible to more easily emphasize important concepts related to design trade-off. In most cases the equations are presented in logarithmic form (the EMC engineer's favorite) so that they directly fit the "limits" of the EMI requirements.

Assume that 1 A of sinusoidal current is flowing at the power frequency. For current flowing in a long single wire, the magnetic field at distance "r" (1 meter in this case) can be simply expressed as dBpT = dBμA − 14. One amp (120 dBμA) produces a 106 dBpT magnetic field, or a whopping 136 dB over the science requirement of -30 dBpT/$\sqrt{Hz}$ at 1 meter distance. Notice that the field intensity is linear with distance. The SPC does not have one wire, but a closely spaced pair with equal (almost) and opposite current flow. However, any common-mode current circulating between the SPC and the PSE through structure will have this "single wire" characteristic. So it only takes 0.16 μA of common mode current to exceed the requirement [−30 dBpT = 20 log(0.16) − 14]. This could be a tough design challenge and so a high permeability solid tube between the SPC and the case of the PSE may be warranted as a conduit for all wiring. If the dimension of the radiating source is comparable in

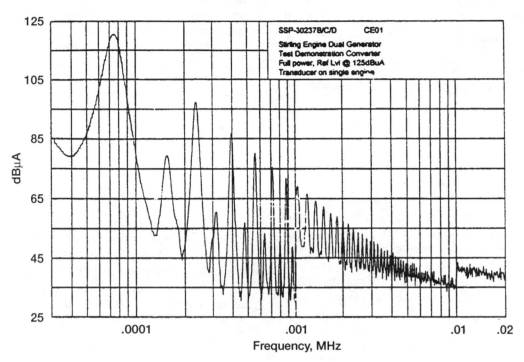

Figure 2.—CE01, 30 Hz-20 kHz, powerline conducted emissions.

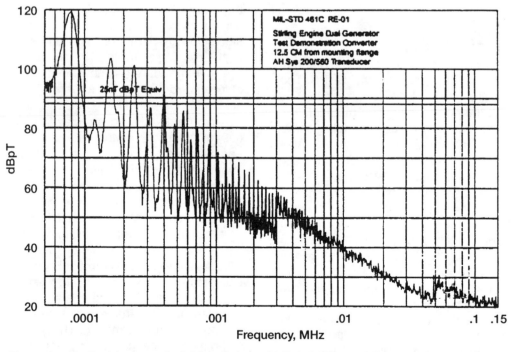

Figure 3.—Radiated magnetic field emissions, 30 Hz-15 kHz at 12.5 cm from convertors.

449

size to the distance at which the field is measured, a current loop may also be an appropriate model.

For two closely spaced wires carrying the same but opposite current, the magnetic field at 1 meter is $dBpT = dB\mu A - 14 - 20\log(\frac{s}{r^2})$, where "s" is the spacing between the wires. Note the field from closely spaced wires varies as $\frac{1}{r^2}$. Assuming the 1 A wires are spaced 5 mm apart, $dBpT = 120\ dB\mu A - 14 - 20\log(0.005)$, or 60 dBpT, still a huge 90 dB over the limit. Twisting of the two wires gives roughly another 12 dB reduction dependent on load impedance and circuit balance. There are other wiring configurations using twisted quads and flat-plate lines that are worth investigating if the interconnecting lines are very long, however another 20 dB or so is all the help one should count on. So there is great advantage of co-locating the PSE at the SPC if at all possible or as mentioned, containing the wiring in a solid tube if co-location is not feasible.

For compact magnetic sources with a closed, tightly coupled magnetic path, like transformers, motors, and the SPC, the magnetic field falls as roughly $\frac{1}{r^3}$ as long as the source is small compared to the measurement distance "r." Models have been developed to calculate this residual flux from the linear alternator by Geng et al. Measurements with the LA alone in the near future would aid this modeling effort.

THE PHYSICS OF SHIELDING MATERIALS

Shielding materials will certainly need to be an integral part of the electromechanical design of the SRPS. The mechanical aspects of the shielding design are equally as important as choosing the correct shielding material. It is worth restating that lack of attention to seams, holes, sharp radius transitions, and wire penetrations can destroy most of the shielding effectiveness of the chosen shielding materials.

The total shielding effectiveness (SE) of a material is the sum of 3 factors: the absorption loss (A), the reflection loss (R), and a multiple reflection loss (B) for thin shields. The multiple reflection loss is of little practical importance if the absorption loss is greater than 10 dB as will be the case in the SRPS design, so that SE = A + R, in dB. As an impinging electromagnetic wave (Ei) encounters a conductive or ferromagnetic material, its amplitude decays exponentially by producing current flow in the material and hence ohmic losses and heating of the material. The resultant exiting wave out of the shield (caused by the current that is actually able to flow completely through the material), (Es) obeys the equation $E_s = E_i^* e^{-\frac{t}{\delta}}$, where t (in meters) is the material thickness, and δ (in meters) is the "skin depth"

or the distance required for an attenuation of 1/e or 37 percent of the impinging value of the field. These expressions can be further simplified to give an absorption loss term $A = 8.7(t/\delta)$ dB, so that the absorption loss through any material is simply 8.7 dB per "skin depth." Engineering charts are available listing skin depths of various materials. However, skin depths are a nonlinear function of frequency for each material. This means the best shielding material (for absorption loss) at low frequency is not the best material for higher frequencies. From a practical point of view that crossover point between high conductivity materials (copper) and high permeability materials (mu-metal) is between 100 kHz and 1 MHz. This means that the shielding material design for the LA magnetic field emissions will be different from that of the power conversion electronics.

The reflection loss term is not as simple to visualize since the loss depends on the impedance of the wave and the media at the shield interface (usually air). The largest reflection (best attenuation) occurs at the first boundary (entering) of the material for electric (high impedance) fields, but at the secondary boundary (exiting) for magnetic (low impedance) fields. In the case of electric field shielding, very thin sheets (foils) provide good SE, however the SE of very thin sheets is much less for magnetic fields. Because of the effect of wave impedance, it is always advantageous to locate the shield as close to the source of electromagnetic energy as possible (in the near field). So, the reflective loss (best in high conductivity materials) varies with frequency, distance from the source, and wave impedance. Wave impedance (high or low compared to free space of 377 Ω) is best intuitively understood if one remembers this relationship: circuits containing high fluctuating currents (dI/dt) produce magnetic fields which are low impedance, and circuits containing high dV/dt compared to the current (dV/dI > 377 Ω), produce electric fields which are high impedance.

In order to take advantage of the best material properties for optimizing both absorption and reflection, copper and magnetic alloy materials may be combined into a sandwich. This combination not only provides maximum reflection at low frequencies where energy absorption is low for both low impedance and high impedance waves, but also increases absorption at high frequencies.

To summarize total shielding effectiveness, the bottom line is shielding low frequency magnetic fields (<100 kHz) requires a thicker high permeability material (mu-metal), and shielding electric fields at all frequencies requires only thin but high conductivity materials (copper). Remember the shield must be a sardine can—no seams or penetrations—for maximum performance. This is the challenge for real world mechanical design. In practice, the actual total SE is limited by the number of thin slots, penetrations, and the shape of the enclosure. Various means can be used to overcome the shield penetrations, like creating "thick holes" by using extension tubes at apertures. Wiring penetrations (and to some extent mechanical fasteners) are by far the most difficult to control since the wires also efficiently conduct coupled energy through the shield, free to re-radiate on the other side. Filtering is the only hope here, and low frequencies are very difficult to filter.

POWER CONDITIONING ELECTRONICS

It is necessary to convert the nominal 90 V, 80 Hz output of the linear alternator to DC voltages useful to the spacecraft. This conversion is accomplished by the PSE. Since the EMI science requirements are most stringent below 150 kHz, the power switching frequency should be at a frequency above 150 kHz, well within today's state-of-the-art. This also puts the unintentional emissions of the switching electronics at a frequency that is much more efficient to filter in terms of filter component size and weight when compared to lower switching frequencies. It is always more effective to provide filtering for wires exiting an electronic circuit, than it is to control the resultant radiation by shielding. The PSE should be in a completely shielded module (sardine can) with filtered input and output wiring. Additional metal partitioning within the box is an effective means to control coupling from the power electronics to the control or instrumentation functions. The PSE should be located as close as possible and preferably integral to, the SPC to reduce the interconnecting lead lengths toward zero.

THE INSTRUMENTATION ISSUE

In the near term development configuration, it will be very difficult to directly verify the "solar probe" magnetic field science requirement of -30 dBpT/$\sqrt{\text{Hz}}$ at 1 meter. Typical EMI loop antennas and flux gate magnetometers are orders of magnitude too insensitive to provide this measurement. Most coil-type magnetic field antennas are simply several turns of wire in an electrostatic shield (so that the electric field is not measured; only the magnetic field). Simply stated, the voltage developed at the terminals of the loop antenna is equal to the number of turns, times the loop area, times the frequency, times the field being measured ($V = 2\pi nAfB$). The typical sensitivity of a top of the line oscilloscope or spectrum analyzer is 0.1 μV, so that at low frequency it is difficult to use an antenna of this type. Also the facility required to reduce the ambient background noise at 60 Hz, to a level that will not saturate the measurement system, may be expensive and bothersome to maintain.

Alternate methods may be used for verification of this requirement that in the end will be more accurate and repeatable. This involves an additive sequence of measurements that will allow direct calculation of the base requirement. Once the radiating sources and rate of decay of their fields have been established, measurements may be made at closer distances. The shielding effectiveness of the pressure vessel itself can be measured, as well as shielded versus unshielded test configurations. Wiring currents can be measured and resultant fields calculated using expressions similar to those developed in this paper, once the circuit and field generating geometries are known. It may well be that the linear alternator is the easiest element of the SRPS to control because of its hermetically sealed configuration, dependant on the selection of materials allowed for the pressure vessel.

Care should be taken in the construction of any test engine fixtures where magnetic field measurements are important, to limit the amount of extraneous magnetic material. This extra material may alter the radiated magnetic flux path in a way that is detrimental to obtaining accurate, representative test data.

The complement of instrumentation should contain a flux gate magnetometer for the DC and low frequency (<100 Hz) magnetic field, and an induction coil sensor for the AC magnetic field. Sources for each type are being explored with candidate types identified that will meet the requirements. In addition it is almost a necessity to have a low frequency (10 Hz to 200 kHz) spectrum analyzer (not just an oscilloscope) capable of resolution bandwidths of 1 Hz or below (as a means of controlling the noise floor of the measurement).

CLOSING REMARK

It is hoped that this simplified view of the EMI design challenge will provide an insight into the physics of these interactions. This paper is primarily intended to sensitize the multidiscipline design team that only through frequent interaction, can the level of EMI control required by the SRPS be maintained throughout the design, and particularly the fabrication process. For this reason, EMI design engineering decisions must be considered beginning with the conceptual design phase of the program.

REFERENCES

Steven M. Geng, Gene E. Schwarze, "A 3-D Magnetic Analysis of a Linear Alternator for a Stirling Power System," NASA/TM—2000-210249, 2000.

Richard Furlong, Richard Shaltens, "Technology Assessment of DOE's 55-We Stirling Technology Demonstrator Convertor (TDC)," NASA/TM—2000-210509, 2000.

Clayton R. Paul, "Introduction to Electromagnetic Compatibility," 1992.

Henry W. Ott, "Noise Reduction Techniques in Electronic Systems," Second Edition, 1988.

Donald R.J. White, "Electromagnetic Interference and Compatibility," Vol. 3–EMI Control Methods and Techniques, 1978.

Proceedings of IECEC'01
36th Intersociety Energy Conversion Engineering Conference
July 29–August 2, 2001, Savannah, Georgia

IECEC2001-CT-41

FABRICATION AND USE OF PARALLEL PLATE REGENERATORS IN THERMOACOUSTIC ENGINES

S. Backhaus and G. W. Swift
Condensed Matter and Thermal Physics Group
Los Alamos National Laboratory
Los Alamos, NM 87545

ABSTRACT

Diagnostic and performance measurements on a thermoacoustic-Stirling heat engine (TASHE) utilizing a parallel-plate regenerator are presented. The diagnostic measurements demonstrate that the spacing between the regenerator plates is very uniform and close to the design spacing of 102 μm. Achieving a high degree of uniformity, which is crucial to the regenerator performance, requires careful fabrication. The details of this process are described. The performance measurements show that the viscous losses in the regenerator are quantitatively understood while unforeseen heat leaks show that the heat transfer properties are only qualitatively understood. By switching from a screen-based regenerator to a parallel-plate regenerator, the power output of the TASHE was nearly doubled with a significant increase in efficiency at the highest acoustic power output.

INTRODUCTION

Regenerators in Stirling or Stirling-like machines are designed to provide sufficient thermal contact to the working gas to minimize loss due to irreversible heat transfer while generating as little viscous loss as possible. Theoretically, parallel-plate regenerators offer a significant performance advantage over screen-based regenerators. In comparing these two types of regenerators, it is useful to look at the ratio $StPr^{2/3}/f$ where St and Pr are the Stanton and Prandtl numbers respectively and f is the friction factor[1]. This ratio becomes larger as heat transfer improves or viscous loss decreases. For steady laminar flow through parallel plates, this ratio is ~0.4 and is independent of Reynolds number. For screen beds with properties typical of Stirling machines and even at a Reynolds number of ~20, this ratio is ~0.1. This difference provides a strong motivation to use the parallel-plate geometry in Stirling regenerators.

However, small variations of the plate spacing can negate any gains from the change of geometry. The viscous flow resistance in the gap between two parallel plates is $\propto 1/h^3$ where h is the plate spacing. Typically h is around 100 μm in engines and 25 μm or smaller in cyrogenic refrigerators. Therefore, variations in h on the order of 10 μm from channel to channel lead to ~30% variations in the flow resistance in an engine and >200% variations in a refrigerator. In a parallel-plate regenerator with flow-channel gaps distributed about some average gap, the wider gaps will garner more of the flow per cross-sectional area than the narrow gaps. In addition, the thermal contact between the gas in the gaps and the plates falls off as 1/h. Therefore, the gaps with the poorest thermal contact have to handle the majority of the heat transfer. This leads to high thermal losses.

Over the past few years, we have developed a thermoacoustic-Stirling heat engine (TASHE) that generates in-phase pressure and velocity oscillations in a regenerator and, therefore, utilizes a Stirling-like thermodynamic cycle[2]. The engine resembles a "gamma-style" Stirling engine where the dynamic properties of the displacer piston are achieved by the inertia of the gas in an "inertance" tube, and the thermal-insulation properties of the displacer piston are achieved by thermally stratified gas motion in a "thermal buffer" tube. Also, the power piston is replaced by the inertia of the gas in an acoustical resonator attached to the TASHE and by deliberate dissipation in a variable acoustic load impedance. We will not describe the TASHE in detail here, but instead refer the reader to Ref. [2]. Recently, we have replaced the TASHE's screen-based regenerator with a parallel-plate regenerator. This article reports on the fabrication and testing of the regenerator as well as the performance of the TASHE while using it.

REGENERATOR FABRICATION

The completed parallel-plate regenerator is a cylinder 88.9 mm in diameter and 73.0-mm long. It is constructed from a stack of alternating stainless-steel sheets shown in Fig. 1; 1,500 sheets are required to complete the regenerator. Both sheet A and sheet B are fabricated by photochemical milling (PCM) of 316L stainless-steel sheet. This alloy was chosen because of its availability and good diffusion bonding properties. Other, lower thermal conductivity metals, such as Inconel 625, could have been used. Sheet A, which is 104-μm thick, has 1.02-mm-wide by 73-mm-long windows etched through it. The windows are spaced 1.27 mm on-center leaving 0.25-mm-wide fingers separating the windows. Along their long dimension, which is also the flow direction, the windows are interrupted at either end and in the middle by 0.25-mm-wide webs that are etched approximately half-way through, i.e. 52-μm deep. The webs provide lateral support for the extremely long, thin fingers, and the half etches provide spacing for the oscillatory flow to enter and exit the windows as well as cross over the middle web. Sheet B, which is 25-μm thick, has 1.02-mm square holes etched though it. When stacked above and below Sheet A, the square holes are centered over the middle set of webs on Sheet A and provide additional area for the flow to cross over the middle web. In a trial PCM run, square notches were also etched at the edges of Sheet B to provide additional room for the entering and exiting flow, but these features made the edges of Sheet B jagged, extremely delicate, and difficult to handle.

Figure 2 shows Section CC, a cross section perpendicular to the flow direction. Sheet B forms the two long walls of many 1.02-mm wide by 104-μm high flow channels. The 1.02-mm width of the channels is set by the width of the PCM'd windows in Sheet A. The crucial, 104-μm height dimension is determined by the accurately rolled thickness of Sheet A. When the A sheets were received from the PCM manufacturer, the variation of the unetched thickness from sheet to sheet and along any one sheet was less than 3 μm. The B sheets showed similar small variations. However, the half-etched regions in the A sheets, shown in Section DD, showed wide variations. On some A sheets, the webs were etched completely through. On others, the half-etched region was 81-μm thick, i.e. only 23 μm etched away.

This wide variation in the half-etch thickness is the main reason why half-etching should not be used to set the important dimensions of the flow channels in a parallel-plate regenerator.

After the PCM process is complete, the next steps are cleaning, sorting, stacking, and diffusion bonding of the stacked sheets. The PCM process is performed on "parent sheets" that contain 9 of the A or B sheets. The cleaning is done before removing the A and B sheets from the parent. For the B sheets, a dilute nitric-acid etch is followed by a de-ionized water rinse and then an isopropyl-alcohol rinse. Then, the sheets are dried

with a heat gun. The A sheets are too delicate to clean in this manner. Water spots are removed with isopropyl alcohol and a cotton swab. Following this, any handling of the sheets is done while wearing latex gloves.

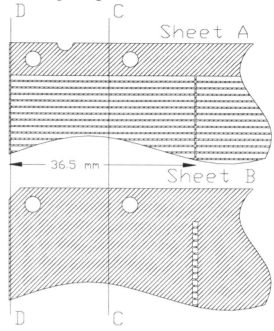

Figure 1. Plan view of the upper left corner of Sheet A and Sheet B. The oscillating flow is from left to right. The hole in Sheet A near the middle web is used to ensure all the A sheets are stacked with the half-etched side up. The area near the top of Sheet A and B is used only for the alignment holes. It is machined away later in the fabrication. Except for the extra hole mentioned above, the sheets are symmetric about the horizontal and vertical centerlines.

During the PCM process, a 0.25- to 0.50-mm gap is etched through around the perimeter of the sheets except in four locations where thin tabs hold the individual sheets in the parent sheet. The individual sheets are removed by cutting the tabs with either a razor blade or surgical scissors. During removal, the individual sheets are sorted into "good," "useable," and "rejects." Rejects include A sheets with torn or etched through webs, A sheets with bent fingers, sheets with incompletely etched windows, or sheets with excessive water spotting. Moderate bends in the A-sheet fingers, which looked like they could be straightened, were deemed "useable." As required to complete the full stacking height, the "usable" sheets were repaired.

An additional detail in the PCM process eases the removal and later stacking of the individual sheets. When cutting the tabs on Sheet B, burrs are raised that would interfere with the stacking. Half circles are etched in Sheet A that, when interleaved with Sheet B, are aligned with the burrs on Sheet B and provide clearance for them. The corners of Sheet B are etched away for a similar reason. Without these details shown in Fig. 1, the burrs would have to be cut flush with the sheets.

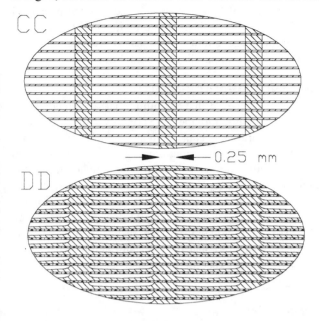

Figure 2. Enlarged view of Sections CC and DD after stacking alternating Sheets A and B. The oscillating flow is into and out of the page. In CC, the rectangular fingers of Sheet A are 104-μm high by 0.25-mm wide. In DD, the half-etched webs of Sheet A are shown. Although they are drawn as if they are very uniform, the half-etch depths show wide variation.

The sheets are then stacked onto a thick 316L plate that holds four 316L registration pins to align the sheets. The pins fit snugly into PCM'd holes near the corners of the individual sheets. Although the registration pins hold the perimeters of the sheets in excellent alignment, the delicate half-etched webs in the A Sheets are often slightly distorted from handling of the sheets leading to misalignment of several of the fingers. Later in the fabrication process, misalignment of a finger would lead to buckling around the misaligned finger and many nearby fingers, resulting in collapsed or irregular flow channels. Therefore, the fingers must be inspected and their alignment individually corrected during the stacking. One technique that was found useful involves stacking up many tens of the sheets on a separate fixture, using a pointed scalpel to align the fingers under a low-power microscope, transferring the short stack onto the main stack, and realigning the interface layers under the microscope. After a sufficient stack height is achieved (95 mm in this case, allowing for machining waste and plastic strain generated during the diffusion bond), a second thick 316L plate is added to the top of the stack.

Next, the stack is vacuum hot pressed by standard techniques[3] to form a diffusion bond between the layers. After fabrication, the regenerator will not be subject to large stresses, so bond strength is not crucial. Avoiding buckling of the 95-mm-high, 0.25-mm-wide columns shown in Fig. 2 is of the highest importance. The Euler formula gives the buckling stress as[4] $\sigma_b = C\pi^2 E/(L/r)^2$ where $E \sim 18 \times 10^6$ psi is the elastic modulus at the bonding temperature, L is the 95-mm height of the stack, and r is the smallest radius of gyration of the column (~ 0.289 W, where W is the 0.25-mm width of a single column). Also, C is a constant between 1 and 4 that depends on the condition at the ends of the column. It should be taken as 1 for a conservative estimate of the buckling stress. During the vacuum hot pressing of this regenerator, the largest applied load never exceeded ½ of the estimated buckling stress. Also, the temperature was ramped up to and down from the bonding temperature slowly to avoid temperature differences between the interior and exterior of the part. Large temperature differences generate large thermal stresses, which would distort the freestanding sections of the B sheets. A post-bond visual inspection of the regenerator showed no buckling of the columns due to overloading or misalignment and no distortion of the B sheets. Also, the interlayer bonds were found to be of sufficient strength.

The bonded stack of sheets is then machined[5] into a cylinder by electric discharge machining (EDM). Within one flow-channel width around the cut, the EDM process distorts the B sheets considerably. For this reason, cuts perpendicular to the flow direction should be minimized. Only clean water should be used during the EDM process and the faces of the regenerator should be masked to minimize the amount of EDM "dust" left inside the flow channels. The resulting cylinder is then pressed into a thin-walled stainless-steel sleeve (0.5-mm wall) to protect it from damage and ease its installation and removal from the TASHE. To obtain a tight fit between the regenerator and the sleeve, the I.D. of the sleeve is machined a few thousandths of an inch undersize and heated while the regenerator is pressed into place. The regenerator should not be "shrunk-fit" into the sleeve, because the thermal shock to the regenerator might distort the freestanding B sheets. To remove any EDM dust that may have been introduced into the flow channels, the regenerator is filled with isopropyl alcohol (surface tension holds the alcohol in the channels), which is then blown out with dry nitrogen. This process is repeated until the alcohol runs out clean. EDM distortion of the flow channels around the O.D. of the regenerator is too severe to allow those channels to be accessed by the oscillating flow in the TASHE. Therefore, the outer layer of channels is blocked at the cold end of the regenerator using Stycast 2850 epoxy. The fractional loss of flow area is only ~4% in a regenerator with this large diameter.

DIAGNOSTIC MEASUREMENTS

Visual inspection of the ends of the completed regenerator (Section DD) shows that the B sheets remained flat and parallel, at least at the ends of the regenerator. A steady-flow pressure-drop test is performed to quantify the uniformity and dimensions of the channels.

To measure the steady-flow pressure drop, the regenerator is sealed into a metal housing using O-rings to ensure no flow can pass around the outside of the regenerator sleeve. The housing is glued to a 1.8-m-long entrance section and a 0.6-m-long exit section. Both are made from PVC pipe with an I.D. slightly larger than the regenerator O.D. A copper screen flow straightener is placed 1 cm from the entrance side of the regenerator. The exit from the test section leads into a 0.8-m-long pipe, which houses a Laminar Flow Element[6] (LFE) used to measure the flow rate. Room temperature, atmospheric pressure, dry argon is supplied at the end cap of the entrance. The temperature of the argon is measured with a type-K thermocouple located between the flow straightener and the regenerator. The pressure drops across the regenerator and LFE are measured with two differential, Bourdon-tube pressure gauges. The LFE and the two pressure gauges are calibrated to NIST-traceable standards.

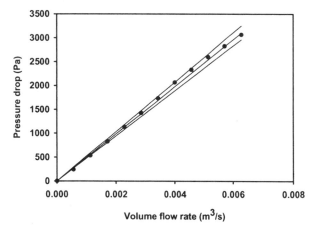

Figure 3. Pressure drop across the parallel-plate regenerator due to steady flow. The circles are the measured data and the upper, middle, and lower lines are the calculated pressure drops for channel heights of 102, 104, and 107 μm respectively.

The pressure drop across the regenerator is measured for flow rates up to 0.007 m^3/s, i.e. 1.7 m/s in the channels. The results are shown as the circles in Fig. 3. For the channel height of 104 μm and width of 1.02 mm, this corresponds to a Reynolds number based on hydraulic diameter Re_{Dh}=18. At this low Reynolds number, the flow in the regenerator is laminar. Also, entrance and exit pressure losses at the ends and middle of the regenerator are estimated to be small. Therefore, the total pressure drop can be calculated from the laminar friction factor[1]. The pressure drop at the ends and middle, where the geometry changes due to the half-etched webs, is calculated assuming the flow in these short regions to be laminar and fully developed. These three short sections contribute about 5% of the total pressure drop, so this

assumption should not significantly affect the calculation of the total pressure drop. The measurements are in closest agreement with a 104-μm channel height, which is the average measured thickness of the A sheets. This result, coupled with the visual inspection of the regenerator, leads us to believe the flow channels are uniform and parallel to about 3 μm.

PERFORMANCE MEASUREMENTS

When we are satisfied that we built what we intended, the regenerator is installed into the TASHE. A complete set of performance measurements is taken in a similar fashion to those reported for the TASHE operating with a screen-based regenerator[2]. In summary, heat is supplied to the hot end of the regenerator by an electric resistance heater, and waste heat is removed from the cold end of the regenerator by water flowing through a shell-and-tube heat exchanger. The power produced by the TASHE, and delivered to the resonator and variable acoustic load, is measured near the resonator entrance. Here, we report on measurements specific to the regenerator.

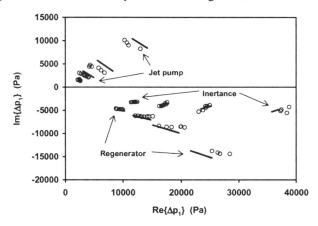

Figure 4. Measured and numerically calculated pressure-drop phasors across the feedback inertance, jet pump, and regenerator. Data are taken at fixed values of $p_{1,c}/p_m$ equal to 0.038, 0.051, 0.069, and 0.10. For the different pressure drops, $p_{1,c}/p_m$ is higher for the data farther from the origin. At each $p_{1,c}/p_m$, T_h is swept from the minimum T_h required to reach that amplitude to a maximum of 725°C by adjusting the variable acoustic load on the TASHE.

With the geometry of the parallel-plate regenerator confirmed by visual inspection and flow testing, the oscillatory pressure drops in the TASHE are measured to confirm our understanding of the acoustic field inside the TASHE. If the pressure drops across the various elements in the loop of the TASHE agree with numerical calculations[7], then the numerical code is making reasonably accurate predictions of the volumetric-flow-rate phasors throughout the TASHE. Figure 4 shows the measured and calculated pressure-drop phasors across the feedback inertance, jet pump, and regenerator. The various groups of data points (open circles) are taken at fixed $p_{1,c}/p_m$ while T_h is varied from the minimum value at which the engine runs at $p_{1,c}/p_m$ to a maximum of 725°C. The solid lines connect the calculated values of the pressure-drop phasors at the lowest and highest T_h for each value of $p_{1,c}/p_m$. Here, p_m is the mean pressure of helium gas (450 psia) inside the TAHSE, $p_{1,c}$ is peak oscillating pressure amplitude measured at the cold end of the regenerator, and T_h is the temperature of the regenerator sleeve at the hot end of the regenerator. For this and all subsequent data, the temperature at the cold end of the regenerator, T_c, is between 20°C and 60°C. At low values of $p_{1,c}/p_m$ the measured pressure-drop phasors are in good agreement with the calculations. At higher $p_{1,c}/p_m$, the ~10% discrepancies are due mainly to our limited knowledge of the acoustic impedance of the jet pump[2].

With some confidence in the numerical predictions of the volumetric-flow-rate phasors, the measured acoustic power flow and heat flows are compared with the numerical predictions. Figure 5 shows the acoustic power flow into the resonator generated by the TASHE for the same values of $p_{1,c}/p_m$ as in Fig. 4. The numerical calculation is in excellent agreement with the data, confirming our understanding of the TASHE acoustics, volumetric flow rates, and the regenerator pressure-drop properties. Also, the acoustic power output of the TASHE utilizing this parallel-plate regenerator is nearly twice that achieved with a screen-based regenerator[2].

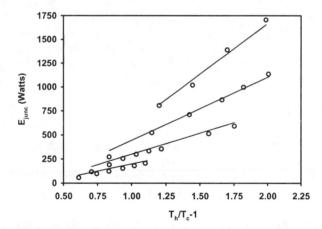

Figure 5. Open circles, measured acoustic power flow into the resonator, E_{junc}, generated by the TASHE. $p_{1,c}/p_m$ is higher for the larger power flows. Acoustic power is measured using a two-microphone technique[8]. For each value of $p_{1,c}/p_m$, E_{junc} is calculated at the highest and lowest value of T_h/T_c-1. The lines connect these values.

An additional data run is taken with the acoustic load set so that the engine will not oscillate even with T_h~725°C. By measuring the heat input to the hot heat exchanger, Q_h, and the heats rejected at the main cold heat exchanger, $Q_{c,1}$, and the secondary cold heat exchanger, $Q_{c,2}$, the heat leaks through the various components are determined. This is described in more detail Ref. [2]. These heat leaks are included in the numerical model of the TASHE.

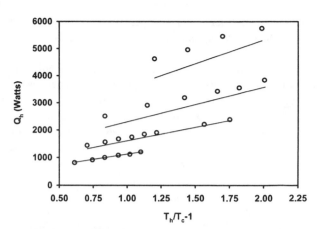

Figure 6. Open circles, measured heat absorbed at the hot heat exchanger vs. T_h/T_c-1 for the same values of $p_{1,c}/p_m$ as in Fig. 4. Higher values of $p_{1,c}/p_m$ correspond to larger Q_h. Q_h is determined by measuring the electric power flowing into the electrically-heated hot heat exchanger. For each $p_{1,c}/p_m$, Q_h is calculated at the highest and lowest values of T_h/T_c-1. The lines connect these calculated values.

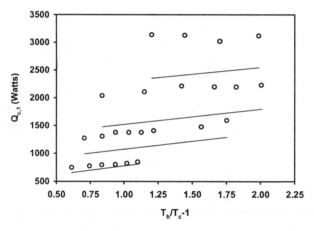

Figure 7. Open circles, measured heat rejected at the main cold heat exchanger vs. T_h/T_c-1 for the same values of $p_{1,c}/p_m$ as in Fig. 4. Higher values of $p_{1,c}/p_m$ correspond to larger $Q_{c,1}$. $Q_{c,1}$ is determined by measuring both the flow rate of cooling water through the main cold heat exchanger and the temperature drop across it. For each $p_{1,c}/p_m$, $Q_{c,1}$ is calculated at the highest and lowest values of T_h/T_c-1. The lines connect these calculated values.

The measured and calculated heat flows Q_h and $Q_{c,1}$ are shown in Figs. 6 and 7. At higher values of $p_{1,c}/p_m$, the measured Q_h is significantly higher than the calculated values. The discrepancy in $Q_{c,1}$ is approximately the same as Q_h. This implies that the extra heat the engine is absorbing at the hot heat exchanger is being rejected at the main cold heat exchanger. There is also a discrepancy between the measured and calculated $Q_{c,2}$, but it is much smaller that the difference in $Q_{c,1}$. All conduction and radiation heat leaks are accounted for by the heat leak measurements. Therefore, the extra heat must be carried by the oscillating gas itself.

One clue to the mysterious heat-transport mechanism is the axial temperature profile in the regenerator. When using the screen-based regenerator[2], the temperature always varied linearly from the cold exchanger to the hot exchanger, i.e. the temperature measured at the axial midpoint of the regenerator was always within a few percent of the average of T_h and T_c. However, the temperature at the axial midpoint of the circumference of the parallel-plate regenerator typically runs 15 to 20% hotter than the average of T_h and T_c. This temperature distribution cannot be explained by either the numerical model or the temperature dependence of the thermal conductivity of the metals used in the regenerator and its pressure vessel. One effect that is known to cause both a temperature distribution that deviates from linear and extra heat transport is a steady flow of gas around the loop of the TASHE, here called streaming[2]. However, streaming around the entire loop would also cause the axial temperature distribution in the buffer tube to deviate from linear. This is not observed. Another possibility is a circulating acoustic streaming loop contained entirely within the regenerator similar to that observed in the pulse tubes of orifice pulse tube refrigerators[9]. The streaming might be driven by the same mechanisms that cause the circulating streaming in pulse tubes or by small, undetected variations in the regenerator plate spacing. Four thermocouples placed around the O.D. of the regenerator sleeve indicate that the axial midpoint temperature does not show a significant angular dependence. Therefore, whatever the mechanism, it must be occurring on a single or few-channel scale.

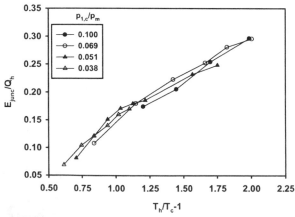

Figure 8. Measured thermal efficiency of the TASHE vs. T_h/T_c-1 for the same values of $p_{1,c}/p_m$ as in Fig. 4.

In spite of the mysterious extra heat leak from the hot heat exchanger and the high thermal conductance of the 316L fingers, the measured TASHE efficiency with a parallel-plate regenerator shown in Fig. 8 is still quite good. By switching from a screen-based to a parallel-plate regenerator, the acoustic power output of the TASHE is nearly doubled with no decrease in maximum efficiency[2]. In fact, the efficiency at $p_{1,c}/p_m$=0.10 has increased from 22% to 30%. This is mainly due to the flow

resistance of the parallel-plate regenerator remaining linear at this amplitude whereas the screen-based regenerator shows significant nonlinear behavior at the higher amplitudes.

CONCLUSIONS

If they are constructed with care and carefully inspected after fabrication, parallel-plate regenerators offer significant advantages over screen-based regenerators. We have attempted to define a procedure for the fabrication and diagnostic testing of parallel-plate regenerators. We have demonstrated that, by following this procedure, the viscous pressure drop of the regenerator is very close to that given by theoretical calculations. This shows that the plate spacing is very uniform and close to the design spacing. However, the heat transfer properties are only qualitatively understood. A small but significant extra source of heat transport through the regenerator exists. The experimental evidence points towards streaming as the source of this heat transport, but the precise mechanism driving the transport remains unknown.

ACKNOWLEDGMENTS

The authors would like to thank everyone at Refrac Systems, Inc. for their expert knowledge and assistance in the cleaning, stacking, and diffusion bonding of the regenerator. We would also like to thank everyone at Micro-Tronics, Inc. for their willingness to take on challenging EDM work, C. Espinoza and D. L. Gardner for their technical assistance in the construction of the TASHE, and the Office of Basic Energy Sciences in the U. S. DOE for financial support.

REFERENCES

[1] W. M. Kays and A. L. London, <u>Compact Heat Exchangers</u>, 3rd edition (McGraw-Hill, New York, 1984).

[2] S. Backhaus and G. W. Swift, "A thermoacoustic-Stirling heat engine-Detailed study," J. Acoust. Soc. Am., **107** (6), 3148-3166, (2000).

[3] Refrac Systems, Inc. 7201 W. Oakland St., Chandler, AZ.

[4] R. J. Roark and W. C. Young, <u>Formulas for Stress and Strain</u>, 5th edition (McGraw-Hill, New York, 1975). pg. 415.

[5] Micro-Tronics, Inc. 2905 S. Potter Dr., Tempe, AZ.

[6] Meriam Instrument, 10920 Madison Ave., Cleveland, OH.

[7] W. C. Ward and G. W. Swift, "Design environment for low amplitude thermoacoustic engines (DeltaE)," J Acoust. Soc. Am., **95**, 3671-3672 (1994). To review or obtain DeltaE, visit http://www.lanl.gov/projects/thermoacoustics.

[8] A. M. Fusco, W. C. Ward, and G. W. Swift, "Two-sensor power measurements in lossy ducts," J. Acoust. Soc. Am. **91**, 2229-2235 (1992).

[9] J. M. Lee, P. Kittel, K. D. Timmerhaus and R. Radebaugh, "Flow patterns intrinsic to the pulse tube refrigerator." In Proc. 7th Int. Cryocooler Conf., Kirtland AFB, NM 87117-5776, Phillips Laboratory, 1993, p. 125.

2001-CT-12

SNAP - A Stirling Numerical Analysis Program with User Viewable and Modifiable Code

Alan Altman
7002 Seward Park Ave. South
Seattle,WA 98118
bn660@scn.org

ABSTRACT

A new second order Stirling Engine thermodynamic design and analysis program for the serious researcher on a budget and the knowledgeable hobbyist has been developed. SNAP - STIRLING NUMERICAL ANALYSIS PROGRAM offers many improvements and enhancements over previously available programs. SNAP is user modifiable so new heat exchanger geometries can be easily added or new constructions modeled. SNAP runs under Excel versions 5.0 and up and is Mac and PC compatible. The code is written and annotated in an easy to follow format. The 'source code' and all formulas are visible.

INTRODUCTION

The writing of SNAP - STIRLING NUMERICAL ANALYSIS PROGRAM began over a year and a half ago to address the need for a reliable, fast running simulation to answer a number of heat exchanger design questions that were not answered in the literature nor easy to answer using existing programs (1,2,3,4). One example is how tubular heaters compare to annular finned heaters in low pressure air engines .

There are very few complete documented design procedures on how to analyze Stirling Engines in the open literature. The one I used was Dr. Martini's Stirling Engine Design Manual editions 1 and 2 (5,6). These equations and overall procedure form the core of SNAP with two very important additions. Beta and gamma arrangements are mapped onto equivalent alpha arrangement and all heater and cooler types are mapped onto the equivalent tubular types. After numerous changes, revisions and expansions it has evolved into a wide ranging program covering all the major engine configurations and heat exchanger types.

SNAP is written in Excel which was a choice based on the belief that if we are going to design 'good' engines a working knowledge of how the different design options affect all the critical loss parameters is necessary. The need to easily add new geometries to the program is important and the code is therefore open and user modifiable.

The program is stable and the temperature loop usually converges in 5 or 10 iterations. The program has built in test procedures to allow non-convergent cases to output enough heat exchanger data to determine the problem. Even the most obstinate cases have been troubleshot with procedures outlined in the documentation.

A number of engines have been simulated and have shown very good correlations at their peak operating power points. Engines like the GPU-3 simulate very well over their entire operating range. Finned low RE, low pressure engines simulate very well up to their peak operating point where burner limitations limit output. SNAP fills the gap: between back of the envelope guesses and expensive commercial software.

DESCRIPTION

The spreadsheet consists of four sheets:

- Main - The first half accepts all the inputs and contains program controls. The second half calculates live and dead volumes, flow losses, basic heat input, other heat inputs including conduction, reheat losses, etc. Finally an iterative loop calculates the internal gas temperatures until the loop converges or the error checking stops it.

- Numerical Simulation - Runs a cycle simulation to calculate basic power, mass flow rates, pressure max/min's, and mass distribution in the expansion and compression spaces. Sinusoidal motion is assumed for all calculations. Different motions can be added easily.

- Output - Shows all the outputs including power, efficiency, losses, dimensionless volume ratios and heat exchanger/regenerator parameters including Reynolds number, mass flow rates, NTU's, etc.

- Burner - Calculates how overall burner/pressure vessel/heater will act as a unit. A bit simplified but a good first approximation. This sheet is for finned annular heaters only and tells you if your burner will supply enough power to the engine and at what gross efficiency.

The basic procedure follows Dr. Martini's isothermal procedure with a number of additions that make the results more understandable and useful for comparing different configurations and engine design optimization.

- The program converts all beta and gamma configurations to the equivalent alpha arrangement allowing ease of computation and comparisons between different engine layouts. The equations used were developed by Dr. Theodor Finkelstein (7) and provide equivalent values of alpha and kappa to beta and lambda for an engine with an identical thermodynamic cycle. Gamma's are also converted to alphas and require an additional compression side dead space.

- A new technique was developed to map expansion and compression exchangers onto a common geometry. This allows comparisons of different exchangers and the addition of new geometries very rapidly. Questions such as "What effect will changing from a tubular heater to a finned heater head have on power and efficiency?" can now be answered quickly.

- Most design cases that do not converge output enough data to immediately pinpoint the problem. Even the most obstinate case has been troubleshot with procedures outlined later. Internal temperatures are tested to prevent invalid numbers from being output.

- Tubular, annular, and finned (milled or cut fins, folded fin material or individual fins) expansion and compression exchangers are included. Regenerator types covered are annular, foil, screen and linear elements. New geometries can be added easily to try out new designs as the program is completely user modifiable.

OUTPUT FORMAT

The following output is taken from a simulation made of Andy Ross's D-60 air engine (figure 1), input data was obtained from (8). The output sheet allows changing the working fluid, pressure, RPM and Alpha or Beta angle of the subject engine. For convenience these four parameters can be varied either in the input section of the main sheet or in the output sheet, like using a 3-way switch.

The following section displays the output including the flow losses of each exchanger and estimated mechanical losses. Heat input is then displayed showing how much heat is needed to derive the calculated output. Internal gas temperatures are listed with the hot and cold design temperatures. Below is a complete panel for the heater, regenerator and cooler giving all the critical numbers to allow design optimization. Reynolds number, net transfer units, heat transfer coefficient, hydraulic radius, surface area, etc. are displayed allowing a rapid analysis of your chosen design and immediate feedback on how changes affect heat transfer and flow losses. The dead space ratio, Beale number, and swept volume as well as some volume ratios are shown to allow quick comparisons to other engine designs.

A lower panel not shown has been added to allow power and efficiency to be graphed versus RPM using Excel's built-in graphing capability.

USING SNAP

SNAP is quite easy to use. Calculation is turned off on your spreadsheet and the engine parameters and operating conditions are entered. Metric units are used throughout. After inputting all your data turn calculation back on and the sheet iterates through the cycle until the internal gas temperatures converge. This is evident from watching the output power.

Stability is assured by testing the heat exchanger temperature drops before continuing calculations. If the temperature drops are too big the output is forced to zero and the output sheet can be examined to find the cause. This will usually be a very low NTU (under .8 or .9) or insufficient surface area for heat exchange. If the output doesn't converge the output sheet will still show the probable cause which frequently are very large flow losses in the heat exchangers, low NTU's or insufficient exchanger surface area. One can simply save a copy of the spreadsheet for the new design and further refine it.

SNAP is not protected as it is intended to be user modifiable. If some types of errors are made (deleting a formula, divide by zero, etc.) and the spreadsheet locks up closing without saving and then re-opening will revert back to a working copy of SNAP. This is also handy to remember as everything in SNAP can be modified and you always want to have an original copy to start with.

SIMULATING HEATER TUBE COUNTS

Numerous simulations have been run using data from as many engines as could be located. Results for Philips type engines are quite good, usually within 10%. These follow the 'standard' curves of increasing power until fluid flow losses start increasing rapidly.

More interesting were the simulations run using low pressure air engines (also within about 10%) such as those constructed by Andy Ross. Andy's engine has laminar flow throughout as can be seen in figure 1. A few other internally finned air engines including some by Sunpower (from the limited data available) have been simulated; they all seem to have laminar flow internally.

A series of simulations comparing tubular heat exchangers to annular finned exchangers (high number tubular equivalents) were run to make an initial assessment lacking in the open literature. The procedure was as follows: a 333cc swept volume beta air engine operating at 100PSI gauge with finned internal cooler and a foil regenerator was used as a starting point. Speed was set at 2000 RPM as flow losses are more apparent at this speed.

A new worksheet was created in SNAP that controlled the heater inputs and displayed all the relevant outputs. The engine output power was then maximized starting with a one tube heater with air as the working fluid (table 1). Only the length of the heater tube(s) and the hydraulic diameter of the tube(s) were varied until the peak was found. The working fluid was then changed to helium with the same geometry and operating conditions.

This was repeated for two tubes then five, ten, twenty up to 10,000. The whole procedure was repeated optimizing on helium then running on air (table 2). A few cases of helium optimized heaters using air required the heater tubes to be 10% longer in order to converge, these are Helium optimized running on AIR at tube counts of 2000, 5000 and 10,000.

The following trends occur in general. Higher tube counts result in higher output power, higher efficiency, lower reduced dead space ratio, lower delta T's across the heater, lower fluid flow losses, and higher heat exchanger net transfer units.

Low tube count heaters optimized on air have fairly good performance when running on helium (figure 2). Low tube count heaters optimized on helium running on air have decidedly inferior performance. At higher tube counts the performance difference is reduced in both scenarios and is somewhat more chaotic as the different configurations go through transition at different times. Kays and London's heat transfer curve for a single round tube with abrupt opening leads was used though SNAP also contains Gnielinski's heat transfer correlation. The differing transitions lead to conditions where higher tube counts lead to lower output locally.

From a practical standpoint tube counts much above 50 are made as equivalent internally finned heaters. The basic equivalent is derived as follows: keep the hydraulic diameter and length the same then calculate the number of tubes required for equal free flow area. A fin array efficiency factor must also be added in to account for non-uniform fin temperatures.

It should be noted that many of the optimized configurations will not be reasonable to build due to fabrication constraints or limited external heat transfer area. Low tube counts simulated are not dissimilar to those used in the part engine concept where Philips used 3 tubes to simulate 1/16th of a 400HP helium engine.

The total time to set up the additional worksheet and run the above series of simulations (26 optimization runs) was about 2 hours.

SENSITIVITY TO TUBE DIAMETERS

The results of the tube count simulations raised the question of how sensitive power output is to non-optimum sized tube diameters and how it varied for helium versus air (table 3 and figure 3). The procedure was similar to the previous study with the exception that the length to diameter ratio was held constant at a near optimum number for the conditions around the maximum power point found during the prior simulations. Hydraulic diameters were varied plus and minus approximately 20% and then 20% again. It appears that air is more sensitive to varying tube sizes than helium particularly when the tube sizes decrease as this increases the fluid flow resistance very rapidly.

SUMMARY

SNAP - STIRLING NUMERICAL ANALYSIS PROGRAM has been developed to allow everyone to have access to a fast and easily modified program that will lead to better Stirling engines. Sample simulations were shown to demonstrate the ease and versatility of using SNAP. It is hoped that SNAP will be used to design new engines further validating the program.

NOMENCLATURE

(primarily from output sheet)

DhC	Cooler Hydraulic Diameter
DhH	Heater Hydraulic Diameter
NTUC	Net Transfer Units in Cooler
NTUH	Net Transfer Units in Heater
NTUR	Net Transfer Units in Regenerator
PMAX	Maximum Cycle Pressure
PMIN	Minimum Cycle Pressure
RE	Reynolds Number
RhR	Regenerator Hydraulic Radius
TC	Calc Cold Gas Temperature in Cooler
TCM	Cooler Metal Temperature
TH	Calc Hot Gas Temperature in Heater
THM	Heater Metal Temperature
VCD	Cold Dead Volume
VD	Cold Live Volume
VHD	Hot Dead Volume
VHL	Hot Live Volume
VR	Regenerator Volume
WCS	Cooler Mass Flow Rate
WHS	Heater Mass Flow Rate
WRS	Regenerator Mass Flow Rate

ACKNOWLEDGMENTS

I would like to thank Rob McConaghy for his unending support, patience and valuable suggestions during the writing and testing of SNAP. I would also like to thank Theodor Finkelstein, Andy Ross and the many others who took the time to talk to me about Stirling engines.

REFERENCES

1) Altman,Alan SNAP- STIRLING NUMERICAL ANALYSIS PROGRAM Stirling Machine World Dec 2000, Los Olivos, California published Brad Ross pp3-5

2) Altman,Alan SNAP- STIRLING NUMERICAL ANALYSIS PROGRAM Stirling Engine News, Volume 4, 2000. Chelsford, England Stirling Engine Society

3) Altman, Alan Stirling Numerical Analysis Program Manual self-published, Seattle WA. July,2000

4) Altman, Alan Stirling Engine Heat Transfer- a Snappy Primer self-published Seattle WA Feb,2001

2) Martini, Dr. William Stirling Engine Design Manual 1st edition DOE/NASA/3152-78/2 NASA CR-135382 April 1978 pp. 107-219

3) Martini, Dr. William Stirling Engine Design Manual 2nd edition DOE/NASA/3194-1 NASA CR-168088 January 1983 pp. 327-349

4) Organ, Allan J. 1992 Thermodynamics & Gas Dynamics of the Stirling Cycle Machine New York, NY, USA Cambridge University Press pp. 95-97

5) Ross, Andy Making Stirling Engines self-published Columbus, Ohio 1993 pp. 56-58

CONTROLS TO CHANGE INPUTS SIMULATION PROGRAM AFTER SETUP						
MODE SWITCH 1>> 2 OR 2>>1 TO CHANGE ACTIVE MODE					2	
ACTIVE MODE 111 IS MAIN AND 222 IS HERE INPUT/OUTPUT SHEET					222	
GAS1>>H2 2>>HE 3>>AIR	3		SELECTED	GAS	**3**	
GAS PRESSURE IN MPa	0.3		SELECTED	Pav in MPa	**0.3**	
RPM of engine	3300		SELECTED	RPM	**3300**	
ALPHA/BETA IN °	90		SELECTED	ALPHA	**90**	
OUTPUTS OF THIS CONFIGURATION			HEAT INPUT REQUIRED			
BASIC POWER OUTPUT	253	watts	BASIC HEAT INPUT		451	watts
REGENERATOR FLOW LOSS	5		REHEAT LOSS		40	
HEATER FLOW LOSS	14		SHUTTLE HEAT CONDUCTION		8	
COOLER FLOW LOSS	13		STATIC HEAT CONDUCTION		158	
INDICATED POWER	221		PUMPING LOSS		22	
MECH. FRICTION	25		TEMP SWING LOSS		3	
BRAKE POWER	**196**	watts	INT TEMP SWING LOSS		0	
			HEATER WINDAGE POWER		-14	
BRAKE EFFICIENCY	**29.36%**		1/2 REGEN. WINDAGE POWER		-3	
INDICATED EFFICIENCY	33.16%		NET HEAT INPUT		**666**	watts
THM	850	°K	TCM	327	°K	PMAX	0.47
TH CALC	842	°K	TC CALC	370	°K	PMIN	0.19
HEATER/REGENERATOR/COOLER DATA			COOLER DUTY- REJECT HEAT		471	
HEATER RE	553		REGEN RE	156		COOLER RE	2280
NTUH	2.95		NTUR	66.50		NTUC	1.07
WHS	8.33	g/sec	WRS	10.07	g/sec	WCS	11.80
DhH	0.08625965	cm	RhR	0.01018786	cm	DhC	0.06804979
Heater Surface	482.29	cm^2	Regen Surface	5250.00	cm^2	Cooler Surface	257.27
L/DhH	44.17		L/RhR	249.32		L/DhC	40.85
u hot gas veloc	2458	cm/sec	u gas velocity	640	cm/sec	u cold gas veloc	2652
Heat Xfer Coef	0.02796383	w/cm^2	Fill Factor	0.11	w/cm^2	Heat Xfer Coef	0.02358514
Burner Surface	102.6						
				Dead Space Ratio		0.55652243	

Volume Ratios normalized to VHL							
VHL	VHD	VR	VCD	VCL		Swept volume	60.16
1.00	0.27	0.49	0.11	1.00		Beale Number	0.20
Volume/Temperature Ratios normalized to VHL/TH							
1.00	0.27	0.76	0.24	2.27			

Figure 1 Output sheet for SNAP simulating Andy Ross's D-60 low pressure alpha engine at maximum power

Optimized for AIR													
Tube Count	1	2	5	10	20	50	100	200	500	1000	2000	5000	10000
Hydraulic Diameter(cm)	4	2.5	1.5	1.1	0.8	0.5	0.4	0.275	0.16	0.12	0.1	0.07	0.05
Length/ Diameter	23	22	23	27	31	30	31	36	50	58	50	36	40
Free Flow Area(sq cm)	13	10	9	10	10	10	13	12	10	11	16	19	20
Brake Power on AIR (watts)	264	601	941	1184	1396	1675	1830	1975	1986	2023	2251	2438	2554
Indicated Efficiency (%)	7	14	19	23	26	29	31	33	33	33	36	39	39
Reynolds Number	157k	118k	72k	45k	29k	17k	11k	7.3k	4.9k	3.3k	1.9k	1.0k	.7k
Net Transfer Units	0.26	0.27	0.32	0.42	0.53	0.57	0.67	0.79	0.8	0.78	0.97	1.14	1.79
Temperature Drop (°C)	153	177	170	131	103	106	87	73	75	78	58	48	21
Heater Flow Loss (watts)	362	417	424	381	346	308	199	220	322	278	132	91	111
Reduced Dead Space Ratio	2.09	1.2	0.83	0.76	0.7	0.55	0.56	0.5	0.44	0.44	0.44	0.4	0.38
Brake Power on HE (watts)	923	1392	1761	1840	1820	1934	2249	2508	2694	2703	2736	2859	2870
Indicated Efficiency (%)	27	33	37	38	37	37	42	45	46	46	47	47	47
Reynolds Number	21k	16k	9k	6k	4k	3k	2k	1k	.6k	.4k	.3k	.15k	.1k
Net Transfer Units	0.59	0.62	0.73	0.78	0.67	0.6	1.03	1.7	3.4	5.6+	6+	6+	6+
Temperature Drop (°C)	83	106	106	100	126	162	79	35	6	1	0	0	0
Heater Flow Loss (watts)	69	75	73	64	56	49	33	42	82	96	65	49	70
Reduced Dead Space Ratio	1.81	1.06	0.74	0.71	0.68	0.54	0.53	0.47	0.42	0.42	0.42	0.38	0.37

Table 1 AIR optimized heater data for different tube counts running on both Air and Helium

Optimized for Helium													
Tube Count	1	2	5	10	20	50	100	200	500	1000	2000	5000	10000
Hydraulic Diameter (cm)	2	1.5	1	0.7	0.5	0.35	0.3	0.2	0.15	0.11	0.09	0.05	0.04
Length/ Diameter	30	30	35	36	40	57	50	50	33	27	22	20	12.5
Free Flow Area (sq cm)	3	4	4	4	4	5	7	6	9	10	13	10	13
Brake Power on HE (watts)	1579	1799	2047	2186	2225	2238	2431	2614	2766	2852	2915	2951	2992
Indicated Efficiency (%)	34	36	39	41	41	41	43	45	46	47	47	48	48
Reynolds Number	34k	22k	12k	8k	6k	4k	2k	1.3k	.7k	.5k	.3k	.2k	.1k
Net Transfer Units	0.71	0.78	1.05	1.12	1.08	1.1	1.41	1.89	2.15	2.45	3.02	3.61	3.31
Temperature Drop (°C)	117	109	76	73	80	76	51	31	24	18	10	5	8
Heater Flow Loss (watts)	173	360	299	292	284	233	103	135	73	68	45	86	52
Reduced Dead Space Ratio	0.58	0.54	0.5	0.45	0.43	0.45	0.46	0.4	0.38	0.36	0.35	0.33	0.33
Brake Power on AIR (watts)	-1133	-290	172	348	545	1004	1687	1724	1918	1692	1700	1712	1799
Indicated Efficiency (%)	0	0	8	11	15	22	31	31	31	27	27	27	28
Reynolds Number	230k	155k	87k	60k	40k	21k	13k	9k	6k	4k	3k	2k	1.2k
Net Transfer Units	0.32	0.35	0.46	0.51	0.63	1.04	1.03	1.09	0.59	0.4	0.36	0.38	0.38
Temperature Drop (°C)	121	133	108	101	82	42	48	46	115	184	204	198	200
Heater Flow Loss (watts)	2900	2225	1827	1798	1740	1384	648	759	334	269	151	228	141
Reduced Dead Space Ratio	0.61	0.57	0.53	0.47	0.44	0.46	0.47	0.42	0.4	0.38	0.38	0.35	0.35

Table 2 Helium optimized heater data for different tube counts running on both Helium and Air

Sensitivity Study of Varying Tube Diameters

AIR with 50 tubes, length/diameter=30						AIR with 1000 tubes, length/diameter=60					
Diameter (cm)	0.3	0.4	**0.5**	0.6	0.7	Diameter (cm)	0.08	0.1	**0.12**	0.14	0.16
Brake Power (watts)	471	1498	**1674**	1597	1419	Brake Power (watts)	1342	1884	**2024**	2019	1948
Heater Flow Loss (watts)	1848	648	**308**	180	120	Heater Flow Loss (watts)	1222	534	**283**	171	115
Temperature Drop (°C)	110	116	**106**	93	81	Temperature Drop (°C)	60	75	**74**	67	55
Reduced Dead Space	0.38	0.44	**0.55**	0.7	0.9	Reduced Dead Space	0.37	0.4	**0.45**	0.51	0.59

Helium with 50 tubes, length/diameter=60						Helium with 1000 tubes, length/diameter=30					
Diameter (cm)	0.25	0.3	**0.35**	0.4	0.45	Diameter (cm)	0.07	0.09	**0.11**	0.13	0.15
Brake Power (watts)	1951	2188	**2239**	2190	2112	Brake Power (watts)	2612	2812	**2852**	2823	2750
Heater Flow Loss (watts)	769	403	**241**	158	114	Heater Flow Loss (watts)	326	140	**73**	43	29
Temperature Drop (°C)	63	72	**69**	65	53	Temperature Drop (°C)	35	21	**13**	9	6
Reduced Dead Space	0.37	0.41	**0.45**	0.52	0.6	Reduced Dead Space	0.33	0.34	**0.38**	0.38	0.42

Table 3 Output power with tube diameters varied around optimum value (**bold**). Power with diameters plus and minus about 20% then another 20% additional. Done for tube counts of 50 and 1000, both Air and Helium

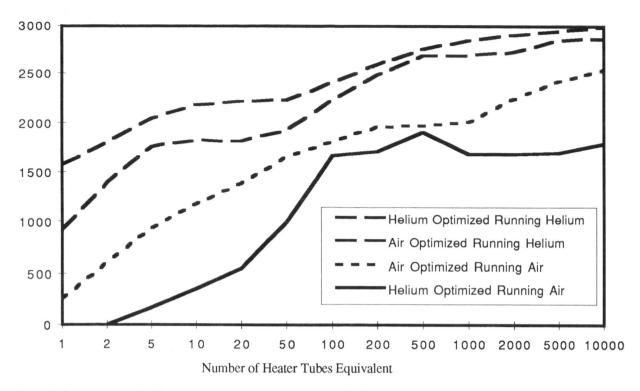

Figure 2 Output Power in Watts versus tube count for Air and Helium Optimized Heaters. Each is run on the other gas once optimized without any other changes

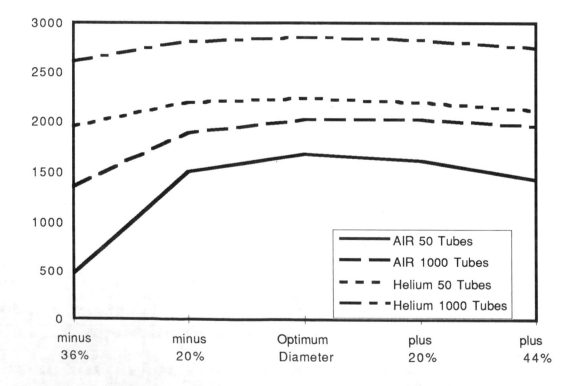

Figure 3 Output power in watts versus tube diameters varied around optimum value. Diameters optimum in center with plus and minus about 20% then another 20% (plus 44% and minus 36%).Taken from data in Table 3 .

IECEC2001-CT-27

STUDY OF TWO-DIMENSIONAL COMPRESSIBLE NON-ACOUSTIC MODELING OF STIRLING MACHINE TYPE COMPONENTS

Roy C. Tew, Jr.
Thermo-Mechanical Systems Branch
Power & On-Board Propulsion Division
NASA Glenn Research Center
Mail Stop 301-2
Cleveland, Ohio 44135-3191
Tel: (216) 433-8471
FAX: (216) 433-8311
e-mail: Roy.Tew@grc.nasa.gov

Mounir B. Ibrahim, Chairman and Professor
Mechanical Engineering Department
Fenn College of Engineering
Cleveland State University
1960 East 24th Street,
Cleveland, Ohio 44115-2425
Tel: (216) 687-2580
Fax: (216) 687-9280
e-mail: m.ibrahim@csuohio.edu

ABSTRACT

A two-dimensional (2-D) computer code was developed for modeling enclosed volumes of gas with oscillating boundaries, such as Stirling machine components. An existing 2-D incompressible flow computer code, CAST, was used as the starting point for the project. CAST was modified to use the compressible non-acoustic Navier-Stokes equations to model an enclosed volume including an oscillating piston. The devices modeled have low Mach numbers and are sufficiently small that the time required for acoustics to propagate across them is negligible. Therefore, acoustics were excluded to enable more time efficient computation. Background information about the project is presented. The compressible non-acoustic flow assumptions are discussed. The governing equations used in the model are presented in transport equation format. A brief description is given of the numerical methods used. Comparisons of code predictions with experimental data are then discussed.

INTRODUCTION

Background

NASA Glenn Research Center (GRC) has been involved in development of Stirling engines for ~25 years. GRC began managing the Stirling Automotive Development Program for the Department of Energy (DOE) in ~1977. The DOE/NASA contractors were first Ford Motor Co. (for one year) and then Mechanical Technology, Inc. (MTI) of Albany, NY. These Stirling automotive engines used hydrogen as the working fluid and were required to operate over demanding automotive driving cycles. This work continued into the early '90's (Ernst and Shaltens [1]). Engines were demonstrated in automobiles and trucks and future development looked promising. However, due to decreases in oil and gasoline prices and improvements in spark ignition engines, the automotive manufacturers did not choose to develop the Stirling technology for the automotive market.

Development of Stirling engines for generation of space auxiliary power began in ~1983 with MTI as the primary contractor. These engines used helium as working fluid, were 25 kWe (12.5 kWe/cylinder) designs with electrical power produced by linear alternators and used gas bearings. Back-to-back cylinders with synchronized pistons damped engine vibrations.

One of the problems encountered in Stirling engine development by GRC and MTI was the limited accuracy of the 1-D engine design computer codes. Performance projections for new designs were usually optimistic. The last major design of a 25 kWe (12.5 kWe/cylinder) Stirling engine for space power was the Component Test Power Converter or CTPC (Dahr [2]). Because of previous power shortfalls with other new engine designs, MTI and NASA chose to design the CTPC with a 20% margin on power. When tested the engine produced slightly in excess of the 25 kWe power goal, meaning that it produced almost 20% less power than predicted by the computer code that was used to design it.

During this period of space power Stirling engine development at MTI (~1983-1993), concerns about 1-D design code accuracy led to fundamental research. A "loss understanding" program was started consisting of grants and small contracts for investigation of thermodynamic losses (Tew and Geng [3]). The participants were hosted by GRC at yearly Stirling Loss Understanding Workshops for several years. The long range goal was to improve the accuracy of 1-D Stirling engine design codes.

The SP-100 (or Space Power-100 kWe) reactor space power system program, for which these 25 kWe space power Stirling designs were intended, ended in 1993. Along with it, funding for the Stirling loss understanding program ended also.

From ~1993-1999, NASA support of Stirling engine development continued at a low level via NASA SBIR funding. Most of this funding was for continued development of free-piston engines at the Stirling Technology Company (STC) of Kennewick, WA based on flexural bearing technology. GRC and CSU also continued low-level development of a multi-D Stirling code via the thesis project reported here.

Now Stirling power convertors are being developed by DOE, NASA and STC as a substitute for the less efficient radioisotope thermoelectric generators for deep-space missions (Thieme, Qui, and White [4]). The success of STC in accumulating years of reliable operation on its free-piston flexural bearing machines (over 6 years on a 10 kWe convertor with no maintenance and no degradation in performance) has been a key factor. In late 1999, a 55 We STC free-piston Stirling was subjected to a number of tests and evaluations to determine its suitability for a deep-space mission (For example, it passed a simulated launch vibration test, while operating). As a result, in 2000, the Stirling radioisotope power system, or RPS (Furlong and Shaltens [5]), was baselined as the advanced power system that may be used on NASA missions such as Europa and Solar Probe missions. It now appears that long-duration Martian rover missions will be the first application of Stirling RPS.

Stirling One-Dimensional Flow Design Codes

Two one-dimensional-flow Stirling machine codes have been in primary use by NASA's Stirling engine contractors in recent years. These are the HFAST code developed by Mechanical Technology, Inc. (Huang [6]) and the GLIMPS code (now Sage, (Gedeon [7])) developed by Gedeon Associates. HFAST was used by MTI in design of the large (25 kWe) Stirling convertors discussed earlier. The GLIMPS code and now it's successor, Sage, have been the primary design codes used by the Stirling Technology Company. Sage is also used by Sunpower, Inc. While these codes have been used to develop excellent engine designs, 1-D codes assume uniform axial flow and are thus believed to be deficient in modeling those interfaces in Stirling engines where significant changes in area take place. Past comparisons of code predictions, such as the GLIMPS/HFAST study done by Geng and Tew [8], showed rough overall performance agreement but differences in predictions of individual losses. It would be of significant value to be able to accurately characterize all major losses so that the design process could do an adequate job of trading off the various losses against each other.

Objective of the Work

The immediate purpose of the 2-D code development reported here was to develop a time-efficient compressible code for study and definition of Stirling machine type cylinder heat transfer/power (or hysteresis) losses; see Tew [9] for more complete documentation. A long range purpose was to provide the basis for development of a 2-D model of a complete Stirling engine that could be used to study interaction of various thermodynamic losses. Even longer range, if computer software/ hardware time efficiencies are increased sufficiently, it may be possible to use a 2-D code (perhaps even a 3-D code) for engine design. The previous IECEC paper by Makhkamov and Ingham[10] suggests such design use of a 2-D code is already underway at the Laboratory for Stirling Engines at the Physical-Technical Institute in Tashkent.

NOMENCLATURE

Latin letters

c_p specific heat at constant pressure

d   denotes differential of variable

D_h hydraulic diameter (=4 x wetted area/wetted perimeter)

$\overline{D}$ viscous force vector

$\overline{f}$ mass force vector (due to gravity, for example)

p   static pressure

P   mean spatial pressure

P_a amplitude of mean spatial pressure

P_o arithmetic mean of max. and min. mean spatial pressures

q_m entropy source due to non-zero mass sources

Q   density of continuously distributed heat sources

R   gas constant

s   entropy per unit mass

t   time

T   temperature

$\overline{\overline{T}}_{ST}$ stress tensor

$\overline{u}$ velocity vector $(= \overline{i}U + \overline{j}V)$

U   axial, x-direction, velocity

V   radial, r-direction, velocity or volume

V_o arithmetic mean of maximum and minimum volumes

$\overline{w} = \xi\rho\overline{u}$ rate of momentum change because of mass sources

$\hat{W}_{loss}$ non-dimensional work or hysteresis loss

Greek letters

γ ratio of specific heats of fluid

ϕ transport quantity per unit mass

λ molecular thermal conductivity

μ absolute viscosity

ρ density, fluid mass per unit volume

ω angular velocity (rad/sec)

ξ mass source strength per unit mass

COMPRESSIBLE NON-ACOUSTIC FLOW

Assumptions

Fedorchenko [11] discusses a number of subsonic initial-boundary-value problems that cannot be solved using the classical theory of incompressible fluid motion which involves the equation $\nabla \cdot \overline{u} = 0$ (where $\overline{u} = \overline{i}U + \overline{j}V$ and the unit vector $\overline{i}$ is the x-direction for both cartesian and axisymmetric coordinates; the unit vector $\overline{j}$ is the y-direction for cartesian coordinates and in the r-direction for axisymmetric coordinates.) Among these problems are: (1) flows in a closed volume initiated by blowing or suction through permeable walls, (2) flows in a closed volume initiated by a moving boundary such as in a piston-cylinder problem, (3) flows with continuously distributed mass sources, and (4) viscous flows with substantial heat fluxes. Fedorchenko notes that application of the most general theory of compressible fluid flow may not be best in such cases because of the difficulties in accurately resolving complex acoustic phenomena and in assigning proper boundary conditions.

Fedorchenko proposes a non-local mathematical model where $\nabla \cdot \overline{u} \neq 0$ in general, for simulation of unsteady subsonic flows in a bounded domain with continuously distributed sources of mass, momentum and entropy, also taking into account the effects of viscosity and conductivity when necessary. The exclusion of sound waves is one of the most important features of the model.

The most general form of Fedorchenko's compressible non-acoustic system of equations for simulation of unsteady subsonic, heat conducting viscous flows are the following forms of the momentum, continuity, energy and state equations:

$$\frac{\partial(\rho\overline{u})}{\partial t} + \nabla \cdot \rho\overline{u}\overline{u} + \nabla p = \rho\overline{f} + \overline{w} + \overline{D} \tag{1}$$

$$\frac{\partial\rho}{\partial t} + \nabla \cdot \rho\overline{u} = \xi\rho \tag{2}$$

$$\frac{\partial s}{\partial t} + \overline{u} \cdot \nabla s = \frac{R}{P}\left[\nabla \cdot \lambda\nabla T + Q\right] + q_m \tag{3}$$

$$F(s, P, \rho) = 0 \tag{4}$$

When $\nabla\mu = 0$ (as it is for constant viscosity problems) then

$$\overline{D} = \mu\Delta\overline{u} + \frac{1}{3}\mu\nabla(\nabla \cdot \overline{u}) \tag{5}$$

Note that the energy equation is written in terms of entropy, s, rather than in terms of internal energy, enthalpy, or temperature.

The simplifications in the Navier-Stokes equations used to eliminate acoustics and arrive at the above system of equations were: (1) Pressure at any time, t, and position, $\bar{r}$, is split into a mean spatial pressure level that varies only with time and a $\Delta(pressure)$ relative to a reference position that varies with position and time:

$$p(\bar{r},t) = P(t) + \Delta p(\bar{r},t) \qquad (6)$$

(2) The pressure appearing in the equation of state is the mean spatial pressure that varies only with time. Therefore, from the ideal gas equation of state

$$\rho = \frac{P(t)}{RT(\bar{r},t)} \qquad (7)$$

So density is a function of mean spatial pressure level and the temperature field. (and is independent of the spatial pressure drop).

Simplification of Equations for the Stirling Problem

For the Stirling piston-cylinder problem of interest here, there are no mass or heat sources distributed within the cylinder volume and the gravity force on the gas is not of interest. Therefore the variables defined above: $\bar{f}$, $\bar{w}$, ξ, Q and q_m are all zero. Also the ideal gas equation of state is sufficiently accurate for the helium gas that is used in most current Stirling engines of interest to NASA. Therefore the equations (1)-(4) reduce to the following set:

$$\frac{\partial(\rho\bar{u})}{\partial t} + \nabla \cdot \rho\bar{u}\bar{u} + \nabla p = \overline{D} \qquad (8)$$

$$\frac{\partial\rho}{\partial t} + \nabla \cdot \rho\bar{u} = 0 \qquad (9)$$

$$\frac{\partial s}{\partial t} + \bar{u} \cdot \nabla s = \frac{R}{P}\left[\nabla \cdot \lambda\nabla T\right] \qquad (10)$$

$$\rho = \frac{P}{RT} \qquad (11)$$

CAST AND MODIFIED CAST CODE

The incompressible flow CAST (Computer Aided Simulation of Turbulent Flow) code, as originally received by Cleveland State University and NASA is documented in Peric and Schuerer [12]. Changes made to CAST in developing the modified CAST compressible non-acoustic code are documented in Tew [9].

Transport Equation Format

The solution technique used in the CAST code is based on a transport equation formulation of the governing equations. A coordinate-free form of the general transport equation is:

$$\underbrace{\frac{\partial(\rho\phi)}{\partial t}}_{\substack{\text{Time rate} \\ \text{of storage}}} + \nabla \cdot \underbrace{\left(\rho\bar{u}\phi - \Gamma_\phi\nabla\phi\right)}_{\substack{\text{convection} \quad \text{diffusion}}} = \underbrace{S_\phi}_{\text{source}} \qquad (12)$$

The text under equation (12) catagorizes the terms in the equation.

Transport quantities, exchange coefficients and source terms for the continuity, momentum and energy equations used in Modified CAST are shown in Table I. Note that the energy equation is in enthalpy format, rather than the entropy format used by Fedorchenko. The turbulent kinetic energy and dissipation rate equations are not shown here due to space limitations. They are defined in Tew [9].

Table I: General Transport Equation - Transported Quantities, Exchange Coefficients and Source Terms for Continuity, Energy and Momentum Equations

Equation Name	Transport Quantity/vol.	Exchange Coefficient	Source Term
Continuity	ρ (mass/vol.)	0	0
Momentum	$\rho\bar{u}$ (momentum/vol.)	μ	$\nabla \cdot \overline{\overline{T}}_{ST,S}$
Energy	ρh (enthalpy/vol.)	$\dfrac{\lambda}{c_p} = \dfrac{\mu}{\text{Pr}}$	$\dfrac{dP}{dt}$

Numerical (Finite-Volume) Methods

The numerical methods in modified CAST are almost the same as those used in CAST (Peric and Scheuerer [12]). An exception is the use of Leibniz's rule (Ferziger and Peric [13]) to account for the effect of the moving boundary (i.e., piston motion). Modifications to the governing equations themselves, to account for the change from incompressible to compressible non-acoustic flow, are discussed in Tew [9]. These included accounting for the non-zero $\nabla \cdot \bar{u}$ terms, adding the time derivative of the mean spatial pressure to the source terms of the energy equation, and using mean spatial pressure in the ideal gas equation so that the density field is a function of the temperature field and the mean spatial pressure. Spatial pressure drop still appears in the momentum equations to help determine the velocity field. Since the density is not affected by spatial pressure, the incompressible SIMPLE algorithm still applies.

The governing equations are solved with a conservative finite-volume method (see Patankar [14]). The basic principle is: (1) Discretize the solution domain by subdividing it into small axisymmetric (or rectangular) control volumes. Locate the numerical grid points in the center of the control volumes. (2) Discretize the transport equations.

Discretization is done by formally integrating the single terms in the equations over a control volume. Application of Leibniz's rule and Gauss's theorem yields an integro-differential equation relating the net increase in the transported quantity per unit time to the convective and diffusive fluxes across the control volume boundaries and to the source (or sink) terms within the control volume. This practice leads to a conservative method because boundary-fluxes leaving one control volume through its right boundary enter the neighboring control volume through its left boundary. Since this principle applies to all control volume faces, the scheme becomes overall conservative. The approach is described in Peric and Schuerer [12] and Tew [9] in more detail.

AVAILABLE DATA FOR CAST COMPARISONS

Rectenwald Computations and Kornhauser Experimental Data

Recktenwald [15] computed heat transfer between the walls and gas inside the cylinder of a reciprocating compressor. A compressor cylinder contains intake and discharge valves, unlike the cylinders of a Stirling engine. Recktenwald used 2-D, unsteady, compressible equations (acoustics included) to simulate the compressor. In order to validate his computer code, he simulated a gas spring for comparison with data generated by Kornhauser and Smith [16, 17]. The comparison between data and experiment was based soley on experimental and simulated values

467

of non-dimensional hysteresis loss for the gas spring over an operating range. It should be noted that a gas spring is essentially a piston-cylinder which has no flow to or from the enclosed volume of the cylinder. Kornhauser [18] reported further details of these gas spring experiments. He also reported on tests made with a modification of the gas spring test rig, to include a heat exchanger mounted on top of the cylinder, such that flow could continuously pass between the cylinder and the heat exchanger as the piston expanded and compressed the gas. This "two-space test rig" operated more like a Stirling machine cylinder than either a gas spring or a compressor.

Kornhauser's gas spring test data was also used as a basis for validation of the 2-D Modified CAST compressible non-acoustic code. Hysteresis losses computed by CAST were compared with the experimental values of Kornhauser and the calculated values of Recktenwald, over a range of gas spring operation. Also, since Recktenwald published plots of his calculated velocity vectors and temperature contours within the gas spring for two operating points, these were compared with similar plots of CAST calculated values for one of the operating points. Thus it was possible to compare the calculations of a compressible model which did account for acoustics in the computations against the compressible non-acoustic calculations of the Modified CAST code. Comparison of CAST and Recktenwald's temperature contours over the cycle for a 10 RPM gas spring showed excellent agreement. Agreement between velocity vector plots also appeared to be very good, although due to some difference in the way the plots were made only a qualitative comparison could be made. These 2-D comparisons are shown in Tew [9].

Gas Spring and Two-Space Test Rig Dimensions

The dimensions of Kornhauser's test rigs are shown in Tables I and II. Due to a limitation of the CAST code in simulating complex geometries, it was necessary to make a slight change in the annular heat exchanger geometry simulated for the two-space test rig. The annular heat exchanger, physically mounted on top of the cylinder, was moved slightly outward so that the outer wall coincided with the cylinder wall, and the heat exchanger volume was maintained the same in order to maintain the same volume ratio. Physical and simulated dimensions are shown in Table III (This small change in geometry likely tended to reduce the difference between test and data results, to be discussed below).

Table II: Gas Spring Dimensions

Physical Quantity	Symbol	Value
Cylinder Bore (Diam.)	D	50.80 mm (2 in.)
Piston Stroke	S	76.2 mm (3 in.)
Volume Ratio	r_v	2.0

Table III: Two-Space Test Rig Dim. (Physical & Simulated)

Physical Quantity	Physical Value	Simulation Value
Cylinder Bore	50.80 mm (2 in)	50.80 mm (2 in)
Piston Stroke	76.20 mm (3 in)	76.20 mm (3 in)
Volume Ratio	2.0	2.0
Annulus O.D.	44.5 mm (1.75 in)	50.80 mm (2 in)
Annulus I.D.	39.4 mm (1.55 in)	46.4 mm (1.83 in)
Annulus Gap	2.5 mm (0.10 in)	2.2 mm (0.09 in)
Annulus Length	445 mm (17.5 in)	445 mm (17.5 in)
Min Pist/Head Clr.	2.9 mm (0.11 in)	2.9 mm (0.11 in)

Both test rigs used helium gas and the walls of the cylinder and heat exchanger were at a constant temperature of approximately 294 K.

RESULTS: COMPUTATIONS VS. DATA

Gas Spring Hysteresis Losses

For a gas spring, the hysteresis loss is the work that is dissipated by the spring per cycle at steady operating conditions; it's also equal to the heat generated in and transferred out of the spring. A good way to compare computational and measured hysteresis losses is via plots of dimensionless work as a function of oscillating flow Peclet number. Dimensionless work and oscillating flow Peclet numbers are defined, respectively, as follows:

$$\hat{W}_{loss} = \frac{\oint P\, dV}{P_o V_o \left(\frac{P_a}{P_o}\right)^2 \left(\frac{\gamma - 1}{\gamma}\right)} \quad (13)$$

$$Pe_\omega = \frac{\rho_o c_p \omega D_h^2}{4\lambda} \quad (14)$$

Recktenwald had previously plotted his calculated dimensionless losses on a plot of Kornhauser's data. The Modified CAST dimensionless losses were superimposed on this plot and the result is shown in Figure 1. This figure shows dimensionless loss as a function of oscillating flow Peclet number. The CAST loss values were plotted at "uncorrected" Peclet numbers as the obviously handwritten "X" symbols. If plotted at the corrected values of the Peclet numbers they would be shifted slightly to the right and would fall on Recktenwald's values (the solid diamonds).

Five of the six pairs (of CAST and Recktenwald values) of calculated dimensionless losses agree well with Kornhauser's data. The one pair of calculated points that did not agree well with the data is the pair shown at the highest Peclet number and corresponds to a 1000 RPM, 1465 kPa gas spring operating condition. Recktenwald [15] duscusses several plausible explanations for disagreement at high Pe_ω.

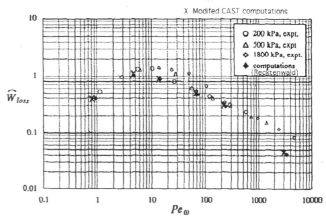

Figure 1: Modified CAST Dimensionless Losses Superimposed on Recktenwald [15] Plot of Recktenwald's Calculated Losses and Kornhauser's [18] Experimentally Derived Losses

Also in the mid-range of the data in the vicinity of Pe_ω=10, where there appears to be a higher and a lower data curve, the data agrees with the lower data curve. Kornhauser [18] found that the higher and lower data "curves" were related to differences between data taken at high-pressure/low-speed (higher curve) and that taken at low-pressure/high speed (lower curve) which had oscillating flow Peclet numbers in the same range. Thus he concluded there was some other dimensionless parameter needed to resolve the data in this range. His experiment also showed that adding fins within the clearance volume of the gas spring suppressed the difference in losses of the two types of data (high-pressure/low-speed and low-pressure/high-speed). See Kornhauser [18] for more discussion.

Comparison of modified CAST and experimental P-V diagrams are shown in Figures 2 and 3 for ~49 and 496 RPM respectively. Agreement is very good. When plotted in Microsoft Excel the curves were smooth. The 'jagged' nature of the curves appeared as an artifact of "pasting" the Excel plots into Microsoft Word.

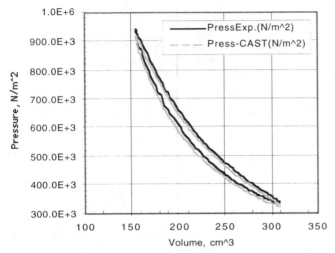

Figure 2: Comparison of Experimental and CAST P-V Diagrams. CAST used 18 x 12 grids, 120 time steps/cycle (Operating Conditions: 48.6 RPM, Mean Pressure = 555.7 kPa (80.6 psia), Wall Temp.=294 K)

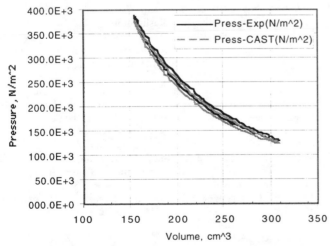

Figure 3: Comparison of Experimental and CAST P-V Diagrams. CAST used 18 x 12 grids, 120 time steps/cycle (Operating Conditions: 495.8 RPM, Mean Pressure=223.3 kPa (32.4 psia), Wall Temp. = 294 K)

Two-Space Test Rig Data and Calculation Comparisons

Figures 4 and 5 show experimental heat exchanger heat transfer per unit area and annulus center-to-wall temperature difference for Kornhauser's [18] two-space rig. Figures 6 and 7 show the corresponding Modified CAST calculated results for comparison with Figs. 4 and 5. The experimental and calculated heat transfer have different signs due to differences in definition of the positive heat transfer direction.

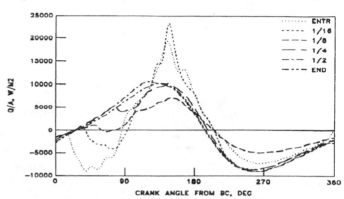

Figure 4: Heat Transfer per Unit Area at Various Positions in Heat Exchanger Relative to Entrance to Cylinder. Kornhauser [18] Exp. Data: Run #12071539, 201.7 RPM, 1.008 MPa Mean Pressure

Figures 4 and 6 show that peak experimental heat transfer/unit area near the entrance is 4000 to 5000 W/m^2 less than the corresponding calculated value; near the end of the heat exchanger the peak experimental value is about 9000 W/m^2 less than the corresponding calculated value. Thus even though the qualitative variations in heat transfer look very similar in the experimental and calculated plots, the quantitative agreement is not very good.

Comparison of the annulus center to wall temperature differences in Figs. 5 and 7 show that the calculated temperature differences are smaller that the experimental values. This is consistent with the calculated heat transfers/area being larger than the experimental values.

Two-dimensional plots of the calculated temperatures, velocities, pressures, etc. are given in Tew [9]. However, there are no experimental values available for comparison.

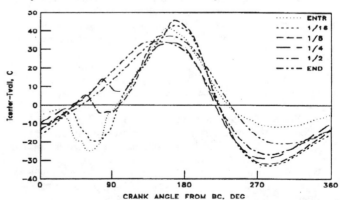

Figure 5: Temperature Difference from Heat Exchanger Center to Wall. Kornhauser [18] Exp. Data: Run #12071539, 201.7 RPM, 1.008 MPa Mean Pressure

469

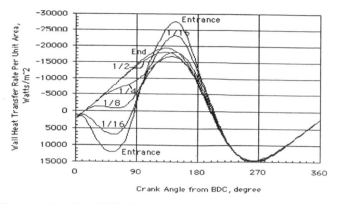

Figure 6: Modified CAST Calculation of Heat Transfer per Unit Area at Various Positions in Heat Exchanger Relative to Entrance to Cylinder (34x20 grids, 120 time steps/cycle). For Comparison to Kornhauser [18] Exp. Data: Run #12071539, 201.7 RPM, 1.008 MPa Mean Pressure.

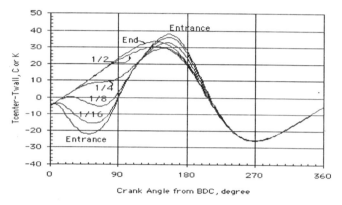

Figure 7: Modified CAST Calculations of Temperature Difference from Heat Exchanger Center to Wall (34x20 grids, 120 time steps/cycle). For Comparison with Kornhauser [18] Exp. Data: Run #12071539, 201.7 RPM, 1.008 MPa Mean Pressure

CONCLUDING REMARKS

The Modified CAST compressible non-acoustic model agreed well with 10 RPM gas spring hysteresis and P-V diagram data, and also with compressible acoustic calculations of two-dimensional velocities and temperatures. However, modified CAST calculations deviated from experimental values of two-space (piston/cylinder-heat exchanger) test rig data, although trends were predicted well.

Recent preliminary comparisons of the commercial CFD-ACE+ code compressible acoustic calculations with CAST, at CSU (Ibrahim, et. al. [19]), for the two-space rig suggests temperature agreement within about 2% for somewhat different grids. Therefore, data and 2-D computations do not agree, for some reason, for the two-space test rig. The reasons will be explored further. New data will likely be required for further multi-D Stirling code validation efforts.

When CAST and CFD-ACE+ calculations were made with the same type of grid and time step size, CFD-ACE+ was about 8% faster than CAST. Although not demonstrated here in comparisons against the highly developed commercial CFD-ACE+ code, Fedorchenko's compressible non-acoustic technique, may have the capability for reductions in simulation time for those transient situations where compressibility must be simulated but acoustics are not important.

REFERENCES

1. Ernst, W. D. and Shaltens, R. K., 1997, "Automotive Stirling Engine Development Project," DOE/NASA/0032-34, NASA CR-190780, MTI Report 91TR15

2. Dahr, M., 1999, "Stirling Space Engine Program," Volume 1—Final Report, NASA CR 19999-209164/VOL1

3. Tew, R. C. and Geng, S. M., 1992, "Overview of NASA Supported Stirling Thermodynamic Loss Research," NASA TM-105690

4. Thieme, L. G.; Qiu, S. and White, M. A., 2000, "Technology Development for a Stirling Radioisotope Power System for Deep Space Missions," NASA/TM-2000-209767 (Prepared for the 34th IECEC, Vancouver, B.C., August 1-5, 1999)

5. Furlong, R. and Shaltens, R., 2000, "Technology Assessment of DOE's 55-We Stirling Technology Demonstrator Convertor (TDC), NASA/TM—2000-210509

6. Huang., S.C., 1993, "HFAST Version 2.0 Analysis Manual," Prepared under NASA Contract Number NAS3-25330 (NASA Glenn Research Center)

7. Gedeon, D., 1999, "Sage User's Guide", 3rd Edition, (Gedeon Associates, 16922 South Canaan Road, Athens, OH 45701)

8. Geng, S. M. and Tew, R. C., 1992, "Comparison of GLIMPS and HFAST Stirling Engine Code Predictions with Experimental Data," NASA TM-105549

9. Tew, R.C. Jr. 2000, *Two-Dimensional Compressible Non-Acoustic Modeling of Stirling Machine Type Components*, Cleveland State University Doctoral Thesis.

10. Makhkamov, K. and Ingram, D.B., 2000, "Theoretical Investigations on the Stirling Engine Working Process," Paper # AIAA-2000-2815, Proceedings of the 35th Intersociety Energy Conversion Engineering Conference, Las Vegas, NV, July 2000

11. Fedorchenko, A. T., 1997, "A Model of Unsteady Subsonic Flow with Acoustics Excluded," J. Fluid Mechanics, Vol. 334, pp. 135-155

12. Peric, M. and Scheuerer, 1989, G., "CAST - A Finite Volume Method for Predicting Two-Dimensional Flow and Heat Transfer Phenomena," GRS - Technische Notiz SRR-89-01

13. Ferziger, J. H. and Peric, M., 1997, *Computational Methods for Fluid Dynamics*, Springer-Verlag

14. Patankar, S. V., 1980, *Numerical Heat Transfer and Fluid Flow*, Hemisphere Ubl. Co., Washington

15. Recktenwald, G. W., 1989, *A Study of Heat Transfer Between the Walls and Gas Inside the Cylinder of a Reciprocating Compressor*, University of Minnesota Ph. D. Thesis

16. Kornhauser, A. A., and Smith, J. L. Jr., 1987, "A Comparison of Cylinder Heat Transfer Expressions Based on Prediction of Gas Spring Hysteresis Loss," in Fluid Flow and Heat Transfer in Reciprocating Machinery, T. Morel, J. E. Dudenhoefer, T. Uzkan, and P. J. Singh (eds.), FED-Vol. 62, HTD-Vol. 93, American Society of Mechanical Engineers, NY

17. Kornhauser, A. A., and Smith, J. L. Jr. 1988, "Heat Transfer During Compression and Expansion," Phase II Progress Report for Oak Ridge National Laboratory Subcontract No. 19x-55915C, Cryogenic Engineering Laboratory, Department of Mechanical Engineering, Massachusetts Institute of Technology, Cambridge, MA

18. Kornhauser, A. A., 1989, *Gas-Wall Heat Transfer During Compression and Expansion*," Massachusetts Institute of Technology S. D. Thesis

19. Ibrahim,M., Tew, R., Zhang, Z., Simon, T., and Gedeon, D., 2001, "CFD Modeling of Free-Piston Stirling Engines", Proceedings of the 36th Intersociety Energy Conversion Engineering Conference"

Proceedings of IECEC'01
36th Intersociety Energy Conversion Engineering Conference
July 29-August 2, 2001, Savannah, Georgia

IECEC 2001-CT-15

A HEAT PUMP USING WATER AND AIR HEAT SOURCES ARRANGED IN A SERIES

Sadasuke Ito,   Yasushi Takano,   Naokatsu Miura
Department of Mechanical Systems Engineering
Kanagawa Institute of Technology, Atsugi, 243-0292, Japan
Fax:81-46-242-6806, Phone:81-46-291-3091, E-mail:ito@sd.kanagawa-it.ac.jp

Yasuo Uchikawa, Kubota Cooperation
Hirakata, 573-8573, Japan

ABSTRACT

The performance of a heat pump which used the ambient air heat-source and a water heat-source arranged in a series in the system was examined experimentally and analytically. In the experiment, a rotary type compressor with the rated capacity of 250 W was used. The analysis on the performance of a heat pump with dual heat-sources was based on the assumption that the performance of the heat pump for each single heat source was already known. It was found that the addition of the air heat-source at a higher temperature than the water heat source increased the COP when the air heat-source was set in the refrigerant loop in the upstream side of the water heat-source. In this case, the analytical results of the COP and evaporation temperature agreed well with the experimental results. The consumption of the heat energy in the water-heat source could be reduced by the assistance of the air heat-source.

INTRODUCTION

In order to reduce the emission of carbon dioxide and thereby to protect against global warming, the effective use of energy such as the efficient use of various types of waste heat and renewable energy should be promoted. A heat pump can produce more heat energy than the energy which is used to run the heat pump. Thus, a heat pump is considered to be one of the representatives of the machines which use energy efficiently.

For heating purposes such as space heating and hot water supply, the ambient air is commonly used as the heat source for a heat pump. The thermal performance of the heat pump can be increased by using water in wells, rivers, and ground as well as water or other liquid, which is heated by solar heat or various kinds of waste heat as the heat source. The interest in heat pumps using these heat sources for practical use has increased recently. However, the supply of such heat would be often unstable in the temperature and in the quantity. Therefore, it would not be easy to use such a heat source singly for evaporating the refrigerant in the evaporator.

In case of a ground source heat pump, Hopkins [1] showed that the ground temperatures around vertical earth heat exchangers decreased when heat was extracted. The amount of heat from the ground can be reduced by using the heat of the ambient air together with ground heat. Rafferty [2] made a capital cost comparison of commercial ground-source heat pump systems. One of the systems used the ambient air heat-source together with the ground heat-source in the loop of the heat transfer medium.

Kurisu [3] used the ambient air heat-source in the refrigerant loop when the temperature of a water heat-source became too low for a heat pump system which was used to melt snow by sprinkling water. Either one of the heat sources was used at a time. Ito [4] proposed to use a heat source together with the ambient air at the same time, which were arranged in parallel. In the study, the heat from water and heat from the ambient air were used for the heat pump. An expansion valve was set at the inlet of each evaporator.

Depending on the conditions, either one or both heat sources were used automatically.

In this study, a heat pump, which used two types of evaporators arranged in a series, was proposed for utilizing a heat source effectively and the ambient air at the same time. Then, the thermal performance of a heat pump, which used water and the air as the heat source, was studied experimentally and analytically. The performance of the heat pump with the dual heat sources were compared with that of the heat pump with the water heat-source or air heat-source.

NOMENCLATURES

COP: coefficient of performance, Eq. (1)

c_p: specific heat, J/(kg·K)

K_e: coefficient, W/K

P: pressure, Pa

H: electric power consumption by compressor, W

Q: heat transferred at evaporator or condenser, W

T: temperature, ℃

Subscripts

a: air

c: condenser

e: evaporator or evaporation

ea: air heat-source or air heat-source evaporator

ew: water heat-source or water heat-source evaporator

o: single evaporator

r: refrigerant

w: water

1: inlet

2: outlet

EXPERIMENT

Figure 1 shows a schematic diagram of a heat pump system for heating water at the condenser. The water in a constant temperature bath and the air in the room in which the system was installed were the heat sources. Either one of the heat sources or both heat sources arranged in a series were used in an experiment. The heat pump, which used the air heat-source singly, was called Type A as shown in Fig. 2(a). The heat pump, which used the water heat-source singly, was called Type W. Type W-A heat pump used the water heat-source in the upstream of the air heat-source as shown in Fig. 2(b). Type A-W heat pump used the air in the upstream of the water heat source. The arrangement of the evaporators in Fig. 1 is the case for Type W-A heat pump.

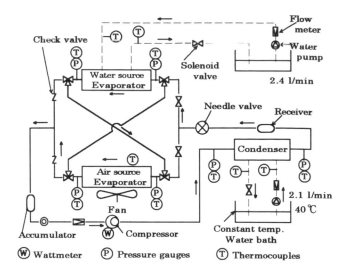

Fig. 1   Schematic of experimental apparatus.

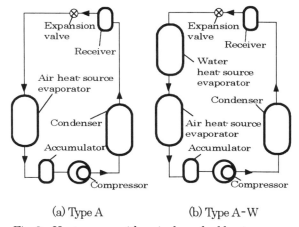

(a) Type A   (b) Type A-W

Fig. 2   Heat pumps with a single or dual heat sources.

Instead of using thermal storage tank to store the hot water, a constant temperature water-bath with a heater and cooler was used to supply the water to the condenser at the temperature of 40 ℃. The rated capacity of the rotary type compressor was 250 W. A needle valve was used as the expansion valve. The degree of superheat of the refrigerant, which was R22, at the outlet of the evaporator was adjusted to 6 ℃ manually. The air heat-source evaporator, which was called Type A evaporator, was a finned tube type. The flow rate of the room air whose temperature was kept constant was chosen to be 4.5 m³/min. The water heat-source evaporator, which was called Type W evaporator, was made of two copper tubes soldered together and rounded cylindrically. One of the copper tubes with the inside diameter of 10.8 mm was for the water flow and the other with inside diameter of 7.9 mm was for the refrigerant flow. The length of each tube was about 11 m. This type of heat exchanger was used for the condenser, too. The flow rates of the water for Type W evaporator and the condenser were chosen to be 2.4 l/min, and 2.1 l/min, respectively. The temperatures and pressures of the refrigerant were measured by thermocouples and Bourdon tube pressure gauges respectively at the location shown in Fig. 1. The temperatures and power consumption of the compressor were measured every 5 minutes. After the temperatures, pressures and degree of superheat became stable in an operation, data were taken for 20 minutes. The temperatures and pressures were averaged in 20 minutes. When the both heat sources were arranged in a series, the temperature of the air was kept constant at 20 ℃.

The coefficient of performance, COP, defined by the following equation was obtained.

$$COP = Q_{c,w} / H = m_{c,w} \cdot c_{p,w} \cdot (T_{c,w,2} - T_{c,w,1}) / H \qquad (1)$$

where $Q_{c,w}$ was the heat gain per unit time at the condenser, H was the electric power consumption of the compressor, $m_{c,w}$ was the mass flow rate of the water, $c_{p,w}$ was the specific heat, $T_{c,w,1}$ and $T_{c,w,2}$ were the temperatures of the water at the inlet and outlet of the condenser, respectively.

ANALYSIS

An analysis is made on the thermal performance of a heat pump using dual heat sources. The thermal performance of Type A and Type W heat pumps is assumed to be already known. The flow rates of the air and water for the heat pump with the dual heat sources are assumed to be the same as those for Type A and Type W heat pumps.

It is assumed that there is no pressure loss in the refrigerant in the evaporators and that the refrigerant in the evaporators absorbs heat at an evaporation temperature, which is the saturated temperature.

Generally, the COP of a heat pump is given by a function of the evaporation temperature, T_e, and condensation temperature, T_c, or the temperature of the heat transfer medium at the condenser. Since the temperature of the water at the condenser is kept 40 ℃, the COP of the heat pump would be given as a function of T_e. Thus,

$$COP = F_1(T_e) \qquad (2)$$

At first, consider Type W heat pump. If there is no heat loss from the compressor, the heat obtained by the water at the condenser, $Q_{c,w}$, is equal to the sum of the heat absorbed by the refrigerant at Type W evaporator and the power supplied to the compressor, H. Thus,

$$Q_{c,w} = Q_{e,w} + H \qquad (3)$$

The heat supplied to the refrigerant at Type W evaporator is given by the previous paper by Ito [1] as follows.

$$Q_{e,w} = K_{ew}(T_{ew,w,1} - T_e) \qquad (4)$$

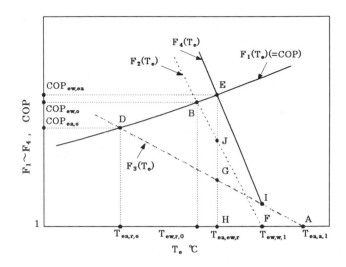

Fig. 3   A diagram for finding COP and the evaporation
temperature of the heat pump with dual heat sources.

where $T_{ew,w,1}$ is the inlet temperature of the water.

From Eq. (3) and Eq. (4), $Q_{c,w}/H$ is given by the following function $F_2(T_e)$.

$$Q_{c,w} / H = F_2(T_e) = (K_{ew} / H) \cdot (T_{ew,w,1} - T_e) + 1 \qquad (5)$$

In case of Type A heat pump, Eq. (6) and Eq. (7) can be obtained.

$$Q_{e,a} = K_{ea}(T_{ea,a,1} - T_e) \qquad (6)$$

$$Q_{c,a} / H = F_3(T_e) = (K_{ea} / H) \cdot (T_{ea,a,1} - T_e) + 1 \qquad (7)$$

Similarly to a single heat source, Eq. (8) and Eq. (9) would be obtained for the heat pump with the dual heat sources as follows.

$$Q_e = Q_{e,a} + Q_{e,w} \qquad (8)$$

$$Q_c = Q_e + H \qquad (9)$$

$$Q_c / H = F_4(T_e) = (K_{ew} / H) \cdot (T_{ew,w,1} - T_e)$$
$$+ (K_{ea} / H) \cdot (T_{ea,a,1} - T_e) + 1 \qquad (10)$$

From Eq. (5), Eq. (7) and Eq. (10), the following equation is derived.

$$\{ F_4(T_e) - 1 \} = \{ F_2(T_e) - 1 \} + \{ F_3(T_e) - 1 \} \qquad (11)$$

The power consumption of the compressor would be given by a function of T_e. Then,

$$H = F_2(T_e) \qquad (12)$$

For given $T_{ew,w,1}$ and $T_{ea,a,1}$, the evaporation temperature for the dual heat sources, $T_{ew,ea,r}$, is the temperature T_e which satisfies Eq. (10). Incidentally, the values of K_{ew} or K_{ew}/H and H for Type W heat pump and K_{ea} or K_{ea}/H for Type A heat pump are supposed to be given as it was assumed.

The coefficient of performance of the heat pump with the dual heat sources, $COP_{ew,ea}$, can be obtained analytically by substituting the evaporation temperature, $T_{ea,ew,r}$, which is obtained by Eq. (11), to Eq. (2).

The heat absorbed from the water heat-source is obtained from Eq. (4) and Eq. (5) and given by

$$Q_{e,w} = \{F_2\,(T_{ew,ea,r}) - 1\}H \qquad (13)$$

Similarly, the heat absorbed from the air heat-source is given by

$$Q_{e,a} = \{F_3\,(T_{ew,ea,r}) - 1\}H \qquad (14)$$

A method to find $COP_{ew,ea}$ and $T_{ea,ew,r}$ is shown graphically by using a figure like Fig. 3. In the figure, the coefficient of performance of Type W heat pump, $COP_{ew,o}$, for the water heat-source at a temperature of $T_{ew,w,1}$ is plotted on the curve $F_1\,(T_e)$ at the point B. The point F is plotted at $COP=1$ and $T_e=T_{ew,w,1}$. Then, the line between the points B and F is $F_2\,(T_e)$. If K_{ew}/H dose not depend on T_e, $F_2\,(T_e)$ is a straight line. Similarly, for Type A heat pump, the line between the points D and A is $F_3\,(T_e)$. It can be understood that the addition of the heights of $F_2\,(T_e)$ and $F_3\,(T_e)$ from the axis of the abscissa gives the height of $F_4\,(T_e)$. The cross point E of the curve $F_4\,(T_e)$ and the curve $F_1\,(T_e)$ gives $COP_{ew,ea}$ and $T_{ea,ew,r}$. The length between the points J and H is equivalent to Q_{ew}/H and the length between E and J or G and H is equivalent to Q_{ea}/H.

RESULTS AND DISCUSSIONS

Figure 4 shows the relation between COP and saturated temperatures at the exits of Type W evaporator, $T_{e,2}$. From these data, $F_1\,(T_e)$ and $F_5\,(T_e)$ were obtained.

The relation between COP and the temperature of the heat sources is shown in Fig. 5 and relation between the saturated temperature at the exit of the evaporator, and the temperature of the heat sources is shown in Fig. 6, respectively. In case of Type W heat pump, the temperature of the water was changed between 8℃ and 20℃. For dual heat sources, the temperature of the air heat source was kept constant at 20℃ and the temperature of the water was changed between 8℃ and 17℃. It was known from these figures that the temperature of the air heat-source was several degrees higher than the water heat source to get the same COP and $T_{e,2}$ when each heat source was used singly. The COP and $T_{e,2}$ were higher for Type W-A heat pump than for Type W heat pump. In this case, the analytical results showed good agreements with the experimental results. The evaporation temperature in case of the single air heat-source at 20℃ was 6.2℃(cf. Fig. 6). Comparing with Type W heat pump, the COP and the evaporation temperature became higher for Type W-A heat pump when the temperature of the water heat-source was higher than the evaporation temperature at 6.2℃. However, the COP and T_e for Type A-W heat pump were about the same as those for Type W heat pump. The addition of the air heat-source did not increase the COP and the evaporation temperature in this case.

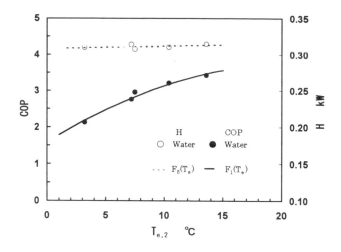

Fig. 4   Variation of COP and power consumption of compressor.

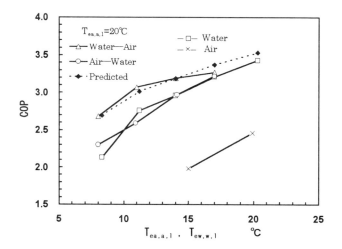

Fig. 5   Relation between COP and the temperature of heat sources.

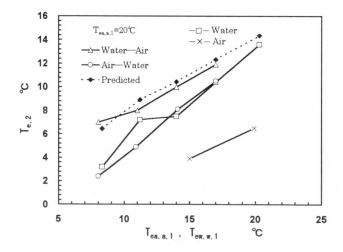

Fig. 6   Relation between the evaporation temperature and the temperature of heat-sources.

474

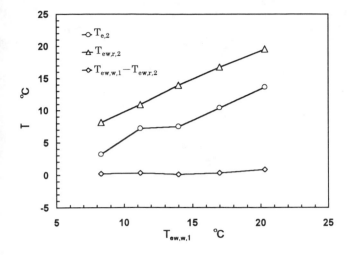

Fig. 7　Variation of the saturated temperature and superheated vapor at he exit of the evaporator of Type W heat pump .

Figure 7 shows the variations of the saturated temperature at the exit of the evaporator, $T_{e,2}$, and the temperature of the superheated vapor of the refrigerant at the exit of Type W evaporator, $T_{ew,r,2}$, for Type W heat pump. Due to the good performance of the evaporator, the temperature of the refrigerant at the exit was raised to the temperature of the heat source. Since the degree of superheat at the evaporator was adjusted to about 6℃, the evaporation temperature became about 6℃ below the temperature of the water heat-source. In this case, the addition of the air heat-source at 20℃ to the system was not possible to increase the evaporation temperature and COP. If the performance of the evaporator had not been good or if the degree of superheat had been much smaller, then the addition of the air heat-source which was set in the upstream of the water heat-source would have increased the evaporation temperature and COP.

The heat absorbed by Type W evaporator and the heat transferred from the refrigerant to the water at the condenser are shown in Fig. 8 for the case of Type W-A heat pump. When the temperature of the water heat-source was about 8℃, the heat transferred from the water heat-source to the refrigerant was about zero. The higher the temperature of the water heat-source, the larger the absorbed heat from the water. The predicted results of the absorbed heat from the water heat-source, $Q_{e,w,p}$, and the heat transferred to the water at the condenser from the refrigerant, $Q_{c,p}$, agreed well with the experimental results.

At $T_{ew,w,1}=12$℃, the heat absorbed by Type W evaporator and the heat absorbed by Type A evaporator became the same. It could be said

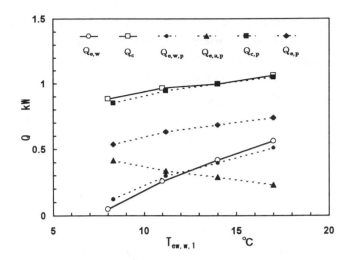

Fig. 8　Heat transferred at the evaporators and condenser for Type W-A heat pump.

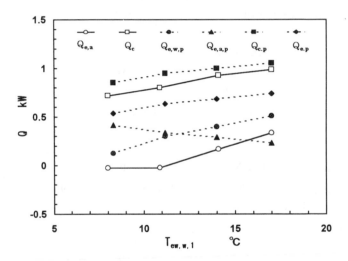

Fig. 9　Heat transferred at the evaporators and condenser for Type A-W heat pump.

that not only the COP increased but also the consumption of the heat in the water heat-source was able to be reduced by the assistance of the air heat-source.

CONCLUSIONS

The performance of Type W evaporator was so good that the evaporation temperature was governed by the degree of superheat at the exit of the evaporator. In this case, the addition of the air heat-source with higher temperature than that of the water heat-source which was set in the upstream of the water heat-source did not increase the evaporation temperature and the COP. The addition of Type A evaporator in the downstream of Type W

475

evaporator increased the COP and evaporation temperature. When the temperature of the water heat-source was higher than the evaporation temperature of Type A heat pump, the evaporation temperature and COP were higher than those for Type A evaporator as well as Type W evaporator. The analytical results agreed well with experimental results for this arrangement of the heat sources. The consumption of heat energy in the water heat-source was able to be reduced by addition of the air heat-source in the system.

REFERENCES

1. Hopkins, R. J., and Burkart, R., 1990, "Earth-Coupled Heat Pumps," Proceedings of the 3rd International Energy Agency Heat Pump Conference, Tokyo, Japan, pp. 411-421.

2. Rafferty, K., 1994, "A Capital Cost Comparison of Commercial Ground-Source Heat Pump Systems," Geothermal Resources Council Transactions, Vol. 18, pp. 387-392.

3. Kurisu,R., Sugiyama, K., Okajima, J., Mitsuoka, F., Sato,M., Sato, M., 1992, "Aqua-air Double Source Heat Pump Chiller Applied for Snow-melting System," Proc. of Japan Society of Mechanical Engineers 2nd Symposium on Environmental Engineering, Kawasaki, Japan, pp. 439-442.

4. Ito, S. and Miura, N., 2000, "Studies of a Heat Pump Using Water and Air Heat Sources in Parallel," Heat Transfer-Asian Research, 29 (6), pp. 473-490.

Proceedings of IECEC'01
36th Intersociety Energy Conversion Engineering Conference
July 29–August 2, 2001, Savannah, Georgia

IECEC2001-CT-17

EXPERIMENTS ON A HEATPUMP-CHILLER WITH SOME HFC REFRIGERANTS

G. Dheeraj*, G. Venkatarathnam*, L.R. Oellrich and S. Srinivasa Murthy*[1]**

*) Refrigeration and Airconditioning Laboratory
Indian Institute of Technology Madras, India

**) Institut für Technische Thermodynamik und
Kältetechnik, Universität Karlsruhe (TH), Germany

ABSTRACT

The performance of a heat pump-chiller has been experimentally studied with the following working fluids: (i) HFC134a (ii) HCFC22 (iii) HFC mixture R507 (iv)HFC mixture R407C. The coefficients of performance, volumetric heating and cooling capacities and exergy efficiency obtained with the above fluids are compared.

INTRODUCTION

Heat pumps are the preferred devices for heating in the temperature range of 50 to 120 °C because of their lower power consumption per unit of heat delivered at these operating conditions. Heat pump-chillers are ideal choice when the difference between the temperatures at which heating and cooling are required is about 50 to 70 °C. Prior to Montreal Protocol ban, the preferred working fluids in high temperature heat pumps were CFC114 and CFC12.

Recently, a number of authors have been studying the performance of systems with different single component as well as mixtures of Hydro-Fluoro-Carbons (HFCs) and Hydro-Carbons (HCs) [1-4]. Most of these studies have largely been on either refrigerators or heat pumps. Very few studies have been reported on the performance of heat pump chillers, in which both useful refrigeration and heating is obtained. In this context, this paper reports the results of experiments on heat pump chillers with different working fluids.

NOMENCLATURE

COP_C	Cooling Coefficient of Performance
COP_H	Heating Coefficient of Performance
COP_O	Overall coefficient of Performance
Q_E	Heat absorbed in the evaporator (kW)
Q_C	Heat rejected in the condenser (kW)
T_a	Ambient temperature (°C)
T_C	Condensing temperature (°C)
T_E	Evaporating temperature (°C)
W	Compression work (kW)
η	Exergy efficiency

HEAT PUMP CHILLER TEST STAND

Figure 1 gives the schematic view of test stand [5] built at the Indian Institute of Technology Madras. The test stand comprises of three different fluid circuits.

- Refrigerant circuit
- Water circuits: (i) Hot water circuit, (ii) Cold water circuit

The evaporator heat load is supplied by a set of ten electrical heaters each of 1.5 kW capacity, and also by two air cooled cooling coils of capacity 5 kW each. All the heaters and cooling coils can be activated individually giving the flexibility for obtaining any desired temperature.

The condenser heat is dissipated to the ambient using a cooling tower. Water is used as the heat transfer fluid in both the condenser and evaporator sides. When the test stand is operated in the sub-zero refrigeration mode, the water in the cold water circuit is replaced by brine solution. Part of the fluid flowing to the cooling tower can be bypassed to achieve the required temperature of water at the condenser inlet.

It is possible to operate the test stand at different evaporating and condensing temperatures by adjusting heat addition and heat dissipation rates. The test stand can be operated in heat pump, refrigeration and heat pump-chiller modes. Different condensing and evaporating temperatures can be achieved by controlling the mass flow rate of the heat transfer fluids, as well as the temperature of heat transfer fluid at condenser and evaporator inlet.

---

[1] Corresponding Author. email: ssmurthy@iitm.ac.in

Compressor

The refrigerant compressor used is a semi-hermetic reciprocating compressor type Bitzer 2DL-3.2. It has a displacement volume of 13.3 m³/hr and operates at 1450 RPM. The maximum allowable working pressure is 25 bar. The compressor is designed for HCFC22 and is suitable for use with alternatives such as HFC134a, other HFCs and mixtures.

Condenser, Evaporator and Internal heat exchanger

All the counter flow heat exchangers are built with enhanced surface tubes donated by Weiland, Germany. The condenser and the internal heat exchanger are of tube-in-tube type. The evaporator has five tubes-inserted-in-single tube type.

The details of the heat exchangers are given in Table 1.

PERFORMANCE WITH DIFFERENT FLUIDS

The performance of the heat pump chiller with and with out the use of glide in the heat exchangers has been discussed elsewhere by the authors [5,6]. In this paper, the performance of a heatpump-chiller with (i) HFC134a (ii) HFC22 (iii) HFC azeotrope R507 and (iv) HFC zeotropic mixture R407C is presented.

The highest condensing temperature at which the system can be operated with each refrigerant is limited by the design considerations of the compressor. In this case the maximum pressure in the compressor cannot exceed 25 bar.

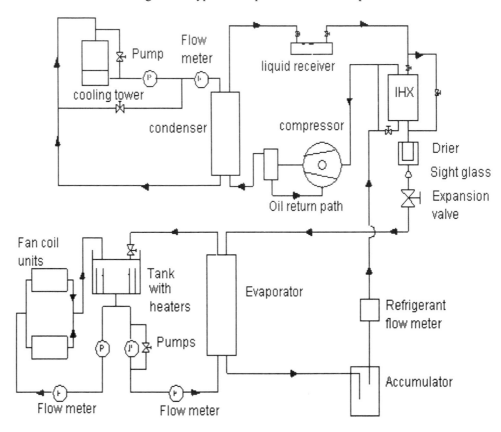

Figure 1: Schematic of the heat pump-chiller test rig

Table 1: Dimensions of the heat exchangers

Type	Length (mm)	Refrigerant (Inside Tube)		Water (Inside Annulus)		No. Of segments in series
		ID (mm)	OD (mm)	ID(mm)	OD (mm)	
Condenser	1100	22.4	25.4	32.7	35.7	9
Evaporator	1400	12.6*	15.6*	50.0	54.0	7
LSHX	1100	22.4	25.4	32.7	35.7	1

* 5 Tubes

Experiments could, therefore, not be conducted at higher condensing temperatures with refrigerants such as HCFC22, HFC mixture R507 etc. In order to compare the performance of zeotropic mixture R407C with that of pure fluids, the reference condensing temperature is taken as the dew point temperature and the reference evaporating temperature is defined as that at the inlet of the evaporator.

In the test rig, the the highest evaporating temperature is limited by the maximum heat that could be added in the evaporator (15kW). All the tests were performed for a superheat of 10 ± 1 K and a subcooling of 2 ± 1 K.

Though heat is rejected at temperatures above the ambient and heat absorbed at temperatures below the ambient, the heating/cooling is useful only at certain temperatures. The overall (heating and cooling) COP is defined to take this into account as given in Table 2.

Figure 2 shows the variation of overall (heating+cooling) COP with condensing and evaporating temperatures. It can be

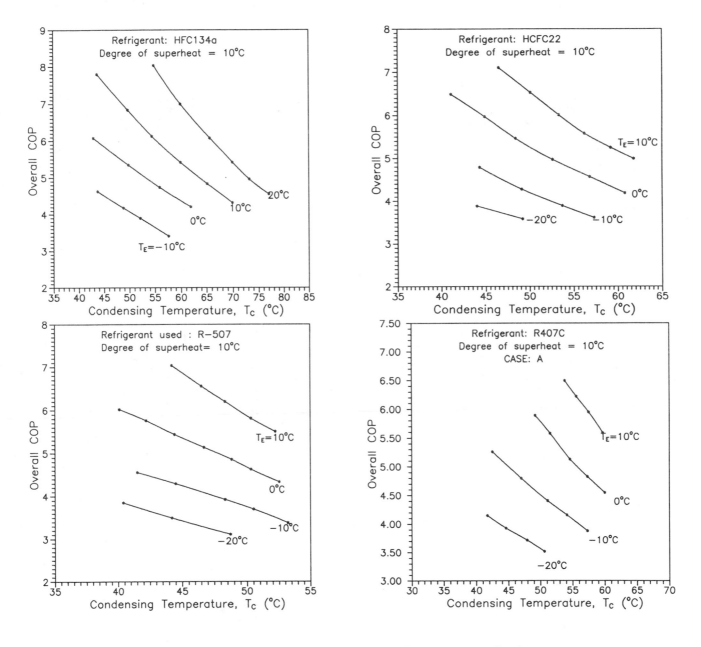

Figure 2: Variation of overall COP (heating + cooling) with condensing and evaporating temperatures.

seen that a heating COP greater than 5 has been achieved with all the fluids tested at evaporating temperatures of -10 °C and above and temperature lifts greater than 50 K. It can also be observed the lift at which a heating COP of about 5 can be achieved is the lowest for R507 and highest for R407C and HFC134a. Because of the lower vapour pressure, it was also possible to reject heat at temperatures as high as 75 °C in the case of HFC134a.

Figure 3 shows the variation of volumetric refrigerating capacity with condensing and evaporating temperatures for the different fluids tested. The volumetric refrigerating capacity of HFC134a is the least of all because of its lower vapour pressure. The results also show that the volumetric refrigerating/heating capacity of R407C and R507 is higher than that of HCFC22 over the range of condensing and evaporating

temperatures studied.

It is evident from Figs. 2 and 3 that when heat rejection needs to be done at temperatures greater than 50 °C, HFC134a becomes the only choice available. Where as at lower condensing temperatures, both R407C and R507 are preferable for use in the heat pump-chiller because of their high volumetric heating capacity. The performance of systems operating with R407C, however, will depend on how the refrigerant glide in the condenser and evaporator are put to use [6].

Figure 4 shows the variation of overall exergy efficiency of the system which is defined as follows:

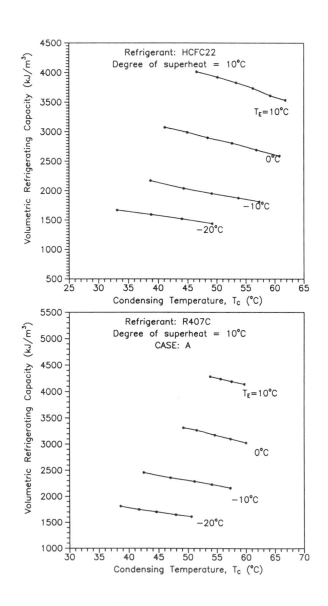

Figure 3: Variation of volumetric refrigerating capacity at different condensing and evaporating temperatures

Table 2: Definition of overall COP

Sl. No.	Evaporating Temperature Range (°C)	Condensing Temperature Range (°C)	Overall COP definition
1	$T_E < 20$	$20\,°C \leq T_C \leq 40$	$COP_O = COP_C = Q_C/W$
2	$T_E < 20$	$T_C > 30$	$COP_H = Q_H/W$; $COP_C = Q_C/W$; $COP_O = (Q_C + Q_H)/W$
3	$20\,°C \leq T_E \leq 40$	$T_C > 40$	$COP_H = COP_O = Q_H /W$

$$\eta = \frac{|Qc|*|1 - Ta\,/\,Tc| + |Qe|*|Ta\,/\,Te - 1|}{|W|}$$

In the case of a conventional heat pump in which the evaporating temperature (T_E) is close to the ambient (T_a), the second term in the numerator of the exergy efficiency expression becomes negligible. Similarly, in the case of a conventional refrigerator in which the condensing temperature (T_C) is close to ambient, the first term in the numerator becomes negligible.

The overall exergy efficiency of the system depends on irreversibilities during heat transfer, pressure drop across different components, the deviation of the cycle from the ideal cycle (superheat horn, isenthalpic expansion), as well as the exergy efficiency of the compressor itself. Since the deviation between the vapour compression cycle and the Carnot cycle increases with an increase in condensing temperature, the overall exergy efficiency of a theoretical vapour compression cycle should decrease monotonically with an increase in condensing temperature. Contrary to that, the results show that the losses in the cycle are somewhat high at very low condensing temperatures or low temperature lifts, resulting in a peak in the exergy curves as seen in Fig. 4. It is not very clear where the losses are occuring. It is felt that it may be due to the performance of the compressor at low lifts or high flow rates. Since the variation of exergy efficiency with temperature lift is not as much for HFC134a compared to other fluids, it may be concluded that the losses seem to be the highest with R134a at low temperature lifts. Here, it should be mentioned that the compressor was originally designed for HCFC22.

From Fig. 4 it is also evident that the maximum exergy efficiency is less than 50% for all the fluids tested at all operating conditions. The exergy efficiency is the highest with R407C mixture. The performance with R407C, however, depends on whether the temperature glides are used beneficially in the heat exchangers or not. During the tests, the flow of heat transfer fluid was regulated such that the temperature approach between the working fluid and the heat transfer fluid at the warm end of the heat exchanger is lower than that at the cold end of the heat exchanger in the case of condenser, and vice versa in the case of evaporator, which corresponds to the „ beneficial use" of glide during condensation and evaporation. It can also be seen that the exergy efficiency is lower with R507 compared to that with HFC134a at corresponding temperatures.

CONCLUSIONS

An overall (heating + cooling) COP greater than 5 has been obtained with all the fluids tested, at temperature lift of about 50 K. For any given overall COP, the temperature lift is the lowest for R507.

With the compressor installed in the test-rig, the overall system exergy efficiency with R134a is higher than that with R507, while the volumetric heating/cooling capacity is much higher with R507 at corresponding temperatures.

ACKNOWLEDGMENTS

This paper is a part of an Indo-German project on „Alternatives to CFCs for heat pump applications" sponsored by DLR and the IIT Madras. The authors acknowledge the support of M/S Bitzer and Weiland, who donated a compressor, and enhanced surface heat exchanger tubes respectively, and Solvay Deutschland, for donating the refrigerants used in this work.

REFERENCES

1. Pannock, J., Didion, D.A., Radermacher, R., (1991) Energetic behaviour of chlorine-free zeotropic refrigerant mixtures in a heat pump or air conditioner: A computer study and practical test, *Deutsche Kaelte Vereins (DKV)_Tagungsbericht*, 18 (2), 125-139
2. Domanski, P.A. and D.A. Didion (1993) Thermodynamic evaluation of R-22 alternative refrigerants and refrigerants mixtures, *ASHRAE Trans.*, 99, Part 2, 636-648
3. Purkayastha, B. and P.K. Bansal (1998) An experimental study on HC290 and a commercial liquefied petroleum gas (LPG) mix as suitable replacement for HCFC22, *Int. J. Refrig.*, 21, 3-17
4. Loi, N.D., Moessner, F and Oellrich, L.R., Ein Versuchsstand zur Untersuchung des Verhaltens von Kältemitteln und gemischen, *Proc. DKV-Jahrestagung*, Bremen, Nov. 1992 II (1), 219-233
5. Dheeraj,G., Experiments on heat pump chiller with different working fluids, M.S. Thesis, Indian Institute of Technology Madras, India, Sept. 2000.
6. Untersuchungen an einem kombinierten Wärmepumpen-Kühler Teststand für gleichzeitige heiz - und Kühlanwendungen, Deutscher Kälte- und Klimatechnischer Verein (DKV) Conference, Bremen, Germany, Nov. 2000.

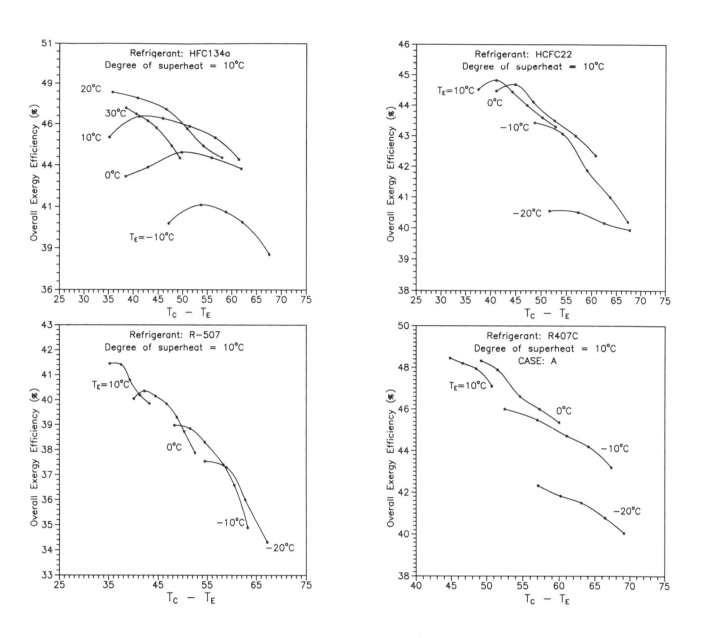

Figure 4: Variation of the exergy efficiency of the heat pump chiller at different condensing and evaporating temperatures.

IECEC2001-CT-18

APPLICATION OF ISOBUTANE - PROPANE MIXTURES FOR HEATING, AND COMBINED HEATING AND COOLING IN VAPOUR COMPRESSION SYSTEMS

Suresh Bhakta Shrestha[*] S. Srinivasa Murthy[@]

Refrigeration & Air-conditioning Laboratory, Department of Mechanical Engineering
Indian Institute of Technology Madras, Chennai - 600 036 (India)

ABSTRACT

Performance of propane and isobutane mixtures in vapour compression systems for heating, and simultaneous heating and cooling applications, is studied. The influence of mixture composition is highlighted leading to the identification of suitable compositions for specific applications. Pressure ratio, pressure difference, compressor discharge temperature, cooling and heating coefficients of performance, volumetric cooling and heating capacities are evaluated. Since temperature glides in evaporator and condensers are inevitable, it is necessary to determine them for possible use in enhancing system performance by matching with the glide of the heat transfer fluid. Based on these parameters, acceptability of hydrocarbon mixtures for the combined heating and cooling, and heating systems are discussed in comparison with those of pure refrigerants such as CFC114, HFC236ea and HFC134a.

INTRODUCTION

Increasing concern on global warming has brought forth the need to search for alternatives for ozone friendly HFCs. Hence, the search is on for environment friendly as well as energy efficient working fluids for various heating and cooling applications. Simultaneous cooling and heating from a compression system has many practical applications in industries like dairy, food processing, pharmaceuticals etc., and can be highly energy efficient. In spite of the flammability, hydrocarbons are being successfully used in domestic refrigerators, freezers, etc by employing newer technologies of design, fabrication, safety and control [1-4]. However, there is a need to investigate the possibility of using hydrocarbon mixtures in applications such as water chilling, air-conditioning, simultaneous heating and cooling, and high temperature process heating. In this context, this paper studies the performance of various compositions of propane and isobutane mixture in vapour compression systems.

NOMENCLATURE

COP = Coefficient of performance (-)
h = specific enthalpy (kJ/kg)
dt = Temperature glide ($^{\circ}$C)
m_r = Refrigerant mass flow rate (kg/s)
P = Pressure (MPa)
PR = Pressure ratio (-)
PD = Pressure difference (MPa)
Q_{vo} = Volumetric cooling capacity (kJ/m^3)
q_e = Refrigeration or cooling effect (kJ/kg)
q_h = Heating effect (kJ/kg)
ρ = Density (kg/m^3)
t = Temperature ($^{\circ}$C)
W = Work of compression (kW)
X = Mass fraction (kg /kg of mixture)

SUBSCRIPTS

c	condenser	crit.	critical
e	evaporator	l	liquid
r	refrigerant	s	saturation
v	vapour	vo	volumetric

[*] On study leave from Mechanical Engineering Department,
Kathmandu University, P O Box No. 6250, Kathmandu, Nepal.
[@] Corresponding Author, e-mail: ssmurthy@iitm.ac.in

THERMODYNAMIC MODEL

The Lorenz cycle for zeotropic refrigerant mixtures is shown in Figure 1. Since propane and isobutane are Type A and Type B refrigerants respectively, their mixtures usually fall into either of the categories depending on concentration and operating temperatures. Moreover, for the mixtures as well as pure fluids belonging to Type B, it is necessary to provide a certain minimum degree of suction superheat to avoid wet compression. The cycle assumes isentropic compression followed by isobaric heat exchange in condenser, isenthalpic expansion in throttle valve and isobaric heat exchange in evaporator. Figure 1 also shows the schematic of the cycle on t-x diagram highlighting temperature glides during phase change in both evaporator and condenser.

Among the performance parameters, COP is a good indicator of energy consumption whereas volumetric efficiency relates to the compressor size. Pressure ratio and discharge temperature significantly influence various design and operational aspects of compressor. Matching of temperature glide of refrigerant mixture with that of the heat transfer fluid is important in order to achieve high heat transfer efficiencies and also to avoid pinch points in the heat exchangers [5-7]. The temperature glides of refrigerant mixtures in both evaporator and condenser are therefore significant parameters. Discharge temperature plays an important role in heat pumps as it governs various factors like viscosity and stability of lubricants, life of motor winding etc. The performance parameters for Type A and B fluids are defined in Table 1.

Normal boiling point and critical point usually offer lower and upper practical limits for vapour compression cycle. Propane is low normal boiling point and low critical point fluid where as isobutane is high boiling point refrigerant with fairly high critical point. Their mixtures have intermediate values of bubble point and critical point. Table 2 shows critical properties for pure refrigerants and select mixtures. Figure 2 shows the saturation curves. In this study, the reference evaporator and condenser temperatures are defined as the dew point temperature at compressor entry and bubble point temperature at the entry to expansion valve respectively [8]. Also, the reference evaporator cooling capacity Q_e is taken as 1 kW.

RESULTS AND DISCUSSION

Performance evaluation is carried out at wideranging evaporator and condenser temperatures as follows:
Combined Heating and Cooling:
t_e = -20°C to +10°C and t_c = 50°C to 100°C
Heating only:
t_e = 20 °C to 60°C and t_c = 50°C to 100°C
Pure and mixture refrigerant properties are computed using REFPROP 6.01 [9].

Figures 3 to 5 show that both pure and mixture hydrocarbons provide better volumetric capacities than CFC114

and HFC 236ea, and the mixtures provide comparable performance with respect to HFC 134a. Among the pure refrigerants, propane maintains the highest volumetric capacities in all temperature ranges. 70% propane mixture closely matches the volumetric capacities of HFC 134a especially for combined heating and cooling. It can be observed that hydrocarbon mixtures provide better volumetric heating capacities when operated at elevated evaporator temperatures. It can be seen from Figure 5 that increasing the propane percentage enhances both the volumetric cooling and heating capacities of the mixtures.

A comparison of above results with Figures 6 to 8 reveals that the highest COPs are registered by those refrigerants which are poor in volumetric capacities. Among pure refrigerants, HC 600a and HFC 236ea yield maximum COPs while HC 290 gives the lowest COP. 50% propane mixture also results in the lowest COP. Figures 6 and 7 also show that COPs of 10% propane compositions are better than those of HFC 134a for both combined and heat pump operations. For heating, 10% propane mixture delivers better COP than pure isobutane and HFC 236ea. Moreover, with increasing condenser temperature, COPs of isobutane and HFC 236ea deteriorate. On the other hand, even mixtures like 50% and 70% propane compositions improve their COPs at increasing condenser temperature from 80°C onwards. Also, the magnitude of COPs particularly for the mixtures for heat pump operation, are higher than sum of the individual COPs for heating and cooling in combined mode. Figure 8 demonstrates that low COPs result at the middle of concentration range.

Figure 9 shows that the highest pressure ratio under all operating conditions is offered by HFC 236ea which has the lowest pressure difference, and the highest value of pressure difference occurs for propane, which exhibits the lowest pressure ratio. Mixtures with 70% to 90% propane compositions mark low pressure ratios among mixtures at all the temperature ranges. The pressure ratios of other compositions and HC 600a match that of HFC 134a. It is evident that high pressure ratios are encountered at high isobutane contents. The pressure difference of most of the compostion is comparable with that of HFC 134a.

From Figures 10 to 14, it is seen that 10% and 90% propane mixtures behave as near azeotropes as the temperature glides in both evaporator and condenser are less than 3K. In general, glides in the condenser are larger than those in the evaporator. The highest temperature glides occur in the vicinity of 50% mass fraction i.e. at equimolar compositions as evident from Figure 14.

Figure 15 reveals that the discharge temperatures for mixtures of 70% to 90% propane compositions are higher than those for pure propane, matching with those of HFC 134a. For the composition upto about 40% propane, discharge temperature remains nearly constant and begins to raise at higher compositions. The discharge temperature variation for different compositions is attributed to slope of saturated vapour curve as upto 40% propane composition, the mixture behaves as Type B and thereafter, it behaves as Type A fluid.

Figure 16 shows the minimum degree of suction superheat required for different compositions of propane-isobutane mixtures to prevent retrograde compression which leads to two phase state at the end of compression process. For most of the temperature range, the mixture compositions that require superheat are from 10% to 30% propane compositions. However, depending on temperature lift as well as evaporation and condensation temperatures, even the 50% propane composition may require superheating.

CONCLUSIONS

Propane offers good volumetric heating and cooling capacities at all temperature ranges. It also offers reasonable COP compared to mixtures except for few compositions with higher isobutane content. However, system pressures exceed those with HFC 134a and HFC 236ea greatly. Propane offers large pressure difference usually more than 1.8 MPa, an allowable limit for reciprocating compressors. It offers discharge temperature comparable with that of HFC 134a. Since it's critical temperature is 96.7°C, it is not suitable for high temperature heat pumps. Therefore, it looks difficult for propane to be retrofitted into systems designed for HFC 134a and HFC 236ea.

Isobutane offers good COPs, particularly at low evaporator and condenser temperatures. At high temperatures, COP decreases drastically. It offers very low volumetric heating capacity and requires high degree of suction superheat. Similarly, the pressure ratio is also very high even though it offers acceptable pressure difference values.

For 0.3 mass fraction mixture, the discharge temperature is low, and pressure difference and pressure ratios are comparable with those of HFC 134a. The mixture requires large degree of suction superheat and the temperature glide in evaporator is less than 3K. Unlike propane, it can be used for high temperature application as it has got fairly high critical temperature of 122.6°C. Another reason marking it suitable for high temperature application is its good COPs at high temperature, which are even more than those of pure isobutane. This composition offers lower system pressures compared to those of HFC 134a.

0.5 mass fraction mixture has favorable pressure characteristics and volumetric heating and cooling capacities. The discharge temperature, pressure ratio and pressure difference are comparable to those of HFC 134a. It offers high temperature glides in both evaporator and condenser, and low COP at low condenser temperatures. However, the COPs improve at higher condenser temperatures. The mixture may require certain degree of suction superheat depending on temperature lift, evaporator and condenser temperature.

0.7 propane mass fraction offers good volumetric heating and cooling capacities and COPs are usually low but improves at high condenser temperature. It does not require suction superheat. Pressure ratio, pressure difference and discharge temperature are close to those of HFC 134a. The temperature glide offered by it is good and may be used for system improvement.

ACKNOWLEDGMENTS

This work presented here is a part of an Indo-German project between IIT Madras and University of Karlsruhe. Authors thank the sponsors. SBS also thanks Kathmandu University, Nepal and IIT Madras for deputation to do this work.

REFERENCES

1. Hammad, M.A., Alsaad, M.A., The use of hydrocarbon mixture as refrigerants in domestic refrigerators, App. Thermal Engng., Vol 19, pp 1181-1189, 1999.

2. Richardson, R.N., Butterworth, J.S., The performance of propane/isobutane mixtures in a vapour compression refrigeration system, Int. J of Refgn., 18, 1, 1995.

3. Bodio, E., Chorowski, M and Wilczek, M., Working parameters of domestic refrigerators filled with propane – butane mixture, Int. J of Refgn., Vol 16, No 5, pp: 353-356, 1993.

4. Venkatarathnam, G., Murthy, S.S., Oellrich, L., Engler, T., Performance of some mixtures as working fluids for high temperature heat pump applications, Proceedings of 20th Int. Congress of Refrgn., IIR/IIF, Sydney, pp: 198 – 204, 1999.

5. Venkatarathnam, G., Mokashi, G., and Srinivasa Murthy, S., Occurrence of Pinch Points in Condensers and Evaporators for Zeotropic Refrigerant Mixtures, Int. J of Refgn., 19, pp 361-368, 1996.

6. Venkatarathnam, G., and Srinivasa Murthy, S., Effect of Mixture Composition on the Occurrence of Pinch Points in Condensers and Evaporators for Zeotropic Refrigerant Mixtures, Int. J of Refgn., 22, pp 205-215, 1999.

7. Venkatarathnam, G., and Srinivasa Murthy, S., Formation of Internal Temperature Pinches in Condensers and Evaporators for Hydrocarbon Refrigerant Mixtures, Paper No. NHT 2000-12051, 34th ASME National Heat Transfer Conf., Pittsburgh, USA, 20-22, August 2000.

8. Mclinden, M.O., Radermacher, R., Methods of comparing the performance of pure and mixed refrigerants in the vapour compression cycle, Int. J of Refgn., Vol 10, , pp: 318 – 325, 1987.

9. NIST Thermodynamic Properties of Refrigerants and Refrigerant Mixtures Database (REFPROP 6.01) Gaithersburg, MD, USA, 1998.

Table 1: Performance parameters

Parameters	Type A fluid	Type B fluid
Cooling effect (kJ/kg)	$q_e = (h_1 - h_5)$	$q_e = (h_2 - h_5)$
Heating effect(kJ/kg)	$q_c = (h_2 - h_4)$	$q_c = (h_3 - h_4)$
Work input(kW)	$W = m_r(h_1 - h_2)$	$W = m_r(h_3 - h_2)$
Pressure ratio	$PR = P2/P1$	$PR = P3/P1$
Pressure difference	$PD = P2 - P1$	$PD = P3 - P1$
Evaporator glide (K)	$dt_e = t_1 - t_5$	$dt_e = t_1 - t_5$
Condenser glide(K)	$dt_c = t_3 - t_4$	$dt_c = t_3 - t_4$
Heating capacity(kJ/m³)	$Q_{vo,h} = \rho_1(h_2 - h_5)$	$Q_{vo,h} = \rho_1(h_3 - h_5)$
Cooling capacity(kJ/m³)	$Q_{vo,c} = \rho_1(h_1 - h_5)$	$Q_{vo,c} = \rho_2(h_2 - h_5)$

Table 2: Physical properties

Pure Refrigerants & Mixtures	Normal Boiling Point (°C)	Critical Properties		
		$t_{crit.}$ (°C)	$P_{crit.}$ (MPa)	$\rho_{crit.}$ (kg/m³)
HC 290	-42.09	96.7	4.248	220.5
HC 600a	-11.61	134.7	3.640	224.4
HFC 134a	-26.07	101.06	4.059	511.9
HFC 236ea	6.19	139.29	3.502	563.0
CFC 114	3.59	145.65	3.257	580.0
0.3/0.7(HC290/HC600a)		122.6	3.910	219.0
0.5/0.5(HC290/HC600a)		114.8	4.042	218.2
0.7/0.3(HC290/HC600a)		107.3	4.144	218.7

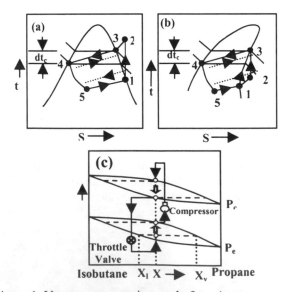

Figure 1: Vapour compression cycle for mixtures
a) Type A fluid b) Type B fluid c) Schematic diagram

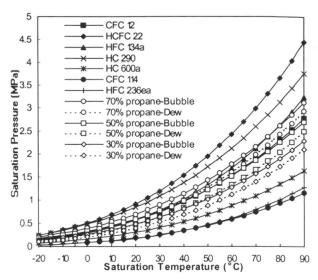

Figure 2: Saturation vapour pressure of various refrigerants

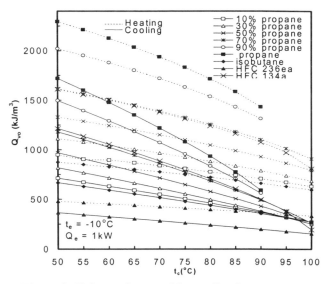

Figure 3: Volumetric capacities vs Condenser temperature

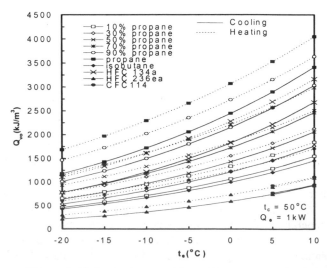

Figure 4: Volumetric capacities vs Evaporator temperature

486

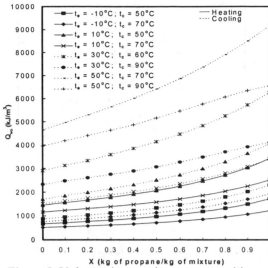

Figure 5: Volumetric capacity vs. composition

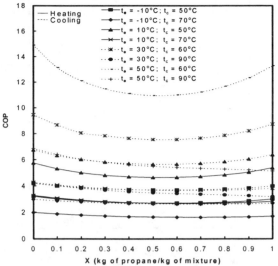

Figure 8: COP vs. composition

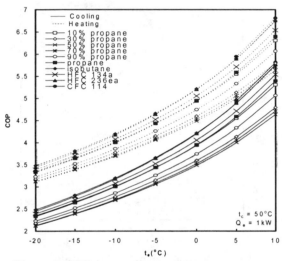

Figure 6: COP vs. condenser temeprature

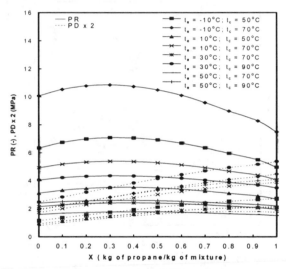

Figure 9: Pr. ratio and pr. Diff. variation

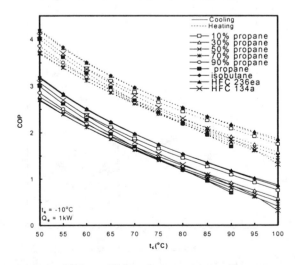

Figure 7: COP vs. Evaporator temperature

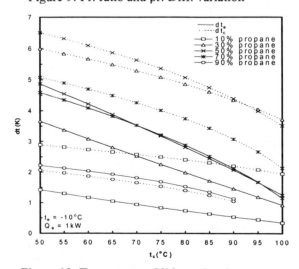

Figure 10: Temperature Glide vs Condenser temp.

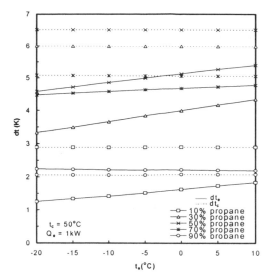

Figure 11: Temperature Glide vs Evaporator temp.

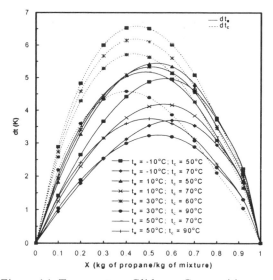

Figure 14: Temperature Glide vs Composition

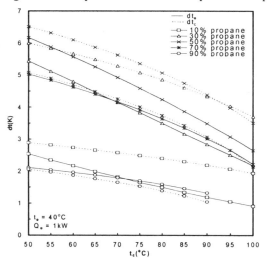

Figure 12: Temperature Glide vs Condenser temp.

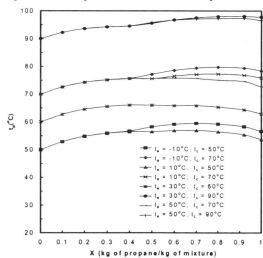

Figure 15: Discharge temp. vs Composition

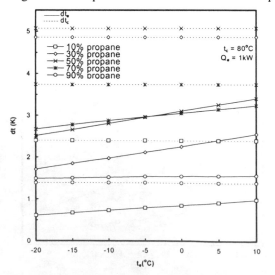

Figure 13: Temperature Glide vs Evaporator temp.

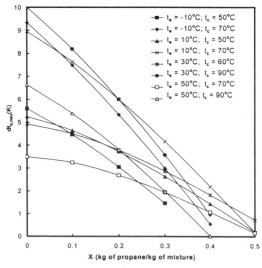

Figure 16: Minimum Superheat vs Composition

Proceedings of IECEC'01
36th Intersociety Energy Conversion Engineering Conference
July 29–August 2, 2001, Savannah, Georgia

IECEC2001-CT-20

A HIGHLY-EFFICIENT HEAT PUMP SYSTEM
WITH SOLAR THERMAL AND SKY RADIATION COOLING

Takeo S. Saitoh
Professor of School of Aeronautics and Space Engineering,
Tohoku University, Sendai 980-8579, Japan.
Phone: +81-22-217-6974
FAX: +81-22-217-6975
E-mail: saitoh@cc.mech.tohoku.ac.jp

Akira Hoshi
Research Associate of School of Aeronautics and Space Engineering,
Tohoku University, Sendai 980-8579, Japan
Phone : +81-22-217-6977
FAX : +81-22-217-6977
E-mail : hoshi@cc.mech.tohoku.ac.jp

Jun Takahashi
Graduate Student, School of Aeronautics and Space Engineering,
Tohoku University, Sendai 980-8579, Japan
Phone : +81-22-217-4067
FAX : +81-22-217-4067
E-mail : ens810@cc.mech.tohoku.ac.jp

ABSTRACT

A unique energy-independent house ("HARBEMAN House"; HARmony BEtween Man And Nature) incorporating sky radiation cooling, solar thermal, and photovoltaic (PV) energies, was built as a residential house owned by one of the authors (TSS) in Sendai, Japan in July 1996. The proposed HARBEMAN house, which meets almost all its energy demands, including space heating and cooling, as well as domestic hot-water, electricity generated by photovoltaic cells and rainwater for a standard Japanese home.

A unique heat pump system with solar thermal / sky radiation cooling installed in the HARBEMAN house is introduced in this paper. The proposed system has a solar collector for collection of heat for space heating and sky radiator and rainwater tank for space cooling. This heat pump is operated with nighttime electricity to heat and cool the main tank water. The system has four operational modes: The first is the sky radiation (SR) cooling mode that is a typical mode in summer. In this mode, the main tank water is cooled using the heat pump connected with the sky radiator. The second is the rainwater tank (RT) mode that cools the main tank water with rainwater as a low temperature reserver. The third is the subtank (ST) mode that is used in spring to cool the main tank water and heat the auxiliary tank water at the same time. The fourth is the winter (WIN) mode. In this mode, the main tank water is heated by the heat pump connected with the auxiliary tank heated by the solar collector. The experimental COP (Coefficient of Performance) of the present solar thermal / sky radiation cooling assisted heat pump system will be shown in detail. The total COP exceeds 7.0 for the SR cooling mode.

INTRODUCTION

The first example which utilizes a large-capacity underground tank for a residence was the MIT Solar house [1] designed by Professor Hoyt C. Hottel in 1939. Later, George O.G. Löf, who was a student of Professor Hottel, built the Löf house [2] in Englewood, Denver in 1957.

Recently, the performance of the low energy houses of the International Energy Agency (IEA) task 13; Solar Heating and Cooling Program was reported [3].

About 20 years ago, one of the authors (TSS) proposed an energy - efficient house incorporating solar thermal and sky radiation, and the fundamental experiments have been conducted by using facilities at the Tohoku University.

Experimental data on long-term heat and cool storage modes were presented in Saitoh[4], Saitoh et al.[5] and Saitoh and Kuwabara[6]. Along with these experiments, theoretical analyses and simulations were done to obtain an optimal design for the proposed house (Saitoh and Ono[7],[8]). An energy -independendent residential house (HARBEMAN house) was built in Sendai, Japan during July, 1996. The monitored results of this house and simulation results are reported (Saitoh and Fujino [9]).

The water of the underground tank is chilled in advance in spring when the ambient temperature and humidity are relatively low. The temperature of the tank is cooled down to 4℃ with the aid of a small-capacity heat pump (0.85kW) that utilizes nighttime electricity. A unique heat pump system with solar thermal / sky radiation cooling installed in the HARBEMAN house is introduced in this paper.

This paper reports the recent experimental results for this heat pump system. Annual variations of water temperature in the underground main tank were also obtained by the measured data.

NOMENCLATURE

COP : Coefficient of performance
P : Electric power consumption
Q : Heat exchanged
S : Area of heat transfer surface
T_H : Average temp of condenser
T_L : Average temp. of evaporator
ΔT : $T_H - T_L$

EXPERIMENTAL RESULTS FOR THE LONG-TERM HEAT STORAGE MODE AND COOL STORAGE MODE

This section gives the experimental results covering the period from August 1996 to February 2001.

Figure 1 shows the average ambient temperature and solar radiation at the site of the house (Kaigamori, Sendai, Japan) from September 1999 to August 2000. The average ambient temperature is slightly lower (1.5 - 2.0℃) than that reported by the local meteorological observatory in Sendai. The annual mean ambient temperature during this period was 12.7℃. The annual solar radiation amounted to 47000 kWh (from Sept.1, 1999 to Sept.1, 2000).

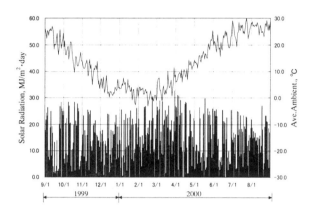

Fig. 1 Average ambient temperature and solar radiation intensity in Sendai

Figure 2 shows the long-term variations in the average water temperature of the main underground tank during the past four years.

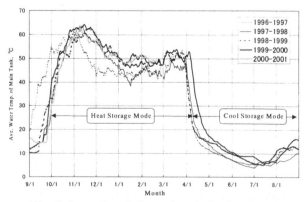

Fig. 2 Annual variation of main tank water temperature

490

First, the long-term thermal energy storage mode extending from September to March is examined. At the beginning of November, 1996 the maximum average water temperature reached 62℃. However owing to a large space heating demand in December, January and February it fell below the critical temperature (50℃) required to provide a sufficient space heating load. In the new heat pump system introduced in this paper, the water of the main tank can be heated using the heat pump. The average water temperature rose again in March with increase of solar radiation, and it was 52.1℃ on March 29, 1997.

PERFORMANCE TEST OF HEAT PUMP SYSTEM

Test equipment and procedure

A schematic of the experimental setup is shown in Fig. 3. In the figure, the operational and functional network in space cooling and heating corresponds to each upper and lower description, respectively. The heat pump system is composed of the compressor, the 4-way valve, and the capillary tube. It becomes for the space cooling mode (full line) when the 4-way valve turned on. The rotary compressor was used for the compressor, and the rotational speed of the compressor was regulated by the inverter.

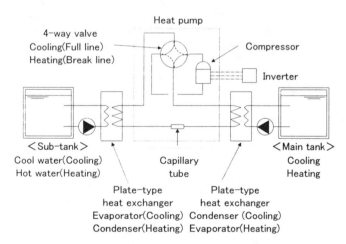

Fig. 3 A schematic of the performance test equipment

The plate-type heat exchangers were used for condenser and evaporator. The plate-type heat exchanger has heating surface of $0.8m^2$. The capacity of the first and secondary side tank was about 100 liters. The first tank temperature was kept constant by heater or cooler. The heat pump system was operated in space cooling and heating mode by changing frequency as a parameter. The inlet and the outlet temperatures of the condenser and the evaporator, and the compressor power were measured. The coefficient of performance (COP) was obtained by the measured data.

Performance of the heat pump system

The results of performance test of the heat pump system were shown in Figs.4 and 5. T_H and T_L are the mean temperatures between inlet and outlet of the condenser and the evaporator, respectively. The COP increases, when the temperature difference (ΔT) between the condenser and evaporator is small. On the other hand, the COP increases in case of low frequency. The inverter is effective for decreasing the power consumption of the compressor. The COP of the present system gives twice or triple better values compared with the conventional heat pump system with air heat exchangers.

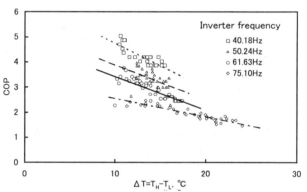

Fig. 4 COP versus ΔT for space cooling mode

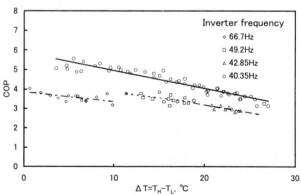

Fig. 5 COP versus ΔT for space heating mode

HIGH-EFFICIENT HEAT PUMP SYSTEM

The HARBEMAN house has already been operated as a natural energy independent house which covers most of energy demand such as space heating and cooling and domestic hot water. The principal purpose of this study is to increase the COP of the entire system by the high-efficient heat pump technology installed into the HARBEMAN house. Introduction of this system is effective for the residence in which the energy demand is big and in the urban area where heat island poses an important environmental issue such as in Tokyo and Osaka.

In this study, the outline of high-efficient heat pump system installed into the HARBEMAN house is described. And, four necessary operational modes in practical use are reported. The specification for the principal components of the high-efficient heat pump system installed is shown in Table 1. The high-efficient heat pump system consists of heat pump (compressor, reversing valve, capillary tube), inverter, and plate-type heat exchanger. The plate-type heat exchangers with excellent heat exchange efficiency were adopted.

Table 1 The principal specifications for the high-efficient heat pump system

Notation	Component	Spec.
P1	Solar collector pump	190W
P2	Sky radiator pump	95W
P6	Pump for heat pump	26W
P11	Pump for rainwater tank	27W
P12	Pump for auxiliary tank	45W
HP	Heat pump	450~1720W (Heating) 365~1530W (Cooling)
HEX	Plate-type heat exchanger	$S=0.55m^2$ $Q=5600kcal/h$
INV	Inverter	$1\phi 100V$, 0.75kW

The summertime operational mode

Sky radiator (SR) mode:

This mode is the most typical operational mode in the summertime. A schematic of the SR mode is shown in Fig.6. Heat from the condenser of the heat pump is rejected to the sky by use of the sky radiator. The heat pump is driven by using nighttime electricity. Since the waste heat is not discharged to the atmosphere, the impact to the environment (for ex. heat island effect) is minor. By undergoing the direct heat exchange

adopting plate-type heat exchangers, heat exchange performance is greatly improved.

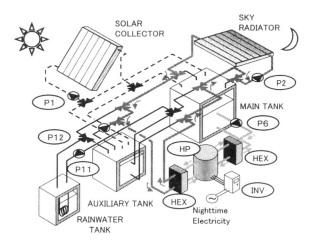

Fig. 6 Sky radiator (SR) mode

Rainwater tank (RT) mode:

This mode is an operational mode, which can deal with rapid cooling in the summertime. A schematic of the RT mode is shown in Fig.7. Heat of the main tank is rejected to the rainwater tank by using the heat pump. The water in the rainwater tank is always kept at low temperature by heat conduction to the surrounding soil. However, only short-time operation is possible, since the capacity (2000 liters) of the rainwater tank is not sufficient. The water in the rainwater tank has been utilized for toilet water, gardening and car wash. It is especially recommended to use the rainwater as a cold heat source as well as water supply.

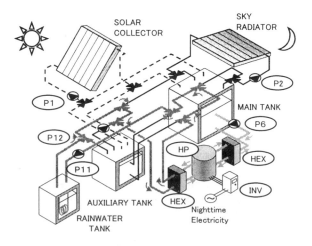

Fig. 7 Rainwater tank (RT) mode

Sub-tank (ST) mode:

This mode is an operational mode in the middle season (spring) in which the water temperature in the main tank is high and the water temperature in the sub-tank is low. A schematic of the ST mode is shown in Fig.8. Heat from the main tank is rejected to the sub-tank by using the heat pump. The notable feature is that both cooling the main tank water and heating the sub-tank water can be simultaneously done. In the past, the water in the sub-tank was heated by using the electric heater with nighttime electricity.

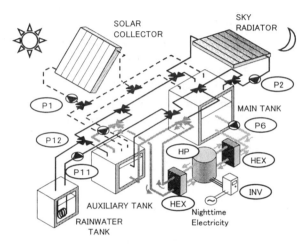

Fig. 8 Sub-tank (ST) mode

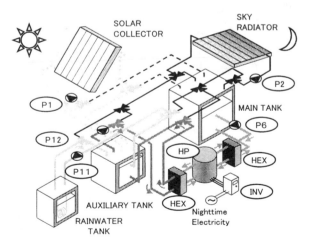

Fig. 9 Winter (WIN) mode

The winter operational mode

Winter (WIN) mode:

This mode is an operational mode in winter. A schematic of the WIN mode is shown in Fig.9. The water in the sub-tank is heated by the solar collector placed on the south roof of the house or by the rainwater tank. By making use of the water in the sub-tank as a heat source, the water in the main tank is heated by the heat pump. The heat pump is driven by nighttime electricity. Since the water temperature in the sub-tank becomes as low as 5∼10℃ in winter, the collector efficiency increases to a great extent (more than 50% in winter). The auxiliary heating of the main tank is possible by the present improvement.

FURTHER IMPROVEMENT OF HEAT PUMP SYSTEM

The variation of the main tank water temperature after the heat pump installation is shown in Fig.10.

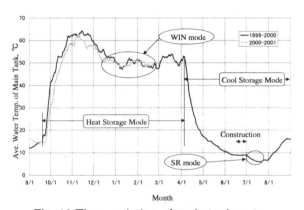

Fig. 10 Time variation of main tank water temperature after the heat pump installation

Since the replacement of the heat pump was finished at the end of June, the main tank water temperature is constant for this period (two weeks). The main tank water temperature decreases again with restarting of the sky radiator (SR) mode that is the most typical operational mode in the summertime. The main tank water temperature is kept over 50℃ which can provide heat for space heating by this WIN mode in bitter cold months in Sendai (January and February). In the past, the water in the main tank was heated by using the auxiliary natural gas powered boiler.

The experimental COP of the solar thermal / sky radiation cooling assisted heat pump system was monitored in detail. Figure 11 shows the experimental COP. The COP amounts to as high as 7.0 for the SR cooling mode, with ΔT being 10℃. And, the COP recorded 6.0 ($\Delta T=12°C$) for the SR cooling mode even in the hottest month (August) in Sendai. For the RT mode, the COP was 6.0 with ΔT was 12°C.

In the future, it is possible to make the total COP to be 9~10, if the performance of the heat pump is further improved, and if the ordinary sky radiator (flat plate type) is replaced with CPC type. Another improvement will be done with having thermal energy storage tank, in which phase-change capsules are filled. The efficient cooling system presented in this paper will be also very promising in urban area like Tokyo to mitigate severe heat island effect in the summertime.

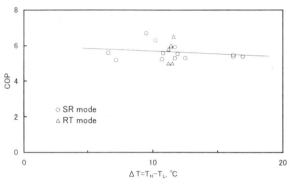

Fig. 11 Experimental COP

CONCLUSIONS

In this paper, a highly-efficient heat pump system with solar thermal and sky radiation cooling has been proposed.

The following conclusions may be drawn from the present study.

1. The outline of the high-efficient heat pump system with solar thermal / sky radiation cooling installed into the HARBEMAN house is described.
2. The long-term variation in the average water temperature of the main underground tank during the past four years including recent data is shown.
3. The performance of the present heat pump system was shown. The COP of the present system gives twice or triple better performance compared with the conventional heat pump system with air heat exchangers.
4. The four operational modes of the high-efficient heat pump system are reported in detail.
5. The variation of the main tank water temperature after the heat pump installation is shown. The main tank water temperature is kept over 50°C which can provide heat for the space heating by the winter (WIN) mode in bitter cold months (January and February) in Sendai.

6. The experimental COP of the solar thermal / sky radiation cooling assisted heat pump system was monitored in detail. The COP amounts to 7.0 for the SR cooling mode.

REFERENCES

1. H.C.Hottel et al., 1942, "The performance of flat-plate solar-heat collectors," Trans. of the ASME, Vol.64, pp.91-104.
2. G.O.G.Löf et al., 1963, "Residential heating with solar heated air — the Colorado Solar House —," ASHRAE Journal.
3. IEA, 1995, Solar Low Energy Houses of IEA Task 13, James & James Science Publishers Ltd., London.
4. T.S.Saitoh, 1984, "Natural Energy Autonomous House with Underground Water Reservoir," Bulletin of JSME, Vol.27, No.266, pp.773-778.
5. T.S.Saitoh, H.Matsuhashi, and T.Ono, 1985, "An energy-independent house combining solar thermal and sky radiation energies," Solar Energy, Vol.35, No.6, pp.541-547.
6. T.S.Saitoh and K.Kuwabara, 1987, "Exerimental performance for long-term skyradiation cooling using prototype sky radiator," ASME-JSME-JSES Solar Conf. Hawaii, pp.512-518.
7. T.S.Saitoh and T.Ono, 1984, "Utilization of seasonal sky radiation energy for space cooling," J. Solar Energy Eng., Vol.106, pp.403-407.
8. T.S.Saitoh and T.Ono, 1984, "Simulative analysis for long-term underground cool storage incorporating sky radiation cooling," J. Solar Energy Eng., Vol.106, pp.493-496.
9. T.S.Saitoh and T.Fujino, 2001, "Advanced energy-efficient house (HARBEMAN house) with solar thermal, photovoltaic, and sky radiation energies (Experimental results)", Solar Energy, Vol.70, No.1, pp.63-77.

IECEC2001-CT-29

DESIGN OF CO$_2$ BASED REFRIGERATOR/FREEZERS

Tony D. Chen
Integrated science and Technology Department
College of Integrated and science and Technology
James Madison University
Harrisonburg, VA 22807
U.S.A.

ABSTRACT

An initial design model was developed for investigating the potential performance and component requirements of a CO$_2$-based refrigerator-freezer using a transcritical cycle. Comparisons were made with results of a second model developed to predict with the performance of a conventional vapor-compression refrigerator-freezer cycle with R-12, R-134a, propane, and isobutane. Capillary tube models were also included to estimate required diameters and lengths.

Predictions for the CO$_2$ cycle with an intercooler and compressor efficiency the same as the conventional cycle gave a 35% lower COP than with R-134a. When a 50% efficient expander was added in place of the intercooler and the CO$_2$ compressor efficiency was boosted by 10 percentage points, the CO$_2$ cycle performance improved to within 10% of the standard design. A major design challenge identified is the need for a CO$_2$ compressor with higher efficiency than the typical refrigerator-freezer design for halocarbons with a displacement almost an order of magnitude smaller but with a 40% reduced pressure ratio. The two hydrocarbons investigated showed similar performance to R-134a but individually showed no compelling reason for use in lieu of R-134a.

INTRODUCTION

One environmentally sound way to address the ChloroFluoroCarbon/ HydroChloroFluoroCarbon (CFC/HCFC) phase out situation is to use substances which are already present in the natural environment and for which distribution systems have been long established. Carbon Dioxide (CO$_2$, R-744) is one of these alternatives. It is also inexpensive and readily available. Its relatively high pressure and moderate molar mass give important advantages in volume and cost of the system when the equipment is properly designed.

Carbon dioxide is a high vapor pressure refrigerant which has a lower normal boiling point and critical point as compared to the conventional refrigerants. The vapor pressure of CO$_2$ is approximately 6 to 7 times higher than that of HCFC-22 (R-22). The high vapor pressure offers the opportunity of reducing heat exchanger size and weight. Due to the lower critical pressure, the CO$_2$ refrigerant cycle becomes a transcritical cycle which has a two-phase evaporation and a supercritical gas cooling process. Because the thermodynamic and transport properties vary widely around the critical point, extra caution must be taken in the selection of property models suitable across this region and in the modeling of a supercritical condenser.

Lorentzen and Pettersen (1993) at the Refrigeration Engineering Department of The Foundation for Scientific and Industrial Research at Norway Institute of Technology (NTH-SINTEF) developed and tested a new, efficient and environmentally safe automobile air conditioning system in 1992. The laboratory prototype CO$_2$ system has been extensively tested, and the results compared with reference data from a standard commercial CFC-12 (R-12) air-conditioning system. A major difference between the CO$_2$ circuit and the standard system is that the receiver is positioned at the evaporator outlet and that an internal heat exchanger is used. A general conclusion from the comparison results is that the COP of the CO$_2$ system is as good as, or even slightly better (from 0 to 25%) than, that of standard R-12 system. the difference is larger at high ambient temperature and low compressor speed, possibly as a result of larger compressor (leakage) losses in the

R-12 system with its sliding vane machine. A reciprocating R-12 compressor could possibly give a slightly better performance at low speed and high pressure differential.

Factors contributing to this improvement in evaporator performance are: larger airside surface within same heat exchanger volume (the refrigerant-side area is, however, much lower than in the R-12 unit); increased air flow due to lower airside pressure drop; better heat transfer characteristics of CO_2; higher refrigerant mass velocity; and, finally, the absence of a superheat zone towards the outlet. The resulting increase in evaporating temperature contributes to an improved COP.

A unique feature of the CO_2 system is that the cooling air flow can be dramatically reduced in counterflow heat exchange as a consequence of the gliding temperature. This opens the possibility of designing the cooler for operation with a small electric air fan, eliminating the problems of poor performance at low driving speed or, in particular, standstill. The reduced face area requirements of this counterflow concept also permit improved aerodynamic car design and more flexible location of the cooler.

A quite different design philosophy was applied for the CO_2 heat exchangers, with an increased number of small-diameter tubes, resulting in a large airside surface, light-weight construction, high-pressure capability, reduced inside volume and reduced airside pressure drop. The reduced refrigerant-side surface is compensated for by the higher inside heat transfer coefficients of CO_2.

Of special significance to the cycle performance, the CO_2 gas cooler approach temperature was measured to be 4 K as compared to 16 K for the R-12 condenser. Further, the compressor isentropic efficiency was about 70% for CO_2 while only 50% for R-12. These were significant advantages for the CO_2 cycle that may not be typical for fully comparable equipment. They are also in much larger sizes than are suitable for residential r/f application.

In contrast, the analysis of Bullock (1997) assumed equal approach temperatures of 4.4 K for the gas cooler versus condenser in his comparison. (The evaporator superheat and condenser subcooling were typical heat pump design values of 5.5 and 8.3 K, respectively.) Under this scenario, he found using a variable-specific heat gas cooler model, that UA levels were nearly equal to a standard heat pump with a counterflow configuration (with nearly twice as large if crossflow). Bullock also assumed cases of equal isentropic compressor efficiency (70%) and a case with a 75% efficient CO_2 compressor.

NOMENCLATURE

A_{cond} : Condenser total heat transfer area (m^2)
A_{evap} : Evaporator total heat transfer area (m^2)

D : Total compressor displacement per revolution (m^3/hr)
F_L : The required fraction of the total evaporator air flow
$\dot{m}_R$: Refrigerant mass flow rate (kg/s)
P_R : Pressure ratio
$\dot{Q}_E$: Evaporator capacity (kW)
Q_L : Ratio of fresh-food-to-freezer load
S : Nominal compressor speed (rpm)
T_{EI} : Evaporator inlet air temperature (^{o}C)
T_{EO} : Evaporator outlet air temperature (^{o}C)
T_H : Fresh-food compartment temperature (^{o}C)
T_L : Freezer compartment temperature (^{o}C)
U_{sub} : Heat transfer coefficient in subcool region (W/m^2.^{o}C)
U_{sup} : Heat transfer coefficient in supheat region (W/m^2.^{o}C)
U_{tp} : Heat transfer coefficient in two-phase region (W/m^2.^{o}C)
$\dot{W}$: Compressor motor input power (kW)
Δh_{actual} : Actual enthalpy change (kJ/kg)
Δh_{isen} : Enthalpy change for an isentropic compression (kJ/kg)
ΔT_E : Airside temperature change across evaporator (K)
ΔT_R : Ratio of ΔT_E to temperature difference between fresh-food and freezer
η_{exp} : Isentropic expander efficiency
η_{isen} : Compressor isentropic efficiency
η_{vol} : Compressor volumetric efficiency
ρ_c : Refrigerant density at the compressor shell inlet (kg/m^3).

MODELING APPROACH

The EES (Engineering Equation Solver) Program

EES is an acronym for Engineering Equation Solver from F-Chart Software (1998). The basic function provided by EES is the solution of a set of algebraic equations. EES can also solve initial value differential equations, do optimization, provide linear and non-linear regression.

There are two major differences between EES and existing numerical equation-solving programs. First, EES automatically identifies and groups equations which must be solved simultaneously. This feature simplifies the process for the user and ensures that the solver will always operate at optimum efficiency. Second, EES provides many built-in mathematical and thermophysical property functions useful for engineering calculations. For example, the steam tables are implemented such that any thermodynamic property can be obtained from a built-in function call in terms of any two other properties. Similar capability is provided for most halocarbon refrigerants, ammonia, methane, carbon dioxide and many other fluids. Air tables are built-in, as are psychrometric functions and JANAF table data for many common gases. Transport properties are also provided for all substances.

Note that one can call REFPROP6 from EES using an interface routine that must be purchased separately. Where they are reasonably accurate, we use properties built-in to EES. Of the refrigerants for this work, only the CO_2 properties near and above the critical point must be tested. EES properties run much faster than calls to REFPROP6 so are preferred if sufficiently accurate. Also, the equation programming is simplified with EES property calls.

EES can also be used to conduct parametric analyses of any desired variable and to generate tables of desired input and output parameters. Tabular results can be pasted to spreadsheets for further analysis if needed.

EES allows the user to enter his or her own functional relationships in three ways. First, a facility for entering and interpolating tabular data is provided so that tabular data can be directly used in the solution of the equation set. Second, the EES language supports user-written functions and procedures, similar to those in Pascal and FORTRAN. The functions and procedures can be saved as library files which are automatically read in when EES is started. Third, compiled functions and procedures, written in a high-level language such as Pascal, C and FORTRAN, can be dynamically-linked into EES using the dynamic link library capability incorporated into the Windows operating system.

Modeling Conditions/Assumptions

The main assumptions made regarding air- and refrigerant-side operating conditions and component performance modeling are as follows:

McLinden and Radermacher (1987) proposed that the airside temperature ranges across the heat exchangers be held constant when comparing cycle performance with different refrigerants. For systems of fixed refrigerating capacity, this gives fixed evaporator airflow rates for all refrigerants and condenser airflow rates that vary directly with the relative condenser heat rejected for each refrigerant. We adopt this approach for the design point analysis here.

To appropriately model a single evaporator refrigerator/freezer (r/f), the mixed air temperature entering the evaporator coil is needed. Usually, however, only the individual fresh food and freezer compartment temperatures, T_H and T_L, are known. An algorithm was derived by Rice and Sand (1992) in the CYCLE-Z model development to calculate the mixed entering air temperature for known compartment temperatures, airside temperature change ΔT_E across the coil, and a given ratio of fresh-food to freezer-load, Q_L. Using mass continuity and energy balances between the air streams for the freezer and fresh-food compartments with the combined airflow across the

evaporator, as shown in Figure 1, the entering evaporator mixed air temperature T_{EI} can be obtained. Once T_{EI} is determined, the exit evaporator temperature T_{EO} is obtained from the specified air ΔT_E.

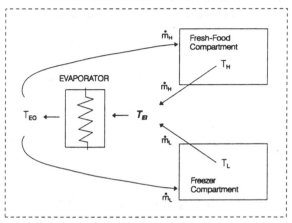

Figure 1. Schematic for determining single-evaporator refrigerator airside temperature (T_{EI}).

The required fraction of the total evaporator air flow, F_L, which serves the freezer section is given by

$$F_L = \frac{1 - \Delta T_R \pm \sqrt{(1 - \Delta T_R)^2 + 4\Delta T_R /(1 + Q_L)}}{2}, \quad (1)$$

where,

$$\Delta T_R = \frac{\Delta T_E}{T_H - T_E}, \quad (2)$$

and Q_L is the ratio of fresh-food-to-freezer load.

Once F_L is determined from the above equation, the evaporator entering air temperature can be calculated from

$$T_{EI} = T_H - F_L(T_H - T_L), \quad (3)$$

and

$$T_{EO} = T_{EI} - \Delta T_E. \quad (4)$$

From Equations 1-4, the appropriate single-evaporator air inlet and exit temperatures under the above assumptions are functions only of the fresh-food and freezer compartment temperatures, the assumed (or calculated) evaporator air-side change across the coil, and the assumed fresh-food-to-freezer ratio.

Conditions and assumptions (Rice and Sand, 1992) are given as follows:

❑ fixed external source and sink inlet and exit temps,
❑ 32.2°C (90°F) suction gas temp obtained through liquid-line subcooling in the high-temperature intercooler,
❑ 2.8°C (5°F) condenser subcooling and evaporator superheat,
❑ a constant compressor overall isentropic efficiency of 0.45 (Rice and Sand 1993),
❑ compressor volumetric efficiency as a function of pressure ratio (Rice and Sand 1993),
❑ compressor can heat loss of 70% of compressor input power (Rice and Sand 1993),
❑ no evaporator or condenser refrigerant pressure drop.

Specified airside conditions selected for the r/f design point calculations are:

❑ 3.3°C (38°F) fresh-food and -15°C (5°F) freezer settings giving in a freezer-coil mixed entering air temperature of -12.8°C (8.9°F) and 5°C (9°F) air glide,
❑ 32.2°C (90°F) condenser air inlet and 40.2°C (104.4°F) air exit temperatures, and
❑ a fixed fresh-food-to freezer heat load ratio F_L of 1/1.

Overall compressor isentropic efficiency is defined as

$$\eta_{isen} = \frac{\dot{m}_R \Delta h_{isen}}{\dot{W}}, \tag{5}$$

where, $\dot{m}_R$ is refrigerant mass flow rate given by $\dot{m}_R = \dot{Q}_E$ /Δh_{evap}, and $\dot{Q}_E$ and Δh_{evap} are evaporator capacity and enthalpy change, respectively; Δh_{isen} is the enthalpy change for an isentropic compression from shell inlet conditions to shell outlet pressure, and $\dot{W}$ is compressor motor input power. It is assumed that a compressor isentropic efficiency of 45% could be achieved at the design pressure ratio for all the considered refrigerants.

Compressor volumetric efficiency is defined as

$$\eta_{vol} = \frac{\dot{m}_R}{S D \rho_c}, \tag{6}$$

where S is the nominal compressor speed in revolution per minute (rpm), D is the total compressor displacement per revolution, and ρ_c is the refrigerant density at the compressor shell inlet. Rice and Sand (1993) tested a 670 Btu/hr (196.3 W) reciprocating r/f compressor at ORNL for three different refrigerants (R-12, R-32/R-124, and R-22/R-141b), and derived the volumetric efficiency from the nine-point calorimeter data. The volumetric efficiency data were seen to group closely together and could be approximated by a single decreasing linear function of pressure ratio of the following form:

$$\eta_{vol} = 0.8061 - 0.029 P_R, \tag{7}$$

where, P_R is the pressure ratio. This equation should be representative for use with HFC refrigerants and is assumed to be reasonably applicable to CO_2 as well.

In this analysis, we will be specifying overall U values for each refrigerant region and total HX external areas. The overall UAs for each heat exchanger (HX) are calculated from the sum of the UAs determined from each refrigerant region as required to balance the heat transfer in each region.

We have assumed typical U values for each refrigerant region for US refrigerator/freezers. These assumed values (Admiraal and Bullard, 1995) are:

U_{tp} = 10.0 Btu/hr.ft².°F (56.8 W/m².°C),
U_{sup} = 6.25 Btu/hr.ft².°F (35.5 W/m².°C), and
U_{sub} = 5.0 Btu/hr.ft².°F (28.4 W/m².°C) for the condenser; and
U_{tp} = 4.0 Btu/hr.ft².°F (22.7 W/m².°C),
U_{sup} = 1.72 Btu/hr.ft².°F (9.77 W/m².°C) for the evaporator.

The assumed HX external areas (Reeves et al., 1992, 1994) are:

A_{cond} = 8.6 ft² (0.799 m²) and A_{evap} = 17.3 ft² (1.607 m²).

The selected HX sizes were applied to a unit with an assumed capacity of 205 W (700 Btu/hr).

For both HXs, the air is assumed to flow over the refrigerant coils in a multirow crosscounterflow arrangement, which for 4 or more rows, is closely approximated by the performance of a simple counterflow arrangement. While some r/f condensers may deviate from simple multirow crosscounterflow, we made this assumption for comparability with a transcritical CO_2 cycle which would require such an arrangement for good HX performance.

For the design point (sizing) calculation, the refrigeration capacity and desired air-side ΔT's are set along with typical values for heat exchanger areas, evaporator superheat, condenser subcooling, and intercooler conditions as given in this section. The required design evaporator airflow is then determined for all refrigerants. The required compressor swept volumes and condenser airflow rates are found for each refrigerant. (Alternatively, one can, for a baseline refrigerant, such as R-12, determine the required condenser airflow. Then for other refrigerants, the airflow can be held fixed, to maintain constant fan power, rather than constant external fluid thermal conditions as done here per the approach of McLinden and Radermacher, 1987.)

CO_2 REFRIGERATOR/FREEZER DESIGN MODEL

A CO_2 r/f design model has been developed and simulations of a CO_2 r/f design model in the EES program have been carried out in the recent study.

Model Description

CO_2 Properties. In developing this model, we first used the built-in EES properties for CO_2 as a real gas. Because of concerns with the accuracy of this representation near the critical point, this version served primarily as a convenient way to test the model for CO_2 over the range of design parametrics of condensing pressure and approach temperature. The primary advantage of the built-in properties is computational speed.

This initial program was then modified to use REFPROP6 CO_2 properties through the EESREFP6 interface. Calculated performance is shown in the following section with both property data sets to show the degree of approximation incurred by the EES properties.

Heat Exchanger Modeling. The major simulation difference compared to the R-12 analysis was that the condenser was modeled with a specified counterflow approach temperature, TD_{Appr}, of 3 K and condenser pressure was studied as a design parameter.

Multi-row cross counterflow condensers and evaporators that approximated counterflow performance were assumed. The evaporator treatment, because it is subcritical, was the same as for the R-12 analysis where we assumed the same values for total external area and two-phase and superheated overall U levels. We used the same specified capacity and air ΔT's, evaporator superheating and low-side intercooler exit temperature to calculate the r/f cycle states, swept compressor volume, required air flows, and system COP.

We have assumed the same external area for the condenser ($A_{Cond} = 0.799$ m^2) and evaporator ($A_{Evap} = 1.607$ m^2) for CO_2 as in the R-12 r/f model. However, because a fixed condenser approach temperature is assumed, the condenser area is only used to estimate the required overall condenser U value consistent with the required UA value.

The explicit calculation of a transcritical condenser performance for a given area was beyond the scope of this analysis. This is because the specific heat of the transcritical fluid varies by more than a factor of 10 as the fluid is cooled (as is suggested by the constant pressure process line in Figure 5b). The standard UA×LMTD analysis requires that the overall U values and the fluid specific heats remain constant over the length of the heat exchanger. A tube-by-tube calculation with an average U and specific heat for each tube would be required to determine an accurate approach temperature due to the widely varying transport property values through the gas cooler. The required UA_{cond} was calculated for the specified approach temperature, but *these values should only be taken in a relative sense* because of the extremely crude approximation of the simple LMTD method for a highly variable-specific heat fluid such as supercritical CO_2.

Compressor. The compressor model for the CO_2 r/f is the same as for R-12 r/f model, with overall isentropic efficiency and compressor can heat loss factor fixed at 0.45 and 0.70, respectively, while the volumetric efficiency model is the same function of pressure ratio. Later we discuss the possibility of a higher isentropic compressor efficiency for CO_2.

Flow Control Modeling. For application to CO_2 in a transcritical cycle, the Bittle, Wolf, and Pate (1998) generalized subcooled correlation is only applicable below the critical pressure where the fluid is subcooled. Above the critical pressure, we used a single-phase flow model as described by Wolf, et al (1995, p. 171) to calculate the tube length L_c required to obtain an isenthalpic pressure drop from the inlet supercritical pressure to 99% of critical pressure. Average refrigerant specific volume v and viscosity μ were used with capillary diameter d_c for this calculation where $\Delta P = (f \times L_c \times V^2)/(2 \times v \times d_c)$ and the friction factor $f = 0.23 \times Re_d^{-0.216}$ where $Re_d = V \times d_c/(v \times \mu)$.

Once the pressure is below the critical value of 7384 kPa, the fluid temperature is calculated from the known pressure and enthalpy, and the amount of subcooling is determined. The generalized subcooled correlation is applied from this point to determine the remaining length needed.

Two EES procedures were written to handle this sequential calculation within EES. With known refrigerant mass flow rate, temperature, and pressure at the expansion device inlet, we calculate, for a specified diameter, the capillary lengths required, in series, for the supercritical and subcritical regions of the expansion. In combination, these calculations determined the required total capillary tube length for the CO_2 r/f throttling device. Alternately, the desired total length can be specified and the model will determine the required capillary tube diameter. The fraction of the total capillary tube length in the transcritical region varies from about one third at 9000 kPa to over one half at 13000 kPa.

CO₂ Refrigerator/Freezer Design Analysis and Parametrics

Figures 2a and 2b show the P-h and T-s diagram, respectively, of the transcritical carbon dioxide refrigeration cycle with intercooling under the assumed operating and design conditions. The assumed condenser operating pressure for the diagrams was 9500 kPa with a 3 K approach temperature. The model calculates the transcritical CO_2 refrigeration conditions at about 1880 kPa suction pressure and -21.4°C evaporating temperature, as shown in Figures 2a-b. EES properties were used for these diagrams.

From the diagrams, one can see that there is a significant amount of lost refrigeration effect and lost expansion work even with the intercooling heat transfer that was accomplished. One can also deduce (from the constant pressure process line in Figure 2b) that the specific heat of the transcritical fluid varies from a low value at the condenser inlet to a maximum near the critical region as the fluid is cooled, and then back to an intermediate value. This wide change in specific heat make the use of a simple LMTD approach inappropriate except for approximate relative assessments of required heat exchanger size or UA level. Because the liquid specific heat for CO_2 is much larger than for the vapor, the cycle does not obtain as much subcooling as one would need to obtain a fuller utilization of the evaporator two-phase region.

Parametrics of condensing pressure and approach temperature were studied to find the optimal pressure and assess the effect on COP of closeness of approach to the condenser inlet air temperature. Results are given for calculations with both EES and REFPROP6 properties. For all cases the refrigerating capacity is held constant at 205 W. Shown are COP, compressor power and required swept volume, refrigerant flow, and estimated UAs for the condenser and the intercooler. The intercooler effectiveness is also given.

Figures 3a and 4a show the calculational differences in terms of condensing pressure and approach temperatures in COP and compressor work between using EES built-in versus REFPROP6 CO_2 properties. In Figure 3a, the condensing pressure is varied while approach temperature is the varied parameter in Figure 4a. The difference in the calculations of COP is as high as 12% when condensing pressure is around 9 Mpa, and condenser approach temperature is in the range from 1 to 3 K. In Figure 3a, the trends in COP between 8.5 and 11 MPa are significantly different with peak COP around 11 Mpa for EES properties versus 9 Mpa with REFPROP6. In Figure 4a, the REFPROP6-calculated COP is seen to increase monotonically with closer approach temperatures with while the EES-COP peaks at 7 K, dips slightly, and then rises only slowly with smaller approach temperatures. These differences were all

caused by discrepancies in the calculation of the required work as the capacities were constant at 205W in all cases.

In Figure 3b, the corresponding required refrigerant mass flow rate and compressor swept volumes are shown. Note that the region of highest COP and minimum power is also close to the minimum in compressor size and flow rate.

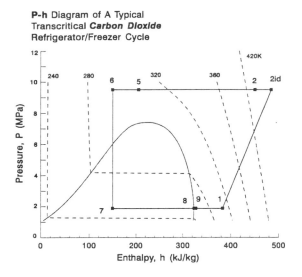

Figure 2a. P-h diagram of a typical transcritical carbon dioxide refrigerator/freezer cycle.

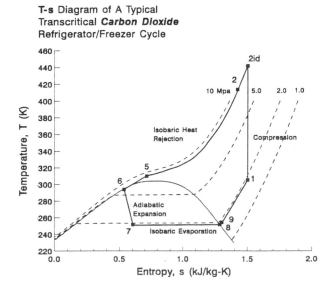

Figure 2b. T-s diagram of a typical transcritical carbon dioxide refrigerator/freezer cycle.

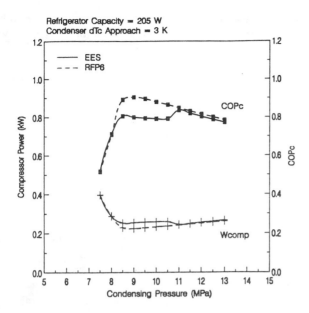

Figure 3a. Comparison of CO_2 refrigerator compressor power and COP as a function of condensing pressure between EES and REFPROP6 properties.

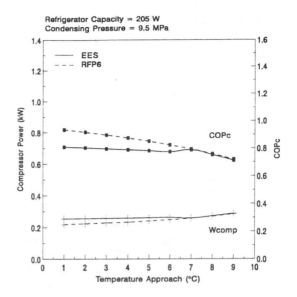

Figure 4a. Comparison of CO_2 refrigerator compressor power and COP as a function of approach temperature between EES and REFPROP6 properties.

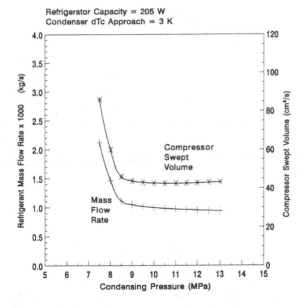

Figure 3b. CO_2 refrigerator refrigerant mass flow rate and compressor swept volume as a function of condensing pressure.

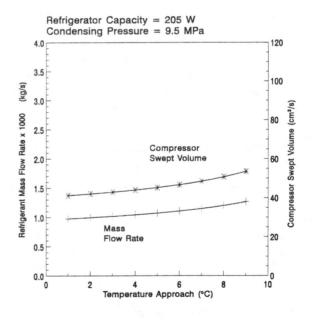

Figure 4b. CO_2 refrigerator refrigerant mass flow rate and compressor swept volume as a function of approach temperature.

COMPARISON OF REFRIGERATOR/FREEZER EFFICIENCY

The initial R-12 r/f model was generalized to run for a range of halocarbon and hydrocarbon refrigerants. This involved setting up calculations in the program for reasonable limiting values for key iteration variables (temperatures/pressures and their associated enthalpies and entropies) for each refrigerant and linking these values to the upper- and lower-limit assignments in EES. Prior to this improvement, it was often a rather tedious and time consuming process to fine tune the model for each new refrigerant considered. In the final version of the model for halocarbon and hydrocarbon refrigerants, the four candidate refrigerants were all evaluated in the same program using the parametric table option. Using this program with EES properties, cycle conditions and performance were calculated for CFC-12 (R-12), HFC-134a (R-134a), propane (R-270), and isobutane (R-600a).

A comparison of P-h and T-s of R-12 and R-134a r/f cycle and the state points of the operating cycle diagrams to those of Figures 2a-b for CO_2 show the large difference in cycle operation with R-134a and propane cycles operating well below their critical pressure and temperatures.

A summary comparison of predicted r/f performance between all refrigerants evaluated in this study is given in Table 1. The R-12 and R-134a results provide CFC- and HFC-baselines to which to compare the three considered natural alternatives. It can first be noted than in moving from R-12 to R-134a, the industry has gone to a refrigerant with nearly the same COP, lower suction pressure, higher pressure ratio, larger compressor requirement, lower mass flow rate, and a longer capillary tube requirement.

Relative to R-134a, propane shows just a slight drop in COP, similar suction pressure, a significantly lower pressure ratio, about a 40% smaller compressor, half the mass flow rate, yet requres capillary tubes twice as long. In contrast, isobutane (R-600a), shows a 3.5% increase in COP but operates at subatmospheric suction pressure and requires twice as large a compressor. Required isobutane mass flow rate is similar to propane but 1/3 the capillary tube length is needed.

By comparison, CO_2 is predicted to have a 37% lower COP and operates at 15 and 9 times higher suction and discharge pressures, respectively, than R-134a. On the positive side, the larger relative rise in suction pressure than discharge results in a 43% smaller pressure ratio. However, because of the high suction density, the required compressor size is only $1/9^{th}$ of that for R-134a. The refrigerant mass flow rate is close to that for R-134a but the capillary tube size is 4-1/2 times longer for the same diameter because primarily of the higher pressure. It should be noted that since supercritical CO_2 is an excellent

cleaning agent, use of smaller than conventional diameter tubes should be possible with CO_2 without major concerns with plugging.

After expansion, the inlet evaporator quality for CO_2 is twice that of R-134a, which indicates a greater loss of expansion work. The amount of subcooling obtained in the intercooler is less than 1/3 of the low-side superheat (51°K) which is less than half of the subcooling obtained with the other fluids. This is because the specific heat of supercritical CO_2 on the high-side of the intercooler is much higher than that of the low-side superheated vapor. (Note also that the inlet temperature to the capillary tube is about 15 °C higher for the CO_2 cycle than for the conventional cases.) This fluid heat capacity mismatch for CO_2 suggests that an expander should be considered as an alternative to a conventional flow control device.

This first-cut comparison between CO_2 and other r/f fluids has assumed the same compressor isentropic efficiency and evaporator performance, assumptions which likely penalize CO_2 systems to a degree. On the other hand, we have assumed a condenser approach temperature of 3 K for CO_2 as compared to about 5 K (actually, a range of 4.9 to 5.4 K as seen from Table 3) for the other fluids. Pettersen (1997) data indicates that 3 °K should be an achievable level. However, this likely implies a larger airside area (although not necessarily volume) for a CO_2 gas cooler than for a subcritical condenser.

We decided to briefly investigate the effect of improved compressor isentropic efficiency and the use of an expander to see how much the performance gap could be narrowed. It was first assumed that the compressor efficiency could be increased from 0.45 to 0.55 because of the lower pressure ratio. Next, two levels of expander isentropic efficiency η_{exp} were studied, 0.25 and 0.50. Isentropic expander efficiency is defined as the actual work obtained from the expander compared to the ideal isentropic work. This is given by :

$$\eta_{exp} = \frac{\Delta h_{exp,actual}}{\Delta h_{exp,isen}} \qquad (8)$$

The amount of intercooling was also varied from a high of 51 K (with the baseline 32.2°C suction temperature) to a low of 9 K. Bullock (1997) has noted that when expanders are used, intercooling may not be beneficial.

The results of the parametric assessment of expander efficiency and intercooling are shown in Figure 5 with the compressor isentropic efficiency at 0.55. With no expander and the baseline value of 51 K, the cycle COP for CO_2 has increased from 0.906 to 1.107, a gain of 18.2% from a compressor efficiency increase of 22.2%. As the intercooling is decreased from the maximum, the COP drops to 1.072 at 9 K, a loss of 3.2% . A parametric of condensing pressure was studied for

Table 1. Comparison of Refrigerator/Freezer Performance Between R-12, R-134a, Propane, Isobutane, and CO_2

Application	Refrigerator/Freezer				
Refrigerant	**R-12**	**R-134a**	**Propane**	**R-600a**	**CO_2**
Suction Pressure [kPa]	142.8	125.1	133.6	68.4	1882.0
Discharge Pressure [kPa]	983.2	1042.2	1398.9	547.7	9000.0
Compressor Pressure Ratio	6.879	8.322	5.984	8.007	4.783
Evaporator Temperature [$^\circ$C]	-21.39	-21.40	-21.40	-21.40	-21.43
Evaporator UA [W/K]	36.04	36.01	35.99	35.98	35.81
Evaporator Inlet Quality	0.156	0.178	0.181	0.178	0.328
Condensing Temperature [$^\circ$C]	40.94	40.88	40.96	41.37	N/A
Condenser Exit Temperature [$^\circ$C]	38.16	38.11	38.18	38.59	35.2
Condenser UA [W/K]	43.73	42.41	42.38	42.63	10.20
Intercooler UA [W/K]	3.53	3.64	3.84	3.99	4.13
Intercooler Effectiveness	0.895	0.896	0.895	0.888	0.944
Condenser Heat Rejected [W]	248.0	248.0	248.0	247.0	273.0
Intercooler Heat Transfer [W]	46.0	50.0	52.0	54.0	58.0
Intercooler Subcooling [$^\circ$K]	32.51	30.60	31.38	32.54	14.95
Capillary Inlet Temperature [$^\circ$C]	5.65	7.51	6.80	6.06	20.25
Capillary Tube Diameter (mm)	0.508	0.508	0.508	0.508	0.508
Capillary Tube Length (m)	**0.832**	**1.357**	**2.652**	**0.980**	**6.294**
Refrigerant Mass Flow Rate [g/s]	1.485	1.160	0.607	0.657	1.050
Compressor Volumetric Efficiency	0.607	0.565	0.633	0.574	0.667
Compressor Swept Volume (cm^3/s)	350.4	398.4	229.0	717.6	43.79
Compressor Power [W]	141.9	143.3	144.1	138.5	226.0
COP	**1.445**	**1.431**	**1.423**	**1.480**	**0.906**

Condenser Entering Air Temp. of 32.2°C and Air Temp. Range of 8°K

Evaporator Entering Air Temp. of −12.8°C and Air Temp. Range of 5°K

Suction Temp. Entering Compressor of 32.2°C

Evaporator Exit Superheat of 2.8°K, Condenser Exit Subcooling of 2.8°K (except for CO_2 where Tc,rfr,exit=Tc,air,in + DT_approach)

Compressor Overall Isentropic Efficiency of 45%, Compressor Shell Heat Loss 70% of Input Power

Evaporator Capacity of 205 W

different expander efficiency levels and 9000 kPa was found to remain optimum for all cases.

When an expander is added in place of the capillary tube, the cycle COP effect from intercooling is seen to be negligible for an efficiency of 25%. When the expander efficiency is increased to 50%, the effect of eliminating the intercooler is positive, from a COP of 1.195 to to 1.243, or a gain of 3.9%. When no expander is present, the increase in refrigerating effect from intercooling slightly outweighs the increased compressor work from the hotter compression gas. With a reasonably efficient expander, the extra amount of lost work from intercooling is enough to offset the net gain from increased refrigerating effect. This is in part because the expander also increases the refrigerating effect by delivering fluid to the evaporator inlet at a lower enthalpy than with isenthalpic expansion.

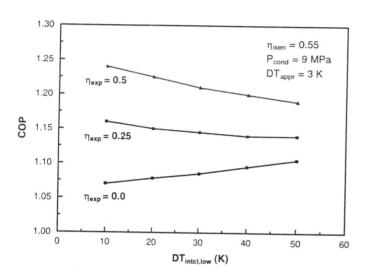

Figure 5. Effects of expander and intercooler on CO_2 R/F Cycle COP.

A maximum COP of 1.264 is obtained at zero intercooling and expander efficiency of 50%. Combined with the 55% efficient compressor, this narrows the performance gap between the transcritical CO_2 cycle and the conventional refrigeration cycle with R-134a from 36.7% to 11.7% for an improvement of 25%. About 44% of this gain is from the addition of the 50% efficient expander in conjunction with elimination of the intercooler.

It should be noted that there are likely to be significant, if not insurmountable, compressor design challenges in achieving a compressor efficiency of 10 percentage points higher than for an R-134a compressor. While the pressure ratio with CO_2 is 43% less, the required compressor size is 1/9th as

large. In the size range of typical r/f reciprocating or rotary refrigerator compressors, those with the higher efficiencies are typically at the larger sizes because of fixed losses which do not scale with smaller sizes. Some other compressor design concept may be needed to approach this efficiency target in such a small compressor size. In contrast, as the density of CO_2 entering the expander is less than half that for R-134a, it is likely to be a lesser challenge to obtain a 50% or higher efficiency expander.

CONCLUSIONS

Refrigerator/freezer design models have been developed for conventional and transcritical cycles. These were implemented using the Engineering Equation Solver program. The conventional version has been run successfully for R-12, R-134a, propane, and isobutane using built-in EES properties. The transcritical cycle uses carbon dioxide as the refrigerant and calls REFPROP6 for all properties as EES properties were found to be inaccurate above the critical region for CO_2.

The conventional model uses EES property data to calculate the performance of a simple refrigeration cycle with heat transfer rate limitations modeled with UA×LMTD. Three refrigerant regions—superheated, two-phase, and subcooled— are included in the condenser model, and two regions—two-phase and superheated—are used in the evaporator. Overall U values are specified for each refrigerant region along with total external HX area. Such an approach should provide reasonable predictions for heat exchanger utilization as evaporator superheat and condenser subcooling vary at different ambient conditions.

The transcritical cycle uses one supercritical region for the high-side gas cooler with a specified approach temperature and an evaporator treatment the same as for the conventional cycle. Parametrics of approach temperature and condensing pressure were conducted and an optimum high-side pressure of 9000 kPa was determined.

Using the two models developed, calculations of system COP, required compressor size, airflow rates, and capillary tube sizes were made at design conditions with specifed refrigerating capacity, evaporator superheat and condenser subcooling, heat exchanger total external area and regional Us, and air-side temperature changes. For these comparisons, the CO_2 cycle used a gas cooler approach temperature of 3 K as compared to about 5 K for the condenser in the conventional cycle.

Comparison of r/f efficiency was made between two halocarbons—R-12 and R-134a and three natural refrigerants— the hydrocarbons propane and isobutane, and CO_2. The analysis shows that isobutane has the highest COP among the considered refrigerants while the r/f COP for CO_2 is about 63% of R-134a. However, an improvement in CO_2 r/f COP to within about 10% of R-134a performance could be accomplished by designing a higher efficiency compressor for CO_2, adding a 50% efficient expander, and eliminating the intercooler.

ACKNOWLEDGMENTS

The author would like to thank Dr. C. K. Rice of Oak Ridge National Laboratory for his great support and supervision during this project under the U. S. Department of Energy agreement No. ERD-97-1589.

REFERENCES

Admiraal, D. M., and C. W. Bullard, 1995, "Experimental Validation of Heat Exchanger Models for Refrigerator/Freezers", *ASHRAE Transactions*, V. 101, Pt.1, pp.34-43.

Bittle, R. R., D. A. Wolf, and M. B. Pate, 1998, "A Generalized Performance Prediction Method for Adiabatic Capillary Tubes", *HVAC&R Research Journal*, January 1998, (RP-762) (4183).

Bullock, C.E., 1997, "Theoretical Performance of Carbon Dioxide in Subcritical and Transcritical Cycles", *Refrigerants for the 21st Century, ASHRAE/NIST Refrigerants Conference*, October 6-7, pp. 20-26.

EES Users Manual, 1998, *EES (Engineering Equation Solver) for Microsoft Windows Operating Systems, F-Chart Software*, 4406 Fox Bluff Rd., Middletown, WI., U.S.A.

Lorentzen, G. and J. Petersen, 1993, "A New, Efficient and Environmentally Benign System for Car Air-Conditioning", *International Journal of Refrigeration*, pp. 4-12, Volume 16, No.1, 1993.

McLinden, M. O. and R. Radermacher, 1987. "Methods for Comparing the Performance of Pure and Mixed Refrigerants in the Vapor Compression Cycle", *Int. J. Ref.*, 10, pp. 318-325.

McLinden, M. O., et al, 1998. "NIST Thermodynamic and Transport Properties of Refrigerants and Refrigerant Mixtures—REFPROP6, Version 6.0, Users' Guide", *NIST Standard Reference Database 23*, U.S. Department of Commerce.

Pettersen, J., 1997, "Experimental Results of Carbon Dioxide in Compression Systems", *Refrigerants for the 21st Century, ASHRAE/NIST Refrigerants Conference*, October 6-7, pp. 27-37.

Reeves, R. N., C. W. Bullard and R. R. Crawford, 1992, Modeling and Experimental Parameter Estimation of a Refrigerator/Freezer System, ACRC TR-9 Report, University of Illinois Urbana-Champaign, January 1992.

Reeves, R. N., C. W. Bullard and R. R. Crawford, 1994, A Measurement of Refrigerator Component Performance, *ASHRAE Transactions*, V. 100, Pt.1.

Rice, C. K. and D. T. Chen, 1999, *Simulation Models of CO_2 Transcritical and Conventional Vapor Compression Cycles for Analysis of Refrigerator/Freezer Application*, ORNL Final Report for the U.S. Department of Energy under Agreement No. ERD-97-1589, March.

Rice, C. K. and J. R. Sand, 1992, *Steady State Performance Potential of the Lorez/Meutzner Cycle for A Refrigerator/Freezer Application*, ORNL Internal Report to DOE, August.

Rice, C. K., L. S. Wright, and P. K. Bansal, 1993, "Thermodynamic Cycle Evaluation Model for R-22 Alternatives in Heat Pumps -- Initial Results and Comparisons", *Proceedings of Heat Pumps in Cold Climates, Second International Technical Conference*, August 16-17, 1993, published January 1994 by Caneta Research, Inc., pp 81-96.

Rice, C. K. and J. R. Sand, 1993, "Compressor Calorimeter Performance of Refrigerant Blend-Comparative Methods and Results for A Refrigerator/Freezer Application", *ASHRAE Transactions*, V. 99, Pt.1, pp. 1447-1466.

Wolf, D. A., R. R. Bittle, and M. B. Pate, 1995, *Adiabatic Capillary Tube Performance with Alternative Refrigerants, RP-762, Final Report*, ASHRAE Research Report.

Proceedings of IECEC'01
36th Intersociety Energy Conversion Engineering Conference
July 29-August 2, 2001, Savannah, Georgia

IECEC2001-CT-24

INVESTIGATION OF THE NIOBIUM/HAFNIUM ALLOY C-103 AS CELL WALL MATERIAL IN AN AMTEC BASED THERMAL TO ELECTRICAL CONVERTER

Daniel P. Kramer
Mound Power Systems
Technologies
BWXT of Ohio, Inc.
Miamisburg, OH 45343

Joseph D. Ruhkamp
Mound Power Systems
Technologies
BWXT of Ohio, Inc.
Miamisburg, OH 45343

James R. McDougal
Mound Power Systems
Technologies
BWXT of Ohio, Inc.
Miamisburg, OH 45343

Dennis C. McNeil
Mound Power Systems
Technologies
BWXT of Ohio, Inc
Miamisburg, OH 45343

Robert A. Booher
Mound Power Systems
Technologies
BWXT of Ohio, Inc
Miamisburg, OH 45343

Edwin I. Howell
Mound Power Systems
Technologies
BWXT of Ohio, Inc
Miamisburg, OH 45343

ABSTRACT

Screening studies were performed on test specimens fabricated out of the niobium/hafnium alloy C-103 that were used in the determination of some of its mechanical and physical properties as they relate to the fabrication/utilization in an AMTEC (Alkali Metal Thermal to Electrical Conversion) cell wall. The investigations were centered in three areas: 1) machinability, 2) weldability, and 3) "inert" atmosphere compatibility. The results show that C-103 can be machined as prototype in-situ ribbed AMTEC cell walls were fabricated out of thin sheet material. Welding studies demonstrated that thin sheet C-103 can be electron beam welded. However, it was determined that the heat treatment of C-103 specimens under an argon atmosphere containing ~2ppm of oxygen did significantly degrade the desired mechanical properties of the alloy.

Keywords: AMTEC, C-103, Electron Beam Welding

INTRODUCTION

The future application of AMTEC technology in nuclear space power systems is ultimately dependent on the converter being robust enough to withstand typical operational environments (temperatures, dynamic loadings, mission duration, etc). One of the technology's critical materials issues centers on the selection of an appropriate cell wall material. The physical and chemical properties of the cell wall must provide sufficient strength and material compatibility to successfully complete anticipated mission requirements. Over the last several years several refractory alloys, most significantly niobium-1%zirconium and molybdenum/rhenium, have been considered for application as the cell wall material in an AMTEC based space power system.[1-4] Recent published performance and stress modeling studies has identified that a two-part AMTEC cell wall that utilizes two niobium refractory alloys, niobium-1%zirconium in combination with the niobium/hafnium alloy C-103, would result in significant performance and strength improvements.[5-7] In the modeling studies the hot-end of the AMTEC cell wall is constructed out of niobium-1%zirconium while the cold-end consists of C-103. The dual alloy AMTEC cell wall results in predicted enhanced thermal to electrical conversion performance. This is principally due to two factors: 1) C-103 has a higher yield strength compared to niobium-1%zirconium at anticipated cold-end temperatures allowing the fabrication of thinner cold-end walls and 2) C-103 has a lower thermal conductivity vs. niobium-1%zirconium. A dual alloy AMTEC cell would in theory yield an increased thermal to electrical conversion efficiency due to reduced parasitic heat losses. Table 1 shows a comparison of the chemical composition, thermal conductivity, ultimate tensile strength, and yield strength of C-103 and niobium-1%zirconium.

Table 1. Nominal chemical compositions (wt.%) and selected properties of niobium-1%zirconium and C-103.

ELEMENT	NIOBIUM-1%ZIRCONIUM	C-103
NIOBIUM	BALANCE	BALANCE
ZIRCONIUM	1%	<0.7%
HAFNIUM	<0.01%	10%
TITANIUM		1%
TANTALUM	<0.02%	<0.5%
TUNGSTEN	<0.05%	<0.5%
PROPERTY		
THERMAL CONDUCTIVITY AT 800°C[8]	59W/m°C	37.4W/m°C
ULTIMATE TENSILE STRENGTH (~25°C)[8]	275Mpa (39.88ksi)	420MPa (60.91ksi)
YIELD STRENGTH (~25°C)[8]	150Mpa (21.75ksi)	296MPa (42.93ksi)

EVALUATION OF C-103 AS A CANDIDATE AMTEC CELL WALL MATERIAL

The present experimental study was initiated to determine various mechanical and physical properties of C-103 as they relate to the fabrication and utilization in an AMTEC cell wall. Specimens for the experiments were fabricated out of 0.38mm (0.015") thick C-103 sheet material obtained from Wah Chang (Albany, OR). Evaluation studies for determining the suitability of C-103 as an AMTEC cell wall material were centered in three areas: 1) machinability, 2) weldability, and 3) "inert" atmosphere compatibility. Machinability is of importance since piece parts would ultimately have to be fabricated. Emphasis in the present work is on the machining of integral ribs in thin sheet material. The second area of interest, weldability, is of importance since the fabrication of dual alloy cells requires the formation of crack-free hermetic welds. During this phase of the investigation samples of C-103 were electron beam welded and some of their properties were evaluated.

Inert atmosphere compatibility is of interest since during the fabrication of an AMTEC based space power system the AMTEC cell walls will be assembled together with the nuclear heat sources while under an argon atmosphere that can contain up to 5ppm of oxygen. This work was initiated to determine the mechanical properties of C-103 after exposure to low concentrations of oxygen for anticipated assembly times and temperatures.

Machining of AMTEC Cell Walls

In order to maximize the thermal efficiency/strength of an AMTEC cell the walls of the cold-end section are typically ribbed. In most cells the wall thickness of the cold-end of the cell is thinner than the wall thickness of the hot-end of the cell. Ribbing increases the stiffness of the cell wall while minimizing heat losses via the application of thinner sheet materials. Ribs have been processed into the cold-end of AMTEC cell walls several different ways: 1) brazing thin ribs of the selected metal directly to the wall, 2) chemical milling, and 3) by conventional milling processes. Figure 1 shows a cold-end ribbed prototype AMTEC cell wall that was machined out of 0.38mm (0.015") thick C-103 sheet stock. The three ~0.13mm (~0.005") thick by 0.95cm (0.375") wide slots were successfully machined using conventional milling processes with carbide tools. Machining of the slots resulted in the "formation" of three 0.38mm (0.015") thick by 0.33cm (0.13") wide in-situ ribs. This work confirmed that C-103 may be machined using relatively standard processes into in-situ ribbed AMTEC cell walls.

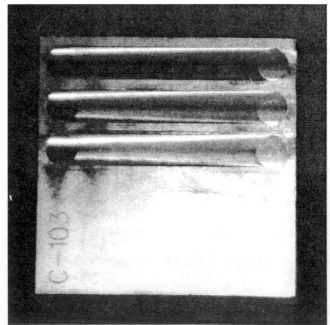

Figure 1. Prototype C-103 cold-end cell wall after conventional machining of three ~0.13mm (~0.005") thick slots resulting in three in-situ ribs.

Preparation of C-103 Weld Samples and Mechanical Test Specimens

C-103 mechanical test specimens and weld samples were obtained from 0.38mm thick sheet stock via Wire Electrical Discharge Machining (WEDM). Figure 2 shows a section of the C-103 sheet after completion of the WEDM process that resulted in the formation of mechanical test specimens and weld "cutouts". The mechanical test specimens are ~7.6cm (3") long with a gauge length of ~2.5cm (~1") and a gauge width of ~0.63cm (0.25"). The grip area is ~1.6cm (0.62") long and ~1.6cm (0.62") wide. After the specimens were fabricated, all of their edges were lightly polished to remove any material that may have been left from the WEDM process.

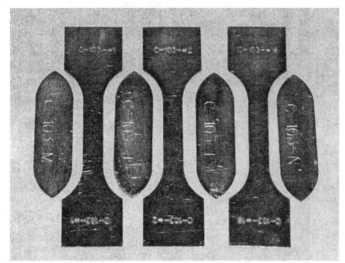

Figure 2. Section of 0.38mm thick sheet of C-103 after WEDM of mechanical test specimens and weld "cutouts".

As shown in Figure 2, when the mechanical test specimens were fabricated out of the sheet stock, small ~1.25cm wide by ~3.8cm long oblong shaped cutouts were also produced. Weld coupons were obtained from the cutouts after their flat edges were ground to yield square weld surfaces. All mechanical test specimens and weld cutouts were cleaned in a liquid vapor degreaser prior to testing.

Electron Beam Welding of C-103

In the fabrication of AMTEC cells it is anticipated that welding of the selected cell wall material will be required. Welding is necessary to obtain cells that are hermetically sealed both to contain the sodium inventory within the cell and to keep the outside environment from penetrating the cell. Welding studies were performed to determine the feasibility of electron beam welding of C-103 to C-103 and C-103 to Nb-1%Zr using previously prepared cutouts.

The electron beam welding studies were performed using a Leybold-Heraeus High Voltage system with an Allen Bradley CNC system. Welding was performed under a vacuum of approximately 10^{-5} torr. Beam voltage, current, diameter, and travel speed were monitored and controlled during the welding process. A defocused beam was used to preheat the C-103/C-103 weld joint piece parts just prior to the formation of the electron beam welds. The parameters developed and used in the formation of the two welds are presented in Table 2. Some of these are similar parameters developed in an earlier study for electron beam welding of niobium-1%zirconium and molybdenum-44.5%rhenium coupons.[4]

All of the cutouts were electron beam welded in a square butt joint configuration. Polished cross-sections of the welds were prepared which showed complete weld penetration. Simple longitudinal and transverse face bend tests (not guided bend tests) were performed on several welded coupons that showed that the welded samples exhibited good ductility. Figure 3 shows an example of a completed electron beam butt-welded C-103/C-103 specimen after a simple 4T longitudinal face bend test.

Table 2. Electron beam welding parameters employed in welding C-103/C-103 and C-103/Nb-1%Zr cutouts.

Parameter	C-103/C-103	C-103/Nb-1%Zr
Beam Voltage	125kilovolts	75kilovolts
Beam Current	3milliamps	3milliamps
Travel Speed	20.3cm/min	38cm/min
Beam Focus	~0.5mm diameter circle	~0.5mm diameter circle

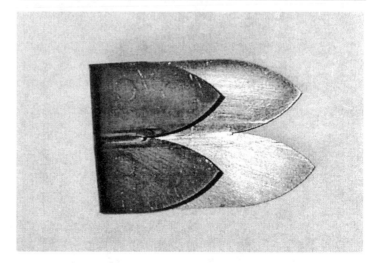

Figure 3. Top view of a simple longitudinal face bent test performed on a C-103/C-103 electron beam welded specimen.

Mechanical test samples were also prepared from C-103/C-103 and C-103/Nb-1%Zr electron beam welded specimens. This was accomplished using the WEDM process to cut mechanical test samples out of several welded coupons as shown in Figure 4. Test samples were ~0.63cm wide by ~2.54cm in length. Additionally, one of the two welded "ends" shown in Figure 4 (bottom) were typically mounted and polished for metallographic examination of the weld.

The results of the tensile tests performed on three electron beam welded C-103/C-103 and two C-103/Nb-1%Zr test samples are presented in Table 3. Due to the short length of the welded coupons it was not possible to obtain percent elongation (%E), so only ultimate tensile strength (UTS) and yield strength (YS) data was obtained. Table 3 also contains mechanical test results obtained on unwelded C-103 "control" 7.6cm (3") long tensile bars as previously shown in Figure 2. All of the results were obtained using an ATS 1620 computer controlled universal testing machine according to ASTM E 8-00. A strain rate of 0.508cm/min (0.2in/min) was used during the tests.

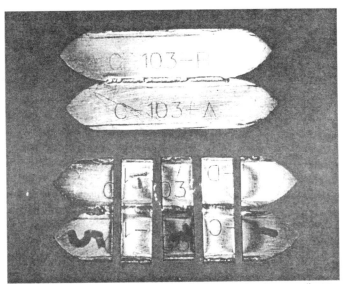

Figure 4. Example of mechanical test specimens (bottom, center) WEDM out of electron beam welded C-103 coupons.

Comparison of the test results show that the unwelded C-103 control specimens and the electron beam welded C-103/C-103 coupons exhibited comparable ultimate tensile and yield strengths. The electron beam welded C-103/Nb-1%Zr coupons exhibited lower but satisfactory ultimate tensile and yield strengths.

C-103 IAAC Atmosphere Compatibility

All of the RTG's (Radioisotope Thermoelectric Generators) which were used to power the Galileo, Ulysses, and Cassini missions were assembled at Mound in the Inert Atmosphere Assembly Chamber or IAAC. The IAAC is ~2 meters wide x

Table 3. Mechanical test results obtained on unwelded C-103 control specimens and on electron beam welded C-103/C-103 and C-103/Nb-1%Zr test specimens.

Sample Number	Type	UTS (psi)	YS (psi)	%E
1	C-103 Control	55,000	41,400	34
2	C-103 Control	56,500	40,800	27
3	C-103 Control	58,500	40,200	27
1	C-103/C-103 EB-Welded	53,500	41,900	
2	C-103/C-103 EB-Welded	57,500	45,000	
3	C-103/C-103 EB-Welded	50,000	41,000	
1	C-103/Nb-1%Zr EB-Welded	41,300	38,500	
2	C-103/Nb-1%Zr EB-Welded	40,900	38,400	

~4.5 meters long x ~3.5 meters high. On one side of the IAAC is a row of mechanical manipulator arms and on the opposite side are two rows of glove ports. During assembly procedures the IAAC contains an argon atmosphere that is closely controlled to contain <15ppm nitrogen, <10ppm moisture, and <5ppm of oxygen. It is anticipated that the final assembly of an AMTEC based power system would likely take place in the IAAC or some similar inert atmosphere chamber. During assembly of a nuclear space power system, the plutonium (238) dioxide fuel, which is contained within the graphitic General Purpose Heat Source (GPHS) modules, would be assembled within the AMTEC converter. This operation may take several days to perform with subsequent heating of the C-103 containing AMTEC cell walls. This work was initiated to determine whether the mechanical/physical properties of C-103 would be detrimentally effected after being exposed to the anticipated IAAC assembly operations.

C-103 mechanical test specimens of the configuration shown in Figure 2 were heat treated within an electrically heated box furnace (hot zone ~10 cm wide x ~9.5 cm high x ~11.5 cm deep). The furnace was located within a flowing argon atmosphere controlled glove box that is ~1 meter wide x ~1 meter high x ~2meters long. The experimental design allowed oxygen concentrations both within the glove box and within the hot zone of the furnace to be monitored using a Vacuum/Atmospheres Oxygen Analyzer Model AO-316-C. Calibration of the oxygen monitor was performed using the recommended calibration procedure and it was verified using an argon/oxygen mixture gas standard with a certified content of 4ppm of oxygen.

The 2 x 2 experimental design required C-103 mechanical test specimens to be heat treated in the glove box furnace at either 873K, (600°C) or 1073K (800°C) for either two different times (25 hours or 150 hours). Furnace temperature and atmosphere oxygen content was monitored during the experiment. During all of the heat treatments the measured oxygen content of the glove box was ~2ppm. After completion of the heat treatments the tensile bars were passed out of the glove box and mechanical tests were performed when possible on intact bars to determine their ultimate tensile strength, yield strength, and percent elongation as a function of each of the four experiment conditions.

The mechanical test matrix was not completed due to the severe embrittlement of the tensile specimens as a function of the experimental parameters. Figure 5 is a photograph of the C-103 tensile bars after heat treatment at 1073K for 25 hours (left two bars) and at 1073K for 150 hours (right two bars) as they appeared coming directly out of the glove box furnace. The 1073K for 150 hours bars were removed from the furnace in pieces and they were reassembled for the photograph. Close examination of Figure 5 reveals that while the 1073K for 25 hours tensile bars were removed from the furnace intact, they were severely warped especially on the surface of the grips.

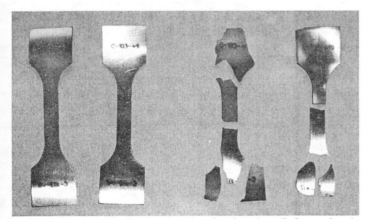

Figure 5. C-103 tensile specimens after heating (left two bars) at 1073K for 25 hours and (right two bars) at 1073K for 150 hours under an inert argon atmosphere containing ~2ppm of oxygen.

An attempt was made to insert the 1073K/25hours bars into the universal testing machine but the grip surfaces readily broke when the grips started to be tightened. Mechanical test data was not obtained from any of the 1073K heat treated specimens. All of the bars could be readily fractured even with a slight hand pressure.

The lower temperature heat treat experiments on C-103 mechanical test bars at 873K were only marginally more successful. Figure 6 shows examples of the C-103 bars heat treated at 873K for 25 hours (left two bars) and 150 hours (right two bars). The tensile bars heat treated for 150 hours were determined to be extremely brittle making placement into the universal testing machine impossible. However, mechanical test data was obtained on the bars heat treated at 873K for 25 hours and the results are presented in Table 4. For comparison purposes mechanical test data obtained on unheated control C-103 bars is also shown in the table. The results of the mechanical tests performed on the 873K/25hr test specimens heat treated under IAAC processing conditions show a marked reduction in ductility.

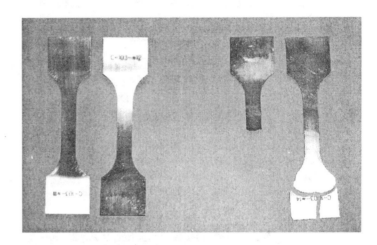

Figure 6. C-103 tensile specimens after heating (left two bars) at 873K for 25 hours and (right two bars) at 873K for 150 hours under an inert argon atmosphere containing ~2ppm of oxygen.

Table 4. Mechanical test results obtained on control specimens and on test specimens heat treated at 873K/25hours under an argon atmosphere containing ~2ppm of oxygen.

Sample Number	Type	UTS (psi)	YS (psi)	%E
1	Control	55,000	41,400	34
2	Control	56,500	40,800	27
3	Control	58,500	40,200	27
1	873K/25hr	26,400	23,700	1.1
2	873K/25hr	32,500	32,200	1.0

CONCLUSIONS

Screening studies for determining the potential of C-103 as cold-end cell wall material in an AMTEC based thermal to electrical converter have been completed. The studies determined that the machinability and weldability of the niobium/hafnium based alloy appears to be adequate for the fabrication of the cold-end section of AMTEC cell walls. However, heat treating C-103 mechanical test specimens under anticipated IAAC processing conditions (~2ppm of oxygen in argon) resulted in the severe embrittlement of the specimens. As test specimens were heat treated for longer times and at higher temperatures the embrittleness became even more severe. The lack of IAAC processing compatibility severely reduces C-103's candidacy as cold-end cell wall material in an AMTEC based space power system.

REFERENCES

[1] Kramer, D.P., Ruhkamp, J.D., McNeil, D.C., Mintz, Jr., G.V., Howell, E.I., 8/1999, "Mechanical Testing Studies on Niobium-1%Zirconium in Association with its Application as Cell Wall Material in an AMTEC Based Radioisotope Space Power System," *Proceedings of the 24th Intersociety Energy Conversion Engineering Conference (IECEC)*, Society of Automotive Engineers, 1999-01-2608.

[2] Kramer, D.P., Ruhkamp, J.D., McNeil, D.C., Howell, E.I., Williams, M.K., McDougal, J.R., and Booher, R.A., 2/2000, "Investigation of Molybdenum-44.5%Rhenium as Cell Wall Material in an AMTEC Based Space Power System," *Proceedings of Space Technology and Applications International Forum (STAIF-2000)*, American Institute of Physics Conference Proceedings 504, pp. 1402-1407.

[3] King, J.C., and El-Genk, M.S., 1/2000, "A Review of Refractory Materials for Vapor Anode Cells," *Proceedings of Space Technology and Applications International Forum (STAIF-2000)*, American Institute of Physics Conference Proceedings 504, pp. 1390-1401.

[4] Kramer, D.P., McDougal, J.R., Booher, R.A., Ruhkamp, J.D., Howell, E.I., and Kwiatkowski, J.J., 7/2000, "Electron Beam and Nd-YAG Laser Welding of Niobium-1%Zirconium and Molybdenum-44.5%Rhenium Thin Sheet Material," *Proceedings of the 35th Intersociety Energy Conversion Engineering Conference & Exhibit (IECEC)*, **2**, AIAA-2000-2971, pp.956-9961.

[5] King, J.C., and El-Genk, M.S., 1/2000, "Analyses of a Nb-1Zr/C-103, Vapor Anode, Multi-Tube AMTEC Cell," *Proceedings of Space Technology and Applications International Forum (STAIF-2000)*, American Institute of Physics Conference Proceedings 504, pp. 1383-1390.

[6] King, J.C., and El-Genk, M.S., 7/2000, "Structural Analyses of a PX-Type, Nb-1Zr/C-103, Vapor-Anode AMTEC Cell," *Proceedings of the 35th Intersociety Energy Conversion Engineering Conference and Exhibit (IECEC)*, **2**, American Institute of Aeronautics and Astronautics, pp.940-949.

[7] King, J.C., and El-Genk, M.S., 10/2000, "*Performance and Stress Analyses of Nb-1Zr/C-103 Multitube, Vapor-Anode AMTEC Cells for Space Power Applications*," Report No. UNM-ISNPS-1-2000, Institute for Space and Nuclear Power Studies, School of Engineering, University of New Mexico, Albuquerque, NM.

[8] Wojcik, C.C., 12/1998, "High-Temperature Niobium Alloys," *Advanced Materials & Processes*, ASM International, **154**, pp. 27-30.

Proceedings of IECEC'01:
36TH Intersociety Energy Conversion Engineering Conference
July 29–August 2, 2001, Savannah, Georgia

IECEC2001-CT-36

AN ANALYSIS OF AN ALKALI METAL THERMAL TO ELECTRIC CONVERTER WITH AN ASYMMETRICALLY GROOVED WALL

Paul E. Hausgen
Spacecraft Component Technology Branch
Space Vehicles Directorate
Air Force Research Laboratory
Albuquerque, NM 87117
paul.hausgen@kirtland.af.mil

James G. Hartley
George W. Woodruff School of Mechanical
Engineering
Georgia Institute of Technology
Atlanta, GA 30332
james.hartley@me.gatech.edu

ABSTRACT

The Alkali Metal Thermal to Electric Converter (AMTEC), conceived in 1962, is a static, direct thermal to electric energy conversion technology. As with any energy conversion technology, the thermal efficiency of the AMTEC is of great importance. Various studies have concluded that the AMTEC could potentially achieve a very high thermal efficiency. To obtain high thermal efficiency, parasitic heat losses within the AMTEC must be minimized. Any attempt to minimize the parasitic heat losses of the AMTEC must involve the effective control of the radiation heat transfer within the AMTEC. This investigation examined the ability of specially designed asymmetric wall grooves to increase conversion efficiency and power output of an AMTEC device by controlling the internal radiation heat transfer. Modeling results showed that the asymmetric wall grooves, with proper groove opening angles and surface specularity, increase conversion efficiency (compared to a flat wall design with no internal heat shields) by 2.3 percentage points (~16.4% increase).

INTRODUCTION

Vining et al. [1] have demonstrated that the Alkali Metal Thermal to Electric Converter (AMTEC) [2] could potentially achieve a very high fraction of Carnot efficiency. For this potential to be realized, parasitic heat losses within the AMTEC must be minimized. Examples of parasitic heat losses in the AMTEC include radiation heat transfer to the condenser, conduction heat transfer to the condenser via the cell wall, and conduction heat transfer through the alkali metal return artery to the condenser. To date, typical methods of reducing parasitic heat transfer have involved placing radiation heat shields between the high temperature region and the low temperature region of the AMTEC [3]. The difficulty with this approach is that it typically impedes not only the radiant heat transfer, but also the flow of the alkali metal working fluid. Impeding the flow of the alkali metal working fluid increases the cathode pressure, which causes a decrease in cell power output. One possible way of controlling the internal radiation heat transfer in a less obtrusive manner is to use geometrically designed surfaces such as asymmetric wall grooves that yield favorable effective directionally dependent radiative properties. The purpose of this investigation was to evaluate the effectiveness of this approach to minimizing the parasitic heat transfer in the AMTEC.

NOMENCLATURE

A_p	surface area of node p [m^2]
B_{ij}	radiation interchange factor from surface i to surface j
F_{ij}	view factor from surface i to surface j
N_R	total number of radiating surface areas
N_s	total number of solid nodes
Q_{bk}^{loss}	net radiant heat transfer rate from surface b that bounds node k [W]
Q_{fk}^{loss}	net radiant heat transfer rate from surface f that bounds node k [W]
S_k	source term in nodal energy balance [W]
T_c	temperature of condenser surface [K]
T_h	temperature of heat input surface [K]
T_p	temperature of node p [K]
β_k	coefficient used in energy balance relationship (β_k=1 if node k has surface b and β_k=0 if node k does not have surface b)

δ_{kp}	Kronecker delta ($\delta_{kp}=1$ if k=p, $\delta_{kp}=0$ if k≠p)
ε_c	emissivity of condenser surface
ε_m	emissivity of surface m
σ	Stefan-Boltzmann constant [W/m²-K⁴]
ψ_k	coefficient used in energy balance relationship ($\psi_k=1$ if node k has surface f and $\psi_k=0$ if node k does not have surface f)
Λ_{kp}	coefficient of nodal temperature p in energy balance relationship for node k

DEVICE CONFIGURATION

An asymmetric wall groove design was chosen to direct more (compared to the current smooth wall) of the thermal radiation coming from the "hot" end of the device (heat input region - β" alumina solid electrolyte (BASE) tubes, BASE tube support plate, etc.) back toward the "hot" end. In effect, the desired result is an increase in the total heat transfer resistance from the "hot" end to the "cold" end (heat rejection region) by re-directing thermal radiation incident on the cylindrical wall that originates in the "hot" end back to the "hot" end instead of allowing it to propagate to the "cold" end. This increase in total thermal resistance from the "hot" end to the "cold" end then decreases the parasitic heat transfer losses and increases the BASE tube and evaporator temperatures for fixed heat input and condenser (heat rejection surface) temperatures. Decreasing the parasitic heat transfer losses and increasing the BASE tube and evaporator temperatures contribute to an increase in conversion efficiency. A fortuitous advantage of the grooved wall is that when the groove shape is present on both the inside and outside of the cylinder, the wall heat conduction path is increased, which contributes to a decrease in the parasitic heat transfer.

The particular AMTEC device configurations that were chosen to be modeled with the asymmetrically grooved wall were the PX-4C and the PX-5A. Both of these cells were fabricated by Advanced Modular Power Systems (AMPS) and tested by the Air Force Research Lab (AFRL), with the experimental results for each of these cells presented in the literature [4,5]. These cells were identical with the exception of the condenser design. The PX-4C had a stainless steel screen wick placed on the condenser and the PX-5A had a micro-machined condenser. The benefit of the micro-machined condenser is that it should lower the condenser emissivity by causing a uniform layer of sodium to form on the condenser [6] (emissivity of stainless steel wick was assumed to be 0.15 and the emissivity of the micro-machined condenser was assumed to be 0.05). Each of the cells were cylindrical, 38 mm in diameter, 96.7 mm long, had 6 BASE tubes that were 40 mm long, a conical evaporator, and were made from stainless steel.

The asymmetric wall groove design chosen for use in the PX-4C and PX-5A was a 60°/30° (θ_1/θ_2) grooved/grooved (inside wall profile/outside wall profile) design (see Figure 1). These grooves were each parallel to the condenser surface and were therefore distinct, as opposed to being continuously

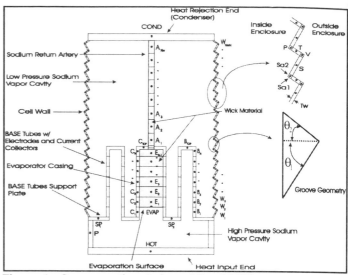

Figure 1. Cross sectional view of the 60°/30° grooved/grooved wall device.

"threaded". This design was chosen because previous analysis by Hausgen [7] of the 60°/30° groove effective reflectance properties (with surfaces 100% specularly reflecting) showed that it had the greatest potential of many θ_1/θ_2 combinations examined to reduce the parasitic axial heat transfer and increase the BASE tube temperature in a cylindrical AMTEC device. Nineteen wall grooves spanned the entire length of the PX-4C and PX-5A low-pressure cavity wall, each with a groove opening width of 4.71 mm. The actual PX-4C and PX-5A designs had cylindrical heat shields on the cell wall near the condenser, but these shields were not included in the devices modeled with asymmetric wall grooves.

MODELING METHODOLOGY

The modeling methodology for the AMTEC devices with asymmetrically grooved walls involved modifying a proven AMTEC device model. This proven model was developed using FORTRAN by the Institute for Space and Nuclear Power Studies at the University of New Mexico and is called AMTEC Performance and Evaluation Analysis Model (APEAM). APEAM is very comprehensive and includes: (1) a sodium flow model, (2) an electrochemical and an electrical circuit model, and (3) a radiation/conduction heat transfer model. Each of these models is coupled with one another due to common dependent variables. Results from APEAM were compared to experimental results obtained for the PX-4C and PX-5A (and other cell designs) and found to be in reasonably good agreement [4,5]. The details of these models are not presented here, but can be found in Tournier and El-Genk [4-5-8-9].

Since APEAM has been validated by experimental data, an effort was made to make only the minimum required modifications to model the asymmetrically grooved walls. These minimum required modifications included (1) altering the energy balance relationship written for each solid node to use radiation interchange factors instead of view factors, (2) changing the wall nodal energy balances to use the thermal conductances of the asymmetrically grooved wall instead of the flat wall thermal conductances, and (3) the emissivity was considered to be constant within APEAM as it solved the coupled fluid flow/electrical/thermal models for the cell power output and heat transfer.

The APEAM thermal model assumes diffuse gray surface behavior and therefore uses diffuse view factors and an enclosure analysis to model the radiation interchange in the AMTEC cell. For the present analysis, radiation interchange factors, calculated using an electromagnetic ray tracing code called NEVADA [10], were used instead of diffuse view factors to allow specular effects to be included. The radiation interchange factor, B_{ij}, is defined as the fraction of radiant energy emitted from surface i that is absorbed by surface j, including all intervening reflections. This is in contrast to the commonly used view factor, F_{ij}, that is defined as the fraction of total energy leaving surface i, both by emission and reflection, that is incident on surface j. The view factor also differs from the radiation interchange factor in that it is only dependent on the geometry while the interchange factor is dependent on both the geometry and the radiative surface properties.

Since APEAM was originally formulated using diffuse view factors to characterize the radiation exchange in the enclosure, it was necessary to modify the nodal energy balance relationship used by Tournier and El-Genk in APEAM. The energy balance relationship for solid node k is given by Tournier and El-Genk [9] as:

$$\sum_{p=1}^{N_s} \Lambda_{kp} T_p - S_k = -\psi_k Q_{fk}^{loss} - \beta_k Q_{bk}^{loss} \qquad (1)$$

The two terms on the right side of Equation 1 represent the net radiant energy loss from two surfaces (f and b) that bound node k. The coefficient ψ_k is equal to one if node k has surface f and zero if it does not, with the same being true for the coefficient β_k for surface b. The net radiant energy loss from a surface is given by Tournier and El-Genk [9] as:

$$[Q^{loss}] = \{M\}[\sigma T_k^4] \qquad (2a)$$

where:

$$M_{kp} = A_k \varepsilon_p \left(R_{kp}^{-1} - \sum_{m=1}^{N_R} F_{km} R_{mp}^{-1} \right) \quad k, p = 1 \; to \; N_R \qquad (2b)$$

and R_{kp}^{-1} is the inverse of:

$$R_{kp} = \delta_{kp} - (1 - \varepsilon_k) F_{kp} \qquad k, p = 1 \; to \; N_R \qquad (2c)$$

As stated earlier, this formulation uses diffuse view factors (F_{kp}) to characterize the radiation exchange between surfaces. For the present analysis, which uses radiation interchange factors, the Q^{loss} term (Equation 2) is modified in the following way:

$$[Q^{loss}] = \{M\}[\sigma T_k^4] \qquad (3a)$$

where:

$$M_{kp} = [\delta_{kp} - B_{pk}] A_p \varepsilon_p \qquad k, p = 1 \; to \; N_R \qquad (3b)$$

This modification to M_{kp} yields the modified energy balance that uses radiation interchange factors instead of diffuse view factors:

$$\sum_{p=1}^{N_s} \Lambda_{kp} T_p - S_k + \sigma \sum_{m=1}^{N_s} [\delta_{km} - B_{mk}] A_m \varepsilon_m T_m^4 = 0$$
$$k = 1 \; to \; N_s \qquad (4)$$

The radiation interchange factor is the fraction of energy emitted by node m absorbed by node k and, for this analysis, includes any radiating surface that bounds node m and k. This approach eliminates the need to include multiple radiation terms in the energy balance relationship for solid nodes that have more than one radiating surface.

NEVADA calculates the radiation interchange factors using a statistical simulation method that tracks the origin and destination of electromagnetic rays emitted from each surface. As a ray bundle is tracked from surface to surface, its energy level is decreased according to the radiation properties of the surfaces it intercepts. Once a ray bundle reaches a minimum allowed energy level, it is no longer traced. The result is that a small amount of energy emitted from a surface is not included in the surface to surface radiation interchange factors. In addition, a small amount of energy emitted from each surface is sometimes "lost" to space. This is likely due to the limited number of significant digits used to define the surface geometries. The loss of emitted energy to these two sources causes the sum of the radiation interchange factors from surface i to all other surfaces to be slightly less than one, but, by conservation of energy, the sum of the radiation interchange factors from surface i to all other surfaces must equal one. Therefore, to insure conservation of energy, the "lost" energy is assumed to be reabsorbed by the emitting surface.

The cross sectional view of the nodal divisions in the device thermal model with the 60°/30° grooved/grooved wall is shown in Figure 1. No nodal divisions, with the exception of the wall, were changed from those originally used by APEAM. The nodal radiating surfaces defined in NEVADA for the cell without the cylindrical wall are shown in Figure 2.

The solid nodes for the 60°/30° grooved/grooved wall are labeled in Figure 1 as "P", "V", "S", and "T". These nodes were circumferential because, due to symmetry, no variation in

Figure 2. Nodal surfaces of the operational device defined in NEVADA without the cylindrical wall.

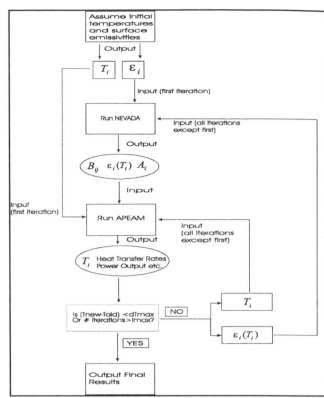

Figure 3. Flow chart of the NEVADA/APEAM coupling.

temperature occurred circumferentially in the cell wall. The surface areas that were associated with the "P" nodes that exchange radiation in the enclosure (Sa1 and Sa2) were assumed to be part of the radiating surface area associated with nodes "S" and "T" (Sa1 to "T" and Sa2 to "S"). This assumption was made to simplify the surface descriptions required in the radiation interchange modeling software (NEVADA). It also allows the wall thickness to be iterated without changing the surface descriptions in the NEVADA model.

The original APEAM formulation included the temperature dependence of the emissivity as it solved the coupled fluid flow, electrical, and thermal models. For this analysis, APEAM was modified to assume a constant emissivity. This was necessary because the present analysis required the use of radiation interchange factors, which are dependent on the surface emissivities, and had to be calculated externally to APEAM. The emissivities used in the energy balance relationship must match the emissivities used in the calculation of the radiation interchange factors for the energy balance relationship to be correct, and therefore could not be changed with temperature in APEAM without recalculating the radiation interchange factors.

To include the temperature dependence of the emissivity, the NEVADA code had to be coupled to the APEAM code. This is because the radiation interchange factors, calculated by NEVADA, depend on the emissivity of the surfaces and the emissivity of the surfaces is dependent on temperature, which is calculated by APEAM. A Visual BASIC program was written to externally iterate APEAM and NEVADA and the process that was followed is graphically shown in Figure 3. This process was repeated until the maximum change in cell temperatures was small from iteration to iteration or a maximum number of iterations had been performed.

The sodium pressure drop from the BASE tubes to the condenser for the device with an asymmetrically grooved wall was assumed to be the same as if the walls were flat. The actual pressure drop should be higher than that associated with a flat wall due to the presence of the grooves on the cylindrical wall. Since the power output of the AMTEC decreases as the BASE tube to condenser pressure drop increases for a fixed condenser temperature, the actual power output will likely be lower than the predicted value.

To ensure that the modifications to APEAM for use of the radiation interchange factors were made correctly, a comparison of results from the coupled NEVADA/APEAM model, which used radiation interchange factors, and the unmodified APEAM model, which used view factors calculated from closed form expressions, was made. The configuration for which this comparison was made was the PX-4C with no internal heat shields. Conversion efficiency, power output, and various device temperatures were compared and excellent agreement was observed [7].

RESULTS AND DISCUSSION

Four cases were examined for both the PX-4C and PX-5A, which included: (1) 60°/30° grooved/grooved cylindrical side wall with all surfaces 100% specularly reflecting except the BASE tubes (Case A), (2) 60°/30° grooved/grooved cylindrical side wall with only the wall 100% specularly reflecting (all other surfaces diffuse), (3) flat cylindrical side wall with internal cylindrical heat shields and all surfaces diffuse (Case C, which was the actual PX-4C and PX-5A configuration) and (4) flat wall device with no internal cylindrical heat shields and all surfaces diffuse (Case D). Cases A, B, and D were modeled using the coupled NEVADA/APEAM model. Case C was modeled using the unmodified APEAM code that used closed form expressions to calculate the diffuse view factors. A comparison of the results for the 60°/30° grooved/grooved wall (Cases A and B) with the flat wall both with (Case C) and without (Case D) internal heat shields demonstrates the performance improvement that the wall grooves provide. For the devices modeled with the wall grooves, no internal shields were present. In all cases examined, emission was assumed to be diffuse, all surfaces were assumed to be gray, and the external cylindrical wall was assumed to be adiabatic.

The wall thickness in the actual PX-4C and PX-5A cells was not uniform along the entire length of the wall from the BASE tube support plate to the condenser surface. While most of the wall was 0.1 mm thick, there was a short span of 7.5 mm in which the wall was 0.175 mm thick and another span of 5 mm in which it was 1 mm thick. However, for the results presented, the wall was assumed to be 0.1 mm thick along the entire length.

The conversion efficiency as a function of resistive load at T_h=1130 K and T_c=565 K is given in Figures 4 and 5 for the PX-4C and the PX-5A, respectively. Figure 4 shows that for the PX-4C design, the 60°/30° grooved/grooved wall with all surfaces 100% specularly reflecting (Case A) outperforms both the device that has a flat wall and internal cylindrical heat shields (Case C) and the device that has a flat wall and no internal heat shields (Case D). The same results were obtained for the PX-5A, but the magnitude of improvement over the device with internal cylindrical heat shields caused by the 100% specularly reflecting 60°/30° grooved/grooved wall was small. However, a significant improvement over the smooth wall with no internal heat shields was demonstrated.

Table 1 gives several calculated parameters at a resistive load near the maximum efficiency point (1.5 Ω) for both the PX-4C and the PX-5A. These calculated parameters include efficiency (η), percent increase in efficiency compared to the flat (smooth) wall with no internal heat shields (%η_{inc}), electrical power output (P_e), parasitic radiation heat transfer rate to the condenser surface (Q_R), parasitic conduction heat transfer rate through the device wall to the condenser (Q_W), and parasitic conduction heat transfer rate through the sodium return artery to the condenser (Q_A). Table 1 shows that both grooved wall cases (Case A and Case B) for the PX-4C and the PX-5A reduce each of the components of the parasitic heat transfer and increase the power output and conversion efficiency compared to a flat wall device with no internal heat shields (Case D). The power output is increased by the wall groove design (compared to a flat wall device with no internal heat shields) by about 0.1-0.2 W due to a slight increase in BASE tube temperature.

It is interesting that the cylindrical wall shields (Case C), as modeled, actually cause an increase (compared to a flat wall with no heat shields – Case D) in the radiative heat transfer to the condenser. However, they decrease the conduction heat transfer through the wall to the condenser by enough to cause a net decrease in parasitic heat transfer. In contrast, the 60°/30°

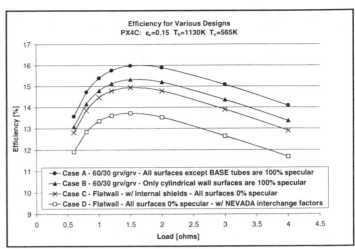

Figure 4. Efficiency for various cases (PX4C - ε_c=0.15, T_h=1130 K, T_c=565 K, adiabatic external wall).

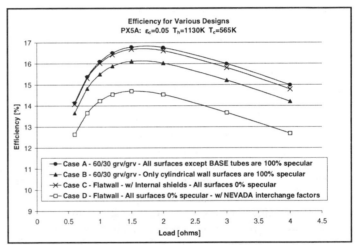

Figure 5. Efficiency for various cases (PX5A - ε_c=0.05, T_h=1130 K, T_c=565 K, adiabatic external wall).

Table 1. Comparison of results for various AMTEC designs (load=1.5 Ω, T_h=1130 K, T_c=565 K, PX-4C and PX-5A, adiabatic external wall).

Load=1.5 Ω : T_h=1130 K : T_c=565 K

PX-4C (ε_c=0.15)	η [%]	%η_{inc}	P_e [W]	Q_R [W]	Q_W [W]	Q_A [W]
CASE A	16.0	16.4	4.8	3.7	6.9	2.8
CASE B	15.3	11.6	4.7	4.1	7.5	2.9
CASE C	14.9	8.9	4.7	6.1	5.5	3.5
CASE D	13.7	------	4.6	5.0	8.9	3.2

PX-5A (ε_c=0.05)	η [%]	%η_{inc}	P_e [W]	Q_R [W]	Q_W [W]	Q_A [W]
CASE A	16.8	14.2	4.9	1.4	7.8	2.9
CASE B	16.1	9.6	4.8	1.6	8.6	3.1
CASE C	16.7	13.5	4.9	2.4	6.3	3.9
CASE D	14.7	------	4.7	1.9	10.2	3.4

grooved/grooved design decreases the radiative heat transfer to the condenser while also reducing each of the other parasitic heat transfer components.

It appears that the lower condenser emissivity present in PX-5A aids the performance of the flat wall with internal heat shields more than the grooved wall designs. Examination of the components of the parasitic heat transfer seems to indicate that this is likely due to the decreased prominence of the radiation heat transfer to the condenser for the case of lower condenser emissivity. The grooved walls do a better job than do the internal cylindrical wall heat shields of reducing the radiation heat transfer to the condenser, while the internal cylindrical wall heat shields do a better job of reducing wall conduction to the condenser. The magnitude of the reduction of radiation heat transfer by the grooved walls is much greater for the higher condenser emissivity.

CONCLUSIONS

While the results of this research show promise for an innovative AMTEC design that could yield increases in conversion efficiency and power output, the actual implementation of it will likely be difficult. To be effective, the grooves must be highly specular, which requires extensive surface preparation. In addition, at the high operating temperatures, atoms tend to migrate at the grain boundaries, thereby changing a polished, highly specular surface to one that is diffuse. This phenomenon was observed in an experimental test [7], but only at the highest temperature locations (~1100 K). This fact is the reason for modeling the grooved geometry with only the walls 100% specularly reflecting (Case B). As can be seen in Figure 4, Figure 5, and Table 1, this situation still yields increased performance over the AMTEC device with a flat wall and no internal heat shields. Fabricating the cylindrical wall with the required groove geometry will also be challenging.

It is clear from the analysis presented that geometrically designed surfaces utilizing specular reflection can significantly affect the performance of AMTEC devices. However, it is likely that the performance of the wall grooves can be increased through further optimization. While supporting research [7] examined the effect of the asymmetric groove opening angles on performance, no investigation has been undertaken to optimize the performance based on the number of wall grooves, device dimensions (such as length and diameter of the cell), and the groove geometry itself (only asymmetric grooves of the type shown in Figure 1 were examined). In addition, the performance improvement caused by the wall grooves may be greater for AMTEC devices that have better performing electrodes than those present in the PX-4C and PX-5A.

AKNOWLEDGEMENTS

The work described in this publication was funded by the United States Air Force under direction of the Space Vehicles Directorate of the Air Force Research Laboratory.

REFERENCES

1. Vining, C. B., Williams, R. M., Underwood, M. L., Ryan, M. A., and Suitor, J. W., 1992, "Reversible Thermodynamic Cycle for AMTEC Power Conversion," *Proceedings of the 27th Intersociety Energy Conversion Engineering Conference*, Vol. 3, pp. 123-127.

2. Cole, T., 2 Sept 1983, "Thermoelectric Energy Conversion with Solid Electrolytes," *Science*, Vol. 221, No. 4614, pp. 915-920.

3. Borkowski, C., Svedberg, R., and Hendricks, T., 1997, "Parasitic Heat Loss Reduction in AMTEC Cells by Heat Shield Optimization," *Proceedings of the 32nd Intersociety Energy Conversion Engineering Conference*, Honolulu, HI, pp. 1130-1135.

4. Tournier, J. M., and El-Genk, M. S., 1998, "AMTEC Performance and Evaluation Analysis Model (APEAM): Comparison with Test Results of PX-4C, PX-5A, and PX-3A Cells," *Proceedings of Space Technology and Applications International Forum (STAIF-98)*, CONF-980103, M. S. El-Genk, ed., American Institute of Physics, New York, NY, AIP Conference Proceeding No. 387, pp. 1552-1563.

5. Tournier, J. M., and El-Genk, M. S., 1998, "Heat Transfer in the Enclosure of a Multitube Alkali-Metal Thermal-to-Electric Converter Cell," *Proceedings of the 7th AIAA/ASME Joint Thermophysics and Heat Transfer Conference*, HTD-Vol. 357-4, Vol. 4, pp. 17-24.

6. Crowley, C. and Izenson, M., 1993, "Condensation of Sodium on a Micro-machined Surface for AMTEC," *Proceedings of the 10th Symposium on Space Nuclear Power and Propulsion*, No. 271, pt. 2, pp. 897-904.

7. Hausgen, P. E., 2000, "A Thermal Analysis of an Alkali Metal Thermal to Electric Converter with Geometrically Designed Interior Surfaces Exhibiting Directionally Dependent Radiative Properties," Ph.D. thesis, Georgia Institute of Technology, Atlanta, GA.

8. Tournier, J. M., El-Genk, M. S., and Huang, L., 1998, "Experimental Investigations, Modeling, and Analyses of High-Temperature Devices for Space Applications", U. S. Air Force Technical Report - AFRL-VS-PS-TR-1009-1108.

9. Tournier, J. M. and El-Genk, M. S., Feb 1999, "Radiation Heat Transfer in Multitube, Alkali-Metal Thermal-to-Electric Converter," *Journal of Heat Transfer*, Vol. 121, No. 1, pp. 239-245.

10. TAC Technologies, Incline Village, Nevada.

Proceedings of IECEC'01
36th Intersociety Energy Conversion Engineering Conference
July 29–August 2, 2001, Savannah, Georgia

IECEC2001-CT-43

ADVANCED AMTEC CONVERTER DEVELOPMENT

Jan E. Pantolin, Robert K. Sievers, Joseph C. Giglio, Daniel R. Tomczak, Quinlan Y. Shuck
and Brian K. Thomas
Advanced Modular Power Systems, Inc.
4370 Varsity Drive
Ann Arbor, Michigan 48108
(734) 677-4260, E-mail: jpantolin@ampsys.com

ABSTRACT

The alkali metal thermal to electric converter (AMTEC) design continues to evolve and improve. Innovations in design have reduced part counts and simplified the assembly process, leading to lower cost and increased reliability. The major design improvement is in the power-producing component of the converter; the β"-Alumina Solid Electrolyte (BASE) tube. This tube incorporates the Internal Self Heat Pipe (ISHP) design that produces more power and is more robust at various power loads, compared to previous designs. The converter discussed in this paper will produce over 5 watts of electrical power. This design is the basis for continued AMTEC development, and is also the converter for small fuel-fired portable demonstration units that produce power to operate small electronic devices.

INTRODUCTION

An alkali metal thermal to electric converter (AMTEC), like the one shown in Figure 1, is a device that efficiently converts heat directly to electricity using the unique characteristics of the alkali metal conducting ceramic, β"-Alumina (Weber, 1974). β"-Alumina is a solid electrolyte (BASE) that conducts sodium ions, but is an electron insulator. An electrochemical potential is generated when sodium is present at two different pressures separated by the electrolyte. In the converters described, here the sodium is recirculated through the converter using a porous stainless steel wicking system that uses capillary forces to transport the sodium from

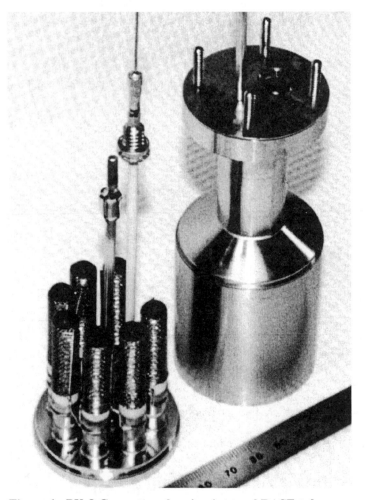

Figure 1. PX-8 Converter showing internal BASE tubes.

the low-pressure region to the high-pressure region. The pressure in each region is dependent on the vapor pressure of sodium at the coldest point within the region. In the converter described here, the sodium in the high-pressure region evaporates from the wick and flows as a vapor to the inner surface of the electrolyte tube where it ionizes and enters the BASE. The freed electrons are captured through a system of electrodes and interconnects. They are used to power a load and then recombined with sodium ions, which were conducted through the electrolyte and are now at the surface of the tube, to form sodium vapor. The vapor then evaporates from the cathode surface and flows to the cold end of the converter where it condenses in the wick structure, from which the cycle can be repeated. The demonstrated efficiencies, the re-circulation of the sodium and the fact that there are no moving parts makes AMTEC conversion attractive for a number of portable and remote power applications.

Development of the first stainless steel multi-tube converters began in 1994. These converters used a thin TiN anode inside the tube. In order to prevent the anode on these series connected tubes from shorting to the converter ground, sodium liquid was not allowed to condense inside the BASE tubes. This led to what was called the "vapor-anode." After some initial component development steps, converters where fabricated and tested. Performance progressed from ~1 watt output in late 1996 for the converter called PX-1, to 5 watts at 850°C in late 1997 for the converter called PX-5 (Borkowski, 1997). The design iterations between PX-1 and PX-5 were made to troubleshoot and solve issues that were revealed with the testing of each successive converter iteration. To culminate this development program, two systems, one using the PX-3 and one using the PX-5 converter, were delivered to the United States Air Force in 1997 for efficiency and life testing. One of the PX-3 converters was then operated at high temperature for over 20,000 hours. A detailed description of the testing and results can be found in Merrill (1999).

A number of issues became clear with the testing of these converters. First, the design, especially the BASE tube, was complex. This complexity led to high cost and low reliability. Second, converters on test would degrade by 10-20% in the first 500 hours, and then an additional 1-2% for every 1000 hours of operation, depending on thermal cycles. Substantial progress had been made in a short time, but there were still some serious issues still to be addressed.

Since the PX-5 and PX-3 system testing, there have been three successive generations. PX-6 increased the number of BASE tubes in the converter from 5 or 6 to 8. This converter was put through the most extensive efficiency test of any converter, producing 16% efficiency at 850°C. PX-7 was the first converter with a "chimney" in the converter wall. PX-8 was the first converter to introduce the internal self-heat pipe (ISHP), changing the BASE tube from the sodium "vapor-anode" to a liquid sodium anode (Sievers 1999). This liquid-anode also acted as a heat pipe to more efficiently transport heat from the hot end into the BASE tube.

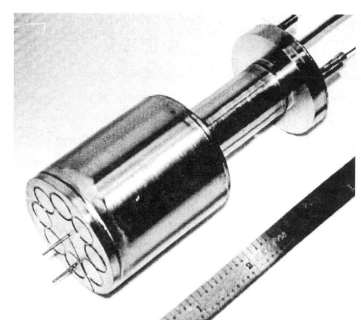

Figure 2. PX-8 Converter after final assembly.

CONVERTER DESIGN

The PX-8 converter design, shown in Figures 1 and 2, is the first 5 W AMTEC converter to use the next generation ISHP BASE tube configuration. It uses the ISHP BASE tube arrangement for improved performance, ease of assembly and fewer components. The old design relied on sodium vapor being supplied on demand to the high pressure side inside the BASE tube. The BASE tubes were required to operate at a higher temperature than the evaporator to avoid sodium condensation. Condensation in the old design could eventually sequester enough sodium to reduce the converter power output by reducing the amount of sodium available and possibly producing and electrical short between BASE tube anodes and ground. Operation of these vapor-anode converters was also more sensitive to temperature distribution.

The ISHP design allows the sodium to condense and collect in the BASE tubes. Only enough sodium collects in the tubes to fill the stainless steel wick material. This greatly reduces the inner current collector system resistance and makes very good electrical contact with the BASE material. The electrical insulator at the open end of the BASE tube assembly operates at a high enough temperature to keep sodium from condensing there and electrically shorting out the tubes. Because of this, the converter is able to operate at zero and low load situations. The ability to operate in this mode is more desirable for consumer and commercial applications.

The AMTEC "Chimney" converter was first conceived for a space power system (Sievers 1994). It has two main design features that contribute to reducing thermal losses and increasing efficiency. One is the reduced converter wall diameter; the other is the converter wall material thickness. Both of these reduce the cross sectional area of the wall, and that in turn reduces the thermal conduction path. Secondary benefits of this design come from the transition cone that joins the larger diameter lower converter wall where the BASE tubes are, to the small diameter "chimney" wall section. This transition cone helps radiate heat back towards the BASE tubes. It also blocks a good portion of the direct thermal radiation from the BASE tube mounting plate, the BASE tubes and other components located in the hottest area of the converter.

This "chimney" design does not rely on any radiation heat shields. This reduces the number of components and eases the assembly of the converter. Although a single, centrally located radiation disc placed in the transition cone section has been used in similar designs, it was omitted for this converter build. It is not clear that it would make a significant difference in performance. It would help retain some of the heat in the BASE tube area, but may also increase the sodium vapor back pressure around the BASE tubes and reducing the overall converter power. The primary reason to use radiation shields is to increase the converter efficiency by several percentage points, but some applications may not warrant the added complexity.

One of the overlooked design elements of an AMTEC converter had been the method of fastening it to a fixture or housing. Previous designs used the threaded electrical feedthrough or the sodium fill port as a means of attaching the converter to a bracket. Usually the threads on the feedthrough were delicate and there was non-uniform contact with the bracket when the nut was tightened. The design used on this converter is just four threaded studs that are welded to the cold end cap. This is the coldest part of the converted and is easily inserted into a mating bracket or housing. The stud locations are uniformly spaced on a bolt–circle. They can easily be changed to accommodate other mating constraints. This mounting arrangement will provide better thermal contact with a mounting surface if more heat rejection is required.

The stepped down converter wall diameter in the "chimney" section creates sodium reservoir issues. If the cold end reservoir is the same diameter as the chimney section, the length of the converter must be greatly increased for an equivalent sodium storage volume, compared to a larger diameter more like the diameter around the BASE tubes. The ISHP BASE tube in this design retains as much as 1 gram of sodium per tube, so a sodium reservoir in the cold end must have the flexibility to deal with large changes in sodium inventory. There must be some sodium left in the cold end when all the BASE tubes are saturated, and be able to retain all the sodium when the BASE tubes are dry. The reservoir diameter on the cold end, is therefore, very close to the larger converter wall diameter and not the chimney diameter.

Once the cold end reservoir volume is determined, the actual design and assembly process is simple. Older converter designs used a same diameter cold end stainless steel fiber pads. In order to hold them in place, they were resistance spot welded to the cold end cap and then sintered in a hydrogen atmosphere furnace. This new design skips the sintering process, the stainless steel fiber pads are installed and compressed in the stepped cold end wall section. A simplified artery to cold end reservoir connection is utilized. Finer stainless steel fiber felt is formed into straps, and the straps are looped around the pads through a center hole. There are several sets of straps that assure that sodium is always wicked towards the artery tip no matter what the converter orientation is.

Several major design changes of this converter have reduced the final assembly and welding time. The BASE tube assemblies require less custom fitting at this stage of the process. At this point, they only require minimal fitting. The BASE tube connections are all resistance spot welds performed on the assembly bench. The converter wall has no radiation shields to get in the way, and the cold end components simply drop into place. Only basic fixturing is required for the final assembly. More upfront engineering helps all the assemblies fit together more quickly.

Initial testing of the PX-8 converter design was performed with electrical heaters for controlled temperature conditions. Under certain temperature conditions, over 5W of peak power is achievable. The output voltage is high enough to power small electronic devices without any step up DC-DC conversion. Figure 3 illustrates the performance as the condenser temperature changes. The peak power varies by 0.7 W with a temperature change of 100°C on the condenser. The lower temperature produces more output power, but may reduce

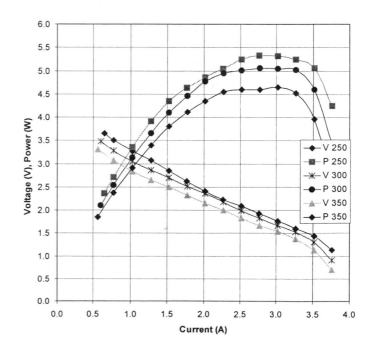

Figure 3. PX-8 power/voltage sensitivity to condenser temperature, with a hot end temperature of 800°C.

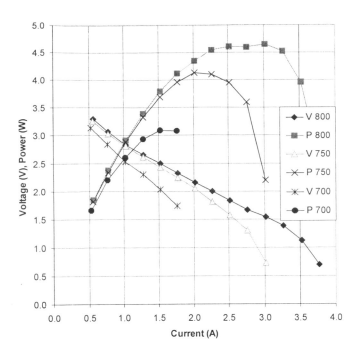

Figure 4. PX-8 power and voltage sensitivity to hot end temperature, with a condenser temperature of 350°C.

overall system efficiency.

The general power output changes dramatically with the hot end temperature, as shown in Figure 4. There are sharp drop offs at the high-current end of these performance curves. Usually the converter has a passive load and set to 0.5 to 1 ampere of current. During this low current the BASE tube wick fills with sodium. Then at high currents, this stored sodium gets used up and the evaporator can't supply enough sodium to keep up with the demand. Eventually, the sodium is depleted and the power output diminishes.

GENERATOR DEVELOPMENT

For AMTEC applications and demonstration purposes, several gas-fired systems, such as the one shown in Figure 5, are being developed. The PX-8 series design is well-suited for small portable power applications and a general test bed for gas fired systems. Current work has produced a preliminary system to evaluate and demonstrate a simple generator. Small electrical devices like a CD player and TV are operated off this system.

Small concentrations of hydrogen can be found in burned fuel products. This hydrogen can diffuse into the converter through its hot stainless steel walls. This reduces the pressure drop across the BASE material, reducing the converter's power output. To combat hydrogen diffusion, gettering material is used in these converter designs. Another way to reduce hydrogen diffusion is by selecting specific outer converter materials. Haynes and Inconel alloys allow less hydrogen to permeate through at high temperatures. It is also a better choice

Figure 5. The 5 watt generator, incorporation the PX-8 converter, operates on propane.

over 300 series stainless steel for oxidation resistance and high temperature exposures.

PX-5 converters have been upgraded with getters and improved hot end materials. These converters are now on test in a propane-fueled system shown in Figure 6. The performance of one of these PX-5 converters is shown in Figure 7. With an initial power of 4 watts at 800°C, this converter has consistently operated within 5% of its initial power for the first 4000 hours of operation. There is some fluctuation in power, but most of this can be attributed to hot end temperature variations due to manual fuel control. There is no sign of the initial degradation so often observed in earlier PX converters, especially the rapid degradation that has typically been observed in the first 500 hours, even when electrical heating is used. Another important aspect of this fuel fired testbed is that

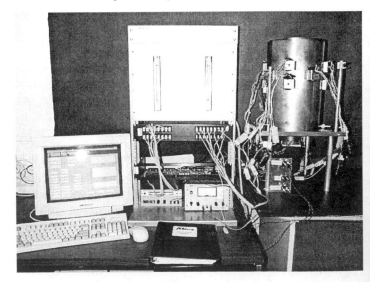

Figure 6. Propane-fueled test bed for AMTEC converters.

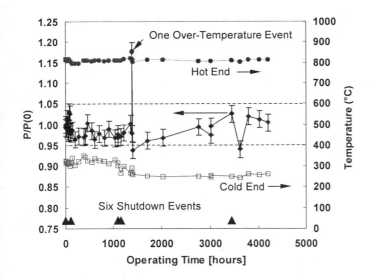

Figure 7. Advanced PX-5 power and temperature history.

no degradation associated with hydrogen diffusion is apparent. Also note that this converter has seen several shutdown and one over-temperature thermal cycles without significant impact on performance.

CONCLUSIONS

The PX-8 converter design promises useful energy conversion that can be scaled up to larger systems. Design changes and improved BASE tube performance have reduced production costs and made it more reliable and longer service life. Recent testing has shown substantial improvements in all aspects of the converter technology. Demonstration units are now feasible and consumer products are closer to becoming reality.

ACKNOWLEDGMENTS

The authors wish to thank the Air Force Phillips Laboratory and the Wright Patterson Air Force Base for their support of the propane-fueled test bed, as well as the Teledyne Brown Energy System Division for the combustion system in that test bed. The authors also wish to thank Dr. Jerry Martin at Mesoscopic Devices, Inc., for the system design and many of the thermal system components for the 5 watt generator.

REFERENCES

Borkowski, C.A., Sievers, R.K. and Hendricks, T.J., "PX Series AMTEC Cell Design, Testing and Analysis," *Proceedings of the 32nd Intersociety Energy Conversion Engineering Conference*, p. 1202 (1997).

Carlson, M.E., Hendricks, T.J., Sievers, R.K., and Svedberg, R.C., "Design of the EPX-1 AMTEC Cell for the Advanced Radioisotope Power System," *Proceedings of the 34th Intersociety Energy Conversion Engineering Conference*, # 2654 (1999).

Hendricks, T.J., Huang, C., and Huang, L., "AMTEC Cell Optimization for Advanced Radioisotope Power System (ARPS) Design," *Proceedings of the 34th Intersociety Energy Conversion Engineering Conference*, # 2655 (1999).

Sievers, R.K., Hunt T.K., Hendricks, T.J., Barkan, A., Siddiqui, S., "AMTEC Internal Self-Heat Pipe", *Proceedings of the 34th Intersociety Energy Conversion Engineering Conference*, p. 2660 (1999).

Sievers, R.K., Pantolin, J.E., Huang, C., "Advanced AMTEC Radioisotope Power Systems for Deep Space Applications" *35th Intersociety Energy Conversion Engineering Conference*, p. 3074 (2000).

Merrill, J.M. and Mayberry, C., "Experimental Investigation of Multi-AMTEC Cell Ground Demonstration Converter Systems Based on PX-3 and PX-5 Series AMTEC Cells," in proceedings of *the 16th Symposium on Space Nuclear Power and Propulsion*, p. 1369 (1999).

Ryan, M.A., Williams, R.M., Homer, M.L., Philips, W.M., Lara, L., and Miller, J., "Lifetime Modeling of TiN Electrodes for AMTEC Cells," *Proceedings of the 15th Symposium on Space Nuclear Power and Propulsion*, p. 1607 (1998).

Schock, A., Noravian, H., Or, C., and Kumar, V., "Recommended OSC Design and Analysis of AMTEC Power System for Outer-Planet Missions," *Proceedings of the 16th Symposium on Space Nuclear Power and Propulsion*, p. 1534 (1999).

Weber, N., "A Thermoelectric Device Based on Beta-Alumina Solid Electrolyte," *Energy Conversion*, 14, 1-8 (1974).

R.K. Sievers, T.K. Hunt, J.F. Ivanenok, III, J.E. Pantolin and D.A. Butkiewicz, "Modual Radioisotope AMTEC Power System," Proceedings of the Tenth International Symposium on Space Nuclear Power, January 1994, pp. 319-324.

Proceedings of IECEC'01
36th Intersociety Energy Conversion Engineering Conference
July 29 - August 2, 2001, Savannah, Georgia

IECEC2001-RE-10

Nighttime and Daytime Electrical Power Production from the Nighttime Solar Cell

Ronald J. Parise
Parise Research Technologies
101 Wendover Road
Suffield, CT 06078
Phone: (860) 668-4599

G. F. Jones
Department of Mechanical Engineering
Villanova University
Villanova, PA 19085
Phone: (610) 519-4985

ABSTRACT

The Nighttime Solar Cell™ works on the principle of a simple thermoelectric generator (TEG) operating in the temperature differential that exists between deep space at an effective temperature of 4K and the surrounding ambient temperature (nominally at 300K). Thus the ambient or surroundings of the device are the source of thermal energy while deep space provides a thermal sink. The cold junction of the TEG is insulated from the surroundings by a vacuum cell, improving its overall effectiveness.

This research is an on-going effort to develop a clean, reliable, safe, inexpensive, alternate source of electric power at night. The model discussed herein investigates the many design parameters that influence electrical power production during daytime operation. Previous research has shown that the 6cm x 6cm cell can produce about 7mW of continuous electrical power from the TEG module during nighttime operation.

Using a model developed for this work, we investigate the operation of the cell during the day without photovoltaic cells present. Therefore the cell converts solar thermal energy into electrical energy by reversing the operation of the TEGs. That is, the nighttime cold junctions now become the daytime hot junctions heated by the sun, and the nighttime hot junctions become the daytime cold junctions cooled by the ambient air. In this mode of operation the model includes thermal radiation from the sun for three selected solar fluxes. These correspond to approximately full-sun, half-sun and quarter-sun exposure during summer and winter operation.

The thermal model has also been expanded to include a twelve-banded model for radiation transmission through the atmosphere for nighttime electrical energy production. The results of the model will provide performance data for material selection and parameter requirements for building a prototype Nighttime Solar Cell™.

INTRODUCTION

The primary drawback to solar (photovoltaic) panels is the lack of nighttime energy production. Costly battery backup is needed if the electrical load powered by the photovoltaics is required at night. The batteries are subject to failure and require maintenance. Solar panel installations are normally in remote locations where service is difficult to provide and often inconvenient.

The Nighttime Solar Cell™ is a unique solid state device that produces electric power at night, similar to daytime solar power; or it can produce electricity by day, with or without solar power accompaniment, or used to augment the operation of solar panels by electric power generation at night (Parise, 1998).

There are many configurations and uses that the cell can assume. The original function of the Nighttime Solar Cell™ is to produce electrical energy both day and night (Parise, 1999). Daytime electrical energy production takes place from one of two methods (or a combination) of direct energy conversion. The first method can be solar cells connected in parallel with the TEGs, converting light into electrical energy when the sun is visible. In this configuration, the TEGs also produce electrical power from the solar thermal energy. The TEGs would produce electricity at night with the photovoltaic cells acting as an electrical conduit.

The second method would not utilize photovoltaic cells but involve the conversion of thermal energy from the sun into electrical energy by reversing the operation of the TEGs. That is, the nighttime cold junctions now become the daytime hot junctions; the sun is the thermal source for operating the system. In this mode of operation, part of the control system that provides electrical power to the load requires circuitry that monitors electrical current flow direction: daytime operation will have a reverse current flow direction from nighttime operation because the junction temperature of the TEGs will reverse (hot junctions become cold junctions and vice versa). This is the primary mode of operation the model considers in this study.

Note that the Nighttime Solar Cell™ could use an alternate energy source (thermal waste stream, ground water, etc.) at a higher temperature than the ambient, if available at the site of the solid state device. This would improve system performance and add to system reliability.

Obviously the Nighttime Solar Cell™ technology is in its infancy with the many uses and designs in which it can be configured. With today's technology, a cell can be manufactured to produce electricity both day and night for low wattage applications for such typical electronic devices as a cell phone, laptop, radio, etc.

THERMAL MODEL

A thermal model was developed from fundamental principles for this work. The model determines the temperature difference between the hot and cold junctions of the TEGs. Figure 1(a) shows the physical configuration and parameters utilized in the model when using deep space as the thermal sink during nighttime operation. The orientation of the cell is such that the cold junction plate (CJP) is parallel to the horizontal. The thermal source, the ambient, supplies energy to the module at T_∞.

Figure 1(a) shows the height of the TEG elements to be L, and the distance from the CJP to the window to be L_a. The TEG junctions are copper and the CJP is aluminum. The thermal conductivities of copper and aluminum are sufficiently large so that the temperature gradients are small through the thickness of each.

Figure 1(b) illustrates the daytime operation of the cell, where the CJP now becomes the hot junction plate (HJP). The HJP is now oriented normal to the solar beam radiation when the sun is at solar noon, and the model accounts for the movement of the sun from horizon to horizon assuming the usual periodic motion of the sun (Duffie and Beckman, 1974).

The CJP (or HJP) is 20cm^2 and the aperture opening is slightly larger to avoid physical or thermal interference. The thickness of the CJP is 5mm to reduce the temperature gradients associated with the fin effect which results from the TEG module being smaller than the CJP.

The surface of the CJP is assumed to be a gray, diffuse surface with $\epsilon_c = \alpha_c$ at temperature T_c. The surface facing the nighttime sky is assumed to have an emissivity, ϵ_c, of 0.90.

Deep space is modeled as a black body at temperature $T_s = 4K$. The model utilizes 12 transmission bands shown in Fig. 2 for radiative transfer through the atmosphere (Hudson, 1969), although less than 1% of the energy occurs below $5\mu m$ at 300K or less. The view factor between the CJP and deep space is assumed to be unity.

The exterior of the window is exposed to the ambient temperature, T_∞, through a specified heat transfer coefficient, $h_w = 10 \ W/m^2K$ corresponding to weak free convection of air at the surface. Temperature variations through the window are neglected.

The TEG elements are bismuth telluride. The specific design configuration of the TEG elements and module are based on standard sizes available from industry. The CJP and cell size are based on the size availability of the ZnSe window (nominal size: 2" x 2").

With solar heating of the CJP during the day, three values of normally incident solar radiation are used in the model: 200 W/sq.m, 500 W/sq.m and 800 W/sq.m. Only the local ambient temperature, of values 300K and 275K as specified in the model, provides the thermal sink for this mode of operation.

As done in the past (Parise and Jones, 2000) the model includes the effect of not having a full vacuum in the cell. That is, there will now be a slight trace of air at the low pressure of about 25 torr. With this air, the only effect that will be considered in the model is between the CJP and the ZnSe window, referred to as the air gap.

Free convection does not occur because the low density of the air results in a Reyleigh number for the air gap less than the critical value (Eaton and Blum, 1975). The air gap is 1cm.

The thermoelectric properties of the Seebeck coefficients (α_n, α_p), thermal conductivity (λ_n, λ_p), and the electrical resistivity (ρ_n, ρ_p) are assumed to be constant. The length of the thermoelectric elements in the direction of heat flow is L.

EQUATION DEVELOPMENT

The thermal model has been developed in a previous study (Parise et al., 1999) and the results will be summarized here.

Radiation Model

The net radiative heat flux on the CJP comes from four sources: the radiosity of the CJP, J_c, the fraction of energy from the night sky that is transmitted through the window, emission from the window, and the fraction of the J_c that is reflected from the window. Thus

$$q_c = (1 - \rho_w)J_c - \tau_w\sigma T_s^4 - \epsilon_w\sigma T_w^4, \quad (1)$$

where the ρ_w, τ_w, ϵ_w refer to the radiative properties of reflectivity, transmissivity and emissivity of the window and T_w is the window temperature.

An energy balance on the window accounts for the convective heat transfer rate at the external surface of the window and the net radiative energy that it absorbs. Therefore

$$h_w(T_\infty - T_w) = 2\epsilon_w\sigma T_w^4 - \epsilon_w\sigma T_s^4$$
$$- \epsilon_w J_c + (k_a/L_a)(T_c - T_w), \quad (2)$$

where h_w is the convective heat transfer coefficient at the surface of the glass. When the effect of air is present, k_a is the thermal conductivity of the air gap in the cell, and T_c is the temperature of the CJP. With no air present, k_a equals zero.

All radiative properties in eqns. (1) and (2) are written as the sum of the contributions from their respective bands. For example, ϵ_w in the term $\epsilon_w\sigma T_s^4$ in eqn. (2) may be written as

$$\epsilon_w = \epsilon_1 F(0,\lambda_1 T_s)$$

$$+ \sum_{i=2}^{11} \epsilon_i F(\lambda_i T_s, \lambda_{i+1} T_s)$$

$$+ \epsilon_{12} F(\lambda_{12}T_s, \infty) \quad (3)$$

where $F(x,y)$ is the blackbody emissive power fraction over the band of λT defined by the values of the first (x) and second (y) arguments in $F(x,y)$ (Dunkle, 1953). In eqn. (1), the radiation properties of J_c are based on the band model at temperature T_c.

Heat Conduction and Thermoelectric Model

A steady-state, quasi one-dimensional heat conduction model with internal energy generation is used (Parise et al., 1999). One boundary condition is the heat flux, q_c, at the CJP, and on the opposite end of the TEG the second boundary condition is the convection heat transfer at the hot junction plate.

The area for heat conduction in the individual thermoelectric elements is A_e. The area ratio, A_r, is equal to A_w/A_e. This is the area parameter used in the development of the conduction model, where A_r is greater than 1.

The relationship between the heat flux at the CJP, q_c, and the temperatures of the hot and cold junctions of the TEG, T_h and T_c, respectively, is written as (Angrist, 1982)

$$q_c \eta A_r (A_p + A_n) = \kappa (T_h - T_c)$$
$$+ (|\alpha_n| + |\alpha_p|) T_c I_{out}$$
$$+ 1/2\, I_{out}^2 R \qquad (4)$$

where the fin efficiency of the CJP, η, is assumed to be unity for convenience, κ is the thermal conductance of a TEG pair,

$$\kappa = (\lambda_p A_p/L_p) + (\lambda_n A_n/L_n),$$

R is the TEG electrical resistance,

$$R = (\rho_n L_n/A_n) + (\rho_p L_p/A_p),$$

A_n is the area of the n-type material, A_p is the area of the p-type material, and L_n, L_p are the lengths of the elements. The electrical current produced by the TEG module, I_{out}, is defined below.

The thermoelectric generator equations will be selected to maximize the thermal efficiency (Culp, 1979; Angrist, 1982) of the module based on the semiconductor material properties and the geometry of the module. Therefore the maximum value for the figure of merit is

$$Z = \frac{(|\alpha_p| + |\alpha_n|)^2}{[(\rho_n \lambda_n)^{1/2} + (\rho_p \lambda_p)^{1/2}]^2}, \qquad (5)$$

where α_p, α_n are the respective Seebeck coefficients, ρ_n, ρ_p are the electrical resistivities and λ_n, λ_p are the thermal conductivities of the materials.

Utilizing the figure of merit, the calculation for the current output of the TEG module is based on optimizing the internal and external resistances of the system (Angrist, 1982). Therefore the equation for the current produced by the module to maximize the thermal efficiency is

$$I_{out} = \frac{(|\alpha_p| + |\alpha_n|)(T_h - T_c)}{R\,[x + 1]}, \qquad (6)$$

where

$$x = [1 + Z(T_h + T_c)/2]^{1/2}.$$

The open circuit voltage for the thermoelectric generator is

$$V_{oc} = (|\alpha_p| + |\alpha_n|)[T_h - T_c]. \qquad (7)$$

The selection of the TEG module will be based on utilizing off-the-shelf or near-off-the-shelf materials. That is, a minimum of modifications to existing tooling will be sought.

Therefore, both n-type and p-type elements are chosen with the same cross-sectional area and length. Typical assembly techniques of copper junctions, ceramic endfaces, etc., are used.

RESULTS

The TEG elements utilized are p- and n-doped bismuth telluride with a 1mm x 1mm square cross-section and a length of 25mm. For these element dimensions, the model is run to optimize the number of junction pairs, based on the power output of the cell. The thermal conductivity is considered constant at 1.35 W/mK, as specified by the manufacturer.

Figure 3 shows the power output of the cell when operated at night as a function of the number of TEG junctions. Twenty-two junctions correspond to a module having 44 TEG elements, the maximum energy output developed for this geometry. With this design, the cell will produce approximately 6.7mW of electrical power with a full vacuum in the cell; 6.3mW for a partial vacuum. Although not shown, the voltages corresponding to these powers are about 0.412 volts at an ambient temperature of 300K and 0.314 volts at 275K.

The model shows that the full vacuum is not justified based on the small improvement in performance when reducing the pressure in the cell below 25 torr.

For the maximum electrical energy production with 22 TEG junctions, the CJP temperature is 253K at 300K ambient temperature and 239K at 275K. These correspond respectively to a 47K and 36K temperature difference across the TEG elements.

Figure 4 illustrates the daily energy produced by the 22-junction module when the device is operated day and night. The daytime performance of the cell is determined, to a large extent, by the number of daylight hours available (summer vs. winter use of the cell) as well as the solar flux (200 W/sq.m, 500 W/sq.m, or 800 W/sq.m). This figure greatly facilitates determining the operation of the cell in a particular geographical location. One only needs to know the expected peak solar flux and number of daylight hours.

For Fig. 4, the nighttime energy production is only a function of the ambient temperature at 300K (summer) or 275K (winter). The daytime energy production is determined from the prescribed solar flux. As shown in the example, for ten hours of daylight in the summer (at 300K ambient), 460 mW-hrs will be produced by the cell. As a comparison, this is about one-fourth the energy produced by a commercially available photovoltaic panel operating over ten hours (C. Crane, 2001), the solar panel obviously not producing any power at night.

Figure 5 shows the temperature of the HJP and the peak power output for the 22-junction cell as a function of incident solar flux for an ambient temperature of 300K. The peak power output indicates the maximum amount of daytime energy that is available from the cell.

DISCUSSION

The 22 TEG junction module design is selected for the prototype. This provides both convenience for the selection of a standard module and utility in the optimum power range of the device. The good performance of the cell with solar heating of the CJP shows the sound practicality of such a power producing device.

The addition of the atmospheric window bands incorporated in the model provides an accurate method for predicting the performance of the cell in various climates.

The period at dawn just when the sun appears on the horizon and in the evening when the sun sets are two transient conditions that must be considered further to determine device performance when the direction of the electric current changes.

Figure 4 can be utilized to predict the performance of the Nighttime Solar Cell[TM] for any location where the solar flux is known. Interpolation between curves provides a simple means for determining device energy production for any locale or sun angle.

CONCLUSIONS

The thermal model shows the performance of the Nighttime Solar Cell[TM] to be satisfactory under several operating conditions and modes of operation, while providing valuable parametric guidelines for the design of a prototype. The electrical power output of the cell, nominally sized at 6cm x 6cm x 3cm, will produce about 7mW of power at night. Four cells connected in series, a 12cm x 12cm panel, will produce about 1.6 volts. This corresponds to a single D-sized battery, with an almost infinite life.

Although the device cannot out-produce a photovoltaic cell in a 24-hour period, the cell still provides sufficient electrical energy production for many applications, with the added advantage of continuous energy production during nighttime operation.

The daytime operation without the use of solar cells can also be achieved successfully. Therefore this new mode of electric power production may be the next source of clean, reliable, safe and inexpensive energy.

ACKNOWLEDGEMENTS

One author would like to acknowledge the efforts of his late son Joseph "Joey" Parise in challenging his father to be more creative and understanding of the laws of nature and their benefits to mankind and the environment.

REFERENCES

Angrist, S.W., 1982, Direct Energy Conversion, Fourth Edition, Allyn and Bacon, Inc., Boston, Massachusetts.

C. Crane Company Catalog, 2001, Radio - Light & Science, 25th Aniversary, No. 12, Fortuna, CA, pp.72-73.

Culp, A.W., Jr., 1979, Principles of Energy Conversion, McGraw-Hill Book Company, New York, NY.

Duffie, J.A. and Beckman, W.A., 1974, Solar Energy Thermal Processes, John Wiley & Sons, New York, NY.

Dunkle, R.V., 1953, Thermal Radiation Tables and Applications, Transaction of the ASME, Paper No. 53-A-220.

Eaton, C.B. and Blum, H.A., 1975, The Use of Moderate Vacuum Environments as a Means of Increasing the Collection Efficiencies and Operating Temperatures of Flat-Plate Solar Collectors, Solar Energy, Vol. 17, p151.

Hudson, R.D., 1969, Infrared System Engineering, Wiley & Sons, New York, NY, pg. 115.

Parise, R.J., 1998, Nighttime Solar Cell™, IECEC98, Colorado Springs, CO, Paper No. IECEC-98-133.

Parise, R.J., Jones, G.F., Strayer, B., 1999, Prototype Nighttime Solar Cell™, Electrical Energy Production from the Night Sky, IECEC99, Vancouver, British Columbia, Canada, Paper No. 1999-01-2566.

Parise, R.J., 1999, "Nighttime Solar Cell", United States Patent No. 5,936,193.

Parise, R.J. and Jones, G.F., 2000, Energy from Deep Space - The Nighttime Solar Cell™ - Electrical Energy Production, IECEC00, Las Vegas, Nevada, Paper No. AIAA-2000-2822.

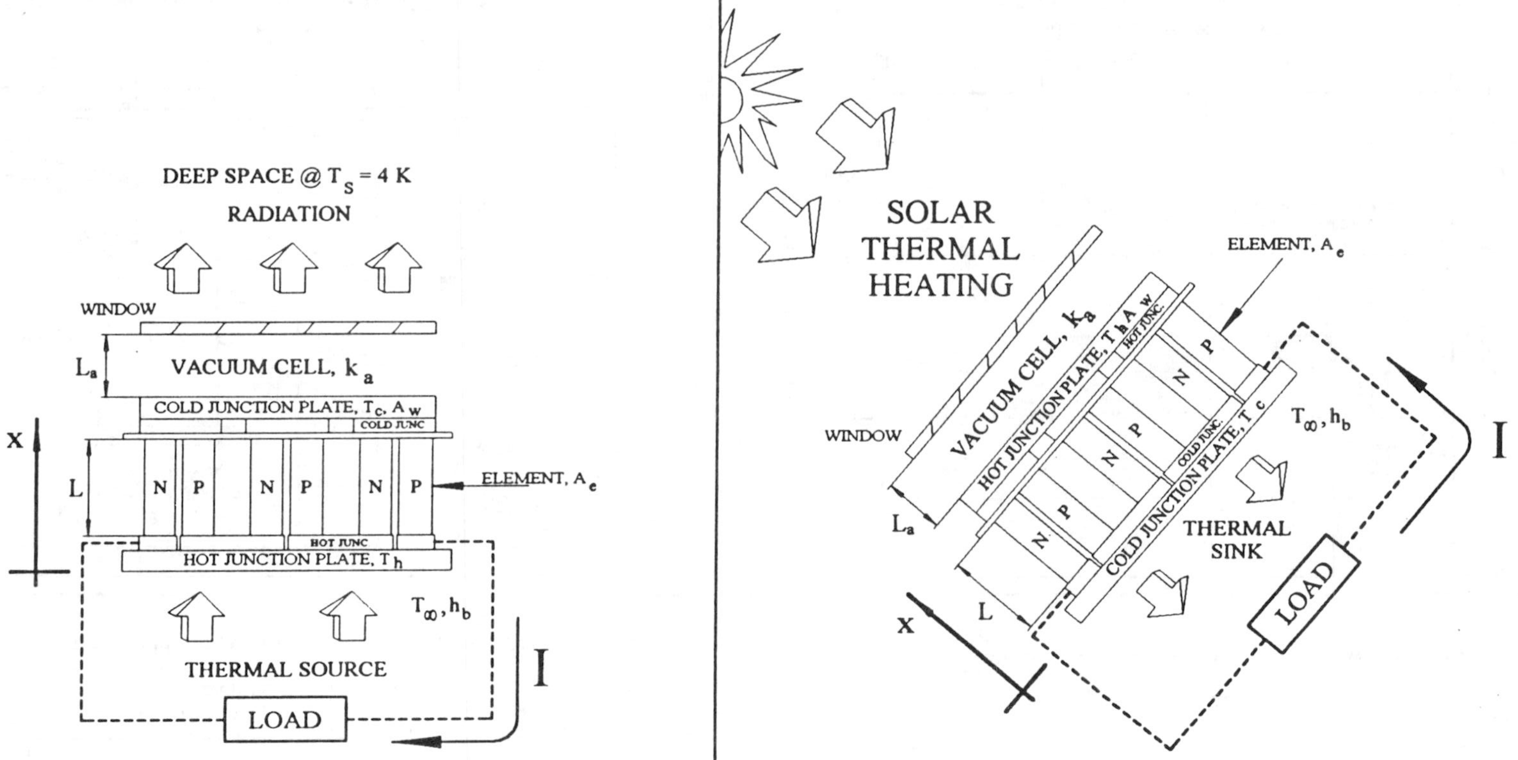

SOLAR THERMAL HEATING

VACUUM CELL, k_a

HOT JUNCTION PLATE, T_h A_w

WINDOW

L_a

x

L

ELEMENT, A_e

COLD JUNCTION PLATE, T_c

N P N P N

HOT JUNC

COLD JUNC

T_∞, h_b

THERMAL SINK

LOAD

I

(b) Daytime Operation
(Normal to Solar Noon
Beam Radiation)

DEEP SPACE @ $T_s = 4$ K
RADIATION

WINDOW

L_a

x

VACUUM CELL, k_a

COLD JUNCTION PLATE, T_c A_w

COLD JUNC

ELEMENT, A_e

N P N P N P

HOT JUNC

HOT JUNCTION PLATE, T_h

L

T_∞, h_b

THERMAL SOURCE

LOAD

I

(a) Nighttime Operation
(Parallel to Horizontal)

Figure 1: Model Configuration.

Parise

531

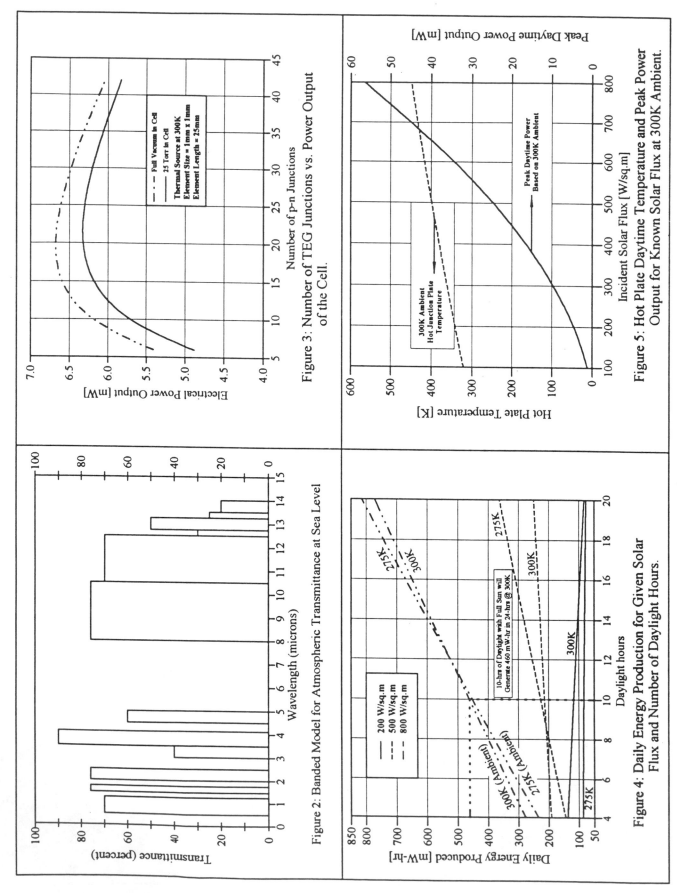

Figure 2: Banded Model for Atmospheric Transmittance at Sea Level

Figure 3: Number of TEG Junctions vs. Power Output of the Cell.

Figure 4: Daily Energy Production for Given Solar Flux and Number of Daylight Hours.

Figure 5: Hot Plate Daytime Temperature and Peak Power Output for Known Solar Flux at 300K Ambient.

532

Proceedings of IECEC'01
36th Intersociety Energy Conversion Engineering Conference
July29- August 2, 2001, Savannah, Georgia

IECEC2001-RE-19

Hybrid PV/Fuel Cell System Design and Simulation

Dr.Thanaa.F.Al-Shater   Dr.Mona.N.Eskandar

Prof. Mohsen.T.El-Hagry

Electronics Research Institute,   National Research Center Bldg.

El-Tahrir Street -Dokki-Giza-Egypt

e.mail : thanaa@eri.sci.eg

ABSTRACT

In this paper, a hybrid Photovoltaic (PV)-fuel cell generation system employing an electrolyzer for hydrogen generation is designed and simulated. The system is applicable for remote areas or isolated loads. Fuzzy regression model (FRM) is applied for maximum power point tracking (MPPT) to extract maximum available solar power from PV arrays under variable insolation conditions. The system incorporates a controller designed to achieve continuous supply power to the load via the PV array or the fuel cell, or both according to the power available from the sun. The simulation results show that the system can run without power shortage for more than four days even in case of zero insolation.

INTRODUCTION

A pilot study to construct a renewable energy system for generating electricity is presented in this paper. It is an extension of a solar generating system that has been erected four years ago, where the maximum power point (MPP) has been calculated using the fuzzy regression model [1]. This extension includes the generation and storage of hydrogen (H_2) to be used in different applications among which is the generation of electricity via fuel cell (FC). This paper includes system simulation and sizing.

This system is designed to overcome the problem of large variances of PV output power under different insolation levels, by integrating the PV power plants with FC through H_2 generation plant. The fuel cell (FC) is an electrochemical device that produces direct current electricity through the reaction of hydrogen and oxygen in the presence of an electrolyte. They are an attractive option for use with intermittent sources of generation, like the PV, because of high efficiency, fast load response, modularity, and fuel flexibility. Their feasibility in coordination with PV systems has been successfully demonstrated for grid-connected applications [2].

The use of electrolysis to produce hydrogen from water proved to be efficient [3]. Additionally, when PV is used with the electrolyzer, it is the cleanest source of hydrogen with no pollutants produced. On the small scale, a PV array coupled to an electrolyzer and H_2 storage tank provides a flexible system, which could be installed in any location with little maintenance.

In this paper a 2.24 kW PV-fuel cell hybrid generation system is designed. An electrolyzer coupled to the PV array is employed for hydrogen production. Maximum power tracking (MPPT) for the PV array is achieved using fuzzy regression. A controller is designed to ensure continuous constant power generation through the day and after sunset via the PV and fuel cell stack. Electrical models for each of the system components are given and integrated with each other. The system is simulated using Matlab software. The system characteristics obtained from simulation are discussed. Similar generation systems to the one proposed in this paper has been conducted before [2,4 -6]. In [2] the performance of a grid connected PV-FC hybrid system was evaluated. Such system differs in the method of analysis and requirements than the stand-alone system proposed in this paper. In [4] the thermal efficiency of a hybrid PV-electrolyzer-FC system was studied. The paper did not study the electrical characteristics of the system. In [5] the operation of a solar regenerative fuel cell test bed facility was described without simulation. In [6] an electrical model of a FC generation system, based on the time constants representation was presented. Such model did not incorporate PV or electrolyzer representation.

SYSTEM DESCRIPTION

The proposed PV-electrolyzer-FC system is shown in Fig.1. The major components of the system are: a polycrystalline PV array Solarex module type MSX-56, a Unipolar Stuart cell electrolyzer, a hydrogen storage tank, proton exchange membrane (PEM) fuel cell stack, and the 72 dc V, 3.1 A load. A control system is employed to monitor the state of the system, and control power and hydrogen flows. The system components are described as follows:

The Unipolar Stuart cell is a high efficiency low maintenance, rugged and reliable cell. Each electrode has a single polarity producing either H_2 (cathode) or O_2 (anode). The electrolyzer consists of a number of cells isolated from one another in separate cell compartments. Cell voltage under

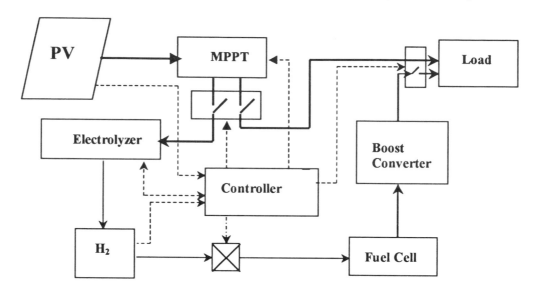

Figure.1 Isolated Hybrid PV-FC Generation System

A. PV Subsystem

The power generation system with a PV array has two application types: a local (isolated) type and an interconnected (grid connected) type. The isolated type system works independent of other power systems, where the local load consumes the electric power from the PV array. The output power of the PV array fluctuates depending on insolation and surface temperature. Hence, a storage system must be used to deliver the required power at low insolation levels and during the night. Fig.2 shows the measured insolation levels (one sun = 1 kw/m²) measured during one year (1996) at El-hammam site (40th Km West of Alexandria – Egypt). The maximum power point (MPP) voltage and current V_m , I_m respectively, are determined on-line using FRM [1]. The fuzzy model input parameters are:

1. The solar insolation incident on the array surface "W".
2. The panel surface temperature "T".

The generated PV power is used to feed the load demand. The power exceeding the load demand is fed to the electrolyzer to generate H_2. The H_2 generated is fed, with the amount of air required, to the FC to generate electrical power.

B. Electrolyzer Subsystem

normal operating conditions is in the range of 1.7-1.9 V_{dc}. Circulation of the electrolyte is facilitated by the H_2 and O_2 gases rising in the channels formed between the respective electrodes and cell separator. The operating temperature of the electrolyzer does not exceed $70^o C$, thus reducing the material constraints. H_2 is directly produced at 99.9% purity. Also the current efficiency is 100%, and hence the hydrogen production rate in Mole/Sec is:

$$X_{H_2} = 5.18 \ e^{-6} \quad I_e \tag{1}$$

Where I_e is the current between electrodes, H_2 is stored at 3 bar in a tank to feed the FC at low insolation levels and hence supply the required load power. An important factor affects the electrolysis process, and should be considered in the system design. This factor takes place after sunset when the electrolyzer current drops to zero, which means that the electrolyzer must be kept under protective voltage in order to prevent the cathodic elements from being excessively attacked by active corrosion. To overcome this defect, the proposed electric storage device is designed to isolate the electrolyte from the electrolysis cell and inject N_2 to the electrolyzer to protect electrodes from corrosion.

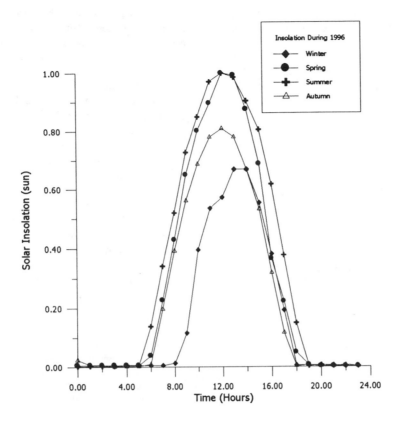

Fig.2 Insolation against Time

C. Fuel Cell Subsystem

Proton Exchange Membrane (PEM) fuel cells stack is used. The PEM uses a polymer membrane as its electrolyte. With such solid polymer electrolyte, electrolyte loss is not an issue with regard to stack life. H_2 produced by the electrolyzer is consumed at the anode, yielding electrons at the anode and producing H_2 ion, which enter the electrolyte. At the cathode, O_2 combines with H_2 ions to produce water, which is rejected from the back of the cathode [7].

In the proposed system, air is used as the oxidant; and cell temperature is $70^{\circ}C$. Current density is designed as 400 mA/cm^2. This leads to a 7.75 cm^2 fuel cell area. At atmospheric pressure, Nernst equation relates the electrical performance of the FC to the state variables [8]:

$$V_o = E_o - \frac{R.T}{zF} \ln \frac{X_{h_2} X_{O_2}^{0.5}}{X_{h_2O}} \qquad (2)$$

Where

V_o = Open circuit reversible cell potential (volt)

E_o = Standard reversible cell potential (volt)

R = Thermo Gas Constant (KJ/Kmol. °K)

T = Absolute Temperature (°K)

z = Charge of the ion

F = Faradays constant (96,485 C/mol)

X_i = Mole fractions of species (g mole)

Under load, cell voltage is affected by ohmic loss. It is also affected by anode polarization, cathode polarization, and temperature. Neglecting polarization losses, the cell voltage under load is:

$$V = V_o - i A R_c - b \log(iA) \qquad (3)$$

Where

i = current density (A/cm^2)
A = cell area (cm^2)
R_c = ohmic resistance (ohm)
b = Tafel Slope

D. The Controller

A controller is designed to control the whole process. The controller functions could be summarized as follows:

(i)   Measuring the insolation and temperature of the PV to determine the point of maximum power and issue commands to the MPPT.

(ii)   Monitoring the state of electrolyzer and issue commands to fill or drain it.

535

(iii)   Monitoring the state of hydrogen tank to estimate the amount of H$_2$ stored and controls the hydrogen valve operation.

(iv)   Moreover it controls the flow of power from PV system (PV + MPPT) to the load and the electrolyzer, as well as the power flow from FC according to the control plan and load demands.

SIMULATION RESULTS

The proposed system is represented by the Simulink toolbox under the Matlab. The output simulation results are shown in Figs 3 and 4. Fig.3 shows the PV output power during the year. This power is calculated using FRM to achieve MPPT. The advantage of FRM is that it does not need extensive calculation or expensive equipment, where only insolation and temperature measurements are needed. These values are used with FRM to determine the MPP and Voltage-Current curve of the PV array. The output from the PV array varies from season to another as shown depending on the insolation level and array surface temperature.

P$_L$ is higher than available PV average power (P$_{av}$). This is given as a controller signal, which allows generation of, needed power via the FC. From sizing equations Egido, M.A. [9], the load capacity that can be supplied from the PV is about 8 to 10 % of the PV maximum power. Therefore, the load capacity is calculated to be 224 w. To achieve the required load voltage of 72 v$_{dc}$, the numbers of the fuel cells needed are 103, connected in series.

Table 1 shows the average insolation hours during different seasons, the panel surface temperature, average insolation levels, the average generated PV power, the H$_2$ generated during insolation hours considering load demand and the equivalent reserve stored energy in kw. The difference between the P$_{av}$ and P$_L$ is used to generate H$_2$. To prove that the proposed system allows the continuous flow of power to the load, the following variables are calculated and indicated in Table1:

1.   The average power generated from the PV array (P$_{av}$).
2.   The PV power exceeding the load demand.
3.   The hydrogen generated using this excess power.

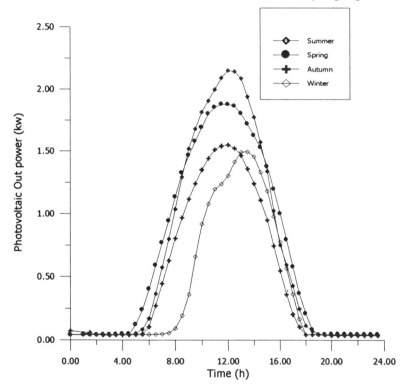

Figure.3 Photovoltaic Output Power against Time

Fig.4 shows the output voltage of the fuel cell against the current density and these data are used to control the FC operation. The output power from the FC is independent on the fluctuations in the insolation level. This is due to the hydrogen storage strategy. The H$_2$ storage strategy is as follows: H$_2$ is generated utilizing the difference between generated power from PV and load demand (P$_L$) during useful insolation hours. It is generated by electrolyzer and stored in the tank until the

The equivalent reserve which is the power exceeding the load demand. This is calculated as follows:

i)   Calculate the total generated PV power during insolation hours.
ii)   Calculate the total load consumption during insolation hours.
iii)   Calculate the excess power stored as hydrogen.

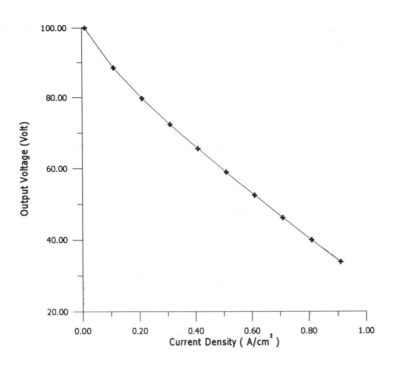

Figure.4 Output voltage against Current Density for PEMFC

iv) Calculate the power generated by the FC at 60% efficiency.

v) Calculate the load power required during zero insolation levels.

vi) The difference between (iv) and (v) is equivalent to the reserve storage power.

The simulation results calculated as described above show that the stored power during winter is equivalent to 0.75 kw which increases to 4.5 kw during summer. In winter the excess power will be useful to supply power during low insolation periods, which may extend to three to four days maximum. In summer the excess hydrogen can be used to supply extra loads such as ventilation or air conditioning.

CONCLUSION

A stand-alone PV-Electrolyzer-Fuel Cell hybrid generation system is designed, sized, and simulated. Fuzzy regression is used for maximum power tracking of the PV array. Electrical integration of the system components is done using Matlab software. A controller is designed to ensure fulfilling the load demands throughout the day. It also allows protection of the electrolyzer electrodes at low insolation levels. The electrolyzer is designed to generate H_2 during the periods when the PV power exceeds the load demands. The generated H_2 is stored in a tank to be utilized at low insolation levels or at night. It is fed to the FC to generate the load power demand. Simulation results proved that the load demand could be supplied continuously up to four days of low insolation.

The feasibility of the system proposed in this paper has been proved. It is a pilot system, which could be extended for higher power ratings.

REFERENCES

[1] T.F.Elshatter, M.T.Elhagree, M.E.Aboueldahab and A.A.Elkousy," Fuzzy Modelling and Simulation of Photovoltaic System", 14th European Photovoltaic Solar Energy Conference, Barcelona, Spain, 30 June-4 July 1997.

[2] Kyoungsoo Ro and Saifur Rahman, "Two-Loop Controller for Maximizing Performance of a Grid-Connected Photovoltaic-Fuel Cell Hybrid Power Plant", IEEE Trans. On Energy Conversion, Vol.13, No.3, Sept.1998, pp.276-281.

[3] Krishna Sapru, Ned T. Stetson, Stanford R.Ovshinsky, Jeffery Yang, Gred Fritz, Matthew Fairlie, and Andrew T.B.Stuart " Development of A small Scale Hydrogen Production-Storage System for Hydrogen Applications", IECEC's 97, pp.1947-1952.

[4] J.Divisek and M. Schwuger, "Hybrid System Solar Powered Water Electrolysis/ Fuel Cells', Renewable Energy Sources Conference,Vol.2, Cairo, Egypt, 1990,681-688.

[5]G.Voecks,N.Rohatgi,D. Jan, S. Moore, M.Warshay, P. Prokopius, H.Edwards, and G.Smith, "Operation of the 25 kW Nasa Lewis Research Center Solar Regenerative Fuel Cell Testbed Facility", IECEC 97 proceedings, pp. 1543-1549.

[6].Yoon-Ho Kim, and Sang-Sun Kim, "An Electrical Modeling and Fuzzy Logic Control of a Fuel Cell Generation System", IEEE Trans. On Energy Conversion, Vol.14, No.2, June 1999, pp.239-244. [7] John H.Hirschenhofer," Fuel Cell Status", IECEC 's 1995, pp.165-170.

[8] Michael D. Lukas, Kwang Y.Lee and Hossein Ghezel-Ayagh," Development of a stack Simulation Model for Control Study on Direct Reforming Molten Carbonate Fuel cell Power Plant", IEEE Trans. On Energy Conversion, Vol.14, No. 4, December 1999, pp.1651-1657.

[9] Egido,M.A. and Lorenzo,E., 1992, " The Sizing of Stand alone PV Systems: a review and a proposed new method", Solar Energy Materials and Solar Cells 26, pp.51-69.

Table 1 Simulation results

	Insolation hrs (hr)	T_{av} (°C)	W_{av} (kw/m^2)	P_{av} (kw/S.H)	H_2 (gmole/hr)	Equivalent reserve (kw)
Winter	8	18.5	0.6	1.1308	29.349	0.75
Autumn	8	45	0.688	1.2207	33.249	1.9
Spring	10	30	0.79	1.4157	39.754	3.8
Summer	10	50	0.856	1.55	44.234	4.6

Proceedings of IECEC'01
36th Intersociety Energy Conversion Engineering Conference
July 29-August 2, 2001, Savannah, Georgia

IECEC2001-RE-05

FUNDAMENTAL STUDY OF SOLAR THERMAL CELL
-AN INFLUENCE OF THE INTERNAL STRUCTURE AND TEMPERATURE ON THE OUTPUT OF THE CELL-

ANDO Yuji and TANAKA Tadayoshi

R. I. of Energy Utilization
National Institute of Advanced Industrial Science and Technology (AIST)
16-1, Onogawa, Tsukuba, Ibaraki 305-8569, JAPAN

ABSTRACT

Solar thermal cell converts low temperature solar thermal energy around 100 °C into electric power. It consists of chemical reactions of 2-propanol dehydrogenation and acetone hydrogenation, and a fuel cell. 2-propanol dehydrogenation proceeds in a reactor with an appropriate catalyst and acetone and hydrogen gas are produced. Acetone and hydrogen gas are supplied to the positive and negative electrode of the fuel cell, respectively. Through the process of acetone hydrogenation at the fuel cell, electricity is generated. In this study, we studied about the influence of the internal structure of the cell and the effect of the temperature in the cell on the output of electricity.

As the main results, (1) We kept the temperature of negative electrode at 60 °C, and increased the temperature of positive electrode from 51 °C to 57 °C. As the temperature in the cell rises, the short circuit current of the cell increases from 24 mA (51 °C) to 39 mA (57 °C). (2) We changed the thickness of the electrode. The thinner the electrode thickness is, the larger the output of the cell becomes. The reason is supposed that most of the reactant (hydrogen gas and acetone) can easily reach the interface of electrode and electrolyte in the cell when thickness of the electrode is thin. (3) When the electrodes are kept in good contact with the polymer electrolyte, the performance of the cell was improved and short circuit current was reached 80 mA.

INTRODUCTION

In recent years, environmental destruction such as forest destruction by acid rain and global warming, which is caused by consumption of large amounts of fossil fuels, has become a very serious problem. Therefore, from the viewpoint of environmental protection, we need to make better use of fossil energy supplies and use clean energy such as solar energy.

Solar energy is clean and inexhaustible energy source and a lot of studies have been conducted. In Japan, it is not effective to operate high temperature solar thermal system above 300 °C because the direct insolation is poor. However, it is not difficult to get low temperature solar thermal energy about 100 °C.

From above point of view, we have proposed "solar thermal cell", which converts low temperature solar thermal energy about 100 °C into electric power and we have conducted fundamental studies about it [1]. It consists of chemical reactions of 2-propanol dehydrogenation (Eq. (1)) and acetone hydrogenation (Eq. (3)) and a fuel cell.

$$(CH_3)_2CHOH \, [l] \rightarrow (CH_3)_2CO \, [l] + H_2 \, [g] \qquad (1)$$
$$\text{2-propanol} \qquad \text{Acetone} \quad \text{Hydrogen}$$
On the negative electrode;
$$H_2 \, [g] \rightarrow 2H^+ + 2e^- \qquad (2)$$
$$\text{Proton} \quad \text{Electron}$$
On the positive electrode;
$$(CH_3)_2CO \, [l] + 2H^+ + 2e^- \rightarrow (CH_3)_2CHOH \, [l] \qquad (3)$$

We have the two concepts for the configuration. One is liquid-phase solar thermal cell and the other is gas-phase solar thermal cell.

Liquid-phase solar thermal cell

Figure 1 shows the concept of liquid-phase solar thermal cell. It is a kind of fuel cell. This cell has the characteristic with non-uniform temperature in the cell. The temperature of both electrodes in the cell is different from each other. The negative electrode is heated by the sun about 100 °C and the positive electrode is cooled by water. 2-propanol is decomposed into acetone and hydrogen, and electron is induced on the negative electrode with an appropriate catalyst (Eq. (1) and Eq. (2)). The proton moves to the positive electrode through the electrolyte and the electron moves to the positive electrode through the outside electric circuit. The acetone is desorbed from the negative electrode and supplied to the positive electrode. The acetone, proton and electron regenerate 2-propanol on the positive electrode with an appropriate catalyst (Eq. (3)) and 2-propanol is supplied to the negative electrode again. Thus solar heat about 100 °C is converted into

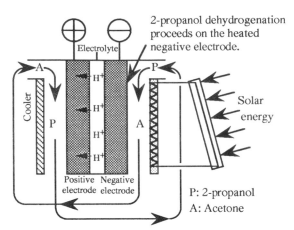

Fig. 1 Liquid-phase Solar Thermal Cell

Hydrogen gas is decomposed into proton and electron on the negative electrode.

Fuel cell

2-propanol dehydrogenation reactor

P: 2-propanol
A: Acetone
H: Hydrogen gas
Fig. 2 Gas-phase Solar Thermal Cell

electric power. "Liquid-phase" is named by the reason that liquid-phase 2-propanol is supplied to the negative electrode. The electromotive force (e.m.f.) of the cell is decided by reaction temperature, catalytic activity, and so on. The e.m.f. of the cell is about 20 mV at 60 °C [2].

Gas-phase solar thermal cell

Figure 2 shows the concept of gas-phase solar thermal cell. It is composed of 2-propanol dehydrogenation reactor and fuel cell. So, its system composition is more complex than that of liquid-phase solar thermal cell. However, temperature in the fuel cell is uniform. So, temperature control is easy. 2-propanol dehydrogenation proceeds in the reactor that is heated by the sun, and acetone and hydrogen gas are produced. Hydrogen gas and acetone is supplied to the negative electrode and the positive electrode of the fuel cell, respectively. Through the process of acetone hydrogenation, electric power is produced at the fuel cell. 2-propanol, which is reaction product, is supplied to 2-propanol dehydrogenation reactor again. "Gas-phase" is named by the reason that hydrogen gas is supplied to the negative electrode. The e.m.f. of the cell is decided by Gibbs-free-energy. So, reaction temperature affects the e.m.f. However, e.m.f. isn't affected by catalytic activity. The e.m.f. of the cell is about 120 mV at 60 °C.

Characteristics of gas-phase and liquid-phase solar thermal cell are summarized in Table 1. As shown in the table, e.m.f. of the gas-phase cell is larger than that of the liquid-phase cell. Therefore, in this study, in order to enhance the efficiency of the cell, we focused on the fuel cell of gas-phase solar thermal cell and examined the effect of the internal structure and temperature on the output of the solar thermal cell.

Table 1 Characteristics of gas-phase and liquid-phase solar thermal cell

	system composition	temperature control	electromotive force
Gas-phase	complex	easy	120 mV
Liquid-phase	simple	difficult	20 mV

EXPERIMENTAL EQUIPMENTS

As an electrode material, a carbon in the form of felt (BET specific surface area: 2500 m²/g, Kuraray Co.; we refer to it as carbon-felt hereafter) and a carbon in the form of cloth (BET specific surface area: 2500 m²/g, Kuraray Co.; we refer to it as carbon-cloth hereafter) were adopted. They were impregnated with aqueous solutions of $RuCl_3$ and/or K_2PtCl_4 at room temperature. The adsorbed metal salts were reduced in the atmosphere of hydrogen gas for 12 hours at 300 °C. Then they were washed with fresh water and were dried up for 10 hours at

50 °C. The ruthenium and platinum supported on carbon-felt and carbon-cloth were used in this study.

Figure 3 shows a cross section of reaction apparatus. The size of electrode area was 2x2 cm^2 and the reaction apparatus was heated with tape heater instead of solar heat. Temperature in the apparatus was measured by a thermocouple and the inside of the apparatus was kept at a fixed temperature with using temperature control system. Nafion® (Du Pont Co.) was used for electrolyte, which is one of polymer electrolytes. Acetone and 2-propanol was mixed with an equal amount of each. The mixture was diluted 10 times with water and supplied to positive electrode. Acetone flow rate was kept at 1.4 mmol/min. Hydrogen gas was supplied to negative electrode at the rate of 4.5 mmol/min. The electric output of the cell was measured with DC voltage current source/monitor (Advantest Co.).

In this study, we examined the effect of the internal structure and temperature on the output of the cell. Figure 4 is the conceptual drawing of the structural difference, which we examined in this study. T1 and T2 are thickness of Teflon® sheet and thickness of an electrode, respectively. From structural restriction of the reaction apparatus, the total thickness of T1 and T2 is constant (8mm). When T1 is small,

few of fuel reaches the interface of electrode and electrolyte (Fig. 4(a)). As T1 becomes larger, most of the fuel reaches the interface (Fig. 4(b, c)). It is supposed that the internal resistance of the cell becomes smaller when most of chemical reaction indicated by Eq. (1), (2) and (3) proceeds around the interface. So, it is supposed that the output of the cell will be improved when the structure of the cell is one shown in Fig. 4(c).

It is also supposed that the internal resistance of the cell becomes smaller when the electrodes are kept in good contact with the electrolyte. The thickness of carbon-felt and carbon-cloth is about 1mm. However, they are so soft that they are easily crushed. The number of carbon-felt or carbon-cloth per 1 millimeter in an electrode is found by the following equation;

$$s = n / T2 \qquad (4)$$

where "s" is the number of carbon-felt or carbon-cloth per 1 millimeter in an electrode and "n" is the number of carbon-felt or carbon-cloth in an electrode. It means that an electrode area is packed tightly with carbon-felt or carbon-cloth when the value of "s" is large. So, it is supposed that the performance of the cell will be improved under the condition of large "s".

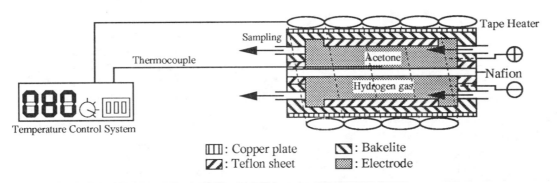

: Copper plate   : Bakelite
: Teflon sheet   : Electrode

Fig. 3   Cross section of reaction apparatus

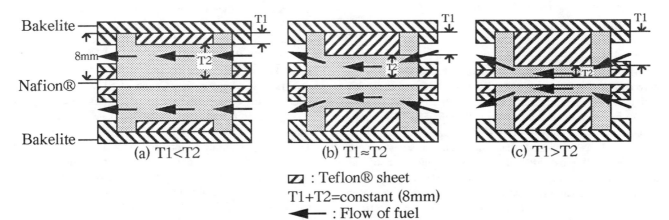

(a) T1<T2   (b) T1≈T2   (c) T1>T2

: Teflon® sheet
T1+T2=constant (8mm)
◄— : Flow of fuel

Fig. 4   Structural difference of the cell and flow of fuel

RESULTS AND DISCUSSION

Figure 5 shows an effect of reaction temperature around positive electrode on I-V characteristics of the solar thermal cell. Temperature around negative electrode was kept at 60 °C and temperature around positive electrode was increased from 51 °C to 57 °C. As temperature around positive electrode increases, the output of the cell also increased. The temperature dependence of the output of the cell was rather large and the short-circuit current was changed from 24mA (51.1 °C) to 39mA (56.9 °C). The reason is supposed that both the chemical reaction rates, which are shown in Eq. (1), (2) and (3), and a proton conductivity of polymer electrolytes, are improved with increasing the temperature in the cell.

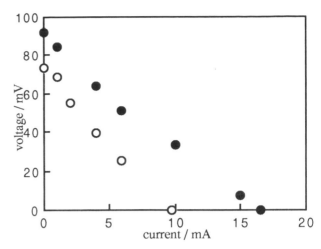

Catalyst : 30 wt% Ru-Pt / carbon-cloth
Temperature around negative electrode : 60℃
Temperature around positive electrode : 50.5-51.5℃
○ : T2=8mm, ● : T2=1mm

Fig. 6 Effect of thickness of an
electrode on I-V characteristics
of the solar thermal cell

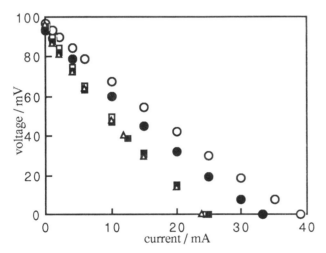

Catalyst : 30 wt% Ru-Pt / carbon-cloth
Temperature around negative electrode : 60℃
Temperature around positive electrode
○ : 56.9℃, ● : 55.3℃, □ : 51.7℃,
■ : 51.4℃, △ : 51.1℃
s=2.0, T2=3 mm

Fig. 5 Effect of reaction temperature
around positive electrode on I-V
characteristics of the solar thermal cell

Figure 6 shows an effect of thickness of an electrode on I-V characteristics of the solar thermal cell. Temperature around negative and positive electrode was kept at 60 °C and at 51 °C, respectively. As the thickness of an electrode was changed from 8mm to 1mm, the output of the cell increased. The short-circuit current was changed from 10mA (T2=8mm) to 17mA (T2=1mm). It is supposed that when the thickness of an electrode is small most of fuel reaches the interface of the electrode and polymer electrolyte, and chemical reactions shown in Eq. (1), (2) and (3) mainly proceeds there. The proton movement from negative electrode to positive electrode becomes easier in this placement, so the output of the cell increased.

Figure 7 shows an effect of "s", which is the number of carbon-felt or carbon-cloth per 1 millimeter, on I-V characteristics of the solar thermal cell. Temperature around negative and positive electrode was kept at 60 °C and at 51 °C, respectively. Figure 7(a) shows the results with 30wt% ruthenium and platinum composite catalyst supported on carbon-cloth. The thickness of an electrode was 8mm, and the number of carbon-cloth per 1 millimeter was changed from 1.5 to 2.0. Figure 7(b) shows the results with using carbon-felt as an electrode material. The thickness of an electrode was 1mm, and "s" was changed from 2.0 to 3.0. In these conditions, as the value of "s" increases, the output of the cell increased. It is supposed that the electrodes are kept in good contact with the polymer electrolyte on the condition of large "s". The internal resistance of the cell becomes smaller in this situation, so the output of the cell increased.

Figure 8 shows an effect of "s" on I-V characteristics of the solar thermal cell. Temperature around negative and positive electrode was kept at 60 °C and at 51 °C, respectively. The thickness of an electrode was 1mm. As an electrode material, 30wt% ruthenium and platinum composite catalyst supported on carbon-cloth was used, and "s" was changed from 1.0 to 4.0. The output of the cell increased with increase in value of "s". The short-circuit current reached 80mA and the maximum electric power reached 1.8mW (Fig. 9) when the value of "s" is 4.0. This value is the highest at present. If an electrode is packed more tightly with carbon-felt or carbon-cloth, the output of the cell is supposed to be improved.

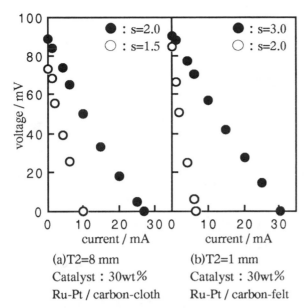

(a)T2=8 mm
Catalyst : 30wt%
Ru-Pt / carbon-cloth

(b)T2=1 mm
Catalyst : 30wt%
Ru-Pt / carbon-felt

Temperature around negative electrode : 60℃
Temperature around positive electrode : 50.5-51.5℃

Fig. 7 Effect of "s" on I-V
characteristics of solar thermal cell

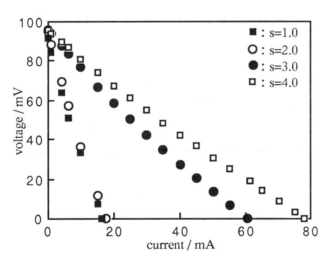

Catalyst : 30wt% Ru-Pt / carbon-cloth
Temperature around negative electrode : 60℃
Temperature around positive electrode : 50.5-51.5℃
T2=1 mm

Fig. 8 Effect of "s" on I-V
characteristics of solar thermal cell

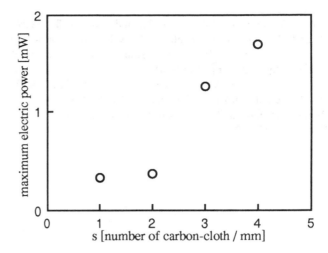

Fig. 9 Relation between "s"
and maximum electric power

CONCLUSIONS

We have proposed "solar thermal cell", which converts low temperature solar thermal energy about 100 °C into electric power with using a fuel cell and chemical reactions of 2-propanol dehydrogenation and acetone hydrogenation.

In this study, in order to enhance the efficiency of the cell, we focused on the fuel cell of gas-phase solar thermal cell and examined the effect of the internal structure and temperature on the output of the solar thermal cell. The following results were obtained from the experiments.

(1) We kept the temperature of negative electrode at 60 °C, and increased the temperature of positive electrode. As the temperature in the cell rises, the short circuit current of the cell increased. The reason is supposed that both the chemical reaction rates and a proton conductivity of polymer electrolytes are improved with increasing the temperature in the cell.

(2) We changed the thickness of the electrode. The thinner the electrode thickness is, the larger the output of the cell becomes. The reason is supposed that when the thickness of an electrode is small, most of fuel reaches the interface of the electrode and polymer electrolyte, and chemical reactions mainly proceeds there. The proton movement from negative electrode to positive electrode becomes easier in this situation.

(3) When the electrodes are kept in good contact with the polymer electrolyte, the performance of the cell was improved and short circuit current was reached 80mA. The reason is supposed that internal resistance of the cell becomes smaller in this situation.

ACKNOWLEDGMENTS

The authors would like to recognize the contributions to this effort by Mr. KAMINAGA Masayuki (Shibaura Institute of Technology).

REFERENCES

1. Ando, Y., Doi, T., Takashima, T., and Tanaka, T., 1997, "Proposal and Fundamental Analysis of Thermally Regenerative Fuel Cell Utilizing Solar Heat," Proceedings of the 32nd ICECE, pp. 1860-1864.
2. Ando, Y., Tanaka, T., and Takashima, T., 1999, "Fundamental Study on Solar Thermal Cell," Proceedings of Renewable and advanced Energy Systems for the 21st Century, RAES99-7725.

Proceedings of IECEC'01
36th Intersociety Energy Conversion Engineering Conference:
Energy Technologies Beyond Traditional Boundaries
July 29 – August 2, 2001, Savannah, Georgia

Paper Number 2001-RE-15

CONVERSION OF SOLAR THERMAL ENERGY: PRACTICAL ENHANCEMENTS AND ENVIRONMENTAL CONSEQUENCES

Ted E. Larsen
Industrial Solar Technology Corporation

E. Kenneth May
Industrial Solar Technology Corporation

ABSTRACT

The technical characteristics of line-focus parabolic trough concentrators (PTC) are described. Major trough developments resulted from the US Department of Energy programs circa 1980. However, during the period since 1980, PTC technology has been proven and major commercialization efforts have taken place. Luz installed over 350 MW of solar thermal power from 1984-1990. Industrial Solar Technology (IST) is the sole surviving US company manufacturing solar PT concentrators and has installed systems from 1985 through the present. The company continues to break new ground to reduce the cost and improve the performance of PTC systems. Examples of PTC applications illustrate the flexibility of the technology to provide heat for commercial and industrial applications, and for electricity generation. PTC systems are already competitive with heat from electricity and liquid fuels, and are approaching parity with natural gas. The current energy crisis appears more permanent than others in the past as energy demand has caught up with supply. At the same time, the potential catastrophic results of global warming are gaining increasing public attention. Parabolic trough technology is an economic alternative to fossil fuels and can have a major impact worldwide in reversing the emission of air pollutants and greenhouse gases.

Keywords: solar energy, parabolic troughs, global warming

INTRODUCTION

Parabolic troughs are a line-focus solar concentrator technology. The troughs are oriented to the sun and reflect parallel beams of light onto a linear receiver located at the focal point of the parabola [Figure 1]. For thermal applications, the receiver is a pipe through which a working fluid is circulated to transfer solar heat to the load. The concentration ratio (ratio of aperture width to receiver diameter) of a parabolic trough concentrator (PTC) ranges from about 30 to 100.

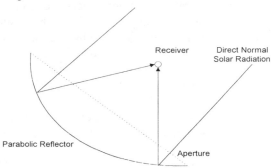

Figure 1: Line-Focus Parabolic Concentrator

It is the ability of parabolic troughs to concentrate solar energy that allows the technology to efficiently and cost-effectively serve thermal loads ranging from 28 C (heating a swimming pool) to 390 C (generating high pressure steam for power production). However, PTC's must continually track the sun to an accuracy of a fraction of a degree. The tracking requirement means that PTC's are most suitable for larger applications ranging from electricity generating systems delivering 50 kW to 160 MW, to industrial applications of no less than 1000 m^2 collector aperture area (600 kW$_{Thermal}$ output at peak conditions), down to commercial applications as small as about 80 m^2 of aperture area (50 kW$_{Thermal}$ peak).

A PTC is typically fabricated from steel or aluminum. Various reflective surfaces are available, including silvered glass, aluminized and silvered plastic films and polished aluminum.

The amount of solar radiation reflected to the receiver [Figure 2] is a function of the optical properties of the concentrator, mainly: the reflectance of the mirror, the optical intercept that determines how much of the reflected light is incident on the receiver, and the transmittance of the glass envelope. Heat loss is a function of the thermal properties of the receiver: the absorptance and emittance of the selective coating applied to the absorber, and whether the annular space between the absorber and the glass envelope is air-filled or evacuated. Increasing the operating temperature of a PTC increases thermal losses from the receiver and hence reduces the efficiency of solar energy collection.

Figure 2: Receiver of PTC

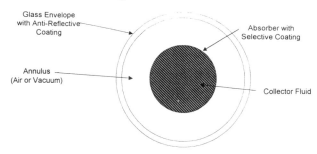

Since heat loss relative to incident solar energy is a function of concentration ratio, the challenge for solar designers is to build concentrators as optically accurate as possible, but at a cost that is economically affordable. The results of two approaches to building PTC's are illustrated in Figure 3. This shows efficiency curves as a function of operating temperature for the IST-PT trough built by IST and for the LS-2 trough built by Luz. Performance equations, produced by Sandia National Laboratories [Dudley and Evans 1995], are illustrated to show that thermal performance is optical efficiency, modified by the angular relationship of the trough to the sun, minus heat loss.

Figure 3. Comparison of Performance of IST-PT and LS-2

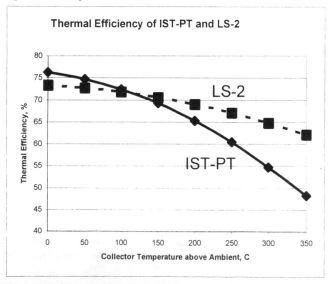

The efficiency of the IST-PT is:
$$\eta = K [76.25 - 0.006836(\Delta T)] - 14.68(\Delta T)/I - 0.1672(\Delta T^2)/I$$
$$K = \cos(Ia) + 0.0003178(Ia) - 0.00003985(Ia)^2$$
Where:

η = Collector efficiency based on a net aperture area of 13.2 m^2, %

ΔT = Average receiver fluid temperature above ambient air temperature, $^\circ$C

K = Incident angle modifier

I = Incident direct normal insolation, W/m^2
Ia = Solar beam incident angle, degrees

The LS-2 concentrator is designed for deployment in utility-scale solar-electric plants. Hence, individual modules are large, measuring 5 m across the aperture and 8 m in length, and its design optimizes operation at temperatures around 370 C. The LS-2 has a concentration ratio of about 70, the receiver is evacuated and the absorber selective coating favors low emittance. In comparison, the IST-PT is targeted at commercial and industrial heating markets, and smaller-scale power plants. Modules measure 2.3 m across the aperture and 6.1 m long, and the design is optimized to deliver heat at temperatures less than 300 C. The IST-PT has a concentration ratio of about 45; it has less shading of the absorber than the LS-2 and a selective surface with higher absorptance so it performs better at lower temperatures

HISTORICAL PERSPECTIVE

Parabolic trough technology has a long history. The most extensive early work was undertaken by Shuman and Boys in Egypt in 1912. They installed 1,200 m^2 of solar collectors to power a 45 kW steam pumping plant. However, the onset of WWI and the discovery of cheap and plentiful supplies of oil curtailed further technical development for many years.

The energy crises in the 1970's led President Nixon to call for US energy independence. However, the next step in the development of PTC's occurred mainly during the Carter era.

DOE PROGRAM CIRCA 1980

This period laid the foundation on which most future developments in PTC technology were based. The U.S. National Laboratories (Sandia in Albuquerque and the National Renewable Energy Laboratory—NREL, in Golden, CO) were active in conducting R&D and managing demonstrations of the technology. Private companies entered the field to meet government demands for technology demonstrations and to join an industry seeking to provide alternatives to conventional fuels.

Some major developments drew on the work of early solar pioneers or transferred products from other fields. The metal finishing industry used research in the field of selective surfaces to produce coatings, such as black chrome and blackened nickel. The lighting industry provided polished aluminum and reflective films adapted to outdoor use as PTC reflectors. Honeywell developed a reliable tracking controller and Corning produced glass tubes with an anti-reflective coating.

The National Laboratories produced developments with immediate practical applications, such as:

    Reinforced concrete foundation designs
    Wind tunnel test data
    Flexible hose designs and testing procedures
    Anti-reflective coatings for glass tubes
    Black chrome with higher temperature capability
    Facilities for PTC performance evaluation

Facilities for accelerated and long-term testing of optical materials

Measurements of heat loss from piping components

The Labs worked with industry to produce new concentrator designs and components. Concentrators were produced from pressed sheet metal, from sagged glass, and by techniques using honeycomb materials and exterior rib supports. Thin and thick glass reflectors were tested along with polished aluminum and reflective films. 3M worked with NREL to develop a high-reflectance, silvered reflective film. Industry produced examples of rotating joints and evacuated tube receivers.

In the drive for energy independence, the DOE goal was to incorporate these technological developments into billions of dollars worth of product that would target the industrial market for thermal energy. EA Mueller [1990] estimated this market at about 6 Quads (10^{15} Btu) for supply temperatures less than 300 C. The electric market was not included in this figure, since conventional wisdom at that time was that other solar technologies: power towers and parabolic dishes, were more suitable for electric power generation. Under the DOE plan, PTC and other renewable energy technologies would be demonstrated and made market ready largely at government expense and then commercialized by industry. The pull of the market was stimulated by tax incentives at the federal and state levels, as well as by projections of continually rising costs for conventional energy.

Numerous demonstrations of PTC technology were undertaken. The last two demonstrations were very large for the time at 5,000 m^2 of collector aperture area and produced 150-psig steam for process use. Much was learned from these demonstration programs, but by and large they were technical and economic failures. All the companies involved left the field, and with the exception of a system built by Suntec in 1978 at the Yuma Proving Grounds in Arizona, none of the demonstration systems are currently operating. In the rush for deployment, too little attention was placed on system reliability and ease of maintenance. The goal of maximum performance was often pursued with little regard for cost. Changing political priorities during the Reagan years, lack of government funding, falling energy prices and the removal of financial incentives almost eradicated the concentrator industry.

CIRCA 1980 – 2000: A PERIOD OF PROGRESS

Industrial Solar Technology from its founding in 1983 and Luz, during a period of explosive growth from 1984 until its bankruptcy in 1992, permanently changed the perception of PTC technology. IST proved that solar parabolic trough technology could be competitive with conventional fuels in niche markets, and demonstrated that large fields of parabolic trough concentrators could be operated in an unattended fashion with a very high degree of reliability. The question was no longer, "Does parabolic trough technology work", but "Is it economic for my application?"

Luz proved that troughs were the most advanced solar thermal technology for the production of utility scale power and that major corporations were willing to invest hundreds of millions of dollars to build over 350 MW of PTC generating plants in the Mojave desert of California.

Engineers Ken May and Randy Gee started IST with the goal of using the lessons learned from the DOE trough program to develop a new generation of trough technology. The guiding principles of this effort were cost-effectiveness, simplicity and reliability.

The most important contribution to the attainment of IST's goals was the use of a patented structural concept to build the parabolic concentrator. Torsional wind loads are the defining criteria in designing a PTC. Previous designs emphasizing performance had sought to limit shading of the reflector by structural members. Torsional rigidity was obtained most commonly by using a large torque tube on the backside of the reflector, or by building the reflector substrate thick enough to impart torsional stiffness. The reflector was cantilevered off the torque tube or applied to the substrate in a manner that did not contribute to the structural integrity of the concentrator. In contrast, the IST design [Figure 4] accepts some shading of the reflector, but uses structural cross-members to convert the entire concentrator into a torque tube of very large cross-section. The size of this cross-section allows the concentrator to be produced in a very rigid and robust form using very thin members. The mass of the IST concentrator is one-fifth or less compared to previous designs. Reduced mass translates into major cost reductions, while front shading has minimal impact on thermal performance.

Figure 4: IST Concentrator Design

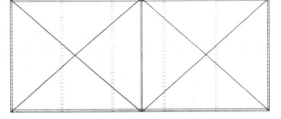

IST Parabolic Trough Concentrator

The IST system includes other important innovations, such as a means to drive up to six parallel rows of collectors with a single drive and control system. Licenses were obtained from Honeywell to manufacture their tracking control system, and from the International Nickel Company to develop their technology for the application of blackened nickel to steel absorber tubes. IST commercialized anti-reflective coating technology developed by Sandia and flexible metal hoses with superior fatigue resistance. These innovations have been introduced into the product and time tested in commercial field installations dating back to 1985. Continuous monitoring has demonstrated the outstanding reliability of these systems, yet they cost a fraction of what was achieved for PTC installations circa 1980 [Figure 5].

IST has been able to achieve gradual cost reductions over time, while continually improving the PTC system in terms of performance and reliability. The projected decrease in the cost of future systems comes from deployment on a larger scale and from R&D leading to a change in concentrator materials from aluminum with plastic or polished aluminum reflectors to thin silvered glass reflectors mounted on steel.

Under a cost-shared contract with NREL, the development of a glass/steel concentrator is proceeding rapidly. The goal is to install PTC systems at a cost less than $200/m^2 while achieving more than a 10% enhancement in performance.

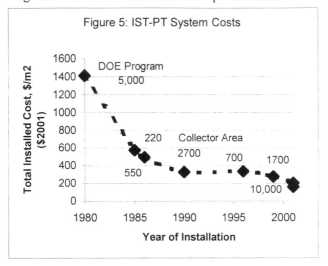

Figure 5: IST-PT System Costs

From 1984–1991 Luz installed nine Solar Electric Generating Systems (SEGS) with a combined capacity of over 350 MW, involving 2.3 million square meters of solar collectors and a cost of over $1.2 billion [Pilkington Solar International, 1996].

Using facilities in Israel, Luz developed and deployed evacuated receivers on a very large scale. To enhance thermal/electric conversion efficiency, they increased PTC operating temperatures to 371 C. Sputtered cermet(ceramic-metal) selective surface coatings were developed for efficient energy collection at high temperatures. Luz established the infrastructure to build very large solar facilities in very short time periods and developed the techniques necessary to operate and maintain them. The systems required special heat transfer fluids, pumps, fittings and rotating joints.

All the SEGS plants are currently operating. They form the backdrop of attempts to install similar systems throughout the world.

NEW MARKET ENTRIES

IST continues to develop its conventional IST-PT system. However, for smaller commercial applications where minimal load and lack of land preclude the installation of the IST-PT, the company has developed the IST-RMT (Roof Mounted Trough). This system is easy to ship and install on flat roofs. RMT modules are 2.4m long and 1.1 m across. The IST-RMT is slightly more costly than the IST-PT, but commercial customers pay higher energy rates than larger industrial users so the additional cost can be justified.

The Europeans led by Spain and Germany, have conducted major research aimed at bringing down the cost of PTC-generated power. One such effort is the development of the Eurotrough [Ciemat & DLR, 1997]. This design is said to be a much improved and less costly version of the LS-3--the final Luz concentrator market entry. The designers of the Eurotrough are seeking US licensees.

Duke Solar is developing the "Power Roof" system. This concept employs the roof as a fixed line-focus concentrator. Reflected sunlight from the roof is focused onto a moving receiver. Their first system, slated to go into operation in the summer of 2001, is designed to provide hot water and air conditioning to an office building in North Carolina. Duke is also active in marketing solar power systems based on the technology left over from Luz.

PTC APPLICATIONS

With the ability to deliver heat up to 370 C, parabolic troughs are a flexible energy source. The following PTC applications are presented to illustrate this flexibility and to show the relative ease with which the technology can be integrated into an energy system. Economic data show the emerging competitiveness of PTC systems with natural gas. Widespread deployment will reduce the cost of PTC installations, save businesses money, reduce fuel imports, increase US energy independence, and have a major impact in reducing air pollution and greenhouse gas emissions.

Hot Water Heating

Large quantities of hot water are used in commercial facilities, such as hotels, laundries and prisons, and in industry particularly food processing: meat packing, bakeries, dairies, corn milling and chicken processing, but also in the chemical and wood products industry. As a low-temperature application, water heating will maximize solar energy delivery. Thermal storage allows energy delivery 24-hours per day.

IST has installed a number of large-scale solar DHW heating systems, including a 1584 m^2 system at the Federal Correctional Institution in Phoenix, Arizona. To illustrate the characteristics of a PTC system, this application is described in detail. A schematic is shown in Figure 6.

The solar field consists of 120 parabolic trough concentrator modules that are arranged in 10 parallel rows. The total system flow of about 100 gpm is divided into five parallel U-loops. The collectors track the sun continually during the day to heat a circulating propylene glycol antifreeze solution. Heat from the solar collectors is transferred through immersed copper coils to two hot water storage tanks with a total volume of 23,000 gallons. In summer, these tanks heat to a limit of approximately 85 C. Incoming domestic water is heated as it flows through a second set of copper coils. Annually, about 70% of the hot water needs of the institution are supplied by solar energy.

Figure 6: Schematic of Solar Water Heating System

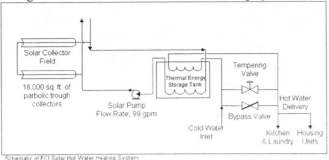

The economics of this installation are favorable because solar energy displaces electricity. If a private company were considering investing in a similar system to produce hot water, at a time when the solar industry had grown to a reasonable rate of production, the following economic conditions might apply:

Total installed system cost:	$240/m^2
System energy delivery: (Sunbelt location)	950 kW$_{Thermal}$/m^2-yr
O&M&I: (increasing with inflation)	2% of capital investment
Inflation rate	2.5%/yr
Fuel escalation rate:	3.5%/yr
Conventional heater efficiency:	
NG & fuel oil	70%
Electricity	95%
Loan term:	10 yr
Loan fraction:	0.4
Loan interest rate:	8%
Corporate tax rate:	
Federal	33%
State	8%
Federal investment credit:	10%
Depreciation:	5 year accelerated

The economics of such an investment compared to three different fuel types are listed in Table 1. Payback is shown in terms of years to recover the capital down payment compared to before- and after-tax savings (revenue less operating costs). The internal rate of return is calculated based on after-tax revenue streams over 10 and 15-year time periods. The projected life of the solar system is 30 years.

Table 1: Economics of Solar Water Heating

Fuel Type	NG	# 2 Fuel Oil	Electricity
Assumed fuel cost	$7/MMBtu	$1.20/gal	6.5 c/kWh
Before tax payback, yr	6	5	2
After tax payback, yr	4	3	2
IROR, 10 years, %	14	24	39
IROR, 15years, %	19	21	37

These figures illustrate that large-scale solar water heating systems are competitive now with electricity and No. 2 fuel oil.

In addition, they are approaching parity with natural gas, even against stringent economic hurdles.

Steam Generation

Process steam is a much larger thermal market than hot water. IST and a joint venture partner are currently installing a solar steam system at a pharmaceutical plant near Cairo, Egypt. 1900 m^2 of solar collectors are being installed to generate 1.6 metric ton/h of steam at a pressure of 8 kg/m^2 (114 psig). The heat from the solar field is converted into steam in an unfired steam generator.

Figure 7: Schematic of PTC Steam Generating System

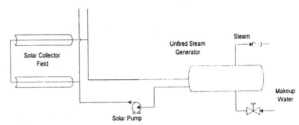

The economics of a steam generating system are less favorable than water heating: the system is more expensive and the delivery of thermal energy will be reduced because system operating temperatures are higher. Using the same economic assumptions as before, at higher system cost and reduced output, in competition with natural gas, the after-tax payback declines to about 6 years, and the 10 and 15-year IROR's are 7% and 13%, respectively.

Steam is a common heating medium. Besides process use, solar steam can be used to desalinate sea water, to power air-conditioning chillers or to run a turbine for mechanical or electric power production.

Air-Conditioning

PTC systems can drive lithium bromide chillers to provide air conditioning. An advantage of chilling systems is that they displace high-priced peak electricity, since demand for air-conditioning coincides with hot sunny weather. The IST-RMT is designed for the commercial market. Electric rates can be high and commercial electric chillers are not as efficient as industrial-scale equipment. Since single stage absorption chillers run on hot water at around 80 – 95 C, it is possible to incorporate hot water thermal storage to deliver cooling at night.

Figure 8: Solar Cooling Using Single Stage Absorption Chiller

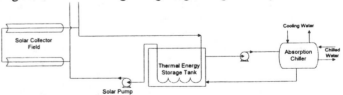

The chiller, storage tank and possibly a cooling tower add to the cost of the cooling system. Using the economic criteria

similar to those described above, for a single-stage chiller with a COP of 0.7, displacing electricity at $0.18/kW used to power a chiller consuming 1.17 kW/ton-hr of cooling, the after-tax payback period is 6 years and the 10 and 15-year IROR is 5% and 12%, respectively. Small-scale, two-stage absorption chillers with a COP of about 1.2 are coming onto the market. They will improve the economics of solar cooling. Even higher COP's and less costly equipment are possible if developments in desiccant systems are successful. Replacing 15-cents/kWh power with such equipment could improve after-tax paybacks to 4 years and 10 year returns to 13%.

Since cooling is often a seasonal load, trough systems can be integrated into buildings to reduce conventional fuel use through the provision of summer time cooling, winter space heating and year-round DHW heating.

Electric Power Generation

Luz pioneered the use of stand-alone PTC systems to generate electricity. These plants generate steam that is run through a conventional turbine. Solar heat is supplemented with a 25% contribution from natural gas. The last plant was built in 1990. It had an output of 80 MW and cost $3440/kW. In 1996, it was projected that a similar plant could be built for $2750/kW and would deliver power at a cost of $0.14/kWh [Pilkington 1996].

The Integrated Solar Combined Cycle System (ISCCS) concept is a combined cycle gas turbine electric generator with an oversized steam turbine. During the day the steam generator is powered about 50% from the solar system and 50% from the gas turbine heat recovery steam generator. By using much of the infrastructure present at the CC plant, the cost of the solar system can be reduced. The estimated cost of power is around $0.10/kWh.

York Research Corporation has developed the Solar Energy Enhanced Combustion Turbine (SEECOT[TM]) concept as another means of integrating solar energy systems into CC plants that are presently favored for new electric generating facilities. Solar energy delivered by PTC's is used to power an absorption chiller to cool the inlet air to the combustion turbine, and can be used to generate steam for injection into the combustion zone of the turbine. Cooling the inlet air stream increases the electric output of a gas turbine on hot days when demand for power peaks. It can also reduce the heat rate. Steam injection increases power output and reduces the conventional heat rate. There are no published estimates for the cost of power resulting from this use of solar energy.

Large-scale solar power plants are selling into competitive markets fueled by natural gas, coal, and nuclear. In much of the world, if electricity is available at remote sites, it is derived from liquid-fueled diesel engines. The cost of power can easily exceed $0.25/kWh. PTC systems could potentially compete with such systems.

In the range of 300 kW to 10 MW, Organic Rankine Cycles have advantages over the use of steam turbines. So far, ORC machines have typically been used for low temperature geothermal applications. However, machines are being developed to use higher temperature energy sources, such as solar or biomass. A reasonable goal for ORC machines at a mature level of production is a cost of $1000/kW for a 1 MW machine that has a thermal-to-electric conversion efficiency of 24%. With such a machine, the installed cost of the solar power plant would be around $3,000/kW. Using the same economic assumptions, but with increased O&M costs and an electric price of 18 cents/kWh, the 10 year IROR is about 12%. At 25 cents/kWh, the before- tax return would be about 5 years and the after-tax IROR is 22%.

ENERGY IN THE 21st CENTURY

We now appear to be entering a period of energy crisis unlike those of the past. Continually rising energy demand has pushed up against energy supply. Energy prices have peaked for the moment, but there is no expectation that oil will again decline to $10 per barrel or natural gas to $2/million Btu. Regardless of energy costs, there is the widely held belief that Global Warming (See Side Bar) is the most serious issue facing the world in the 21st century. Governments, the United States notwithstanding, have undertaken initiatives that will drive the widespread deployment of renewable energy technologies. The crisis in California is an illustration that the old assumptions regarding the reliability of the energy supply are no longer valid. Driven by the demands of the market, large multi-national companies have become suppliers of wind power and photovoltaics. These same forces will come to drive the solar thermal market and PTC's in particular, although at this point the technology is little understood and the potential not yet appreciated

CONCLUSIONS

The applications of PTC technology are broad and this has significant implications for the energy and environmental crisis that we are facing in the new century. Parabolic trough technology has been proven as an effective means of converting solar energy into heat at temperatures up to 370 C. There are no technical barriers to the introduction of PTC's for applications including water heating, steam generation, air-conditioning and electric power generation As this survey has shown, multi-faceted development efforts in material and operations technology have produced significant improvements in cost, performance and reliability, and continue to do so.

Rising energy costs are rendering PTC technology competitive with conventional fuels as a source of heat and power. The potential to displace fossil fuel burning with solar energy using parabolic troughs in sunny parts of the globe (such as the south-western United States) is enormous.

Global warming is real and the potential ramifications are catastrophic. Parabolic trough collector systems can make a major contribution to reversing this trend. Ultimately, economic forces will drive the technology into the market place. Commitments by individuals and governments who take a longer-term economic perspective and recognize the environmental benefits to be gained will accelerate the pace of deployment.

REFERENCES

Ciemat & DLR: Plataforma Solar de Almeria, Annual Technical Report 1997 July 1998.

Dudley, V. E. and Evans, L. R. Test Results: Industrial Solar Technology Parabolic Trough Solar Collector, SAND94-1117, Sandia National Laboratories, Nov. 1995.

Mueller, E. A.: Estimating the Potential for Solar Thermal Applications in the Industrial Process Heat Market 1990 – 2030, Nov. 1990.

Pilkington Solar International: Status Report on Solar Thermal Power Plants, 1996.

SIDE BAR

Heat energy traveling from the Sun to Earth penetrates our atmosphere with very little difficulty as visible radiation at wavelengths ranging from about 3 to 7 micrometers. Once absorbed by the earth, however, the wavelengths of this energy are converted to the far infrared, peaking at about 12 micrometers. If heat imported from the sun on an annual basis is not re-radiated to space, the earth's average temperature will increase. Because gases such as carbon dioxide strongly absorb radiation in the 12-micrometer range, the continual release of these gases into the atmosphere prevents an ever-increasing portion of the re-radiated heat from penetrating the atmosphere and returning to space.

The polar regions of Earth are hovering close to the freezing point, which means that much of their ice is only a few degrees away from melting. Scientists predict a rise in ocean levels due to melting ice; and there is no rational reason to believe that increased temperatures due to increased carbon dioxide concentrations in the atmosphere won't cause melting. When viewed from the standpoint of the ocean currents which transport heat from the equator to the northern latitudes harboring much of Europe, the question that must be asked is, what would happen if this current flow shuts down? The evidence strongly suggests (1) that climactic changes have taken place, some quite sudden, because of disruption in North Atlantic currents with consequent drastic cooling in northern Europe.

Helping to distribute heat throughout the globe, moderate depth currents of the Atlantic Ocean travel north along Africa and Spain, gathering heat as they flow through the equator, and in doing so acting as a significant heat sink for those regions. Continuing their flow northward they arrive in the vicinity of Iceland, where these warm, highly saline currents begin to encounter colder and therefore more dense water. Forced to the surface, they release an enormous amount of heat to the surrounding regions. Having thus cooled and gained greater density than the waters over which they pass, the currents sink downward at deep-water formation sites, and flow back south at greater depths than on the trip northward, only to repeat the cycle after excursions into the Pacific Ocean.

There is strong scientific evidence (2) that, at the time of the small Ice Age 11,000 years ago, these deep-water formation sites were diluted by a precipitous flow of fresh water from a glacial reservoir called Lake Agassiz, which covered most of what is now the Province of Manitoba. Consisting of water from melting glaciers, held on the east by an ice dam and on the west by higher elevations of land, Lake Agassiz grew to a depth of about 75 feet. As warming continued, the ice dam to the east was eventually breached, and the vast reservoir of fresh water flowed down the St. Lawrence River into the North Atlantic. Buoyed to the upper levels of the salt water, it was able to adversely affect the deep-water formation sites of the Atlantic currents, reducing their density at the surface, and thus preventing them from plunging downward to complete their cycle. Over a very short period of time the northern reaches of Europe froze over, and the equatorial regions lost a major heat sink.

Given the fact that there is enough frozen fresh water on Greenland alone to equal the volume of water that flowed from Lake Agassiz, and the further fact that Greenland is in closer proximity to the deep water formation sites, the probability of melting temperatures in regions north of the Arctic Circle carries with it the possibility of catastrophic rather than moderate environmental change. If an argument were needed for control of greenhouse gas emissions, this would seem to be an effective one. The question is whether or not we as a society and as a species are capable of rational action to prevent it.

(1) Wallace S. Broecker, Thermohaline Circulation, the Achilles Heel of Our Climate System: Will Man-Made CO_2 Upset the Current Balance?, Science, Nov 28, 1997.

(2) W. Broecker and G. Denton, What Drives Glacial Cycles?, Scientific American, January 1990.

Proceedings of IECEC'01
36th Intersociety Energy Conversion Engineering Conference
July 29-August 2, 2001, Savannah, Georgia

IECEC2001-RE-13

COOLING POWER ENHANCEMENT OF CPC (COMPOUND PARABOLIC CONCENTRATOR) SKY RADIATOR

Takeo S. SAITOH
Professor of School of Aeronautics and Space Engineering,
Tohoku University, Sendai 980-8579, Japan
Phone : +81-22-217-6974
FAX　 : +81-22-217-6975
E-mail : saitoh@cc.mech.tohoku.ac.jp

Koichi TATSUO
Graduate student, School of Aeronautics and Space Engineering,
Tohoku University, Sendai 980-8579, Japan

ABSTRACT

A compound parabolic concentrator (CPC) type sky radiator was proposed to greatly enhance the sky radiation cooling performance.

The authors have demonstrated that the CPC sky radiator has by far the better performance for cooling power than the flat plate type. However, the reflector contour of the previous CPC sky radiator was originally designed as the solar collector, not as the proper sky radiator. Therefore, in this article, improved CPC sky radiator with an improved reflector was proposed.

The results demonstrate that the new type CPC sky radiator shows much better performance in cooling power than the conventional one.

INTRODUCTION

Global warming due to carbon dioxide (CO_2) emissions has been posing an important environmental issue.

For instance, the earth's annual mean temperature in 1998 was 14.46 degree Celsius, which is 1.2 degree Celsius higher than the past average value.

On the other hand, urban warming rate in the past fifty years was about ten times higher than the global warming rate. According to the most recent computer projection, the air temperature at the center of Tokyo at end of July of 2030 would exceed 40 degree Celsius if business goes as usual[1].

Motivated mainly by the above fact, the authors have proposed a sky radiator to supply cooling loads by virtue of sky radiation cooling, which will be very promising among natural energies from the standpoint of reduction of CO_2 emissions, as well as mitigating thermal pollution (heat island) due to heat emissions in urban areas [2]-[7].

The authors have studied two types of the sky radiator, i.e. flat plate type and CPC type, and revealed that the CPC sky radiator has by far the better performance in cooling power than the flat plate one. However, the reflector shape of the previous CPC sky radiator was originally designed as the solar collector, not as the sky radiator.

In this paper, the new CPC sky radiator with excellent cooling power will be designed. And the fundamental experimental cooling performance for this improved CPC sky radiator will be shown.

NOMENCLATURE

R_{sky} : amount of atmospheric radiation
P_{SR} : cooling power
Q : heat quantity
T_c : radiator surface temperature
A : aperture
d : height of full CPC
d_1 : height of truncation

Greek symbols

λ : wave length
σ : Stefan-Boltzmann's constant
θ_a : acceptance angle
Δ : difference

SKY RADIATION COOLING

Figure 1 shows the radiative energy balance emitted from the black body radiator. It was assumed that the radiating surface is perfectly thermally-insulated, and convection heat transfer and heat conduction are negligible. A dotted line shows radiation emitted from a black body with the same temperature as the ambient temperature. Line A shows sky radiation spectrum, and line B radiation emitted from the sky radiator. Sky radiation outside the wave length range of 8~13 μm conforms to blackbody radiation. On the other hand, sky radiation within 8~13 μm range is smaller than blackbody radiation. This relatively transparent range is called the "atmospheric window". Sky radiation cooling power is calculated from the difference between lines A and B. As the radiator surface is cooled, cooling power increases because sky radiation is larger than radiation emitted from the radiating surface. And finally an equilibrium temperature below ambient temperature is established.

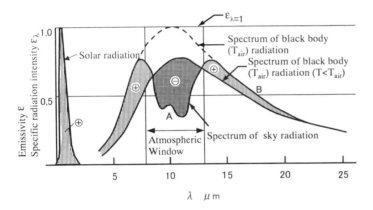

Fig. 1 Energy balance of black body radiator

Figure 2 shows the wavelength distribution and angle distribution of sky radiation measured in Colorado and Florida by Bell et al[8][9]. As zenith angle becomes wider, sky radiation intensity increases because radiation path becomes longer.

Figure 3 shows the wavelength distribution of sky radiation from zenithal direction measured by Berdahl and Martin[10]. As water vapor content increases, sky radiation intensity increases in the atmospheric window range, the atmospheric window becomes opaque.

Therefore, in case of high ambient temperature, high relative humidity, and high cloudiness, the equilibrium temperature does not decrease below ambient temperature because of opaque atmospheric window and convective heat transfer.

It is well known that the utilization of the sky radiation cooling in the Japanese climate is almost impossible. For this reason, the development of the sky radiator with good cooling power even under high ambient temperature and high relative humidity is greatly needed.

As the black body radiator absorbs radiation of all wavelength distribution, heat exchange must be limited within the atmospheric window to gain the lowest possible equilibrium temperature. Therefore, a selective absorption film was considered to improve optical characteristics of the radiator surface. The selective absorption film has a high reflectivity within the atmospheric window, and low emissivity outside the atmospheric window. This improvement provides the lower equilibrium temperature than the black body radiator.

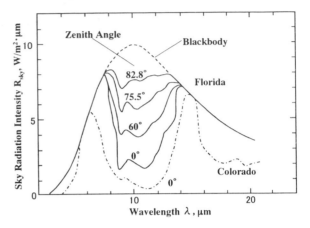

Fig. 2 Relation between distribution of sky radiation intensity and zenith angle[8][9]

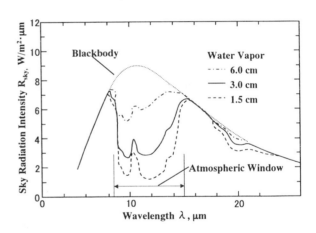

Fig. 3 Relation between wavelength distribution of the sky radiation intensity and water vapor[10]

But the maximum cooling power cannot exceed that of the black body radiator because radiation balance is limited within the atmospheric window even if ideal selective absorption film were used. It is strongly influenced by relative humidity and zenith angle. For this reason, the cooling performance could not be improved as long as the flat plate sky radiator is utilized.

CPC SKY RADIATOR

To overcome above difficulties, the CPC sky radiator was proposed. The CPC sky radiator has a parameter acceptance angle (θ_a) as shown in Figure 4, and the incident sky radiation within the acceptance angle is absorbed by the sky radiator surface but the incident sky radiation outside acceptance angle is reflected by the reflector. On the other hand, almost all of the radiation emitted from the radiating surface are rejected to the sky. Therefore, exchange of radiation between the sky and the radiator is done only at limited zenithal aperture, and as the sky radiation intensity decreases, the radiation emitted from the radiator surface increases.

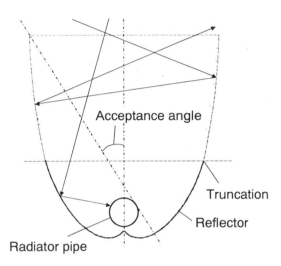

Fig. 4 Illustrative view of CPC trough

The CPC sky radiator gives much better performance than the conventional flat plate sky radiator with selective emission film. Because it can curtail absorption of sky radiation extending for a wide wavelength distribution and it is not influenced by the strong sky radiation from the direction of earth's surface due to high relative humidity.

OPTIMAL DESIGN SIMULATION FOR CPC RADIATOR

The cooling power of CPC sky radiator, P_{SR} is represented by the next equation with consideration of the thermal balance at the radiator surface.

$$P_{SR} = (Q_{out} + Q_{in})/A$$

Where, Q_{out} means the amount of thermal radiation from radiator pipe, and represents σT_c^4 with using temperature of the radiator T_c. Q_{in} means the amount of incident sky radiation to the radiator pipe. And A indicates CPC aperture area that is effective area.

When a thermal balance for the CPC sky radiator is made, radiative heat transfer between each constituent element occurs on a very complex aspect. The ray trace method was used to obtain the optical characteristics for given configurations of the CPC sky radiator.

With adopting 50 points on CPC cover film and 100 points on the radiator pipe and the reflector and tracing total 36 rays that are radiated with each 5 degrees interval in tangent direction, the radiative heat transfer exchanges between constituent elements were calculated. Figure 5 shows the illustrative example of the ray trace in case of acceptance angle 30°. And Figure 6 shows simulation models for the CPC sky radiators.

Fig. 5 Illustration of the ray trace (in case of 30°)

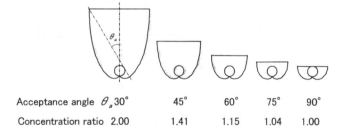

Acceptance angle θ_a	30°	45°	60°	75°	90°
Concentration ratio	2.00	1.41	1.15	1.04	1.00

Fig. 6 Simulation models of CPC sky radiators

COOLING POWER SIMULATION

Cooling power simulation was conducted with assumptions that sky radiation is radiated equally from whole sky and the CPC cover penetrates it equally. Further, optical properties of each constituent element were established approximately shown as Table 1.

The simulation results are shown in Figure 7. Where, a vertical line means fraction of cooling power with respect to conventional CPC sky radiator. Weather conditions were assumed that the air temperature is 21 °C and the relative humidity 80 %, which correspond to a summer condition.

Table 1 Optical properties of each constituent element

	emmitance absorptance	reflectance	Trans-mittance	Reflective form
Radiator pipe	1	0	0	diffusive
Reflector	0	1	0	mirroring
Cover	0	0	1	diffusive

Cooling power simulation based on ray tracing with considering optical characteristics of the CPC is conducted. Figure 7 shows a relation between acceptance angle and cooling power without truncation.

It is seen that the CPC sky radiator has a peak of the cooling power near $\theta_a = 60°$ and be superior to the conventional one in cooling power by around 20%. This is due to the fact of large acceptance angle, and small concentration ratio and so on.

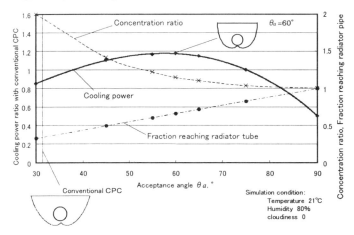

Fig. 7 Cooling power simulation results

without truncation

Furthermore, moderate truncation is effective in order to reduce concentration ratio and manufacturing costs.

Figure 8 shows that relation between fraction of truncation and the cooling power without truncation in case of $\theta_a = 60°$. Here, fraction of truncation is defined as d_1/d shown as in Figure 9. In short, fraction of truncation is 0 in case of no truncation, and 1 in case that the CPC is truncated at the same height as the top of radiator pipe.

As a result of simulation, even if the CPC is truncated to the extent of fraction of truncation 0.5, the cooling power will be coincident within error of 2%.

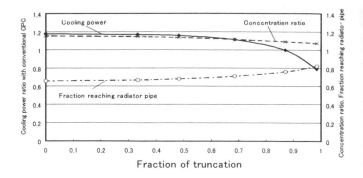

Fig. 8 Effect of truncation

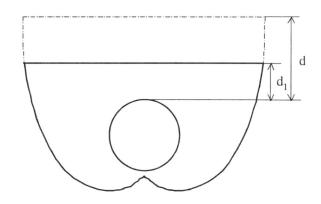

Fig. 9 Fraction of truncation

EXPERIMENT OF COOLING POWER

The authors manufactured an improved CPC sky radiator based on the optimal design simulation and validated the simulation results through cooling power experiment.

Figure 10 shows illustration of experimental setup for the improved CPC sky radiator. The CPC sky radiator consists of the copper pipe through which working fluid flows, the CPC reflector, and the cover. Since the aperture area ratio between improved CPC and conventional CPC is 5 : 7, it was determined that the number of improved CPC trough is 7, and the number of conventional CPC trough is 5 in order to specify the same aperture area. As both of the radiator length is 1600 mm, module aperture area is 0.96m². The back of the sky radiator is thermally insulated by glasswool and urethane foam. The sides are also thermally insulated by urethane foam. Polyethylene film (20~25μm thick) with a high transmittance (0.9) for infrared radiation is used as a cover to reduce heat loss due to convection.

Table 2 shows thermophysical properties of each material.

Table 2 Thermophysical properties*

Material	Specific gravity kg/m²	Specific heat kJ/(kg·K)	Thermal conductivity W/(m·K)
Water	999.8	4.213	6.562
ethylene gricohol	1060	3.798	0.256
aluminum	2690.7	0.8808	235.93
copper	8917.6	0.3806	401.22
urethane foam	33.26	1.129	0.034
glass-wool	16	0.0502	0.00246

*Reference Temp.: 0°C

EXPERIMENTAL RESULTS

Figure 11 shows a photograph of the experimental apparatus. To compare the cooling power, the conventional CPC sky radiator was installed next to the improved CPC sky radiator. Experiment was conducted at clear night at the roof of the Mechanical Engineering Building of the Tohoku University. Antifreeze agent (concentration : 22.5%wt, specific heat : 3.798 kJ/kg) is used as a working fluid. To keep the flow rate fixed, the working fluid is circulated only by pressure head of the upper tank. The working fluid through the radiator flows to the lower tank. A needle valve is used to adjust the flow rate. The working fluid in the lower tank is pumped up to the upper tank with a magnet pump. And the working fluid returns back to the sky radiator again. To keep the head of the upper tank constant, an overflow is attached. The both of the flow rate are 1 ℓ/min.

Figure 12 shows time variations of temperature and amount of atmospheric radiation under an actual weather condition. The average temperature was −4.2°C, amount of atmospheric radiation was comparatively stable to be 242 W/m². Furthermore, cloudiness was low and the wind was weak.

Figure 13 shows the experimental result that is plotted by adopting Δ T, i.e. temperature difference between inlet temperature and outlet temperature and cooling power.
As a consequence of this experiment, it was clarified that the improved CPC sky radiator is superior to the conventional CPC sky radiator in cooling power at least by 30%.

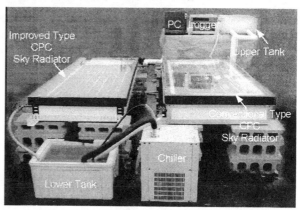

Fig. 11 Photo of the experimental apparatus

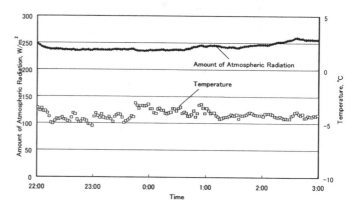

Figure 12 Time variations of temperature and amount of atmospheric radiation

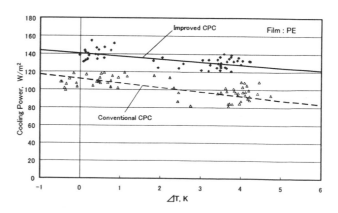

Figure 13 Experimental cooling power

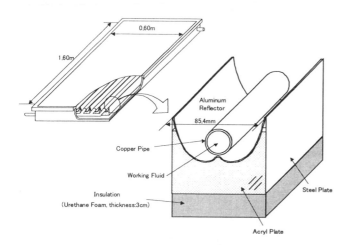

Fig. 10 Illustration of improved CPC sky radiator

CONCLUSION

From the present study, the following conclusions may be drawn:

(1) The optimal design for the CPC sky radiator was presented by considering optical characteristics of the CPC based on ray tracing.

(2) The CPC sky radiator has a cooling power peak at θ_a =60° and is superior to the conventional CPC in cooling power by around 20%.

(3) Even if the CPC is truncated to fraction of truncation 0.5, the cooling power loss will be less than 2%.

(4) Improved CPC sky radiator was manufactured and the fundamental experimental results were obtained for the cooling performance.

(5) As a consequence of this experiment, it was clarified that the improved CPC sky radiator is superior to conventional CPC sky radiator in cooling power by at least 30%.

REFERENCES

(1) T. S. Saitoh, "Heat Island", Kodansha Publ. Co., Ltd (1997).
(2) T.S. Saitoh, H. Matsuhashi and T. Ono, "An Energy-Independent House Combining Solar Thermal and Sky radiation Energies", Solar Energy, vol.135, No.6 (1985), pp.541-547.
(3) T.S. Saitoh and T. Ono, "Utilization of Seasonal Sky Radiation Energy for Space Cooling", Trans. ASME Journal of Solar Energy Engineering, vol.106 (1984), pp.403-407.
(4) T.S. Saitoh and T. Ono, "Simulative Analysis for Long-Term Underground Cool Storage Incorporating Sky Radiation Cooling", Trans. ASME Journal of Solar Energy Engineering, vol. 106(1984)
(5) T.S. Saitoh, "Natural Energy Autonomous House with Underground Water Reservoir", Bulletin of JSME, 27-226 (1984), pp.773-778.
(6) T. S. Saitoh amd T. Marushima, "Research about sky radiation cooling system", Journal of JSES, vol.24, No.1, 46-55 (1998).
(7) T. S. Saitoh and T. Fujino, "An energy-sufficient house(HARBEMAN house) combining solar thermal photovoltaic and sky radiation energies"(1st Report: operational results), Journal of JSES, vol.24, No.6, 35-42 (1998).
(8) E. E. Bell, L. Einsner, J. Young and R. A. Oetjen, "Spectral Radiance of Sky and Terrain at Wavelength between 1 and 20 Micron 2 Sky Measurements", J. Opt Soc. Am., 50 (1950), pp.1313-1320.
(9) B. Landro and P. G. McCormic, "Effect of Surface Characteristic and Atmospheric Condition on Radiative Heat Loss to a Clear Sky", Int. J. Heat Mass Transfer, 23 (1980), pp.613-620.
(10) M. Martin and P. Berdahl, "Summery of Results from Spectral and Angular Sky Radiation Measurement Program", Solar Energy, vol.33, No.3/4 (1984), pp.241-252.
(11) T.S. Saitoh, Japanese Patent (2000-1).
(12) T.S. Saitoh, Japanese Patent (2000-2).
(13) T.S. Saitoh, Japanese Patent (2000-3).

Proceedings of IECEC'01
36th Intersociety Energy Conversion Engineering Conference
July 29--August 2, 2001, Savannah, Georgia

2001-RE-16

A NUMERICAL HEAT TRANSFER STUDY FOR A STORAGE TANK OF A HEAT-PIPE SOLAR HOT WATER SYSTEM

Ru Yang and Zhi-Hao Yang
Department of Mechanical and Electro-Mechanical Engineering
National Sun Yat-Sen University
Kaohsiung, Taiwan 804

ABSTRACT

For a heat-pipe solar collector hot water system, one of the major design problems is addressed on the heat transfer between the heat-pipe and the storage tank. In this paper, a natural convection heat transfer model is considered in the water storage tank with a load taking on a wall. Governing equations are transformed into vorticity-stream equations. Gauss-Seidel method with a finite-difference implicit numerical scheme is applied. The temperature of the heat-pipe surface in the storage tank is considered as uniform that corresponding to a condensation saturation condition. It is shown that the heat transfer rate increase with the Rayleigh number, the heat-pipe length, the heat-pipe thickness and the angle of inclination, and decrease slightly with the height location of the heat-pipe. The results may provide design information for related heat transfer considerations.

INTRODUCTION

Natural convection heat transfer in an enclosure has been studied for many different applications [1-10]. However, natural convection inside a cavity with a portion of a heating block intruded into one of the side wall and cooling load taking place on another side wall has not yet been studied previously. This problem is motivated by the design for a storage tank for the heat-pipe solar collector hot water system. The evaporation section of the heat-pipe is located in solar collector while the condensation section is located in water storage tank. Latent heat is transferred into water inside the tank during the condensation process, therefore the surface of the heat-pipe inside the tank can be considered as a constant-temperature heating surface. The load is considered taking place on a side wall with a constant cold temperature.

NOMENCLATURE

Ar   aspect ratio, $=W/H$
C   specific heat
d   thickness of heat-pipe
g   gravity
Gr   Grashof number, $=g\beta(T_h-T_c)H^3/\nu^2$
H   tank height
h   distance between heat-pipe and bottom wall
h_c convection coefficient
K   thermal conductivity
L   heat-pipe length in tank
Nu_x local Nusselt number in x direction
Nu_y local Nusselt number in y direction
$\overline{Nu}$ average Nusselt number
P   pressure
Pr   Prandtl number, $=\nu/\alpha$
q'' heat flux
Ra   Rayleigh number, $=g\beta(T_h-T_c)H^3/\nu\alpha$
t   time
T   temperature
T_c temperature of load
T_h temperature of heat-pipe
u   x velocity
v   y velocity
W   width of tank

Greek letters
α thermal diffusivity
β thermal expansion coefficient
ν viscosity
θ angle of inclination
ρ density

ω   vorticity
ψ   stream function

MATHEMATICAL MODEL

The problem is to study the natural convection heat transfer inside a water storage tank with a heating block attached on a side wall and a cooling load taken on the other side wall. A simplified two dimensional problem is considered in this study. Figure 1 is the schematic of the storage tank which has width W, height H, heat-pipe thickness d, heat-pipe length inside the tank L, distance between heat-pipe and tank bottom wall h, and system inclination angle θ. The suitable governing equations for incompressible flow with constant properties are:

Continuity:

$$\frac{\partial u}{\partial x} + \frac{\partial v}{\partial y} = 0 \tag{1}$$

x-momentum:

$$\rho(\frac{\partial u}{\partial t} + u\frac{\partial u}{\partial x} + v\frac{\partial u}{\partial y}) = -\frac{\partial p}{\partial x} + \mu\nabla^2 u - \rho'g\sin\theta \tag{2}$$

y-momentum:

$$\rho(\frac{\partial v}{\partial t} + u\frac{\partial v}{\partial x} + v\frac{\partial v}{\partial y}) = -\frac{\partial p}{\partial y} + \mu\nabla^2 v - \rho'g\cos\theta \tag{3}$$

Energy:

$$\rho c(\frac{\partial T}{\partial t} + u\frac{\partial T}{\partial x} + v\frac{\partial T}{\partial y}) = k\nabla^2 T \tag{4}$$

Boundary conditions are:

No-slip conditions on all walls.
$T=T_h$ at heat-pipe walls as constant temperature condensation inside the heat-pipe.
$T=T_c$ at x=W as a constant temperature load.
Insulation on walls x=0, y=0 and y=H.

By taking curl of the momentum equation, the vorticity equation can be obtained

$$\rho(\frac{\partial \omega}{\partial t} + u\frac{\partial \omega}{\partial x} + v\frac{\partial \omega}{\partial y}) = \mu(\frac{\partial^2 \omega}{\partial x^2} + \frac{\partial^2 \omega}{\partial y^2}) - g(\frac{\partial \rho'}{\partial x}\cos\theta - \frac{\partial \rho'}{\partial y}\sin\theta) \tag{5}$$

where the vorticity, ω, is defined as

$$\omega = \frac{\partial v}{\partial x} - \frac{\partial u}{\partial y} \tag{6}$$

The buoyancy force is modeled by Boussinesq approximation, therefore

$$\rho = \rho[1 - \beta(T - T_c)]$$

A stream function, Ψ, is defined to satisfy the continuity equation as

$$u = \frac{\partial \psi}{\partial y}, \qquad v = -\frac{\partial \psi}{\partial x} \tag{7}$$

If the variables are non-dimensionalized by

$$\bar{x} = \frac{x}{H}, \qquad \bar{y} = \frac{y}{H}, \qquad Pr = \frac{\nu}{\alpha}$$

$$\bar{u} = \frac{u}{\alpha/H}, \qquad \bar{v} = \frac{v}{\alpha/H}, \qquad \bar{\omega} = \frac{H^2}{\alpha}\omega$$

$$\bar{t} = \frac{t}{H^2/\alpha}, \qquad \bar{\psi} = \frac{\psi}{\alpha}, \qquad \bar{T} = \frac{T - T_c}{T_h - T_c}$$

$$Gr = \frac{g\beta(T_h - T_c)H^3}{\nu^2}$$

$$Ra = Gr \times Pr = \frac{g\beta(T_h - T_c)H^3}{\nu\alpha}$$

then the governing equations become

$$\frac{\partial \bar{\omega}}{\partial \bar{t}} + \frac{\partial \bar{\Psi}}{\partial \bar{y}}\frac{\partial \bar{\omega}}{\partial \bar{x}} - \frac{\partial \bar{\Psi}}{\partial \bar{x}}\frac{\partial \bar{\omega}}{\partial \bar{y}} = Pr\left(\frac{\partial^2 \bar{\omega}}{\partial \bar{x}^2} + \frac{\partial^2 \bar{\omega}}{\partial \bar{y}^2}\right)$$
$$+ RaPr\left(\frac{\partial \bar{T}}{\partial \bar{x}}\cos\theta - \frac{\partial \bar{T}}{\partial \bar{y}}\sin\theta\right) \tag{8}$$

$$\bar{\omega} = -\frac{\partial^2 \bar{\psi}}{\partial \bar{x}^2} - \frac{\partial^2 \bar{\psi}}{\partial \bar{y}^2} \tag{9}$$

$$\frac{\partial \bar{T}}{\partial \bar{t}} + \frac{\partial \bar{\Psi}}{\partial \bar{y}}\frac{\partial \bar{T}}{\partial \bar{x}} - \frac{\partial \bar{\Psi}}{\partial \bar{x}}\frac{\partial \bar{T}}{\partial \bar{y}} = \frac{\partial^2 \bar{T}}{\partial \bar{x}^2} + \frac{\partial^2 \bar{T}}{\partial \bar{y}^2} \tag{10}$$

Boundary conditions become

1. $\bar{u} = \bar{v} = \bar{\psi} = 0$ at $\bar{x} = 0$、$\bar{x} = \frac{W}{H}$、$\bar{y} = 0$、$\bar{y} = 1$

2. $\bar{T} = 1$ on heat-pipe surfaces

3. $\bar{T} = 0$ on the wall $\bar{x} = \frac{W}{H}$

4. $\frac{\partial \bar{T}}{\partial \bar{n}} = 0$ at $\bar{x} = 0$, $\bar{y} = 0$,   and $\bar{y} = 1$

and initial condition become

$$\bar{t} = 0, \quad \bar{u} = \bar{v} = \bar{T} = 0$$

ANALYSIS

The convective heat transfer coefficient, h_c, is defined by

$$q = h_c \cdot (T_h - T_c)$$

Therefore, local dimensionless heat transfer coefficient on the right wall of the heat-pipe can be written as

$$Nu_{\bar{y}} = \left(\frac{\partial \bar{T}}{\partial \bar{x}} \right)_{\bar{x} = \frac{L}{H}} \tag{11}$$

and local dimensionless heat transfer coefficient on the upper and lower surfaces of the heat-pipe can be written as

$$Nu_{\bar{x}} = \left(\frac{\partial \bar{T}}{\partial \bar{y}} \right)_{\bar{y} = \frac{h}{H}, \frac{(h+d)}{H}} \tag{12}$$

The average dimensionless heat transfer coefficient on the right wall of the heat-pipe is

$$\overline{Nu}_{\bar{y}} = \frac{H}{d} \int_{h/H}^{(h+d)/H} Nu_{\bar{y}} d\bar{y} \tag{13}$$

The average dimensionless heat transfer coefficient on the upper and lower surfaces of the heat-pipe is

$$\overline{Nu}_{\bar{x}} = \frac{H}{L} \int_0^{L/H} Nu_{\bar{x}} d\bar{x} \tag{14}$$

The average Nusselt number is then

$$\overline{Nu} = \frac{1}{2L+d} \left[L \left(\overline{Nu}_{\bar{x},upper} + \overline{Nu}_{\bar{x},lower} \right) + d\, \overline{Nu}_{\bar{y}} \right] \tag{15}$$

NUMERICAL METHOD

A fully implicit finite difference scheme is employed to discretize the equations. Then, a Gauss-Seidel numerical iterative method is applied to solve the problem.

Grid number of 81×81 was found to be adequate for Ar=1 after several tests for different grid sizes. For other aspect ratios, same grid size was also found to be suitable.

RESULTS AND DISCUSSIONS

There are seven parameters that control the problem. They are the Rayleigh number (Ra), the ratio of heat-pipe length to tank height (L/H), the ratio of heat-pipe thickness to tank height (d/H), the ratio of heat-pipe height location to tank height (h/H), the angle of inclination (θ) of the system, the Prandtl number (Pr), and the ratio of tank width to tank height (W/H). Except W/H was fixed to be 1 and Pr was set to 7 (water) in this study, the individual effect of each parameter on the flow and heat transfer for the system was studied. The base case with Ra=10^7, L/H=1/2, d/H=1/20, h/H=1/2, and θ=0° was chosen for parametric study. The effect of Rayleigh number is studied in the range of $10^3 \leq Ra \leq 10^7$. The effect of L/H is studied for four different values:1/8, 1/4, 1/2 and 3/4. d/H effect is studied for

values of 1/80, 1/40, 1/20 and 1/10. h/H effect is studied for values of 0, 1/4, 1/2 and 1/4. The inclination angle effect is studied in the range of 0°≤θ≤60°. The results for two cases of each parameter are presented in the form of streamlines and isotherms.

Figures 2 and 3 show the results for Ra=10^3 and 10^7, respectively. It can be seen that the convection effect is very weak and conduction dominate the transfer process for Ra=10^3. A larger Rayleigh number corresponding to a higher natural convection driving force induces a stronger convection motion. It can be observed in Fig. 3 that a thermal boundary layer is formed on the cooled surface for Ra=10^7. Figure 4 illustrates the effect of Rayleigh number on $\overline{Nu}$. $\overline{Nu}$ increases with increasing Rayleigh number.

Heat-pipe length has first major influence on the size of heating area, and second major influence on flow field geometry. Figures 5 and 6 illustrate the results for L/H=1/8 and 3/4 respectively. Convection motion is more prominent for longer heat-pipe length inside the tank especially for the upper portion of the tank. Figure 7 shows that $\overline{Nu}$ increases with increasing heat-pipe length.

Figures 8 and 9 show the effect of heat-pipe thickness on the problem for d/H=1/10 and 1/80 respectively. It is shown that the thicker the heat-pipe thickness, the larger the heat transfer heating area, and therefore, the larger the heat transfer rate. Flow in the lower portion of the tank is stronger that increases the convection rate for larger heat-pipe thickness. Figure 10 illustrates that $\overline{Nu}$ increases with increasing heat-pipe thickness.

The above results all show that the flow and heat transfer are strong in the upper portion of the tank and weak in the lower portion of the tank for a low inclination angle. Therefore, decrease the area of upper flow portion by increase the height location of the heat ;pipe results in a lower transfer rate. The results are shown in Figs. 11 and 12 for h/H =0 and 3/4 respectively. Figure 13 shows that $\overline{Nu}$ decreases with increasing heat-pipe height location. However, the effect is not significant especially for h/H less than 0.5.

An increase of inclination angle results in a stronger convection in the lower portion of the tank and, in turn, results in a stronger overall convection rate. Figures 14 and 15 illustrate the results for θ=0° and 60° respectively. Figur an shows that $\overline{Nu}$ increases with increasing inclination angle for the ranges of 0°≤θ≤20° and 40°≤θ≤60°. The reason that 20°≤θ≤40° or θ > 60° has no significant effect on $\overline{Nu}$ is not clear and remains to be studied.

CONCLUSIONS

The effects of the Rayleigh number (Ra), the ratio of heat-pipe length to tank height (L/H), the ratio of heat-pipe thickness to tank height (d/H), the ratio of heat-pipe height location to tank height (h/H) and the angle of inclination (θ) of the system on natural convection heat transfer in a water storage tank with a heat-pipe intruded in through one of the side wall and cooling load taken on another side wall have been studied by a numerical method. Concluding results are:

1. $\overline{Nu}$ increases with increasing Rayleigh number for $10^3 \leq Ra \leq 10^7$.

2. $\overline{Nu}$ increases with increasing heat-pipe length for $1/8 \leq L/H \leq 3/4$.

3. $\overline{Nu}$ increases with increasing heat-pipe thickness for $1/80 \leq d/H \leq 1/10$.

4. $\overline{Nu}$ decreases with increasing heat-pipe height location for $h/H > 0.5$. The effect is not significant, especially for $h/H < 0.5$.

5. $\overline{Nu}$ increases with increasing inclination angle for the ranges of $0° \leq \theta \leq 20°$ and $40° \leq \theta \leq 60°$, but no significant effect for $20° \leq \theta \leq 40°$ or $\theta > 60°$.

The present results can provide reference information for the design of a heat-pipe solar hot water system as well as for the heat transfer taking place in the same geometry considered in this study.

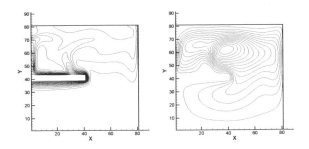

Fig. 3 Streamlines and isotherms for Ra=10^7.

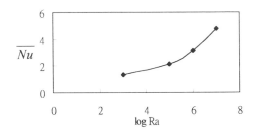

Fig. 4 Effect of Rayleigh number on $\overline{Nu}$.

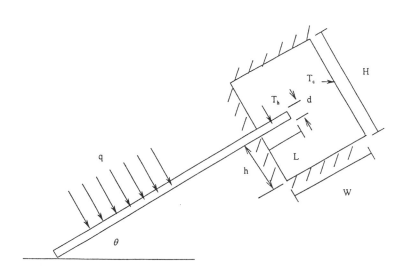

Fig. 1 Schematic of the problem.

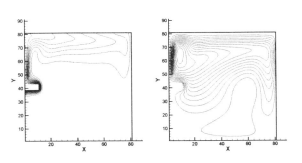

Fig. 5 Streamlines and isotherms for L/H=1/8.

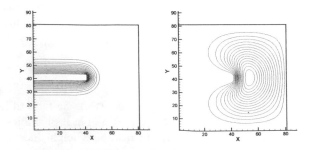

Fig. 2 Streamlines and isotherms for Ra=10^3.

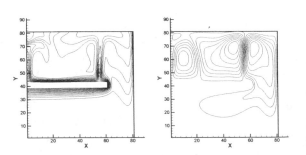

Fig. 6 Streamlines and isotherms for L/H=3/4.

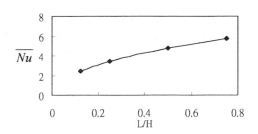

Fig. 7 Effect of heat-pipe length on $\overline{Nu}$.

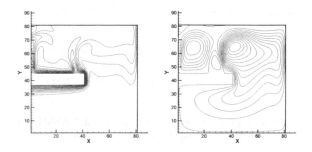

Fig. 8 Streamlines and isotherms for d/H=1/10.

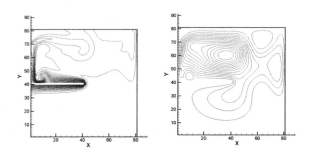

Fig. 9 Streamlines and isotherms for d/H=1/80.

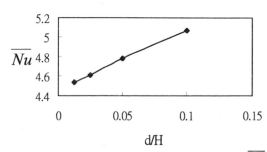

Fig. 10 Effect of heat-pipe thickness on $\overline{Nu}$.

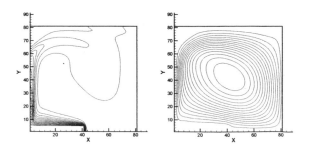

Fig. 11 Streamlines and isotherms for h/H=0.

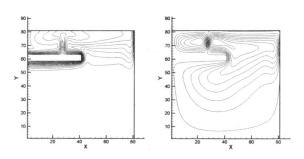

Fig. 12 Streamlines and isotherms for h/H=3/4.

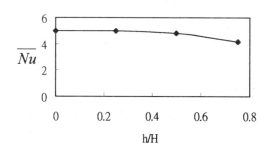

Fig. 13 Effect of heat-pipe location on $\overline{Nu}$.

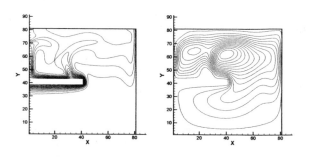

Fig. 14 Streamlines and isotherms for θ=0°.

563

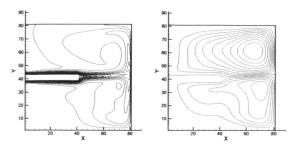

Fig. 15 Streamlines and isotherms for θ=60°.

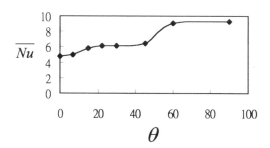

Fig. 16 Effect of inclination angle on $\overline{Nu}$.

ACKNOWLEDGMENTS

This study was sponsored by the National Science Council, Taiwan, ROC under the contract NSC89- 2212-E-110-052.

REFERENCES

[1]Eckert, E. R. G. and Carlson, W. O., 1961, "Natural convection in an air layer enclosed between two vertical plates with different temperature," International Journal of Heat and Mass Transfer, Vol. 2, pp. 106-120.

[2]Cormack, D.E., Leal, L. G. and Imberger, J., 1974 "Natural convection in a shallow cavity with differentially heated end walls, Part 1," Journal of Fluid Mechanics, Vol. 65, pp. 209-229.

[3]Cormack, D. E., Leal, L. G. and Seinfeld, J. H., 1974, "Natural convection in a shallow cavity with differentially heated end walls, Part 2," Journal of Fluid Mechanics, Vol. 65, pp. 231-246.

[4]Simpkins, P. G., and Dudderal, T. D., 1981, "Convection in rectangular cavities with differentially heated end walls," Journal of Fluid Mechanics, Vol.110, pp. 433-456.

[5]Inaba, H., and Fukuda, T., 1984, "Natural convection in an incline square cavity in regions of density inversion of water,"

Journal of Fluid Mechanics, Vol. 142, pp. 363-381.

[6]Chenet, K. S., Ho, J. R. and Humphrey, J. C., 1987, "Steady, two-dimensional, natural convection in rectangular enclosuers with differently heated walls," ASME Journal of Heat Transfer, Vol. 109, pp. 400-406.

[7]Aydin, O., Unal, A., and Ayhan, T., 1999, "Natural convection in rectangular enclosures heated from one side and cooled from the ceiling," International Journal of Heat and Mass Transfer, Vol. 42, pp.2345-2355.

[8]Ismail, K.A.R., and Scalon, V.L., 2000, "A finite element free convection model for the side wall heated cavity," International Journal of Heat and Mass Transfer, Vol. 43, pp. 1373-1389.

[9]Baytas, A. C., 2000, "Entropy generation for natural convection in an inclined porous cavity," International Journal of Heat and Mass Transfer, Vol. 43, pp. 2089-2099.

[10]Ismail, K. A. R., and Adogderah, M. M., 1998, "Performance of a heat-pipe solar collector," ASME Journal of Solar Energy Engineering, Vol. 120, pp. 51-59.

Proceedings of IECEC'01
36th Intersociety Energy Conversion Engineering Conference
July 29-August 2, 2001, Savannah, Georgia

IECEC2001-RE-26

RECENT PROGRESS IN HEAT-PIPE SOLAR RECEIVERS

James B. Moreno
Sandia National Laboratories
Albuquerque, NM 87185-0703
jbmoren@sandia.gov

Marcos A. Modesto-Beato
Sandia National Laboratories
Albuquerque, NM 87185-0703
mamodes@sandia.gov

Charles E. Andraka
Sandia National Laboratories
Albuquerque, NM 87185-0703
ceandra@sandia.gov

Steven K. Showalter
Sandia National Laboratories
Albuquerque, NM 87185-0703
skshowa@sandia.gov

Scott Rawlinson
Sandia National Laboratories
Albuquerque, NM 87185-0703
ksrawli@sandia.gov

Timothy A. Moss
Sandia National Laboratories
Albuquerque, NM 87185-0703
tamoss@sandia.gov

Mark Mehos
National Renewable Energy
Laboratory
Golden, CO 80401
mark_mehos@nrel.gov

Vladimir Baturkin[*]
National Technical University of
Ukraine (KPI)
Kyiv, Ukraine 020147
baturkin@carrier.kiev.ua

ABSTRACT

Metal-felt wicks have demonstrated excellent heat-transfer performance in high-temperature liquid-sodium heat-pipe solar receivers. Their practical implementation has been delayed by efforts to overcome liquid-sodium corrosion, wick fragility, and unexplained hot spots. Recent results indicate substantial progress on all three fronts. With regard to corrosion and wick fragility: a small-scale heat pipe has passed 5000 hours of high-flux testing. In addition, an alternative-wick project has produced new designs that should be more robust than our current designs, with heat-transfer ability nearly as good. There has also been progress in understanding hot spots: a new technique confirms the existence of wall-wick bond flaws. We have identified simple modifications that will make our wick flaw tolerant. In other efforts, we have developed a method to measure relative permeability for liquids in a wick partially filled with gas, which offers many advantages over our present suite of permeability measurements, porosimetry measurements, and relative-permeability models. We are also beginning short-term testing of new wick designs, and the incorporation of these designs into full-scale receivers as well as into a long-term gas-fired heat-pipe test. These developments are described in the following paper.

Keywords: solar, receiver, heat-pipe, Stirling

INTRODUCTION

The addition of sodium heat pipe receivers to solar dish/Stirling systems offers significant power, lifetime, and operational enhancements [1]. The development of dish/Stirling systems is being pursued by the United States Department of Energy's Concentrating Solar Program, as well as by other governments, and by industry here and abroad [2]. This interest is driven primarily by dish/Stirling's high efficiency. The dish/Stirling world record is 29.4%, peak net sunlight to electric conversion [3]. This was achieved with a directly-illuminated receiver (DIR), i.e., the Stirling engine heater-head tubes were heated directly by concentrated solar radiation. In dish/Stirling tests at Sandia Laboratories, we have demonstrated that replacing the DIR with a heat-pipe receiver on a multi-cylinder engine can increase the system power output by 20% [1].

The challenges in developing sodium heat pipe solar receivers have been: first, to extend the existing heat-pipe technology to the very large size and power capacity required; second, to demonstrate the heat pipe's performance advantages; third, to demonstrate practical lifetimes; and fourth, to demonstrate acceptable cost. We are now in the third stage. In all stages, the central issue is the capillary structure (wick). For our application, the best-known wicks, made from screens or

[*]Representing the heat-pipe team at KPI

565

sintered powder metals, although durable, are marginal in power capacity. Metal felt wicks, comparatively unknown, offered much more comfortable heat-transfer safety factors. We demonstrated these safety factors in previous years, on full-scale receivers, thereby completing stage 2.

Our present challenge is to demonstrate durability. Unlike wicks made from screens or sintered powder metal, our metal felt wicks are extremely porous -- greater than 96%. This produces the high permeability and small pore sizes needed for our power and size requirements. But the low solid fraction also makes the wick vulnerable to damage by corrosion and by the mechanical load imposed by capillary forces. We have observed both effects. In addition, we have seen the development over time of isolated warm or hot spots on some heat pipes. We reviewed our work in these areas several years ago [4]. Our progress since then is the primary focus of this paper.

Our efforts to minimize corrosion have concentrated on removing oxygen from the heat pipe. Previously, we showed that this could be done effectively using sodium, in a process analogous to vapor degreasing [5]. More recently, we have explored the possibility of a simpler and quicker technique that uses very high temperature vacuum baking. Here, we report encouraging details of a just-completed 5000-hour sub-scale test.

Our work on wick mechanical strength has proceeded by two approaches: buttressing the strength of our present wick, using an exoskeleton [4], and exploring alternate metal felts. An exoskeleton was used in the afore-mentioned 5000-hour test. Early x-ray diagnostics indicated that the exoskeleton was separating, albeit slowly, from the wick in this sub-scale test. We have seen similar separations after short-term full-scale tests. As a result, we began efforts to improve the wick-exoskeleton bond strength, which we report on here. A notable feature of our effort on wick strength is the involvement of the Heat Pipe Laboratory of the National Technical University of Ukraine (KPI). This involvement was established under the DOE's Initiatives for Proliferation Prevention [6], and contracted through the National Renewable Energy Laboratory (Golden, CO). KPI has proposed and fabricated metal-felt designs that use larger diameter fibers for increased robustness. Our models indicate that some of these designs approach the power capacity of our present best, at least in some applications. Details are now being worked out on how to fabricate one of these designs on a full-scale dome-shaped absorber. Sub-scale limit tests on these wicks are just beginning, and are described here.

Developing an understanding of absorber hot spots has been more problematic. Since corrosion has been brought under control, this behavior has been observed sporadically, and only in on-sun tests, which, because of cost and time constraints, are infrequent. Recently, we were able to show for the first time that wick flaws exist, which, at least under some conditions, could cause hot spots. Here we describe our flaw-detection method, positive results, and plans for flaw-tolerant designs.

Finally, we wish to describe a novel technique we have developed to measure two-phase liquid permeability in a metal felt wick, under simulated receiver conditions. At present, our wick model estimates this permeability using a theoretical dependence on vapor fraction in the wick [7]. Compounding the uncertainty, it obtains the vapor fraction from a linearized version of the pore-size distribution determined by mercury porosimetry, and the local computed vapor-liquid pressure difference. Our new technique provides a direct, measured relationship between the two-phase liquid permeability and the vapor-liquid pressure difference.

HEAT PIPE RECEIVERS

The dish/Stirling heat pipe receiver is a vacuum vessel with a solar absorber (heated surface) at one end, and Stirling-engine heater-head tubes at the other. The vacuum vessel contains a small pool of sodium. Inside the receiver, the absorber is coated with a capillary structure (wick), which distributes sodium from the pool over the entire heated area. The sodium is evaporated from the wick, condenses on the heater-head tubes, and is returned to the pool by gravity. At the design point, the sodium vapor is nearly isobaric and saturated, so the system is close to isothermal. The operating temperature is typically 750 C. At this temperature, sodium is the heat-transfer fluid of choice, because its vapor pressure is sub-atmospheric, yet its vapor density and latent heat of vaporization are high enough to support the required heat fluxes.

In a directly-illuminated receiver, the heater tubes are not isothermal, because of non-uniform illumination. The peak temperatures on the tubes limit the average temperature of the tubes and thus the overall heat transfer and system operating temperature. This in turn limits the power and efficiency, compared to the isothermal operation of a heat-pipe receiver. We have confirmed this in our back-to-back tests of a directly-illuminated receiver and a heat-pipe receiver [1]. The heat pipe's isothermal operation also has the potential to extend the lifetime of the heater-head tubes. Another advantage is that the increased thermal mass of heat pipes can allow system startup to be delayed to higher temperatures, easing starter-motor and battery requirements. And finally, heat-pipe receivers offer much greater design flexibility for hybridization, since the solar and flame-heated surfaces can be optimized independently [8].

The central issue in heat-pipe receivers is the wick/solar absorber. The wick is a porous structure performing many functions. It is a pump, a sodium conduit, and, together with the sodium, a heat conductor. The concave interfaces (menisci) that form on its pores support the pressure difference between the liquid sodium within and the vapor pressure without. This enables the wick to retain sodium to some height above the pool. For the simplified case of a spherical-segment meniscus, the pressure difference is given by the Laplace bubble-equilibrium equation [9],

$$\Delta P \equiv P_V - P_L = \frac{2\sigma}{r} \qquad (1)$$

where r is the meniscus radius of curvature, and σ is the surface tension. Heat, applied to the absorber, conducts through the wick and sodium to the menisci, where evaporation occurs. Evaporative depletion decreases r, which provides the pressure increment required to move sodium up the wick, against viscous forces. As the power is increased, or as the wick height is extended, the vapor-liquid pressure difference increases, causing r to decrease. Eventually, r is too small to be supported by the pore, and vapor intrudes into the pore. This explains the limit on various power/lift-height combinations. If the contact angle is zero (a good approximation for all the liquids considered here), then

$$\Delta P_{max} = \frac{2\sigma}{R_p} \qquad (2)$$

where R_p is the pore radius (or in the case of non-circular pores, the effective pore radius). Decreasing the pore size has limited benefit – for example, in a bundle of capillary tubes, the viscous pressure drop would increase faster than the capillary pressure ($1/R_p^2$ vs $1/R_p$)[10]. Increasing the wick thickness may also help, by reducing the viscous pressure loss, but this is limited by the concomitant increase in absorber temperature (because of heat conduction) and risk of nucleation of bubbles that could become trapped in the wick. Thus, to further improve performance, the structure itself must be fundamentally changed. This is what led us to metal felt wicks.

Our wicks are fabricated from commercial metal-felt blankets that are draped over the heated surface, trimmed to fit, and sintered in place in a vacuum furnace. Our present design uses 3 layers of Bekaert's Bekipor® WB 8/300 felt (8 micron diameter fibers, each layer 300 grams/m² areal density), sintered while compressed to a thickness of 3.75 mm. Unlike our first wicks, which were made from layers of screen or from sintered powder metals, felt metal wicks are extremely porous, providing a structure with very small pores and at the same time, having greatly reduced surface area on which viscous forces act. This translates into higher power/lift combinations. In addition, felt wicks have a distribution of pore sizes, which means that, as the vapor-liquid pressure difference increases, the largest pores fill with vapor first. Vapor begins to form within the wick, escaping through the largest pores, while the smaller pores continue to transport liquid. The result is lower temperature at the absorber and therefore less likelihood of trapped-bubble formation, and gradual rather than abrupt dryout: the wick is tolerant of and even benefits from vapor formation within.

CORROSION REDUCTION

Although our felt wicks provided the superior power capacity we expected, they suffered early failures because of corrosion [5] The attack was on the wick, not the vacuum vessel, and in some cases resulted in perforation of the wick, with consequent loss of heat transfer and localized hot spots on the absorber. We have used felts produced by Bekaert Corporation from Type 316L stainless steel fibers, either 4 or 8 microns in diameter. The cause of the corrosion is believed to be oxygen contamination of the sodium [5]. The contamination comes mainly from the native oxide on the fibers that make up the felt, and is exacerbated by the small size of the fibers [5]. The problem is also aggravated by the concentration and deposition of contaminants at flow sinks on the wick, where liquid flow ends, and evaporation leaves the contaminants behind [5].

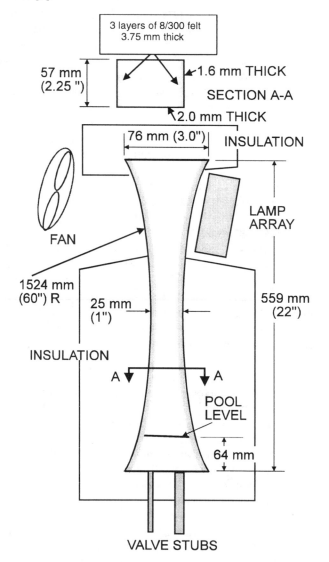

Figure 1. Subscale heat pipe & test rig.

We have considered several approaches to reducing corrosion. The primary ones are: using materials that are more corrosion resistant, and reducing the oxygen content of the heat pipe. Among alternative materials, nickel is available as felt, and there is some indication that it is insensitive to oxygen content [11]. Unfortunately, the high-temperature strength of nickel is poor: the ASME allowable design stress for nickel at 650C is only about 8 MPa, versus about 12MPa for Inconel 600, and 50MPa for 316L [12]. In a subscale test using nickel, we observed re-crystallization of the fibers into nodules.

Meanwhile, we have demonstrated that cleaning our heat pipes using a sodium-vapor refluxing technique practically eliminates corrosion damage [5]. Using this approach, we have seen no new cases of wick perforation in our long-term sub-scale tests. Now we have turned our attention to devising a simpler and faster cleaning procedure. Very-high-temperature vacuum baking was identified as a possibility, based on vacuum-brazing technology. Our latest sub-scale heat pipe was cleaned by simply vacuum baking for two hours at 950C. This was accomplished by heating the device in an air oven, while maintaining internal vacuum with a turbo-molecular pump station. The heat pipe was then tested as illustrated in Figure 1. We use quartz lamps to produce incident fluxes on the order of 60 watts/cm^2, and throughput powers of about 2600 watts. Approximately 5000 hours of operation at a vapor temperature of 750C have been successfully completed. While these results are encouraging, a definitive verdict awaits metallurgical examination, which has just begun.

WICK STRENGTHENING

In early full-scale and sub-scale tests, we found that our felt wicks became compressed in thickness after only tens of hours. In sub-scale tests, we have observed this by periodic x-ray examinations. The wicks became progressively thinner, especially in the heated zone. Compressed wicks have smaller pores, and, as mentioned previously, increased flow resistance (decreased permeability). This reduces their safety factors for power capacity

We have been pursuing a number of possible remedies to wick compression. The first is to support the free surface of the wick with an exoskeleton. Our exoskeleton is formed from perforated sheet metal (PM). It is used to compress the felt to the desired thickness during sintering, and in the process, becomes sintered to the wick. Figure 2 shows an example of this design. Figures 3a and 3b show x-rays of a sub-scale heat pipe after 286 and 1085 hours of testing. This heat pipe was prepared by grit blasting the PM and the absorber, and the PM/wick/absorber was sintered at 1050C for 2 hours. It is apparent that the wick is slowly separating from the PM, and compressing. Nevertheless, this same heat pipe has successfully completed 5000 hours of testing. Previous tests without the PM failed after fewer than 2000 hours when the wick compressed sufficiently to limit liquid flow to the heated zone. In the last

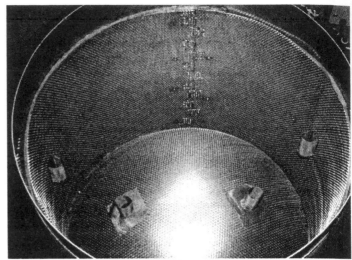

Figure 2 Full-scale absorber dome and hybrid cylinder with exoskeleton/wick.

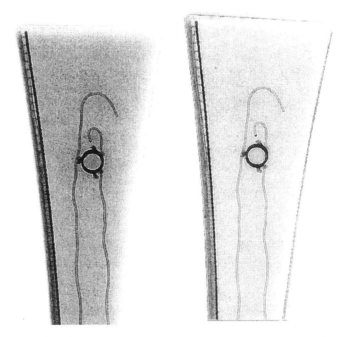

Figure 3. X-rays of subscale heat pipe at 286 hours (left) and 1085 hours (right). Periodic structure on left side is the perforated metal. Objects in the center are thermocouples and a spacer. Separation of the wick is evidenced by the light area between it and the perforated metal, near the top left side of each image.

year, we have explored strengthening the PM/wick bond. We have fabricated tensile-test specimens of the PM/ wick/ substrate assembly, prepared in various ways. Figure 4 shows results obtained in exploring the effects of sinter temperature, time, and grit blasting. The best results were obtained without grit blasting, sintering for four hours at 1100C. We are implementing these results in our latest receivers. We are also

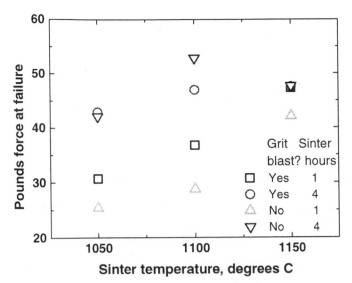

Figure 4. Data on wick/wall bond optimization.

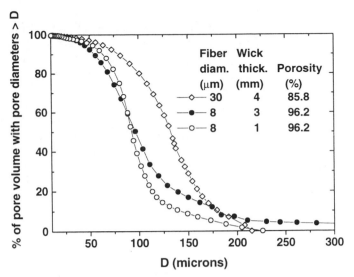

Figure 5. Wick pore distributions

exploring enhancing the wick/substrate bond by depositing nickel on the absorber and the PM.

We have also considered alternative materials and structures, obtaining samples of felt from Bekaert made of Hastalloy HR, 600-series Inconel and FeCrAlloy, as well as a sample of knitted metal fabric. We have characterized some of these for pore size and permeability, but none have been incorporated into a heat pipe yet.

A different approach to dimensional stability has been taken at the Heat Pipe Laboratory of the National Technical University of Ukraine (KPI). They have produced sintered felt wicks using 30-micron diameter fibers of the Soviet steel 03X18H9T-VI and more recently, 316L. Obviously, these wicks can be much stiffer than 8/300 felt – fiber stiffness varies as the third power of the diameter [13]. They can also be more permeable – felt permeability varies as the square of the wire diameter, for a given fiber arrangement and porosity [14]. Less favorably, we expect these larger-diameter fibers to shift the pore-size distribution towards larger values. Our measurements confirm the much-greater stiffness of these wicks, as well as the increased permeability and pore size. Here, we focus on one of the most promising variants: 30-micron fibers, 3-mm long, with porosity $\varepsilon \approx 0.86$. Its pore size distribution is compared with our 8/300 design, in Figure 5. The permeability of both wicks is approximately 250 square microns. We used our wick-performance computer model [7] to compare the thermal performances. Figure 6 shows the results. The calculations were for a receiver dome having a 457-mm rim diameter and a 244-mm radius of curvature, with condensate returned by gravity to the central area of the wick. The KPI wick is 4-mm thick. We assumed the flux distribution of Sandia's Test Bed Concentrator [15]. The results are presented as maximum computed wick vapor fraction versus receiver throughput

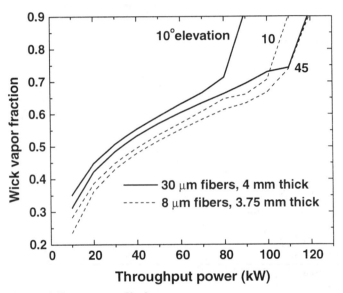

Figure 6. Dryout predictions.

power, for sun elevations of 10° and 45°. They show that the KPI wick is nearly equivalent to the 8/300 wick in power capacity at 45°, and about 20% lower at 10°. The KPI wick's combination of power capacity and wick robustness makes it a leading candidate for further testing. Short-term testing (1000 hours) in our quartz-lamp test rig is just getting underway, and work is beginning on how to apply this wick to our domed absorbers.

HOT SPOT INVESTIGATIONS

During our early problems with wick corrosion, we experienced hot spots in both sub-scale and full-scale tests. Since corrosion was brought under control, hot spots have developed in only a small number of the full-scale tests. For example, in the last two full-scale tests, only one hot spot developed. The two tests used identical receivers, having 8/300 wicks and exoskeletons. In the first, we operated the receiver for a total of 50 hours, with gentle (low thermal flux) sunrise starts. After about 10 hours, a warm spot developed, but did not worsen. The temperature excursion was less than 10 C above adjacent areas, as determined with a solar-blind infrared camera. Post-test examination revealed a small copper lump imbedded in the wick. This was the result of copper (associated with the manufacture of the felt fibers) being dissolved by the sodium and deposited at a flow sink. A similar warm spot and copper lump also developed in the second receiver. Because the temperature excursion is small and stable, and its cause understood, the problem is not regarded as serious.

In the second test, the receiver was treated gently for the first 20 hours, and then was subjected to a series of full-power frozen-sodium starts. On the 48th start, a hot spot developed just above the pool. Hot spots are a serious problem, because they are tens of degrees C above adjacent areas, and because they worsen with time. Before we removed the sodium from the receiver, we examined the absorber with ultrasound and a developmental thermographics technique [16]. Both of these techniques have the potential to detect abnormalities in the wick. There were no indications that the area of the hot spot differed from the rest of the wick. The sodium was evaporated from the receiver under vacuum, and a 130-mm diameter disc including the hot spot was cut out of the absorber/wick. It was potted in epoxy, sectioned and polished. Examination revealed nothing to distinguish any one section from the others.

One of the theories advanced to explain hot spots is that a sodium-vapor bubble nucleates at the absorber/wick interface, preventing liquid sodium from reaching the absorber. A simple heat-conduction model shows that a disc-shaped vapor bubble 6 mm in diameter would cause a temperature excursion similar to what we observed [17]. But could such a bubble form? To begin, it would require a nucleation site (a pre-existing free surface in the sodium), and vapor pressure high enough to begin to inflate a bubble [18]. The probability of a suitable nucleation site is difficult to establish. Micro-cracks, naturally found in metals, could serve as such. The presence of non-condensable gas, possibly diffused from the metal, would enhance nucleation [19]. Actual magnitudes are difficult to quantify. Even if nucleation occurred, vapor should only penetrate the largest pores in the wick. The occurrence of dryout would depend on the distribution and connectivity of large pores. A flaw, such as a delamination or a cluster of large pores, would seem necessary. To shed further light on this question, we built the apparatus shown in Figure 7, using another section out of our most recent receiver. The idea was to use a non-wetting solder (Silvabrite 100) as a surrogate for sodium vapor, forcing it into the absorber/wick interface region. The solder is then frozen, carefully controlling the solidification wave so that the solder remains in place. We chose the solder pressure as follows. The incident flux at the hot spot was about 36 watts/cm², which a

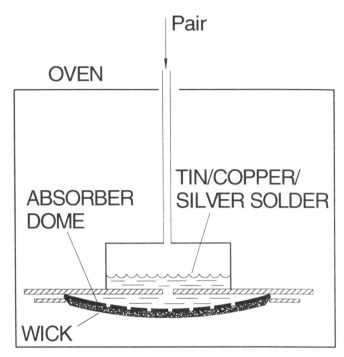

Figure 7. Solder intrusion apparatus.

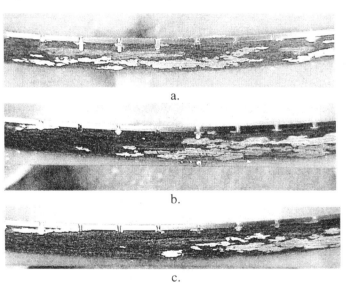

a.

b.

c.

Figure 8. Solder intrusion results. Orientation as in Figure 7. Absorber blanketing evident in (b), 1st hole on left, and (c), 2nd hole on left.

simple conduction calculation indicated would produce 775 C at the wick/absorber interface (for 750 C receiver vapor temperature). This corresponds to a liquid/vapor pressure difference of 8.5 kPa. Scaling this by the solder/sodium surface tension ratio, the solder pressure should be 40kPa. In reality, solder leaked through the wick at about 16 kPa (we expected 21 kPa, based on bubble point measurements), and we were limited to 10 kPa during freezing to avoid continued leakage. A few of the potted/sectioned/polished results are shown in Figure 8.

They demonstrate that the wick is far from homogeneous. The intrusion in the mid-plane of the wick would not be a problem: heat-pipe transport would occur across these volumes. But the ~6-mm intrusions next to the absorber would be a problem, because nucleation there would dry out the absorber. It remains to be shown if nucleation could occur. But even if not, these flaws would be a problem, if normally-expected vapor "fingers", extending inward from the free surface, reach them. The source of the flaws remains uncertain. They may have formed as a result of the frozen starts, the initial fabrication, or both. We believe our wick could be made flaw-tolerant, by underlying it with a simple structure that could keep the absorber wetted at all times. This structure would only need to supply sodium over flaw-size distances. Possibilities include a compacted layer of felt, or inexpensively photo-etched micro-channels. This development is now underway.

TWO-PHASE PERMEABILITY

Our wick-performance computer model includes vapor intrusion into the wick [7]. It models the effect of the vapor fraction α on the liquid permeability k_l with the theoretical dependence

$$k_l = k(1-\alpha)^3 \qquad (3)$$

where k is the single-phase permeability. It calculates α from the vapor/liquid pressure difference, by calculating the radius r of the smallest vapor-filled pore from Eq. 2, and using that to calculate the vapor fraction α, from a linear approximation to Figure 5. This methodology has only been validated indirectly, in terms of the overall accuracy of the model. We recently developed a direct determination of k_l as a function of the vapor/liquid pressure difference. The apparatus we used is illustrated in Figure 9. We used air as a surrogate for sodium vapor, and methanol in place of sodium. The wick internal gauge pressure is

$$p(z) = \rho g h_{out} - \rho g [h_{out} - h_{in}]\frac{z}{z_{max}} \qquad (4)$$

where ρ is methanol density and g is gravitational acceleration. Choosing $h_{out} = h_{in}$ makes the wick internal pressure uniform. When $P_{air} = \rho g h_{in}$, the methanol is just prevented from flooding the wick free surface. Increasing P_{air} causes air to intrude into the wick, slowing the flow of methanol. We measure the flow rate with a beaker and stop watch, for various settings of P_{air}. The permeability is calculated from [20]

$$k_l = \frac{Qv}{Ag} \qquad (5)$$

where Q is flow rate, ν is kinematic viscosity, and A is wick cross sectional area. The pressure difference $P_{air} - \rho g h_{in}$ should be equivalent (in percent intrusion) to a sodium vapor/liquid pressure difference $\sigma_{sodium}/\sigma_{methanol}$ times as large. Our measurement (Figure 10), indicates that the permeability model can be significantly improved by using the real rather than the

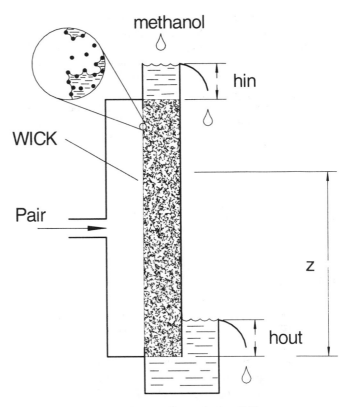

Figure 9. Schematic of two-phase permeability apparatus.

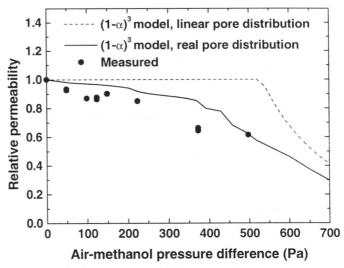

Figure 10. Relative permeability comparison.

linearized pore size distribution. These results should be regarded as preliminary.

SUMMARY

A simple alternative to sodium reflux cleaning, vacuum baking for 2 hours at 950 C, has enabled a sub-scale heat pipe with a metal-felt wick to survive 5000 hours of testing. Exoskeletons show promise of dimensionally stabilizing metal felt wicks, but further sinter optimization is required. Felt wicks with 30-micron fibers have projected power capacity almost as good as those made with 8 micron fibers, while being much more robust. A solder intrusion method has revealed the existence of wick flaws that could explain observed hot spots. Flaw-tolerant designs involving wick sub-layers are being investigated. Two-phase liquid permeability has been measured using a new method that mimics actual service conditions, suggesting improvements to our wick model.

ACKNOWLEDGMENTS

Sandia National Laboratories is a multi-program laboratory operated by Sandia Corporation, a Lockheed Martin Company, for the U.S. Department of Energy under Contract DE-AC04-94-AL85000.

REFERENCES

[1] Andraka, C.E., Rawlinson, K.S., Moss, T.A., Adkins, D.R., Moreno, J.B., Gallup, D.R., Cordeiro, P.G., Johansson, S., 1996, "Solar Heat Pipe Testing of the Stirling Thermal Motors 4-120 Stirling Engine," Proceedings, 31st Intersociety Energy Conversion Engineering Conference, Washington DC.

[2] Grasse, W., ed., 2000, "Solar PACES Annual Report", International Energy Agency, Paris, France, p.2.9.

[3] Washom, B., 1984, "Parabolic Dish Stirling Module Development and Test Results," Paper No. 849516, Proceedings, 19th Intersociety Energy Conversion Engineering Conference, San Francisco, CA.

[4] Adkins D.R., Andraka C.E., Moreno J.B., Rawlinson K.S., Showalter S.K., Moss T.A, 1999, "Heat Pipe Solar Receiver Development Activities At Sandia National Laboratories", Proceedings, ASME SED Conference, Maui HI.

[5] Adkins D.R., Rawlinson K.S., Andraka C.E., Showalter S.K., Moreno J.B., Moss T.A., Cordeiro P.G., 1998, "An Investigation of Corrosion in Liquid-Metal Heat Pipes", Proceedings, American Society of Mechanical Engineers International Mechanical Engineering Congress and Expositions, Anaheim, California.

[6] Gottemoeller, R., 1999, "Program Strategy: Initiatives for Proliferation Prevention", Department of Energy, Washington, D.C.

[7] Andraka C.E., 1999, "Solar Heat-Pipe Receiver Wick Modeling", Proceedings, ASME SED Conference, Maui HI.

[8] Moreno, James, Rawlinson, Scott, Andraka, Charles, Mehos, Mark, Bohn, Mark S., and Corey, John, 1999, "Dish Stirling Hybrid Receiver Sub-Scale Tests and Full-Scale Design", Proceedings, 34th Intersociety Energy Conversion Engineering Conference, Vancouver, BC.

[9] Dwyer, O.E., 1976, "Boiling Liquid Metal Heat Transfer", American Nuclear Society, Hinsdale, IL, p.5.

[10] Schlicting, H., 1960, Boundary Layer Theory", McGraw-Hill Book Company, New York, p. 11.

[11] Foust, O.J., ed., 1972, "Sodium NaK Engineering Handbook, Vol. 1", Gordon & Breach, Science Publishers Inc., New York.

[12] Haynes International, 1995, "Haynes 230 Alloy", Haynes International, Kokomo, IN, p. 7.

[13] Avalone, E. A., and Baumeister, B., 1987, "Mark's Standard Handbook for Mechanical Engineers", McGraw-Hill Book Co., New York, p. 5.3.

[14] Jackson, G. W., and James, D. F., 1986, "The Permeability of Fibrous Porous Media", The Canadian Journal of Chemical Engineering, **64**,pp. 364-374.

[15] Moreno, J.B., Andraka, C.E., Moss, T.A., Cordeiro, P.G., Dudley, V.E., and Rawlinson, K.S., 1994, "On-Sun Test Results from Second-Generation and Advanced-Concepts Alkali-Metal Pool-Boiler Receivers", Sandia National Laboratories Report SAND93-1251.

[16] Valley, M., Roach, D., Dorrell, L, and Mullis, T., 1998, "Evaluation of Commercial Thermography Systems for Quantitative Composite Inspection Applications," Second Joint NASA/FAA/DoD Conference on Aging Aircraft.

[17] Rawlinson, K. S., 2000, private communication.

[18] Dwyer, p. 13.

[19] Dwyer, p. 8.

[20] Bear, J., 1972, "Dynamics of Fluids in Porous Media", American Elsevier Publishing Co., New York, pg120.

Proceedings of IECEC'01
36th **Intersociety Energy Conversion Engineering Conference**
July 29-August 2, 2001, Savannah, Georgia

IECEC2001-RE-06

HYDROGEN FROM SOLAR ENERGY: AN OVERVIEW OF THEORY AND CURRENT TECHNOLOGICAL STATUS

Paul A. Erickson / University of Florida

D. Yogi Goswami / University of Florida

ABSTRACT

Hydrogen production from solar energy sources and water becomes a critical technology in shifting to a renewable energy economy. While the shift to such an economy is still a few years away, it is important to review the current methods and state of the art of hydrogen production by solar means. This paper reviews the theory and practice of current hydrogen production methods by use of solar energy and water. Such a review gives an overall perspective of recent developments in the technology. Key areas are thus identified for further development and research. This review also allows those unfamiliar with solar hydrogen production to become acquainted with the principles, literature and current technological status of the water splitting process.

The following methods of water splitting are reviewed and discussed: electrolysis by solar derived power, photoelectrochemical and photochemical hydrogen production, direct thermal decomposition of water, thermochemical cycles for hydrogen production, and biological hydrogen production. The accomplishments in each research area are discussed.

INTRODUCTION

Solar produced hydrogen is an enticing research topic because it allows conversion of cheap and abundant resources (sunshine and water) to a practical and widely usable fuel (hydrogen). Although solar energy is diffuse when compared to other energy sources the sun provides an unlimited energy source without adverse environmental impact.

Hydrogen becomes a practical means of storing and effectively concentrating solar energy with minimal environmental damage. The potential direct environmental impacts of using hydrogen are limited to high temperature reactions when hydrogen is burned with air producing a potential for nitrogen oxide formation. Hydrogen can be used in conventional thermal energy systems or it can be reacted with oxygen in a fuel cell to directly produce electricity. The reaction of hydrogen and oxygen produces water

which is the original source for hydrogen by the methods to be discussed.

However enticing, the practical means of producing hydrogen from solar energy and water is difficult, complex, and currently very expensive. Several methods have been discovered to produce hydrogen from splitting water into its hydrogen and oxygen components. The methods included in this review are electrolysis by solar derived electrical power, photoelectrochemical and photochemical methods, thermal decomposition of water, thermochemical hydrogen production, and biological hydrogen production. It is the objective of this paper to provide an overview of each method and report on the current status of solar to hydrogen production. Key areas of interest are identified.

THEORETICAL WATER SPLITTING

The water splitting reaction is shown in equation 1:

$$H_2O + Energy \rightarrow H_2 + \frac{1}{2}O_2 \qquad (1)$$

The change in enthalpy (ΔH) can come from a change in Gibbs free energy (ΔG) or a change in entropy (ΔS) at a temperature (T) as shown in equation 2.

$$\Delta H = \Delta G + T\Delta S \qquad (2)$$

For the water-splitting reaction the change in enthalpy at 25° C is 285.58 kJ/mol. This is the theoretical minimum energy required to break water down into its components. In practice the energy required is higher due to irreversibilities in the process and the fact that high rates of H_2 production are desired. There are other practical limitations and irreversibilities associated with the concentration and harvesting of solar power, inefficiencies and irreversibilities in each individual method, and the pressurizing and/or liquefying for storage of the hydrogen product. The energy to drive the water splitting reaction in the methods discussed ultimately derives from solar energy but may come from intermediate electrical, chemical or thermal sources. The main concept behind each hydrogen production method discussed is to utilize intermediate steps to drive the water splitting reaction by way of electrical, chemical, thermal or a combination of such energies derived from solar flux. Ideally the path from solar to hydrogen energy would be as direct as possible to avoid irreversiblities associated with multiple steps but direct paths often are not plausible given material and economic constraints.

The methods or paths presented each have various limitations and benefits associated with them due to economics, thermodynamics, practical material usage, current experience and knowledge, etc.

EFFICIENCIES AND COST TARGETS OF SOLAR HYDROGEN PRODUCTION

Table 1 Efficiencies and Estimated Cost of Hydrogen Production

Method	Theoretical Efficiency	Achieved Efficiency	Estimated Cost	DOE goal
PV electrolysis	27.8% (Pyle et. al., 1996)	13% Calculated for Multi-junction PV (Bolton, 1996)	41$/GJ (Padro and Putsche, 1999)	
Solar heat engine and electrolysis	36% (Williams, 1980)	18.8% Calculated for Dish Stirling (Pyle et. al., 1996)	46 $/GJ projected Power Tower (Glatzmaier et. al., 1998)	
Photoelectro-chemical	31% Single Photosystem (Bolton, 1996)	12.4% Concentrated light 7.8% Natural light (Khaselev and Turner, 1998)	24 $/GJ projected (DOE 1999)	9-14 $/GJ (DOE, 1999)
Thermal Decomposition	40% (Kogan et. al. 2000)	2.1% (Pyle et. al. 1996)	??	
Biological	31% (Bolton, 1996)	11% Transient only (Bolton, 1996)	??	
Thermochemical cycles	~40% (Wendt, 1988)	18% From Heat only (Bockris, 1975)	??	
Fossil Based Hydrogen Production	N/A	N/A	10$/GJ (Sherif et. al. 1999)	6-8 $/GJ (DOE, 1999)

As reported by Pyle et. al. (1996), the National Renewable Energy Laboratory (NREL) has established a goal of 25% for thermal efficiency of converting solar energy into hydrogen. However, because the energy source (solar flux) is basically free, a more useful goal rather than thermal efficiency is the overall hydrogen production cost. With a low efficiency system, one is forced to collect more solar power in order to produce a given rate of hydrogen flow. However, if the low efficiency system is inexpensively enlarged such a system may achieve the cost goal and become more practical before a highly efficient and more expensive system. Current goals established by the U.S. Department of Energy for hydrogen production from renewable resources are 9-14$ per gigajoule hydrogen (hhv) (DOE, 1999). The efficiencies and estimated costs of each method are found in Table 1. Currently the methods of electrolysis with photovoltaic and solar heat engine derived electricity are both the most mature technologies and also have the highest achieved efficiencies. In general for solar hydrogen production, costs come from capital equipment rather than from actual harvesting of the solar source (fuel costs). Capital costs are normally high for solar derived power systems; thus the bulk of the lifetime cost must be paid before power (or hydrogen) is produced. However, as energy prices go up and reflect the true cost of fossil fuels including the externalities of social, environmental, and health costs associated with them, solar derived power begins to become cost competitive.

ELECTROLYSIS WITH SOLAR DERIVED ELECTRICITY

Electrolysis uses electrical energy to separate H_2 from the host molecule (H_2O). The electrical energy can be derived from any electrical source. A schematic for a simplified polymer or acid electrolyte electrolysis system is shown in Figure 1. For this type of electrolyzer, electrical energy is used to drive the electrons to the anode where the protons and electrons form hydrogen. This process is essentially reversed when electrical energy is produced in a solid polymer or acid type fuel cell. Although solid polymer electrolyzers are becoming more

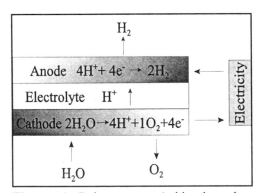

Figure 1 Polymer or Acid electrolyte electrolysis process

available (Friedland and Speranza, 1999) the most widespread practice of electrolysis uses electrolyzers with an alkali electrolyte.

As presented by Ohta (1979), the electrolysis proceeds for the alkaline electrolyzer as follows. Cathodic and anodic reactions for the alkaline type electrolyzers are found in equations 3 and 4 respectively.

$$2H_2O + 2e^- \rightarrow 2OH^- + H_2 \qquad (3)$$

$$2OH^- \rightarrow H_2O + \frac{1}{2}O_2 + 2e^- \qquad (4)$$

Equilibrium potentials at both the cathode (E_c) and the anode (E_a) at 1 atm pressure for hydrogen and oxygen and 25°C are written as equations 5 and 6.

$$E_c = -0.828 - 0.059\log a_{OH^-} \qquad (5)$$

$$E_a = 0.401 - 0.059\log a_{OH^-} \qquad (6)$$

Where a is the activity of the hydroxyl radical at the respective cathode and anode. The difference of the anode and the cathode equilibrium potentials yields a theoretical decomposition voltage of 1.229 V. The corresponding electrical energy (Gibbs free energy) required for this reversible case at 25° C is 236.96 kJ/mol. This value is noticeably less than the 285.58 kJ/mol needed to drive the water splitting reaction so the balance of energy (48.62 kJ/mol) to complete the reaction would be absorbed from the surroundings. However, practical electrolysis at 25° C does not follow the theoretical case without a thermal gradient but requires the full amount of driving energy (285.58 kJ/mol) from the electricity. The voltage required to drive the practical process is 1.48 volts. This voltage is referred to as the thermoneutral voltage, or where water can be split isothermally at 25°C. Actual used voltages (1.75V-2.05V) are higher than this value (known as overpotential) for electrolyzers because of losses and high rates of hydrogen production required. As explained by Linkous et. al. (1995), addition of heat to the electrolysis process can drive the reversible voltage down and thus minimize the electrical energy required. Although thermodynamically the process requires the same amount of energy, heat is a larger source of the enthalpy change at high temperatures. Because heat can be made more efficiently than electricity from solar sources, overall cost may be minimized by such high temperature electrolysis application (Linkous et. al., 1995). While the required change in enthalpy remains at 285.58 kJ/mol the electrical power represented by the Gibbs free energy is offset by the increase of entropy by higher temperature. Figure 2 shows theoretical regions of where the hydrogen is produced from heat and electricity, and a region where waste heat is also generated. The practical demonstrated cell efficiencies of electrical to hydrogen energy conversion range from 60% to more recently 92% (Dutta, 1990). Recent advances in electrolysis minimize the voltage or overpotential by use of novel catalysts, higher temperature, or hydrocarbon gases to keep the catalyst activation energy low (Pham, 2000).

Electricity for electrolysis can come from any source but the electrical power is typically low voltage dc which is ideally suited for photovoltaic application. Hydrogen from solar energy through photovoltaics and electrolysis is usually limited by the low

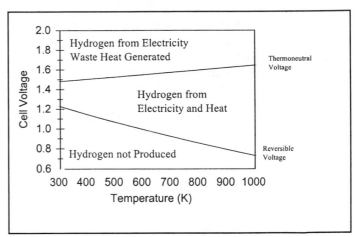

Figure 2 Thermoneutral and Reversible Voltages vs. Temperature (adapted from Divisek, 1990)

efficiencies of the photovoltaic conversion. The photovoltaic quantum effect can only use a portion of the available radiation spectrum due to band gap limitation, which results in low conversion efficiencies of the solar source to electrical energy. As multijunction cells are developed which can use more of the spectrum, overall conversion efficiencies increase as well. Although photovoltaic conversion of sunlight to electricity is relatively inefficient, the concept of producing hydrogen with photovoltaics has been proven and the technology is established. Demonstrations of some solar to hydrogen conversion systems using photovoltaics around the world are briefly outlined by Sherif et. al. (1999). These are:
- the Solar-wasserstoff-Bayern pilot plant in Nuenburg vorm Wald in Germany,
- the HYSOLAR project in Saudi Arabia,
- a system at Schatz Energy Center, Humboldt State University in Arcata, California,
- a system at the Helsinki University of Technology in Helsinki, Finland,
- and a system at INTA Energy Laboratory in Huelva, Spain.

The electricity for electrolysis can also come from heat engines using solar sources including those using the Rankine, Brayton, and Stirling cycles. Of special interest are the Power Tower systems developed and systems using parabolic dish reflectors and a Stirling engine combined with a generator. These power systems produce electrical energy from solar heat which can then be converted to hydrogen through electrolysis. Methods using the Stirling cycle and electrolysis have been calculated to have solar to hydrogen efficiencies of 18.8% (Pyle et. al., 1996) and the use of such systems has been proven (Friedland and Speranza, 1999).

Further development of solar electrolysis systems needs to take place in several areas. Some of the most important are:
- cost reduction of current methods (PV and electrolyzers),
- continued demonstrations of solar powered high temperature electrolysis (via solar heat in combination solar derived electricity),

- methods for decreasing overpotential in electrolyzers,
- increasing the efficiency of photovoltaic electricity (i.e. multijunction cells),
- demonstration of conventional power generation cycles using solar heat to produce energy for electrolysis.

PHOTOELECTROCHEMICAL AND PHOTOCHEMICAL METHODS

Systems that combine the photovoltaic effect and electrolysis in a solution are known as photoelectrochemical cells. Research interest in photoelectrochemical cells is high because these are essentially a marriage and miniaturization of the two well-known processes: photovoltaic electricity production and electrolysis. These cells seem to have potential for being a simple, low cost option for solar hydrogen production. The landmark publication for solar produced hydrogen (Fujishima and Honda,1972) was essentially one of these systems. Research interest in these methods is high as reflected by review articles in previous years (Getoff, 1990; Amouyal, 1995; Bolton, 1996) and breakthroughs continue to push the efficiencies higher and the costs lower (DOE, 1999). In these systems a light absorbing semiconductor acts as the anode or cathode of an electrochemical cell which electrolyzes water. Physical separation of the product stream occurs as there is a natural separation of the anode and cathode where oxygen and hydrogen are evolved. Figure 3 is a simplified schematic of the process (adapted from Miller and Rocheleau, 2000). The research efforts into photoelectrochemical methods are investigating different geometric designs. Some of these designs utilize physical separation of the cathode and anode (electrical wire connection) with both reduction and oxidation steps using light energy in a dual bed approach (Linkous and Slattery, 2000). Research groups are also investigating novel catalyst and semiconductor materials to maximize efficiency.

Although optimal cell designs are still being developed and investigated, photoelectrochemical cells have demonstrated that hydrogen can be produced at over 10% efficiency using

concentrated light for long periods of time (Khaselev and Turner, 1998). Near term efficiencies using photoelectrochemical methods with a solar source should achieve 15% (Miller and Rocheleau, 2000). Difficulties still being addressed with these methods are semi-conductor stability, and interfacial kinetics (Bansal et. al., 2000), and optimization of material properties (transparency, conductivity, resistance to KOH) (Miller and Rocheleau, 1999). Photoelectrochemical cells are currently being developed by research groups including universities and the National Renewable Energy Laboratory (DOE,1999).

Photochemical systems are similar to but are distinguished from photoelectrochemical cells by the difference that in photochemical systems sunlight is absorbed by isolated molecules in solution. Hydrogen and oxygen are evolved on the same particle rather than at a separated cathode and anode. The particle must act as the bandgap material and the electrolyzer.

Because water does not absorb the spectrum of light necessary to initiate water splitting, a photosensitizer must be used to absorb the light energy. In addition the sensitizer must be linked to a catalyst or catalysts to allow the reduction and oxidation steps (electrolysis) to occur. Figure 4 is a minimal scheme for a photochemical system. The bandgap material (photosensitizer) is typically used in connection with a relay compound and catalysts which help the reaction to proceed in the preferred direction (Harriman and West, 1982).

Several bandgap materials are known to have enough energy to split water in a single photon arrangement (Connolly, 1981). However some of these materials may not be suitable for photocatalysis because they will breakdown when used in a aqueous solution. Other more stable materials such as $SrTiO_3$

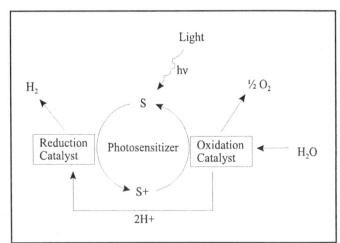

Figure 4 Minimal scheme for photochemical splitting of water.

have such a large bandgap that they only use a small portion of the solar spectrum for water splitting resulting in unacceptable efficiency. Work with photosensitizers for improving the spectral response of the bandgap materials is noted by Mallinson et. al.

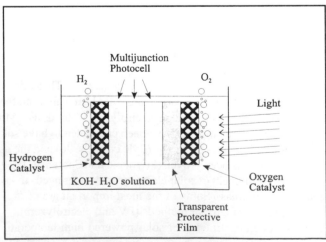

Figure 3 Representative photoelectrochemical process schematic (adapted from Miller and Rocheleau, 2000).

(1999). In principle the smallest bandgap material with good stability would be preferred because it would be able to use a larger portion of the available light spectrum resulting in a higher efficiency.

Demonstrations of this technology, and studies investigating catalysis in these systems (Mallison et. al, 1999) have taken place recently. However, photochemical hydrogen production systems have the following difficulties as outlined in the review by Bolton (1996).

- Photodegradation reactions which destroy the photosensitizer.
- Slow diffusion of excited state molecules to the catalyst sites.
- Back electron transfer.
- Separation of an explosive H_2/O_2 mixture.

While efficient hydrogen production from water may not be currently practical using photochemical methods, research continues in photocatalysis, because of other applications such as photodegradation of pollutants (Goswami and Blake, 1996). Interestingly, photodegradation of pollutants can yield hydrogen as a byproduct (Sartoretti et. al., 1999).

THERMAL DECOMPOSITION

Another process of producing hydrogen from water is to use thermal energy to break water into its components. Such a process is known as thermal decomposition and when solar thermal energy is used the process is known as Hydrogen by Solar Thermal Water Splitting (HSTWS) (Kogan, 1997). As shown in table 2, equilibrium dictates that water will disassociate at elevated temperatures. The products of the disassociation process are H_2O, H_2, OH, O, H, and O_2. Low pressure also aids such disassociation.

Table 2 Equilibrium Mole Fractions of a Water Mixture at Different Pressures and Temperatures

Temperature	2000 K		
Pressure	1 atm	0.1 atm	0.02 atm
H	0.0001	0.0006	0.0016
H_2	0.0058	0.0124	0.0209
H_2O	0.9896	0.9773	0.9607
O	0.0000	0.0002	0.0004
OH	0.0021	0.0042	0.0075
O_2	0.0024	0.0052	0.0088
Temperature	2500 K		
Pressure	1 atm	0.1 atm	0.02 atm
H	0.0052	0.0230	0.0627
H_2	0.0428	0.0843	0.1259
H_2O	0.9110	0.8059	0.6685
O	0.0018	0.0081	0.0226
OH	0.0233	0.0464	0.0705
O_2	0.0160	0.0322	0.0497
Temperature	3000 K		
Pressure	1 atm	0.1 atm	0.02 atm
H	0.0578	0.2114	0.4153
H_2	0.1352	0.1811	0.1398
H_2O	0.6440	0.3239	0.1021
O	0.0244	0.0914	0.1866
OH	0.0914	0.1260	0.1010
O_2	0.0470	0.0062	0.0552

Theoretical efficiency is a function of the level of water disassociation so at extremely low pressure and high temperature theoretical efficiency (neglecting reradiation losses) has been calculated as high as 80% (Kogan et. al., 2000). However, such extremes in temperature and pressure are not feasible. At more practical pressures and temperatures reported to be within the state of the art, Kogan et. al. (2000) calculate 40% theoretical efficiency at 2 mb and 2500 K.

While efficiency increases with higher temperature, some disassociation does occur at temperatures below 2000 K. As long as a feasible method exists for removing the small amount of hydrogen and oxygen from the reaction, disassociation at reasonable temperatures (<2000K) may yield practical amounts of product.

Although the overall process is conceptually simple, the practice of harvesting hydrogen from the product stream in this method is difficult. The main difficulties are in finding materials that will withstand the corrosive environment of disassociated water (radicals in products), and separation of the hydrogen from the product stream. Some schemes for quenching and low temperature separation have been studied (Fletcher and Moen, 1977) but any quenching process introduces a high level of irreversibility (Glatzmaier et. al., 1998) and creates an explosive mixture. A more feasible idea seems to be that of separating the hydrogen at high temperature using semi-porous membranes. This idea has been demonstrated to work with the HSTWS process as reported by Kogan (1998) and by Naito and Arashi (1995). However, sintering and clogging of the porous membrane structure has also been documented which degrades the hydrogen production over time. Although with low efficiency, another separation technique proven with this method is that of using a separator nozzle (Pyle et. al., 1996). The efficient separation of the product stream is currently a technological hurdle for the further development of the thermal decomposition method.

Aside from the materials and separation problems the attainment of temperatures above 2000 K with a solar source is also difficult. According to Kogan et. al. (2000) concentration ratios of 10000 suns are necessary for a practical HSTWS process. Such concentration ratios have been successfully demonstrated with a solar tower and secondary concentrator (Yogev et. al., 1998). Partial thermal decomposition at lower temperatures does however show promise in conjunction with other types of water splitting methods such as electrolysis (Padin et. al., 2000).

THERMOCHEMICAL CYCLES

Direct thermal decomposition is not the only thermal method to break down water. Several chemical cycles driven by thermal energy can be used to decompose water at lower and thus more practical temperatures. These cycles constitute an overall oxidation reduction reaction and require at least four steps although in theory it may be possible to combine these steps into two individual reactions. These are a water consumption step, oxygen production step, hydrogen production step, and a material regeneration step. The basic idea behind these cycles can be shown by the following general equations for a four-step cycle.

General Thermochemical Cyclical Process

$AB + H_2O \pm Heat \rightarrow AH_2 + BO$	Water Consumption
$2BO \pm Heat \rightarrow O_2 + 2B$	Oxygen Production
$AH_2 \pm Heat \rightarrow A + H_2$	Hydrogen Production
$A + B \pm Heat \rightarrow AB$	Material Regeneration

Individual steps may be endothermic or exothermic but the overall cycle would be endothermic as the driver for the overall reaction is thermal energy. The individual reactions may take place at different temperatures. With some cycles one can utilize the endothermic and exothermic nature in different steps to take advantage of the cyclical nature of solar energy harvest. Hybrid cycles (using both thermal and electrical energy) are also possible which would use a work input to drive or enhance some of the reactions.

At a minimum, the reduction oxidation reaction requires one compound or element that can easily change its oxidation state. Thus group 7 elements (chlorine, bromine) and metals (iron, zinc, mercury) are used in many proposed cycles. Also the change in Gibbs free energy for each step should approach zero for the temperature considered (Norbeck et. al., 1996). This allows for recycling of the intermediate products and reduces accumulation of a stable compound or element. Build up and consumption of intermediate compounds or elements will also occur when any reaction does not react to completion, which is the case for nearly all reactions.

One of the earlier thermochemical processes developed by deBeni and Marchetti in 1970 is shown (Casper, 1978).

Mark I Cyclical Process

$CaBr_2 + 2H_2O \rightarrow Ca(OH)_2 + 2HBr$	730°C
$Hg + 2HBr \rightarrow HgBr_2 + H_2$	250°C
$HgBr_2 + Ca(OH)_2 \rightarrow CaBr_2 + H_2O + HgO$	200°C
$HgO \rightarrow Hg + \frac{1}{2} O_2$	600°C

As one can see this process is a multi-step thermal reaction that consumes 1 mol of H_2O producing 1 mol H_2 and ½ mol O_2. Wendt (1988) estimates that 2000 - 3000 different theoretical cycles have been proposed and evaluated. Some of these processes have been physically constructed and continue to be developed and evaluated (Glatzmaier et .al., 1998; Steinfeld et. al., 1999; Sakurai et. al., 1996; Tamaura et. al., 1999; Sturzenegger and Nuesch, 1999). Methods for evaluating these thermochemical cycles are given by Casper (1978). Theoretical efficiencies are estimated at 40% (Wendt, 1988) but are essentially Carnot limited as discussed by Casper (1978). Multiple steps also contribute to lower efficiencies. For individual reaction yields of 94% in a four step cycle, the resultant overall reaction yield is 78% (Williams, 1980). This can result in significant build up and depletion of some intermediate compounds. Practical implementation of cycles has been limited by separation of hydrogen from the product mixture, toxicity and corrosion by the intermediate chemical compounds, matching reaction rates of the individual steps, and integration of the solar energy heat source. Research continues in this field as catalyst

application (Kirillov, 1999) and the above difficulties are addressed.

BIOLOGICAL METHODS

In 1942, certain types of algae were observed to produce hydrogen in their metabolic activities using light energy and similar hydrogen producing photosynthetic bacteria were found in 1949. Since then several researchers have debated how exactly such hydrogen production takes place. Figure 5 shows a representative schematic of such a system. Note that this schematic is essentially the same type of system as is found with photocatalytic methods except that enzymes and biological photosensitizers and relay compounds are used. Thus some of the same limitations of photochemical methods apply to biological methods. Catalysts and engineered relay compounds may be used in conjunction with the biological systems and thus increase the efficiency of the biological methods (Willner and Willner, 1988).

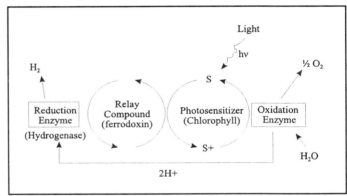

Figure 5 Simplified schematic for biophotolysis (adapted from Miyake et. al., 1999).

Research in biological hydrogen production is accelerating. Recently a system has been developed that uses a metabolic switch (sulfur deprivation) to cycle algal cells between a photosynthetic growth phase and a hydrogen production phase (Melis et. al., 2000).

Photobiological hydrogen technology is not yet fully investigated and has been shown to have promise, but certain hurdles presently must be overcome. Because oxygen is produced along with the hydrogen, the technology must overcome the limitation of oxygen sensitivity of the hydrogen-evolving enzyme. Other systems are being evaluated that use spatial or temporal separation of the hydrogen and oxygen production steps (Ghiradi et. al., 2000). Other major research projects in this field as discussed by Greenbaum and Lee (1999) are linearizing the light saturation curves of photosynthesis, and constructing real world bioreactors that can work against the back pressure of the produced gases. Research projects are underway to screen for naturally occurring organisms that are more tolerant of oxygen, and by genetically engineering new forms of the organisms that can sustain hydrogen production in the presence of oxygen (Ghiradi et. al., 2000).

CONCLUSIONS

Table 3 shows the status of the reviewed methods along with the significant hurdles these methods must overcome to become practical. Shown also are the author's explanatory notes and comments about the methods.

Table 3 Summary of the Solar to Hydrogen production methods.

Method	Status	Technological Hurdles	Notes
PV electrolysis	Commercial	PV efficiency and cost	High temp electrolysis may double efficiency.
Solar driven Heat engine w/ electrolysis	Demonstrated, approaching commercial	Further demonstrations with conventional thermal cycles, cost.	Efficiencies and production costs warrant further investigation and demonstrations.
Direct Thermal Decomposition	Research, transient H_2 production demonstrated	High efficiency in steady state, Materials, Solar Concentrators, Separation of Products	Efficient separation of high temperature products necessary.
Thermochemical Cycles	Research, laboratory cycles demonstrated	Toxicity with intermediate steps. Build up of byproducts. High temperature and corrosion resistant materials.	High theoretical efficiencies, practical issues (see hurdles) seem to be holding back the technology.
Photoelectro-chemical and Photochemical	research, H_2 production demonstrated	Improvement of bandgap materials and catalysts, cell design	A major research topic. Potential for simple lowcost systems.
Biological	transient H_2 production demonstrated	Steady state production, light saturation of H_2 production	Combinations of biological and engineered systems needed.

Research in hydrogen production is accelerating in many areas including all of those reviewed. Currently the only commercial method is the use of electrolysis with photovoltaics. However, because of the quantum nature of photovoltaics, theoretical efficiencies (light to hydrogen) are limited to under 31%. This same low efficiency limits other photoelectrochemical and photochemical and biological methods. Thermal decomposition and thermochemical schemes have higher theoretical limits but are currently less practical because of materials, toxicity, corrosion, and separation limitations. Potential exists for employing several portions of separate methods in a hybrid fashion to increase efficiencies and provide practical implementation

REFERENCES

Amouyal, E. (1995). Photochemical production of hydrogen and oxygen from water: A review and state of art. *Solar Energy Materials and Solar Cells*, Vol. 38, p. 249-276.

Bansal, A., O. Khaselev, and J. Turner (2000). Proceedings of the 2000 DOE Hydrogen Program Review NREL/CP-570-28890.

Bockris, J. (1975). Energy: The solar-hydrogen alternative. John Wiley & Sons: New York.

Bolton, J. (1996). "Solar Photoproduction of Hydrogen: A Review", *Solar Energy,* Vol. 57 n. 1, p. 37-50.

Casper M. (Ed.) (1978). Hydrogen Manufacture by Electrolysis Thermal Decomposition and Unusual Techniques Noyes Data Corporation: Park Ridge, New Jersey.

Connolly, J. S. (Ed.) (1981). Photochemical conversion and storage of solar energy. Academic Press: New York.

Divisek, J. in Wendt, H (editor) (1990). Electrochemical Hydrogen Technologies, Elsevier: London.

DOE (1999) "Report to Congress on the Status and Progress of the DOE Hydrogen Program" U. S. Dept. of Energy, available at http://www.eren.doe.gov/hydrogen/pdfs/bk28423.pdf

Dutta, S. (1990) "Technology assessment of advanced electrolytic hydrogen production" *Int. J. Hydrogen Energy*, Vol. 15 n. 6, p.379-386 .

Fletcher E., and R. Moen (1977) "Hydrogen and Oxygen from Water" *Science,* Vol. 197, p.1050-1056.

Friedland R. and A. Speranza, (1999) "Integrated Renewable Hydrogen Utility System" Proceedings of the 1999 U.S. DOE Hydrogen Program Review. NREL/CP-570-26938.

Fujishima, A., and K. Honda (1972) "Electrochemical photolysis of water at a semiconductor electrode" *Nature*, Vol. 238 n. 5358, p. 37-8.

Getoff, N. (1990) "Photoelectrochemical and photocatalytic methods of hydrogen production: a short review" *Int. J. Hydrogen Energy,* Vol 15 n. 6, p. 407-418.

Ghiradi, M, L. Zhang, J. Lee, T. Flynn, M. Seibert, E. Greenbaum, and A. Melis, (2000) Microalgae: a green source of renewable H_2, *Trends in Biotechnology*, Vol. 18 n.12, p. 506-511.

Glatzmaier, G., D. Blake, and S. Showalter. (1998) "Assessment of Methods for Hydrogen Production Using Concentrated Solar Energy" Natl. Ren. En. Lab. NREL/TP-570-23629.

Greenbaum E. and J. Lee (1999), "Photosynthetic Hydrogen and Oxygen production by Green Algae" U.S. Dept. of Energy CONF-9706179–1.

Goswami, D. Y., and D. M. Blake "Cleaning up with sunshine" *Mechanical Engineering,* Vol.118 n.8, p 56-59.

Harriman, A. and M. A. West, (Eds.). (1982). Photogeneration of hydrogen. Academic Press : New York.

Khaselev, O., and J. Turner, (1998) "Monolithic photovoltaic-photoelectrochemical device for hydrogen production via water splitting" *Science* Vol. 280 n. 5362, p. 425-427.

Kogan A., (1997) "Direct thermal splitting of water and on-site separation of the products-I Theoretical evaluation of hydrogen yield" *Int. J. Hydrogen Energy* Vol, 22 n. 5, p. 481-486.

Kogan A., (1998) "Direct thermal splitting of water and on-site separation of the products-II Experimental feasibility study" *Int. J. Hydrogen Energy* Vol 23 n. 2, p. 89-98.

Kogan A., E Spiegler, and M. Wolfstein, (2000) "Direct thermal splitting of water and on-site separation of the products. III Improvement of reactor efficiency by steam entrainment" *Int. J. Hydrogen Energy*, Vol. 25, p.739-745.

Kirillov, V. A. (1999). "Catalyst application in solar thermochemistry" *Solar Energy*, Vol. 66 n.2, p.143-149.

Linkous C., and D. Slattery, (2000) "Solar photocatalytic hydrogen production from water using a dual bed photosystem" Proceedings of the 2000 U.S. DOE Hydrogen Program Review, NREL/CP-570-28890.

Linkous, C., R. Anderson, and R. Kopitzke (1995) "Development of Solid Electrolytes for Water Electrolysis at Intermediate Temperatures" U. S. Dept. of Energy, DOE/AL/85802–T2-Pt.3.

Mallinson, R., D. Resasco, K. Nicholas, and L. Lobban, "Novel Catalytic Approaches for Hydrogen Production (Photocatalytic Water Splitting)" U.S. Dept. of Energy, DOE/GO/10141--F.

Melis A., L. Zhang, M. Forestier, M. L. Ghirardi, and M. Seibert (2000) "Sustained Photobiological Hydrogen Gas Production upon Reversible Inactivation of Oxygen Evolution in the Green Algae Chlamydomonas reinhardtii" *Plant Physiology*, Vol. 122, p. 127-136.

Miller, E., and R. Rocheleau, (2000) "Photoelectrochemical Hydrogen Production" Proceedings of the 2000 U.S. DOE Hydrogen Program Review, NREL/CP-570-28890.

Miller, E., and R. Rocheleau, (1999) "Photoelectrochemical Hydrogen Production" Proceedings of the 1999 U.S. DOE Hydrogen Program Review, NREL/CP-570-26938.

Miyake, J., Miyake, M., and Asada, Y. (1999). Biotechnological hydrogen production: Research for efficient light energy conversion. *Journal of Biotechnology*, 70, 89-101.

Naito, H., and H Arashi (1995) "Hydrogen production from direct water splitting at high temperatures using a ZrO_2-TiO_2-Y_2O_3 membrane" *Solid State Ionics*, Vol 79, p. 366-370.

Norbeck, J., J. Heffel, T. Durbin, B. Tabbara, J. Bowden, M. Montano (1996) Hydrogen Fuel for Surface Transportation Society of Automotive Engineers, Inc: Warrendale, Pennsylvania.

Ohta, T. (Ed.). (1979). Solar-hydrogen energy systems. Pergamon Press: New York.

Padin, J., T. Veziroglu, and A. Shanin (2000) "Hybrid solar high-temperature hydrogen production system" *Int. J. Hydrogen Energy* Vol 25, p 295-317.

Padro, C. and V. Putsche (1999), "Survey of the Economics of Hydrogen Technologies" Natl. Ren. En. Lab., NREL/TP-570-27079.

Pham, A. (2000) "High Efficiency Steam Electrolyzer" Proceedings of the 2000 DOE Hydrogen Program Review NREL/CP-570-28890.

Pyle, W., M.. Hayes, and A. Spivak (1996) "Direct Solar-Thermal Hydrogen Production from Water using Nozzle/Skimmer and Glow Discharge" Proceedings of the 31st Intersociety Energy Conversion Engineering Conference, IECEC 96535.

Sakurai, M., E. Bilgen, A. Tsutsumi, and K. Yoshida. (1996) "Solar UT-3 Thermochemical Cycle for Hydrogen Production" *Solar Energy*, Vol. 57 n. 1, p. 51-58.

Sartoretti, C., M. Ulmann, J. Augustynski, and C. Linkous (1999). Photoproduction of Hydrogen in Non-Oxygen-Evolving Systems International Energy Agency Hydrogen Programme Annex 10.

Sherif, S., T. Veriziglo and F. Barbir, in Webster, J. G. (Ed.). (1999). "Hydrogen Energy Systems" p.370-402 Wiley Encyclopedia of Electrical and Electronics Engineering Volume 9. John Wiley & Sons: New York.

Steinfeld, A., Sanders, S., and Palumbo, R. (1999). "Design aspects of solar thermochemical engineering–a case study: Two-step water-splitting cycle using the Fe_3O_4/FeO redox system" *Solar Energy*, Vol. 65 n.1, p. 43-53.

Sturzenegger M., and P. Nuesch (1999). "Efficiency analysis for a Manganese-oxide-based thermochemical cycle" *Energy,* 1999, p. 959-970.

Tamaura, Y., Y. Ueda, J. Matsunami, N. Hasegawa, M. Nezuka, T. Sano, and M. Tsuji (1999). "Solar Hydrogen Production Using Ferrites" *Solar Energy*, Vol. 65 n. 1, p. 55-57.

Wendt, H. in Wendt, H (editor) (1990) Electrochemical Hydrogen Technologies, Elsevier: London.

Williams, L. (1980). Hydrogen Power Permagon Press, New York.

Willner I. and B. Willner,(1988) "Solar Hydrogen Production through Photobiological, Photochemical and Photoelectrochemical Assemblies" *Int. J. Hydrogen Energy,* Vol. 13 n. 10, p. 593-604.

Yogev, A., A Kribus, M. Epstein and A. Kogan, (1998) " Solar tower reflector systems : A new approach for high-temperature solar plants" *Int. J. Hydrogen Energy,* Vol. 23 n. 4, p. 239-245.

2001-RE-03

ON-BOARD HYDROGEN STORAGE FOR FUEL CELL VEHICLE

Thomas Vernersson
Div. Applied Electrochemistry
Dept. Chemical Engineering & Technology
Royal Institute of Technology
Stockholm, Sweden

Kristina Johansson
Div. Energy Processes
Dept. Chemical Engineering & Technology
Royal Institute of Technology
Stockholm, Sweden

Per Alvfors
Div. Energy Processes
Dept. Chemical Engineering & Technology
Royal Institute of Technology
Stockholm, Sweden

ABSTRACT

Methods for onboard storage of hydrogen were evaluated for use in a fuel cell vehicle. Compressed hydrogen gas and cryogenic liquid hydrogen seem to be the two most viable options. Both these storage options were modelled, for storage of 5 kg hydrogen, to be implemented in an automotive fuel cell system simulation model.

Hydrogen discharge was simulated for different values of cell stack operating pressure and temperature, using a constant rate of hydrogen release, and the power requirement for heating of the hydrogen to fuel cell stack operating temperature was calculated. The calculations show that compressed gaseous hydrogen storage requires a heating capacity of 0.72 – 1 kW for stack operating temperatures of 343 – 368 K. In the case of liquid hydrogen storage, heating demand for vaporisation and heating of the fuel was calculated to between 10 and 13 kW for stack operating temperatures of 343 – 368 K. The fuel cell stack produces surplus heat that can be used for fuel heating. Calculations show that the heat content of the cooling medium is sufficient to heat the fuel stream to approximately 20 K below stack temperature, with temperature differences in heat exchangers being the limiting factor. The radiator / compartment heating and humidifier will also extract heat from the cooling medium. However, to reach system temperature an auxiliary heat source will be required. This could be in the form of an electrical heater or a hydrogen burner. Also, for liquid hydrogen storage, a power demand arises for maintaining operating pressure inside the storage vessel during hydrogen release. This was calculated to between 13 and 28 W for the fuel cell stack operating conditions simulated, and this power demand can be supplied by directing a stream of released and heated hydrogen through a coil running inside the storage vessel.

INTRODUCTION

Compressed hydrogen gas and liquefied hydrogen are the options pursued by major car manufacturers of the world, (together with methanol storage for onboard reforming, which lies outside the scope of this study). The advantages that compressed gas and cryogenic liquid storage offer in rapid refuelling rate, simplicity of hydrogen discharge, and charge / discharge cycle stability are significant. Depending on vehicle type and application, either of these storage methods could be preferred. Naturally, cost is an issue with these methods as well but manufacturers of graphite fibres and pressure vessels have expressed optimism about lowering costs in large-scale production.

Here, a direct, *i. e.* fuelled on pure hydrogen, fuel cell vehicle is studied. The automotive polymer fuel cell system consists of a 50 kW_e fuel cell stack, compressor, humidifiers and condensers, see Figure 1. The system is operated at pressures between 1.5 to 5 bar(a) and temperatures in the interval of 70- 95 °C.

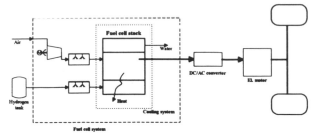

Figure 1. Schematic presentation of the PEM fuel cell system.

The requirements on hydrogen storage systems include weight and volume limitations, energy efficient charge-discharge cycle and cycling stability, low cost and simple operating principles. Ideally, the hydrogen storage system would work at near ambient temperature and have a high hydrogen storage capacity in a small volume. Mechanical stability is desired from a safety point of view and low cost in purchase and maintenance is also desirable.

NOMENCLATURE

a	[J m^3 mol^{-2}]	van der Waal constant of hydrogen, 0,0247 [J m^3 mol^{-2}] (Wark, 1988)
C_v	[J kg^{-1} K^{-1}]	heat capacity at constant pressure
C_p	[J kg^{-1} K^{-1}]-	heat capacity at constant volume
k_v	[W m^{-1} K^{-1}]	conductivity of vessel material
m	[kg]	mass of hydrogen
mf	[kg s^{-1}]	mass flow of hydrogen
m_v	[kg]	mass of vessel material
M	[kg mol^{-1}]	molecular weight of hydrogen
p	[N m^{-2}]	pressure
Q	[W]	heat transfer to the vessel from surroundings
t	[s]	time
T	[K]	temperature
T_s	[K]	fuel cell system operating temperature
u	[J kg^{-1}]	internal energy
v	[m^3 kg^{-1}]	specific volume
V	[m^3]	vessel volume
V_g	[m^3]	gaseous hydrogen volume in vessel
V_l	[m^3]	liquid hydrogen volume in vessel
Z		compressibility factor of the hydrogen gas
π_T	[N m^{-2}]	internal pressure of a gas
ρ	[kg m^{-3}]	density of hydrogen

1 BACKGROUND

1.1 Compressed hydrogen gas

The simplest form of storage and utilisation of hydrogen in a fuel cell powered vehicle is compressed hydrogen gas. A rapid refuelling capability together with uncomplicated discharge and excellent dormancy characteristics all speak in favour of compressed hydrogen gas storage. Development of tanks for hydrogen containment has had great influence on making compressed gas storage a more attractive option. The U.S. Department of Energy has stated that the DOE 2000 goal for compressed hydrogen storage systems is to have a hydrogen density of 12% by weight. This goal seems achievable with high

performance tanks made from graphite fibre material. (Mitlitsky *et al.*, 1999).

Despite the progresses made in tank design and construction, poor volumetric energy density is still a drawback of compressed hydrogen storage as well as the high pressure in terms of safety aspects. Looking at storage volume, estimates show that in a rather small, 1150 kg, 50 kW$_e$ fuel cell vehicle, 5 kg of compressed gas would provide a driving range of 553 km, the vehicle being operated in the New European Drive Cycle, NEDC. A single 34.5 MPa cylinder designed to store this amount of compressed hydrogen gas would take up a volume of approximately 279 litres (it is assumed that a 25% extra volume must be added to the hydrogen volume). A high pressure tank of 690 bar, for the same amount of hydrogen (and the same range), would take up a volume of 160 litres. Conformable tank storage, where segments of cylinders are fitted together, looks promising in structural flexibility of compressed hydrogen storage and can provide up to 50% more storage within available vehicle space compared to cylindrical tanks (Kunz & Golde, 1999).

Estimates of future, large-scale, tank production costs, where cost of graphite fibres is of central importance, have been studied by Mitlitsky *et al.* (1999) and it has been shown that the energy storage cost for compressed hydrogen would be very expensive compared to that of gasoline (Jung, 1999). Conformable tanks, albeit more expensive than cylindrical tanks per unit, are capable of having the same installed cost. (Kunz & Golde, 1999).

1.2 Liquid hydrogen

Liquid hydrogen is a very energy dense fuel with energy to mass ratio of three times that of gasoline. Liquid hydrogen, due to the low evaporation temperature, faces the problem of boil-off losses. Storage in insulated high-pressure vessels could reduce the problem and enable filling with either liquid hydrogen or ambient temperature compressed gas, depending on need and / or availability of fuel (Aceves *et al.* 1998). The reduced boil-off losses from an insulated pressure vessel, being an obvious economical advantage, could also be seen as a safety issue since avoiding venting of large volumes of hydrogen is clearly desirable. On the other hand, with a vehicle that is regularly used the need for high-pressure storage becomes unnecessary and a smaller and cheaper vessel with a much lower maximum pressure could just as well be employed. For a vehicle used regularly, lower pressure storage would probably be a better choice. One drawback is the cost, since liquefaction of hydrogen is a very energy intensive and lengthy process, where up to 40% of the heating value of the hydrogen is spent (Pettersson & Hjortsberg, 1999).

1.3 Metal hydrides

Metal hydrides are solid alloys capable of reversibly storing hydrogen at ambient temperature and at a pressure far lower than that required for compressed hydrogen gas storage or liquid hydrogen storage. Absorption pressures for metal hydrides is normally in the range of 30 - 55 bar and desorption pressures lower than 10 bar (Arthur D. Little, 1995; Pettersson & Hjortsberg, 1999; T-Raissi et al., 1996). Hydrogen storage in metal hydride form offers the advantage of high safety and is a compact, solid state storage, which gives positional flexibility. Losses from long storage time are very low and the hydrogen fed to the fuel cell is pure, thus reducing risks of poisoning the catalyst. However, none of the present day hydride materials offer a favourable combination of all of the criteria mentioned above. Common problems include slow hydrogen release at low temperatures, low gravimetric hydrogen storage density, high cost, high weight, and sensitivity to poisoning (Arthur D. Little, 1995; Daugherty et al. 1998; Sapru, 1998; Thomas et al., 1995, 1996, and 1998). Both Mazda and Toyota have constructed prototype vehicles employing metal hydrides for fuel storage.

1.4   Conclusions on storage method

Together with compressed hydrogen gas, liquefied hydrogen storage seems like the most viable option for automotive applications and they both have advantages and disadvantages depending on proposed application. Due to high weight and cost, metal hydrides storage appears to be the method with the least scope of future use. The slow refuelling rates also speak against hydride storage in other than public transport or similar fleet type vehicular applications. Liquid and compressed gas storage differ mainly in the complexity of hydrogen discharge and storage volume. Liquefied hydrogen holds an advantage over compressed gas when it comes to storage vessel size and weight as shown in Table 1.

Table 1. Comparison of liquid and gaseous hydrogen storage.

	Liquid storage	Gas storage
storage capacity	5 kg	5 kg
design pressure	0.8 MPa	35 MPa
internal volume	0.095 m^3	0.22 m^3
external volume	0.125 m^3	0.28 m^3
vessel mass (empty)	23 kg	37 kg

2.   MODELLING

2.1   Introduction

Mathematical models have been created in order to evaluate the applicability of the two simplest storage alternatives, compressed gas and liquid hydrogen. The thermodynamic calculations describing the storage vessel and hydrogen release are used in conjunction with the fuel cell model to estimate the amount of energy that has to be drawn from the fuel cell and/or exchanged from its exhaust gas to supply fuel of the required pressure and temperature to the system. NOTE: All thermodynamic and physical property data of hydrogen that are used for the calculations and simulations are collected from (R. D. McCarty, 1975).

2.2 Compressed hydrogen tank

The fuel tank to hold compressed hydrogen gas is based on the type described by Mitlitsky, Weisberg & Myers (1999). This vessel is a graphite fibre (T100G) cylinder with a low permeability polymer liner (high-density polyethylene HDPE). The suggested tank is dimensioned to hold 5 kg of hydrogen at 300 K under a maximum storage pressure of 35 MPa. The relevant data of the proposed tank are given below. (Mitlitsky et al., 1999; Perry & Green, 1997; Brandrup & Immergut, 1989)

Table 2. Data on tank for gaseous hydrogen

Safety factor[1]		2.25
Vessel mass (empty)	kg	37
Internal volume	m^3	0.22
External volume	m^3	0.28
Internal diameter	m	0.55
Inner surface area	m^2	2.0
HDPE liner thickness	m	0.0050
T1000G fibre density	g/cm^3	1.82
HDPE liner density	g/cm^3	0.95

Thermodynamic data of tank materials: (at 300 K)
- Conductivity of graphite 2.0 W/m,K
- Conductivity of HDPE 0.3 W/m,K
- Heat capacity of graphite $2.673 + 0.002617T - 116900/T^2$ cal/K,mol
- Specific heat of HDPE 1.916 J/g,K

Hydrogen gas is released at system pressure and heated in a heat exchanger to system temperature. The possibility of using the fuel cell cooling medium as the heat source is investigated. Tank temperature together with temperature of extracted hydrogen is calculated in order to determine heat demand for bringing the hydrogen gas to operating temperature before admitting it to the fuel cell. If the fuel cell cooling medium is used as heat source, the program enables calculation of the auxiliary heat demand, which will be necessary since the cooling medium can not be used for heating of the hydrogen fuel all the way up to system temperature.

2.3 Liquid hydrogen tank

In the case of liquid hydrogen storage, the fuel tank is modelled as a vacuum insulated (Multi Layer Vacuum Super Insulation, MLVSI) vessel with a maximum operating pressure, or venting pressure, of 0.8 MPa. Storage vessels for these pressures have been described and tested by Aceves et al. (1997, 1998 & 1999) and the results from these works serve as basis for this study. A schematic of the liquid hydrogen storage vessel and fuel discharge is seen in Figure 2 (Peschka, 1992).

Hydrogen is extracted at system pressure and passed through a heat exchanger, where the cooling medium from the fuel cell could be used as heat source. When the vehicle has been dormant for a period of time and pressure build-up has occurred

---

[1] Safety factor is defined as burst pressure over design pressure.

in the tank, hydrogen is first released in gaseous form to lower the pressure. After vessel pressure reaches a predefined level, liquid hydrogen is instead extracted to the heat exchanger, where it is vaporised and heated to system temperature before entering the fuel cell system. A pressure management heating-coil, where part of the heated fuel stream is passed through, is needed in order to maintain operating pressure in the vessel when hydrogen is being released during longer periods of operation (Michel *et al.*, 1998). The hydrogen circulated in the pressure management heating-coil is re-heated in a second heat exchanger.

Heat demand for vaporisation and bringing of hydrogen to system temperature is calculated. As for compressed hydrogen storage, auxiliary heat demand can be calculated when cooling medium is used as heat source. The heat demand for maintaining the desired vessel operating pressure during hydrogen release is also calculated.

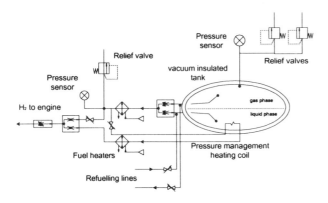

Figure 2. *Schematic of liquid hydrogen storage and discharge from a vacuum insulated tank (based on Peschka, 1992).*

As mentioned earlier, the modelled tank is chosen as a 0.8 MPa storage vessel with a filling pressure of 0.3 MPa. The tank volume is calculated from the EIPH 2000 guidelines on maximum filling levels in liquid hydrogen vessels. (EIHP, 2000). The relevant material data for this tank are taken from Aceves *et al.* (1998) and Peschka, (1992) and are as follow.

Table 3. Data on tank for liquid hydrogen

Design pressure	MPa	0.8
Aluminium mass	kg	11
Internal volume	m³	0.095
External volume	m³	0.125
Internal diameter	m	0.42
Inner surface area	m²	1.1
Insulation thickness	m	0.02
Conductivity of MLVSI[2]	W/m,K	10^{-5}

---

[2] Thermal conductivity of vessel insulation is assumed to be independent of internal and external temperature.

2.3 Thermodynamic analysis of compressed hydrogen gas storage

The following assumptions are made to simplify the calculations:

- the kinetic and potential energies are neglected in all energy balances.
- both temperature and pressure are assumed to be uniform within the vessel. This is an assumption that has been confirmed for small vessels of the size required for light-duty vehicles (Aceves *et al.*, 1998).

The energy balance over the pressure vessel, when applying the first law of thermodynamics (Aceves *et al.* 1998) is:

$$m\frac{du}{dt} + m_v\frac{d(C_{p,v}T)}{dt} = Q + \left(\frac{p}{\rho}\right)\frac{dm}{dt} \qquad (1)$$

The left-hand side of the equation represents the rate of change in internal energy of the hydrogen gas contained in the vessel and that of the vessel material. Heat transfer into the vessel, Q, is positive and the last term on the right-hand side, commonly known as flow work, represents a cooling effect on the vessel and its content due to mass flow out from the vessel. In order to obtain vessel temperature, the change in internal energy is expressed as:

$$du = \pi_T dv + C_v dT \qquad (2)$$

Internal pressure itself is a function of specific volume:

$$\pi_T = \frac{a}{(vM)^2} \qquad (3)$$

This gives the following expression for rate of change in internal energy:

$$\frac{du}{dt} = C_v\frac{dT}{dt} - \frac{a}{VM^2}\frac{dm}{dt} \qquad (4)$$

As the perfect gas law is not valid due to the high storage pressure, the compressibility factor, Z, is introduced as a compensation for the gas molecules own volume in the calculations. Since heat transfer into the vessel results in only minor change in temperature, Z is expressed as a function of vessel pressure at 300 K. Combining the modified gas law with equations (1) and (4) gives the following expression for the energy balance over the pressure vessel (heat capacity of the vessel material is assumed to be constant):

$$mC_v\frac{dT}{dt} - \frac{a}{VM^2}m\frac{dm}{dt} + m_vC_{p,v}\frac{dT}{dt} = Q + \frac{RTZ}{M}\frac{dm}{dt} \qquad (5)$$

This equation is rearranged and solved stepwise over time for temperature with a defined rate of mass release. Mass release can either be constant over time or a vector formed from driving scheme data.

Hydrogen gas is released from the pressure vessel at a defined system pressure. In order to know the heat requirement to bring the hydrogen gas up to system temperature, it is necessary to calculate the temperature of the outgoing gas. Extraction of hydrogen through the release valve is an isenthalpic process. The release temperature is calculated from two relations that describe the change in internal energy of the hydrogen gas released from the pressure vessel.

With the outside pressure known, *i.e.* the system pressure, and with the perfect gas law valid outside the vessel, the isenthalpic process is expressed as a function of temperature.

$$u_2 - u_1 = p_1 v_1 - p_2 v_2 = p_1 v_1 - \frac{RT_2}{M} \quad (6)$$

The same expression integrated, taking the internal pressure in consideration:

$$u_2 - u_1 = \left(\frac{a}{v_1 M^2} - \frac{ap_2}{MRT_2} \right) + C_v (T_2 - T_1) \quad (7)$$

The heat capacity of hydrogen can be seen as constant in the calculations since the difference between highest pressure and lowest pressure is only 2.5%. With equations (6) = (7) and solving this for T_2, yields one solution that is close to zero, which is discarded, and one solution where temperature is around 300. This solution seems likely, noting that the Joule – Thompson factor, *i.e.* the change in temperature with pressure at constant enthalpy, of hydrogen is negative at these pressures and temperatures (McCarty, 1975). A reduction in pressure at constant enthalpy thus results in increased temperature.

2.4 Thermodynamic analysis of liquid hydrogen storage

In the case of an insulated pressure vessel for liquid hydrogen storage, the fuel will normally be released in liquid form. Liquid hydrogen is extracted at system pressure, vaporised and heated to system temperature in the heat exchanger/auxiliary heater. Longer periods of rest, typically a few days, will result in pressure build-up inside the vessel that necessitates gaseous hydrogen release (Aceves *et al.*, 1998). A vessel pressure of 0.1 MPa above system pressure is recognised as the limit above which hydrogen is released in gaseous form. The equations 6 and 7 described above for compressed gas storage are then used. They are slightly modified in the Matlab program to account for the temperature dependence of the specific heat of hydrogen and the different construction materials of the tank, considering that the release temperature from liquid storage will be around 20 – 30 K.

Vessel temperature and temperature of released hydrogen is simply taken as the equilibrium temperature at the vessel pressure. The vessel pressure is held at 0.1 MPa above system pressure by the pressure management system. The heat required for pressure management is taken from the hydrogen that has been discharged from the vessel and vaporised in the heat exchanger. This heat demand is dependent on the mass flow rate of hydrogen from the vessel, because of the need of compensating the pressure drop due to increasing gas volume in the tank. The heat demand is obtained from the condition that vessel pressure during liquid hydrogen discharge remains constant. At this low pressure and temperature, gas pressure in the storage vessel is expressed with the perfect gas law.

The gas volume will increase when liquid hydrogen is released from the vessel, and is expressed as:

$$V_g = V_{g0} + \frac{1}{\rho_l} (m_f + m_v) \quad (11)$$

Solving for the mass of hydrogen vaporised from the liquid phase, "m_v", gives:

$$m_v = \frac{\dfrac{m_{g0} m_f}{\rho_l}}{V_{g0} - \dfrac{m_{g0}}{\rho_l}} \quad (8)$$

This is the mass of hydrogen that has to be vaporised in order to compensate for the release of liquid hydrogen, and enables the calculation of the heat demand.

3. RESULTS AND DISCUSSION

Below, the results from simulation for compressed gas an liquid hydrogen storage are displayed. For each storage method, the full (100%) load and part (50%) load cases are tested. For a 50 kW_e fuel cell stack at 0.75V, this means a mass flow of hydrogen of 0.8 g/s at full load and 0.4 g/s at part load.

3.1 Compressed hydrogen gas storage

With compressed gaseous hydrogen storage, the simulation of fuel discharge over time is carried out for the two cases 100% and 50% fuel cell stack load. In these simulations the storage vessel temperature, the temperature of released hydrogen, and the heating power requirement for bringing hydrogen to system temperature, is calculated for the simplest case of a constant rate of hydrogen discharge. Vessel and released hydrogen temperature during operation with 50 % and 100 % stack load is displayed in Figures 3 and 4 respectively. The heating power requirement is displayed in Figures 5 and 6.

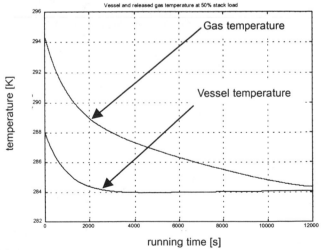

Figure 3. Vessel temperature and temperature of released hydrogen at 50% stack load.

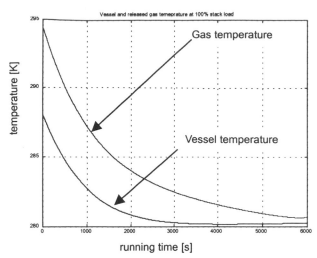

Figure 4. Vessel temperature and temperature of released hydrogen at 100% stack load.

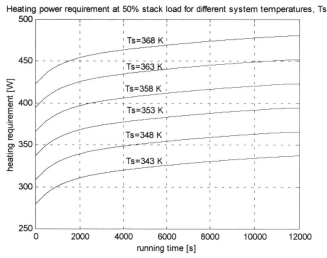

Figure 5. Heat requirement for fuel pre-heating at 50% stack load.

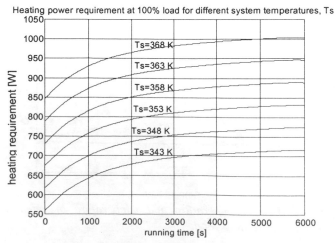

Figure 6. Heat requirement for fuel pre-heating at 100% stack load.

The simulation made with constant hydrogen release rate shows that vessel temperature decreases during release of hydrogen and consequent decrease in pressure (Figures 3 and 4). Equilibrium is reached when the heat conducted through the vessel wall makes up for the energy loss that hydrogen release constitutes. As a consequence, the temperature of the released gas, although being higher than that of the vessel, will also decrease over time of operation thus calling for a need of fuel heating, increasing with time of cell stack operation (Figures 5 & 6). The stack operating pressure did only have a minor influence over heating requirement in compressed gas storage.

It is seen in Figure 6 that at 100% fuel cell stack load a heating power requirement of between 720 and 1000 W is necessary to provide fuel of adequate temperature to the fuel cell system. It was suggested earlier in this work that cell stack cooling medium could be used to provide the heat and there is certainly a possibility to utilise the waste heat to some extent for this purpose. However, to reach the desired temperature, an auxiliary heat source would be required. This could be solved either through an electrical heater or through burning of hydrogen. Using an electrical heater to heat the fuel might seem like wasting produced energy. On the other hand, implementing a hydrogen burner introduces extra cost, space and weight requirements to the vehicle. A burner would perhaps be an option for fuel cell vehicles, operated on reformed fuels, where there already is a burner for the reforming processes. This is, however, something that has to be decided from a system design viewpoint.

3.2 Liquid hydrogen storage

The simulations in the case of liquid hydrogen storage are performed under the same conditions as the compressed gas simulations. The heat requirement for bringing hydrogen to system temperature as well as the heat requirement for maintaining operating pressure in the vessel is calculated. The power requirement for vaporisation and heating of hydrogen to system temperature together with the power requirement of the pressure management system at different system conditions is displayed in Figures 7 & 8. This power requirement constitutes the necessary capacity of the heat exchangers.

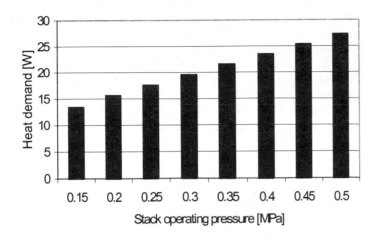

Figure 7. Heat requirement for pressure management in liquid hydrogen tank.

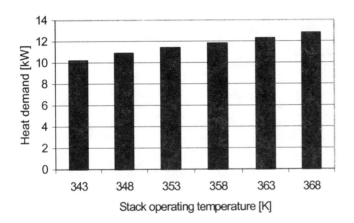

Figure 8. Heat requirement for pre-heating of fuel released from liquid hydrogen tank.

Regarding the pressure management system, the proposed mode of functioning was through redirection of a part of the released and heated hydrogen stream into a heating coil, running within the storage vessel. There, heat transfer from the fuel gas stream to the liquid hydrogen evaporates a sufficient amount of hydrogen to keep the desired operating pressure. It should be noted that this work focuses on calculating the power demand, and does not deal in detail with the design of the pressure management system and its control. The simulation shows that a very modest heating demand exists for pressure management, some 27 W at the most severe (in terms of pressure management power demand) system conditions of 0.5 MPa pressure. For a less complicated design, this energy could possibly be added from a heating coil or similar utilising ambient temperature air as heat source, thus simplifying the fuel management and control.

Under the "worst" conditions, at the lowest investigated system pressure and lowest investigated system temperature, the power requirement for fuel heating is quite significant, 12.8 kW. This seems like a really large amount of energy and a drawback of liquid hydrogen usage, seeing that the fuel cell stack has a power output of 50 kW. However, most of this heat is added at a very low temperature (boiling point of hydrogen is 22 K at 0.15 MPa), which means that heat could be drawn from virtually anywhere, the ambient temperature surrounding air sufficing a long way. Also, waste heat from the fuel cell stack can be used for fuel heating purpose. But just as in the case of compressed gaseous hydrogen storage, it will be necessary to include some sort of auxiliary heating in the system design.

8 CONCLUSIONS

For both storage options, some extra heating device will be needed to reach the fuel cell system operating temperature in the interval of 343 – 368 K. While in the case of compressed hydrogen, the heat demand is quite modest at full load, between 0.72 and 1 kW, the need of heat in the liquid hydrogen case is larger, up to 12.8 kW. The fuel cell stack produces surplus heat that can be used for fuel heating. Calculations show that the heat content of the cooling medium is sufficient to heat the fuel stream to approximately 20 K below stack temperature, with temperature differences in heat exchangers being the limiting factor. The radiator and/or compartment heating and humidifier will also draw heat from this source. Also, for liquid hydrogen storage, a power demand arises for maintaining operating pressure inside the storage vessel during hydrogen release. This was calculated to between 13 and 28 W for the fuel cell stack operating conditions simulated, and this power demand can be supplied by directing a stream of released and heated hydrogen through a coil running inside the storage vessel.

REFERENCES

Ahn C. C., Ye Y., Ratnakumar B. V., Witham C., Bowman R. C., Fultz b., 1998, "Hydrogen desorption and adsorption measurements on graphite nanofibres", Applied Physics Letters, Vol. 73, No. 23, Published on web by American institute of physics.

Aceves S. M., Berry G. D., 1997, "Thermodynamics of insulated pressure vessels for vehicular hydrogen storage", Proceedings of the ASME Advanced Energy Systems Division – 1997, pp 155 – 163, Dallas, USA.

Aceves S. M. & Garcia-Villazana O., 1998, "Hydrogen storage in insulated pressure vessels", Proceedings of the 1998 U.S. DOE Hydrogen Program Review, pp 503 – 518, Alexandria, USA.

Aceves S. M., Martinez-Frias J., Garcia-Villazana O., 1999 (1), "Analysis and experimental testing of insulated pressure vessels for automotive hydrogen storage", Proceedings of the 1999 U.S. DOE Hydrogen Program Review, Published on Web by U.S DOE at http://www.eren.doe.gov/hydrogen/, accessed April 2000.

Aceves S. M., Martinez-Frias J., Garcia-Villazana O., 1999 (2), "Evaluation of insulated pressure vessels for cryogenic hydrogen storage", Proceedings of the ASME Advanced Energy Systems Division – 1999, pp 355 – 364, Nashville, USA.

Arthur D. Little, 1995, Multi fuel reformers for fuel cells used in transportation – Assessment of storage technologies, Final report prepared for the U.S Department of Energy, Cambridge, USA.

Brandrup J. & Immergut E. H., 1989, "Polymer Handbook", 3rd Ed., John Wiley and Sons, Inc., USA.

Daugherty M. A., Prenger F. C., Daney D. E., Hill D. D., Edeskuty J. F., 1995, A comparison of hydrogen vehicle storage options using the EPA urban driving schedule, CEC/ICMC 1995 cryogenic engineering conference and international cryogenic materials conference, Columbus, USA.

European Integrated Hydrogen Project, 2000 (1), "Uniform Provisions Concerning the Approval of [1]Specific components of motor vehicles using compressed hydrogen; [2]Vehicles with regard to the installation of specific components for the use of compressed hydrogen." EIHP, Draft for Regulation, Revision 7, Published on Web at http://www.hyweb.de/eihp, accessed 2000-08-31.

European Integrated Hydrogen Project, 2000 (2), "Uniform Provisions Concerning the Approval of [1]Specific components of motor vehicles using liquid hydrogen; [2]Vehicles with regard to the installation of specific components for the use of liquid hydrogen." EIHP, Draft for Regulation, Revision 10, Published on Web at http://www.hyweb.de/eihp, accessed 2000-08-31.

Jung P., 1999, "Technical and economic assessment of hydrogen and methanol powered fuel cell vehicles", Master of Science Thesis, Chalmers University of Technology, Gothenburg, Sweden.

Kunz R. K., Golde R. P., 1999, "High-pressure conformable hydrogen storage for fuel cell vehicles", Proceedings of the 1999 U.S DOE Hydrogen Program Review, Published on Web by U.S DOE at http://www.eren.doe.gov/hydrogen/, accessed April 2000.

McCarty R.D., 1975, "Hydrogen Properties", Hydrogen: its Technology and Implications, Vol. 3, CRC Press, Cleveland, Ohio, USA.

Michel F., Fieseler H., Meyer G., Theissen F., 1998, "On-Board Equipment for Liquid Hydrogen Vehicles", International Journal of Hydrogen Energy, Vol. 23, No. 3, pp. 191 – 199, Elsevier Science Ltd, Great Britain.

Mitlitsky F., Weisberg A. H., Myers B., 1999, "Vehicular hydrogen storage using lightweight tanks (regenerative fuel cell systems)", Proceedings of the 1999 U.S DOE Hydrogen Program Review, Published on Web by U.S DOE at http://www.eren.doe.gov/hydrogen/, accessed April 2000.

Perry R.H. & Green D.W., 1997, "Perry's Chemical Engineers' Handbook", 7th Ed., McGraw – Hill, Inc., USA.

Peschka W., 1992, "Liquid Hydrogen, Fuel of the Future", Springer Verlag, Vienna Austria.

Petterson J. & Hjortsberg O., Volvo Technological Development Co., 1999, Hydrogen storage alternatives – a technological and economic assessment, Stockholm, Sweden.

Sapru K., Ming L., Stetson N. T., Evans J., 1998, "Improved Mg-based alloys for hydrogen storage", Proceedings of the 1998 U.S DOE Hydrogen Program Review, pp 433 – 447, Alexandria, USA.

T – Raissi A., Banerjee A., Sheinkopf K., 1996, "Metal hydride storage requirements for transportation applications", Cocoa, USA.

Thomas G. J., Guthrie S. E., Bauer W., 1995, "Lightweight hydride storage materials", Proceedings of the 1995 U.S DOE Hydrogen Program Review, pp 543 – 550, Coral Gables, USA.

Thomas G. J., Guthrie S. E.,, Bauer W., Yang N. Y. C., 1996, "Hydride development for hydrogen storage", Proceedings of the 1996 U.S DOE Hydrogen Program Review, pp 807 – 818, Miami, USA.

Thomas G. J., Guthrie S. E., 1998, "Hydrogen storage development", Proceedings of the 1998 U.S DOE Hydrogen Program Review, pp 419 – 431, Alexandria, USA.

Proceedings of IECEC'01
36th Intersociety Energy Conversion Engineering Conference
July 29–August 2, 2001, Savannah, Georgia

IECEC2001-RE-07

PREMIXED COMBUSTION OF HYDROGEN/METHANE MIXTURES IN A POROUS MEDIUM BURNER

Chung-jen Tseng
Department of Mechanical Engineering
National Central University, Chungli 32054
Taiwan, R.O.C.

ABSTRACT

The premixed combustion of hydrogen-methane mixtures in porous medium burners is investigated numerically. The numerical model considers solid and gas phase energy equations separately, and includes nongray radiation transport for the solid phase. GRI-Mech 2.11 is used for the detailed chemical kinetic model. The lean limit of pure methane can be extended from $\phi = 0.53$ for free flame to $\phi = 0.4$ for porous medium flame. Adding hydrogen in the fuel, the lean limit can be further reduced to $\phi = 0.35$. The flame speeds of porous ceramic burner flames are several times as that of free flames. For $\phi = 1.0$, the free flame speed of methane is about 40 cm/s whereas that of porous burner flame is 87 cm/s approximately. Adding hydrogen to the fuel, the flame speed is increased to 103 cm/s when the fraction of hydrogen in the fuel, g, equals to 0.3. Increasing g to 0.6 raises the flame speed to 142 cm/s. The CO and NO_x emissions increase slightly as the hydrogen fraction in the fuel is raised.

INTRODUCTION

Recently, developing high efficiency, low pollution combustion technologies has become a very important subject due to worsening global environments and depleting oil reserves. In order to obtain more efficient combustion systems with low pollutant emissions, several methods have been proposed. Many of them use various heat transfer mechanisms to recirculate the product enthalpy to the pre-combustion zone. A detailed review of those techniques is given by Weinberg [1]. A simple and practical way to recirculate the enthalpy from the products is by embedding the flame within a highly porous material. Takeno and Sato [2] first proposed inserting a porous solid of high conductivity into the flame to conduct the post-flame enthalpy to aid in preheating the fresh mixture. By choosing the porosity of the solid, they could control the heat transfer coefficient between the solid and the gas. They observed that the flame structure was altered significantly by the presence of the solid and that the region of stable flames extended beyond the normal lean limit. However, they did not identify the importance of the radiation transport on the system.

Echigo and coworkers [3,4] presented a rigorous model for multi-mode heat transfer, Arrhenius-type one-step reaction kinetics, and radiative transfer for absorbing/emitting media. However, they did not consider the scattering effects.

More recently, various researchers [5-8] have used reticulated dodecahedral structure ceramic foam as the heat-recirculating medium. Combustion within a highly porous material is a very promising technique for efficient combustion systems with low pollutant emissions. A detailed review of the combustion of hydrocarbon fuels within porous inert media was presented by Howell et al. [9].

It has been found that the combustion within porous media has many advantages over conventional burners. These include, first, higher burning speeds and volumetric energy release rates, which allows a more compact burner design. Second, it is possible to maintain combustion to much lower lean limits within the porous material because the incandescent porous ceramic acts as a means of maintaining stable combustion during any minor fluctuations in temperature of the flame, and the hot ceramic maintains an ignition source within the flame zone. Third, the ability to maintain a very lean flame means that combustion can occur at a lower temperature. This low combustion temperature has a further advantage. For lean combustion, thermal NO_x is the primary source for NO_x production [10]. That is, the chemistry of production of NO_x is very temperature dependent, NO_x production can be very low at low flame temperatures. Thus, porous medium burners can be operated with low NO_X pollutant outputs.

Most previous work on porous medium burners used methane as the fuel. The present study focuses on using hydrogen-methane mixtures as the fuel. Hydrogen has become a very promising energy carrier due to its cleanness and abundance. Effects of hydrogen on the burner performance are investigated. A numerical simulation has been done on the premixed combustion of hydrogen-methane mixtures in porous medium burners. The numerical model considers solid and gas phase energy equations separately, and includes nongray radiation transport for the solid phase. GRI-Mech 2.11[11] is employed for the detailed chemical kinetic model. The range of equivalence ratio ϕ, covered in this work is from 1.0 (stoichiometric) down to the lean limit.

GOVERNING EQUATIONS

The geometry of the problem is shown in Fig. 1. Because the porosity of the porous medium used in this type of burner is typically very large (0.85 and above), the pressure loss is very small. And usually, good insulation is placed around the burner wall, so radial-direction heat loss may be neglected. Experimental investigations of porous medium burners [7, 8] showed that these burners maintain a one-dimensional flame profile. So a steady one-dimensional laminar flow model is used in this work. Due to the high extinction coefficient of porous ceramics, radiative transport is localized in space, and radiative propagation is assumed one-dimensional in the axial direction. Gas radiation from the fuel/air combustion mixture is considered to be negligible. The gas radiation is insignificant compared with solid radiation. The porous ceramic emits, absorbs, and scatters radiation in local thermal equilibrium, and it acts as a homogeneous medium.

Under the above assumptions, a set of differential equations can be obtained.

The gas-phase continuity equation:

$$\frac{d}{dx}(\rho u) = 0 \qquad (1)$$

The gas-phase species conservation equation:

$$\rho A u \frac{dY_k}{dx} + \frac{d}{dx}(\rho A Y_k V_k) = A \dot{\omega}_k W_k, \quad k = 1, 2, \ldots, K \qquad (2)$$

where $\dot{\omega}_k$ is the production rate of the k-th species.
The i-th chemical reaction is of the general form

$$\sum_{k=1}^{K} v'_{ki} x_k \leftrightarrow \sum_{k=1}^{K} v''_{ki} x_k \quad , \qquad (3)$$

$$\dot{\omega}_k = \sum_{i=1}^{I} (v''_{ki} - v'_{ki}) \left(\kappa_{fi} \prod_{k=1}^{K} [x_k]^{v'} - \kappa_{ri} \prod_{k=1}^{K} [x_k]^{v''} \right), \qquad (4)$$

$$\kappa_{fi} = A_i T^{\beta_i} \exp\left(-\frac{E_i}{R_c T}\right), \qquad (5)$$

where κ_{fi} is the forward rate constant for reaction i and κ_{ri} is the reverse rate constant.

Convection is included by solving separate energy equations for the solid and the gas, and coupling them through the convective heat transfer coefficient.

The gas phase energy equation:

$$\rho_g A u c_{pg} \frac{dT_g}{dx} = \frac{d}{dx}\left(k_g A \frac{dT_g}{dx}\right) - A \sum_{k=1}^{K} \rho_g Y_k V_k c_{pk} \frac{dT_g}{dx}$$
$$- A \sum_{k=1}^{K} \dot{\omega}_k h_k W_k - A H_{gs}(T_g - T_s), \qquad (6)$$

in which the term on the left hand side is the advection of enthalpy by the main flow; the first term on the right hand side is the conduction term; the second term represents the energy transfer by interspecies diffusion; the third term is the energy released or removed by the chemical reactions; the fourth term represents the convection heat transfer between the gas and the solid.

The ideal gas equation of state is also employed:

$$\rho = \frac{\overline{W}p}{R_c T}. \qquad (7)$$

The solid-phase energy equation:

$$\frac{d}{dx}\left(k_s \frac{dT_s}{dx}\right) = \frac{d}{dx}(q_r) - H_{gs}(T_g - T_s), \qquad (8)$$

where H_{gs} is the volumetric heat transfer coefficient which is obtained from an experimental correlation reported by Younis and Viskanta [12]. The divergence of radiative heat flux can be obtained by solving the radiative transfer equation (RTE). The RTE can be written as

$$\frac{dI_\lambda}{ds} = -(K_{a\lambda} + K_{s\lambda})I_\lambda + K_{a\lambda} I_{b\lambda}$$
$$+ \frac{K_{s\lambda}}{4\pi} \int_{4\pi} I_\lambda(s, \Omega') p_\lambda(\Omega', \Omega) d\Omega' \qquad (9)$$

where $K_{a\lambda}$ is the spectral absorption coefficient, $K_{s\lambda}$ the spectral scattering coefficient, $I_{b\lambda}$ the spectral blackbody radiation, and p_λ the spectral scattering phase function.

The phase function is modeled by the modified Henyey-Greenstein phase function given by the following expression:

$$p(\theta, \lambda) = f_{iso,\lambda} + (1 - f_{iso,\lambda}) \frac{(1 - g_{hg,\lambda}^2)}{(1 + g_{hg,\lambda}^2 - 2g_{hg,\lambda} \cos\theta)^{1.5}}, \qquad (10)$$

where $-1 \leq g_{hg,\lambda} \leq 1$, $g_{hg,\lambda}$ is the asymmetry factor, $f_{iso,\lambda}$ is the isotropic scattering parameter, and θ is the angle between incoming and scattered radiation beams. The parameters $K_{a\lambda}$, $K_{s\lambda}$, $f_{iso,\lambda}$, and $g_{hg,\lambda}$ are obtained from a work by Hendricks [13].

At the inlet, it is assumed that the fuel-air mixture enters at a known room temperature T_o. A value of 25 °C is used for T_o in this study. The species concentrations are assumed to be at prescribed values at the inlet. It is also assumed that radiation exchanges with a blackbody at room temperature T_o.

The boundary conditions for species and gas temperature at the exit are obtained by appying Eqs. (2) and (6) to the exit control volume. The boundary conditions for the solid temperature at both ends are obtained by applying energy balance at the inlet and exit boundaries. The exit radiation boundary condition used is $\pi I_b(T_e)$, which exchanges radiation with a blackbody cavity at the burner exit temperature. Mathematically, the boundary conditions can be expressed as:

at the inlet,

$$T_g = T_o, \; Y = Y_{k,o}, \; q^+ = \pi I_b(T_o), \; k_s \frac{dT_s}{dx}\bigg|_{x=0} = \left[\sigma T_s^4 - \sigma T_o^4\right], (10)$$

and at the exit,

$$\frac{dY_k}{dx} = 0, \; q^- = \pi I_b(T_e), \; -k_s \frac{dT_s}{dx}\bigg|_{x=L} = \left[\sigma T_s^4 - \sigma T_o^4\right]. \quad (11)$$

METHOD OF SOLUTION

The gas phase and solid phase equations are discretized by the finite difference method, and solved as a two-point boundary value problem by a modified damped Newton method with adaptive grid [14]. The computer code allows for the use of multi-step detailed chemical kinetics (CHEMKIN II, [15]), and the TRANFIT subroutine [16] for accurate determination of the transport properties of the gas. GRI-Mech 2.11 is employed for the detailed chemical kinetic model. GRI-Mech is an optimized detailed chemical reaction mechanism capable of the best representation of natural gas flames and ignition. It is a compilation of 279 elementary chemical reactions and associated rate coefficient expressions and thermochemical parameters for the 49 species involved in them. The RTE is solved by using the discrete-ordinates method [17, 18]. Iteration is required because the equations are coupled. For all cases, a relative convergence of 10^{-4} is used.

RESULTS AND DISCUSSION

Figure 2 depicts the temperature profiles of the solid and the gas stream in the porous medium burner for $\phi = 1.0$ and g = 0 and 0.6, where g represents the hydrogen mole fraction in the fuel gas. Note that only the later portion of the burner is shown in the figure for better resolution. The gas temperature at the post-flame region is several hundred degrees higher than the solid temperature. Enthalpy is transferred from the gas to the solid via convection. This enthalpy, in turn, is transported from the post-flame region to the pre-flame region by both solid conduction and radiation. Since the solid temperature in the pre-flame region is higher than the gas temperature, the enthalpy is then transferred from the solid back to the gas in the pre-flame region to

preheat the gas, causing the burning rate to be a lot higher than for free burning laminar flames. The temperature profiles near the exit drop due radiation loss to the ambient. For hydrogen-added flame, the gas temperature in the postflame region remains relatively constant and is approximately 100 K higher than pure methane flame.

Figure 3 shows the effects of hydrogen addition on the maximum flame temperature at various equivalence ratios. The maximum flame temperatures for the corresponding free flames are also shown for comparison. Placing a porous ceramic block in the burner increases the maximum flame temperature, especially in the lower equivalence ratio region. The lean limit is extended to $\phi = 0.4$. Addition of hydrogen further lowered the lean limit to $\phi = 0.35$. However, it does not seem to have large effects on the maximum flame temperature. Although hydrogen has higher adiabatic flame temperatures than methane, it has also higher conductivity. The heat transfer coefficient between the solid and the gas phases also increases due to higher flame speed (see next figure). As a result, the maximum flame temperature does not change much when adding hydrogen in the porous medium burners.

Figure 4 depicts the effects of hydrogen addition on the flame speed at various equivalence ratios. Flames in the porous medium burner have higher flame speed than that of free flames. For $\phi = 1.0$, the flame speed increases from 40 cm/s to 87 cm/s approximately. This is due to the preheating effect that is brought about by the enthalpy feedback mechanism. Adding hydrogen further raises the flame speed. For $\phi = 1.0$, the flame speed increases to 103 cm/s for g = 0.3 and 142 cm/s for g =0.6. Several factors account for this increase. First, hydrocarbon combustion suffers from the relatively slow $CO \rightarrow CO_2$ reaction step, which is absent in H_2 combustion. The reaction kinetics for hydrogen is therefore very fast. Second, hydrogen atoms act as chain carriers to promote the reaction, and hence the propagation. Third, hydrogen has much higher thermal and mass diffusivities than hydrocarbon fuels, so it helps to transport energy out of the flame zone.

The emissions characteristics of the burner are presented in figures 5 to 7. In these figures, the emission concentrations are all corrected to 3% O_2. Figure 5 shows the CO emission vs. the equivalence ratio for several hydrogen fractions. The CO concentration for porous medium burner is slightly lower than for free flame for the same equivalence ratio. Noting that, for the same equivalence ratio, the porous medium burner flame has higher flame speed and thus the energy release rate, the difference in concentration is actually more significant. The decrease in CO concentration in tne porous medium burner is possibly due to the temperature drop near the exit, where energy is lost to the ambient by radiation. However, as the hydrogen fraction in the fuel is raised, the CO emission increases. The residence time is very short such that CO is not equilibrated and exits at levels

591

somewhere between equilibrium at peak temperature and exhaust temperature. The CO_2 emissions shown in figure 6 show similar trend as CO_2 concentration increases with hydrogen fraction.

Figure 7 presents the NO_x emission vs. the equivalence ratio for several hydrogen fractions. For fuels containing no nitrogen, NO_x is produced by three mechanisms: the thermal or Zeldovich mechanism, the prompt or Fenimore mechanism, and the N_2O-intermediate mechanism [19]. The N_2O-intermediate mechanism is important where the total NO formation rate is relatively low [10]. The thermal NO_x mechanism is very temperature dependent and only important for temperatures over 1800 K. Because this process is rather slow, most thermal NO is formed in the postflame region. For the Fenimore prompt NO scheme, CH radicals first react with N_2 to form amines or cyano compounds. The amines and cyano compounds are then converted to intermediate compounds that ultimately form NO. Increasing the hydrogen fraction would thus reduces prompt NO by cutting the supply of CH radicals. Non-equilibrium O and OH can accelerate the rate of thermal NO mechanism. This is sometimes considered as part of the prompt-NO mechanism. From figure 7, the NO_x emission increases with H_2 fraction. This indicates the non-equilibrium mechanism plays an important role in porous medium burner because hydrogen itself is a catalyst that promotes O and OH production.

CONCLUSIONS

A numerical simulation has been done on the premixed combustion of hydrogen-methane mixtures in porous medium burners. The numerical model considers solid and gas phase energy equations separately, and includes nongray radiation transport for the solid phase. GRI-Mech 2.11 is employed for the detailed chemical kinetic model. The range of equivalence ratio ϕ, covered in this work is from 1.0 (stoichiometric) down to the lean limit.

Based on the study, the following conclusions can be drawn.

1. The lean limit of pure methane can be extended from $\phi = 0.53$ for free flame to $\phi = 0.4$ for porous medium flame. Adding hydrogen in the fuel, the lean limit can be further reduced to $\phi = 0.35$ approximately.
2. The flame speeds of porous medium burner flames are several times as that of free flames. For $\phi = 1.0$, the free flame speed of methane is about 40 cm/s whereas that of porous burner flame is 87 cm/s. Adding hydrogen to the fuel, the flame speed is increased to 103 cm/s for g = 0.3. Increasing g to 0.6 raises the flame speed to 142 cm/s.
3. The CO emission increases as the hydrogen fraction in the fuel is raised, possibly due to the residence time is very short such that CO is not equilibrated and exits at levels somewhere between equilibrium at peak temperature and exhaust temperature.

4. The NO_x emission from porous medium burner increases with hydrogen fraction, indicating the non-equilibrium mechanism plays an important role in porous medium burner.

ACKNOWLEDGMENTS

The author acknowledges the support for the research from the National Science Council of Taiwan under Grant NSC89-2212-E008-046.

REFERENCES

1. Weinberg, F. J., 1986, "Combustion in Heat-Recirculating Burners," *Advanced Combustion Methods*, Edited by F. J. Weinberg, Academic Press, New York, pp. 183-236.
2. Takeno, T., and Sato, K., 1979, "An Excess Enthalpy Theory," *Combustion Science and Technology*, Vol. 20, pp. 73-84.
3. Echigo, R., Yoshizawa, Y., Hanamura, K., and Tomimura, T., 1986, "Analytical and Experimental Studies on Radiation Propagation in Porous Media with Internal Heat Generation," *Proceedings of Eighth Int. Heat Transfer Conference*, pp. 827.
4. Yoshizawa, Y., Sasaki, K., Echigo, R., 1988, "Analytical Study of the Structure of Radiation Controlled Flame," *International Journal of Heat and Mass Transfer*, **31**, No. 2, pp. 311-319.
5. Chen, Y. K., Matthews, R. D., and Howell, J. R., 1987, "The Effect of Radiation on the Structure of Premixed Flame Within a Highly Porous Inert Medium" 1987 ASME Winter Annual Meeting, ASME HTD-81.
6. Hsu, P.-F., Evans, W. D., and Howell, J. R., 1993, "Experimental and Numerical Study of Premixed Combustion Within Non homogeneous Porous Ceramics," *Combustion Science and Technology*, **90**, pp. 149-172.
7. Evans, W. D., 1991, "Experimental Stability Limits of Methane Combustion Inside a Porous Ceramic Matrix," M.S. Thesis, Mechanical Engineering Department, The University of Texas at Austin, Austin, Texas.
8. Chaffin, C., Koenig, M., Koeroghlian, M., Matthews, R. D., Hall, M. J. , Nichols, S. P. and Lim, I-G., 1991, "Experimental Investigation of Premixed Combustion Within Highly Porous Inert Media," *Proceedings of ASME/JSME Thermal Engineering Joint Conference*, **4**, pp. 219-224.
9. Howell, J. R., Hall, M. J., and Ellzey, J. L., 1996, "Combustion of Hydrocarbon Fuels within Porous Inert Media," *Progress in Energy and Combustion Science*, **22**, pp. 121-145.
10. Miller, J. A., and Bowman, C. T., 1989, "Mechanism and Modeling of Nitrogen Chemistry in Combustion," *Progress in Energy and Combustion Science*, **15**, pp. 287-338.

11. Bowman, C.T., Hanson, R.K., Davidson, D.F., Gardiner, W.C., Jr., Lissianski, V., Smith, G. P., Golden, D. M., Frenklach, M., and Goldenberg, M., http://www.me.berkeley.edu/gri_mech/.

12. Younis, L. B., and Viskanta, R., 1993, "Experimental Determination of the Volumetric Heat Transfer Coefficient Between Stream of Air and Ceramic Foam," *International Journal of Heat and Mass Transfer*, **36**, No. 6, pp. 1425-1434.

13. Hendricks, T. J., 1994, "Thermal Radiative Properties and Modeling of Reticulated Porous Ceramics," Ph.D. Dissertation, Mechanical Engineering Department, The University of Texas at Austin.

14. Kee, R. J., Grcar, J. F., Smooke, M. D., and Miller, J. A., 1985, "A Fortran Program for Modeling Steady Laminar One-dimensional Premixed Flames," Sandia National Lab., SAND85-8240.

15. Kee, R. J., Rupley, F. M., and Miller, J. A., 1989, "CHEMKIN II: A Fortran Chemical Kinetics Package for the Analysis of Gas Phase Chemical Kinetics," Sandia National Lab., SAND89-8009.

16. Kee, R. J., Dixon-Lewis, G., Warnatz, J., Coltrin, M. E., and Miller, J. A., 1986, "A Fortran Computer Code Package for the Evaluation of Gas Phase Multicomponent Transport Properties," Sandia National Lab., SAND86-8246.

17. Carlson, B.G. and Lathrop, K.D., 1968, "Transport Theory - The Method of Discrete Ordinates," *Computing Methods in Reactor Physics*, edited by Greenspan, Kelber, and Okrent, Gordon and Breach, New York.

18. Fiveland, W.A., 1984, "Discrete-Ordinates Solutions of the Radiative Transport Equation for Rectangular Enclosures," *Journal of Heat Transfer*, **106**, pp. 699-706.

19. Turns, S.R., 2000, An Introduction to Combustion, 2nd ed., McGraw-Hill, New York, Ch. 5.

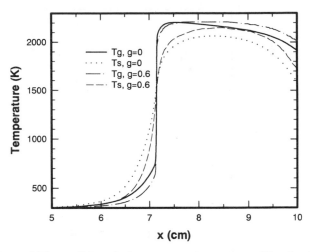

Figure 2 The solid and the gas temperature profiles in the porous medium burner. ϕ =1.

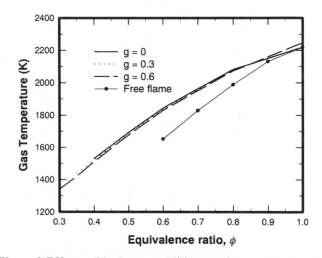

Figure 3 Effects of hydrogen addition on the gas temperature profiles in the porous medium burner.

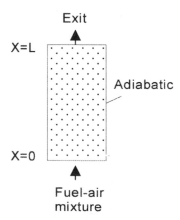

Figure 1 Schematic of the geometry of the problem.

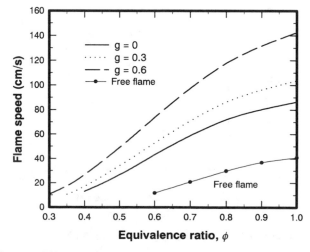

Figure 4 Effects of hydrogen addition on the flame speed.

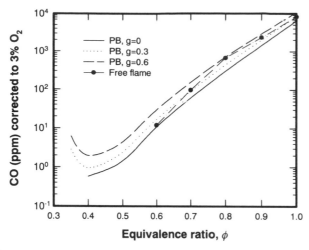

Figure 5 Effects of hydrogen addition on CO emission.

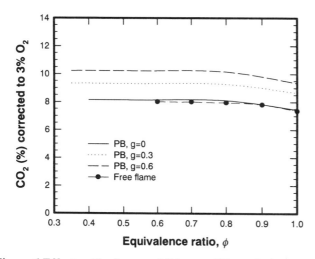

Figure 6 Effects of hydrogen addition on CO_2 emission.

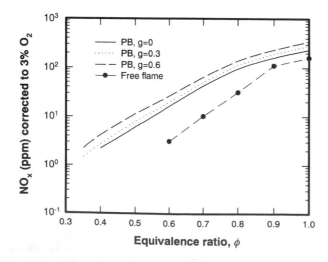

Figure 7 Effects of hydrogen addition on NOx emission.

Proceedings of IECEC'01
36th Intersociety Energy Conversion Engineering Conference
July 29–August 2, 2001, Savannah, Georgia

IECEC2001-RE-04

BIOMASS BRIQUETTING TECHNOLOGY WITH PREHEATING

Sk Ahad Ali and Md Nawsher Ali Moral
Department of Mechanical Engineering
Bangladesh Institute of Technology (BIT), Khulna, Bangladesh

S. C. Bhattacharya
Energy Technology Program
Asian Institute of Technology, Bangkok, Thailand

ABSTRACT

This paper presents a technical and economical cross evaluation of the different biomass residues (rice husk and saw dust) compacted by briquetting machine using preheater. The study also presents the information to date on biomass briquetting technology in Bangladesh, configurations of preheating technology and experimental results on briquettes produced with or without preheating option. This study shows the growing trend of biomass densification as an alternative fuel in the rural areas of Bangladesh. The quality of the briquettes product and the combustion characteristics of the briquettes product from preheated material were better compared to conventional briquettes. Biomass densification with preheating can help overcome some of the technical and economical barriers that inhabit the wider use of waste agricultural and forestry residues.

1. INTRODUCTION

With ever-increasing civilization of modern society the demand for energy is also increasing. But the mankind is always facing fuel crisis as well as combustion technology crisis. Utilization of agriculture and forestry residues is often difficult because of their uneven and troublesome characteristics. This drawback can be overcome by means of densification of the residues into a product of high density. The term densification refers to the process of compaction of residues into a product of higher bulk density than the original raw materials and regular shape. Biomass is an economically and environmentally attractive fuel, but it often difficult to collect, transport, store and use. Densifying biomass to a specific gravity of 1.0 to 1.2 gm/cm^3 or more eliminates most of these problems and producing a uniform, clean, and stable fuel.

The biomass briquetting technology has been developed in two distinct directions Europe and the United States pursued and perfected the reciprocating ram/piston press while Japan has independently invented and developed the screw press technology. But the screw pressed briquettes are far superior to ram press solid brequittes in terms of their storability and combustibility. In the screw press technology which is used for briquetting of biomass has a screw to compress the agricultural and forestry residues such as rice husk, saw dust, baggage, wood wastes etc. The two major impediments for the smooth working of the screw press identified were the high wear of the screw and comparatively large specific power consumption. Extensive damage of the screw is observed for the most abrasive material rice husk. The effect of wear, which is extremely expensive, can be repaired by means of welding. Many developing countries produce huge agro residue but they are used inefficiently causing extensive pollution to environment. Apart from the problem of transportation, storage and their handling, the direct burning of loose biomass in conventional grate is associated with very low thermal efficiency and wide scale of air pollution. In rice husk for burning loose condition about more than 40% ash is occurred. This technology can help the expanding the use of biomass energy uses such as for the industries that are dependent on wood, coal and lignite and also can help in develop the fuel situation in rural areas.

The wearing problems could be reduced to a great extent by preheating the feedstock prior to densification (Hislop, 1986, Aqa and Bhattacharya, 1992). Preheating can reduce the power consumption and pressure required for briquetting and the wear of the screw. Nowadays, there exists two technology options for increasing the life span of the screw and enhance the briquette production: preheating and hard facing technology. Due to hard facing technology may be only life span of screw will be increase and the maintenance cost will be reduced. With a preheater for heating the feed material to a suitable temperature before it supplied into the briquetting machine can expect many advantages such as (a) reduction of resistance of biomass to pressing process (with high temperature biomass become soften and not elastic) what results in a lower press and energy requirement and (b) the briquettes have better quality, easy to transport and storage. Therefore, it is very useful to pay attention to the study of preheating systems. The briquetting machines in combination with preheater will definitely find out the large market for their product in many developing counties, particularly in Bangladesh, where the forest is degraded day by day.

2. COUNTRY PROSPECTIVE OF BRIQUETTING TECHNOLOGY IN BANGLADESH

Bangladesh is surrounded by India in the east, north and west, by Myanmar in the south east and by Bay of Bengal in the south. It has a total area of about 144,000 km^2 and population about 116.8 million of whom about 17% live in urban areas (ADB 1995). As noted by Islam (1993), the biomass fuels used in Bangladesh are mainly in the form of agricultural residues, animal dung and wood fuels. Any significant increase in the energy services provided by biomass fuels will only be possible through either surplus the

Division	No. of Machines	Price of the rice husk (per 40kg)	Price of briquetting (per 40kg)
Sylhet	250	8	45
Chittagong	150	30	80
Rajshahi	180	35	85
Khulna	175	33	90
Barisal	60	35	85
Dhaka	35	40	90

Table 1. Distribution of the Briquetting Machine in Bangladesh1[1]

biomass generated by improvements in efficient utilization of these fuels or addition biomass produced by plantation. About 1.31 million hectors of denuded forest exist in the country. Biomass briquetting is a relatively recent development in Bangladesh. The briquetting machines are of heated die screw press type (capacity = approx. 78 kg/hour). The technology appears to have been developed by the local entrepreneurs without any support from the government or donor agencies. Once briquettes were accepted by users, the technology quickly spread over in the last few years. Currently about 850 briquetting machines are operating in the country, which is shown in Table 1. Bangladesh Institute of Technology (BIT), Khulna has also improved by surface hardening with tungsten carbide alloy by micro flow super jet spray. The screw of the local briquetting machine is about 6-7 hours, but in BIT, Khulna are found screw life 10-12 hours by micro flow alloy spray. Also a preheater have been designed to preheat the biomass-using conveyor surrounded by the electric heater.

So far briquetting technology in Bangladesh has been limited to only one raw material, namely rice husk. Among the total rice production 70 -75% are used as parboiling rice and the rest are unboiled. Only unboiled rice husk is used for producing briquettes as it contains lignin, which helps to compact the product. About 19,48,498 metric tons of rice husk may be used for producing briquettes. But only a portion of the rice husk is used for producing briquettes, rest of these is used for burning and a small portion is unused. The expected market of the briquettes in Bangladesh may be: already the restaurants, hotels, small scale food, tea stall, agro-product processing, small boilers industrial, the brick industries and a great number of households have been interested to use it. But there is limited production of briquettes for not availability technological improvement. So, all these listed briquettes consumers would ensure the future prospects of biomass briquetting in Bangladesh. For this the development of the briquetting technology for Bangladesh is basic need.

3. BRIQUETTING TECHNOLOGY

The biomass briquetting is the process of densification from a lower density of the raw biomass into a higher bulk density of the product by compressed the loose raw materials. Normally the screw press and ram press technology are used for briquetting. In the screw press technology, which is used for briquetting of biomass, has a screw to compress the material through a taper die. This technology has been in use for so many years in Japan and European countries for briquetting some easy material, i.e. saw dust, soft wood and rice husk etc. The only moving part of the machine which is a screw rotating at a speed of 600 rpm is prone to more damage than the stationary contact part between because of wear caused by the abrasive behavior of the biomass. A long run of the machine without changing the screw is expected when the quality of raw material used is very poor, i.e. without any associated dust or inherent presence of ash or if the screw will suffer damage because of its lack of sustenance to high temperature and pressure condition. This screw is designed such to convey the material and to compress it partially before it reaches the die. The flights of the screw are more prone to damage because of their smaller surface area. The wear is done due to mechanical friction of biomass with the screw surface.

Bhattacharya et al. (1993) have analyzed the potential use of densified rice husk and sawdust in Thailand from an energy viewpoint. A survey has found a number of densifying machines lying idle due to lack of market for the products. An experimental investigation have found that water hyacinth which produces rather fragile briquettes on its own probably because of its low lignin contains (8%) can be made to produce denser briquettes by mixing it with rice husk (lignin contents 40%). Bhattacharya et al. (1984) densified the rice husk in a heated die extruder. Rice was mixed with water hyacinth at different ratios. The quality test carried out for the products showed that rice husk alone formed better briquettes than the mixture.

3.1 Briquetting Technology with Preheatering

Thermal pretreatment virtually eliminates these problems. If the biomass is heated for a period of temperature at $180\text{-}200^0C$, the following changes occur. The physical and most chemically bonded water is removed. The cellulose fibers soften and some of the lignins and a very small amount of the biomass material turn to charcoal. The energy requirements for the compaction are reduced by upto 50% as the material fibers are softer and lignin's it as a lubricant. Relaxation of the compressed material does not occur as the hot material is plastic not elastic. Mild steel die can replace the alloy steel dies due to lower wear and the reduced pressure required to compact the material. Once the briquette is cooled, the lignin acts as a binder and provides a water resistant surface coating. The lignin also increases the briquettes durability and ensures that degradation during transport is minimum. Since the biomass has started to decompose and since it is very dry (approx. 5% moisture) it is very easy to ignite.

[1] Source: Field Survey Report, BIT, Khulna, 1997

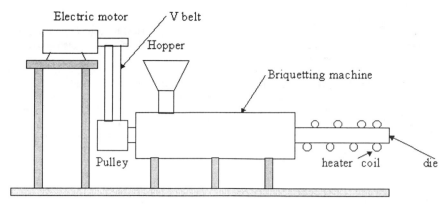

Figure 1. Block diagram of the briquetting machine

Joseph and Hislop (1985) suggested that grinding or preheating the raw materials might reduce the power required for briquetting, allowing higher quality briquette for a given energy input, lower wear of dies or a combination of these. Although pre-heating requires extra thermal energy for complete drying and to heat the biomass a higher temperature, these are offset by lower briquetting power requirement, possible reduction in die/screw wear due to improved lubricate of the biomass at increased temperature and increased fuel value due to complete water removal and pre-pyrolysis. Rahman (1989) carried out an experimental work to examine the technical feasibility of a rotary drum heater and of a vertical electric heater to preheat the raw material for densification. Using the same rotary drum, Aqa and Bhattacharya (1992) studied the effect of varying the die temperature and the raw material preheat temperature on the energy consumption for sawdust densification using a heated die screw press. A significant amount of energy could be saved by densifying sawdust preheated to a suitable temperature. The energy input to the briquetting machine motor, die heaters and overall system were reduced by 54,

30.6 and 40.2% respectively in case of sawdust preheated to 115°C. The decrease in the electric energy requirement per kg of sawdust allows operation of the briquetting machine at higher throughput with the existing motor.

3.2 **Briquetting Machine**

From the construction point of view the briquetting machine looks quite simple whose major parts are the driving motor, die and screw which is shown in Figure 1. Pulley and belt are used for transmission of power from the motor to the screw. The die is stationary and warped around with electric heaters. When the temperature reached at certain limit, the motor is started which rotates the screw by pulley mechanism and the raw material (rice husk) is feed into the hopper of the briquetting machine. The screw conveys the raw material and also compresses. As the heater is set outside the die, the temperature melt the lignin of rice husk and acts as a binder. The die is set as taper. When the screw rotates the material compresses and the density increases. Die and screw are available in different lengths, short and long. The mismatching of the die and screw may create a problem.

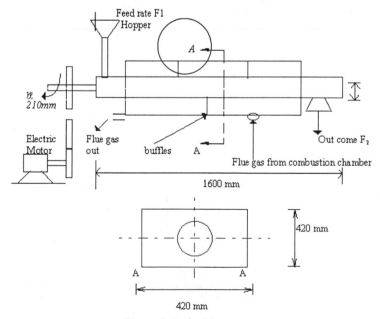

Figure 2. Preheating system

3.3 Biomass Preheater

The preheating system is designed to heat the biomass indirectly. The biomass is feed into the preheater by the hopper. The screw conveyor is used to convey the biomass with agitate which is surrounded by the cylinder. The screw is designed to match the output of the heater with the input of the briquetting machine. The output rate of the preheater can be controlled by the rotational speed of the screw. The screw is driven by a motor with pulley mechanism. The preheater was heated by the flue gas coming out from combustion chamber in which producer gas from a gasifier burnt. The flue gas is supplied at the counter flow into the annular space between the cylinder and outer shell. The buffles is used to increase the heat transfer rate. By the heat transfer principle, the biomass absorb heat from the flue gases. The heated biomass is supplied into the hopper of the briquetting machine. Figure 2 shows the details of the preheating system.

4. EXPERIMENTAL RESULTS AND DISCUSSIONS
4.1 Cold Run without Preheater

All densification experiments were performed with a heated die briquetting machine available in the energy park of Energy Technology division. Before carrying out an experimental study of briquetting technology with preheating option, some preliminary tests on the briquetting machine and the preheater had conducted separately under cold condition in order to find out the range of operation for the whole system. Rice husk and saw dust were used as raw materials for the study. A temperature controller unit with probe attached to the external surface of the die is mounted on the machine. For the experimental procedure the switch of heater for the heated die was on and the temperature also set at $250^{0}C$ for rice husk. When the temperature reached $250^{0}C$, the indicator shown and the switch off automatically. Then the motor is started which rotates the screw by pulley mechanism and the raw material is feed into the hopper of the briquetting machine. The screw conveys the raw material and also compresses. As the heater is set outside the die, the temperature melt the lignin of rice husk and acts as a binder. The die is set as taper. When the screw rotates the material compresses and the density increases. For the sawdust, the setting the temperature $350^{0}C$ as the sawdust contains large amount moisture. The energy consumption for the die and heater is calculated by knowing the meter readings separately set two meters for measuring the power.

In conventional briquetting rice husk and sawdust at the ambient temperature was feed to the briquetting machine. In sawdust at the die temperature $250^{0}C$ and $350^{0}C$, the quality of the product is not good as the sawdust contains large amount moisture. When the die temperature was set $300^{0}C$, the product quality was good, the outer surface of the product was not crack. Only increase of high die temperature the color of the briquette was a little bit darker and their surface was partly pyrolyzed. For the rice husk at die temperature $250^{0}C$, the quality of the product was good, the outer surface of the product was dark brown and their no jammed on running. But the sawdust was jammed and the outer surface of the product was crack.

4.2 Cold Run with Preheater

An amount of 2-3 kg of biomass was initially fed into the preheater in order to facilitate the uniform distribution of the biomass inside the preheater pipe, the screw was manually rotated in both directions: clockwise and anti-clockwise for 5-10 minutes. The raw material was then fed manually into the preheater through the feeding hopper one the electric motor started. The manually feeding was made smoothly to avoid biomass jam inside the preheater. Three speeds are set foe rice husk 150 rpm, 200 rpm and 300 rpm and two speeds are set 100 rpm and 200 rpm for sawdust. Figure 3 and 4 show the output of the preheat verses motor speed and corresponding mathematical correlation between preheater output rate and motor rpm of the preheater screw. From the graph and correlations ($y = - 0.0041x2 + 3.794x - 142.22$ for rice husk and $y = 0.49x + 124$ for saw dust, where y is the output rate and x is the corresponding motor speed of the preheater screw) between the output rate of the motor of the preheater screw could find for setting any output according to the briquetting machine. From the experimental observations, for sawdust, it was very difficult to run the preheater at a speed of 300 rpm for the electric motor because the biomass was jammed inside the preheater. This fact attributed to the relatively wide particle size distribution of the sawdust used.

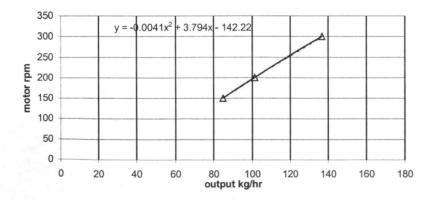

Figure 3. The variation of the preheater output corresponding motor speed of the preheater screw

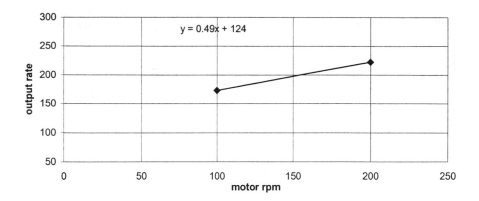

Figure 4. Variation of the preheated saw dust output from the preheater corresponding to the motor speed.

4.3 Hot Run with Preheater

The gasifier was initially charged with 10 kg of charcoal to guarantee the time period of operation (approximately 3 hours) for the whole system. From the preliminary test on the gasifier itself, a fixed value of air flow rate of 80 lit/min was chosen in terms of the stability in working of the gasifier, also an initial value of the mixing air of 60 lit/min was chosen to dilute and eliminate the accumulation of the producer gas during the start-up process. Temperature inside gasifier in the combustion chamber and in the mixing path was measured by three K-type thermal couples. After 10 minutes, the producer gas was then ignited in the combustion chamber using one small heater located inside the chamber. The combustion air was initially chosen as 70-80 lit/min. Once the flame was stable, this heater is taken out from the combustion zone, and the temperatures at different locations of the whole system were then recorded for every five minutes. When the temperature of the flue gas in the mixing path reached at 650^0C, an amount of 2 kg of rice husk was manually fed into the preheater. By rotating the shaft of the preheater in both clockwise and anti clockwise directions, the biomass was uniformly distributed inside the preheater. When the temperature of the flue gas at the outlet of the preheater was about 70-80^{0}C, the shaft of the preheater was then rotated by the electric motor, and the feeding of raw biomass was made smoothly. Temperature of the biomass at the outlet of the preheater was traced using one K-type thermal couple. Given the same speed of the shaft, this temperature could be controlied by increasing or decreasing the mixing airflow rate in the mixing path.

The effect of the mixing airflow rate on temperature of the preheated biomass can be easily identified from Figure 5. The preheated biomass temperature and the exhaust temperature of the flue gas increases with the increase of the flow rate of the mixing air. The energy consumed to the system has saved 7.5% of the total energy consumed by using preheater on the system. The energy input to the system could probably decrease further if the production rate was increased at 300^0c and 350^0C and the production capacity increased and

briquettes surface smooth and without crack. Their color was dark brown at 350^{0}C and brown at 300^{0}C. At the sawdust of 125^{0}C and die temperature of 250^{0}C, the densified process could be carried out smoothly and without any problem but only 250^{0}C without preheating saw dust briquettes is not so good. The result from the preheated sawdust densification has shown input to the motor decreases and the entire system reduces at the sawdust temperature 125^{0}C. As the production rate increased energy consumption rate is reduce, therefore it is better to run the machine at full capacity to achieve less energy input per ton of briquettes production, energy loss due to friction in the different bearing of the machine and heat loss due to the ambient from the die remain unchanged.

5. CONCLUSIONS

The study found for rice husk and saw dust that the temperature of heated die should be in the range from 250 - 350^{0}C in cold run condition. The appropriate die temperature depends on the moisture content of the raw biomass. At the die temperature lower than 300^{0}C - 350^{0}C briquettes from sawdust are not so good quality. The most suitable die temperature for densification of sawdust appears to be 300^{0}C and for rice husk 250^{0}C. At the same rpm of screw of the preheater, the output rate of sawdust was always higher than that of rice husk. This results from the difference in size distribution of sawdust and rice husk. The results of experimental study on hot run condition with a preheater have showed the advantages compared to the case without preheater. Preheating briquettes appearance and quality were good. They were strong enough and did not break immediately in to pieces while emerging from the die and partly carbonized (outer surface) that facilitate the ignition, transportation and of cause they could be supplied to the areas very far from the production base where the fuel is in a shortage. The preheater can significantly improve the quality of the briquettes, save energy on the electric motor as well as whole system. On the other hand, they become less sensible in contact with water and more sustained more convenient to use and acceptable by the users.

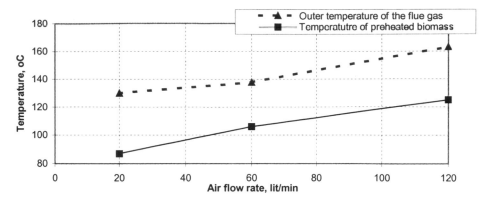

Figure 5. Variation of the outer temperature of the flue gas and preheated biomass corresponding to the airflow rate

The energy input to the densification process was reduced significantly when the briquetting machine was operated at higher throughput capacity. The concept of preheating of agro residues was found to be successful in reducing the power consumption and increasing the production rate and may increasing the screw life due to lower pressure on the screw. In economic point, energy saving in briquette production by the system with preheater could reduce the price of the briquettes that encourage again the people in using this fuel and develops the market. By briquetting technology with preheating option the loose biomass then becomes a competitive fuel to be commercialization in the market.

6. REFERENCES

1. Aqa, S. and Bhattacharya S.C., 1992 " Densification of Preheated Saw Dust for Energy Conservation, Energy, Vol. 17, No. 6, pp. 575-578.

2. Asian Development Bank (ADB), 1994, Energy Indicators of Developing Member Countries of ADB.

3. Bangladesh Energy Planning Project (BEPP), Vol. IV, Rural Energy and Biomass Supply, 1987, Planning Commission, Government of the Peoples Republic of Bangladesh.

4. Bhattacharya, S.C. and Yeasmin, H., 1984, Effect Of Densification Pressure And Temperature On The Properties Of Densified Biomass, Asian Institute Of Technology, Bangkok.

5. Bhattacharya, S. C., 1993, State-of-the-Art of Utilizing Residues and Other Types of Biomass as an Energy Source; RERIC International Energy Journal; Vol. 15, no. 1, pp. 1-21.

6. Hislop, D., 1986,, Residue Briquetting in the Gambia, Appropriate Technology, Vol.13, No. 2, pp. 22-23.

7. Islam, M. N., 1993, ed. by K.V. Ramani, M.N. Islam and A. K. N. Reddy, Bangladesh Country Report in Rural Energy Systems in the Asia Pacific Development Centre.

8. Joseph and Hislop, 1984, Small Scale Briquetting Project 470, Intermediate Technology Development Group, Rugby, UK.

9. Rahman, M., 1989, A Study of Biomass Torrefaction and Preheated Biomass Densification, A Research study, AIT, Bangkok.

ACKNOWLEDGEMENT

[This study is the part of a special program on Biomass Briquetting at Asian Institute of Technology, Thailand. And the programme is funded by the Swedish International Development Cooperation Agency (SIDA)]

Proceedings of IECEC'01
36th Intersociety Energy Conversion Engineering Conference
July 29-August 2, 2001, Savannah, Georgia

2001-RE-17

EXPERIMENTAL STUDY OF COMBUSTION CHARACTERISTICS IN A RICE HUSK FIRED VORTEX COMBUSTOR

Pongjet Promvonge
Mechanical Engineering Department, Faculty of Engineering,
King Mongkut's Institute of Technology Ladkrabang,
Bangkok 10520, Thailand
Email: kppongje@kmitl.ac.th

Kulthorn Silapabanleng
Former Associate Professor
Mechanical Engineering Department, Faculty of Engineering,
Chulalongkorn University,
Bangkok 10330, Thailand

ABSTRACT

This paper presents the study of combustion characteristics of rice husk fuel in an annular vortex combustor (VC). The temperature distributions for selected locations inside the combustor and the fly ash and smoke from its flue gas were measured and observed respectively. Measurements were made by setting a constant mass flow rate of rice husk to be 0.2 kg/min and by varying the equivalence ratio, Φ, to be 0.8, 1.0 and 1.2. To study the effect of feeding secondary air on flame stability, three values of the ratio of volumetric flow rates of the secondary air to the total air, (λ), were used and set to be 0.0, 0.2 and 0.4. The experiment shows the maximum temperature of about 1000° C in the upper chamber with less smoke of flue gas. Besides, emissions and the sizes of flyash particles from the exhaust stack can be controlled by the flow rate ratios of the secondary air to the total air, λ. The VC shows an excellent performance, low emissions, high stabilization and ease of operation in firing the fine rice husk.

Keywords: annular vortex combustor, rice husk, biomass.

INTRODUCTION

The conventional sources of energy have been depleting at an alarming rate and the price of conventional energy is going up. Thus, the focus on alternative renewable sources of energy has been increasing and biomass is one of them that has been getting continued and increased attention. In general, paddy rice is one of the mostly produced crops throughout the world. Since it is not possible to take the whole of it as food, about 22% by weight of this paddy rice is generated as a waste, known as rice husk. It is estimated that over 60 million tons of rice husks are generated each year worldwide. Thailand generates 4.4-4.6 million tons of rice husks annually, which their thermal potentials are equivalent to 1.46-1.53 million tons of crude oil.

About 10% of rice husk are utilized as a source of heat energy in Thailand. It is creating waste management problem, especially in the rice milling sites. Therefore, an attempt to energy recovery from this rice husk waste by combustion technique may be worthwhile. With this the burning of rice husk by an annular vortex combustor (VC) can be considered.

There have been many reports published on various combustors using biomass materials or coals as a fuel. Singh et al. [1] designed a cyclonic rice husk furnace for drying a ton of paddy and its moisture was reduced from 35% to 14%d.b. Different furnace efficiencies were found for various rice husk feed and air flow rates. Tumambing [2] also investigated a rice husk furnace for drying paddy. Xuan et al. [3] studied two types of husk furnaces. One was a furnace with inclined grate and cylindrical combustion chamber with heat exchanging in the furnace upper part. Inlet air entered at the lower part of inclined grate and burnt the rice husk on grate. The other was a pneumatic-fed furnace (vortex type) including combustion chamber and rice husk feed system. Rice husk was fed into the combustion chamber with primary air and burnt and fell to the lower part of the furnace. The secondary air entered tangentially the chamber to create the vortex flow with a view to eliminating dust from flue gas. Soponronnarit et al [4] applied a rice husk fired furnace to a fluidized bed paddy dryer. An extensive study of a vortex combustor for burning dry ultrafine coal and coal water fuel was studied by Nieh and Fu [5]. A vortex combustor similar to [5] but using rice husk fuel instead was experimentally investigated by Promvonge et al. [6]. For above combustors burning rice husk fuel, effects of ash inside the combustors on combustion efficiency including its problem in elimination were reported.

This article deals with a preliminary study of combustion characteristics and temperature distributions in a vortex combustor burning rice husk fuel. Effect of feeding secondary air on flame stability, temperature control in the combustor, emissions, ashes and smokes from the exhaust stack are studied and observed for a guideline in design and improving the performance of the combustor.

Table 1 Composition of fine rice husk

Carbon	38.0 %
Hydrogen	5.70 %
Oxygen	41.6 %
Nitrogen	0.69 %
Sulfur	0.06 %
Moisture	10.3 %
Ash	14.0 %
Density, kg/m^3	100.00
Gross heat of combustion, kcal/kg	3,580.00
Stoichiometric air	4.850 kg/kg fuel

EXPERIMENTAL SETUP

The rice husk particles were milled and grinded, sieved up to about 1.3-mm size (between 1.19 and 1.41 mm), and stored in the laboratory under dry condition (10.3% of moisture content). The arrangement of experimental system of the combustor is shown in Fig. 1 below. The combustor was a concentric cylindrical pipe made of steel with castable refractory cement lining as insulation while the exhaust center pipe was made of stainless steel. It is 0.755-m high and 0.20-m

inside diameter (D) with multiple injection nozzles of 0.005-m diameter each for the secondary air (Q_2) as depicted in Fig. 2. The exhaust center pipe is 0.078 m in diameter.

The combustor was operated over a temperature range from 600 – 1000° C. A blower and two compressors were used for providing both primary and secondary combustion airs. Rice husk fuel was fed through a screw feeder and injected to the bottom chamber by pneumatic conveying via the primary air (Q_1). Start up process was commenced by heating up the VC with LPG torch inserted at the lower air nozzle slot. The preheating took about 20 minutes for the chamber to raise its temperature to be about 400°C. Then feeding commenced through the hopper, slowly with the rice husk until 0.2 kg/min and kept constant. When the temperature in the chamber reached 700° C, stop of preheating with LPG was made. A thick, black smoke was seen at the beginning and slowly thinned out by adjusting the air/fuel (A/F) ratio. Excess air at various levels for equivalence ratio, defined as $\Phi = (A/F)_{stoi}/(A/F)_{actual}$, of 0.8, 1.0 and 1.2 was tried for each of λ. The temperatures were monitored at various selected locations with chromel-alumel (type K) thermocouples while volumetric flow rates of primary and secondary airs were measured by using orifice meters. Flue gas emissions were measured by a gas analyzer. All data collection was taken at steady state condition.

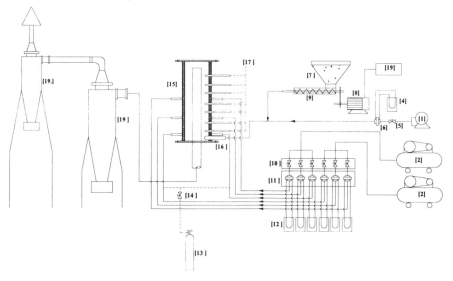

1. Blower	2. Air compressor1	3. Air compressor2	4. Manometer	5. Needle valve	6. Orifice plate
7. Hopper	8. Motor	9. Screw feeder	10. Needle valve	11. Orifice plate	12. Manometer
13. L.P.G. supply	14. Burner	15. Vortex combustor	16.Primary air nozzle	17. Indicator	18. Inverter
19. Cyclone					

Fig. 1 Schematic diagram of experimental setup of combustion system.

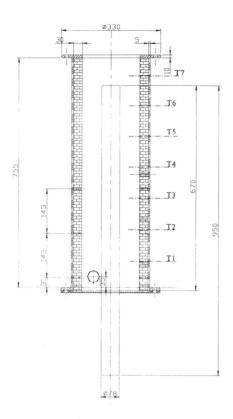

Fig. 2 Configuration of vortex combustor (unit in mm.)

RESULTS AND DISCUSSION

The VC was originally designed to accommodate various types of coal fuels as sources of heat. The VC responded significantly to the various fuels as expected. For rice husks, the level of equivalence ratio (or excess air) was varied to optimize the temperature inside the VC. The measurements of temperature were made at seven axial locations namely, x/D = 0.5, 1.0, 1.5, 2.0, 2.5, 3.0 and 3.5.

Effects of Equivalence Ratios

The radial profiles of temperature inside the VC for $\Phi = 1.2$ (rich), $\Phi = 1.0$ (stoich.) and $\Phi = 0.8$ (lean) are presented in Figs. 3a, 3b and 3c respectively. The feed-rate of fine rice husk with 10.3% moisture content was 0.2 kg/min and kept constant throughout.

For $\Phi = 1.2$ and $\lambda = 0$, Fig. 3a, it is found that the temperature profiles for all locations generally are not uniform. The temperature in the vicinity of the center pipe is higher than that near the chamber walls. Maximum temperature of about 800° C takes place in the top of chamber while lower temperature can be seen near the chamber wall. Unburned fuel and black smoke emission from the exhaust stack were observed for this case. This indicates the incomplete combustion (high CO concentration) or combustion reaction takes place very slowly.

Fig. 3b shows the temperature distributions for $\Phi = 1.0$. It can be seen that (for $\lambda = 0$) the temperature profiles are nearly uniform for all stations except at the top chamber. The maximum temperature above 800° C is found in the top annular space whereas the minimum temperature in the vicinity of the chamber walls at early station. This shows insufficient turbulence at the bottom chamber, which was due to too many rice husks, resulting in inadequate mixing and consequently the presence of fuel rich zones within the bottom space. This case yielded a less black smoke emission from the exit pipe.

The temperature distribution for $\Phi = 0.8$ is depicted in Fig. 3c. It is worth noting that the temperature profiles similar to the case for $\Phi = 1.0$, are nearly uniform for all stations. The peak temperature of some 800° C is also seen in the top annular space while lowest temperature at the bottom one. The peak one for every station is found in the middle between the chamber and center pipe walls. The maximum temperature for this case has the lower value than the case for $\Phi = 1.0$, as expected, since an increase in the excess air leads to decreasing temperature in the combustion system. White and thin smoke emission and small size of flyash particles are visible. This points out that complete combustion takes place for this case.

Effects of Volumetric flow rate ratios of Secondary to total Airs (λ)

The ratio of volumetric flow rate of the primary air to the secondary air, defined as $\lambda = Q_2/(Q_1+Q_2)$, is an indication of the strength of the vortex of the flow. The increase in value of λ results in the higher tangential velocity component (strong vortex). This means that the impact of swirling phenomenon of the secondary air on the flow and temperature fields becomes more pronounced as the value of λ is increased. Comparison of the temperature distributions for $\lambda = 0.0$, 0.2 and 0.4 for various equivalence ratios is also presented in Fig. 3.

For $\Phi = 1.2$, it is obvious that adjusting of $\lambda = 0.4$ leads to a flatter and higher temperature distribution curve than that of $\lambda = 0.0$ and 0.2. The use of a value of λ can improve substantially the temperature distributions in the VC as can be seen in Fig. 3a. This indicates that for the rich mixture, λ should be used to help increase combustion efficiency.

When $\Phi = 1.0$, it is interesting to note that the influence of λ on the temperature profiles shows slightly improvement on the first three stations at the bottom while significant improvement due to λ on the temperature profiles is found at other stations as illustrated in Fig. 3b. Again, use of λ leads to higher combustion temperature in the VC at about 20%.

As $\Phi = 1.2$, it is worth noting that effect of λ on the temperature distributions is seen at early stations while slightly influence on the top part of the VC as exhibited in Fig. 3c. Also, use of λ shows better combustion temperature obtained.

It is concluded that for all values of Φ, use of λ results in significant improvement of temperature profiles and yields higher temperature of combustion in the VC. Besides, the increased swirling intensity leads to an abatement of particle elutriation, and to enlargement of the region of recirculating zone.

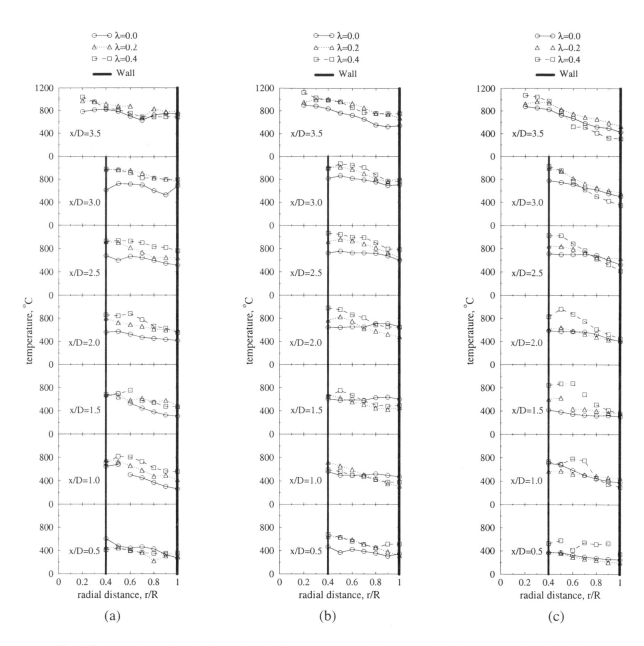

Fig. 3 Temperature distributions inside the combustor, (a) Φ = 1.2, (b) Φ = 1.0, and (c) Φ = 0.8.

Emissions of exhaust gas

The flue gas emissions of CO, CO_2, NO_x and O_2 for various equivalence ratios and different λ are studied and presented in Figs. 4a, 4b 4c and 4d, respectively.

For CO emissions in Fig. 4a, it is interesting to note that the emissions of CO are reduced substantially when Φ is decreased (or lean mixture). The formation of CO is very low once λ is introduced for the three equivalence ratio cases. The emissions of CO_2 can be also reduced by increasing the excess air (or decreasing Φ). Nevertheless, the application of λ leads to the reduction of CO_2 as can be seen in Fig. 4b.

For NO_x emissions, Fig. 4c shows that lower equivalence ratio favours the formation of NO_x. Also, use of λ results in higher NO_x emissions. This can be attributed to higher temperature of combustion in the VC. The level of O_2 emissions is presented in Fig. 4d. It is found that the O_2 level increases for lower equivalence ratio and using λ.

CONCLUSIONS

The measurements of temperature distributions and flue gas emissions have been conducted during the combustion of fine rice husk in the VC. The conclusions derived from the results of these experiments are as follows: combustion of rice husk in the

VC with equivalence ratio between 0.8 and 1.0 yields better combustion efficiency, flame stabilization, low emission and reliable furnace condition. For all equivalence ratios, use of secondary air has a significant effect on the temperature distributions inside and helps reduce emissions and the size of flyash particle.

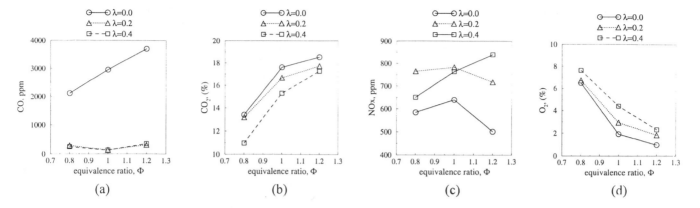

Fig. 4 Emissions of flue gas, (a) CO, (b) CO$_2$, (c) NO$_x$, and (d) O$_2$.

ACKNOWLEDGEMENTS

The authors would like to acknowledge with appreciation, the Thailand Research Fund (TRF) for the financial support and the Mechanical Engineering Department, Mahanakorn University of Technology (MUT) for their support in this effort and the MUT undergraduate students for data collection.

REFERENCES

[1]. Singh, R., Maherhweri, R.C., and Ojha, T.P., 1980, "Development of Husk Fired Furnace", The British Society for Research in Agricultural Engineering, India, pp. 109-120

[2]. Tumambing, J.A., 1984, "Testing and Evaluation of Rice Hull-Fed Furnaces for Grain Drying", National Post Harvest Institute for Research and Extension, pp. 197-214.

[3]. Xuan, N.V., Vinh, T., Anh, P.T. and Hien, P., 1995, "Development of Rice Husk Furnaces for Grain Drying", Proceeding of the International Conference on Grain Drying in Asia, 17-20 October, Bangkok, Thailand, pp.109-120.

[4]. Soponronnarit, S., Prachayawarakorn, S. and Wangi, M., 1996, "Comercial Fluidized Bed Paddy Dryer", Proceedings of 10th International Drying Symposium (IDS'96), Krakow, Poland, 30 July-2 August, Vol. A, pp. 638-644.

[5]. Nieh, S. and Fu, T.T., 1988, "Development of a Non-slagging Vortex Combustor (VC) for Space/Water Heating Applications", Proceeding of The 5th International Coal Conference, Pittsburgh, pp. 761-768.

[6]. Promvonge, P., Vongsarnpigoon, L. and Piriyarungroj, N., 2000, "A Low Emission Annular Vortex Combustor Firing Rice Husk Fuel: Part II – Experimental Investigation", Proceeding of The First Regional Conference on Energy Technology Towards a Clean Environment, 1st – 2nd December, Chiang Mai, Thailand, pp. 279-284.

Proceedings of IECEC'01

36th Intersociety Energy Conversion Engineering Conference

July 29-August 2, 2001, Savannah, Georgia

IECEC2001-RE-01

WIND FARM SIZING AND COST-SCREENING

Mukund R. Patel, PhD, PE

U.S. Merchant Marine Academy

Kings Point, New York 11024

Patelm@usmma.edu

ABSTRACT

The wind speed having a Weibull probability distribution and the wind power having the cubed relation with the instantaneous wind speed make the wind farm sizing and cost estimating a complex process. A simplified process is presented for estimating a community wind farm size and the energy cost based on the following approaches:

- The root mean cubed velocity derived from the generally available average speed at a proposed site. This approach adopts the results of a previously published paper by the author.
- The probabilistic consideration on the total community load using the National Electrical Code® load factors as applicable for a given number of customers to be served with electric power.
- Cost screening look-up chart developed for quick estimates of the energy cost at a site having certain average wind speed and certain capital cost of the equipment.

The data presented in the paper would enable the engineers and community planners to estimate the wind farm size and the energy cost using the generally available average wind speed at a proposed site.

1.0 COMMUNITY LOAD ESTIMATE

In today's homes, numerous electrical appliances connected to the power distribution lines require a peak demand of typically 20 kVA during some time period of the day. The peak demand on the power distribution station serving the community of homes and small businesses, however, is less than the arithmetic sum because they are collectively non-coincidental or diversified. For this reason, the peak-pooled demand at the distribution station gets reduced by a factor, which depends on the number of customers connected to the substation. The National Electrical Code (NEC®) gives the pooled demand factor as a function of the number of customers connected to the power distribution station (Table-1).

Table-1: NEC® demand factors (Adapted from the National Electrical Code® Handbook, 9th Edition, 2000).

Number of Dwellings	Demand Factor
3	0.45
10	0.43
15	0.40
20	0.38
25	0.35
30	0.33
40	0.28
50	0.26
>62	0.23

Thus, a community of more than 62 customers, each having 20-kVA peak load, poses a pooled peak demand of only 20x0.23, or approximately 5 kVA per customer at the central power station level.

2.0 WIND SPEED DISTRIBUTION

The wind speed is the most critical data needed to appraise the power potential of a candidate site. The wind speed variations over the period can be described by the following Weibull probability distribution $h(v)$ as a function of speed having two parameters, the shape parameter 'k' and the scale parameter 'c',

$$h(v) = \left(\frac{k}{c}\right)\left(\frac{v}{c}\right)^{(k-1)} e^{-\left(\frac{v}{c}\right)^k} \text{ for } 0 < v < \infty \qquad (1)$$

where h(v) is in hours per year per meter/second.

The actual wind speed distributions measured at numerous sites show a value of k approximately 2 as seen in Figure-1[1]. The Weibull distribution with k=2 is specifically known as the Rayleigh distribution. The Rayleigh distribution is then a simple and yet accurate representation of the wind speed with just one parameter, the scale parameter 'c'.

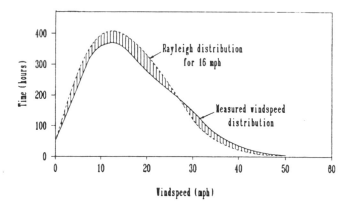

Figure-1. Measured wind speed distribution at most sites approximates the Rayleigh distribution (Weibull distribution with parameter k=2)

Letting λ=1/c, we can write the Rayleigh distribution function as

$$h(v) = 2\lambda^2 \cdot V \cdot e^{-(\lambda V)^2}$$

Hours per year per m/s (2)

Since the wind power is related with the wind speed cubed, the concept of the 'root mean-cube (RMC)' speed has been developed by the author[2]. The energy collected over the year is the integral of h·v³·dv over 8760 hours of the year. We therefore define the 'root-mean-cube speed', V_{rmc}, as

$$V_{rmc} = \sqrt[3]{\frac{1}{8760}\int_0^\infty h \cdot v^3 \cdot dv} \qquad (3)$$

All sites having the same average wind speed do not generate the same average electrical power. However, all sites having the same RMC speed would have the same annual power generation. The RMC speed concept, therefore, offers a great simplicity in assessing the annual energy potential of a wind site.

Using V_{rmc} in the classical wind power equation gives the annual average power, denoted by P_{rmc},

$$P_{rmc} = \frac{1}{2}\rho \cdot A \cdot C_p \cdot V_{rmc}^3 \text{ Watts} \qquad (4)$$

where ρ = air density of the sweeping air (kg/m³)
A = swept area of the blades (m²)
C_p = power coefficient of the rotor.

An analytical expression for the RMC speed for the Rayleigh probability distribution is shown to be[2]

$$V_{rmc}^3 = \frac{6\sqrt{\pi}}{8}\left(\frac{V_{mean}}{0.9}\right)^3 \qquad (5)$$

or $V_{rmc} = 1.222\ V_{mean}$ (6)

Thus, Equation 6 relates the RMC speed with the commonly reported mean speed. The coefficient 1.222 is easy for a wind power engineer to remember as it coincides with the standard air density value in kg/m³.

The RMC wind speed in turn when used in the classical wind power equation determines the annual wind power potential of a site where the annual average wind speed is known. That is,

$$P_{rmc} = \frac{1}{2}\rho \cdot C_p \cdot (1.222 \cdot V_{mean})^3$$

Watts/meter² (7)

The theoretical maximum value of the power coefficient is 0.59. Modern high efficiency rotors have C_p approaching 0.45. The air density is 1.225 kg/m³ at sea level, one atmospheric pressure and 20 degrees Celsius. The mechanical power output of the rotor driving the electrical generators is therefore

$$P_{out} = \frac{1}{2} \times 1.225 \times 0.45 \times (1.222 \cdot V_{mean})^3 = 0.5 V_{mean}^3$$

watts/meter² (8)

Equation 8 is a simple expression for the annual average power output of a wind turbine using the generally available annual average mean speed.

3.0 TURBINE SIZE AND TOWER SPACING

When installing a cluster of machines in a wind farm, certain spacing between the wind towers must be maintained to optimize the power cropping. The spacing depends on the terrain, the wind direction, the speed and the turbine size. The optimum spacing is found in rows 8 to 12 rotor diameters apart in the wind direction and 1.5 to 3 rotor diameters apart in the crosswind direction as shown in Figure-2. The average number of machines in wind farms varies greatly, ranging from several to hundreds depending on the required power capacity.

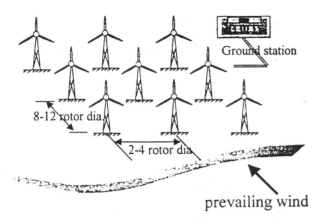

Figuure-2: Optimized tower spacing in wind farm located in a flat terrain.

When the land area is limited or is at premium price, one optimization study that must be conducted in an early stage of the wind farm design is to determine the number of turbines, their size and the spacing for extracting the maximum energy from the farm. The engineering trades in such study are:

- Larger turbines cost less per MW capacity and occupy less land area
- Fewer large machines can reduce the MWh energy crop per year, as downtime of one machine would have larger impact on the energy output.
- If the wind farm is also interconnected with the utility grid, the wind power fluctuations and electrical transients on fewer large machines would cost more in electrical filtering of the power and voltage fluctuations, or would degrade the quality of power, inviting penalty from the grid.

The optimization study takes into account the above trades. Additionally, it must include the effect of tower height that goes with the turbine diameter, the available standard ratings, cost at the time of procurement, and the wind speed. The wake interaction and tower shadow are ignored for simplicity.

Such optimization leads to a site specific number and size of the wind turbines that will minimize the energy cost.

4.0 COST SCREENING CHART

As with any conventional project, the profitability is measured by the profitability index, defined as

$$PI = \frac{Present\ worth\ of\ future\ revenues - Initial\ project\ cost}{Initial\ project\ cost}$$

By definition, the Profitability Index of zero gives the break-even point.

The profitability obviously depends on the price at which the wind farm can sell the energy it produces. In turn, it depends on the prevailing market price the utilities are charging to the area customers. The nation wide average electricity prices in the year 2000 was about 12 cents / kWh in the USA, and 9 pence/kWh (about 14 US cents / kWh) in the United Kingdom.

Inputs to the wind farm profitability analysis include the following.

- Anticipated energy impinging the site, that is the wind speed at hub height
- Initial capital cost of installing the farm
- Cost of capital, usually the interest rate on the loans
- Expected economic life of the plant
- Operating and maintenance cost, and
- Average selling price of the energy generated from the farm.

A detailed multivariable profitability analysis with the above parameters is always required before making financial investments. Potential investors can make initial profitability assessment using easy-to-use profitability screening chart below.

The profitability screening chart above is based on the cost of capital (discount rate) of 8 percent, the project life of 15 years, the total initial cost of 1100 $ per kW capacity, and the operating and maintenance cost of 3 percent of the initial project cost. The chart takes into account the inflation at a constant rate, i.e. the costs and revenues rising at the same rate. The taxes on sales, if any, must be deducted from the average selling price before it is entered in reading the chart.

An example for using the chart follows.

At a site with 7.5 m/s wind speed at hub height, the profitability index would be zero (a break-even point) if the energy can be sold at 0.069 $/kWh. If the energy price is 0.085 $/kWh, the profitability index would be 0.30, generally an attractive value for private investors. At a site with only 7 m/s wind speed, the same profitability can be achieved if the energy can be sold at 0.097 $/kWh.

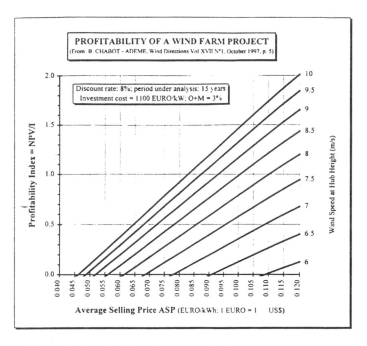

Figure-3: Profitability chart for wind farms. (Adapted from B. Chabot[3], ADME, France, and Wind Directions, Magazine of the European Wind Energy Association, London, October 1997.)

5.0 WIND FARM SIZING EXAMPLE

The following example illustrates the use of the simple sizing approach and the profitability index chart presented in the paper.

- A small town community consists of 5000 homes and small businesses to be served with electricity produced by the proposed wind farm. Each customer on average has the connected load of 20 kVA of individual peak demand.

- The average wind speed at the proposed site is 9 meters per second (20 miles per hour).

For this community, the design estimates proceeds in the following steps.

1. Using Table-1, the load factor for the community is 0.23.

2. The pooled peak demand for the central power distribution station is 5000 x 20 x 0.23, or say 25,000 kVA.

3. Typical power factor for a community of this size is about o.8 lagging. The efficiency of the electrical generator, power control electronics and power distribution wires from the turbines to the customers is typically 70- 80 percent with an average value of 75 percent. Therefore, the required power output of the wind farm is 25,000 x 0.80 / 0.75 = 26,670 kilowatts or 27 Megawatts.

4. Selecting the turbine number and size is a trade between the unit cost and the reliability. Candidate wind turbines for this farm may fall in the size range of 500 kW to 1500 kW. We choose the mid-range of 1000 kW, or 1 Megawatts output capacity of each turbine. Therefore, the number of turbines required to be working on average during the year is 27 / 1 = 27 turbines.

5. We add 10 percent margin for downtime during routine repairs and maintenance, Thus the number of turbines to be installed is 27 x 1.1, which gives 30 turbines each of 1 MW electrical output capacity.

6. Using Equation 8, the required blade diameter for the 1 MW wind turbine is determined to be 59 meters. The farm layout is then selected to have 6 turbines facing the prevailing wind in 5 rows. In accordance with Figure-2, we allow 10 rotor diameters between the rows in the prevailing wind direction and 3 rotor diameters between the columns in the crosswind direction. This requires the land area of 2950 x 1062 meters, amounting to 3,133,000 square meters or 774 acres. Allowing and extra 3 percent for the service buildings and the power distribution station, the total land area required for the wind farm is 800 acres.

For pricing the electricity delivered to the customers from the farm, we make the following assumptions:

- Land is provided free by the township.

- The wind farm is to be operated at economic break-even point (no-profit, no loss), or in terms of Figuure-3, at the profitability Index of zero, which represents the break-even point.

- Township borrows money for the project at 8 percent annual interest rate.

- The annual operating and maintenance cost of the equipment and the farm is 3 percent of the total capital cost per year.

For 9 m/s wind speed on the right hand side vertical scale, and the profitability index of zero on the left-hand side vertical scale in Figurue-3, we read the average selling price of 5.2 cents/kWh, which is about one-half the price paid by an average U.S. customer to the power utility company.

6.0 GRID CONNECTION OR ENERGY STORAGE

The wind farm size determined in Section 5.0 would generate the average power just to meet the community demand. The actual electrical output will vary in the cubed relation of the wind speed. For example, if the wind is at 8 m/s on a given day, the farm would generate only 70 percent of what is needed, and about 50 percent at 7 m/s wind speed. Therefore, on 50 percent of the days in the year, the electrical output would not exactly match with the demand. For the remaining 50 percent of the days, the output would exceed the demand. If the availability of power to all the customers is to be high near 100 percent, the following options must be considered:

- Interconnecting with the area grid. This would allow to draw power from the grid when the wind speed is below average, and to feed power back to the grid when wind speed is above average.

- Energy storage plant, such as the electrochemical battery, superconducting coil, flywheel, etc[4, 5].

- Diesel generator of needed size, which will burn fuel only when the wind speed is below average. At high wind, the excess power potential is wasted by deflecting the blades away from the optimum position.

- Increase the farm size to accommodate a much greater number of turbines than required to generate the 27 MW of power at average wind speed.

Each option will add significantly into the project capital cost and the resulting energy price to the customers. However, interconnecting with the grid is generally the most cost-effective approach, if available.

REFERENCES

1. Elliott, D. L., Holladay, C. G., Barchet, W. R., Foote, H. P. and Sandusky, W. F., "Wind Energy Resources Atlas of the United States", DOE / Pacific Northwest Laboratory Report No. DE-86004442, April 1991.

2. Patel, M. R., "Weibull distribution and RMC speed for sizing wind energy potential", Paper No.182, The Proceedings of the Intersociety Energy Conversions Engineering Conferences, Vancouver, Canada, August 1999.

3. Chabot, B., "L'analyse economique de l'energie ecolienne, Revue de l'Energie", Report No. 485, ADME, France, February 1997.

4. Patel, M. R., Wind and Solar Power Systems (Book), CRC Press LLC, Boca Raton, FL, 1999, p. 46.

5. Dorf, R. C., Technology, Humans and Society, Towards a Sustainable World (Book), Section 16.4, Academic Press, 2001.

CONTACT

Mukund R. Patel, PhD, PE, is a research and development engineer with over 35 years of experience in the power industry. He has served as Principal Power System Engineer at the General Electric Company, Fellow Engineer at the Westinghouse Research Center, Senior Staff Engineer at Lockheed Martin Astro Space, and the 3M McKnight Distinguished Visiting Professor at the University of Minnesota, Duluth. Presently, he is a professor at the U.S. Merchant marine Academy, Kings Point, New York.
Dr. Patel earned his MSEE and PhD degrees from the Rensselaer Polytechnic Institute, Troy, NY, MSIE from the University of Pittsburgh, and ME from Gujarat University, India. He is a Fellow of the Institution of Mechanical Engineers (UK), Senior Member of the IEEE, Registered Electrical Engineer in PA, Chartered Mechanical Engineer in UK, and a member of Eta kappa Nu, Tau Beta Pi, Sigma Xi and Omega Rho. Dr. Patel has presented and published over 40 papers at national and international conferences, holds several patents, and has earned NASA recognition for exceptional contribution to the UARS power system design. He is active in teaching and consulting. He can be contacted at patelm@usmma.edu.

Proceedings of IECEC'01
36th Intersociety Energy Conversion Engineering Conference
July 29-August 2, 2001, Savannah, Georgia

IECEC2001-RE-02

ASSESSING POTENTIAL WIND TURBINE SITES USING COMPUTATIONAL FLUID DYNAMICS

Marc P. LeBlanc
CANMET Energy Technology Center
1 Haanel Drive, Nepean,
Ontario, K1A 1M1. Canada

Gordon D. Stubley   **George A. Davidson**
Department of Mechanical Engineering
University of Waterloo, Waterloo
Ontario, N2L 3G1. Canada

ABSTRACT

The use of computational fluid dynamics to assess potential wind turbine sites found in complex terrain regions less than 10 km by 10 km is demonstrated and discussed. A three-dimensional numerical model was developed to implement recent advances in unstructured, finite-volume discretization methods and in solution algorithms. The result is a flexible modeling tool that can adequately and cost effectively predict wind flow patterns such as flow separation, recirculation zones, and turbulent wakes. To validate the numerical model, simulation results are compared with experimental data from the Askervein Hill study. Good agreement is found for fractional speed-up and turbulent kinetic energy profiles for both the windward and leeward sides of the hill.

NOMENCLATURE

$C_\mu, C_{\varepsilon 1}, C_{\varepsilon 2}$	Turbulence model coefficients
k	Turbulent kinetic energy
P	Mean dynamic pressure
$\boldsymbol{P}$	Mean shear production of k
S	Mean speed
U_i	Mean velocity components
x_i	Cartesian coordinates

Greek

μ_T	Turbulent viscosity
$\sigma_k, \sigma_\varepsilon$	Turbulence model constants
ρ	Local air density
ε	Dissipation rate turbulent kinetic energy

INTRODUCTION AND BACKGROUND

The demand for installations of wind energy conversion systems is expected to grow between 50 000 and 70 000 MW over the next 10 years in North America and Europe. With this anticipated increase and the sensitivity of the available power to local wind speed, it is important to accurately assess the wind resource at possible wind turbine sites. Several commercially available simulation models (*e.g.* WAsP and MS-Micro) are currently used to assess potential sites for wind power by simulating airflow over relatively large areas, i.e. greater than 10 km^2. These tools accurately model flows that remain attached and have weak vertical accelerations such as the flows over low, shallow terrain features. Even though these tools have been used to simulate flow over complex terrain features such as hills and ridges over areas less than 10 km^2, they are not capable of predicting important wind patterns such as flow separation, recirculation zones, and turbulent wakes. These wind patterns can significantly affect the performance of wind turbines by altering favorable flow speed-ups over hills and ridges and by creating undesired regions of high turbulence.

To predict these important wind patterns in complex terrain, steady-state three-dimensional (3D) numerical models based on computational fluid dynamics (CFD) technology that are traditionally used for engineering calculations in power generation and other industries can potentially be used. Surprisingly, only a small number of 3D CFD studies have been performed to predict the wind patterns over complex terrain, *e.g.* [1-3]. This lack of effort is most likely due to the high computational requirements traditionally associated with 3D CFD models. However, recent advances in computer hardware and numerical algorithms have opened up the possibility of making 3D CFD-based models cost effective.

The purpose of this paper is to demonstrate and discuss the use of CFD methods for assessing potential wind turbine sites found in complex terrain regions less than 10 *km* by 10 *km*. First, the modeling approach of the CFD model developed for this study is presented. Second, simulation results are compared with experimental data from the Askervein Hill study [4]. This comparison provides a realistic example of modeling atmospheric flows. Third, suggestions, as well as the strong and weak points, for applying the model for assessing wind turbine siting are discussed.

MODELING APPROACH

In the present study, an unstructured, steady-state 3D, finite volume CFD code was developed to simulate the flow over complex hilly terrain. An overview of the equations of motion, turbulence model, boundary conditions and CFD modeling approach used in the current model is given in this section. For implementation details, the reader is referred to [5].

Equations of Motion

When considering atmospheric conditions at a potential wind site, the flow of air can be considered as an incompressible fully turbulent (i.e. negligible molecular transport) flow of an ideal gas. For the present study, the Coriolis force due to the earth's rotation is negligible since this effect is only important when considering distances greater than 100 *km*. Thermal effects, which are typically weak for windy conditions, were also neglected. Based on these assumptions, the steady-state equations of motion for incompressible, turbulent flow are given by the Reynolds Averaged Navier-Stokes (RANS) equations as[1]:

$$\frac{\partial(\rho U_j)}{\partial x_j} = 0 \tag{1}$$

$$\frac{\partial(\rho U_j U_i)}{\partial x_j} = \frac{\partial}{\partial x_j}\left(\mu_T\left(\frac{\partial U_i}{\partial x_j} + \frac{\partial U_j}{\partial x_i}\right)\right) - \frac{\partial P}{\partial x_j} \tag{2}$$

where x_i, U_i, ρ and P represent the spatial Cartesian coordinates, mean velocity components, local air density and mean dynamic pressure, respectively. As discussed in the following section, the parameter μ_T represents an effective dynamic turbulent viscosity, which is used to model turbulence.

Turbulence Model

A large number of turbulence models have been developed for the RANS equations to model the turbulent transport and mixing of atmospheric flows [6]. However, two-equation models seem to provide the best balance between computational effort and an accurate description of the turbulent structures developing over hilly terrain [2]. Two-equation models use a molecular diffusion analogy to model turbulent stresses by the effective turbulent viscosity, μ_T, term of Eq. (2).

The standard form of the popular two-equation k-ε model was used for this study based on its potential shown in earlier studies [1-3]. For this model, μ_T is expressed as:

$$\mu_T = \rho C_\mu \frac{k^2}{\varepsilon} \tag{3}$$

where C_μ, k and ε are an empirical constant, the turbulent kinetic energy per unit mass and the dissipation rate of turbulent kinetic energy, respectively. The transport of k and ε is modeled using two conservation equations. The steady-state form of these two equations is given by:

$$\frac{\partial(\rho U_j k)}{\partial x_j} = \frac{\partial}{\partial x_j}\left(\frac{\mu_T}{\sigma_k}\frac{\partial k}{\partial x_j}\right) + P - \rho\varepsilon \tag{4}$$

$$\frac{\partial(\rho U_j \varepsilon)}{\partial x_j} = \frac{\partial}{\partial x_j}\left(\frac{\mu_T}{\sigma_\varepsilon}\frac{\partial \varepsilon}{\partial x_j}\right) + \frac{\varepsilon}{k}(C_{\varepsilon 1}P - \rho C_{\varepsilon 2}\varepsilon) \tag{5}$$

where the mean shear production of k, P, is defined as :

$$P = \mu_T\left(\frac{\partial U_i}{\partial x_j} + \frac{\partial U_j}{\partial x_i}\right)\frac{\partial U_i}{\partial x_j} \tag{6}$$

The coefficient values for the standard k-ε model given by Launder and Spalding [7] are $C_\mu = 0.09$, $C_{\varepsilon 1} = 1.44$, $C_{\varepsilon 2} = 1.92$, $\sigma_k = 1.0$ and $\sigma_\varepsilon = 1.30$.

As suggested by several researchers, see Raithby *et al.* [1], a value of 0.09 for C_μ is thought to be too high for atmospheric boundary layer flows. Based on measurements from several field studies performed over a wide range of surface roughness conditions [8] and assuming an equilibrium turbulent shear flow, a C_μ value of 0.033 is calculated. This value, first applied by Raithby *et al.* [1], is used for the current study. All other coefficients are set to their standard values.

Boundary Conditions

To solve the equation set defined by Eqs (1), (2), (4) and (5), boundary conditions must be specified. As shown in Fig. 1, five different domain boundaries can be identified for a potential site: inflow, outflow, surface, top and lateral (sides).

The inflow boundary is the only boundary used to specify the flow entering the domain (or box) defined over the region of interest. Consequently, this boundary must be aligned with the dominant flow direction shown in Fig. 1. To set the fluxes across the inflow boundary, the boundary values of U_i, k and ε are specified by the solution of a 1D CFD model. The conditions for the 1D model are selected to obtain vertical profiles that correspond with available meteorological data at the potential site.

---

[1] The Einstein summation is used throughout this paper.

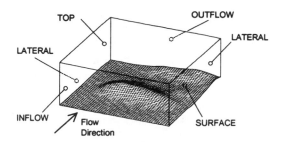

Figure 1: Computational Domain and Boundaries

The outflow boundary is the only flow exit of the solution domain. To leave the upstream flow features as undisturbed as possible, the mass, momentum, k and ε fluxes across the boundary are left unchanged. It is important to place the outflow far enough downstream from any possible recirculation zones to avoid numerical instabilities.

To represent the strong interaction between the surface and airflow at the surface boundary, a wall function based approach as suggested by Launder and Spalding [7] is implemented. Wall functions use a Log-Law velocity profile to relate the turbulent surface shear stress to the mean speed above the surface.

The top boundary is typically placed 500-1000 m above the surface as to not affect the underlying flow. At this elevation, an inviscid free stream flow can be assumed. This condition is imposed at the top boundary by setting the vertical velocity to zero. The normal gradients of horizontal velocities, k and ε are also set to zero.

Since the flow is perpendicular to the inflow and outflow boundaries, both lateral (side) boundaries are parallel with the dominant flow direction. It is therefore a reasonable assumption to restrict the flow of air through these boundaries. To impose this no-flow restriction, a symmetry condition is used. It is important to place the lateral boundaries sufficiently far away from terrain features. This precaution ensures that the lateral boundaries are not channeling the flow in the dominant flow direction.

Computational Mesh

Once the equations and boundaries to be modeled are defined, the first step for all CFD simulations is to generate a computational mesh by subdividing the domain shown in Fig. 1 into control volumes. For wind turbine siting, this region is typically a 500 to 1000 m deep layer above the terrain of the potential site. A computational mesh for a 2D ridge is shown in Fig. 2. At each control volume of the mesh, values of U_i, P, k and ε are calculated.

Figure 2: Example of a Computational Mesh

The more control volumes that are used to represent the flow region, the more accurate the CFD model becomes. However, as the number of control volume increase so does the incurred computational cost. To minimize these costs and to ensure an adequate representation of developing flow patterns, the distribution of the mesh is typically altered by increasing the resolution near the terrain surface as shown by the vertical mesh distribution of Fig. 2.

Discretization of Equations

Once the computational mesh has been generated, the second step is to approximate or discretize the equations of motion and boundary conditions that describe the behavior of the flow over each control volume. For the current CFD model, the finite-volume method [9] is used for the discretization. This method has been proven quite effective at modeling a wide range of small-scale engineering flows found in the power generation, processing, manufacturing, aeronautical, and aerospace fields.

To ensure a balance of geometrical flexibility, accuracy and computational cost, a modified 3D steady-state version of a recently developed unstructured second-order finite-volume method is applied to current model [10].

Multigrid Linear Solver

To solve the algebraic equations set formed by the discretization process, the 3D iterative linear solver developed by Raw [11] is used. This robust solver incorporates several recent advances in algebraic multigrid methods and has been extensively tested through its use in CFX-TASCflow©, a commercial CFD software package.

MODEL VALIDATION

The 3D CFD model, described in the previous section, is validated by comparing the simulated airflow over the Askervein hill to that measured during a multinational field study, [4]. This study is selected because it contains extensive wind speed and turbulence data. Furthermore, the measurements from the Askervein study have been used extensively as a benchmark to assess the predictions of earlier numerical models, *e.g.* [1,13,14].

Description of Askervein Hill Study and Data

The field measurements from the Askervein Hill study were taken in 1983 at Askervein, a 116 m high hill on the west coast of South Uist, one of the islands of the Outer-Hebrides in Scotland. The topography of the Askervein hill and surrounding area is given in Fig. 3.

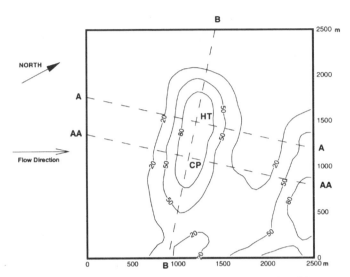

Figure 3: Topography and Measurement Positions of the Askervein Hill Study
Elevation contour lines are in meters

The experimental data selected for this comparison is tabulated as MF03-D, TU03-B and TK03 in the report of Taylor and Teunissen [4]. This set of data was obtained during the period of 14:00 to 17:00 British Summer Time on October 3, 1983. Towers equipped with propeller, cup, and sonic anemometers were used to collect the experimental data. These towers were sited at the hilltop (HT), at the center point (CP), and along the center point cross-ridge line (line A-A on Fig. 3), the hilltop cross-ridge line (line AA-AA), and the ridge line (line B-B). All these locations were used for measuring the flow 10 m above the surface. Vertical profiles up to 60 m above the surface were also obtained on the HT and CP towers as well as at a reference (RS) tower located on ground 2.8 km upwind from CP.

Setup of the Numerical Simulation

The computational mesh used for the flow simulation over Askervein is derived from the elevation data presented in Fig. 3 as provided by Taylor [15]. As a first step, the data is flooded to a minimum of 10 m above sea level. This flooding removes slightly uneven terrain around the hill at elevations between 8 and 10 m above sea-level.

A 2.5 km by 2.5 km horizontal area centered near the hilltop is then chosen to define the inflow, outflow and lateral boundaries of the computational domain, see Fig. 3. Note that

the inflow of the computational mesh is aligned to be perpendicular to the dominant flow direction. The vertical extent of the domain is fixed at an elevation of 700 m thus defining the top boundary. A uniform distribution of 50 by 50 control volumes is used for the horizontal mesh while an expanding mesh distribution of 30 control volumes is used for the vertical part of the mesh giving a total mesh size of 75 000 control volumes.

To specify the inflow boundary condition profiles, the numerical results of the 1D version of the CFD model are used. The vertical profiles of wind speed, S, and k generated by the 1D model are consistent with the experimental RS vertical profile data.

Based on the topographical data provided Taylor [15], a surface roughness length of 0.03 m with no displacement height is applied over the entire surface boundary. These values, which represent grass, are consistent with several previous studies [1,14].

Computational Cost

The Askervein simulation required only 2 hours of CPU time on a modest 300 MHz SUN workstation.

Comparison of Numerical Results to Measured Data

When comparing the numerical results to the measured data, good agreements is found for the fractional speed-up[2] and the turbulent kinetic energy. The profiles along the A-A line and above HT are shown in Figs. 4 and 5, respectively. Similar results are also found for the profiles along AA-AA and B-B lines and above CP. For complete set of comparison data the reader is referred to [5].

The numerical predictions presented in Figs. 4 and 5 also demonstrate that wind flow patterns such as the hilltop speed-up, flow separation, recirculation zone, and turbulent wake are well represented over the Askervein hill. The hilltop speed-up can be seen by the high fractional speed-up coinciding with the hilltop in Fig. 4. The vertical profile of the speed-up is also well represented in Fig. 5. A leeward side flow separation, recirculation zone, and turbulent wake can also be seen in Fig. 4 by the fractional speed-up dip and turbulent kinetic energy increase found beyond the hilltop. Models such as WaSP or MS-Micro cannot predict these abrupt flow features which will affect the predicted hilltop speed-up.

One of the notable features of the current model is that the kinetic energy levels in the recirculating wake at the lee of the hill are better predicted than those of an earlier study attempting to capture this pattern [1]. This improvement is likely due to the higher resolution of the computational mesh.

---

[2] The fractional speed-up is given as $S/S_{rs}-1$ where S and S_{rs} are the local and reference site speeds at the same height above the surface.

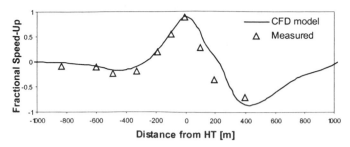

(a) Fractional Speed-Up

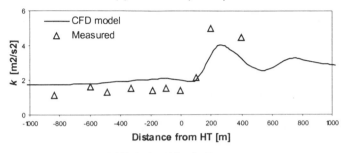

(b) Turbulent Kinetic Energy

Figure 4: Horizontal Profiles along the A-A Line 10 _m_ Above Surface
Dominant flow direction is from left to right.

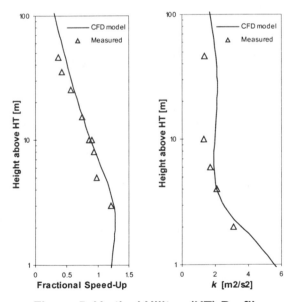

Figure 5: Vertical Hilltop (HT) Profiles

CFD AND SITING WIND TURBINES

The example shown in the previous section demonstrates the validity of applying CFD technology for adequately modeling wind patterns at a potential wind turbine site. By visualizing the data generated by the simulation, desirable flow speed-up areas or available power regions as well as undesirable recirculation and high turbulence regions can be

identified to properly place the wind turbines. Continuing with the Askervein simulation as an example, the contour plots of Fig. 6 presents the available power[3] over the hill. This figure clearly identifies the hilltop region as the highest area of available power. An undesirably high turbulence region is shown in Fig. 7 on the leeward side of the hill.

Figures similar to Fig. 6 and 7 can be generated for a range of meteorological conditions relevant to the potential site. These conditions could include variations in the direction, speed and vertical profile of the dominant flow. Since a geometrically flexible, unstructured discretization method is applied in the current CFD model, it is also feasible to introduce one or more wind turbine type obstacle(s) in the flow simulation. Although this type of study will increase the computational resources required, this approach would be very useful for assessing the effects of the wind turbine(s) on the developing flow patterns.

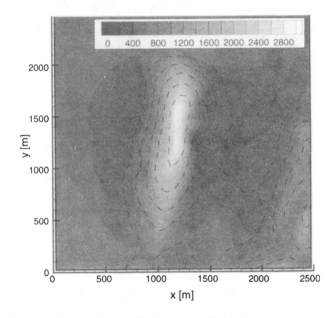

Figure 6: Contour Plots of Available Power 50 _m_ above surface
The units for the available power are W/m²

The only main drawback associated with using CFD technology for assessing potential wind turbine sites is obtaining detailed topographical and meteorological data, especially from remote areas. However, this problem is common to all types of atmospheric modeling efforts and, in most cases, the required information can be derived from available topographical maps and existing meteorological measurements.

---

[3] The available power is calculated based on the local density, ρ, and wind speed, S, as $\frac{1}{2} \rho S^3$. Units are in W/m².

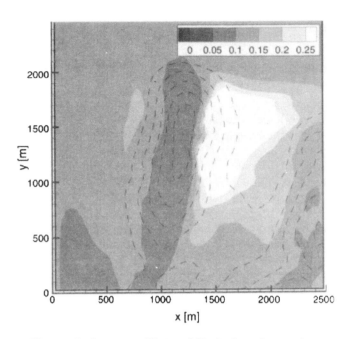

Figure 7: Contour Plots of Turbulent Intensity 10 *m* above Surface

It is important to note that the CFD model used in this study is tailored to only represent steady-state flows and neglects thermal effects. If prevalent weather conditions for a potential site are typically very sunny with calm winds (bad idea for wind turbines but great for solar panels!), the current model will not accurately represent the dominant flow features. In the event that transient and thermal effects are required, these features can be implemented in future versions of the CFD model.

CONCLUSION

Models based on CFD technology can provide a useful modeling tool for assessing potential wind turbine sites. CFD models are particular useful for visualizing regions of available wind power and high turbulence. With recent advances in computer hardware and numerical algorithms, CFD simulations can be performed at minimal computational costs thus providing an effective tool to the wind power industry.

ACKNOWLEDGMENTS

The National Science and Engineering Research Council of Canada is gratefully acknowledged for funding this study. The authors would like to thank Dr. P. A. Taylor at York University for providing the Askervein field data as well as AEA Technology for permission to use their multigrid solver.

REFERENCES

[1] Raithby, G. D., Stubley, G. D., and Taylor, P. A., 1987, "The Askervein Hill Project: A Finite Control Volume Prediction of Three-Dimensional Flows over the Hill," Boundary-Layer Meteorology, **39**, pp. 247-267.

[2] Apsley, D. D., 1995, "Numerical Modelling of Neutral and Stably Stratified Flow and Dispersion in Complex Terrain," Ph.D. thesis, University of Surrey, Surrey, UK.

[3] Utnes, T., and Eidsvik, K. J., 1996, "Turbulent Flows over Mountainous Terrain Modelled by the Reynolds Equations," Boundary-Layer Meteorology, **19**, pp. 393-416.

[4] Taylor, P. A., and Teunissen, H. W., 1985, "The Askervein Hill Project: Report on the Sept./Oct. 1983, Main Field Experiment," Research Report: MSRB-84-6, Atmospheric Environment Service, Downsview, ON.

[5] LeBlanc, M. P., 2000, "The Numerical Prediction of Neutrally Stratified Atmospheric Boundary Layer Flows over Hills," M.A.Sc thesis, University of Waterloo, Waterloo, ON.

[6] Hurley, P. J., 1997, "An Evaluation of Several Turbulence Schemes for the Prediction of Mean and Turbulent Fields in Complex Terrain," Boundary-Layer Meteor., **83**, pp. 43-73.

[7] Launder, B. E., and Spalding, D. B., 1974, "The Numerical Computation of Turbulent Flows," Computer Methods in Applied Mechanics and Engineering, **3**, pp. 269-289.

[8] Panofsky, H. A., and Dutton, J. A., 1984, *Atmospheric Turbulence: Models and Methods for Engineering Applications*, John Wiley and Sons, NY.

[9] Ferziger, J. H., and Peric, M., 1996, *Computational Methods for Fluid Dynamics*, Springer-Verlag, Germany.

[10] Zwart, P. J., and Raithby, G. D., 1998, "An Integrated Space-Time Finite Volume Method for Moving Boundary Problems", Numerical Heat Transfer: Part B, **34**, pp. 257-270.

[11] Raw, M., 1996, "Robustness of Coupled Algebraic Multigrid for the Navier-Stokes Equations," AIAA Paper 96-0297.

[13] Zeman, O., and Jensen, N. O., 1987, "Modification of Turbulence Characteristics in Flow over Hills," Quarterly Journal of the Royal Meteorological Society, **113**, pp. 55-80.

[14] Walmsley, J. L. and Taylor, P. A., 1996, "Boundary-Layer Flow over Topography: Impacts of the Askervein Study," Boundary-Layer Meteorology, **78**, pp. 291-320.

Proceedings of IECEC'01
36th Intersociety Energy Conversion Engineering Conference
July 29 - August 2, 2001, Savannah, Georgia
IECEC2001-RE-08

HIERARCHICAL TECHNIQUE FOR HYDROPOWER PLANTS AT THE NEW VALEY OF EGYPT AND TOSHKA AREA.

Prof. Faten H.Fahmy
Co-head of PV Dept.
Electronics Research Institute
Cairo, Egypt
e-mail: faten@eri.sci.eg

ABSTRACT

The South Egypt development project is a great opportunity to deepen and revive the Egyptian roots taking into consideration the regional balance and Arab African Integration.

Hence, when Egypt speaks of itself, through the giant project to develop the south (Toshka and New Valley), it is interpreting into living reality the century-long dream of the Egyptian people through the tranitional 5-years plan between the 20th Century and the beginning of the 21st Century.

This paper presents a new algorithm to solve a deterministic optimization control problem of new eight hydro power plants are connected in series on the stream of El Sheikh Zayed Canal with it's four branches that located at Toshka & New Valley Area.

The works shows two different techniques depending on the nature of the system to maximiz the electrical generated energy of the new hydrop power plants Once, the system is solved globaly by applying conventional technique and the other, it is divided into subsystems and applying the hierarchical technique. This demonstrates the effectiveness of the hierarchical approach in handling such large scale problem.

Keywords
Deterministic control problem, Hierarchical Technique, Hydro power plant, El Shiekh Zayed Canal.

1. INTRODUCTION

As a result of civilization, development and the increasing number of population in the world, several problems have been created which need a solution and consequently attracted the attention of many scientists in the different branches of science. One of the major and important problem which faces most of the societies is the increasing demand of energy, specially the electric energy. While at the same time, it is expected to have in the near future a shortage in natural resources, such as petroleum normally used for electric power generation.

It is therefore necessary to increase the efficiency of the electric power generation from the other resources such as solar energy hydropower generation,... etc, in order to satisfy the increasing demand of communities. In Egypt, the same problem exists, the demand of electric energy increases rapidly and the demand must be satisfied to implement the different projects required for the economic development.

The research in this paper is directed towards the medium term hydropower plant operating strategy problem which is highly affected by the dynamic behaviour of the suggested hydro reservoir chain system of El sheikh Zayed Canal, [1].

The promotion of the utilization of the River water embodies great and wide hopes to extend new branches and arteries to expand the circle of development in Egypt. Consequently, the Sheikh Zayed Canal which carries water from Lake Nasser to the South Egypt development project shall wide the circle of development to reach a broader, [2].

The objective of this research is to develop efficient control techniques capable of handling multiarea hydropower plants. These techniques are suitable to solve the deterministic formulation of the optimization problem of the cascaded hydro power plants [3] on El Sheikh Zayed Canal, [1]. Also, this paper will presents the optimal solution of system operation in both cases of global sens and a multilevel structure.

2. THE SYSTEM DESCRIPTION

Toshka is one of major development projects in south of Egypt. particularly, we bring into focus the important problems of designing optimal control structures for open loop opertion of the system.

The system under consideration can be expressed by the mass balance equation governing the operation of each reservoir:

$$X_{ji} = X_{j,i-1} + y_{ji} - U_{ji} - Losses \qquad (1)$$

3. THE OPTIMIZATION PROBLEM

The system is subjected to various physical and operational constraints:
Hydraulic Coupling

$$y_{j+1,i} = U_{ji} - Losses \qquad (2)$$

There is also an important constraint regulating the total release from first reservoir according to certain legal construction. This constraint may expressed as:

$$\sum_{i=1}^{n} u_{1i} = Const \tag{3}$$

The water conservation principle for each reservoir

$$X_{ji} \leq X_{j,i-1} + y_{ji} - U_{ji} \tag{4}$$

The Boundry limtis

$$\overline{X}_{ji} \leq X_{ji} \leq X_{ji} \tag{5}$$

$$O \leq U_{ji} \leq U_{ji} \tag{6}$$

While the objective function is to maximize the generated electrical energy of the hydraulic power system during total horizon time T divided into n equal intervals.

$$\text{Maximize} \quad F = \sum_{j=1}^{m} \sum_{i=1}^{n} E_{ji} \tag{7}$$

$$E_{ji} = \alpha_j . \Delta X_{ji} U_{ji} - \beta_j U_{ji}^2 - \gamma_j X_{ji}^2 \tag{8}$$

Thus the optimization problem is maximization as the total electric energy of system in (7-8), subject to the set of operational and physical constraints (1) to (6).

4. THE CONVENTIONAL DETERMINISTIC OPTIMIZATION TECHNIQUE

As shown before, the optimization control problem is expressed in the form of linear optimal control problem with quadratic objective function.

The quadratic programming is the suitable one to solve this dynamic problem to optimize it and obtain the optimal values of X_{ji}, $U_{ji}, \lambda_{ji}, \mu_{ji}$

Lagrange equation:

$$L = F + \lambda_{ji} \{equality\ constrants\} \tag{9}$$

$$+ M_{ji} \begin{Bmatrix} inequality \\ constra\ int\ s \end{Bmatrix}$$

$$= \sum_{j=1}^{m} \sum_{i=1}^{n} \{\alpha_{ji}\left(X_{ji} - X_{j+1,j}\right) - \beta_j U_{ij}^2 - \gamma_j X_{ji}^2$$

$$+ \lambda_{ij}\left(X_{ij} - X_{j,i-1} - y_{ji} + U_{ji} + W_{ij}\right)$$

$$+ \lambda'_{ji}\left(u_{1,i} - const\right)$$

$$+ \lambda''_{ji}\left(y_{j+1,i} - U_{ji} + W_{ji}\right)$$

$$+ M_{j,i}\left(X_i, -X_{ji}\right)$$

$$+ M_{ji}\left(\overline{X}_{ji} - X_{ji}\right)$$

$$+ M''_{ji}\left(U_{ji} - U_{ji}\right)$$

$$+ M'''_{ji}\left(\overline{U}_{ji} - U_{ji}\right)$$

5. HIERARICHAL TECHNIQUE

Complexity, is generally recognized in the interaction among model parameters, decision variables and the existence of many conflicting goals and objectives [3].

Thus, the successful operation of multilevel system is best described via two basic processes decomposition or infimals generation and coordination or overall objective synthesis, [4].

Maximize:

For Simplicity, this complexe, system, it can be divided into n=8 subsystems as shown in Fig(1).

The optimization problem of each subsystem (i=1,...) during intervals (j=1,...n) can be represented by the following equations:

$$F_{ji} = \sum_{i=1}^{n} \left[\alpha_j\left(X_{ji} - X_{j,i-1}\right)U_{ji} - \beta_j U_{ji}^2 - \gamma_j X_{ji}^2\right] \tag{10}$$

$$I=1,...n \qquad j=1,...m$$

Subject to

$$X_{ij} = X_{j,i-1} + y_{ji} - U_{ji} - W_{ji} \tag{11}$$

$$y_{j+1,i} = U_{ji} - W_{ji}$$

$$y_{1,j} = cons\tan t$$

$$X_{ij} \leq X_{ij} \leq \overline{X}_{ij}$$

$$U_{ij} \leq U_{ij} \leq U_{ij}$$

As shown before, the Lagrange equation of each subsystem m is composed of the objective function and the equality and inequality constrants

$$i.e \; L_{ij} = \sum_{i=1}^{n} \left[\alpha_j \left(X_{ji} - X_{j,i-1} \right) U_{ji} - \beta_j \, U_{ji}^2 - \gamma_j \, X_{ji}^2 \right]$$

(12)

$$+ \lambda_{ij} \cdot \left(X_{ij} - X_{j,i-1} - y_{ij} + U_{ij} + W_{ij}' \right)$$

$$+ \lambda_{ij} \left(y_{i,j} - const \right)$$

$$j = 1$$

$$+ \lambda_{i,j}'' \left(y_{j+1,i} - U_{j,i} + W_{ij}' \right)$$

$$+ \mu_{ij} \left(X_{ij} - X_{-ij} \right)$$

$$+ \mu_{ij}' \left(\overline{X}_{ij} - X_{ij} \right)$$

$$+ \mu_{ij}'' \left(U_{ij} - U_{ij} \right)$$

$$+ \mu_{ij}''' \left(\overline{U}_{ij} - U_{ij} \right)$$

where λ_{ij} and μ_{ij} are the Lagrange and Kuhn Tuke multipliers of the equality and inequality constraint such that:

$$\lambda_{ij} \geq 0 \quad , \quad \mu_{ij} \geq 0 \qquad (13)$$

By diffirintating the Lagrange equation (12) w.r.t the system variables $\left(X_{ij}, \, U_{ij}, \right)$ and Lagrange multipliers $(\lambda_{ij}, \mu_{ij}, \mu_{ij}', \mu_{ij}'', \mu_{ij}''')$ the following equations can be obtained:

$$\alpha_{ij} - 2\gamma_{ij} \, X_{ij} + \lambda_{ij} + \mu_{ij} - \mu_{ij}' = 0 \qquad (14)$$

$$- \lambda_{ij} - \alpha_{j,i-1} = 0 \quad \therefore \lambda_{ij} = \alpha_{j,i-1}$$

$$- 2\beta_{ij} \, u_{ij} + \lambda_{ij} + \lambda_{ij}'' + \mu_{ij}'' - \mu_{ij}''' = 0$$

$$X_{ij} - X_{j,i-1} - y_{ij} + u_{ij} + w_{ij} = 0$$

$$y_{i,1} - const = 0$$

$$\therefore \; y_{i,1} = const$$

$$y_{j+1,i} - u_{ij} + w_{ij} = 0$$

$$X_{i,j} - \overline{X}_{i,j} = 0$$

$$\therefore \quad X_{i,j} = \overline{X}_{i,j}$$

$$u_{i,j} = \overline{u}_{ij}$$

By solving the set of equations (-) and applying the quadratic programming, the optimal values of variables and the global optimal solution can be obtaind.

6. RESULTS

All the optimal values are shown in figures the lagrange multipliers λ varries according to the size of reservoire and it's content, while λ' and λ'' have the same optimal values.

Also, it is noted that, μ, μ', μ'' and μ''' are equal for the same power station.

The designed parameters of the power plants α, β and γ are varried linearly against the position "location and distance" of the designed hydropower station.

The solution of the control problem gives the optimal values of all state and control variables H,X and U, taking inconsideration their boundry values.

Finally, the figures indicate also the relation between the generated electrical power of the new and developed hydro power plants that located at El Sheikh Zayed Canal.

The first power station is designed as the most huge one w.r.t. it's size, content, head and it's water release.

For that, this reservoir is considerd as the main inflow to the other hydropowers chain.

The highest values of electrical energy is generated from the first designed station.

7. CONCLUSION

In this dissertation, the optimal operating policy of a hydropower plant chain connected with a complete power network, El Sheihk Zayed canal with it's four branches, has been studied. The deterministic formulation of the problem leads to a complex optimization problem with quadratic cost function and set of inequality constraints to be solved. Moreover due to the complexity and the distributed nature of the problem, a new multilevel algorithm has been developed using the quadratic programming approach at the lower level and the prediction principle to update the coordinating variables at the higher level.

The new optimization problem of real case of designed hydropower generation at the new giant developed projects in Egypt (El Sheihk Zayed Canal) had been solved in both cases as global and as decomposed system, using the nonlinear programming techniques and the concepts of large scale system theory, the resulted optimization problem has been solved .

Through numerical Simulation, it has been shown that the multilevel approach is more efficient than the global one, since it needs less computational time and memory storage.

8. RFERENCES

[1] F.H.Fahmy, March 2000, "A New Cascaded Hydropower plants on El-Sheikh Zayed in The New Valley in Egypt", EE7' 2000, Cario, Egypt, 11-13.

[2] Dr. Diaa El Din, A., May 1998 "Southern Egypt Development project", Arab Republic of Equpt Ministry of Public works and water Resources National Water Research Center, Cairo, Egypt.

[3] M.G.Singh, S.Drew and J.F.Coals, July 1975, "Comparison of practical Hierarchical Control Methods for Interconnected Dynamical Systems", Automatica, vol.11, pp.331-350.

[4] M.D.Mesarovic and D.Macko, 1969, Foundations for a Scientific Theory of Herarchical Systems in Hierarchical Structures, edited by L.L Whyte and D.G.Wilson, Nit, American Elsevier.

List of Sympols

$X_{j,i}, X_{j,i-1}$ The content of reservoir of station j at end of period $i, i-1 \ (m^3)$

u_{ji} The release from station j at period $i, (m^3 / time)$

E_{ji}, P_{ji} The hydro electric energy and power of station j at period i

W   Blobal water losses of station j at each period $i (m^3 / time)$

y_{ji} The water input to station j at each period $i (m^3 / time)$

$\overline{H}_{ij}, H_{ij}$ The boundry limits of water head of station j at time i (m^3)

$\overline{X}_{ij}, X_{ij}$ The tooundry limits of reservoir content (m^3)

$\overline{U}_{ij}, U_{ij}$ The bourdry limits of water release $(m^3 / time)$

$\overline{P}_{ij}, P_{ij}$ The boundry limits of electrical power (watt)

$\alpha_j, \beta_j, \gamma_j$ Specific constants of reservoir j

$\lambda, \lambda', \lambda''$ Lagrange multipliers

μ, μ', μ'', μ''' Kuhn Tuker multipliers

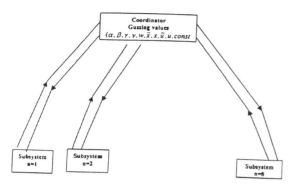

Fig.(1) The hierarchical technique to solve the decomposition
System (n=8 subsystems).

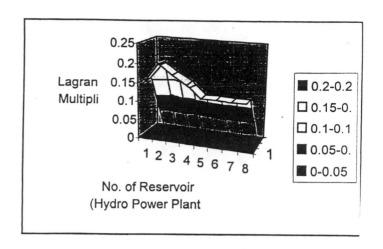

Fig. (2) Lagrang Multiplier of the Optimal Values

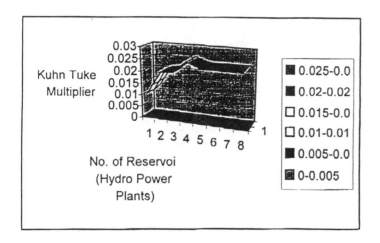

Fig. (3) Optimal Value of Kuhn Tuker Multiplier

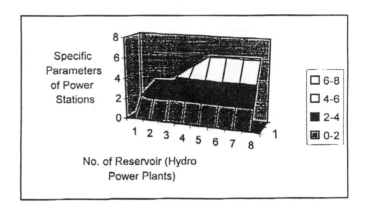

Fig. (4) Specific Parameters of Power Stations

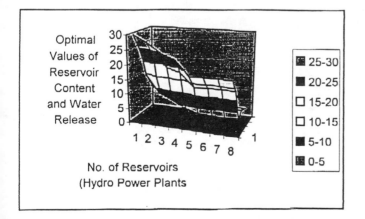

ig. (5) Optimal Values of Reservoir Content and Water Release

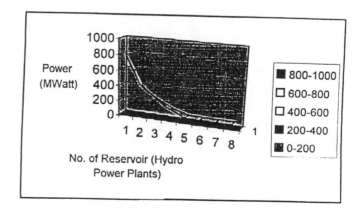

Fig. (6) Optimal Values of Generation Power

Proceedings of IECEC'01
36th Intersociety Energy Conversion Engineering Conference
July 29 - August 2, 2001, Savannah, Georgia
IECEC2001-RE-09

A Multilevel Stochastic and Optimization
Approach of Clean Energy Plants Chain Operation

Prof. Faten H.Fahmy
Electronic Research Institute,
Cairo, Egypt
e-mail: faten@eri.sci.eg

ABSTRACT

In Egypt, the demand of electric energy increases rapidly to satisfy the demand to implement the projects required for the economic development specially that in New Valley and Toshka. For the near future, we presented in previous published paper, a design of eight hydropower plants on El Shiehk Zayed Canal to generat neccessary electric energy. Therefore, it is naturally to try to maximize the electric power generated from these eight stations along the Canal basin, while satisfying the constraints imposed on the system in order to fulfill water requirements. Moreover, due to the randomness of the various problem variables it can also be seen that the problem modeling should incorporate stochastic terms. Thus, the problem is rendered difficult to solve and requires new tools for solution. The problem is transformed into an approximate deterministic one using chance constraint approach. Two different techniques are applied, Global technique and Multilevel approach. The two level technique is so easy to apply. The results show that the two techniques yield almost the same answer yet they differ in their Computational burdens (cpu and memory). The hierarchical approach requires 62% of the memory needed in the global approach. The execution time is about 28% of global technique.

Keywords
Hierarchical Technique, hydropower plants, water release, stochastic problem, chance constraint technique.

1. STACHASTIC OPTIMIZATION PROBLEM FORMULATION

In some cases, one may wish to identify a system with total ignorance of the process. While in others, like the present case, a considerable knowledge of the nature of the system may be available, but the specific values of the system parameters may be unkown or to be identified. The mathematical representation has to be verified against the real data through a calibration process. It must be pointed out, that in general the model must be detailed enough to be reasonably representative of the real system, yet sufficiently simple to keep the necessary calculations from being intricate.

As the variables of the system $(X_{ji} \, and \, U_{ji})$ are random in nature around certain mean values and the optimization problem has correspondingly to be formulated as stochastic nonlinear programming problem. However, Some difficulties may arise in estimating the means and the variances of these random variables.

The control problem can be expressed by the objective function which is the centercore of any optimization problem (maximization of global electrical energy of eight hydro power stations) while, the system constraints are of two Categories. The equality constraints define the hydraulic coupling between the stations and the inequality constraints define the feasible region for each station [1].

2. THE STOCHASTIC NONLINEAR PROBLEM

The basic idea used in solving any stochastic problem into an equivalent deterministic problem by applying chance constraint [2]. Each of random variable can be expressed by it's mean value and it's variance.

The resulting equivalent deterministic problem can be solved by using the suitable techniques.

THE EQUIVALENT DETERMINISTIC PROBLEM
a- The objective function:

$$F(y) = K_1 + K_2 \, \sigma_{\neq} \tag{1}$$

where K_1 and $K_2 \geq 0$ and their numerical values indicate the relative importance for $\quad and \, \sigma_{\neq}$.

b- The Constraints

As $X_{ji} \, and \, U_{ji}$ are random variables, the constraints will also be random

$$\int_0^\infty f_{g_{ji}} \left(g_{ji} \right) dg_{ji} \geq P_j \tag{2}$$

$$F(y) = K_1 \qquad + K_2 \sigma_* \qquad (3)$$

where

$$\sum_{i=1}^{n} \sum_{j=1}^{m} [\alpha_j (\overline{X}_{ji} - \overline{X}_{j,i-1}) \overline{y}$$

$$- \overline{\beta}_j \, \overline{U}_{ji}^2 - \gamma_j \, \overline{X}_{ji}^2]$$

$$\sigma = \sum_{i=1}^{n} \sum_{j=1}^{m} [(\alpha_j u_{ji} - 2\gamma_{ji} X_{ji})^2 \sigma_{xji}^2$$

$$+ (-\alpha_j u_{ji})^2 \, \sigma_{x_{j,i-1}}^2 + \{ (\alpha_j (X_{ji} - X_{j,i-1})$$

$$- 2\beta_j \, u_{ji} \}^2 \, \sigma_{uji}^2 \}^2]^{1/2}$$

b- The Constraints:

$$\overline{g} - \varphi_j (P_j) \left[\sum_{i=1}^{n} \sum_{j=1}^{m} \left(\frac{\partial g_i}{\partial y_j} \right)^2 \sigma_j^2 \right]^{1/2} \geq 0 \qquad (4)$$

$$g_1 : \overline{X}_{ji} - \overline{X}_{j,i-1} - \overline{u}_{i,j-1} + \overline{u}_{ji} + W_{ji}' - \varphi_j (P_j)$$

$$[\alpha_{xij}^2 + \sigma_{xj,i-1}^2 + \sigma_{ui,j-1}^2 + \sigma_{uj,i}^2]^{1/2} \geq 0$$

$$g_2 : \overline{u}_{1i} - Const - \varphi_j (P_j) [\sigma_{u1i}^2]^{1/2} \geq 0$$

$$g_3 : \overline{y}_{j,i} - \overline{u}_{j,i} - \varphi_j (P_j) [\sigma_{ui}^2]^{1/2} \geq 0$$

$$g_4 : \overline{X}_{ij} - \overline{X}_{ij_{min}} - \phi_j (P_j) [(-1.0)^2 \, \sigma_{xj}^2]^{1/2} \geq 0$$

$$g_5 : \overline{X}_{ij_{max}} - \overline{X}_{ij} - \phi_j (P_j) [(1.0)^2 \, \sigma_{xji}^2]^{1/2} \geq 0$$

$$g_6 : \overline{u}_{ij} - \phi_j (P_j) [(-1.0)^2 \, \sigma_{uij}^2]^{1/2} \geq 0$$

$$g_7 : \overline{u}_{ij_{max}} - \overline{u}_{ij} - \phi_j (P_j) [(1.0)^2 \, \sigma_{uij}^2]^{1/2} \geq 0$$

4. THE OPTIMIZATION TECHNIQUE

a- The Conventional Technique

I Lagrange Equation

Invoking The above objective function, the equality and inequality constraints, the Lagrange equation of the formulation can be written as:

$$L = \sum_{i-1}^{n} \sum_{j=1}^{m} \left\{ \alpha_j (\overline{X}_{ji} - \overline{X}_{j,i-1}) \overline{u}_{ji} - \beta_j \, \overline{u}_{ij}^2 - \gamma_j \overline{X}_{ij}^2 \right] \qquad (5)$$

$$+ [(\alpha_j u_{ij} - 2\gamma_{ij} X_{ij})^2 \sigma_{xij}^2 + (-\alpha_j u_{ij})^2 \, \sigma_{x_{j,i-1}}^2$$

$$[(\alpha_j (X_{ji} - X_{j,i-1}) - 2\beta_j \, u_{ij})^2 \sigma_{uij}^2]^{1/2}$$

$$+ \sum_{i=1}^{n} \sum_{j=1}^{m} [\lambda_{ij} (\overline{X}_{ij} - \overline{X}_{j,i-1} - \overline{u}_{i,j-1} + \overline{u}_{i,j} + W_{ij}$$

$$- \phi_j (P_j) [\sigma_{xij}^2 + \sigma_{X_{j,i-1}}^2 + \sigma_{u_{i,j-1}}^2 + \sigma_{u_{i,j}}^2]^{1/2})$$

$$+ \lambda_{ij}' (\overline{u}_{1i} - Const - \phi_j (P_j) [\sigma_{u1i}^2]^{1/2})$$

$$+ \lambda_{ij}'' (y_{i,j} - \overline{u}_{ij} - \phi_j (P_j) [\sigma_{uij}^2]^{1/2})$$

$$+ \mu_{ij} (\overline{X}_{ij} - \overline{X}_{ij_{min}} - \phi_j (P_j) [\sigma_{xij}^2]^{1/2})$$

$$+ \mu_{ij}' (\overline{X}_{ij_{max}} - \overline{X}_{ij} - \phi_j (P_j) [\sigma_{xij}^2]^{1/2}$$

$$+ \mu_{ij}'' (\overline{u}_{ij} - \phi_j (P_j) [\sigma_{uij}^2]^{1/2})$$

$$+ \mu_{ij}''' (\overline{u}_{ij_{max}} - \overline{u}_{ij} - \phi_j (P_j) [\sigma_{uij}^2]^{1/2})]$$

As X_{ij} are state variables, U_{ij} are the control variables, $(\lambda_{ij}, \lambda_{ij}', \lambda_{ij}'')$ are Lagrange multipliers and $(\mu_{ij}, \mu_{ij}', \mu_{ij}'', \mu_{ij}''')$ are Kuhn Tuker multipliers .

II The Necessary Conditions

Differentiating Lagrange equation L w.r.t the variables $X_{ij}, U_{ij}, \lambda_{ij}$ and μ_{ij}

$$\overline{X}_{ij} - \overline{X}_{ji-1} - \overline{U}_{ij-1} - \overline{U}_{ij} + W_{ij} - \phi_j(P_j)\left[\sigma^2_{xij}\right.$$

(6)

$$\left. + \sigma^2_{X_{ji-1}} + \sigma^2_{U_{ij-1}} + \sigma^2_{uij}\right]^{1/2} = 0$$

$$\overline{U}_{1i} - Const - \phi_j(P_j)\left[\sigma^2_{u1i}\right]^{1/2} = 0$$

$$\overline{y}_{ij} - \overline{U}_{ij} - \phi_j(P_j)\left[\sigma^2_{uij}\right]^{1/2} = 0$$

$$\overline{X}_{ij} - \overline{X}_{ij_{min}} - \phi_j(P_j)\left[\sigma^2_{X_{ij}}\right]^{1/2} = 0$$

$$\overline{X}_{ij_{max}} - \overline{X}_{ij} - \phi_j(P_j)\left[\sigma^2_{X_{ij}}\right]^{1/2} = 0$$

$$\overline{U}_{ij} - \phi_j(P_j)\left[\sigma^2_{uij}\right]^{1/2} = 0$$

$$\overline{U}_{ij_{max}} - \overline{U}_{ij} - \phi_j(P_j)\left[\sigma_{uij}\right]^{1/2} = 0$$

$$\alpha_j u_{ji} - 2\gamma_j X_{ij} + 2\left(\alpha_j u_{ij} - 2\gamma_{ij} X_{ij}\right).2\gamma_{ij} \sigma^2_{xij}$$

$$+\alpha_j \sigma_{uij} + \lambda_{ij}\left[\left(\overline{X}_{ij} - \overline{X}_{j,i-1} - \overline{U}_{i,j-1} + \overline{U}_{ij} + W_{ij}\right)\right.$$

$$\left. - \phi_j(P_j)\left[\sigma^2_{xij} + \sigma^2_{X_{j,i-1}} + \sigma^2_{U_{i,j-1}} + \sigma^2_{i,j}\right]^{1/2}\right)$$

$$+ \mu_{ij} + \mu'_{ij} + \mu''_{ij} + \mu'''_{ij}$$

b- THE PROPOSED HIERARCHICAL STOCHASTIC OPTIMIZATION APPROACH

Theoretically, this optimization problem can be solved by the conventional control theory techniques, Nevertheless of the problem increases, it is rendered impractical to obtain a solution due to the computational burden memory limitations and the numerical instability of the solution algorithms.

The hierarchical approach is applied, using the decompostion-coordination technique in two levels. The first level is the coordinator, while in the second level is the set of subproblems.

The Lagrangian equation must be decomposed between the subproblems and the variables must be distributed between the two levels.

The Lagrange multiplier λ and the inflow for the station y are the coordinator variables. The content of each reservoir x, release from each station u and the kuhn Tuker variable μ are constitutes of the lower level variables

Lagrangian equation of each subproblem
 j=1,....m:8

$$L = K_1 \sum_{i-1}^{n}\left[\alpha_j\left(X_{ji} - X_{j,i-1}\right)U_{ij} - \beta_j U_{ij}^2 - \gamma_j X_{ij}^2\right]$$

(7)

$$+ K_2 \sum_{i=1}^{n}\left\{\left(\alpha_j u_{ij} - 2\gamma_j X_{ij}\right)^2 \sigma^2_{X_{ij}} + \left(\alpha_j u_{ij}\right)^2 \sigma_{X_{j,i-1}}\right.$$

$$\left. + \left(\alpha_j\left(X_{ij} - X_{ji-1}\right) - 2\beta u_{ij}\right\}^{1/2}\right.$$

$$+ \sum_{i=1}^{n}\left\{\lambda_{ij}\left(\overline{U}_{ji} - Const.\phi_j(P_j)\left[\sigma^2_{uij}\right]^{1/2}\right.\right.$$

$$\left. + \lambda'_{ij}\left(y_{ij} - u_{ij} - \phi_j(P_j)\left[\sigma^2_{uij}\right]^{1/2}\right)\right.$$

$$+ \lambda''_{ij}\left(X_{ji-1} - y_{ij} - W_{ij} + u_{ij} - \phi_j(P_j)\left[\sigma^2_{X_{ij}}\right.\right.$$

$$\left.\left. + \sigma^2_{X_{ji-1}} + \sigma^2_{y_{i,j}} + \sigma^2_{U_{ij}}\right]^{1/2}\right\}$$

Subject to boundry limits

$$\underline{X}_{ij} \leq X_{ij} \leq \overline{X}_{ij}$$

(8)

$$U_{ij} \leq U_{ij} \leq \overline{U}_{ij}$$

while the necessary condition can be obtained by differintiating the Lagrange equation w.r.t the coordinator and decomposition variables.

5. RESULTS ANALYSIS

It can be seen from the previous section that the problem under consideration can be solved using two approaches: the global technique and the hierarchical technique. The two approaches yield almost the same answer, yet they differ in their computational burdens. The hierarchical approach requires about 62% of the memory needed in the global approach. While, the execution time is about 28% of global technique this demonstrates the effectiveness of the hierarchical approach in handling such large-scale problem. Due to the relative sizes of reservoirs (X,H), it can also be seen that the first reservoir generates the highest amount of energy in comparison with the rest of the system. Also, the water release from the first reservoir has the highest value as it is considerd as the main input flow to the rest of other hydro plant chain. Finally, it can be pointed out that the results are satisfactory and correspond well with actual conditions.

6. CONCLUSION

This work study a very complicated problem in stochastic nature as it composed of several random variables. For that, it can not be solved easily using conventional estimation theory and stochastic control theory. In order to overcome this difficulty, the problem has been discretized in subproblems. Then, by applying the stochastic programming approach which is mainly based on the chance constraint theory, the stochastic problem has been transfordd into nonlinear dterministic problem with strong nonlinearties. Using nonlinear programming techniques and the concepts of large scale system theory, the resulted optimization problem has been solved.

Two different techniques have successfully been proposed in this research in order to solve stochastic large scale multiarea hydropower system chain operation at El shiekh Zayed Canal.

For simplicity to solve stochastic control problem, it must be transformed into deterministic one by applying chance constraint approach, depending on the average and variance of state and control variables.

Through numerical simulation, it has been shown that the multilevel approach is more efficient than the global one, since it needs less computational time and memory storage.

7. REFERENCES

1- F.H.Fahmy, 11-13 March 2000, "A new Cascaded Hydro powe plants on El Shiekh Zayed in the New Valley in Egyp EE7'2000, Cairo, Egypt.
2- M.E.E-Hawary, G.S.Charistenesen, 1979, Optimal Economic Operation of Electrical Power Systems, An Introduction, Facult of Engineering and Applied Science, Memorial University o New found land, st. John, New foundland, Canada.
3- M.Aoki, 1971, Aggregation in Optimization Methods for Larg Scale System with Application New York, Ch-5, pp. 191-232.

LIST OF SYMBOLS

y_{ji}	Inflow for the station j during period $i(m^3/time)$
X_{ji}	The Content of reservoir of station
U_{ji}	The release from station
E_{ji}	The hydroelectric energy of
W	Global losses for each
$\alpha_j, \beta_j, \gamma_i$	specific constants of reservoir j
(y)	mean value of F(y)
ϕ	standard deviation of F(y)
var	variance of F(y)
$f(g_{ji})$	Probability density function of random variable g_{ji}
$\overline{g}_j$	meanvalue of constraint $g(y_j)$
σ_{g_i}	Standard deviation of g_j
$\phi_j(P_j)$	The value of the standard normal variate corresponding to the probability P_j
λ, μ	The Lagrange and Kuhn Tuker multiplie

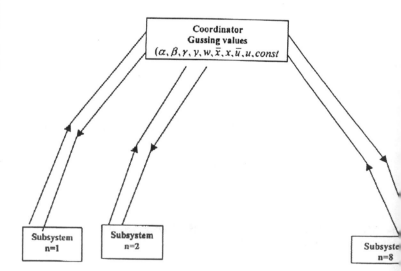

Fig.(1) The hierarchical technique to solve the decomposition System (n=8 subsystems).

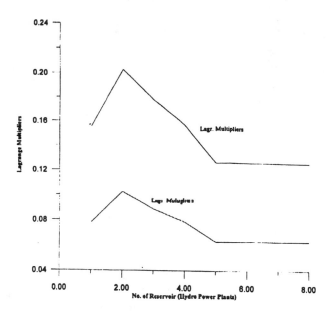

Fig. (2) Lagrang Multiplier of the Optimal Values

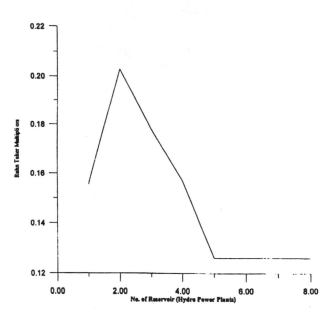

Fig. (3) Optimal Value of Kuhn Tuker Multiplier

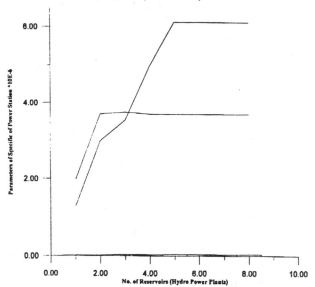

Fig. (4) Specific Parameters of Power Stations

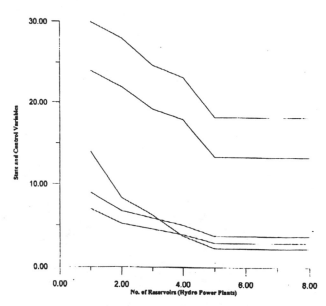

Fig. (5) Optimal Values of Reservoir Content and Water Release

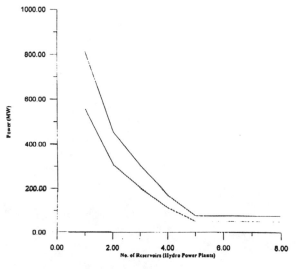

Fig. (6) Optimal Values of Generation Power

629

Proceedings of IECEC'01
36th Intersociety Energy Conversion Engineering Conference
July 29—August 2, 2001, Savannah, Georgia

IECEC2001-RE-12

FULL-SPECTRUM SOLAR ENERGY SYSTEMS FOR USE IN COMMERCIAL BUILDINGS

Jeff D. Muhs and D. Duncan Earl
Oak Ridge National Laboratory

ABSTRACT

This paper describes a systems-level analyses of full-spectrum solar energy systems for use in commercial buildings. We describe and analyze a new approach that more efficiently uses different portions of the solar spectrum simultaneously for multiple end-use applications such as hybrid solar lighting and distributed power generation, e.g., combined solar light and power. Our work suggests that full-spectrum solar energy systems could achieve nonrenewable energy displacement efficiencies of greater than 75%, commercial quantity costs of under \$2/Wp, and displaced electricity costs under 10 cents/kWh during peak demand periods. Our work also predicts the chromaticity of distributed, non-diffuse sunlight will be between 4500 to 5000 °K depending on several factors.

INTRODUCTION

Throughout the 1900s, use of the sun as a source of energy has evolved considerably. Early in the century, the sun was the primary source of interior light for buildings during the day. Eventually, however, the cost, convenience, and performance of electric lamps improved and the sun was displaced as our primary method of lighting building interiors. This, in turn, revolutionized the way we design buildings, particularly commercial buildings, making them minimally dependent on natural daylight. As a result, lighting now represents the single largest consumer of electricity in commercial buildings[1].

During and after the oil embargo of the 1970s, renewed interest in using solar energy emerged with advancements in daylighting systems, hot water heaters, photovoltaics, etc. Today, daylighting approaches are designed to overcome earlier shortcomings related to glare, spatial and temporal variability, difficulty of spatial control and excessive illuminance. In doing so, however, they waste a significant portion of the visible light that is available by shading, attenuating, and or diffusing the dominant portion of daylight, i.e., direct sunlight which represents over 85% of the light reaching the earth on a typical sunny day[2]. Further, they do not use the remaining half of energy resident in the solar spectrum (mainly infrared radiation between 0.7 and 1.8 um), add to building heat gain, require significant architectural modifications, and are not easily reconfigured. Previous attempts to use sunlight directly for interior lighting via fresnel lens collectors, reflective light-pipes, and fiber-optic bundles have been plagued by significant losses in the collection and distribution system, ineffective use of nonvisible solar radiation, and a lack of integration with collocated electric lighting systems required to supplement solar lighting on cloudy days and at night.

Similar deficiencies exist in photovoltaics, solar thermal electric systems, and solar hot water heaters. For example, the conversion efficiency of traditional silicon-based solar cells in the ultraviolet & short wavelength visible region of the solar spectrum is low and the solar energy residing beyond ~1.1 um is essentially wasted. To overcome this and address other economic barriers, one approach has been to develop utility-scale PV and solar thermal concentrators. The rationale being that the cell area and, consequently, the cell cost can be reduced by approximately the same amount as the desired concentration ratio. Unfortunately, this cost-savings is typically offset by the added cost and complexity of the required solar concentrator and tracking system.

This work is the outgrowth of the fundamental premise that different portions of the solar energy spectrum are inherently more effective at displacing nonrenewable energy when used for different end-use purposes (such as lighting and distributed power generation). Further, the development of more flexible and architecturally-compatible hybrid solar lighting systems that are integrated with conventional electric lighting will ultimately lead to more widespread use of sunlight when compared to previous topside daylighting systems.

MODELING OF COLLECTOR SYSTEM DESIGN

We developed several Z-MAX optical design models for novel hybrid solar concentrators that efficiently collect, separate, and distribute the visible portion of sunlight for interior lighting purposes while simultaneously generating electricity from the infrared portion of the spectrum using new GaSb infrared thermophotovoltaics. Using a segmented secondary UV cold mirror, the visible portion of sunlight is reflected and focused onto several 12 mm large-core optical fibers that transport visible light into buildings while transmitting the UV and IR wavelengths to IR-TPV's. One of the preferred designs is illustrated in Fig. 1.

Figure 2. *Reflective Coating being applied to 1.5 m primary mirror.*

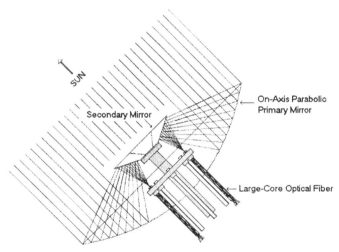

Figure 1. *Collector Design*

This design was selected to accommodate the use of two different commercially-available primary mirrors: a) a polished 1.2 m diameter glass mirror with an enhanced aluminum 92% reflective coating and b) a steel 1.5 m diameter mirror (originally a satellite dish) with an 85% adhesive-backed first-surface aluminum reflective coating, see Fig. 2. The secondary mirror consists of eight segmented mirror surfaces each focusing light onto separate optical fibers. The dual-axis tracking system selected for initial prototypes was a modified Array Technologies gear drive azimuth tracker with a linear actuator elevation tracker. Fig. 3 illustrates the expected spot-size of visible sunlight focused on the ends of the optical fibers given the above design parameters using the 1.2 meter mirror considering expected errors in mirror surface quality. Note that the location of the focused spot will be dependent on the accuracy of the two-axis tracking system.

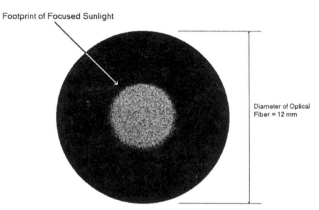

Figure 3. *Footprint of Focused Sunlight on Optical Fiber*

MODELING OF SYSTEM-LEVEL LOSSES AND COLOR SHIFT

Using commercially-available MatLab software, custom-designed visual basic code developed by RPI/LRC and ORNL, and solar resource data provided by NREL, we determined the average optical losses in the system to be 51% for a lighting system that includes 10 m of optical fiber and a hybrid luminaire (discussed later). A CIE 1931 chromaticity diagram illustrating chromaticity values and associated color shift caused by the above-described hybrid solar lighting system is illustrated in Fig. 4. This data was derived using direct sunlight in Knoxville, TN at 12:30 p.m. in mid-January 2001 using the above collector design and a 10 m length of 3M large-core optical fiber.

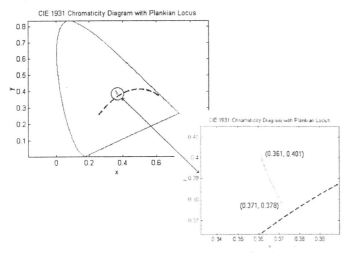

Figure 4. Chromaticity Values, x-y Coordinates

The Color Correlated Temperature (CCT) of solar light before it enters the optical fiber is 4273°K with a Color Rendering Index (CRI) of 98. After transmission through 10 meters of 3M Fiber, the CCT is 4670°K with a CRI of 84. This output can be compared with commercially available fluorescent tubes such as the 32W Sylvania Octron® T-8 Tube which has a CCT of 4100°K and a CRI of 85. Assuming a 0.1 m collection area, 7750 lumens will be delivered to the fiber end-face. Of this, our model predicts 5750 lumens will be delivered through the optical system to the desired work-plane. The overall efficiency improves somewhat (6470 lumens delivered) when using through 6 meters of optical fiber rather than 10 meters.

We also developed several Z-MAX optical design models for new light fixtures capable of blending distributed sunlight and electric illuminants in new "hybrid" luminaries, see Fig. 5.

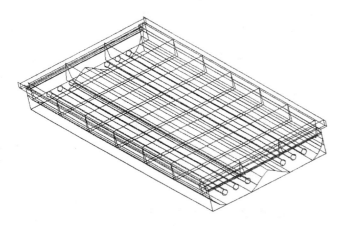

Figure 5. ZEMAX model of Hybrid Luminaire

These resulted in the development of initial prototypes, one of which is shown in Fig. 6. In this case, cylindrical acrylic rods were placed between conventional 32W T8 fluorescent lamps and light from two large-core optical fibers was piped into the rods (one from each end) to distribute the natural light similar to the way collocated electric lamps distribute light. Results of this work were reported by Earl et al[3].

Figure 6. Prototype Hybrid Luminaire

SYSTEMS-LEVEL COST AND PERFORMANCE PROJECTIONS

The electrical energy displacement efficiency of the system in a single story application was determined to be over 75%. At first glance, this might seem unreasonably high. However, included in the performance summary are the following considerations: 1) the sunlight is filtered, the visible portion (~450 W/m^2 of 1000W/m^2) used for displacing much less efficient electric light and the near-IR radiation (~400 W/m^2) used to generate electricity using ideally-suited infrared thermophotovoltaics, 2) the luminous efficacy of the displaced electric light (63 lm/W) is significantly lower than filtered sunlight (~100 lm/W) and includes the luminous efficacy of the lamp/ballast (~90 lm/W) as well as the luminaire efficacy (70%); and 3) the elimination of excess heat generated by electric lights in sunbelt regions reduces subsequent HVAC loads by ~ 15%.

Based on this design scenario and associated cost values for each component, a systems-level cost and performance model was developed. The installed system cost for a single-story application was estimated to be ~$3,200 in commercial quantities. This assumes cost reductions similar to those found in other solar technologies during their early years of commercialization. Although an optimization routine has yet to be developed, we used a 2-m^2 collector illuminating approximately 14 hybrid luminaires covering close to 1000 ft^2 of floor space. Peak electrical power displacement in buildings was estimated to be ~ $1.65/W$_p$. The cost of displaced electricity over a 20 year lifetime was estimated to be between 5 and 13 cents/kWh.

COMPARISONS WITH ALTERNATIVES

Alternative # 1: Advanced Electric Lighting

Because hybrid lighting systems require the use of state-of-the-art electric lamps when sunlight is not available, their cost is additive. As such it is not fully appropriate to compare them directly. However, in a "head-to-head" comparative analysis, the estimated additive cost of installed hybrid solar lighting systems (a clean energy alternative) in terms of ¢/kwh displaced (5 – 13 cents/kWh) is typically lower than the cost of running electric lighting systems in a deregulated market considering time of day rates (10 – 15 cents/kWh) during peak demand periods on hot, sunny days.

Alternative # 2: Conventional Topside Daylighting

A complete study of all types of topside daylighting is not warranted for the purposes of a comparative analysis with adaptive full-spectrum solar energy systems. We limit this discussion to skylights, generally accepted as the most cost-effective form of conventional topside daylighting. On average, incident sunlight does not enter skylights normal to the horizontal plane. Depending on the type and configuration of skylight, light transmission varies dramatically and is attenuated significantly. This is due to several factors but is predominately determined by the efficiency of the light well and glare control media. The typical transmittance of state-of-the-art tubular, domed skylights varies widely, depending on lighting requirements, but for commercial applications is typically well under 50%.

Comparatively speaking, several other factors must also be considered. First, the coefficient of utilization (CU) of a single 1-m2 tubular skylight will inherently be much lower than a system that distributes light from the same square meter to six or more luminaires. Assuming that the room cavity ratio and other room parameters are identical, the CU of the more distributed hybrid system is significantly better. If the single 1-m^2 skylight were replaced by ~6 much smaller skylights, the two systems CUs would compare equally, yet the cost of the skylights would increase prohibitively.

Skylights are typically not designed based on the maximum amount of light that can be supplied but rather designed to approximate that which is produced by the electric lighting system when the total exterior illuminance is 3000 footcandles. This reduces over-illumination and glare. Because of this, all light produced by skylights beyond this value is typically wasted. As such, preliminary estimates suggest that on average, depending on location, approximately 30% of the total visible light emerging from skylights on a sunny day is excess light not used to displace electric lighting. Conventional skylights are also plagued by problems associated with heat gain and do not harvest non-visible light. Finally, conventional skylights are not easily reconfigured during floor-space renovations common in today's commercial marketplace. Once all factors are considered, the simple payback (typically >8

years) and energy end-use efficiency of even the best topside daylighting systems is considerably worse than projected adaptive, full spectrum solar energy systems.

Alternative #3: Solar Electric Technologies

To date, the United States has invested billions of dollars in systems capable of converting solar energy into electricity. The most relevant examples include solar PV modules and solar thermal technologies. The advantages of these systems are obvious. First, PV modules require no moving parts to convert sunlight into direct-current electricity, and they can be conveniently used for any electrically-powered end use. Unfortunately, these advantages come with a steep price in terms of overall efficiency. For example, commercial solid-state semiconductor PV modules typically have a total conversion efficiency of < 15%. Solar thermal systems typically have a conversion efficiency somewhat higher (< 25%), depending on system design and complexity. Further, losses attributed to electric power transmission/distribution (~8%) and dc-ac power conversion (10 - 15%) further reduce the overall efficacy of conventional solar technologies. Because of these and other reasons, conventional solar technologies have not displaced significant quantities of nonrenewable energy and are expected to be used in the United States for residential and commercial buildings, peak power shaving, and intermediate daytime load reduction. PV modules currently sell for between $3 - $5/$W_p$. The projected peak performance of adaptive full spectrum solar energy systems ($3,200 per 1,940 W_p or $1.65/$W_p$) have the immediate potential to more than double the affordability of solar energy when compared to these solar technologies.

CONCLUSIONS

This paper has described a systems-level analyses of full-spectrum solar energy systems for use in commercial buildings. We analyzed a new approach that more efficiently uses different portions of the solar spectrum simultaneously for multiple end-use applications such as hybrid solar lighting and distributed power generation, e.g., combined solar light and power. Our work suggests that full-spectrum solar energy systems could achieve nonrenewable energy displacement efficiencies of greater than 75%, commercial quantity costs of under $2/Wp, and displaced electricity costs under 10 cents/kWh during peak demand periods. Our work also predicts the chromaticity of distributed sunlight will be between 4500 to 5000 °K depending on several factors. Initial comparisons suggest that spectrum solar energy systems will outperform other renewable energy alternatives including topside daylighting, photovoltaics, and solar thermal electric systems.

ACKNOWLEDGMENTS

The authors wish to thank the U.S. Department of Energy Office of Energy Efficiency and Renewable Energy and the Tennessee Valley Authority for funding of this work. The authors also wish to thank Dr. N. Narendran and Nishantha Maliaygoda from RPI/LRC for the development of portions of the MatLab chromaticity model and J.K Jordan, D.L. Beshears, and L.K. Laymance from ORNL for their technical and administrative support.

REFERENCES

1. National Laboratory Directors for the U.S. Department of Energy, "Technology Opportunities: to Reduce U.S. Greenhouse Gas Emissions, and "Technology Opportunities: to Reduce U.S. Greenhouse Gas Emissions: Appendix B; Technology Pathways," October 1997

2. F.A. Barnes et al., Electro-Optics Handbook: Technical Series EOH-11, RCA Corporation, 1974, p. 70.

3. D.D. Earl et al, "Preliminary Results of Luminaire Designs for Hybrid Solar Lighting Systems", The Solar Forum 201, Washington, D.C. April 21-25, 2001. (in print).

Proceedings of IECEC'01
36th Intersociety Energy Conversion Engineering Conference
July 29 – August 2, 2001, Savannah, Georgia

IECEC2001-RE-22

A STUDY OF SEMITRANSPARENT CYINDRICAL RADIATION SHIELDS

Sophie Guers, Marta Slanik, David A. Scott, and B. Rabi Baliga
Department of Mechanical Engineering
McGill University
817 Sherbrooke Street West
Montreal, Quebec H3A 2K6
Canada

ABSTRACT

An experimental and analytical study of heat transfer through semitransparent cylindrical radiation shields is presented in this paper. The experimental work is aimed at obtaining data the establishes the effectiveness of single and multiple cylindrical shields with relatively high transmissivity (almost transparent) to thermal radiation with wavelength in the range $0.1\ \mu m \leq \lambda \leq 3.0\ \mu m$, and relatively low transmissivity (almost opaque) for wavelengths longer than $3.0\ \mu m$. The theoretical part of this work is aimed at determining the accuracy of predictions obtained using a very simple diffuse-gray enclosure theory, coupled with the assumption that the shields are opaque to the thermal radiation emitted by surfaces with temperatures in the range $9\ ^\circ C$ to $85\ ^\circ C$. The results show that this diffuse-gray/opaque-shield theory overestimates the effectiveness of the shields. However, the maximum deviation of the predictions from the experimental data is less than 38 % for all cases considered.

Keywords: Semitransparent cylindrical radiation shields, experimental data, diffuse-gray enclosure, opaque-shield model.

INTRODUCTION

In energy conversion, thermal, solar, building, and agricultural engineering, and in heat transfer experiments, it is often necessary to use materials that are essentially transparent to thermal radiation with wavelength in the range $0.1\ \mu m \leq \lambda \leq 3.0\ \mu m$, and almost opaque for wavelengths longer than $3.0\ \mu m$. Examples include the following: observation windows used in combustion chambers, boilers, and furnaces; cover plates and radiation shields used in solar collectors; plate glass windows used in residential and commercial buildings; materials used in the construction of greenhouses; and insulation of experimental set-ups that allow visualization of fluid flow and heat transfer phenomena. Details of these applications can be found in the works Sparrow and Cess (1970), Siegel and Howell (1972), Duffie and Beckman (1974), Fuschillo (1974), Modest (1993), Incropera and De Witt (1996), Holman (1997), Kreith and Bohn (2001), and Lagana (1996). In all these applications, the materials of interest may be viewed as semitransparent radiation shields with the aforementioned transmission characteristics. Experimental data and a very simple analytical method that would potentially aid in the design of such semi-transparent radiation shields are the focus of this work.

An experimental and analytical study of heat transfer through semitransparent cylindrical radiation shields is presented in this paper. The specific objectives of this work are the following: (i) conduct experiments and provide data that establish the effectiveness of single and multiple semitransparent cylindrical shields in limiting radiation heat transfer from a hot inner cylinder (diameter 0.0381 m and length 0.2743 m), with uniform temperatures in the range $30\ ^\circ C$ to $85\ ^\circ C$, to a concentric cold outer cylinder (diameter 0.1397 m and length 0.2743 m) maintained at a uniform temperature of about $9.5\ ^\circ C$; and (ii) determine the accuracy of predictions obtained using a very simple diffuse-gray enclosure theory, in which it is assumed that the shields are essentially opaque to the thermal radiation emitted by the aforementioned inner and outer cylinders. Concentric cylindrical radiation shields made of fused quartz, Plexiglas (polymethylmethacrylate or acrylic), and Lexan (polycarbonate resin) were studied in this work.

NOMENCLATURE

C Thermal conductance [W/$^\circ$C]

D Diameter [m]

F_{jk} Radiation shape factor from surface j to surface k

G	Irradiation or incident radiation flux [W/m^2]
I	Current input to heater [A]
J	Radiosity [W/m^2]
k_{sh}	Thermal conductivity of shield material [W/m-^{o}C]
k_{eff}	An effective thermal conductivity [W/m-^{o}C]
L	Effective length of the apparatus and shields [m]
P	Input power to the heater [W]
q	Heat transfer rate [W]
q_{cond}	Conduction heat transfer rate [W]
q_{rad}	Radiation heat transfer rate [W]
r	Radius [m]
$R_{th,cond}$	Thermal conduction resistance [^{o}C/W]
$R_{th,rad}$	Thermal radiative resistance [m^{-2}]
S	Heat conduction shape factor [m]
T	Temperature [^{o}C]
T_{abs}	Absolute temperature [K]
V	Voltage input to heater [V]
ε	Total hemispherical emissivity of a surface
λ	Wavelength of radiation [μm]

EXPERIMENTAL SET-UP

A schematic representation of the overall experimental set-up is shown in figure 1.

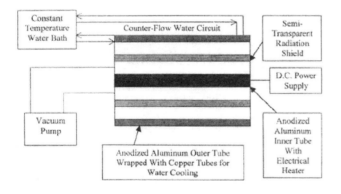

Figure 1: Schematic representation of the experimental apparatus, shown here with one semitransparent shield.

The concentric inner and outer tubes were made of anodized (black) aluminum. The outer diameter of the inner tube was 1.5 inches or 0.0381 m, and its effective length was 10.8 inches or 0.2743 m. The inner diameter of the outer tube was 5.5 inches or 0.1397 m, and it too had an effective length of 10.8 inches or 0.2743 m.

The outer tube was tightly wrapped on its outer surface with two ¼-inch copper tubes. A high thermal conductivity paste (Omegatherm 101) was used to minimize thermal contact resistance between the surfaces of the copper tubes and the surface of the outer tube. Water from a constant temperature bath (Neslab RTE-221) was circulated through the copper tubes, in a counter-flow path, to maintain the outer tube at a desired uniform temperature (about 9.5 ^{o}C in this work). The

inner tube was fitted with an electrical cartridge heater. This heater was supplied with dc power from a laboratory grade power supply (Xantrex XKW 300-3.5), and the input power was adjusted to achieve steady, essentially uniform, inner tube temperatures in the range 30 ^{o}C to 85 ^{o}C. The voltage and current inputs to the heater were measured using two high-accuracy digital multimeters (Hewlett Packard HP 3478A). These multimeters were set for voltage and current measurements with resolutions of ± 1 mV and ± 0.001 mA, respectively.

In the apparatus, removable endplates made of Delrin (3/4 inch or 0.01905 m thick) were used to seal the openings between the concentric inner and outer tubes. Sliding joints, made air tight with suitable o-rings, were used between the endplates and the tubes. The endplates were designed to accommodate one, two, or three concentric cylindrical radiation shields in the annular space between the inner and outer anodized aluminum tubes. Aluminum foil was used on the inner surface of the Delrin endplates to reduce the emissivity. A photograph of the anodized aluminum inner and other tubes, the copper tubes (which were later wrapped around the outer tube), Delrin endplates, and some related fittings is given in figure 2.

During the experiments, the air in the space between the inner and outer tubes was evacuated, typically to an absolute pressure of less than 10 mbar, using a vacuum pump (Edwards High Vacuum Pump E2M8). Vacuum grease (Dow Corning high vacuum grease) was used to coat all o-rings and sliding joint surfaces to achieve and maintain the desired vacuum conditions.

Figure 2: Photograph of the anodized inner and outer tubes, copper tubing, Delrin endplates, and some related fittings.

Temperatures at 12 locations on the outer tube and 12 locations on the inner tube were measured using 30-guage chromel-constant (Type E) thermocouples, and an electronic digital readout (Omega 410A1B-OE-C). With this set-up, the temperature measurements were determined to be accurate to within ± 0.1 ^{o}C, using a certified (National Research Council, Ottawa, Canada) digital quartz thermometer (Hewlett Packard 2804A) as a secondary reference.

The complete apparatus, consisting of the assembly of inner and outer tubes, semitransparent cylindrical radiation shield(s), and the endplates, was wrapped with about two layers of cotton and hydrogel (baby diapers) to insulate it and prevent moisture buildup on hot humid days, an idea borrowed from the work of Elkouh (1996). The entire apparatus was then inserted in a large plastic box, filled with silica gel and paper insulation. A photograph of the overall experimental set-up is given in figure 3.

Figure 3: Photograph of overall experimental set-up.

As was stated earlier, cylindrical radiation shields made of three different materials, Plexiglas, fused quartz, and Lexan, were studied. Three different Plexiglas tubes, three different quartz tubes, and one Lexan tubes were used. These seven radiation shields are denoted as Shields # 1-7. Their inner and outer diameters are given in table 1. A photograph of the apparatus loaded with the three quartz shields, and the Plexiglas and the Lexan shields outside, is shown in figure 4.

Table 1: Details of the radiation shields used in this study

Material	Shield #	Inside Diameter	Outside Diameter
Plexiglas	1	0.04400 m	0.05057 m
	2	0.08800 m	0.09510 m
	3	0.10740 m	0.11360 m
Fused Quartz	4	0.04610 m	0.05052 m
	5	0.08030 m	0.08550 m
	6	0.11000 m	0.11480 m
Lexan	7	0.08935 m	0.09450 m

THEORETICAL CONSIDERATIONS

As was mentioned in the previous sections, with regard to radiation calculations, attention in this paper is limited to a very simple diffuse-gray enclosure theory, based on the assumptions that the radiation shields considered here are essentially opaque to thermal radiation emitted by surfaces at temperatures

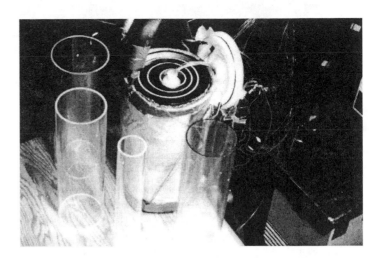

Figure 4: Photograph of the test section loaded with the three quartz shields, and the three Plexiglas and the one Lexan shields outside.

between 30 °C and 85 °C. More sophisticated models that consider semi-transparent mediums, and specularly and diffusely reflecting enclosures are available, for example, in the works of Sparrow et al. (1962), Perlmutter and Siegel (1963), Sparrow and Cess (1970), Fuschillo (1974), Viskanta et al. (1978), Timoshenko and Trenev (1986), Modest (1993), and Holman (1997), but the intention in this work is to explore the possibilities offered by the diffuse-gray/opaque-shield theory.

Heat Transfer in the Empty Apparatus (No Shields) and Estimation of Conduction Losses

Heat transfer in the empty apparatus, without any shields, is considered first, in order to develop a method to estimate the portion of the heater power input that is transmitted by heat conduction from the inner to the outer tubes. This portion of the heater power input will be referred to here as conduction losses, as it occurs because of heat conduction through the Delrin endplates and the insulation packed around the apparatus.

Consider the schematic of the empty apparatus given in figure 5.

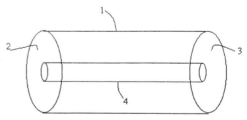

Figure 5: Schematic of the empty apparatus.

The inside surface of the outer tube is denote as 1, the outer surface of the inner tube is denote by 2, and 2 and 3 indicate the inner surfaces of the Delrin endplates. Each of these four surfaces is assumed to be isothermal and diffuse gray. It is also assumed that the irradiation, G, and radiosity, J, for each of these surfaces are uniform, and surfaces 2 and 3 are considered to be re-radiating surfaces (Holman, 1997). With these assumptions, and the temperatures T_1 and T_4 as inputs, a simple diffuse-gray enclosure theory can be used to calculate the rate of radiative heat transfer, q_{rad}, from the hot inner tube to the relative cold outer tube. Details of this enclosure theory can be found in the works of Incropera and De Witt (1996) or Holman (1997). The total hemispherical emissivities of the surfaces of the anodized aluminum tubes and the radiation shields, and the thermal conductivity of the shield materials considered here, are given later in this paper. All required shape factors, F_{jk}, and surface and space radiative resistances, $R_{th,rad}$, were computed using well-established practices (Sparrow and Cess, 1970).

It is assumed that the rate of heat loss from the plastic box, which is used to house the apparatus, to the ambient is negligible compared to the power input to the heater. Thus, under steady-state conditions, the power input to the heater embedded inside the inner tube is equal to the heat loss from the inner tube to the outer tube by radiation and conduction:

$$P = q_{rad} + q_{cond} \qquad (1)$$

Furthermore, the rate of conduction heat transfer, q_{cond}, may be expressed in terms of the temperature difference ($T_4 - T_1$), a heat conduction shape factor, S, and an effective thermal conductivity, k_{eff}. S and k_{eff} characterize the Delrin endplates and the insulation packed around the apparatus. Thus,

$$q_{cond} = S k_{eff} (T_4 - T_1) = C(T_4 - T_1) \qquad (2)$$

where C is an effective heat conductance, given by

$$C = S k_{eff} \qquad (3)$$

In the experiments, the voltage and current inputs to the heater, V and I, respectively, and the temperatures T_4 and T_1, were measured. These measurements were used to calculate the power input to the heater ($P = VI$), and the simple diffuse-gray enclosure theory was used to calculate q_{rad}. With these calculated values, the effective heat conductance, C, for the apparatus can be determined as follows:

$$C = q_{cond} /(T_4 - T_1) = (P - q_{rad})/(T_4 - T_1) \qquad (4)$$

In this manner, C, was determined for the empty apparatus, and this value of C was assumed to apply in experiments with the radiation shields.

Heat Transfer in the Apparatus Loaded With The Shields

The radiative heat transfer from the inner to the outer tubes goes through a sequence of finite-length annular spaces, each

bounded by portions of the Delrin endplates. The calculation of radiation heat transfer in each of the annular spaces was again handled using the simple diffuse-gray enclosure theory, as described in the previous section. However, the heat transmission from one such annular space to the next one involves heat conduction through the wall of the intervening radiation shield. One-dimensional radial heat conduction was assumed inside the walls of the radiation shields. Thus, for any one shield, with $T_{in,sh}$ and $T_{out,sh}$ denoting the steady-state temperatures of its inside and outside surfaces, respectively, $r_{in,sh}$ and $r_{out,sh}$ the corresponding radii, and k_{sh} the thermal conductivity of the material, the rate of heat conduction through the shield, $q_{cond,sh}$, is given by the following equation:

$$q_{cond,sh} = (T_{in,sh} - T_{out,sh}) / R_{th,cond,sh}$$
$$= (T_{in} - T_{out}) / [\ln(r_{out,sh} / r_{in,sh}) / (2\pi k_{sh} L)] \qquad (5)$$

For each experimental run with the shields, with the measured temperatures of the inner and outer anodized aluminum tubes, T_{inner} and T_{outer}, respectively, as inputs, the equations yielded by the diffuse-gray enclosure/opaque-shield theory and the equations that describe heat conduction through the shields, such as equation (5), were solved iteratively. In this manner, the rates of radiation heat transfer from the inner to the outer tubes, $q_{rad,calc}$, for each of the experimental runs with the shields were calculated.

The accuracy of the $q_{rad,calc}$ values yielded by diffuse-gray/opaque-shield theory was checked by comparing them with those deduced from the experimental data, $q_{rad,expt}$:

$$q_{rad,expt} = P - q_{cond} = P - C(T_{inner} - T_{outer}) \qquad (6)$$

For each experimental run with the shields, the percentage deviation of the calculated values from the experimental values was determined using the following equation:

$$\%Dev = 100(q_{rad,expt} - q_{rad,calc}) / q_{rad,expt} \qquad (7)$$

EMISSIVITY AND THERMAL CONDUCTIVITY DATA

Anodized Aluminum

In the published literature, the total hemispherical emissivity values for anodized aluminum range from 0.76 to 0.9. After a careful examination of the consistency of the calculated effective heat conductance values, C, for the apparatus, it was found that the emissivity of the surfaces of the anodized aluminum inner and outer tubes is best given by the following equation:

$$\varepsilon_{al} = 0.82 - \{(0.82 - 0.76)/100\}(T_{abs} - 300) \qquad (8)$$

The ε_{al} values of 0.82 and 0.76 that correspond to T_{abs} values of 300 K and 400 K, respectively, were obtained from the comprehensive works of Touloukian and Ho (1972).

The thermal conductivity of the anodized aluminum tubes was a moot point, because it is so high ($k_{al} > 200$ W/m-°C) compared to that of the shields ($k_{sh} < 1.4$ W/m-°C). Thus, in each experimental run, the inner and outer anodized aluminum tubes were essentially isothermal.

Shield Materials

Again, there is a fair bit of variation in both the emissivity and the thermal conductivity data for the shield materials used in this work. After consulting many different sources of such data, for example, the web site and the on-line technical help facility of the General Electric Corporation, Incropera and De Witt (1996), and Touloukian and Ho (1972), and examining the available data in the context of the consistency of calculated results deduced from the experimental data with the shields, the values given in table 2 were assumed to apply in this work.

Table 2: Emissivity and thermal conductivity values for the shield materials.

Shield Material	ε_{sh}	k_{sh} [W/m-°C]
Plexiglas	0.88	0.18
Fused Quartz	0.90	1.40
Lexan	0.88	0.19

RESULTS, DISCUSSIONS, AND CONCLUSIONS

In each experimental run, once the set-up was completed, the input voltage and current were set, and the desired vacuum conditions were achieved, at least 24 hours were allowed to elapse before any readings were taken. Data from numerous preliminary experiments indicated that this amount of time was sufficient to achieve steady-state conditions. Data and deduced results from a total of four different experimental runs with no shields, and 25 different experimental runs with one or multiple shields, are presented in tables 3 and 4, respectively. In these tables, T_{inner} and T_{outer} refer to the average temperatures of the inner and outer anodized aluminum tubes, respectively. These average values were obtained by calculating the mean of data from not only many different thermocouples but also three different sets of data, for each run, with at least one hour between successive sets. The input voltage, V, and input current, I, to the heater are also similarly averaged values, for each run.

The very consistent values of C in table 3 establish that proposed method of estimating heat conduction losses in the apparatus works well. An average value of C = 0.04968 W/°C was adopted as characteristic of the apparatus, and used in processing the results obtained from the experimental runs with one or multiple shields.

Table 3: Data from experimental runs without shields (empty apparatus) and the corresponding deduced values of C.

Run No.	V [V]	I [A]	T_{inner} [°C]	T_{outer} [°C]	C [W/°C]
1	20.77	0.1336	24.20	9.65	0.049929
2	34.08	0.2188	46.01	9.63	0.049610
3	42.28	0.2710	62.50	9.66	0.049681
4	41.96	0.2691	61.80	9.54	0.049504

Table 4: Data from the experimental runs with one or multiple shields and corresponding deduced results.

Run No.	V [V]	I [A]	T_{inner} [°C]	T_{outer} [°C]	q_{rad} Calc. [W]	q_{rad} Expt. [W]	% Dev
5	34.18	0.2183	48.96	9.73	3.41	5.51	38
6	38.11	0.2432	57.45	9.72	4.31	6.90	38
7	42.10	0.2683	67.52	9.70	5.47	8.42	35
8	34.09	0.2147	51.33	9.59	4.64	5.25	12
9	42.02	0.2669	70.38	9.57	7.37	8.19	10
10	33.95	0.2177	48.67	9.64	4.53	5.45	17
11	38.06	0.2438	57.33	9.60	5.76	6.91	17
12	42.06	0.2692	66.25	9.66	7.11	8.51	16
13	34.00	0.2169	51.01	9.63	3.84	5.32	28
14	42.09	0.2676	71.98	9.59	6.38	8.17	22
15	38.11	0.2428	60.42	9.66	4.92	6.73	27
16	34.08	0.2186	55.48	9.68	5.22	5.17	0.9
17	42.06	0.2693	75.14	9.66	8.17	8.07	1.2
18	33.91	0.2174	50.64	9.63	4.95	5.33	0.7
19	38.03	0.2435	59.77	9.68	6.31	6.77	6.9
20	41.90	0.2681	68.76	9.63	7.75	8.30	6.5
21	34.03	0.2183	53.48	9.69	4.96	5.25	5.7
22	42.03	0.2691	72.19	9.69	7.70	8.21	6.2
23	34.05	0.2182	57.43	9.68	3.49	5.06	31
24	38.08	0.2437	68.08	9.74	4.47	6.38	30
25	42.03	0.2685	81.32	9.71	5.82	7.73	25
26	34.10	0.2186	59.88	9.73	3.89	4.96	22
27	42.02	0.2687	83.57	9.74	6.41	7.62	16
28	34.05	0.2176	62.73	9.72	3.42	4.78	28
29	38.03	0.242	74.96	9.69	4.45	5.95	25

Notes:
One-Shield Runs: Runs # 5-7 with Shield # 1; Runs # 8-9 with Shield # 2; Runs # 10-12 with Shield # 3; Runs # 13-15 with Shield # 4; Runs # 16-17 with Shield # 5; Runs # 18-20 with Shield # 6; and Runs # 21-22 with Shield # 7.
Two-Shield Runs: Runs # 23-25 with Shields # 1 and 2; and Runs # 26-27 with Shields # 4 and 5.
Three-Shield Runs: Runs # 28-29 with Shields # 1, 2, and 3

The results presented in table 4 show that the simple diffuse-gray/opaque-shield theory provides results that deviate from the experimental results by less than 38 % in all cases considered here. The one-shield runs show that the high percentage deviations occur for shields with the lowest inside diameters (Shields # 1 and 4), and the low percentage deviations are obtained with shields of intermediate diameters (Shields # 2, 5, and 7). The two-shield results, obtained with combinations of the small- and intermediate-diameter shields (Shields # 1 and 2, and Shields # 4 and 5), are a bit less accurate than the three-shield results.

The reasons for why the predicted one-shield results for the small-diameter shields are the worst and those for the intermediate-diameter shields are the best are not clear. It appears as if the diffuse-gray/opaque-shield theory applies best to the intermediate-diameter shields, is not too bad for the large-diameter shields, and not very valid for the small-diameter shields, but to determine the exact causes of this behavior it would be necessary to formulate and apply a semitransparent shield theory, perhaps also accounting for specular and diffuse reflections, following, for example, the proposals of Perlmutter and Siegel (1963), Sparrow et al. (1962), Viskanta et al. (1978), Timoshenko and Trenev (1986), Modest (1993), and Holman (1997). The predicted results for shields made with fused quartz are consistently the most accurate, indicating that the diffuse-gray/opaque-shield theory applies to this material better than it does to Plexiglas and Lexan.

ACKNOWLEDGMENTS

The authors gratefully acknowledge financial support of this work by the Natural Sciences and Engineering Research Council (NSERC) of Canada. One of the authors, Sophie Guers, worked on this project under the auspices of a student-exchange program between Ecole Centrale, Nantes, France, and McGill University, Montreal, Quebec, Canada. Mr. Emile Aboumansour, a B.Eng. Honours student at McGill University, assisted in the assembly of the apparatus in some of the experimental runs.

REFERENCES

Duffie, J.A. and Beckman, W.A. (1974), *Solar Energy Thermal Process*, John Wiley & Sons, New York.

Elkouh, N. (1996), "Laminar Natural Convection and Interfacial Heat Flux Distributions in Pure Water-Ice Systems", *Ph.D. Thesis*, Department of Mechanical Engineering, McGill University, Montreal, Quebec, Canada.

Fuschillo, N. (1974), "Semi-Transparent Solar Collector Window Systems", *Solar Energy*, Vol. 177, pp. 159-165.

Holman, J.P. (1997), *Heat Transfer*, 8th Edition, McGraw-Hill Book Co., New York.

Incropera, F.P. and De Witt, D.P. (1996), *Fundamentals of Heat and Mass Transfer*, 4th Edition, John Wiley & Sons, New York.

Kreith, F. and Bohn, M.S. (2001), *Principles of Heat Transfer*, 6th Edition, Brooks/Cole, Pacific Grove, California.

Lagana, A. (1996), "Mixed Convection Heat Transfer in Vertical, Horizontal, and Inclined Pipes", *M.Eng. Thesis*, department of Mechanical Engineering, McGill University, Montreal, Quebec, Canada.

Modest, M.F. (1993), *Radiative Heat Transfer*, McGraw-Hill Book Co., New York.

Perlmutter, M. and Siegel, R. (1963), "Effect of Specularly Reflecting Gray Surface on Thermal Radiation Through a Tube and From Its Heated Wall", *ASME Journal of Heat Transfer*, Vol. 85, pp. 55-62.

Siegel, R. and Howell, J.R. (1972), *Thermal Radiation Heat Transfer*, McGraw-Hill Book Co., New York.

Sparrow, E.M. and Cess R.D. (1970), *Radiation Heat Transfer*, Revised Edition, Brooks/Cole Publishing Co., Belmont, California.

Timoshenko, V.P. and Trenev, M.G. (1986), "A Method for Evaluating Heat Transfer in Multilayered Semitransparent Materials", *Heat Transfer – Soviet Research*, Vol. 18, pp. 44-57.

Touloukian, Y.S. and Ho, C.Y. Eds. (1972), *Thermophysical Properties of Matter*, Vols. 1-9, Plenum Press, New York.

Viskanta, R., Siebers, D.L., and Taylor, R.P. (1978), "Radiation Characteristics of Multiple-Plate Glass Systems", *Int. J. Heat Mass Transfer*, Vol. 21, pp. 815-818.

IECEC2001-RE-23

MODELING AND PERFORMANCE COMPARISON OF AN INTERNALLY COOLED LIQUID DESICCANT ABSORBER OPERATING IN SEVERAL HYBRID COOLING AND DEHUMIDIFICATION MODES

Arshad Y. Khan
Department of Mechanical Engineering
University of Puerto Rico
Mayaguez, PR 00681

ABSTRACT

Conventional air cooling and dehumidification systems are based on electric driven vapor-compression cycle. These systems have achieved an optimum level of energy efficiency for sensible load dominated air conditioning applications. However, for high latent load applications, the energy efficiency ratio of these systems tends to degrade rapidly due to the inherent coupling between the sensible and the latent loads in their operation. Therefore electric driven vapor-compression technology is more energy intensive in high humidity regions. A well-designed hybrid liquid desiccant system is capable of accomplishing the required cooling and dehumidification at a higher energy efficiency ratio for medium to high latent load applications. This paper describes the model development and discusses the performance predictions of an internally cooled liquid desiccant absorber (*ICLDA*) operating in several cooling and dehumidification modes. These modes include when part of the absorber is allowed to perform as a conventional cooling coil or an evaporative cooling coil. The cooling and dehumidification performance in several different modes is predicted and compared with one another for a variety of coolant inlet temperature and its flow rate in this work. The performance comparison revealed that under similar sensible and latent load conditions, two hybrid liquid desiccant cooling and dehumidification modes could outperform a conventional cooling coil regardless of the coolant inlet temperature or its flow rate.

INTRODUCTION

Air cooling and dehumidification for comfort and/or process applications in high humidity regions requires the air conditioning system to operate at high latent heat ratios.

Furthermore, to improve indoor air quality and to address public health issues arising from sick building syndrome, more outside air for ventilation is needed. Supply air containing larger fraction of outside air tends to further increase the operating latent heat ratio for the air conditioning system. Traditionally, systems based on the conventional electric driven vapor-compression cycle are used for air cooling and dehumidification. These systems have achieved an optimum level of energy efficiency for sensible load dominated air conditioning applications. However, for high latent load applications, the energy efficiency ratio of these systems tends to degrade rapidly due to the inherent coupling between the sensible and the latent loads in their operation. Typically, under these operating conditions a vapor compression system performs longer duty cycles and is required to over cool the air in order to meet the latent load.

Theoretically, the de-coupling of latent and sensible loads provides the opportunity to design a system that can remove both the sensible and latent load at their optimum energy efficiency ratios. Systems based on advanced cooling and dehumidification concepts exhibit the characteristic where the sensible and latent loads can be removed independent of each other.

In order to de-couple the latent load from the sensible load, research efforts in the past have considered systems based on solid or liquid desiccant cycle to perform dehumidification of air. These systems have the characteristic to remove latent load independent of the sensible load, thus eliminating the possibility of reheat situations for high latent load applications. Also, desiccant systems can utilize low-grade thermal energy at lower temperatures, which make them more attractive for integrating renewable energy sources. However, the fact that a desiccant system also tends to have a low energy efficiency ratio has overshadowed some of its inherent advantages.

To de-couple the latent load from the sensible load, another potential technology currently gaining rapid attention is the use of combined or hybrid desiccant cycles. These cycles use a desiccant system to bring the humidity to within the comfort range along with a sensible cooling loop to bring the temperature within the comfort range. The purpose is to have the desiccant cycle operate within its most efficient range along with the cooling cycle to operate at higher evaporator temperature to accomplish the sensible load removal task. It is expected that this may result in an appreciable increase in both the energy efficiency ratios of the desiccant as well as the cooling system. The energy efficiency ratio of the whole system can be further improved if part or all the heat rejection from the cooling cycle can be used for desiccant regeneration. Previous studies dealing with the use of hybrid liquid desiccant absorbers that were refrigerant cooled, evaporatively cooled, and water cooled are available in the literature [1-3]. These studies have revealed that a hybrid system may have good potential in terms of de-coupling the latent load from the sensible load.

This paper describes the model development and discusses the performance predictions of an internally cooled hybrid liquid desiccant absorber (*ICLDA*) operating in three different modes. The first mode is the *CC-ICLDA* mode. In this mode only a section of the internally cooled liquid desiccant absorber located at the process air downstream direction is sprayed with the desiccant solution while the remaining absorber operates as a conventional cooling coil. The second mode is the *ICLDA-CC* mode. In this mode the conventional cooling coil section of the desiccant absorber is located at the process air downstream direction. The third mode is a modification of the second mode and it is referred to as a modified internally cooled liquid desiccant absorber (*MICLDA*) in this work.

Modified-ICLDA system

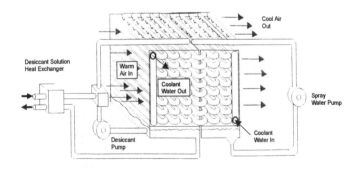

Fig. 1: Schematic diagram of a *MICLDA* system.

In this mode the conventional cooling coil section of the absorber is sprayed with pure water to accomplish evaporative cooling of the process air. The schematic diagram of the internally cooled liquid desiccant absorber operating in the third mode is presented in Figure 1. Due to space limitations schematic of only the third mode is included here. However, it is easy to imagine that the first and the second modes as being internally cooled liquid desiccant absorbers having only a down-stream or up-steam portion that is sprayed with a desiccant solution while the remaining absorber is kept dried.

MODEL DEVELOPMENT

The *ICLDA* can be considered as a spray film fin tube heat exchanger. Essentially, it is a multi-row coil with a desiccant solution spray nozzle header located at the top of the coil as shown schematically in Figure 1. Solution is sprayed on the tube bundle where it comes in direct contact with the process air. As seen form this figure air and desiccant solution as well as the coolant water and the desiccant solution are in a cross-flow arrangement. Also, the process air and the coolant water are in an overall counter-flow arrangement.

A similar approach as used by Khan [3] to develop a model of heat and mass transfer in a desiccant absorber was utilized by considering a differential control volume in an *ICLDA* as shown in Figure 2. The following assumptions were made to develop the steady state heat and mass transfer model presented in this work.

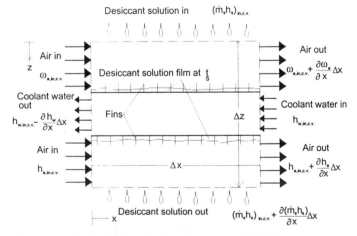

Fig. 2: A differential control volume in an *ICLDA*.

- Dry air and water vapor are treated as a non-reacting mixture of ideal gases.
- Air and water flow directions are in an overall counter-flow configuration, as the local cross-flow effects in each control volume are neglected. The assumption is valid, as the performance of a cross-flow heat exchanger approaches that of a counter-flow heat exchanger when the number of tubes passes is greater than four.

- Air and the coolant water are in a cross-flow configuration with the lithium chloride solution as the solution is sprayed at the top of the absorber unit and it flows in a downward direction.
- Entrainment of solution by the air is negligibly small.
- The airside heat and mass transfer coefficients include the effect of fins at the outer surface.
- Solution is in equilibrium with the local coil surface temperature. Additionally, the change in temperature and concentration of desiccant solution over the thickness of its film are negligibly small. This assumption is especially valid as solution forms a thin liquid film on the absorber surface.
- In each control volume, all the properties are assumed to be uniform.

After making the above listed assumptions, the steady state conservation equations for the control volume for each flow stream in its respective flow direction can be written as presented in the following section.

The change in air enthalpy, air humidity, and coolant water temperature in the airflow direction can be written as follows.

$$\frac{\partial h_a}{\partial x} = -\frac{NTU_o}{DX}\left[\left(h_s - h_{s,t_s}\right)+\left(\omega_a - \omega_{s,t_s}\right)\left(\frac{1}{Le_o}-1\right)h_{v,t_s}\right] \quad (1)$$

$$\frac{\partial \omega_a}{\partial x} = -\frac{NTU_o}{DX\,Le_o}\left(\omega_a - \omega_{s,t_s}\right) \quad (2)$$

$$\frac{\partial t_w}{\partial x} = \frac{NTU_i}{DX}\left(t_w - t_s\right) \quad (3)$$

Where

$$NTU_o = \frac{h_{c,o}A_o}{\dot{m}_a C_{p,m}}, \quad NTU_i = \frac{h_{c,i}A_i}{\dot{m}_w C_{p,w}}, \text{ and}$$

$$Le_o = \frac{h_{c,o}}{h_D C_{p,m}} \quad (4)$$

The conservation of mass for the dry air, coolant water, and the desiccant solution for the control volume shown in Figure 2 can be expressed as given in Equations (5) and (6).

$$d\dot{m}_a = d\dot{m}_w = 0 \quad (5)$$

$$\dot{m}_a \frac{\partial \omega_a}{\partial x} + C_1 \frac{\partial \dot{m}_{sol}}{\partial z} = 0 \quad (6)$$

Finally, the conservation of energy equation for the control volume can be written as:

$$\dot{m}_w \frac{\partial C_{p,w}t_w}{\partial x} - \dot{m}_a \frac{\partial h_a}{\partial x} - C_1 \frac{\partial h_{sol}\dot{m}_{sol}}{\partial z} = 0 \quad (7)$$

Where,

$$C_1 = \left(DZ/DX\right) \quad (8)$$

Equations (1) through (3) along with the conservation of mass and energy Equations (6) and (7) constitute the set of equations that can be solved for a given value of NTU_o, NTU_i, Le_o, water, air and solution mass flow rates, and the appropriate boundary conditions to predict the steady state cooling and dehumidification performance of an $ICLDA$. These equations are also valid and can be used to predict the performance of the absorber section that is operating in a conventional cooling coil mode by setting the desiccant mass flow rate to zero. In the case of evaporative cooling coil section the solution mass flow is replaced by the spray water mass flow rate in the above equations.

METHOD OF SOLUTION

Starting with a slice of absorber in the solution flow direction and with an assumed solution temperature and given solution concentration, the slice was divided into a number of control volumes in the air flow direction to solve Equations (1), (2), (3), (6), and (7). Since the flow of water and the air in a counter-flow direction required that the condition of water be known in advance in order to begin the computation in the airflow direction, an iterative scheme was used. The scheme begins with assuming the water exit temperature and then proceeding with forward calculation of tube surface temperature, air humidity ratio, and air enthalpy in each control volume as air flows toward the absorber exit. Because the equilibrium humidity ratio in Equations (1) and (2) depends on the tube surface temperature, another iteration was required within each control volume to obtain the correct local tube surface temperature that satisfies the conservation of mass and energy equations for the control volume. The procedure for the given absorber slice was repeated until the calculated water temperature converges to the actual entering water temperature within 0.1% of the given water inlet temperature. After a successful convergence is achieved, the solution scheme then proceeds to the next slice in the solution flow direction and continues until all the slices in the solution flow direction are exhausted. Another iterative loop was devised at this point to calculate the actual solution temperature based on the overall energy balance criterion for the solution loop in the absorber. Convergence is declared when the summation of all contributing terms in the overall energy balance in the solution loop converges to a value less than |1.0E-05|.

645

In order to increase numerical stability and to minimize the number of control volumes, piece-wise linear approximation over a control volume was used for all the variables involved in the algebraic equations corresponding to Equations (1), (2), (3), (6), and (7). Furthermore, to ensure convergence, a hybrid under relaxation technique was employed for all the iterative loops in the computational procedure.

To find the enthalpy of lithium-chloride water solution, specific heat capacity relation reported by Uemura [4] along with the procedure suggested by Himmelblau [5] were used. Also the pressure of water vapor in equilibrium with lithium-chloride solution was obtained by the equations suggested by Uemura [4]. The relationship for moist air enthalpy as reported by ASHRAE [6] was also utilized in the solution.

DISCUSSION OF RESULTS

In this section the results from the cooling and dehumidification performance study of the three different modes, that is, the *CC-ICLDA*, *ICLDA-CC*, and *MICLDA* are presented. The model described earlier was used to predict the cooling and dehumidification performance for these three modes under the same operating conditions and input parameters. Two additional modes that is when the entire absorber is sprayed with the desiccant solution and when the entire absorber operates as a conventional cooling coil were also included as base modes for comparison purposes. The air bulk humidity distribution in its flow direction for the above mentioned five modes is plotted in Figure 3. As seen from this figure, the *CC-ICLDA* and *ICLDA-CC* modes show better dehumidification performance characteristics than a conventional cooling coil. The ability to dehumidify the air first followed by a controlled humidification to swap some latent load with sensible load is also seen from this figure for the *MICLDA* mode.

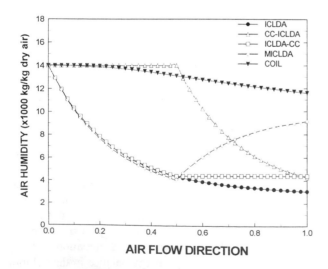

Fig. 3: Air bulk humidity ratio variation in its flow direction for different operating modes studied in this work.

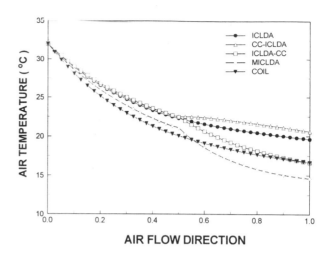

Fig. 4: Air bulk temperature variation in its flow direction for different operating modes studied in this work

Despite the humidification of air in this mode the exit air bulk humidity is still less than the exit bulk humidity for the conventional cooling coil.

The air bulk temperature distribution in its flow direction for different operating modes is plotted in Figure 4. As seen from this figure, the *ICLDA-CC* mode matches the sensible cooling performance of a conventional cooling coil while the *MICLDA* mode provides the best sensible cooling. Based on the results shown in Figure 4, it should be concluded that the controlled humidification in the evaporative cooling section of the absorber provides an effective means to cool the air while keeping the air humidity in check. Also, as seen from this figure the *CC-ICLDA* mode exhibit a poor cooling performance as very little cooling is accomplished in the *ICLDA* section.

The model was used to predict the air exit humidity as a function of coolant inlet temperature for the five modes studied in this work. The averaged air exit humidity along with the corresponding averaged air exit temperature for different operating modes are plotted in Figures 5 and 6 respectively. Figures 5 and 6 can be used to study the cooling and dehumidification performance behavior of the five modes considered in this study. For all the cases reported in Figures 5 and 6 the coolant inlet temperature was allowed to vary from 5°C to 20°C while the inlet air condition was maintained at 32°C dry bulb temperature and 0.014 kg/kg humidity ratio.

The results from Figures 5 and 6 can be used to compare the cooling and dehumidification performance of different options. Also these figures can be used to study what if scenarios dealing with the latent and sensible load removal characteristics when the *ICLDA* operates in different modes. Finally it can be seen from these figures that for a given inlet coolant temperature *ICLDA-CC* and *MICLDA* modes remove more latent and sensible loads than a conventional cooling coil